DATE DUE		
JAN 8 1982		
FEB 2 9 1984		
MAR 2 1 1984		
APR 9 1984		
MAR 2 7 1995		
April 11		
APR 29 1996		

ALSO BY ALLAN CHASE

The Biological Imperatives:
Health, Politics, and Human Survival (*1972*)

Shadow of a Hero (*1949*)

The Five Arrows (*1944*)

Falange: The Axis Secret Army in the Americas (*1943*)

THE LEGACY OF MALTHUS

ALLAN CHASE

THE LEGACY OF MALTHUS

The Social Costs of the New Scientific Racism

ALFRED A. KNOPF / NEW YORK / 1977

THIS IS A BORZOI BOOK
PUBLISHED BY ALFRED A. KNOPF, INC.

Copyright © 1975, 1976 by Allan Chase
All rights reserved under International and Pan-American Copyright Conventions.
Published in the United States by Alfred·A. Knopf, Inc., New York, and
simultaneously in Canada by Random House of Canada Limited, Toronto.
Distributed by Random House, Inc., New York.

Since this page cannot legibly accommodate all copyright acknowledgments,
they may be found on pages viii–ix.

Part of chapter 9 was published in slightly different form as "The Great
Pellagra Cover-Up" in Psychology Today, February 1975.

Library of Congress Cataloging in Publication Data

Chase, Allan, [date]
    The legacy of Malthus: the social costs of the new scientific racism.

    Bibliography: p.
    Includes index.
    1. Race discrimination—History.   2.   Eugenics—
History.   3.   Genetics—History.   I.   Title.
HT1521.C43   1976   301.45'1'042   76–13726
ISBN 0–394–48045–7

Manufactured in the United States of America

First Edition

67675

To Karen Elizabeth Schachter,
my granddaughter,
who is entitled to know why
her generation was cheated of the
full measure of so many of the
world's scientific treasures,
this book is dedicated with
all of my love.

Spring 1945: emerging from the nightmare, the world discovers the camps, the death factories. The senseless horror, the debasement: the absolute reign of evil. Victory tastes of ashes.

Yes, it is possible to defile life and creation and feel no remorse. To tend one's garden and water one's flowers but two steps away from barbed wire. To experiment with monstrous mutations and still believe in the soul and immortality. To go on vacation, be enthralled by the beauty of a landscape, make children laugh—and still fulfill regularly, day in and day out, the duties of a killer.

There was, then, a technique, a science of murder, complete with specialized laboratories, business meetings and progress charts. Those engaged in its practice did not belong to a gutter society of misfits, nor could they be dismissed as just a collection of rabble. Many held degrees in philosophy, sociology, biology, general medicine, psychiatry and the fine arts. There were lawyers among them. And—unthinkable but true—theologians. And aristocrats.

—ELIE WIESEL, in *One Generation After,* 1970

PERMISSIONS ACKNOWLEDGMENTS

Grateful acknowledgment is made to the following for permission to reprint previously published material:

American Association for the Advancement of Science: Excerpts from the following articles, which appeared in the specified issues of *Science,* "Rudolf Virchow's Anthropological Work," by F. Boas, Vol. 16, p. 441, 19 Sept. 1902; "Review of *Passing of the Great Race,*" by F. A. Woods, Vol. 48, pp. 419–20, 25 Oct. 1918; "Review of *America's Greatest Problem,*" by B. G. Wilder, Vol. 42, p. 768, 26 Nov. 1915; "Annual Meeting of the American Genetic Association," by P. Popenoe, Vol. 42, pp. 391–96, 24 Sept. 1915; "The Fruits, Prospects and Lessons of Recent Biological Science," by C. W. Eliot, Vol. 42, pp. 919–31, 31 Dec. 1915; "The Competitive Exclusion Principle," by G. Hardin, Vol. 131, pp. 1292–97, 29 April 1960; "Heritability of Intelligence," by P. Bohannan, Vol. 182, p. 115, 12 Oct. 1973; "Science and the Race Problem: A Report of the AAAS Committee on Science in the Promotion of Human Welfare," Vol. 142, pp. 558–61, 1 Nov. 1963; "In Defense of the Kallikak Study," by H. S. Goddard, Vol. 95, pp. 574–76, 5 June 1942; "Obituary of F. A. Woods," by C. B. Davenport, Vol. 90, pp. 607–08, 29 Dec. 1939; "Parenthood: Right or Privilege?," by G. Hardin, Vol. 169, p. 427, 31 July 1970; "The Races of Europe," by F. Boas, Vol. 10, pp. 292–96, 1 Sept. 1899; "The Growth of Children," by F. Boas, Vol. 20, pp. 256–57, 281–82, 351–52, 23 Dec. 1892, May 6, 20, 1892; "Growth of Children of the Same Race Under Different Environmental Conditions," by W. W. Gruelich, Vol. 127, pp. 515–16, 7 May 1958. Copyright © 1960, 1963, 1970, 1973 by the American Association for the Advancement of Science.

American Medical Association: Excerpts from pp. 1, 12, 13, 15, 18, 19, 20, 21, 41, and 43 of *Mental Retardation: A Handbook for the Primary Physician,* 1964 ed.

Atheneum Publishers: Excerpts from pp. 63 and 212 of *The Social Contract,* by Robert Ardrey. Copyright © 1970 by Robert Ardrey.

Ballantine Books, a Division of Random House, Inc.: Excerpts from pp. 15, 16, 166, and 195 of *The Population Bomb,* by Paul Erhlich. Copyright © 1968, 1971 by Paul R. Ehrlich.

Blackwell Scientific Publications, Ltd.: Figures 4, 6, and 10 and a brief quote from p. 8 of *An Introduction to Social Medicine,* 2nd ed., by T. McKeown and C. R. Lowe.

The Bobbs-Merrill Company, Inc.: Brief excerpt from *Trail to Light,* by R. L. Parsons. Copyright 1943, renewed 1970 by R. L. Parsons.

Columbia University Press: Excerpts from pp. 43–46, 56–58, also Tables 22, 23, 24, 25, and 31 of *Negro Intelligence and Selective Migration,* by Otto Klineberg (1935). Excerpt from p. 129 of *Science and the Concept of Race,* by M. Mead et al. (1968; paperback edition, 1969). Excerpt from pp. 8–9 of *Mental Retardation and Its Social Dimensions,* by Margaret Adams (1971).

Curtis Brown Ltd.: Excerpts from pp. 48, 72–75, 77, 145, 215, 218, 228, and 282–83 of *Road to Survival* by William Vogt. Copyright 1948 by William Vogt. Published by William Sloan Associates.

Doubleday & Company, Inc., and The Hutchinson Publishing Group, Ltd.: Figure 2.1 from *Cells: Their Structure and Function,* by E. H. Mercer, Rev. Ed. Copyright © 1962 by The American Museum of Natural History. Copyright © 1961, 1965 by E. H. Mercer.

W. H. Freeman and Company: Excerpts from pp. 608, 610, 611, 612, 613, 618, 619, and 620 of *Biology: Its Human Implications,* by Garrett Hardin. Copyright 1949 by Garrett Hardin.

Grune & Stratton, Inc.: Excerpts from pp. 504, 505, and 512 from a selection by Dr. Eisenberg that appears in *Psychopathology of Mental Development,* ed. by Joseph Zubin and G. Jervis. Reprinted by permission of the publisher and the author.

Harper & Row Publishers, Inc.: Table entitled "NIT Score and Length of New York Residence" from *Race Differences*, by Otto Klineberg. Copyright 1935 by Harper & Brothers; renewed 1963 by Otto Klineberg.

Harvard Educational Review: Excerpts from pp. 19, 45, 61, 74, 87–88, 95, 114, and 117 of *How Much Can We Boost IQ and Scholastic Achievement?*, by A. R. Jensen (1969).

Harcourt Brace Jovanovich, Inc.: Excerpts from pp. 6 and 7 of *The Biology of Race*, by James C. King (1971). Excerpt from pp. 194–95 of *On Aggression*, by Konrad Lorenz (1966). Excerpts from pp. 357–61 of *Hunger Fighters*, by Paul de Kruif (1955).

Lewis B. Holmes, M.D., and Hugo W. Moser, M.D.: Table, "Clinical Evaluation of 1378 Individuals at the Walter E. Fernald State School, Waltham, Mass." Reprinted from *Mental Retardation: An Atlas of Associated Physical Abnormalities*, ed. by Holmes and Moser (Macmillan, 1972).

The Houghton Library: Selected material from the Immigration Restriction League papers, housed at the Houghton Library, Cambridge, Mass.

Alfred A. Knopf, Inc.: Excerpt from p. 490, "Nature and Measurement of Intelligence," by Read Tuddenham, from *Psychology in the Making*, ed. by Leo Postman. Copyright © 1962 by Alfred A. Knopf, Inc.

Longman Group Ltd.: Excerpts from pp. 79–93 and Figures 14, 15, 17, and 18 of *From Birth to Seven*, ed. by R. Davie, N. Butler, and H. Goldstein. Reprinted by permission of the publisher and the National Children's Bureau.

The New York Times: Map entitled "Population Change, 1960–1970," *The New York Times*, 14 February 1971. Chart entitled "Birth, Fertility Rates at a New Low in U.S.," *Times*, 16 April 1974. Chart entitled "Blacks in the South," *Times*, 23 June 1974. Copyright © 1971, 1974 by The New York Times Company.

Princeton University Press: Text excerpts from pp. 182–83, 190, 191–92, and 207–08, Figures 33 and 36, and Table 33 from *A Study of American Intelligence*, by C. C. Brigham (1923).

Random House, Inc.: Excerpt from p. 4 of *One Generation After*, by Elie Wiesel. Copyright © 1965, 1967, 1970 by Elie Wiesel.

Charles Scribner's Sons: Excerpts from pp. 43–44 of *The Plague Killers*, by Greer Williams (1969). Excerpts from pp. 47, 554–55, and 195 and figure from p. 161 of *Is America Safe for Democracy?*, by William McDougall (1921).

Stanford University Press: Excerpts from pp. 14, 15, and 16 of *The Gifted Group at Mid-Life*, Vol. V of *Genetic Studies of Genius*, by Lewis M. Terman and Melita H. Oden (1959).

State of Mississippi, Dept. of Archives and History: Front page of the Nov. 1, 1915, *Jackson Daily News*, housed in the Archives.

University of Chicago Press: Six lines from "To Malthus" from *Perspectives in Biology and Medicine*, ed. by Richard Landau, p. 225. Copyright © 1966 by University of Chicago Press.

D. Van Nostrand Company: Figure 19 from *General Psychology*, 2d Ed., by Henry E. Garrett (American Book Company, 1961).

Viking Penguin, Inc.: Excerpts from pp. 38, 99, and 102 and Tables on pp. 39 and 104 of *Education and Income: Inequalities in Our Public Schools*, by Patricia Cayo Sexton. Copyright © 1961 by Patricia Cayo Sexton.

CONTENTS

PART THREE
**THE PHOENIX: THE RISE OF THE NEW
SCIENTIFIC RACISM**

PART FOUR
THE LEGITIMATE SCIENCES' ANSWER TO THE OLD AND NEW SCIENTIFIC RACISM—ONCE AND FOR ALL?

PREFACE

This book deals with the life-enhancing effects of the legitimate biological, behavioral, and biomedical sciences—and with the malignant effects of the pseudosciences on individuals and societies, in our own times and since the dawn of the Industrial Revolution. As such, it is a book about the growing and realistic hope for the human condition derived from the discoveries of the legitimate sciences—and also about the defeatist dogmas of scientific racism, which for the better part of two centuries have so perverted and lied about the discoveries of the legitimate life sciences that "society reaps at this moment but a small fraction of the advantage which current knowledge has the power to confer."[1]

Scientific racism is, essentially, the perversion of scientific and historical facts to create the myth of two distinct races of humankind. The first of these "races" is, in all countries, a small elite whose members are healthy, wealthy (generally by inheritance), and educable. The other "race" consists of the far larger populations of the world who are vulnerable, poor or non-wealthy, and allegedly uneducable by virtue of hereditarily inferior brains.

In the teachings of scientific racism, most of the human race's physiological ailments, anatomical defects, behavioral disorders, and—above all else—the complex of socioeconomic afflictions called poverty are classified as being caused by the inferior hereditary or genetic endowments of people and races.

Historically, these core pseudogenetic myths of scientific racism have provided venal legislators and leaders with "scientific" rationales for doing nothing or next to nothing about the prevention of scores of well-understood impediments to proper physical and mental development of entire nations. Those preventable and therefore social plagues range from the common infectious diseases (such as German measles) that cause congenital eye, ear, skeletal, and brain disorders, to tuberculosis, illiteracy, low birth weight, malnutrition, and other well-defined causes of infant underdevelopment and death.

Far from being aimed at ethnic, social, and racial minorities, scientific racism has from its early-nineteenth-century origins been directed at the majorities of the populations of England, France, Germany, the United States, and other industrial nations. The original victims of scientific racism were as white, as Anglo-Saxon, as Protestant as was the noted British political economist who gave scientific racism the first of its historically devastating pseudolaws of demography and biology. In our own times, the white, Anglo-Saxon, and Protestant people who do the hardest work on the land, in industry, and in commerce remain, numerically, the chief targets and victims of the new scientific racism in America.

Coupled, as it often is and has been, with the much older forms of gut racism based on religious, racial, and ethnic bigotry, scientific racism invariably exacerbates the already agonizing traumas of plain old-fashioned racism for all minorities from Auschwitz and Belfast to Boston and Birmingham. Nevertheless, bigotry is not one of the functions of scientific racism; it is merely a later adjunct in the furtherance of the basic socioeconomic functions of scientific racism. This is the testimony of the history of science in its interactions with society.

I am neither a scientist nor a professional historian. By profession, I am a writer.

The differences between working scientists or physicians who treat living patients and writers who write about their scientific and clinical work are quite distinct. An extraordinary number of biological scientists—including the bacteriologists Pasteur and Zinsser, the Nobel laureates in medicine Nicolle and Wald, and the pathologist and microbiologist Lewis Thomas—have written about biology and its role in human affairs as well as or better than any professional science writer. However, our species has yet to produce a single lay writer on science who, without any advanced education and professional laboratory training in one of the biological sciences, was able to do research and make significant findings in any of the sciences dealing with the genetics, the biology, the chemistry, or the physics of human life. Nor, for that matter, have scientists trained in one discipline ever been able to do any research in another scientific discipline without first taking years of university and/or laboratory training in the second science.

To be sure, neither Louis Pasteur, a successful chemist, nor Gregor Mendel, a failed science teacher, had enjoyed any professional education or training in bacteriology or genetics before working in these sciences. That, however, was because before they applied their previous training in chemistry and the life sciences to their historic field and laboratory studies of the functions of microorganisms in fermentation and disease, and the hereditary characteristics of hybrid peas, the very sciences of bacteriology and genetics did not exist. They and their contemporary pioneers of what became modern microbiology and biochemical genetics made possible, and actually essential, the prescribed courses of college, graduate school, and postdoctoral field and laboratory work that are today required to produce professional and scientifically competent microbiologists and geneticists.

After World War II, some professional scientists and their lay literary followers helped bring about the ongoing revival of the long since scientifically disproven Victorian and pre-Victorian myths about the innate or genetic physiological, anatomical, mental, and moral superiority or inferiority of human beings born into the right or wrong social classes, ethnic groups, cultures, or races. As it happens, none of these scientists or lay journalists writing on human genetics has had any professional or Pasteur-Mendel type of exhaustive self-training in any of the sciences most directly concerned with human heredity, human brain development, or human physiology. A Nobel prize in physics, for example, no more makes a professional electrical engineer an expert in human population genetics, mentation, and ethics than an American Academy of Motion Pictures Arts and Sciences grand prize—an Oscar—makes a professional actor an expert in electrical engineering or solid-state physics.

This is not to say that actors and other people in the performing arts lack the inborn or hereditary mental capacity to be educated and to have constructive careers in any of the legitimate sciences. Max von Pettenkofer, the hygienist whose application of the water, sewage, and housing reforms of the English Sanitary Movement in his own nation helped reduce the general death rate of Munich by 33 percent between 1873 and 1900, once worked as a professional actor. The late Gordon M. Tomkins, one of the world's leading authorities on the role of hormones in the activation and deactivation of the more than 100,000 genes in the human body, was a saxophonist in the Stan Kenton jazz band before becoming a professor of biochemistry at the University of California medical center in San Francisco. Tomkins, however, did not effect this change of careers by suddenly deciding that he knew as much biochemistry as any professional biochemist; first he had to complete all the years of training that help define the enormous differences between a life scientist and a saxophone player.

In the absence of equally demanding professional education and training in genetics, when it comes to such purely scientific matters as the genetics of human mentality, brain anatomy, physiology, and behavior, all commentators are laymen. This rule applies in equal measure to the presidential speech writer who writes about the genetics of intelligence in a magazine read by millions of Americans;[2] the Nobel laureate in physics who holds forth on such subjects on television talk shows; the professor of educational psychology who writes articles and books on the alleged differences in brain structure and the "heritability" of intelligence in different human races; and, of course, the writer on the history and sociology of science who comments on the literary efforts of his fellow amateurs in genetics.

When scientifically qualified people such as Tomkins speak or write on such subjects as the actions of genes in our cells, they speak with the authority of the replicable findings of their trained scientific observations of people and their controlled and often highly sophisticated experiments with laboratory animals. That is, they speak in the great scientific tradition of Virchow, Pasteur, and Goldberger—rather than in the equally distinct intellectual tradition of, say, Edward Freeman, the Victorian-era Teutonist and professor of history at Oxford. As we shall see in chapter 4, Professor Freeman made a speaking tour of America in 1881, delighting college, university, learned society, and lyceum lecture audiences with his famous mot: "The best remedy for whatever is amiss in America would be if every Irishman should kill a negro and be hanged for it." Even then scientific racism could well be termed the opium of the educated classes.

An extremely influential twentieth-century inheritor of this brand of scientific racism was the erstwhile teacher of agriculture at the North Missouri State Normal School, Harry Hamilton Laughlin (Sc.D., Princeton). As recounted in chapter 13, it was Dr. Laughlin, in his capacity as supervisor of the Eugenics Record Office, Cold Spring Harbor, New York, who convinced all but one of the nine Justices of the U.S. Supreme Court that bad heredity had created a malignant subpopulation, categorized in Laughlin's "Scientific Analysis" for the courts as "the shiftless, ignorant and worthless class of anti-social whites of the South," whose sheer existence gravely imperiled the biological integrity of the entire American human gene pool. Laughlin's "scientific" analysis of what he called the hereditary "social inadequacy" of a poor southern white girl named Carrie Buck helped convince the high court that Virginia and all other states had the constitutional right to subject citizens who failed to inherit wealth and/or achieve high IQ test scores to compulsory "eugenic" surgical sterilization—on the eugenic grounds that "pauperism" (poverty) begets "pauperism" and that illiteracy in the parents of one generation biologically predetermines illiteracy in all of their descendant generations unto the end of time.

While science writers untrained in the sciences lack the professional and moral qualifications to pass judgment on the intrinsic scientific worth of *new* scientific concepts, they do have the professional and the social obligation to utilize the judgmental capacities that the hindsights of history provide to anyone who can read the literature of science. Properly studied and understood, history—particularly the history of science and its interactions with the culture in which it is done—is a major weapon in the serious science writer's arsenal. Two examples of the value of historical knowledge and perspective, particularly in terms of how they aid in the identification of the real and the fake in the biomedical and behavioral sciences, are:

1. In the summer of 1975, while visiting relatives in western North Carolina, I acquired Volume IV of a four-volume work published in 1914 on the prevention and treatment of most diseases known to man.[3] Written by W. Grant Hague, M.D., a noted New York physician and member of the faculty of Columbia University's College of Physicians and Surgeons, the book made various statements about preventive medicine that even today represent the best thinking of the health professions. However, on page 567, in a section on a then inevitably fatal disease, pernicious anemia, Dr. Hague wrote: "For this condition arsenic is the one remedy useful."

About the only value this poison had in the treatment of this blood dyscrasia was that, since no known remedy existed as of 1913, large enough doses of arsenic probably shortened the period of suffering between the onset of pernicious anemia and the death of its victim. But in 1914, when Hague's book was published, lay science writers had no way of proving this.

Within a dozen years, however, Drs. George R. Minot, George H. Whipple, and William P. Murphy discovered that animal livers contained some as yet unidentified active principle that kept people with pernicious anemia free of all symptoms of this disease for as long as they ate massive amounts of liver. By 1926 Minot's Harvard faculty colleague, the physical chemist Edwin J. Cohen, had isolated and purified most of the anti-pernicious-anemia fraction present in animal liver, and this was soon made available as the extract called Fraction G. Further scientific analysis of Fraction G revealed that its active principle was cyanocobalamin—vitamin B_{12}. By 1975, as a result of these biomedical research findings, a confirmed diagnosis of pernicious anemia was no longer tantamount to a death sentence: pernicious anemia had long since become a chronic condition that is kept completely neutralized with a monthly injection of vitamin B_{12}.

So that if, today, a professor of medicine were to write that "arsenic is the one remedy useful" for the treatment of pernicious anemia, any professional science writer would have the historical hindsight *and the scientific authority* to dismiss this claim as being totally erroneous. More than that, every concerned science writer would also have the professional and moral duty to denounce this bit of misinformation in all communications media in order to prevent its acceptance from causing the preventable death of persons who now develop pernicious anemia.

2. Our nation has not always been as fortunate in the reception of scientific truths as we were when the findings of Minot and his colleagues (for which they won a Nobel prize in 1934) turned the once invariably fatal disease of pernicious anemia into an inconvenience that rarely if ever kills. Before World War II, for example, when the orthodox eugenicist Dr. Henry Herbert Goddard translated from the French and adapted for American use the intelligence tests of Alfred Binet and Théodore Simon, his administration of the first of these American-style "Binet tests" to steerage immigrants at Ellis Island in 1912 showed that, *according to the scores they made on these tests,* more than 80 percent of all Jewish, Italian, Hungarian, Russian, Polish, and other non-Nordic people tested were feebleminded defectives. These "scientific" findings were quickly used to lend scholarly credibility to the political crusade then being waged jointly by highly educated scientific racists and barely literate gut racists to end all "non-Nordic" immigration to this country (see especially chapters 11 and 12).

To Goddard, who even before his use of these so-called intelligence tests accepted the eugenic dogma that such groups were hereditarily inferior to the affluent Nordic Protestants of Europe, the results of his Ellis Island tests offered the ultimate scientific confirmation of what the eugenics propagandists had been writing about so-called inferior and superior races and social classes since 1869. (Eugenics; Malthusianism; Spencer's Social Darwinism, which was neither social nor evolutionary; and Teutonism, later to be rechristened Nordicism, were the four major cults of the scientific racism that developed, for historically explicable reasons, in the wake of the contiguous Agricultural and Industrial revolutions of the eighteenth and nineteenth centuries.)

At about the same time that Goddard was "measuring" what he truly believed to be the innate intelligence of bone-poor Southern and Eastern European immigrants in New York Harbor, his fellow Clark University alumnus and friend, the clinical psychologist J. E. Wallace Wallin, gave the same Goddard mental tests to wealthy and successful landowners, businessmen, farmers, merchants, and their wives in his native Iowa. Wallin had grown up among these people, and he knew them to be mentally competent, resourceful human beings, as well as the parents of college professors, doctors, and business executives. These rural Iowans, most of them Protestants of Northern European (Nordic) ancestry, did even worse on the Goddard intelligence tests than did the non-Nordic Jews, Italians, and other Catholics Goddard had tested at Ellis Island.

To Wallin, these results did not suggest that Iowa Nordics were mentally inferior to other European strains. What they did suggest to Wallin was that, as far as accurately measuring human mental acuity, let alone hereditary individual or "racial" intelligence, the Goddard and other American-style "Binet tests" were not worth the powder to blow them to hell.

As we shall see in later chapters, the scientific conclusions of Wallin were rejected by the greater society—while the scientistic claims of Goddard, Terman, Yerkes, and other eugenicists were accepted at face value. As a result, when the Army Alpha and Beta mental tests, adapted by Goddard, Lewis M. Terman, Robert M. Yerkes, and other eugenicists from their own prewar "Binet tests," were administered to close to two million recruits of World War I between 1917 and 1919, they "proved" to most educated Americans that the average American adult male had the innate mentality of a thirteen-and-a-half-year-old child. For longer than the generation that followed, all American educational, social, and foreign policies were to be conditioned by the historical reality that most of the legislators who fabricated public policies, and the teachers who trained these lawmakers from kindergarten through graduate school, had accepted as a scientific truth this eugenical interpretation of the psychological and anthropological implications of the Army intelligence tests of World War I.

Clearly, in the presence of the ongoing revival of the IQ-test-"validated" myths of genetically limited and fixed racial intelligence created by Goddard and others before World War I, one of the most useful functions a responsible writer on science can perform for his country is to rescue from the forgotten but abundantly documented records of the history of mental testing some essential elements of the real story of the rise and scientific destruction of these racial intelligence myths between 1910 and 1940. This I hope I have done, to the best of my ability, in various chapters of this book.

Today, as before, scientific truths remain the best answers to the barrage of scientific errors, half-truths, myths, and even deliberate perversions of both science and history that the present revival of the dogmas of eugenics and related forms of scientific racism has once again inflicted upon us. If this book takes issue—as it strongly does—with the words and claims of the revivalists of the dogmas of the old-time eugenics zealots who held poverty and its pathologies to be caused by the supposedly inferior genes in the blood of the poor whites of the bucolic South and the industrial North, it counters the resurrection of these scientific falsehoods not with the impassioned rhetoric of the advocate but, rather, with the hard, objective, and universally replicable data of the legitimate biological, behavioral, and anthropological sciences.

Since I will be presenting the findings of the legitimate sciences, I will from time to time, in the interest of clarity, use some of the time-saving and convenient data-display devices of the sciences (and the New York Stock Exchange and the sports pages of our daily newspapers) where tables, graphs, and charts save both time and space for reader and writer alike.

What will not be found in this book are any of the differential equations and other mathematical displays many scientists find convenient to use when communicating their findings or concepts to other scientists. I do not feel there is any justification for using such constructs in this book. Moreover, while many of these mathematical displays might be perfectly good arithmetic, they are just as capable of being bad science as were the tables, graphs, pedigree trees, and curves of distribution used to lend credibility to the scientistic myths of hereditary pellagra, genetic intelligence, and biologically transmitted "hereditary pauperism" that have cost this nation so dearly in this tragic century. (See pages 491–92.)

Along with the history and the lexicon of science, the other major asset available to the men and women who write about science and society is the advice, counsel, and patient guidance of the qualified scientists who work in the specific fields of science that the lay writer elects to write about. There is nothing very new or novel about this practice, as Abraham Flexner demonstrated in 1910, in his now historic study, *Medical Education in the United States and Canada.*

For most of the hundred years ending around 1920, nearly all American medical education was grossly inferior to English, French, German, and other European medical training in every respect. The European schools demanded, as an entrance requirement, premedical education that included college-level text and laboratory work in biology, chemistry, and physics; the American schools, most of them commercial ventures unaffiliated with hospitals or universities, did not even require a high school diploma. Whereas European medical students were given years of laboratory training in the life sciences and clinical training under the tutelage of superbly educated physicians in hospital wards and clinics, the average American medical school (even those with some links to our universities) offered neither laboratory nor clinical training. The American schools, which took as little as one year to graduate "doctors," offered merely lectures and quizzes by doctors graduated from the same or identical diploma mills.

Until 1893, when the opening of the new Johns Hopkins University Medical School offered a European-type medical education, with laboratory and ward training in scientific medicine, to educationally qualified students, even the average American university-affiliated medical schools provided little in the way of modern education. After the Johns Hopkins reforms, very few university-affiliated or independent commercial medical schools even attempted to match the educational and scientific advances introduced at Johns Hopkins.

It is now a matter of history that it was the report made by Flexner for the Carnegie Foundation for the Advancement of Teaching on the entrance requirements, the scientific and clinical training, the hospital affiliations, and the intellectual standards in the then over 150 American and

Canadian medical schools that helped most in accelerating the spread of Johns Hopkins-type standards throughout American medical education.

Flexner was neither a physician nor a natural scientist but a graduate in the classics whose previous adult career had been spent as a teacher of high school Greek in Louisville, Kentucky, and as the director of a tutorial school in the same city. He did, however, have the advice and guidance of top biomedical scientists and educators, notably the pioneer American cellular pathologist and microbiologist William H. Welch, the founding dean of the Johns Hopkins Medical School, and of Welch's protégé, Dr. Simon Flexner (Abraham's brother), bacteriologist and later president of the Rockefeller Institute for Medical Research.[4]

It was experts such as Welch and Simon Flexner who made Abraham Flexner aware of the guidelines for proper medical education, and who also exposed him to the source materials of the short history of American medical education with which his report opened. Essentially, he was instructed in how, when he visited a school, to determine for himself whether it was a scientifically advanced center of modern medical education or a commercial diploma mill.

After this orientation, his scientific consultants did not have to tell Flexner what he soon discovered for himself: as far as the world's resources in the biomedical arts and sciences were concerned, "society reaps at this moment but a small fraction of the advantage which current knowledge has the power to confer."[5] Nor did Flexner have to be a scientist himself to learn that "the overwhelming importance of preventive medicine, sanitation, and public health indicates that in modern life the medical profession is an organ differentiated by society for its own highest purposes, not a business to be exploited by individuals according to their own fancy."[6]

In the preparation of this report, Flexner functioned as a serious science writer doing a responsible job of investigative reporting on science and charlatanism in the schools which then produced the vast majority of America's doctors, medical professors, and government health officials. His excellent performance did not turn Abraham Flexner into a medical scientist professionally capable of making a scientific judgment or a clinical diagnosis. It did, however, serve the greater society by making other intelligent laymen understand the hope for the human condition, created by medical science, that was being destroyed by the very nature of American medical education as of 1910. It also enabled Flexner to set a proper example for all other authors who, since 1910, have written about the natural sciences and their applications in human affairs.

Allan Chase,
New York, *February 1, 1976*

ACKNOWLEDGMENTS

There is no adequate way for me to express my gratitude to the many fine scientists and their lay co-workers in the United States, England, Denmark, Canada, and Australia, who, during the half dozen years that this book was in the making, gave of their time, their personal data, their literature files, and their wisdom in the various stages of research, writing, rewriting, and editing. To thank each and every one of these talented and concerned men and women by name would fill more pages than any reader would have the patience to spare. At the same time, I cannot see this book go to press without thanking, by name, a few of the individuals whose aid was far and above the dimensions that a total stranger to most of them had any right to expect when first I sought their help.

Four scientific workers—the biochemist, cell biologist, and historian of biology Alfred E. Mirsky, Professor and Member, The Rockefeller University, New York; the psychiatrist and social biologist Benjamin Pasamanick, Aubrey and Hilda Lewis Professor of Social Psychiatry at the New York School of Psychiatry and, until 1976, Associate Commissioner for Research, New York State Department of Mental Hygiene, Albany; the behavior geneticist and psychologist Jerry Hirsch, Professor of Psychology and Professor of Zoology, University of Illinois-Urbana; and the educational and developmental psychologist Joseph McVicker Hunt, Professor of Psychology, University of Illinois-Urbana—graciously involved themselves in many aspects of the work. Over the years they plied me with new published and unpublished information, called my attention to new findings in the sciences dealing with human growth and development, read and reread many of the chapters of manuscript, and offered endless personal and professional encouragement.

Dr. Mirsky died on June 19, 1974, before he could finish his own book on the realities and myths of human inheritance and development. During the four years that I knew him, and learned from him, he often called my attention to materials he had already collected for his unfinished book and presented me with countless photocopies of many of them for my use in this book. Readers familiar with modern cell biology and biochemistry will recognize at a glance where Alfred Mirsky's guidance and help were most productive; his knowledge of and insights into the history of science and society have left their equally clear imprints on this entire book.

As a young visiting scientist, he spent a year in the Caltech laboratory of the Nobel prize-winning geneticist Thomas Hunt Morgan. In the privacy of this laboratory, Morgan—whose pre-World War I work in chromosomal genetics shattered the last of the scientific pretensions of Galtonian genetics—

often amused Mirsky by poking fun at the continuing efforts of the eugenicists to pose as legitimate genetical scientists. However, when Mirsky asked him why he never expressed his real feeling about the scientific worthlessness of eugenics and its dogmas in the cold print of a scientific journal, Morgan (a lifelong personal friend of Charles Benedict Davenport, Henry Fairfield Osborn, and other hard-core eugenicists) answered that he could never do such a thing because "it would hurt too many old friends."

I have never met a kinder person, or one more considerate of other people's dignity, than Alfred Mirsky. Yet, when he examined scientific or assertedly scientific statements in what Franz Boas had termed "that cold enthusiasm for truth that enables [scientists] to be always clearly conscious of the sharp line between attractive theory and the observation that has been secured by hard and earnest work," Mirsky never put his personal or academic ties above his deeper obligations as a man of science. A case in point was his essay review, "Genetics and Human Affairs," of Sir Julian Huxley's eugenics tract *Essays of a Humanist,* in the October 1964 *Scientific American:* the now classic scientific dissection of the dogmas of eugenics, IQ-testing myths and all.

Most of the scientific workers named below helped me obtain and understand their research and clinical studies described at many points in the book. Others, whose specific scientific work is not cited as such in the book, supplied me with basic and corroborative data on the work in the life, behavioral, and social sciences cited in the text. Many of these scientists reviewed, for accuracy of fact and interpretation, various chapters or segments of the manuscript. Others, here and in England, took the time to explain to me, in interviews, letters, and reprints, their eugenics concepts despite the fact that each of them knew we were not even in remote agreement. My thanks are offered to:

The late Dr. Edward A. Ackerman, Carnegie Institution of Washington, Washington, D.C. *Dr. Alexander Alland, Jr.,* Associate Professor of Anthropology, Columbia University, New York. *Jay P. Arena, M.D.,* Professor of Pediatrics, Duke University Medical Center, Durham, North Carolina. *Alan Berg,* Population and Nutrition Projects Department, International Bank for Reconstruction and Development, Washington, D.C. *The late Herbert G. Birch, Ph.D., M.D.,* Professor of Pediatrics, Albert Einstein College of Medicine, New York. *Dr. Donald J. Bogue,* Director, Community and Family Study Center, University of Chicago. *Roy E. Brown, M.D., M.P.H.,* Associate Professor, Department of Community Medicine, Mount Sinai Medical Center, New York. *The late Dr. Bacon F. Chow,* Professor of Biochemistry, School of Hygiene and Public Health, The Johns Hopkins University, Baltimore, Maryland. *Dr. James Coleman,* Associate Professor, Department of Sociology, The Johns Hopkins University, Baltimore, Maryland. *Louis Z. Cooper, M.D.,* Professor of Pediatrics, Columbia University College of Physicians and Surgeons, and Director of Pediatric Service, Roosevelt Hospital, New York. *Dr. Francis H. S. Crick,* MRC Laboratory of Molecular Biology, Cambridge, England. *E. Perry Crump, M.D.,* Chairman, Department of Pediatrics, Meharry Medical College, Nashville, Tennessee. *Dr. C. H. Doy,* Genetics Department, Research School of Biological Sciences, The Australian National University, Canberra, Australia. *Dr. Paul R. Ehrlich,* Professor of Biology, Stanford University, Stanford, California. *Leon Eisenberg, M.D.,* Professor of Psychiatry, Harvard Medical School, and Chief of Psychiatry, Massachusetts General Hospital, Boston,

Massachusetts. *Samuel S. Epstein, M.D.,* Swetland Professor of Environmental Health and Human Biology, Case Western Reserve University, Cleveland, Ohio. *Dr. John P. Ertl,* President, Neural Models, Ltd., Toronto, and former Director, Center of Cybernetics Studies, University of Ottawa, Canada. *Dr. Jill Forrest,* Research Fellow, Children's Medical Research Foundation, Sydney, Australia. *Dr. Robert Friedrichs,* Professor and Chairperson, Department of Sociology, Williams College, Williamstown, Massachusetts. *Dr. Mark Golden,* Assistant Professor of Psychiatry, Albert Einstein College of Medicine, New York. *Dr. Rae Goodell,* Massachusetts Institute of Technology, Cambridge, Massachusetts. *George Grier,* Vice-President, Washington Center for Metropolitan Studies, Washington, D.C. *Dr. Florence Halpern,* Clinical Associate Professor of Psychology, New York University School of Medicine. *Dr. Garrett Hardin,* Professor of Biology, University of California, Santa Barbara. *Janet B. Hardy, M.D.,* Director, The Johns Hopkins Collaborative Perinatal Study, The Johns Hopkins Hospital, Baltimore, Maryland. *Dr. Marvin Harris,* Professor of Anthropology, Columbia University, New York. *Dr. Rick Heber,* Professor of Psychology, University of Wisconsin, Madison. *Caroline Hoffman,* Rehabilitation Research and Training Center in Mental Retardation, University of Wisconsin, Madison. *Dr. Liam Hudson,* Professor of Educational Psychology, University of Edinburgh, Scotland. *Dr. Leon Kamin,* Professor and Chairman, Department of Psychology, Princeton University, Princeton, New Jersey. *Dr. Clarence J. Karier,* Professor and Chairman, Department of Educational Policy Studies, University of Illinois, Urbana. *David M. Kessner, M.D.,* Project Director, Health Services Research Study, Institute of Medicine–National Academy of Sciences, Washington, D.C. *Dr. James C. King,* Professor of Genetics, New York University School of Medicine. *Dr. Michael C. Latham,* Professor of International Nutrition, Graduate School of Nutrition, Cornell University Ithaca, New York. *Dr. David Layzer,* Professor of Astronomy, Harvard University, Cambridge, Massachusetts. *Howard Lewis,* Director, Office of Information, National Academy of Sciences, Washington, D.C. *Dr. Michael Lewis,* Professor and Director, Infant Laboratory, Educational Testing Service, Princeton, New Jersey. *Jean D. Lockhart, M.D.,* Director, Department of Committees, American Academy of Pediatrics, Evanston, Illinois. *Charles Upton Lowe, M.D.,* Chairman, Committee on Nutrition, American Academy of Pediatrics, Evanston, Illinois. *Professor Martin L. Maehr,* Department of Educational Psychology, University of Illinois, Urbana. *Drs. Lida and Paul Mattman,* Wayne State University, Detroit, Michigan. *Dr. George W. Mayeske,* Research Psychologist, U.S. Office of Education, Washington, D.C. *Dr. Joseph L. Melnick,* Professor and Chairman, Department of Virology and Epidemiology, Baylor University College of Medicine, Houston, Texas. *Dr. Jane R. Mercer,* Professor and Chairman, Department of Sociology, University of California, Riverside. *Professor Jan Mohr,* University Institute of Medical Genetics, Copenhagen, Denmark. *Walter F. Mondale,* U.S. Senator (D.-Minnesota). *Joseph Mori,* Medical Editor, the National Foundation, White Plains, New York. *Richard L. Naeye, M.D.,* Professor and Chairman, Department of Pathology, Pennsylvania State University Medical School, Hershey, Pennsylvania. *Magna Norgaard,* Head of Department of Information and Statistics, Mothers' Aid Centers, Copenhagen, Denmark. *Erik Olaf-Hansen,* Medical Editor, *Politiken,* Copenhagen, Denmark. *Frederick Osborn,* former Secretary, American Eugenics Society, Garrison, New York. *Donald Pinkel, M.D.,* St. Jude Children's Research Hospital, Memphis, Tennessee. *Dr. T. E. Reed,* Professor of Zoology and Professor of Anthropology, University of Toronto, Canada. *Dr. Steven Rose,* Professor of Biology and Head of the Brain Research Group, The Open University, England. *Dr. Mark R. Rosenzweig,* Professor of Psychology, University of California, Berkeley. *Arthur M. Sackler, M.D.,* International Publisher, *Medical Tribune,* New York. *Dr. Peggy Sanday,* Associate Professor, Department of Anthropology, University of Pennsylvania, Philadelphia. *Dr. Sandra Scarr-Salapatek,* Associate Professor, College of Education, University of Minnesota, Minneapolis. *Professor Earl S. Schaefer,* Duke University, Durham, North Carolina. *Dr. William Shockley,* Alexander M. Poniatoff Professor of Engineering Science, Stanford University, Stanford, California. *Dr. Richard J. C. Stewart,* London School of Hygiene and Tropical Medicine; Clinical Nutrition and Metabolism Unit, Hospital for Tropical Diseases, London, England. *Dr. Henry L. Sultz,* Professor of Epidemiology, Department of Social and Preventive Medicine, State University of New York, Buffalo. *George Tarjan, M.D.,* Professor of Psychiatry, Schools of Medicine and Public Health; Program Director, Mental Retardation, Neuropsychiatric Institute, University of California, Los Angeles. *Dr. Ethel Tobach,* Curator, Department of Animal Behavior, American Museum of Natural History, New York. *The late Gordon M. Tomkins,* Professor of Biochemistry, University of California School of Medicine, San Francisco. *Richard E. Wexler,* Special Assistant Attorney General, State of Minnesota, St. Paul. *Myron Winick, M.D.,* Director, Institute of Human Nutrition, Columbia University College of Physicians and Surgeons, New York. *Paul Zee, M.D., Ph.D.,* St. Jude Children's Research Hospital, Memphis, Tennessee.

Mr. George Salomon, Senior Editor, Publications Division, American Jewish Committee, New York, was kind enough to translate Eugen Fischer's introduction to the 1937 German edition of Madison Grant's *Conquest of a Continent,* quoted on pages 343–44, and to check out the working translation (previously obtained through the kindness of Dr. Leon Eisenberg) of the article Konrad Lorenz wrote in the *Zeitschrift für agnewandt Psychologie und Charakterkunde* of February 1940, now quoted on page 349. Neil Schachter, M.D., Assistant Professor of Medicine, Yale University Medical School, and director of the respiratory therapy department, Yale–New Haven Hospital, translated from the French various writings of the craniometrist and eugenicist Count Vacher de Lapouge, quoted on pages 95–96 and 265. Dr. Schachter also reviewed all of the chapters dealing with experimental and clinical medicine. Other scientific and clinical specialists reviewed the same pages, but only Neil Schachter turned me into a grandfather in the course of this book's preparation. The guidance and moral support of my friend Israel Katz were of constant help through the years. My wife, Martha, did some of the thankless and tiring library research. Two very old friends, Monica McCall, my agent, and Angus Cameron, my editor, kept both book and author afloat in stormy and becalmed waters; few writers have ever had more helpful friends or more constructive colleagues.

It was Angus Cameron who suggested the term "The New Enclosures" used in chapter 17. We were both helped by the skills and contagious enthusiasm of Jack Lynch, the copy editor; Sally Rogers, the production editor; and Bobbie Bristol, assistant to Mr. Cameron.

I again want to thank Professor Marvin Harris, my sponsor, and the Trustees of Columbia University for the designation of Visiting Scholar, which gave me priceless access to the vast resources of the university's general and specialized medical, biological, psychological, educational, scientific, business, law, and social-work libraries. I am also most grateful to the Harvard University libraries, particularly the Houghton Library, where I was able to study the voluminous papers of the Immigration Restriction League and its principals; the Charles Scribner's Sons Archives at the Princeton University Library; the Medical School and Sterling Memorial libraries at Yale University; the Mount Sinai Medical Center libraries; the New York Academy of Medicine Library; the Library of the American Philosophical Society, Philadelphia, where I had unlimited access to the papers of the Eugenics Record Office, the letters of William Bateson, and the letters and papers of Franz Boas; the library and archives of the American Museum of Natural History; the New York Public Library; the Library of Congress, Washington, D.C.; and the new and magnificent National Library of Medicine, Bethesda, Maryland. In each of these institutions, the helpfulness and the unfailing courtesy of every staff member I dealt with over the years made the tasks of research that much more gratifying.

Whatever errors might in time be discovered in this book are my own, for which I apologize to the individuals and institutions who helped me gather the materials, and to the readers, who always deserve better.

PART ONE

The Functions and Social Costs of Scientific Racism

1 | The Genesis and Functions of Scientific Racism

> The lower class individual lives in the slum and sees little or no reason to complain. He does not care how dirty and dilapidated his housing is either inside or out, nor does he mind the inadequacy of such public facilities as schools, parks, libraries: indeed, where such things exist, he destroys them by acts of vandalism if he can. Features that make the slum repellent to others actually please him. He finds it satisfying in several ways.
>
> —from *The Unheavenly City* (1970), by EDWARD C. BANFIELD, Harvard urbanologist and adviser on urban affairs to President Nixon

> We can take the people out of the slums but we can't take the slums out of the people.
>
> —U.S. SENATOR ROBERT BYRD (D., W.Va.), Assistant Majority Leader, U.S. Senate, and Ku Klux Klan organizer in the 1940's[1]

Few measurements of national well-being are more accurate indicators of the entire human condition of any country than are the percentage of its new-born babies who die during their first year of life and the average life expectancies of its population. As of this writing, the infant mortality rates of sixteen other industrial nations are lower than in the United States. Although a poor seventeenth in the world in the level of its infant mortality rate, the United States is an incredible twenty-seventh among all nations in its male life-expectancy rate. Right now, the males of twenty-six other countries can anticipate living longer than the males of the world's richest nation.[2]

There are a number of well-documented reasons for this gap between the well-being of the people of the United States and that of the people of many other industrial but less affluent nations. These reasons include the biological and cultural inadequacies of our present governmental public health or preventive medicine and health care delivery systems; the shortcomings of our present educational systems; and our life-shortening ratio of one registered motor vehicle for every 1.6 human beings. There is also a little known and less understood factor that for the better part of this century has had a major role in the defective structuring and/or administration of our programs for environmental, public, and family health, education, and welfare. It is called scientific racism, and it remains a formidable determinant of the quality of all American life.

This book is about the origins, functions, and continuing human costs of scientific racism.

Brought into being exactly twenty-two years after the first battles of the American Revolution of 1776, scientific racism had only one function. That function was to preserve the type of social arrangements that had prevailed before the writing of the Constitution of the new United States of America made the establishment of equal justice and the promotion of "the general welfare" two of the basic functions of the state itself.

What the authors of the newborn American republic's constitution meant by the societal obligation to "promote the general welfare" ranged from the making of good health, a decent education, and human dignity the birthright of all newborn children to making certain that the total environment in which all classes of people lived, worked, and raised their families be kept biologically and mentally fit for human life.

The purposes of scientific racism, on the other hand, were to blame the hereditary endowments "in the blood" of the ill, the uneducated, and the poor for their chronic (and eminently environmental) infectious and hunger diseases, for their illiteracy, and above all else for their poverty. To the scientific racists, the poor and the near poor of all nations were—and still are—held to be a race apart, "a definite race of chronic pauper stocks."

In the countinghouse philosophy of scientific racism, the physiological, cultural, and economic woes of the poor and the nonaffluent "middling classes" were scientifically ordained by Nature, and therefore neither preventable nor reversible. As a result, the fiscal responsibility of the state was to avoid spending a farthing of public treasury funds on such naturally foredoomed efforts as programs of environmental and clinical health care for all citizens; and free education for the children of people who worked in mines, mills, or fields for a living; and legally mandated minimum sanitary standards for the human habitations, and the water and sewage systems, of the new industrial towns and cities; and laws to prevent the carnage of mine and factory accidents; and, among various other social acts to promote the general welfare, laws mandating minimum wages and maximum hours.

Scientific racism, like the steam-powered railroads and factories—and the massive environmental pollution produced by the coal they burned for energy—was new to the human experience, all of them being uniquely artifacts of the Industrial Revolution. Unlike the old gut racism—that is, the blind, unreasoning, xenophobic, and bigoted types of racism based on racial, skin-color, religious, political, geographic, linguistic, and cultural differences —scientific racism was not based on emotion and hate, but on love. What the creators of scientific racism loved, however, was money.

Miners and mill hands who humbly petitioned for living (higher) wages, or free education for their children, or lower rents for their hovels, were, clearly, acting contrary to the immutable Laws of Nature. However, since these humble poor were, by virtue of the hereditary provisions of the same scientific laws, *born* too unintelligent to understand that what they were asking for was the unpardonable waste of money, they were more to be pitied than scorned. Hatred was reserved for those traitors to their class and their wealth, such as the Utilitarian philosopher Jeremy Bentham, and the millionaire cotton-mill owner and socialist Robert Owen, and the addled bands of Quaker, Unitarian, and other Christian reformers, who took to print and pulpit to advocate precisely such unscientific squandering of sacred wealth.

Created in England, the birthplace of the Industrial Revolution, the original "enemies list" of scientific racism consisted solely of people who were as white, as Protestant, and infinitely more Anglo-Saxon than the members of the Hanoverian Succession who then, as now, occupied the throne of England. To be sure, many of the early proponents of scientific racism were also social anti-Semites, anti-Catholics, and white supremacists. It is even true, as we shall see in some detail in later chapters, that in due time the old-fashioned American gut racists were to make tragically effective use of the high-toned literature and lexicon of scientific racism in such non-economic efforts as the forging of the 1924 barriers against further Italian, Jewish, and other non-Nordic immigration to the United States. Nevertheless, regardless of the frequent and synergistic collaborations, over two centuries, between scientific racism and the forces of the much older xenophobic racism, scientific racism has been dedicated to very different goals for very different reasons.

Now, as in 1798, scientific racism remains color-blind and free of all racial, religious, and cultural biases. It is not concerned with people but, simply, with what is known as the maximization of profits and the minimization of taxes on these profits—particularly when these taxes are earmarked for promoting the health, education, and general welfare of the men and women whose labors make such profits possible in the first place.

Scientific racism did not, obviously, invent the reluctance to pay living wages and taxes for the promotion of the general welfare of an entire nation's people. It did, however, appear at a critical turn in history when the classic Scriptural excuses for greed, selfishness, and poverty were fast losing their traditional credibility. Scientific racism supplanted Scripture as the fount of "scientific" rationales for do-nothingism in terms of the promotion of the general welfare of the greatest numbers of the people of the rapidly industrializing nations of Europe and, shortly, North America.

Since its inception, scientific racism has not, of course, won all of its wars for the preservation of the life styles of the *ancien régime*. Despite the efforts of the old scientific racists (those who flourished between 1798 and the Nazi *Götterdämmerung* of 1945) universal and compulsory education was to become the rule and not the exception in Europe and North America by the first quarter of the twentieth century. Many life-redeeming environmental sanitary reforms were achieved, such as the national network of clean-water and sewage systems that were to end the recurring epidemics of Asiatic cholera that had started to plague all of England in 1831. These clean-water systems ended epidemic cholera in England by 1866—or eighteen years before Koch's rediscovery of the cholera bacillus was announced in 1884.[3] In most industrializing nations, the sanitary and civic reformers did win the passage of governmental health and fire regulations that made urban housing safer than it had been during the opening decades of the Industrial Revolution. Minimum-wage laws did raise wages to more viable levels, and the once standard eighty-hour workweek was reduced by law—with more than a little prodding by the trade unions—to biologically safer limits. In Europe and America, these social advances did reduce, in meaningful proportions, the prevalence of such preventable diseases as tuberculosis to considerably lower levels than had prevailed when the Industrial Revolution first changed the living environment of the toilers.

Nevertheless, all such governmental social programs for the betterment of human life were subjected to the constant opposition and sabotage of the greedy, whose taxes had to pay for them. Armed with the propaganda and the biological pseudolaws of scientific racism, the powerful opponents of state promotion of the general or human welfare were able to delay the enactment of many of the new environmental and social reform laws. When they were unable to prevent passage of such legislation, they often succeeded in sabotaging their administration and/or achieving the addition of budgetary and enforcement restrictions that, for generations, turned many of the first health, education, and general welfare advances into little more than dead letters on the code books of the law.

Although the biomedical, genetic, and behavioral sciences have, for nearly two centuries, exposed the pseudogenetic dogmas of scientific racism as utterly contrary to the realities of human biology, the social and political influence of the new scientific racism (the post-World War II version) is greater than any exerted by the extremely influential old scientific racism.

This is true not only in relation to social programs designed to promote the general welfare. It is equally true in such antisocial efforts as the forced sterilization of the poor in this country. Not since the early theoretician of scientific racism and founder of its eugenics wing, Sir Francis Galton, first called for the sterilization of the world's less affluent and less fortunate people in 1891 has this demand been as well heeded as it has been in the United States. Here, for the past five years, we have been experiencing a growing epidemic of forced sterilization of the poor that quite possibly dwarfs even the orgy of compulsory eugenic sterilization inflicted upon its own citizens by Nazi Germany between 1933 and 1945.[4]

Despite the lavish use of it by various regressive cabals in history, scientific racism itself was never a conspiracy. It has been, rather, a self-perpetuating system of perverting the facts of science—and of nature—to make these altered facts justify the status quo during two centuries of vast social changes. As a general rule, the statesmen, professors, journalists, and clergymen who, since 1798, accepted its claims and based their value judgments on them were not conspirators against the commonweal. Through the years, their numbers have included monarchists and republicans, socialists and fascists, liberals and reactionaries, anti-Semites and Jews, clergymen and atheists. Many of them were even innocents, men and women of goodwill who considered themselves to be just and moral people but who were too well educated (in everything but, alas, biology) to ignore the counterfeit laws of "human biological inheritance" broadcast by the scientific racists.

Far from having developed into a conspiracy in our own times, scientific racism stands as a classical example of the life-wrecking powers of a single bad idea. That idea, the central dogma of scientific racism, is the pseudogenetic simplism that everything about a person's condition in life—from his socioeconomic status and his educational achievement to his life span and the quality of his health—is immutably preformed in the genes he inherits from his parents at the moment of his conception. As William Shockley, a Nobel laureate in physics and the president (and founder) of the Foundation for Research and Education on Eugenics and Dysgenics (FREED), likes to put it, in the great Crap Game called Life, "babies too often get an unfair shake from a badly loaded parental genetic dice cup."[5]

THE FOUNDING FATHER OF SCIENTIFIC RACISM

The founding father of scientific racism, Thomas Malthus (1766–1834), the first professor of political economy in British university history, spelled out its purpose in the sixth edition (1826) of his famous *Essay on the Principle of Population:*

> We are bound in justice and honour formally to disclaim the *right* of the poor to support.
> To this end, I should propose a regulation to be made, declaring that no child born from any marriage, taking place after the expiration of a year from the date of the law, and no illegitimate child born two years from the same date, should ever be entitled to parish assistance. . . .
> The infant is, comparatively speaking, of little value to society, as others will immediately supply its place.

If governmental aid to the poor and to helpless infants roused Malthus' ire, every social measure that in any way improved the health or the minds of the population was, in his view, an even greater crime. In Book IV, Chapter 5, of the same edition of his *magnum opus,* Malthus wrote:

> All children born, beyond what would be required to keep up the population to this [desired] level, must necessarily perish, unless room be made for them by the deaths of grown persons.

Therefore, Malthus concluded:

> . . . we should facilitate, instead of foolishly and vainly endeavouring to impede, the operations of nature in producing this mortality; and if we dread the too frequent visitation of the horrid form of famine, we should sedulously encourage the other forms of destruction, which we compel nature to use. Instead of recommending cleanliness to the poor, we should encourage contrary habits. In our towns we should make the streets narrower, crowd more people into the houses, and court the return of the plague. In the country, we should build our villages near stagnant pools, and particularly encourage settlements in all marshy and unwholesome situations. *But above all, we should reprobate specific remedies for ravaging diseases; and those benevolent, but much mistaken men, who have thought they were doing a service to mankind by projecting schemes for the total extirpation of particular disorders* [italics added].

To Malthus, any measures that eased the lot of the greatest numbers of people—from sanitary reform and medical care to birth control and, above all else, higher wages—were not only immoral and unpatriotic but also against the laws of God and Nature. According to Malthus, all such social instruments for the improvement of individual, family, and public health shared the common and fatal "tendency to remove a necessary stimulus to industry."

The functions of scientific racism have not changed since the publication of the first edition of *An Essay on the Principle of Population* in 1798. These functions are:

1. To preserve what Malthus saw as *the* "necessary stimulus to industry": poverty. It was from poverty that cheap help and child labor flowed to the landed gentry and the new class of mine and factory owners for whom Malthus spoke.

2. To prevent the passage of all social legislation that threatened to diminish poverty by improving in any way the lot of most people. Such "bad" laws included those that provided welfare and medical assistance to keep the poor, the aged, and the sick alive; free public schools for all children; and sanitary, housing, and public hygienic standards and community services that enabled people born into poverty to achieve at least the hope of upward mobility via better living conditions.

Whenever it was not possible to prevent the enactment of any social legislation, the Malthusians and later scientific racists learned to settle for the perversion and/or the sabotage of potentially beneficial statutes—such as, to cite a classic example, the Poor Law Act of 1834, which Malthus and his disciples in and out of Parliament perverted from a law to bring relief to the victims of poverty into an instrument that conferred virtual serfdom upon them. The well-intended British Act to Extend and Make Compulsory the Practice of Vaccination, passed in 1853, was turned into a dead letter by the failure to vote any funds for the manufacture or administration of the smallpox vaccine. And when, in 1871, the British Parliament and local governments appropriated 20,000 pounds sterling for "providing the people of England with gratuitous vaccination," the *British Medical Journal* observed with dismay that, at the same time, "in Ireland, only 400 pounds per year is spent on the same object," a bit of *Realpolitik* that proved particularly deadly during the smallpox epidemics of 1871, 1872, and 1873.[6]

The prevention of the enactment of medically realistic vaccination laws, or the evisceration of such weak vaccination laws that manage to find their way to state or federal codes, remains to this very day one of the prime activities of the new scientific racism. If this rich nation does not have, as do England, Denmark, China, the Netherlands, and many other modern nations, national programs that guarantee every newborn child all of the immunizations against all of the major viral and bacterial infectious diseases for which safe vaccines are now available, we can thank the generations of the teaching of the pseudohereditary theories of disease as legitimate science in the colleges and universities that produced this century's Presidents and lawmakers, as well as the teachers and writers who influenced their thinking. Malthus had written, in his second edition (1803), that, according to "the laws of nature, which are the laws of God" any man unable to find work had to understand that "at nature's mighty feast there is no vacant cover for him" and that any people who out of "compassion" lend any aid to such undeserving poor endanger the greater society.

The vaccination of the children of those poor people who were doomed, even before birth, to carry the "bad blood" that made such immunizations a sheer waste of public funds, was therefore very poor public policy. And productive of higher taxes to boot.

Greed, of course, preceded scientific racism by many millennia. With or without Malthus and his contributions to Western civilization, people of means would still hate to pay the taxes required to fund effective legislation for human welfare. However, by demeaning the intrinsic worth to society of most newborn children and their families, as well as by denigrating the value to the greater society of any and all human-welfare legislation, scientific racism and its propagandists have for two centuries served successions of venal men and pinchpenny governments. They performed this function by providing "scientific" excuses for not undertaking any measures that would protect their populations from easily preventable diseases and disorders of body and mind.

With the advent of two of scientific racism's major English prophets and theoreticians, Herbert Spencer, the creator of the intellectual aberration called Social Darwinism, and Sir Francis Galton, the inventor of eugenics, two additional functions were added to the portfolio of world scientific racism. Spencer and Galton institutionalized scientific racism by the since common techniques of counterfeiting biological laws and by misinterpreting and mis-using legitimate scientific discoveries and methods to serve ends that, in reality, bore no relation whatsoever to the sciences.

What had been just plain larceny and slumlordism, before Spencer's Social Darwinism sanctified greed and aggression as merely the "survival of the fittest," now became the inevitable and biologically ordained fulfillment of evolutionary destinies. Spencer held it to be contrary to the "general truths" of biology for any society to enact laws mandating safety standards for human housing, clean-water systems, effective sewage systems, mine and factory safety regulations, and minimum-wage and working-hours regulations. Such social legislation, according to Spencer, acted against his perversions of the Darwinian laws of natural selection by "the artificial preservation of those least able to take care of themselves."

Spencer's nineteenth-century American admirers included the railroad tycoon James J. Hill, who defended the ruthlessness of the new railroad monopolies by writing that "the fortunes of railroad companies are determined by the law of the survival of the fittest." In this he was echoed by John D. Rockefeller, Sr., who told a Sunday-school class that "the growth of a large business is merely a survival of the fittest . . . merely the working-out of a law of nature and a law of God." Andrew Carnegie wrote that in Spencer's writings he had finally "found the truth of evolution."[7]

Spencer's twentieth-century American devotees have ranged from the ultraliberal sociologist Edward A. Ross to the apostle of Rugged Individualism, the one-term President Herbert C. Hoover. In our own times, University of Pennsylvania (formerly Harvard) professor Edward C. Banfield, one of President Nixon's senior advisers on the Model Cities Program enacted to aid those victims of urban poverty "least able to take care of themselves," is possibly the foremost contemporary exponent of the views long associated with Social Darwinism.

What had been ill-mannered and even immoral and un-Christian bigotry before the coming of Galton's eugenics now became race hygiene and the preservation of "the race" against the awesome specter of what was known in the nineteenth century as "racial degeneration" and is better known today as "genetic enslavement." The roster of scientific disciplines and discoveries perverted by the eugenicists to justify the aggressions of old-fashioned Irish baiting, anti-Semitism, and Jim Crow range from nineteenth-century anthropometry to physiology, Darwinian evolution, Mendelian genetics, Binet's mental tests, the morbidity and morality statistics on infectious and deficiency diseases, and such socially useful contemporary studies as the 1966 study "Equality of Educational Opportunity" under the principal authorship of Professor James S. Coleman of Johns Hopkins University.

Three classic twentieth-century uses of such eugenic "scholarship" to influence legislation and social practices in America and England will be described in detail elsewhere in this book. They were, in brief:

1. The employment, by a U.S. Congress whose staff included an official Expert Eugenics Agent, of the Army intelligence test scores of World War I as the ultimate "scientific" verification of the hereditary mental inferiority of the Jews, Italians, Poles, Hungarians, Spaniards, and other non-Anglo-Saxon Protestant racial and ethnic groups. The further immigration of such genetically inferior groups was subsequently restricted to minuscule token quotas by the U.S. Immigration Act of 1924. Less than a decade later, millions of the Jews and non-Jews barred from sanctuary in this country by these 1924 quotas were to be trapped in Nazi-occupied Europe and exterminated in the race hygiene (the German word for eugenics is *Rassenhygiene*) camps from Auschwitz and Buchenwald to Dachau to Treblinka. (See chapters 12 and 15.)

2. The frank attempt by Galton's anointed heir and prophet, the mathematician and socialist Karl Pearson, to force Parliament to enact similar anti-Semitic immigration barriers. The vehicle in which Pearson made this effort was the first issue of his new *Annals of Eugenics,* in October 1925, which opened with his exhaustive eugenics study, "The Problem of Alien Immigration into Great Britain, Illustrated by an Examination of Russian and Polish Jewish Children." In this investigation, Pearson found in the IQ test scores and the anthropometric measurements of the relative roundness or longness of the heads of the children of Jewish immigrants living in London all the mathematical and scientific proof he needed to validate Galton's 1884 judgments that the accursed Jews were indeed a hereditary race of national parasites not worthy of being allowed to live in England.

3. The revival of the moribund myths of racial intelligence in 1954, after the Supreme Court held the denial of educational equality to black children to be unconstitutional. The racial intelligence myths were revived at that time by the educated southern gut racists who mounted a still growing and on the whole successful attack on the court-ordered desegregation of public schools in America. The southern revivalists did not, like the Ku Klux Klan, take such actions against educational equality in the name of white

supremacy. On the contrary, they screamed for the blood of Chief Justice Earl Warren and his "Black Monday" Court because "science" had proven— via the ever so scientific IQ test scores—that the black children were by genetic endowment the intellectual inferiors of the white children.

In addition to the preservation of poverty and the conversions of old-fashioned greed and bigotry into popular modern "sciences," scientific racism has also from its inception had an ultimate and historically obvious function. This function is to perform the same violence to all legitimate life and behavioral sciences that, as described in Gresham's Law, bad money inflicts upon good money in the economy. This crowning function of pseudoscience was noted well over a century ago by one of Thomas Malthus' arch political enemies, the sanitary reformer Sir Edwin Chadwick.

Over four decades before Pasteur and Koch made the first of the discoveries of the bacterial and viral disease agents that lurked in stagnant water, crowded tenements, backyard accumulations of human excrement and garbage, and other pungent sources of environmental filth, the barrister Chadwick and many British physicians, nurses, and civic reformers saw in such filth the sources of *preventable* contagious and epidemic diseases of urban and rural life in a rapidly industrializing England. Yet even then there were physicians, statesmen, and authors such as Malthus who attacked Chadwick's proposals to make the industrializing urban environments safe for human life as futile and fatuous wastes of public tax monies that would in no way reduce fevers and contagions such as cholera, typhoid, typhus, tuberculosis, scarlatina (scarlet fever), measles, and smallpox.

In his famous report to Parliament in 1842, *The Sanitary Condition of the Labouring Classes,* Chadwick noted that the effects of such pseudomedical controversies over the causes of "fever" were "prejudicial in diverting attention from the *practical means of prevention"* (Chadwick's emphasis). As Chadwick complained, "the great preventives, drainage, street and house cleaning by means of supplies of [clean] water and improved sewerage, and especially the introduction of cheaper and more efficient modes of removing all noxious refuse from the towns," were all needlessly jeopardized, sabotaged, and even aborted by such pseudoscientific obstacles.

During our own times, this employment of the pseudoscientific arguments and data against all of the techniques and tools that the legitimate sciences have since 1842 developed and continue to develop as "practical means of prevention" of literally millions of cases of diseases of body and mind, emerges beyond all doubt as the most serious of the real and pressing dangers inherent in the philosophy and practice of all forms of scientific racism.

WHO IS VICTIMIZED BY SCIENTIFIC RACISM?

Since 1954, possibly because of the way they managed to revive the IQ test score controversy of the 1920's, the spokesmen for eugenics and other forms of scientific racism have succeeded in making it seem as if the whole problem of scientific racism is one that can be stated in very simple and self-limiting terms. To wit, that it presents the decent people of our society with the purely moral problem of the speed at which we should make up to our nonwhite minority for the traumas of chattel slavery and the handicaps of Jim Crow segregation and inferior education that followed the abolition of slavery.

As we shall see in this book, the issues raised by the new wave of scientific racism are not of a nature that can be either prevented or resolved by the busing of black children to formerly all-white public schools. The basic dangers of the new scientific racism are, actually, directed against the physiological and mental health of *all* Americans—starting with that majority of white, Anglo-Saxon, Protestant (WASP) Americans who happen to have been born too poor to be able to give to their children the same adequate levels of immunization, clean air and water, and other modalities of preventive pediatric medicine that are now restricted, for economic reasons, to an affluent *minority* of American families.

It is obviously high time that most of us became aware of the social and political history, as well as the functions, of scientific racism. For only through such knowledge can we begin to understand just who among us are affected by scientific racism. More importantly, perhaps, we have to understand *how* scientific racism affects the lives of millions of us.

History provides many leads to the answers to such questions. For example, the success of the scientific racists of the 1920's in convincing the Congress and the President of the United States of the validity of their pseudogenetic interpretations of the Ellis Island and Army induction center intelligence test scores of European immigrants between 1913 and 1919 resulted in much more than the anti-Jewish, anti-Italian, and anti-Catholic immigration quotas of 1924. Like a spreading cancer, this success also led to over a half century of basing our local and state educational systems on the same myth that IQ test scores measured *hereditary* capacities for learning. According to this myth three out of every four American schoolchildren were by genetic endowment simply incapable of absorbing any education beyond the last grade of elementary school. (See chapter 11.)

Black children deprived of equal opportunity to attend the best public schools in their communities are far, far from being the only educational victims of scientific racism in America. As we shall see in this book, all children, most of them white, educated in the public elementary and high schools of America since 1920 have paid the intellectual penalty of being served in school systems funded by legislators who themselves were taught in their universities that 75 percent of all students in this country are genetically ineducable beyond the eighth grade. And who believe in their hearts that most of the tax money spent on the education of three in every four public elementary and high school students of all races and ethnic backgrounds is appropriated only to court the votes of ethnic and liberal pressure groups.

All children, the majority of them white and WASP, whose eyes, ears, limbs, nervous systems, and brains are permanently damaged and disabled by prenatal and preschool exposure to rubella and measles viruses, *for which effective vaccines already exist,* are full-fledged victims of scientific racism.

Every one of the children suffering from preventable infectious, deficiency, and other environmental diseases that permanently handicap children is a certified victim of the new scientific racism. (As of 1972, according to the National Institute of Neurological Diseases and Stroke, these overwhelmingly avoidable disorders included 5,000,000 cases of mental retardation; 1,500,000 cases of epilepsy; 760,000 cases of auditory handicap; 345,000 cases of visual handicap; 555,000 cases of cerebral palsy; and the speech disorders suffered by "two to five percent of all children.") The National Advisory Neurological Diseases and Stroke Council estimates "that approximately 20 million individuals in the United States have handicaps or defects which fall within this general category."[8] Every child, every mother, every adult who dies for lack of a sufficient supply of doctors in this nation, is and will continue for decades to be a victim of the new scientific racism.

Every human being who—because of *preventable* diseases and disorders of our modern environment—is rendered unemployable, and chronically in need of public welfare agency assistance for survival, is and will continue to be a needlessly handicapped victim of scientific racism.

Every taxpayer forced to meet the increased medical, custodial, and social costs of keeping such needlessly unemployable victims of preventable disabilities alive is and for decades to come will remain a victim of the new scientific racism.

Every American who, after a lifetime of hard work and the weekly payment of social security taxes, is warehoused in one of the thousands of highly lucrative extermination *Lager* called nursing homes for senior citizens because he or she was born and lived in a society with grossly inadequate and inhuman social legislation, is a victim of the direct impact of all types of scientific racism on the value systems of every generation of American lawmakers, teachers, and mass-media managers born in the twentieth century.

In short, then, what scientific racism is all about is nothing less than the total quality, and duration, of *all* American life, white and nonwhite.

Of all branches of scientific racism that arose since Malthus started it all, few have been more damaging to human life—or offer more real and immediate dangers to contemporary America—than the cult of eugenics founded by Sir Francis Galton.

GALTON OF THE GONADS

The heir, at twenty-two, to a comfortable fortune, Francis Galton (1822–1911) could hardly be blamed for confusing nepotism with biological heredity. Galton's first and most famous book, *Hereditary Genius* (1869), was an actuarial study of prominent men in government, religion, commerce, and the arts which proved, with redundant statistics and dozens of quaint notions about human development, what every adult has always known. To wit, that the children of bankers and generals and cabinet ministers are statistically much more likely to find their way into the professions and the corridors of political and economic power than are the children of charwomen, peasants, and ditch diggers.

To Galton, whose ignorance of socioeconomics was matched only by his painful ignorance of human biology—an ignorance made all the more appalling because Galton was in his fourth year of medical school when he came into his inheritance and promptly quit school—the reasons for the well-known tendency of the children of the mighty to take over their ranks were purely "in the blood."

At the time Galton wrote his book, the mounting sanitary and medical benefits of the Industrial Revolution had been enjoyed, for over a generation, by the more affluent elements of British society. These health benefits were, in turn, directly responsible for the steep and swift decline in the infant- and maternal-mortality rates of the affluent families. Then, as now, falling infant-mortality rates were followed by a sharp and proportionate decline in the live-birth rates of the fortunate families who were the first to benefit from the new theories and practical applications of environmental hygiene.

Galton mistook this decline in the birth rates of the affluent families in British society as evidence of the "sterility" that he claimed reduces the birth rates of people of superior blood who move from the country to the city. At the same time, Galton saw the lag in the decline of the birth rates of the lower classes—a lag related directly to the fact that the depressed classes had as yet received almost none of the sanitary and medical benefits of the Agricultural and Industrial revolutions—as a population explosion of inferior strains of white Englishmen that threatened to swamp what Galton saw as the most suitable of all strains of the Anglo-Saxon race: his own. The specter of the more suitable children Galton never got around to fathering being dominated by the corporeal but, in Galton's opinion, hereditarily unsuitable children of the nation's miners and mill hands, drovers and servants, gnawed at the eminent Victorian until he came up with his famous solution to this "problem."

The present name for this solution did not come as easily to Galton as did its nature. As he ultimately defined the "eugenic questions" of the "race," they were:

> . . . questions bearing on what is termed, in Greek, *eugenes,* namely, good in stock, hereditarily endowed with noble qualities. This, and the allied words, *eugeneia,* etc., are equally applicable to men, brutes, and plants. We greatly want a brief word to express the science of improving the stock, which is by no means confined to questions of judicious mating, but which, especially in the case of man, takes cognisance of all the influences that tend in however remote a degree *to give the more suitable races or strains of blood a better chance of prevailing speedily over the less suitable* [emphasis added] than they otherwise would have had. The word *eugenics* would sufficiently express the idea.[9]

Thus, from the Greek word for "wellborn," Galton coined the word "eugenic," meaning pertaining to racial improvement by boosting the birth rate of the wellborn to the levels where they *speedily* prevailed over the less suitable strains or socially less wellborn classes.

While he was at this lexicographic crossroads, Galton also created the word for the opposite of "eugenic." The word was "kakogenic," derived from the Greek word *kakos,* meaning "bad." "Kakogenic" (or "cacogenic" or its much more commonly used synonym, "dysgenic") means "of low birth and tending towards, or productive of, racial degeneration."[10]

With these two words, Galton now brought into being the cult of eugenics, which today is recognized as having approximately the same relationship to the legitimate biological science of genetics that astrology bears to astronomy, or numerology to mathematics. "He had in view," wrote Karl Pearson, Galton's foremost interpreter, "eugenics not only as a science, not only as an art, but also as a national creed, amounting, indeed, to a religious faith."[11]

As a science, eugenics would deal with the factors, as yet admittedly unknown to Galton, that in his view tended to improve or impair superior racial breeding stock in our species.

As a religion, eugenics was to provide the moral and spiritual motivation to encourage increased fecundity in families of Anglo-Saxon, noble, wellborn, affluent (Galton always equated fat bank balances with the noblest of all human qualities), and thus superior human breeding stock.

To the palpably class-conscious Francis Galton, only the breeding successes of the eugenics movement could prevent the superior hereditary qualities of the "race" from being overwhelmed by the rising tides of equally white, equally Anglo-Saxon, equally Protestant Englishmen of inferior heredity and bank balances. As the high priest and theologian of this Victorian racist cult, Galton even established its very "scientific" scale of racial values.

In this scale, all men were graded by letters. The highest grade was X, the lowest A, the grades representing what Galton deemed to be the noble and bad qualities of human beings. In this mathematicized scheme of values, the higher the grade, the rarer its incidence. For example, Galton wrote, whereas "C implies a selection of 1 in 16, or somewhat more than the natural abilities possessed by the average foreman of common juries, D is as 1 in 64— a degree of ability that is sure to make a man successful in life." The V-class man appears once in every 300 births, while the individuals born into the ultimate X class were so rare that Galton decided to "avoid giving any more exact definition of X than as a value considerably rarer than V."[12]

Since in Galton's theology black people were as an entire race grossly inferior to even the lowliest of any white people, Galton added that "classes E and F of the negro may roughly be considered as the equivalent of our C and D—a result [sic] which again points to the conclusion that the average intellectual standard of the negro race is some two grades below our own."[13]

Galton did not need any IQ test scores to tell him this much about the intelligence of the blacks. Nor, for that matter, did the founder of eugenics need any IQ test score data on October 17, 1884, when he made the discovery that "the Jews are specialised for a *parasitical* [Galton's emphasis] existence upon other nations, and that there is need of evidence that they are capable of fulfilling the varied duties of a civilised nation by themselves."[14]

Although, as early as 1912, the fervent adherents of Galton's eugenic creed in England were to include both the future Prime Minister Winston Churchill and Harold Laski, who was to be the chairman of the Labour Party in the election that turned Churchill out of office in 1945, the two countries in which the eugenics movement was to have its greatest influence were the United States and Nazi Germany, where eugenics became an integral part of the "national creed, amounting, indeed, to a religious faith."

It is, in fact, almost impossible to estimate the enormous influence that Galton's brand of scientific racism was to have in this century on the educated and patrician classes. American Presidents proved to be particularly receptive to the dogmas and political programs of the eugenicists.

An early American convert to the eugenics creed, President Theodore Roosevelt, sent a most revealing communication to the Committee to Study and to Report on the Best Practical Means of Cutting Off the Defective Germ-Plasm in the American Population, a committee organized by the American Breeders' Association's Eugenics Section. In this letter of January 14, 1913, the twenty-sixth President of the United States declared, in words that clearly paraphrased Galton's dicta of 1883, that "it is obvious that if in the future racial qualities are to be improved, the improving must be wrought mainly by favoring the fecundity of the worthy types . . . At present, we do just the reverse. There is no check to the fecundity of those who are subnormal. . . ."

Between the presidencies of Theodore and Franklin Roosevelt, we were led by Presidents who, guided by the preachments of Galton and his astonishingly influential American disciples, advanced, sponsored, signed into law, and preserved legislation and policies aimed at the ultimate extinction of humanity's "less suitable races or strains," such as the Jews, Italians, Hungarians, Poles, Russians, and other non-Nordic immigrants; the native-born white Anglo-Saxon Protestant poor; and the even poorer native-born black Americans. These legislative acts of commission and omission ranged from immigration restriction and compulsory eugenic sterilization laws to the failure to support and enact critically needed public health programs such as universal and compulsory vaccination measures.

Few societal actions demonstrate the direct influence of Galton's brand of scientific racism on twentieth-century governments more tragically than do the seven decades of forced sterilization of men, women, and children for the crime of being born poor in America. It was the eugenics movement that, in the United States and later in Nazi Germany, prompted state and national governments to make sterilization their weapon of choice against what the scientific racists called "the menace of racial pollution."

Galton's lifelong obsession with the gonads of other people led him to call for some new means of checking the fertility of what, in an address to the Seventh International Congress of Hygiene and Demography in 1891, he described in the phrase "overgrowth of population."

His disciple Pearson wrote: *Galton is here foreshadowing the sterilisation of those sections of the community of small civic worth,* which has since become a pressing question of practical politics" (italics added).[15]

In 1907, Galton's gonadal obsession became so pressing a practical political question in Indiana that it led to the passage of the world's first compulsory sterilization law. Under this law, the state of Indiana voted itself the authority to make compulsory the sterilization of all "confirmed criminals, idiots, rapists, and imbeciles" who were confined in state institutions "entrusted with the care" (*sic*) of such people. Subsequently, between the Indiana compulsory sterilization law and today, some thirty sovereign states and the colony of Puerto Rico passed equally Draconian forced sterilization laws.

Most of the compulsory state laws were based on the Model Eugenical Sterilization Law drafted primarily by Harry H. Laughlin, superintendent of Davenport's Eugenics Record Office and co-editor of the *Eugenical News*. The model gelding bill was written some years before it was published in Laughlin's book *Eugenical Sterilization in the United States* in 1922, at which time he was also the official Expert Eugenics Agent of the U.S. House of Representatives Committee on Immigration and Naturalization. The Model Eugenical Sterilization Law called for every state to appoint a State Eugenicist, who would direct the compulsory sterilization of all people he judged to be members of the "socially inadequate classes."

According to the model law, "the socially inadequate classes, *regardless of etiology or prognosis,* are the following: (1) Feeble-minded; (2) Insane (including the psychopathic); (3) Criminalistic (including the delinquent and wayward); (4) Epileptic; (5) Inebriate (including drug-habitués); (6) Diseased (including the *tuberculous,* the syphilitic, the leprous, and others with chronic, *infectious,* and legally segregable diseases); (7) Blind (including those with seriously impaired vision): (8) Deaf (including those with seriously impaired hearing); (9) Deformed (including the crippled); and (10) Dependent (including *orphans,* ne'er-do-wells, the *homeless,* tramps, and *paupers*)" (italics added).[16]

Not every state law included all of these conditions as reasons for eugenic sterilization, but all of them included epilepsy, "feeblemindedness," and "criminality" among the offenses calling for forced sterilization. Between 1907 and 1964, a total of 63,678 people suffered compulsory sterilization in the thirty states and one colony that had passed such laws. Of this number, 33,374, or 52.4 percent of the total, were sterilized against their will for being adjudged feebleminded or mentally retarded, which in most of these states was defined as having an IQ test score of 70 or lower. These victims of Galton's obsessive fantasies represented, however, the smallest part of the actual number of Americans who have in this century been subjected to forced eugenic sterilization operations by state and federal agencies.[17]

As Federal District Judge Gerhard Gesell declared in 1974 in a case brought on behalf of poor victims of involuntary sterilization performed in hospitals and clinics participating in federally funded family-planning programs alone: "Over the last few years, an estimated 100,000 to 150,000 low-income persons have been sterilized annually under federally funded programs."[18]

This, of course, equals the rates at which poor people were subjected to compulsory sterilization in Nazi Germany, where two million Germans were sterilized as social inadequates during the twelve years of the Third Reich. The German Sterilization Act of 1933 was derived quite openly from Laughlin's Model Eugenical Sterilization Law.

Judge Gesell, in his landmark opinion of 1974, said:

Although Congress has been insistent that all family planning programs function on a purely voluntary basis, there is uncontroverted evidence in the record that minors and other incompetents have been sterilized with federal funds and that an indefinite number of *poor people have been improperly coerced into accepting a sterilization operation under the threat that various federally supported welfare benefits would be withdrawn unless they submitted to irreversible sterilization.* Patients receiving Medicaid assistance at childbirth are evidently the most frequent targets of this pressure, as the experience of plaintiffs Waters and Walker illustrate. Mrs. Waters was actually refused medical assistance by her attending physician unless she submitted to a tubal ligation after the birth. [Italics added.][19]

Judge Gesell, who knows his history well, observed that "the dividing line between family planning and eugenics is murky," and ruled that the inadequate provisions to protect patients are "both illegal and arbitrary because they authorize involuntary sterilizations, without statutory or constitutional justification."

THE CURRENT EPIDEMIC OF FORCED STERILIZATIONS

The compulsory sterilizations denounced by Judge Gesell do not begin to include all of the involuntary gonadal surgery presently being committed against equally poor people by nonfederal state and voluntary agencies. In 1974, the Association for Voluntary Sterilization, Inc., estimated that 936,000 American people—538,000 (57 percent) males, and 398,000 females—had been surgically sterilized during 1973. This represented a drop of 166,000 from the total of 1,102,000 sterilizations in 1972—but a jump of 13 percent in the *estimated* number of females sterilized. If what a high government official involved in these matters told me off the record proves correct, possibly another 250,000 sterilizations are disguised in the hospital records as hysterectomies in which female sterilization is an unavoidable side effect.[20]

In ethical medical practice, a hysterectomy—the removal of the uterus —is performed only when its tissues become damaged, diseased, or malignant, and not to achieve sexual sterilization. Medically, therefore, to perform a hysterectomy in order to sterilize a female is, observed the president of the American College of Surgeons, Dr. Charles McLaughlin, "like killing a mouse with a cannon."[21] It is also much more lethal than simple sterilization operations to women, since at present some 12,000 deaths a year occur among women receiving hysterectomies.[22] Nevertheless, since the current revival of Galton's nineteenth-century crusade for surgical sterilization was launched, the hysteria unleashed by the new Malthusian population extremists has caused the annual number of hysterectomies to rise, so that this operation is second only to the appendectomy as the most frequently performed surgical procedure in America.

The chief victims of medically needless hysterectomies are, predictably, Galton's original targets: the lower-middle-class and working-class families who cannot afford the costs of proper medical care, and the unemployed. By

1975, according to *The New York Times:* "In New York and other major cities, a hysterectomy which renders a patient sterile costs up to $800, while a tubal ligation [the tying off of the Fallopian tubes], which does the same thing, pays only $250 to the surgeon, increasing the motivation to do the more expensive operation. Medicare, Medicaid and other health plans for the poor and the affluent both will reimburse a surgeon up to 90 percent for the costs of any sterilization procedure, and sometimes will allow nothing for abortion. As a consequence, 'hyster-sterilizations'—so common among some groups of indigent blacks that they are referred to as 'Mississippi appendectomies'—are increasingly popular among surgeons, despite the risks."[23]

The officially reported hospital sterilization operations represent about half of the actual sterilizations performed each year in America. Most vasec-tomies—male sterilizations—are performed under local anesthesia in doctors' offices, rather than in hospital operating rooms.

The actual number of Americans who are each year sterilized voluntarily, or under threats of government reprisals, is not known. In his 1973 annual report, Dr. H. Curtis Wood, Jr., the medical consultant for the Association for Voluntary Sterilization, Inc., estimated that, between 1960 and 1972, "the increase in sterilizations in the United States [was] from around 100,000 per year to 1 or possibly even 2 million a year." In 1974, Robert E. McGarrah, Jr., the attorney for the Public Citizen's Health Research Council of Wash-ington, D.C., wrote that "at present rates, a total of two million people un-dergo surgical sterilization each year and the federal government alone is estimated to have paid for 100,000–150,000 low income sterilizations an-nually."[24]

Given the hard data that are available, it is quite possible that the esti-mate of two million surgical sterilizations per year is on the very conservative side. However, in the absence of easily obtainable factual documentation, it must remain only an estimate.

Not all sterilizations performed in the United States are involuntary. In 1974, federal agencies estimated, according to *The New York Times,* that "at least one American couple in six in the main child-bearing years—age 20 to 39—has had a sterilization operation" for purposes of contraception.[25] On the other hand, a survey of surgical sterilizations in American and Canadian teaching hospitals published in 1970 by Johan W. Eliot and his associates[26] showed that what is known in obstetrical circles as the "Package Deal"—a variation of the form of compulsory sterilization denounced by Judge Gesell in the cases of "plaintiffs Waters and Walker"—is now endemic in North American hospitals. As Eliot et al. described it: "Some women desiring an abortion *were required to have simultaneous sterilization as a condition of approval of the abortion* in from one-third to two-thirds of these teaching hospitals in different regions of the country. This practice was most common in the Mountain States, the Far West, and Canada, and lowest in New England and the Plains States. In all, 53.6 percent of teaching hospitals made this requirement for some of their patients." (Italics added.)

Betty Sarvis and Hyman Rodman, commenting on the above statement in their book *The Abortion Controversy* (1973), write: "Considering the illegality of the requirement, it is likely that an even higher percentage of hospitals sometimes insist upon sterilization as a condition of abortion. . . . The large number of hospitals acknowledging that they sometimes practice this form of compulsory sterilization indicates that it is not uncommon and that it has been thoroughly rationalized within the medical profession."

A federal measure aimed at forcing more poor people to have "voluntary" sterilization went into effect on January 10, 1975. As of that date, under the revised Code of Federal Regulations governing Section XIX (Medicaid) of the Social Security Act, the national government would now provide a flat 90 percent of the medical costs of surgical sterilizations suffered by poor or medically indigent people under Medicaid in all fifty states—but would, henceforth, provide only matching funds of the costs of abortions (which do not sterilize women) performed on people eligible for Medicaid. This meant that, depending on their current rates of assistance to their citizens, the individual states would be reimbursed for from 50 to 81 percent of the costs of hospital abortions. Enacted in a time of rising inflation and swelling unemployment, this new federal regulation gave local health agencies and hospitals compelling economic incentives to manipulate Medicaid patients to choose "voluntary" sterilization as their method of family planning.[27]

From all present information, it is not unreasonable to fear that upward of half of the more than one million sterilization operations performed upon American men and women yearly are quite possibly involuntary. In the absence of a full-scale congressional investigation, with an adequate research staff properly funded and armed with full subpoena power, and whose hearings are open to the mass media and the public, the actual numbers of compulsory sterilizations committed against helpless people in the United States will remain a grim and tragic secret from the rest of us—whose taxes happen to pay for a majority of these forced mutilations.

THE HISTORICAL ROOTS OF TODAY'S FORCED STERILIZATIONS

Whatever the actual number of the victims of involuntary and irreversible sterilizations (probably at least 200,000 Americans per year), there can be no doubt that they owe their surgical mutilations to Sir Francis Galton's paranoiac obsession with the gonads of the poor.

The First International Congress of Eugenics, held at the University of London in 1912—whose vice-presidents included the Right Honorable Winston Churchill, First Lord of the Admiralty; Dr. Charles B. Davenport, director of the Eugenics Record Office and secretary of the American Breeders' Association; Dr. Charles W. Eliot, president-emeritus of Harvard University; Dr. David Starr-Jordan, president of Stanford University; and Gifford Pinchot, a future governor of Pennsylvania—announced as one of its goals the "Prevention of the propagation of the unfit by segregation and sterilization."

The Second International Congress of Eugenics, held in New York in 1921, whose sponsoring committee included Herbert Hoover, then Secretary of Commerce, as well as the presidents of Clark University, Smith College, and the Carnegie Institution of Washington, advanced the cause of the compulsory gelding of the poor in many ways.

The program of that Second International Congress of Eugenics not only included a paper, "The Present Status of Eugenical Sterilization in the United States," by Harry H. Laughlin. It also boasted a highly influential paper by E. J. Lidbetter, entitled "Pedigrees of Pauper Stocks," in which he presented the results of his eugenic study of the poor-white subpopulation in the East End of London whose status was described by the term "pauperism."

What Lidbetter discovered among the white Nordic poor of London was, he and his eugenics peers agreed, "scientific" proof of Galton's postulate that "pauperism" was the product of "bad blood." His conclusions, as presented to the International Congress of Eugenics, included the "findings":

"1. That there is in existence *a definite race of chronic pauper stocks* (italics added). . . .

"2. That modern methods of public and private charity tend to encourage the increase of this class. . . .

"5. That the reduction of this class may be brought about by a due observance of the laws of heredity . . . and that reduction may become progressive in proportion as our knowledge grows."[28]

In short, the poor of all nations constitute separate and distinct *races* of chronic pauper stocks, who are by heredity mental and physiological degenerates, and who have to be wiped out by steadily expanding programs of forced gonadal surgery.

At the Third International Congress of Eugenics, also held in New York, in 1932, the visiting president of the British Eugenics Society, Sir Bernard Mallett, in his oration to the faithful, "The Reduction of the Fecundity of the Socially Inadequate," called for the enforced extermination of what he described as this "definite race of chronic paupers, a race parasitic upon the community, breeding in and through successive generations." The poor, Sir Bernard claimed, as a group are those whose "anti-social characteristics are the result, mainly, of inferior heredity, and its fertility is higher than that of any other social element."

Therefore, concluded the head of the British Eugenics Society, "it is only to sterilization that we can look to limit the fertility of mental defectives and of those classes composing the Social Problem Group [i.e., the 'definite race of chronic paupers']."[29]

This call for the forced sterilization of the poor and the rapidly escalating numbers of the world's unemployed—the Third International Congress of Eugenics was held in Year Three of the Great Depression—was taken up by the solvent physicians, professors, and businessmen in attendance. Dr. Theodore Russell Robie, of the Essex County Mental Hygiene Clinic in New Jersey, in his presentation, "Selective Sterilization for Race Culture,"[30] called, for example, for the sterilization of, at the very minimum, the 14,000,000 Americans who had racked up low intelligence test scores since World War I. In a blazing peroration compounded of equal elements of Malthusian demography, Galton's gonadal nightmares, the new IQ test score mystique, and

Spencer's Social Darwinism, Dr. Robie told the assembled true believers and unemployed college graduates with nothing else to do in 1932:

> . . . there are those who believe that our population has already attained a greater number than is necessary for the efficient functioning of the race as a whole. Certainly our present picture of millions of unemployed would point to the belief that this suggestion is a reasonable one. It would undoubtedly be found, if such research was possible, that a major portion of this vast army of unemployed are social inadequates, and in many cases mental defectives, who might have been spared the misery they are now facing if they had never been born. It would certainly be understandable how many of them would prefer not to have been born, if they could have known what was in store for them on this earth where the struggle for existence and the urge toward the survival of the fittest makes it necessary for all those who would survive to possess a native [genetic] endowment of at least average intelligence.

Speaker after speaker, including Lena Sadler, an irascible Chicago surgeon, and the California millionaire Eugene S. Gosney, president of the Human Betterment Foundation of Pasadena, echoed Dr. Robie's call for the compulsory sterilization of the millions of Americans whom the raging Depression had thrown out of work since the Wall Street crash of 1929.[31]

Nor were these merely the ravings of some lunatic fringe of the greater society. Drs. Robie and Mallett and Sadler spoke for the *majority* of the educated people of this country in 1932. Two generations of the teaching of Social Darwinism and eugenics as legitimate sciences in most leading American universities had, by 1932, turned Galton's gonadal obsession into the political solution of choice for thousands of *educated* Americans, many of them in high political, editorial, and academic office. Many of the college graduates of that era who had been thoroughly brainwashed by such "science" during the critical years in which their lifelong value systems had been formed were, evidently, also the teachers of the succeeding generations of legislators, opinion molders—and voting citizens.

Of all the noble professions, that of medicine seemed to be particularly susceptible to the punitive dogmas of eugenical sterilization of the poor. A 1972 study made by the Planned Parenthood–World Population society[32] showed that, in different urban and rural regions of the country, from 30 to 52 percent of all doctors polled advocated that mothers on welfare who became pregnant should be forced to accept sterilization as a condition of being allowed to remain on the public assistance rolls. In this belief the doctors as a group were in accord with the thinking of the general public.

In 1965, for example, some thirty-three years after Drs. Mallett and Robie and Sadler had called for the sterilization of the poor, the unemployed, and the unfortunate, the national Gallup Poll put the following question to Americans from coast to coast: "Sometimes unwed mothers on relief continue to have illegitimate children and get relief money for each new child born. What do you think should be done in the case of these women? How about the children?" One in every five Americans replied that the solution was to "sterilize the women."[33]

By 1971, the editors of the *Philadelphia Inquirer,* in the wake of the public impact of the IQ test score and genetic enslavement claims of Professor Arthur R. Jensen, conducted a phone poll on the question: "Should the U.S. Encourage Sterilization among Low I.Q. Groups?" A whopping 69.2 percent of the Philadelphians polled favored the forced sterilization of fellow Americans with low IQ test scores.[34]

NOT WITH A SCREAM BUT A SHRUG

If these Gallup and *Philadelphia Inquirer* polls of modern times are even close to being statistical approximations of national attitudes, they also explain why Judge Gesell's historic denunciation, on March 15, 1974, of the fact that, in federally funded birth control programs, "an indefinite number of *poor people have been improperly coerced into accepting a sterilization operation under the threat that various federally supported welfare benefits would be withdrawn unless they submitted to irreversible sterilization"* (italics added), was immediately followed not by national cries of shame and outrage but, instead, by a national shrug and yawn. By precisely the same kind of shrugs and yawns with which the good, decent, educated, and solvent Germans greeted the daily newspaper accounts of the decisions of the *Erbgesundheitsgerichte*—the notorious Nazi Eugenics Courts—which sentenced an average of 165,000 Germans a year to be sterilized against their will for the crime of being what their Nazi judges termed socially inadequate.

Clearly, three generations of the solemn teaching of the essentially religious dogmas of Galtonian eugenics as a legitimate science are reflected in the fact that polls taken in this nation in 1965 and 1971 showed that between 20 and 70 percent of the respondents are convinced that forced surgical sterilization is the only feasible solution to some of our most pressing problems. These problems include the physiological and mental difficulties of that one third or more of our national population who lack the income presently required to meet the ever soaring costs of the minimum amounts of food, medical care, and cultural amenities scientifically proven to be the basic prerequisites for the proper development of their inherited or genetic potentials for physical and mental health.

With this growing public acceptance of the myths of eugenic sterilization, the scientific racists have already helped dim the viable hopes for the human condition abundant in the vast reservoir of biomedical and other scientific discoveries about human potentials and development that are now part of our national heritage. For the greater social dangers of the forced sterilization movement extend far beyond the human prices paid by its immediate victims. After all, poor people are still the only sources of cheap help and even cheaper child labor—two commodities still in great demand. Historically, starting with Malthus and Spencer, the prophets of scientific racism have always been among the fiercest defenders of low wages and the employment of very young children in agriculture, mining, and manufacturing.

Much more devastating than the actual sterilization of a few hundred thousand sacrificial welfare mothers is the more widespread political objective of the compulsory sterilization movement. This is, of course, to brainwash the nation's teachers, doctors, government officials, and taxpayers into believing that the *only* practical way to cope with the mounting welfare costs of the infectious, deficiency, and mental disorders of poverty is and remains the forced surgical sterilization of the poor.

When, in the minds of our educated classes, the sterilization of the poor becomes the only acceptable alternative to the meaningful environmental and family health measures proposed since early in the nineteenth century by the English Sanitary Reformers (our first environmentalists), and the living wages sought by the trade unions, and the universal free education for all children sought by social and educational reformers for over two centuries, it will be at precisely this point that the scientific racists will have won total victory in their long battle for the minds of the men and women who determine American priorities and policies. The acceptance of the get-'em-in-the-gonads Final Solution to the eminently environmental and social causes of the biological and intellectual pathologies of poverty also acts to make solvent and literate taxpayers feel that every penny of their tax dollars spent on such "fatuities" as food for the starving, shelter for the homeless, and nationwide free vaccinations against the common causes of infant mortality and eye, ear, heart, and brain defects, is all part of the continuing tribute we pay for the inability of vote-hungry politicians to say no to the "bleeding-heart liberals."

The roll of the victims of compulsory eugenic sterilization therefore only begins with the 200,000-plus poor people subjected to irreversible surgery on their gonads every year in America. In 1977, as during every International Congress of Eugenics since 1912, the compulsory sterilization of the victims of the preventable and nongenetic diseases and disorders of poverty—from tuberculosis and nutritional anemia to low IQ test scores and other sequelae of biocultural deprivations—remains the Final Solution of scientific racism for the environmental afflictions of society. We forget, at mortal peril, that the knives that each year inflict involuntary and irreversible sterilization on our poor also inflict equally lasting and even more painful injury upon the rest of us.

The surgeons' knives that act out our educated fantasies of aggression on the sexuality of the poor are tempered in the same forges which turned out the knives that have killed every political effort in our times to achieve passage of a medically adequate universal vaccination law. The continued absence of precisely such a law in our national and state public health codes guarantees that, every year, thousands of children with minds damaged by medically preventable infections (such as rubella, measles, whooping cough, and diphtheria) will get the low IQ test scores that make them the victims of compulsory sterilization.

The gelders' knives, with their pseudo-solutions to the real and soluble biomedical problems of our society, also guarantee that for the rest of this century hundreds of millions of Americans will continue to die, *long before their genetically allotted time,* for want of the medical and social modalities that the legitimate sciences continue to develop as scientific and moral alternatives to the sterilization of the "socially inadequate classes."

2 Pseudoscience Deprives Nations of the Health Benefits of the Real Sciences

> The fact that of all occupations of females that of servant shows the highest death rate for consumption does not imply that this occupation is extra-hazardous to the lungs or to body-resistance rather than that servants are largely Irish (who as a nation lack resistance to tuberculosis) or that they are below the average in mental and physical development, incuding disease resistance.
>
> —CHARLES BENEDICT DAVENPORT, PH.D., Director, Eugenics Record Office, in "Euthenics and Eugenics," *The Popular Science Monthly,* January 1911, p. 19

The quacks who purport to treat all human afflictions with their useless nostrums and devices kill thousands of victims each year by depriving them of the legitimate medical help available for hundreds of diseases, including even some forms of cancer, that *can* be cured if professionally diagnosed and treated early enough. Some quacks peddle potions and gadgets. Others peddle pseudoscientific notions that deprive their victims of the useful knowledge of legitimate scientific findings that can either prevent or cure otherwise fatal and crippling diseases.

The quacks who peddle worthless cures for profit kill thousands of people. The quacks who peddle pseudoscientific myths about the causes and cures of medically and socially preventable disorders kill and maim millions.

The only difference between both groups of quacks is in their motivation. The peddlers of fake "cures" do it, in most instances, to make a dishonest if comfortable living. They know they are crooks, and would never dream of using any of their own "remedies." The peddlers of pseudosciences spread their propaganda because, as a rule, they are true believers in eugenics and/or related other branches of scientific racism.

A very practical insight into exactly how much more deadly to the health of millions of us the eugenics zealots are than the quick-buck quacks is provided by a single biosocial problem that now affects over six million sufferers and their families: mental retardation.

In 1959, the American Association on Mental Deficiency (AAMD) adopted, as the official definition of mental retardation, the formulation written by Professor Rick Heber.[1]

It reads: "Mental retardation refers to subaverage general intellectual functioning which originates during the developmental period and is associated with impairment in adaptive behavior."

In plainer terms, this means that mental retardation is by working definition a behavioral condition in which intellectual development, as measured by an IQ test score of below 70, *is associated with the inability to hold a job, drive a car, raise a crop, run a household, care for children, travel to and from one's job, and in general manage to care for oneself in his or her given environment.* Contrary to what many well-educated people who are not educated or trained in clinical psychology or psychiatry like to believe, a low IQ test score in persons with *normal adaptive behavior* is in no way to be considered a sign of mental retardation—no more than the well-known fact that students in rural schools achieve lower IQ test scores than children of urban schools suggests that country children are genetically stupider than city children. Bad politics, not bad genes, make bad rural schools.

Professor Heber's colleague Harvey A. Stevens, director of the Bureau for Mental Retardation of the state of Wisconsin, wrote, in the opening chapter of the now standard book he and Heber edited on the epidemiology, genetics, pathologies, environmental causes, and nature of mental retardation:

"The [AAMD] definition recognized that mental retardation is now viewed as a *reversible* condition. This is a departure from the classical and historical concept of 'once mentally retarded, always mentally retarded.' It is a term describing the *current* status of the individual in regard to his intellectual functioning as well as his adaptive behavior." (Italics added.)

True mental retardation—that is, subaverage intellectual and adaptive functioning—has, as of 1975, over two hundred known causes. Research into mental retardation in modern times has been so extensive that this represents an increase of 100 percent in the number of known causes in the decade since Stevens and Heber edited *Mental Retardation: A Review of Research* (1964).[2]

Margaret Adams, in her *Mental Retardation and Its Social Dimensions* (1971), listed ten basic categories of the causes of mental retardation, as formulated by the American Association on Mental Deficiency and the American Psychiatric Association. Where any given category of causes is listed by only one of these professional societies that deal with mental retardation, it is so indicated by the initials of that group:

1. Diseases and conditions due to infection.
2. Diseases and conditions due to intoxication.
3. Diseases and conditions due to trauma or physical agent.
4. Diseases and conditions due to disorder of metabolism, growth, and nutrition.
5. Diseases and conditions due to new growth (AAMD).
6. Conditions associated with gross brain disease, postnatal (APA).
7. Diseases and conditions due to unknown prenatal influence.
8. Conditions associated with chromosome abnormality (APA).
9. Conditions associated with prematurity (APA).
10. Diseases and conditions due to unknown or uncertain causes with structural reaction alone manifest (AAMD).[3]

In their encyclopedic medical text, *Mental Retardation: An Atlas of Associated Physical Abnormalities* (1972), Harvard Medical School professors Lewis B. Holmes and Hugo W. Moser and their collaborators have presented an interesting summary of clinical evaluations of 1,378 patients at the Walter E. Fernald State School in Waverley, Massachusetts, one of the nation's oldest institutions for the mentally retarded. Dr. Holmes serves as the clinical geneticist at this institution; Dr. Moser is also its assistant director as well as the co-director of the Eunice Kennedy Shriver Center at the school.

Their summary was presented in the following table:

Clinical Evaluations of 1,378 Individuals at Walter E. Fernald State School, Waverley, Massachusetts

Disease Category	Number of Patients		% of Surveyed Population
	IQ <50	*IQ >50*	
1. Metabolic and endocrine diseases	38	5	3.1
2. Progressive diseases of the nervous system	5	7	0.9
3. Acquired conditions	278	79	25.9
4. Chromosomal abnormalities	247	10	18.7
5. Central nervous system abnormalities	49	16	4.7
6. Multiple congenital deformities	64	16	5.8
7. Neurocutaneous diseases	4	0	0.3
8. Psychosis	7	6	0.9
9. Mentally retarded, cause unknown	385	156	39.3
10. Not mentally retarded	0	6	0.4
Total	1,077	301	100.0

What becomes immediately apparent here is that, in this representative population of institutionalized patients, the physicians, pediatricians, psychiatrists, neurologists, psychologists, and other clinical specialists who examined them had to admit that in 39.3 percent—the largest single category—the cause of their mental retardation was unknown. Not environmental. Not genetic. Simply *unknown*.[4]

The next largest category, the 25.9 percent of the patients with mental retardation caused by acquired conditions—that is, infections, nutritional deficiencies, radiation, traumas, and other environmental insults to the brain, the nervous system, the glands, the eyes, the ears, and other organs—were patients whose mental retardation was clearly nongenetic in etiology (causation).

The patients in the next largest category, the 18.7 percent with chromosomal abnormalities—as well as the smaller groups of patients with metabolic and endocrine disorders, multiple congenital deformities, central-nervous-system abnormalities, progressive diseases of the nervous system, and psychoses—could as easily be classified as victims of prenatal and postnatal infections, radiation, and chronic emotional stresses, as well as people with genetic or inborn errors of metabolic, endocrine, and neurological development. As the noted British experts Leon Crome and Jan Stern observed in the second edition of their widely used medical text, *Pathology of Mental Retardation* (1972):

> The rigid distinction between inherited and acquired disease has lost much of its former meaning. Although the final outcome may be vastly different, fundamentally similar causes operate at all stages of ontogenesis [the complete developmental history of an individual]. Ionizing radia-

tion may, for example, induce mutation in germ cells or their precursors to cause, say, phenylketonuria or tuberous sclerosis [the first a congenital, the second a postnatal cause of mental retardation] in the offspring; acting on the embryo *in utero* it is responsible for microencephaly [abnormal smallness of the head], while in the adult it may lead to anaemia, sterility, or cancer. Chromosomal abnormality may be associated with Down's disease [mongolism] if it arises in the germ cells or early zygote, and with some forms of leukaemia if it arises postnatally.[5]

An earlier study at the Fernald school by two other Harvard professors, the neurologist David C. Poskanzer and the pediatrician Maria Z. Salam, showed that many of the biological acquired conditions that cause mental retardation are, in reality, as much social as physiological. Their report, *Mental Retardation Related to Infectious Disease in Patients at the Walter E. Fernald State School* (1966), showed that in the 157 patients with mental retardation due to infectious diseases—6.4 percent of the total patient population of 2,440—"the largest single group of cases, 64, is related to bacterial meningitis. . . . Encephalopathy [inflammation of the brain] due to pertussis [whooping cough] accounted for nine cases . . . and central nervous system effects of maternal rubella [i.e., German measles infection during pregnancy], four cases."[6]

What was socially and politically most interesting, however, was their discovery that:

> Encephalopathy secondary to pertussis was not found in patients under the age of 15. This may reflect a fall in reported cases of pertussis in Massachusetts from an average of 8,800 cases per year in 1930–1939 to 300 per year in 1960–1964. This remarkable reduction in pertussis cases *despite* rising population may be *related to the widespread use of pertussis vaccine*. It seems likely that vaccination of older siblings has a protective effect on infants *under the age of 1 year, the group most likely to suffer neurological complications following an episode of pertussis.* [Italics added.][7]

This clearly suggests that, on this planet, all communities—from villages to nations—whose leaders really wish to eliminate various known biological causes of mental retardation can (and, indeed, often do) mandate vaccination and other public health programs that help prevent such infectious and deficiency diseases from damaging fetal, infant, and child brains and organ systems in all people living within their political borders. (See chapters 23–25.)

THE PREVENTION OF MENTAL RETARDATION

Elsewhere in this book, and particularly in the four closing chapters, some of the biological systems and mechanisms involved in the gene-environment interactions that make us each as tall or as short, as smart or as stupid, as healthy or as sickly as we are will be examined in some detail. At this point, however, let us consider merely four known causes of mental retardation, since, as a group, they are quite possibly responsible for a considerable proportion of the six million cases of mental retardation now recorded by our health agencies.

These four known causes of mental retardation are:

1. Low birth weight.
2. Iron-deficiency anemia.
3. Visual disorders, including legal blindness.
4. Hearing disorders, including legal deafness.

Low birth weight is, by definition, a weight at birth of below 2,500 grams —about 5½ pounds. The American Academy of Pediatrics defines prematurity in terms of birth weight: "A premature infant is one who weighs 2,500 gm. (5½ lb.) or less at birth regardless of the period of gestation," a definition quite similar to that of the World Health Organization. However, while most premature babies—that is, infants born before the full nine-month term of gestation—are low-birth-weight infants, millions of full-term babies, particularly those born to malnourished mothers, are born weighing less than 5½ pounds. These full-term low-birth-weight babies are often classified as immature, rather than premature, babies.

For an exploration of the complexes of interacting biological and neurological reasons why low birth weight, whether because of prematurity or immaturity, is one of the most reliable predictors of mental retardation known to science, see chapters 23–26. Here it is necessary to cite only one famous and well-verified reason: the brains, like the bodies, of low-birth-weight babies are grossly underdeveloped. Low birth weight can nearly always be prevented by adequate nourishment of the mother during her pregnancy.

As Professor Donald B. Cheek and his collaborators at Johns Hopkins University have demonstrated, "the human has one third of his eventual number of brain cells at birth." Professor Jack Tizard, of the University of London, noted in a recent article[8] that the human brain is, at birth, already about 25 percent, and by six months nearly 50 percent, of its mature weight. The whole body, by contrast, is at birth only 5 percent of its young adult weight, and it is not until a child is ten years of age that it attains 50 percent of its young adult weight. The human brain, as every doctor knows, reaches more than 90 percent of its final, mature weight, by the time the child is only six years old.

All of these uncontestable biological facts about brain growth and development have a very simple and incontrovertible implication: by the time a child enters school at the age of six, the biological history of his brain growth has a direct and predictable bearing on his subsequent mental development. The brain is, after all, the one biological prerequisite for intelligence that not even the most orthodox of eugenicists could deny.

Iron-deficiency anemia, also known as nutritional anemia, is a deficiency of hemoglobin, the oxygen-carrying pigment of red blood cells. It can be caused by excess bleeding and by the side effects of malaria and other diseases that act on the blood. Usually, however, nutritional anemia is caused by the lack of sufficient food, particularly of those foods rich in iron, such as, to quote the latest federal nutrition and examination survey, "liver and other organ meats, dark-green leafy vegetables, dried fruits, whole grain and enriched cereals and cereal products, and molasses."

A brain deprived of its biologically established requirements for red blood cells and the iron and oxygen they deliver to the brain cells cannot function as efficiently as it is genetically capable of functioning. In old people with brain cells deprived of oxygen because of atherosclerosis and other artery-clogging disorders that impede blood circulation, such oxygen deficits are associated with many of the mental conditions of senility. In fetuses carried by mothers suffering from iron-deficiency anemia, the resulting lack of oxygen has predictable and adverse effects on prenatal brain growth and development. In infants and preschool children, the oxygen deficits of nutritional anemia in their growing brains have predictable quantitative and qualitative effects on their learning capacities.

Both low birth weight and iron-deficiency anemia are caused, essentially, by a chronic lack of sufficient food. Since the basic reason for such malnutrition is, among the world's working and nonworking poor, a chronic lack of enough money to pay for sufficient food, the low birth weights and nutritional anemias associated with millions of cases of mental retardation are not only environmental but also purely societal in origin. These social causes for the low birth weights in infants born to chronically malnourished families also guarantee that the low-birth-weight infants will remain biologically malnourished during the first six years of life while the human brain attains 90 percent of its mature weight.

An insight into how widespread these underlying environmental and *totally preventable* causes of mental retardation are in our society can be derived from the findings of the U.S. Public Health Service's *First Health and Nutrition Examination Survey.* Concerning deficiencies in nutrient iron, this study, which used both biochemical and dietary intake data, revealed:

"This dietary deficiency occurs at all age levels and is not limited to persons in the below poverty level group. The biochemical iron deficiency is more prevalent at the younger age groups, particularly in children of ages 1–5 years."

The federal survey also showed that, among whites and nonwhites, "approximately 95% of the children of ages 1–5 years, and females of ages 18–44 years [i.e., of child-bearing and lactating ages] in both race and income groups [i.e., below and at or just above the poverty level], had iron intakes below the [biologically safe] standard." What these data indicate, among other facts of modern American life, is that the cruel and continuing fiscal inflation caused by our decades of involvement in the Vietnamese, Laotian, and Cambodian wars has pushed the prices of iron-bearing foods beyond the purchasing powers of millions of the working and taxpaying poor, as well as of the welfare families at the bottom of our income scale.

In terms of the physiological and mental development of preschool children, it is of interest to learn, from this same government survey, that "one-third of the Negroes in both income groups had calcium intakes below standard." And that, when it came to vitamins A and C, so essential in the maintenance of both normal vision and normal genetic resistance against infections, "for white and Negro children ages 1–5 years in both income groups the percentage of persons who had intakes of Vitamin A below the standard ranged from 37 to 52 percent," while the corresponding "range of percentage for Vitamin C was from 43 to 58 percent below standard."

Although in most nutrient categories malnutrition was higher among the blacks than among the whites, this did not mean that all whites were well off. In fact, the national survey discovered, "about 73 percent of the white females ages 18–44 years in the low income group had intakes of Vitamins A and C below the standards. More than half in this age group, without regard to race or income level, had intakes of both vitamins below the standards."[9]

Needless to say, these low vitamin levels did not represent familial or genetic vitamin deficiencies. They represented only the rising dollar costs of liver, whole milk and whole milk products, and dark-green leafy vegetables rich in vitamin A, and the equally inflationary retail prices of the citrus fruits, tomatoes, strawberries, cantaloupes, raw cabbage, and green peppers in which dietary vitamin C is found.

The other two causes of considerable mental retardation that we are reviewing here are:

Visual disorders—from mild strabismus (squint) and nearsightedness to near and total blindness, and

Hearing disorders—from barely measurable but nevertheless handicapping hearing loss to total deafness.

The etiological relationships of visual and hearing deficiencies to low IQ test scores, and low classroom achievement test scores, are quite obvious. In order for the children of all parents, be they geniuses or dunces, to learn, they must first be able to see well enough to read all school books and blackboard lessons, and also to hear well enough to be aware of what the teacher and the other children in the class have to say about any and all subjects. In order for Johnny to be able to provide the right answer on any test form, he first has to be able to see well enough to read the test question.

The relationships between low IQ and classroom test scores and visual and hearing disabilities are, however, not linear. Thus, a 30 to 40 percent visual or hearing loss can add up to a near *total* learning loss—particularly if the hearing losses are in the upper registers.

According to the U.S. National Health Survey, nearly 10 percent of the 24 million American schoolchildren in the 6–11-year age group, or "an estimated 2.2 million" in the entire population, "have a disease condition or other abnormality in one or both eyes."

The same National Health Survey reported that "about 20 percent of an estimated 4.8 million children aged 6–11 years were found to have some abnormality in at least one of their eardrums."

In children of affluent families, most eye and ear disorders are detected in the course of routine periodic pediatric examinations years before the children are of school age, and are corrected with or without the aid of eyeglasses and hearing aids. In children of nonaffluent families, such eye and ear handicaps—unless they are very gross—are generally not detected and corrected even after the handicapped children enter school at age six. These preventable and/or correctable eye and ear handicaps play major roles in the failures of poor children of all races to keep up with their school work.

Among the major causes for children to be born with mild to total visual and hearing defects are the infections of the mother, during pregnancy, with the viruses of rubella and/or mumps and with many bacteria and fungi of diseases common in our environment. Similarly, many of the visual and hearing disorders children acquire between birth and their first day of school are caused by postnatal infections by the same pathogens.

The mothers of these children can, prior to their conception, be immunized against many of these microbial agents. All children can be vaccinated against many of the commonest agents of congenital blindness and deafness. The American Academy of Pediatrics, in fact, recommends that during the first year of life every newborn child should be immunized, with safe and available vaccines, against: rubella, measles, mumps, polio, diphtheria, pertussis, and tetanus.

In families affluent enough to be able to afford birth-to-puberty preventive pediatric care for every newborn child, such vaccinations are now given routinely during the first year of every infant's life. However, as the government's latest U.S. Immunization Survey[10] shows, in this country, of all children between the ages of one and four:

43 percent have not been vaccinated against rubella;
38 percent have not been vaccinated against measles;
37 percent have not been vaccinated against polio; and
24 percent have not been immunized against diphtheria, whooping cough, and tetanus.

This comes to an average of 35 percent of all children between the ages of one and four who are not vaccinated against the commonest preventable infectious-disease causes of mental retardation. These unimmunized children represent a revealing measurement of the actual proportion of all American families with incomes below and above the official poverty levels who are too medically indigent to be able to afford the costs of having their children vaccinated against rubella, measles, polio, diphtheria, whooping cough, and tetanus during the years when all children are at the greatest risk of being permanently damaged, deformed, or killed by these medically preventable infectious diseases.

The so-called compulsory immunization laws on the books of thirty states call for proof of vaccination against one or two of these diseases as a precondition for admission to schools at age six. They are, for the most part, so casually enforced as to be nearly meaningless medically.

All of the hard medical data collected by the federal Center for Disease Control, which conducts the U.S. Immunization Surveys, and by other government agencies on many of the two-hundred other known causes of mental retardation, show that "familial mental retardation," like "familial tuberculosis" and "hereditary pellagra," is not a clinical reality but an artifact of antic statistics. As the U.S. Senate Committee on Nutrition and Human

Needs reported in 1974, some 75 to 85 percent of the approximately 150,000 mentally defective and retarded children born in this country every year are born in poverty—the classic and of course environmental underlying cause of most low birth weight, most nutritional anemias, most prematurity, and most of the family failures to vaccinate their newborn children against the infectious diseases associated with vision and hearing losses.[11]

Dr. Ronald W. Conley, a high official in the U.S. Rehabilitation Services Administration, and the economist on the President's Committee on Mental Retardation from 1968 to 1971, spells out just how much our failure, as a nation, to wipe out the known preventable causes of mental retardation costs us in human as well as in dollar terms in his extremely significant book *The Economics of Mental Retardation* (1973). When Conley analyzed all the available data he found that in our wealthy nation "the children of the poor are about thirteen times more likely to be retarded than the children of the middle and upper classes."[12]

Nor should it come as a surprise to learn, as Conley did, that America's blacks, who as a group are considerably less affluent than our whites, are "six to seven times more likely to have IQ's below 70 than whites."[13]

These data reflect the probability that, far from being genetically ordained, most cases of mental retardation among the white and the nonwhite poor who make up the majority of our mentally retarded population are socially and medically preventable. In current medical practice, let alone in principle, there is much—from vaccinations and nutritional supplements to extended obstetrical and pediatric care—that we can already do to prevent scores of the major known biological causes of mental retardation.

Conley has also estimated the dollar costs, to all of us who pay taxes and retail prices based on current taxes, of continuing to base our national and state public health policies on the ancient and disproven eugenical premise that mental retardation is primarily genetic, and as such not preventable by any social or clinical measures:

"The benefits of prevention [of mental retardation] are large. For each case of severe retardation among males that is averted, the undiscounted total gain to society is almost $900,000 (1970 dollars). For an 18-year-old adult in 1970, this would have a present value of over $200,000.

"Prevention is important. If all groups in society had the same percentage of persons with IQs below 50 as upper- and middle-class white children, the prevalence of this level of retardation would decrease by almost 80%. In 1970 this would have meant an increase of about $800 million in resources available to improve living standards."[14]

One of the most insidious reasons why this nation to this day has no serious programs for the protection of the poor and the near poor—at least one third of our entire population—from the socially costly ravages of *preventable* mental retardation has been the continuing twentieth-century successes of America's eugenicists in convincing the nonscientists who formulate public value systems and public health policies that such interventions against "the laws of God and nature" are liberal illusions that can never overcome the baleful "realities" of the genetic predetermination of IQ test scores and other aspects of human mentality.

As two of the leading American eugenicists, Drs. Sheldon Reed and Elizabeth Reed, stated the classic Galtonian case against initiating social and medical programs for the prevention of mental retardation in their well-known book *Mental Retardation: A Family Study* (1965): "Many people think of mental retardation as a kind of mysterious blight that could be eliminated if we had a more liberal government, more special classes, earlier testing for phenylketonuria or some other kind of social action by some organization."[15]

The Victorian doctors and politicians who, more than a century ago, mocked the claims of Sir Edwin Chadwick, and Drs. John Snow and John Simon, that the squalor and overcrowding of urban slums, fetid privies, putrid water, rotting garbage, and other environmental filth bore the causes of cholera and other contagious diseases, could not have put the case for environmental and medical do-nothingism more plainly.

GET THEM IN THE GONADS

The conclusions of Sheldon Reed, a geneticist, and his wife, Elizabeth, a plant physiologist, have been cited again and again, since their book came out in 1965, by various eugenically oriented authors as scientific "evidence" of the at least 80 percent hereditary origins of mental retardation. These authors have included such of the Reeds' peers as their University of Minnesota colleague Professor Irving I. Gottesman; Professor Arthur R. Jensen, of Berkeley; Professor Dwight J. Ingle, of the University of Chicago; and the Nobel laureate in physics, the nonbiologist, nonbehavioral scientist William Shockley, of Stanford. These and other like-minded professors have been citing the Reeds' book as what they consider to be a scientific basis for their shared conviction that "five out of every six retarded individuals, some five million of the six million retardates in the United States," are genetic defectives who each "have a retarded parent or a normal parent who has a retarded sibling."[16]

Sheldon Reed, director of the Dight Institute for Human Genetics of the University of Minnesota, is also a former president of the Society for Human Genetics (1956). Elizabeth Reed spent the years 1940 through 1966 as a research associate at the Dight Institute for Human Genetics, where her field of research was, according to her biography in *American Men of Science,* "inheritance of mental deficiency." To the lay reader, such credentials are most impressive, and members of Congress and state legislatures, as well as people who edit and report news of science for television, radio, and printed news media, can hardly be faulted for taking the Reeds' 719-page book at face value as a serious work of scientific scholarship, written by two authors clinically qualified to deal with the etiology, the epidemiology, the nature, and the solutions to the manifold problems of mental retardation.

The deflationary facts are, however, that for all of their undoubted training, experience, and high competence in other fields of biology, neither

Sheldon nor Elizabeth Reed has had any more clinical training and clinical experience in the serious study of the pathology and treatment of the well-defined clinical condition called mental retardation than that possessed by any of the electrical engineers, retired airline executives, educational psychologists, urbanologists, and other rank amateurs in psychiatry and clinical psychology who have blandly accepted as valid the Reeds' professional competence to make psychiatric diagnoses and judgments about mental retardation.

This lack of clinical training in the field of the clinical disorder that their book is supposed to deal with is made embarrassingly apparent in their introduction, where in two extraordinary paragraphs the Reeds manage to negate everything that every professionally trained clinical psychologist, psychiatrist, and educational specialist who actually worked with patients and students who were mentally retarded has put forward in the way of a definition of mental retardation itself since the turn of the century.

> Our definition of mental retardation is a simple one. The only persons to be classified as mentally retarded in our study are those who scored below 70 on standard intelligence [IQ] tests, or those so classified by careful subjective judgment. However, this definition might be objected to on the grounds that it does not include the social performance of the subject. . . .
> . . . even though our definition of mental retardation is rather narrow, it agrees with other types of evaluation of mental status, such as the reaction of the general public to a person.[17]

This "simple"—in fact, simplistic—definition of mental retardation not only runs directly counter to the definition adopted by the American Association on Mental Deficiency cited above. It is also no secret to all psychiatrists and psychologists that the use of an individual's IQ test score as the *sole* diagnostic parameter in the clinical diagnosis of the clinical disorder called mental retardation (known before the post-World War I era as "feeble-mindedness") had been thoroughly discredited in the scientific community for well over a half century. Certainly since the American clinical psychologists J. E. Wallace Wallin and Mary Campbell, working independently of one another between 1913 and 1915, discovered that the Goddard-Binet and other standard IQ tests then in wide use by schools, hospitals, and government agencies to certify and institutionalize thousands of people as being clinically feebleminded were at the same time also labeling some extremely competent, socially and economically well-adapted, and self-made rich farmers and businessmen in rural Iowa—and the most important elected and appointed public officials in the city of Chicago—as being *both* feebleminded and morons who "could not pass the ten-year-old [IQ] test."

In our own times, a famous study begun in Riverside, California, over a decade ago by the University of California sociologist Jane E. Mercer has revealed that 75 percent of the Mexican-American and black-American children classified as being mentally retarded solely on the basis of their IQ test scores—*and who were subsequently placed in psychologically deforming school classes for the mentally retarded as a result*—were of perfectly normal mentality. According to Dr. Mercer, these pseudo-retardates would never have been mislabeled as "mentally retarded" if their adaptive behavior and other equally significant behavioral and sociocultural variables had been taken into account in their diagnoses.

It was as far back as 1917, in the first issue of *Mental Hygiene,* that the grand old man of American research on mental retardation, Walter Elmore Fernald, M.D., associate professor of mental diseases at Tufts, lecturer on the mental diseases of children at the Harvard University Graduate School of Education, and, since 1887, superintendent of the Massachusetts State School for the Feebleminded, at Waverley, Massachusetts (since renamed the Walter E. Fernald State School), published his classic article, "Standardized Fields of Inquiry for Clinical Studies of Borderline Defectives." In this article, as every graduate student in psychiatry knows today, the great psychiatrist and educator listed not one but *ten* "Fields of Inquiry" that, in his professional opinion based on over a quarter of a century of clinical experience in the treatment and education of mentally retarded human beings, had to be investigated to provide even a "working basis for individual case study" in the diagnosis of mental retardation.

These minimal diagnostic parameters, Fernald wrote, were:

1. Physical examination.
2. Family history.
3. Personal and developmental history.
4. School progress.
5. Examination in school work.
6. Practical knowledge and general information.
7. Social history and reactions.
8. Economic efficiency.
9. Moral reactions.
10. Mental examination.

The physical and mental examinations Fernald called for in his institution, and in psychiatric practice in general, were to be made by qualified physicians, psychiatrists, and clinical psychologists—not by amateurs, or social workers, or "Binet testers," or even by the then still fashionable "eugenics field workers."

There was simply no room for IQ tests in these guidelines for the professional, clinical diagnosis of mental retardation. "Although [Fernald] used the Goddard and Terman revisions of the Binet-Simon tests, he was not convinced that psychological tests gave an accurate picture of the entire personality of the child."[18]

In his famous 1917 *Mental Hygiene* article, Fernald cited Wallin's contemporary attack on the use of IQ tests to label, and consign to institutions and classes for the dull-witted, thousands of children in America, quoting Wallin as saying that "I assert boldly that from one-tenth to one-half of the children in special public school classes for mental defectives are not at all feebleminded." Fernald prefaced this with the clinical observation that, in the diagnosis of mental defects of all degrees of severity, "the intelligence test findings in these cases are of value as additional evidence but they are not conclusive and should be correlated with sufficient evidence in the other possible fields of inquiry."

In 1908, before Fernald wrote this article (and, for that matter, before either Sheldon or Elizabeth Reed was born), the French authors of the first of the modern intelligence tests, Drs. Alfred Binet and Théodore Simon, produced what emerges in our times as an equally firm warning against using intelligence tests alone as the sole measurement of mental normality or mental retardation. Writing in *L'Année Psychologique,* the journal Binet had edited since 1894, they addressed themselves to the question of differentiating, clinically, between "that which separates moronity from the normal state." This, they recognized, was a highly complicated matter: ". . . we do not consider it fixed but variable according to circumstances. The most general formula that one can adopt is this: an individual is normal when he is able to conduct himself in life without need of the guardianship of another, and is able to perform work sufficiently remunerative to supply his personal needs. . . ."

This, of course, describes what modern behavioral and social scientists term an individual's "adaptive behavior," or his ability to function independently in his environment. Therefore, the two French psychologists continued, in their 1908 article: ". . . a peasant, normal in ordinary surroundings of the field, may be considered a moron in the city. In a word, retardation is a term relative to a number of circumstances which must be taken into account in order to judge each particular case."

Some fifty-six years after the IQ test originators Simon and Binet published their cautionary essay in *L'Année Psychologique,* and nearly a half century after the psychiatrist Fernald published his diagnostic criteria for mental retardation in 1917, a panel of 175 psychiatrists, psychologists, pediatricians, educators, psychiatric social workers and other "American experts in various aspects of mental retardation" met in Chicago from April 9 through April 11, 1964, under the auspices of the Council on Mental Health and the Committee on Maternal and Child Care of the American Medical Association, to draw up a report on the causes, the prevention, and the management of mental retardation. The conclusions of the panel were published in the January 18, 1965, issue of the *Journal of the American Medical Association* and later that year republished by the AMA as a soft-cover book, *Mental Retardation: A Handbook for the Primary Physician.* Of the Reeds' sole "objective" test for mental retardation, the IQ test score, the authors of this handbook declared:

"The physician must remember that the IQ is *not* a diagnosis but one more useful piece of data to be used in arriving at a diagnosis [of mental competence or retardation]. . . .

"Too rigidly used, the IQ may become the determining factor in diagnosis, evaluation, and management with the concomitant danger of committing a child, if even for a short time, to a program above or below his potentialities."

Far from defining mental retardation as a serious condition that can be diagnosed merely by a child's IQ test score, the AMA professional handbook opened by noting: "The identification and diagnosis of mental retardation can prove one of the most challenging problems confronting the primary physician. *Over 200 causes of retardation have been identified,* yet in most cases the physician can make no specific etiologic diagnosis. He can define retardation only in terms of functional characteristics, of significant impairments in intellectual functioning and in the social adaptation of the individual" (italics added).

By far the longest chapter of this handbook for family physicians was Chapter 2, devoted to the *prevention* of mental retardation. While the book did not ignore the role of such genetic causes of mental retardation as phenylketonuria and other probably inborn errors of metabolism, it also listed many of the *known* preventable and/or medically or socially reversible causes of mental retardation. These nongenetic causes included: maternal infections with pathogenic bacteria, viruses, fungi, and other parasites that damage the brain and other neurological organs and systems of the developing fetus; the Pandora's box of maternal medications, ranging from antibiotics and quinine to excessive doses of drugs such as aspirin and vitamin K, that cause brain, eye, ear, and other neurosensory defects in the developing fetus; excessive exposure of the fetus to X rays; intrauterine growth retardation in babies with low birth weight, here defined as under four pounds, seven ounces; postnatal infectious and other common preventable disorders, particularly during the first year of life; accidents and poisons; distorted patterns of maternal care; institutionalization—"A child resident for long periods in an institution may display the effects of sensory deprivation as well as other effects of abnormal mothering"; and familial-cultural-social deprivation— "Retardation in the lower socioeconomic groups is frequently associated with inadequate prenatal care, e.g., prematurity, the toxemias, and high perinatal mortality."[19]

No modern observer, however conservative his personal politics, has ever accused the American Medical Association of either liberal "naïve environmentalism" or radical "creeping egalitarianism." That the AMA's handbook on mental retardation for the family physician reflected the developmental conclusions of twentieth-century physicians, psychologists, and biologists from Binet and Fernald to Crome and Stern—and, in fact, directly contradicted the eugenical or pseudogenetic postulates of the new scientific racism—was not a reflection of the political postures of the AMA or of its panel of 175 experts (127 of whom were M.D.'s and whose non-clinical cohort included "Sheldon C. Reed, Ph.D., Minneapolis"). It spoke, rather, for the quality of the scientific education and the professional training, as well as for the clinical experience with mental retardation, of most of the members of this panel. If it reflected any social or historical condition at all, it was the fact that by 1964 our professional experts in mental retardation had obviously come a long way since the decades when the perverted thinking of Galton and his equally extremist followers had turned the new IQ tests into instruments for the mass misdiagnosis of perfectly competent and intelligent children as being genetically "feebleminded," and of their subsequent compulsory institutionalization and/or equally compulsory sterilization.

All of this suggests, therefore, that in any consideration of the didactic and statistical arguments the Reeds mounted in their book against the scientific claims that millions of cases of mental retardation are *not* genetic, and that they *can* be (or could have been) prevented by early, adequate, and chiefly public, health measures, the first thing that must be borne in mind is that their book is *not* about mental retardation at all. Not, that is, as mental retardation is defined officially by the American Association on Mental Deficiency.

The Reeds' book, on the contrary, is about what amateurs in clinical psychology and psychiatry like themselves define as "mental retardation." Their famous book, *Mental Retardation: A Family Study,* is not about mental retardation but merely about people "who score below 70 on standard intelligence tests, or those so classified by careful subjective judgment."

The makers of the "careful subjective judgment" of mental retardation in the Reeds' book, the Misses Sadie Deavitt and Marie Curial, were the trained "eugenics field workers" who, between 1911 and 1918—working initially without the benefit of IQ tests—made the eugenical field studies of the families of 549 inmates of the State School and Hospital for the Feebleminded and Epileptics at Faribault, Minnesota, that constituted the start and the core of the "data" of the Reeds' study.

As the Reeds inform us in their introduction, after these two eugenics field workers were assigned to the Faribault institution, their salaries were paid by the Eugenics Record Office (headed by Charles B. Davenport and Harry H. Laughlin), in Cold Spring Harbor, New York, and their project was further subsidized by a $25,000 grant from the Minnesota State Legislature for a scientific study of familial feeblemindedness.[20]

These visiting eugenics field workers, neither of them trained in the professional and clinical diagnosis of mental retardation or any other mental or physical disorder, proceeded to gather the usual Eugenics Record Office–style collections of family memories, church and civic records, verified and unverified neighborhood gossip about the living and the dead, and other careful subjective judgments. All of these items, combined with their own untrained personal observations, were used to compile eugenical "family pedigrees" in the form of pseudogenetic pedigree charts, statistics, and miscellaneous notes.

The eugenics field worker phase of the Minnesota project ended in 1917, prior to which IQ tests were introduced into the study by Drs. Maud Merrill and Frederick Kuhlmann. The study was resumed in 1949 by the Reeds, with the assistance of grants from various private foundations, as well as from state and federal agencies. "The material Miss Deavitt and Miss Curial collected," write the Reeds, "gave information on three and sometimes four generations which included the patient's grandparents, parents, aunts and uncles, siblings, first cousins and occasionally children or nieces and nephews, including, of course, records on the spouses as well."

Through questionnaires mailed to the descendants of 289 probands—a proband, or propositus, is the original person diagnosed as having a mental or physical disorder whose descendants and relatives are studied in genetic investigations—the Reeds were able, they write, to assemble the data on over 80,000 persons on which their book is based. Various state, county, and educational agencies opened their supposedly confidential files on those probands and their 80,000 Minnesota relatives for the Reeds.

Not only did school officials make the IQ test scores of 80,000 citizens of Minnesota available to the Reeds, but the state psychological testing bureau was equally generous with the clinical records of the same people. The Reeds claim, on page 10 of their book, that no harm came to any of the 80,000 human beings whose IQ and psychological testing records, so crucial to this study, were made available to them by these public agencies. "Had individual permission to release IQ values been required," they concede, "it would not have been possible to complete the project."[21]

Whether or not, as the Reeds claim, "no person was harmed in any way" because government agencies turned over—*without the informed and written consent and permission of the 80,000 individuals involved*—confidential data as private as their mental test scores to investigators who were neither medically nor psychiatrically nor otherwise clinically licensed to deal with such personal records, raises ethical and even legal questions in the mind of at least one author. Personal clinical records (as the Messrs. Nixon, Erlichman, Liddy, and Hunt learned in 1973) are legally inviolate—particularly when, as in the Reeds' project, the records consist of IQ test scores on which, only too often, clinical diagnoses of mental retardation are made.

Thanks to Charles Benedict Davenport and other of our century's earlier eugenics zealots, twenty-seven states still have laws permitting the forced sterilization of inmates of mental hospitals and other state institutions. In nearly all of these states, mental retardation is still a legal cause of involuntary surgical sterilization. In the State of Minnesota, where the Reeds made their study of the IQ test score records of some 80,000 citizens, Minnesota Statute 256.07, entitled "Sterilization of Feebleminded Persons; Consent to Operation," deals with persons "committed as feebleminded" to the "guardianship" of the state's Commissioner of Public Welfare. According to Richard E. Wexler, special assistant attorney general, this sterilization law "was enacted in 1925 and has remained in force since then in the form in which it was initially enacted."[22]

After 1925, 2,111 human beings, most of them women, were sterilized under this law by January 1, 1943. Between 1943 and the end of 1962, another 239 Minnesotans "committed as feebleminded" to the care of the state were, in consequence, subjected to forced sterilization under this law. Between 1963 and 1968, there were no mandated surgical sterilizations of Minnesotans found to be "feebleminded." In 1969, according to Mr. Wexler, one individual was sterilized "pursuant to section 256.07"; none in 1970; three in 1971; five in 1972; and ten in 1973.

The Reeds' report on their project raises a few scientific as well as ethical questions. According to their Chapter 2, "Organization of the Project," they confined their investigation to three sets of materials. The first was the eugenical family reports collected by the two ladies from Davenport's Eugenics Record Office between 1911 and 1917. The second body of data consisted of the mailed-in answers to the question forms the Reeds sent to the descendants and other relatives of their 289 pre-World War I probands. The third was the collection of IQ test scores kindly provided to them by various Minnesota government agencies. By their own description, therefore, their project was patently *not* a psychiatric and/or psychological study of 80,000 living human beings but, rather, a study of multiples of 80,000 bits and pieces of paper by and about these human beings.

Just as Karl Pearson's 1925 study of flesh-and-blood children of Jewish immigrants in London confirmed for him all of Galton's 1884 subjective judgments about the hereditary parasitic nature of the entire Jewish people, so too did the Reeds' 1965 study of the questionnaires and IQ test scores of 80,000 strangers confirm for them all of the postulates about the causes of mental retardation that Galton's foremost American prophet, Charles Benedict Davenport, had published as early as 1911.[23]

Over half a century before the publication of the Reeds' book, and writing, moreover, as did Galton, without the guidance of a single IQ test score, Davenport had concluded:

". . . a child is born an imbecile, and neither the best of nutrition, the most scrupulous cleanliness, the purest air and sunshine, nor the best of physical and mental training will make anything else out of him. Imbecility can not be cured; in most of its forms it is a necessary result of the nature of parental mating. It is a defect due to a patent or latent defect in both of the paternal germ plasms. The imbecile is an imbecile for the same reason that a blue-eyed person is blue-eyed."[24]

In the preface to their book, the Reeds hailed Davenport as the possessor of "one of the most brilliant minds of the early day geneticists" whose enthusiasm for "his cause—the importance of the gene to mankind" had subjected poor Davenport to unfair attacks by liberals, egalitarians, and other homozygous recessive types. As the Reeds sadly noted, "the vilification which [Davenport] received is still the usual reward for crusaders."

It was, therefore, not too surprising that, a half century after Davenport wrote that no social improvements affecting the nutrition, hygiene, physical and mental care of people labeled as imbeciles by subjective judgments (and/or IQ test scores) could make them any less idiotic than their germ plasms (genes) had ordained them to be, the Reeds' orthodox eugenic pedigree chart and statistical analysis of the IQ test scores of their 289 probands of 1911 and their 80,000 descendants and other relatives should have led them in 1965 to conclusions that mirrored almost exactly those published by Davenport in 1911.

The Reeds concluded that bad genes were involved in at least five out of every six retardates in these 80,000 white Americans. This they reported in only 81 pages of text and a 638-page appendix consisting of the summaries of the question blanks answered by the people who responded to the Reeds and hundreds of pedigree charts and tables on each proband and his or her kinfolk. Nothing quite equal to this vast array of the famous "black charts" of the halcyon days of Goddard and Davenport had been seen in American scientific and nonscientific literature since 1917, when Davenport used batches of them to document his pseudogenetic claim that pellagra—which Goldberger had by 1914 *proven* to be a deficiency disease caused by malnutrition —was a hereditary disease found only in white people who lacked the genes required to resist the undiscovered microbe supposed to cause pellagra.

Just as Davenport, in 1917, had proved with the same types of pedigree charts and tables that poor people with pellagra had one or more ancestors and living relatives who had also suffered from this classic poverty disease, the Reeds proved that pellagra was not the only poverty disorder to be endemic in the families of the poor. And, of course, again like Davenport and Galton, the Reeds concluded from these pedigree charts and tables of IQ test scores that mental retardation was transmitted by the genes of retarded parents to the chromosomes of their equally retarded children.

Only one year before the publication of the Reeds' book on mental retardation and its 80 percent likelihood of being genetic, Harvey A. Stevens revealed himself to be considerably less certain about the causes of mental retardation. In the first chapter of *Mental Retardation,* the massive review of research into the nature, etiology, genetics, and epidemiology of mental retardation that Stevens and Heber had edited, Stevens wrote: "With present techniques of diagnosis it is possible to make a positive and precise identification of the cause of mental retardation in only 15 to 20 percent of all cases. *Our present state of knowledge does not permit definitive diagnoses in the remaining 80 to 85 percent"* (italics added).[25]

Stevens, however, had spent all of his professional life working with flesh and blood mentally retarded human beings, and not with bloodless questionnaire blanks and IQ test scores obtained without the permission of the individuals tested. The Reeds, uninhibited by such cautionary professional experiences, did not hesitate to leap to the conclusion that "the transmission of mental retardation from parent to child is by far the most important single factor in the persistence of this single misfortune . . ." (p. 48).

Thirty pages later, these modern disciples of Galton and Davenport urged that what they term the "transmission of mental retardation" could no longer be ignored. They charged that "this problem has been largely ignored on the assumption that if our social agencies function better, that if everyone's environment improved sufficiently, then mental retardation would cease to be a major problem."

The Reeds did concede knowledge of the fact that diseases of pregnancy, infancy, and childhood, including polio, encephalitis, meningitis, scarlet fever, measles, pneumonia, and whooping cough—many of these preventable infectious diseases of chronic or long-term duration—had been reported to have been followed by mental retardation in some of their 80,000 "subjects." They even had a few modern-sounding things to say against poverty, the socioeconomic condition in which most of their "subjects" lived.

However, like Davenport, they were far from convinced that "the best of nutrition, the most scrupulous cleanliness, the purest air and sunshine" could ever have prevented mental retardation in five out of every six mental retardates. As they put the problem on their pages 63–64, when it comes to what they categorize as the cultural-familial type of mental retardation, the basic question is "whether social conditions were bad because of the parents or were the parents retarded because of cultural deprivation in the families into which the parents were born."

This was a purely rhetorical question, which the Reeds answered with the announcement that in this nation "we have surprisingly little residue of our 2 to 3 percent of the population which is retarded and which could be ascribed as primarily the result of social deprivation."

In their closing chapter, "Some Implications of the Study," the Reeds offer only one solution to the national problem of mental retardation. It is, predictably, the Final Solution for the menace of genetic enslavement carried in the genes of the "definite race of chronic pauper stocks" that the eugenics movement has been peddling since the late nineteenth century: the sterilization of the defenseless poor.

To be sure, the Reeds call for the *voluntary* sterilization of the people with low IQ test scores. But as Judge Gesell's landmark opinion of 1974 reminded us, such surgical sterilizations of the poor are, and in our society will always have to be, about as "voluntary" as the gonadal surgery currently suffered by mothers on welfare as society's price for not removing them from the welfare rolls—or as the sterilizations now being accepted under the extortionate "Package Deal" by poor pregnant women as the precondition for therapeutic abortions in most teaching hospitals.

The Reeds' massive "study of mental retardation," then, turns out to be just another eugenic sterilization tract, in the stereotyped tradition of all eugenic writings of this century, from Davenport and Madison Grant to Lena Sadler and Theodore Russell Robie.

To read the Reeds' words on the genetics and the gonadal transmissibility of low IQ test scores and poverty itself is to be, again, reminded of how very little Galton's true-believing followers have learned of the biology, chemistry, and cytology of human heredity, variation, growth, and development since 1883, when Galton the ex-medical student wrote not only that epilepsy was a genetically transmitted hereditary disease—which it never was—but also and much less pardonably, that epilepsy was as well a major cause of insanity and violent crime. To go through the Reeds' black charts and number tables on the genetic transmission of mental retardation is to be transported, in time, to January 1911, when the new issue of *The Popular Science Monthly* reached the stands—the issue in whose lead article, "Euthenics and Eugenics," Charles Benedict Davenport dealt with both the bacteriology and the sociology of the eminently nongenetic infectious disease called tuberculosis or consumption. Wrote the director of the Eugenics Record Office:

> It is an incomplete statement to say that the tubercle bacillus is the cause of tuberculosis, or alcohol the cause of delirium tremens. Experience proves it, for not all drunkards have delirium and not all that harbor the tubercle bacillus die of consumption—else we should all die of that disease. . . .

The fact that of all occupations of females that of servant shows the highest death rate from consumption does not imply that this occupation is extra-hazardous to the lungs or to body-resistance rather than that *servants are largely Irish (who as a nation lack resistance to tuberculosis)* or that they are below the average in mental and physical development, including disease resistance [italics added].

Because of Galton and Davenport, and all who echoed their pseudo-scientific dogmas in college lecture rooms and legislative chambers, in pulpits and in mass-media editorials, for over a century, scores of thousands of poor people who because of traumas or postnatal sickness acquired epilepsy were to be locked away like wild beasts for life in institutions for the feebleminded and epileptics. And millions of poor, hard-working Irish families were to suffer needless deaths from tuberculosis (and other preventable poverty diseases) because of the absence, in their life environments, of social and public health legislation dealing with the control or elimination of the well-known contributory social causes of tuberculosis.

THE GROWING AND COSTLY GAP BETWEEN SCIENTIFIC RACISM AND SCIENTIFIC PROGRESS

However, whereas in 1883 and 1911 the hereditarian views of Galton on epilepsy and Davenport on tuberculosis were probably shared by most of their peers, things have changed for the better in our times. Whatever the judgments of the educated but uninformed people outside the life and behavioral sciences, in 1965 the Reeds' view of the validity of IQ test scores as measurements of genetic mentality had become the *minority* view among their peers in the sciences that deal with the genetics and the development of human mental capacities.

By 1970, for example, after Professor Arthur R. Jensen of Berkeley achieved instant world fame by publishing pretty much the same conclusions about the heritability of intelligence that had previously been published by the Reeds, by Davenport, by Pearson, and by Sir Francis Galton, much water had passed under many bridges. In that year, the prize-winning American sociologist Dr. Robert W. Friedrichs (now chairman of the Department of Sociology at Williams College) polled a representative sampling of the membership of the American Psychological Association regarding their evaluation of Professor Arthur Jensen's conclusion that "it [is] a not unreasonable hypothesis that genetic factors are strongly implicated in the average Negro-white intelligence difference." Far from accepting the "Jensen thesis" as either fact or good science, the response of the cross-sectional sample of 526 APA members revealed:

(a) Over two-thirds (68%) either disagreed or tended to disagree;

(b) those disagreeing were, on average, over five years younger than those agreeing;

(c) those professionally resident in the Deep South (Alabama and Mississippi) were significantly more receptive to the Jensen thesis than the overall national sample; and

(d) [understandably, in view of the historic ties between the anti-Semitic quotas of the Immigration Act of 1924 and the six million Jews annihilated in the Nazi holocaust of 1933–45] those tentatively identifiable by name as "ethnically Jewish" were significantly less receptive to the Jensen thesis than the overall national sample.

After an analysis of the responses to his questionnaire, Dr. Friedrichs concluded that "the 'Jensen thesis' is opposed by the great majority of Jensen's peers in the United States (even when evaluations were solicited with a guarantee of personal anonymity) and that, whatever role may be played by heredity in differentiating the intellectual achievements of blacks and whites, it is quite clear that subtle environmental differences are most assuredly prognostic of intellectual—even scientific—judgment."[26]

The reason why the majority of the psychologists polled by Professor Friedrichs in 1970 disagreed with Jensen was *not* because the APA members are now liberal or socialist whereas in 1920 they had been centrist and conservative, as all too many influential but painfully uninformed commentators suggest. Nor can the change in the stated conclusions of the non-Jewish members of the American Psychological Association be attributed—as I myself have heard some cocktail-party analysts maintain—to the "fact" that as Christians they have a collective guilt feeling about the historical associations between IQ testing in America and the dead of Auschwitz and Belsen, and that they therefore permit naïve and radical or liberal egalitarian sentiments to block their expressions of their real and objective scientific evaluations of Jensen's IQ-test-score-derived "genetic data."

As we saw in the instances of the ultrarightist eugenicist Charles Benedict Davenport and the socialist eugenicist Karl Pearson, let alone in the persecution by the Marxist-Leninist Joseph Stalin and his ideologues of the Soviet geneticists who gave the lie to the pseudogenetics of the malevolent amateur in genetics Trofim Lysenko, political convictions and scientific judgments follow no such pseudopsychoanalytical rules. When it comes to a profound ignorance of the hard and abundantly available data of our civilization's life and behavioral sciences, history reveals that liberals and conservatives, radicals and reactionaries, skeptics and true believers are—like Judy O'Grady and the Colonel's lady—all sisters under the skin. Ignorance of science transcends all known political philosophies.

What had of course happened, in the sixty years between the publication of Davenport's master work, *Heredity in Relation to Eugenics* in 1911 and Professor Friedrichs' presentation of his poll of Jensen's peers' evaluations of Jensen's eugenic thesis at the meeting of the American Sociological Association in New Orleans in the summer of 1972, was the onset of a still growing hurricane of new scientific knowledge about human genetics, about the etiology and epidemiology of mental retardation, about the developmental biology of the human brain and the hundreds of environmental factors that can enhance or impede brain development, and about the prevention of hundreds of formerly unavoidable body- and mind-wrecking diseases and disorders. The professional psychologists who answered Friedrichs' question about Jensen's eugenic thesis were, apparently, more aware of the discrete scientific findings about the processes of human growth and development in health and disease that have been made since 1911 than were the Reeds and Jensen.

Unfortunately, unlike the majority of the professional psychologists polled by Professor Friedrichs, far too many of the educated citizens of this generation have not been equally aware that the confirmed findings of the legitimate sciences have long since destroyed the validity of eugenics and other once fashionable forms of scientific racism in the minds of all but a tiny, and for the most part aging, minority of our scientists. The public and family health effects of this lag in the delivery of scientific knowledge to the body politic are, alas, quite predictable.

One of the historic reasons why the writings of the Reeds and their eugenic-minded cohorts in the sciences have had more influence on state and national legislatures than have the writings and the hard data of Heber and Stevens, Conley, Pasamanick, and Knobloch, and all serious physicians and psychiatrists since Walter E. Fernald is, of course, purely economic. The prevention of mental retardation—by (1) improving the sociobiological and cultural environments of pregnant mothers and preschool children; (2) making certain that *all* newborn children are immunized during their first year of life against rubella, diphtheria, tetanus, whooping cough, polio, measles, mumps (and, as new experimental vaccines against cytomegalovirus and the streptococcus of rheumatic fever become available, against these pathogens), and other causes of mental and physical retardation; and (3) guaranteeing every pregnant woman the prenatal, obstetrical, and postnatal medical, nursing, and social care she and her developing child biologically require—calls for very costly public and family health programs. The billions of tax dollars such programs require make them far more costly, *at the outset,* than the forced sterilization of that 2 to 3 percent of our population—some six million people—currently diagnosed as being mentally retarded.

This kind of "economy," however, is as illusory as were the pinchpenny social policies that, earlier in this century, turned millions of potential American wage earners and taxpayers into chronic invalids supported by government welfare agencies because they contracted wholly preventable cases of tuberculosis, pellagra, and other *socially preventable,* degenerative, infectious and deficiency diseases. Such "savings" of the taxpayers' dollars are as costly to a nation as were the cash savings to England when, in 1871, Parliament allotted only 400 pounds sterling for smallpox vaccination programs in Ireland—on the eve of the infinitely more expensive smallpox epidemics of 1871, 1872, and 1873, each of which could have been completely averted by a more sensible vaccination appropriation in 1871.

For over a century, scientific studies of the human body's interdependent systems, from the cells and skeletal structures to the brain and the various immune mechanisms, have taught our physicians that not only is a sound mind indeed predictably associated with a sound body, but also that an environment most conducive to the optimal bodily health of newborn infants is invariably associated with lower infant and child and adult mortality rates.

As every doctor knows, the socially and medically preventable causes of mental retardation in children—starting with the lack of food, the lack of vaccinations, and the lack of medical care—are also among the leading causes of low birth weights and of fetal and infant deaths in this world. "Data from the National Center for Health Statistics reveal that families with an annual income of less than $2,000 have four times as many heart conditions, six times as much mental and nervous disorders, six times as much arthritis and rheumatic fever, and almost eight times as many visual defects as those in the higher income brackets."[27]

False social economies, like the false correlations of the effects of poverty diseases with the genes of the poor, offer only transient comfort to the people who feel it is more important to balance the national budget than the national diet. The congressional economists have figured out, quite convincingly, that the dollar costs of ending hunger in America would come to a staggering $3 billion a year. This also happens, however, to be 40 percent less than the $5 billion loss to the economy caused by the effects of malnutrition on health, productivity, and education.[28]

The $1.25 spent on the dose of rubella vaccine that immunizes a human fetus *in utero* against the brain damage caused by this virus also protects the same fetus against the other harm done by the same virus, from partial or total blindness and deafness to widespread and irreversible psychomotor defects. There is good scientific evidence, in recent years, not only that "a virus infection both before and after birth may cause diabetes," but also that, according to researchers at the Children's Medical Research Foundation in Sydney, Australia, "probably no infection is as likely to cause [diabetes] as congenital rubella." (See chapter 23.)

The $1.25 spent to immunize a newborn infant against the measles virus protects the child not only against eye damage caused by this virus, but also against the crippling and often fatal brain inflammations caused by the same measles virus. Recent research findings strongly suggest that multiple sclerosis is a delayed sequel of childhood measles infection, and that the best way to prevent multiple sclerosis might be to vaccinate every infant against measles.[29]

The few pennies that a community spends on the DTP (diphtheria, tetanus, pertussis) three-in-one vaccine not only protects the infant against whooping cough and the pertussis encephalopathy (brain inflammation) that is a known cause of mental retardation, but also protects the same child against what were formerly two of the most common causes of infant and child deaths in America.

The forced sterilizations of the poor might call for smaller *initial* costs to the greater society than the dollar costs of universal family vaccination, medical care, and nutritional supplement programs. However cheap the immediate dollar costs of the compulsory sterilizations the eugenicists have long proposed as the "scientific" alternative to public health programs (programs that the Reeds sneeringly dismissed with their statement that the real problem of eliminating mental retardation "has been largely ignored on the assumption that if our social agencies functioned better, that if everyone's environment improved sufficiently, then mental retardation would cease to be a major problem"), their ultimate dollar costs are quite predictable. The long-term reliance upon the sterilizers' knives as an alternative to meaningful public health programs guarantees that the proportion of mentally retarded people will rise, not fall, in our total population. And that this nation will continue to have to waste billions of dollars on coping with the immediate clinical effects, and the chronic social sequelae, of the preventable infectious, deficiency, and mental diseases and disorders that can be neither prevented nor cured by the compulsory surgical sterilization of the poor and their helpless children in this nation.

Bad science, like heroin and cocaine, always delivers infinitely more permanent damage than transient euphoria.

3 History Provides Clues to Healthy Alternatives to the Social Effects of Scientific Racism

> . . . the ordinary struggle for existence under the bad sanitary conditions of our towns seems to me to spoil and not to improve our breed. It selects those who are able to withstand zymotic [infectious] diseases and impure and insufficient food, but such are not necessarily foremost in the qualities which make a nation great. On the contrary, it is the classes of a coarser organization who seem to be, on the whole, most favoured under this principle of selection, and who survive to become the parents of the next generation.
>
> —SIR FRANCIS GALTON, in "Hereditary Improvement," *Fraser's Magazine,* January 1873

> The public health is the foundation on which repose the happiness of the people and the power of a country. The care of the public health is the first duty of a statesman.
>
> —PRIME MINISTER BENJAMIN DISRAELI, in a speech to Parliament, 1875

As we have already seen in microcosm, in the matter of mental retardation, a working knowledge of the findings of the legitimate life and behavioral sciences provides all of us with practical public health antidotes to the real and pressing health hazards of the pseudo-epidemiology of the new scientific racism. Vaccination of all newborn children—and particularly of the future mothers of children—is still a far healthier and more economical policy for a nation interested in survival than is the barbaric involuntary sterilization of the poor, the sick, the friendless.

Similarly, a basic knowledge of at least the outlines of the histories of the real and the pseudosciences as they interacted and continue to interact with the social and political history of our civilization becomes a prerequisite of responsible modern citizenship. As we shall see in succeeding chapters, the IQ test score and forced sterilization gambits that have marked the post-Hitler era of the new scientific racism have been far from the only political and social weapons of scientific racism since the prime of "T. R. Malthus, Strewer of the Seed which reached its Harvest in the Ideas of . . . Francis Galton."[1]

Before the American eugenics movement latched on to and perverted the Binet intelligence tests (see chapter 10) there was, to name but one of the previous well-used pseudoscientific ploys of scientific racism, *craniology,* with its scale of cephalic-index measurements. In this system the "races" with long heads—particularly if the long-headed or *dolichocephalic* folk had blue eyes, blond hair, fair skin, and spoke an Aryan mother tongue—were designated the intellectual and moral superiors of the "races" with rounder or *brachycephalic* heads, particularly those brachycephalics who had brown eyes, dark hair, off-white skin, and spoke a Latin, Semitic, or Grecian mother tongue.

Craniology was preceded by a variety of pseudo-anatomical "scientific" studies of the brains of superior and inferior races. In these studies, biased American and European physicians and equally racist amateurs solemnly reported finding structural and functional anatomical differences between the brains of the superior races (to which, of course, all of these pseudo-anatomists belonged) and the brains of the inferior races (which invariably proved to be of Oriental, African, and, if white, non-Nordic origins).

Before these brain studies there was *phrenology,* or the mapping of the head for the bumps of wisdom, stupidity, wealth, poverty, and other hereditary traits revealed—by these external cranial features—to skilled professional phrenologists. The roster of European and American notables who took this precursor of craniometry seriously, and had their own heads examined, included U.S. President James A. Garfield, the abolitionist John Brown, Dr. G. Stanley Hall (the psychologist under whom Goddard and Terman, the fathers of American-style IQ testing, took their doctorates), the philosopher Auguste Comte, and the poet Walt Whitman.

John D. Davies notes that Queen Victoria and Prince Albert called in the famous phrenologist George Combe to "examine the heads of their large brood, and Senator Charles Sumner had him inspect his own battered cranium after the famous beating by Preston Brooks. Benjamin Moran, Henry Adams' adversary in the London Embassy, wrote out phrenological analyses of his colleagues, as did Otto von Bismarck; Karl Marx always judged the mental qualities of a stranger from the shape of his head, as did Baudelaire, Balzac, and George Eliot. Henry George was accomplished enough to make an examination of himself. A random catalogue of 19th century figures who accepted the rule-of-thumb aspect of phrenology could be expanded almost indefinitely."[2]

Before phrenology there was *physiognomy,* in which the inborn character traits of a person were revealed by his or her facial angles and related "scientific measurements." And, of course, before physiognomy there was the precursor of them all, *preformation,* about which we will read in a later chapter. (See pages 85–90.)

Most of these scientistic precursors of IQ test scores as scientific measurements of inherited individual and racial traits and characteristics have long since been forgotten. But some of them, notably craniometry, persisted well into the IQ testing phase of scientific racism.

As we shall see in some detail in later chapters, the IQ testing myths of scientific racism did not arrive until the eve of World War I. These myths held that not only were individual *and racial* intelligence limitations preformed in the genetic endowments inherited by all newborn babies, but also that since intelligence was wholly genetic, it was fixed at its preformed levels for the life of all people. This portion of the great IQ testing myths had existed long before the so-called intelligence tests developed by American believers in eugenics such as Henry H. Goddard, Lewis M. Terman, and Robert M. Yerkes gave the scientific racists a statisticized tool for validating the myths of fixed individual and racial intelligence. They also gave budget-minded legislators a powerful excuse for not spending enough money on schools and libraries for that majority of the American population "revealed" by the intelligence test scores of World War I's military recruits to be, by heredity, genetically incapable of completing more than an eighth-grade education. This, of course, was particularly true in an America in which, according to Harvard's professor of psychology William McDougall, the "innate [i.e., genetic] capacity for intellectual growth is the predominant factor in determining the distribution of intelligence in adults, and that the amount and kind of education is a factor of subordinate importance."[3]

Although hard-core eugenicists such as McDougall clung to their faith in this myth, two decades of more scientific analyses of IQ testing had rendered it scientifically obsolete by the time America was drawn into World War II in 1941. The modern revival of the myths of genetically fixed intelligence and the accuracy of IQ testing in the measurement of this fixed intelligence, which began in 1954, had its origins in purely political, rather than scientific, developments.

This revival has, once again, made IQ test scores appear to be measurements of known mental and genetic factors in the eyes of those who help create our public value systems, as well as Presidents and legislatures who remain in office by passing legislation that institutionalizes the values of the majority of our voters. Scientifically, this restoration of the IQ test myths to the heights to which people such as McDougall, and institutions such as the Congress of the United States of America, had raised them more than a half century ago is equivalent to the successful revival of the Flat Earth Theory a century or more after Galileo and Newton.

Now that, as we shall see in later chapters, the IQ test scores are, for the second era in our history, again being employed for the *misdiagnosis* of perfectly normal individuals, who are capable of being educated, of earning their own living and paying taxes, as mental retardates who are uneducable, unemployable, and chronic welfare clients, the tragedy of these victims of IQ testing has also become our tax increases. We cannot right this wrong to the misbranded—let alone end this needless drain of our tax dollars—without knowing at least some of the key historical events that gave the American perverters of the perfectly harmless Binet tests this dangerous power over our lives and taxes.

Some of these highlights in the long history of psychological testing and its effects on the laws governing education, public health, family health, immigration, and other aspects of human life will be reviewed in later chapters, notably chapters 10–13, as well as other chapters, such as 14, 21, 25, and 26, in which we will witness the confrontations between the legitimate life and behavioral sciences and the IQ testing myths. Before getting to those chapters, however, and in view of the tragic successes the old scientific racists had in using the IQ test scores of immigrant military recruits of World War I as scientific evidence that Jews, Italians, Poles, Hungarians, Russians, Greeks, Spaniards, Portuguese, and other non-Nordics were genetically too stupid to ever be educated, we are going to take a quick look at a modern book on the genetics of intelligence by a prominent true believer in the scientific worth of IQ testing.

PROFESSOR INGLE'S DISCOVERY OF AN
EVOLUTIONARY MIRACLE—OR TWO

It was not an intellectual argument that the post-World War I exploitation of the military intelligence test scores helped the old scientific racists win: it was, rather, the hearts and minds of the Congress which passed, and the President who signed into law, the anti-Semitic, anti-Italian, and anti-non-Nordic racial quotas of the U.S. Immigration Act of 1924, in response to this final "scientific" proof of what night-riding Nativists and Brahmin Harvard Teutonists had been saying since the early and middle years of the nineteenth century.

A half century after the IQ testing myths helped close the gates of America to the Jews and other Europeans Hitler and the Nazis were already promising to annihilate once they took power, there have been some interesting changes of human targets among the eugenics true believers.

Dwight J. Ingle, the recently retired University of Chicago professor of physiology and editor of the learned journal *Perspectives in Biology and Medicine,* yields to no man in the fervor of his faith in the scientific reliability of IQ test scores as measurements of genetic intelligence. A follower of Professor Arthur R. Jensen, who was kind enough to review the manuscript of Dr. Ingle's 1973 book, *Who Should Have Children?,* Ingle is one of the few modern eugenicists who has very respectable qualifications and training in human biology.

The thesis of Ingle's book is fully in the orthodox eugenics tradition: Ingle proposes that with *or without* the consent of the nation's achievers of low IQ test scores, the time has come for this nation to protect itself against the menace of "genetic enslavement" by sterilizing them.

Of course, there no longer being any hordes of low-IQ-testing immigrant Jews and Italians to threaten us with genetic enslavement, the eugenics movement had by 1973 fallen back upon the venerable Undeserving Poor who had so terrified Malthus, the "definite race of chronic pauper stocks" whose extermination had been urged at the last two International Congresses of Eugenics. In the orthodox eugenics tradition, Ingle writes that "the majority of steady welfare clients are below average intelligence and have a greater-than-average number of *inherited* diseases" (italics added). (A medical claim that Ingle leaves *totally undocumented* with any of the abundantly available health statistics on inherited diseases issued by federal agencies and the World Health Organization.)

Since, because of the new peril called the "Population Explosion," Ingle predicts that it is soon "going to be necessary to *force* people to limit family size" (italics added), the retired professor of physiology feels we should therefore start solving what he and other eugenicists see as the nation's dual problem of rising welfare budgets and "overpopulation" by sterilizing all Americans with low IQ test scores lest their fecundity enslave us high-IQ-test-scoring types genetically.

Up to this point Ingle's book, like the Reeds' volume, whose data on "80,000 cases" he cites as scientific proof of his claims, proved to be just another in the tiresome flow of compulsory sterilization tracts issued since 1900 by Galton's truest believers. However, only eight pages after writing that "genetic factors count for at least twice as much [i.e., at least 50 percent] for variation in IQ as do environmental factors," Ingle then proceeds to deal with the Jews and their IQ test scores in terms that the great founders and prophets of eugenics, from Galton and Pearson to Davenport and Madison Grant, would most certainly have denounced to all the faithful as infamous heresy. For on page 58 of his book—forty-nine years after the imposition of anti-Semitic immigration quotas based in lethal measure on the low Binet and Army intelligence test scores of Jewish immigrants between 1913 and 1919, and twenty-eight years after the Nazi holocaust in which millions of the Jews, Italians, Poles, Hungarians, and other low intelligence test achievers excluded by the same 1924 racial quotas were murdered—Professor Ingle writes:

> Jewish children in the United States have an average IQ about eight points *above* the national average. They generally do well in school and jobs. Jews have contributed a great deal to discovery and creativity. And yet in the recent past many of them were subjected to cruel discrimination and were forced to live in slums. Many of them attended poorly equipped and staffed schools. Is there something in the culture of the Jews that encourages them to work hard to overcome environmental handicaps? Indeed there is! When I was talking with a Jewish friend about this, he said, "Don't you know that Jewish mothers put *honey* on books?" [Italics added.]

Far from being excommunicated from the eugenics faith as a Semitophile, Ingle was instead hailed as the author of a great book by the eugenics faithful.

More than the Jewish attitude toward books was involved in this rocketlike climb of Jewish intelligence test scores since 1913, however. In order for the eugenic, or pseudogenetic, theory of intelligence as a more than 80 percent heritable factor to hold any water, two ethnic groups in our population would have had to undergo some striking mutations, on scales of magnitude and numbers never previously matched in the history of human evolution, since 1919, when the Army intelligence test scores showed the immigrant Jewish draftees to have a mental age of just a little over eleven years.

For starters, the genes of our five million Jews would have had to have mutated qualitatively and quantitatively to the point where every Jewish chromosome now contained the requisite assortment of the genes supposed to be responsible for high IQ test scores, the same high IQ genes that were formerly the exclusive racial trait of the Nordics. Secondly, the chromosomes of the formerly intellectually superior Nordics would now, because of an equally miraculous mass mutation, have to contain those complexes of the genes for *low* IQ test scores that, between 1913 and 1919, were, by scientific test, confined solely to the genomes of the immigrant Jews, Italians, Poles, Hungarians, Russians, and native American blacks.

The mechanisms of variation, natural selection, and evolution, having evolved Michelangelos and Einsteins from Neanderthal hominids, are of course among nature's greatest wonders. However, until now there has been no suggestion that these mechanisms, which formerly required scores of thousands of years to evolve poets and physicists from cave-dwelling savages, have themselves been altered, so that they now produce such miracles in forty-nine years or less.

There is, of course, a nongenetic, non-honey-on-books explanation for the climb in the intelligence test scores of the Jews in America since 1920. This has to do with such purely environmental factors as being born in America and educated in American schools, and hearing American radio and television news broadcasts, *before* taking American-style intelligence tests. And with the quite predictable sociobiological benefits of upward socioeconomic mobility on the growth and brain development of these American-born children.

Ingle's nondiscovery of two evolutionary miracles that never happened underscores a very serious reality about the power of the pseudogenetic IQ myth in our times: the fact that IQ test scores are still taken as seriously as they are by the nonscientists who govern this nation, edit its news, teach its teachers and toddlers and make up the majority of its voters is tragic testament to the lasting impact of the dogmas of scientific racism. The tragedy of the current controversies over the percentages of "heritability" of intelligence, and the validity of IQ test scores per se as measurements of mental retardation or genius, is that they all obscure the nature or even the reality of what we call "intelligence" itself.

Most of our better psychiatrists and nonmedical behavioral scientists have long since rejected the notion that there even exists any single discrete biological or mental trait than can properly be labeled as "intelligence." For example, the psychologist John Ertl, director of the Center of Cybernetic Studies at the University of Ottawa and developer of a neural-efficiency analyzer that combines electroencephalographs and computers to measure brain functions, was quoted as declaring: "Intelligence is a concept equivalent to truth and beauty. I don't really know what it is, but I do know what it is not. It's not the score of an I.Q. test, and it is not what our equipment measures."[4] Dr. Ertl said this after his brain-wave measurements had proven that large numbers of children who had been labeled as retarded on the basis of their IQ test scores were, actually, quite bright. (See pages 607–11.)

The anthropologist Paul Bohannan, of Northwestern University, probably spoke for most of our better behavioral, medical, biological, and anthropological scientists in a letter sent to the magazine *Science*[5] concerning the long obsolete concepts of Harvard Psychology Department chairman Richard Herrnstein, when he wrote:

> The question that *should* be asked is, Why do serious students of human behavior fool around with a dated idea like "intelligence"? It is possible, of course, to measure *performance;* it is possible to deal with *perception* (either physiologically or insofar as it is turned into *behavior*); it is even possible to deal with hidden *values* and *assumptions*. But is "intelligence" an adequate concept for summarizing all that?
>
> The fact that the results of IQ tests can be statisticized makes mat-

ters worse—*it gives the figures something of the quality of scientific "data" and thereby implies a "reality" that the [IQ test score] figures do not have.*

Obviously, behavioral scientists badly need summarizing concepts or just shorthand terms with which to bring together some of the things they measure. But just as obviously, "intelligence" is a culture-bound western European idea that has been given far more scientific weight than it can bear. [Italics added.]

In modern times, Intelligence, as a scientific construct, is, like Phlogiston and the Ptolemaic Universe before it, a concept whose day has come and gone. (See pages 505–09.)

FROM RACE SUICIDE TO RACE HYGIENE

If the eugenicists of the post-World War II era took the high Jewish IQ test scores of our era not as a refutation of their seminal heritability-of-intelligence dogma but as added evidence of its validity, their attitude toward the birth control movement shows an even more startling capacity for adapting to changed historical conditions.

The original natal policies of Sir Francis Galton were based, primarily, as Galton disciple President Theodore Roosevelt put it in 1913, on improving "the future racial qualities" by "favoring the fecundity of the worthy types." What the eugenicists of Galton's generation wanted most of all was to increase the birth rates of "worthy types" of people like themselves.

Teddy Roosevelt and the other well-heeled eugenics followers, who saw with what great enthusiasm the turn-of-the-century families of the rich (and, as such, worthy) took copious advantage of newly introduced contraceptive devices and techniques to limit their own family sizes, denounced birth control as "race suicide."

Until World War II, only a few more sophisticated scientific racists, such as Lothrop Stoddard, Edward M. East, and Guy Irving Burch, saw the birth control movement as an answer to their elitist and racist dreams. Most eugenicists put down the birth control movement (started as a product of pre-World War I urban radicalism) as a dangerous and uncouth conspiracy of socialists, liberals, Jews, anarchists, crackpots, Jew lovers, racial degeneraters, and otherwise dysgenic hyphenated Americans.

Margaret Sanger had coined the term "birth control," and created her movement with two stated goals: (1) protecting families from the economic and emotional burdens of having unwanted children; (2) preserving poor pregnant women from the agony of dying of the toxemias of self-induced or neighborhood amateur abortions. When Mrs. Sanger opened America's first birth control clinic, in 1915, it was set up in a Brooklyn slum neighborhood peopled so largely by poor immigrant Italian and Jewish families that the handbills she distributed were printed in three languages—Italian, Yiddish, and English. This "sentimental softness" toward poor mothers of "the less suitable races and strains" was easily explained by types like Davenport and Madison Grant. In 1915 Mrs. Sanger was, after all, the health columnist for the *New York Socialist Daily Call,* as well as the editor and publisher of the ultraleftist magazine *The Woman Rebel.*

But by 1925 Mrs. Sanger had long since become a true-believing convert to eugenics, Social Darwinism, and neo-Malthusianism. As early as 1919, in the May issue of her *Birth Control Review* (the monthly successor to *The Woman Rebel*), she had written: "More children from the fit, less from the unfit—that is the chief issue of birth control." Despite her conversion to Galton's civic religion, Mrs. Sanger's formal invitation to the dean of American scientific racism, Dr. Charles Benedict Davenport, to participate as a vice-president in that year's Birth Control Conference, was coolly but firmly declined. As Davenport stated the position of the hierarchy of the eugenics cult in his reply of February 13, 1925:

"For one thing, the confusion of eugenics (which in its application to humans is qualitative) with birth control (which as set forth by most of its propagandists is quantitative) is, or was considerable, and the association of the director of the Eugenics Record Office [Davenport] with the Birth Control Conference would only serve to confuse the distinction. I trust, therefore, you will appreciate my reason for not wishing to appear as a supporter of the Birth Control League or of the conference."[6]

After World War II, however, there was a pronounced shift in postures. In the United States—*despite the fact that the American birth and fertility rates went into continuous and historically unprecedented and still very much ongoing and record-breaking decline* starting in 1955—thousands of literate, educated, and influential people began talking, writing, and waxing paranoid about what was termed, for various reasons wholly unrelated to the verified biological facts of the matter, the American Population Explosion. (See chapters 16–17.)

Out of this new "dancing mania" about the totally nonexistent American Population Explosion rose new mass movements, such as Zero Population Growth, Inc., dedicated to saving us from this non-menace. Unlike the hardcore scientific racists who work up storms about the ultra-fecundity of the hereditary defectives and paupers—meaning most of our blacks plus the numerically greater population of white poor and near poor—many of the most active and vocal partisans of Zero Population Growth (ZPG) and other Population Bomb crusades are not racists. Some of them, in fact, have proven themselves to be active opponents of the old-fashioned gut racism.

It is merely that they have not the faintest suspicion of the fact that the entire postwar Population Explosion concept and the *Population Control* Movement (as opposed to the clearly nonracist and *individual birth control* movement of 1915) that it has fathered are now essential ingredients of the new scientific racism. "Where historically the birth control movement had worked to protect the poor from the ravages of unwanted pregnancies, the new population control movement emerged to protect the nonpoor from the fiscal and social ravages of the unwanted poor."[7]

While many card-carrying ZPG members, who dutifully distribute Population Bomb literature and proudly wear inane lapel buttons reading PEOPLE POLLUTE, are not racists, it is also next to impossible to find an American racist—scientific or old-time unvarnished Ku Klux Klan variety—who is not at the same time a thoroughgoing believer in the Population Explosion, and an ardent supporter of compulsory birth control, sterilization, and even mandatory abortions for the less suitable races and strains.

It has been far from a coincidence that, since the Population Explosion mania took hold, the bulk of the poor people subjected in ever increasing numbers to involuntary sterilization operations were rendered barren in the operating rooms of federally supported birth control programs all over the United States. In his 1974 report, *Preliminary data on Sterilization of Minors and Others Legally Incapable of Consenting,* prepared for the Department of Health, Education, and Welfare, Carl W. Tyler, Jr., M.D., chief of the Family Planning Evaluation Branch of the Center for Disease Control, revealed:

> Blacks younger than age 21 underwent sterilization with greater frequency than did whites in the same age group.
> Welfare recipients under age 21 underwent sterilization more often than did those not receiving welfare, regardless of race.

To the eugenicists and other believers in (or dupes of) scientific racism, the Population Bomb hysteria became just another scientistic rationale for the forced surgical sterilization of the poor—*regardless* of color, ethnic origin, or religion. And, of course, the spurious menace of "overpopulation" provided very convenient rationales for not funding existing public health programs designed to keep infants alive and for vetoing congressional proposals for initiating new programs aimed at meeting presently unmet needs of human health.

ON THE LIFE-ENHANCING ALTERNATIVES THE LEGITIMATE SCIENCES OFFER TO SCIENTIFIC RACISM

Were we, as a nation or a species, without viable, practical, life-enhancing alternatives to the gonadal surgeons' knives as the primary tools of human and social betterment, the compulsory eugenic sterilizations of the poor, now becoming endemic in this country, would still be the moral abominations the war-crimes trials of the Nazi gelders found them to be at Nuremberg. Our life and behavioral sciences, however, have for over a century combined to give us not only moral but also very practical and demonstrably wealth-incrementing alternatives to the genocidal gonadal programs of the scientific racists.

Human history has been more than a parade of fools and knaves. Victorian England, for example, contained among its educated people leaders very unlike the palpably class-conscious agitator for the advancement of the "superior races and strains'" of mankind, Sir Francis Galton. There were, for example, the prominent physicians never mentioned in any of Galton's writings, men like Drs. Neil Arnott, Joseph Toynbee, and Southwood Smith, whose work on the government's Royal Commission on the Health of Towns resulted in the historic reports of 1844 and 1845. Documented scientific reports which, as Chadwick observed, "brought scrofula [tuberculosis] home to defective ventilation."[8]

The publication and wide distribution of these Parliamentary Reports by serious British physicians led to the formation of a number of large public health associations. These included the Association for Promoting Cleanliness among the Poor; the Society for the Improvement of the Conditions of the Labouring Classes; and the highly influential federation of scores of municipal, county, and regional public health associations all over England, the Health of Towns Association. In the minutes of its first public meeting, on December 11, 1844 (twenty-five years *before* the publication of the Sacred Writ of the eugenics cult, *Hereditary Genius*), the Health of Towns Association announced that it had been formed "for the purpose of diffusing among the people the information obtained by recent enquiries, as to the physical and moral evils that result from the present defective sewerage, drainage, supply of water, air and light, and construction of dwelling houses: and also for the purpose of assisting the legislature to carry into practical operation any effectual and general measure of relief, by preparing the public mind for the changes."[9]

Galton not only blamed the declining birth rates of the wealthy families on the adverse effects of urban living on the fertility of the superior races and strains of humanity. He also blamed the urbanization of swiftly industrializing England for what he equally erroneously thought was the soaring birth rates of the less suitable strains of white Englishmen. In 1873, for example, in an article in the January issue of *Fraser's Magazine* called "Hereditary Improvement," Galton exhibited as great a talent for missing the meaning of his half cousin Charles Darwin's theory of variation, natural selection, and evolution as he did for the great public health movements of England that, in a generation, had caused an unprecedented decline in infant mortality in the first segment of the British population to benefit from these gains in environmental biology: the upper classes. As doctors well knew in Galton's times, a decline in infant mortality in any population is invariably followed by a decline in the live birth rates. Galton, however, had a truly Olympian capacity for rising above the vulgar laws of human biology. Thus, in *Hereditary Improvement,* he wrote:

> . . . the ordinary struggle for existence under the bad sanitary conditions of our towns seems to me to spoil and not to improve our breed. It selects those who are able to withstand zymotic [infectious] diseases and impure and insufficient food, but such are not necessarily foremost in the qualities which make a nation great. On the contrary, it is the classes of a coarser organization who seem to be, on the whole, most favoured under this principle of selection, and who survive to become parents of the next generation. Visitors to Ireland after the potato famine generally remarked that the Irish type of face seemed to have become more prognathous, that is, more like the negro in the protrusion of the lower jaw; the interpretation of which was that the men who survived the starvation and other deadly accidents of that horrible time were

more generally of a low or coarse organization. So again, in every malarious country, the traveller is pained by the sight of the miserable individuals who inhabit it. These have the pre-eminent gift of being able to survive fever, and therefore, by the law [*sic*] of economy of structure, are apt to be deficient in every quality less useful to the exceptional circumstances of their life.

The Victorian physicians, lawyers, and statesmen of the Health of Towns Association and other voluntary agencies of the nineteenth century's great Sanitary Movement, on the other hand, knew that the poor whose labors in mine and mill made the wealth of the nation (and the fortune that Galton had inherited) were far less protected against the *preventable* infectious and deficiency diseases of humankind than were people born into the more affluent families, such as their own. The Sanitary Reformers, unlike Galton, knew very well that, far from being "able to withstand zymotic diseases and impure and *insufficient* food," the hard-working, underpaid, badly housed, and—to the Galtons of this world—less suitable strains of English people who dug the coal and manned the machines and tended the furnaces that were making the nation great had death rates that were soaring because of the insufficient food and the "physical and moral evils that result from the present defective sewerage, drainage, supply of water, air and light, and construction of dwelling houses [i.e., slum tenements]" that came with the new urbanization of the Industrial Revolution.

Where Galton called for a frontal attack on the fertility of the new urban poor, the Sanitary Reformers mounted demands for legislative and social actions against the new industrialized environmental conditions that bred tuberculosis, malaria, epilepsy, and, among other preventable afflictions, mental retardation. Nor was this vast network of public health societies an expression of what the scientific racists of two centuries have categorized as "naïve liberalism."

The gadfly of the Health of Towns Association, and author of many of its principal resolutions for over two decades, was Sir Edwin Chadwick, a fiscal conservative who looked upon labor unions as conspiracies against progress, and upon their leaders as dangerous demagogues. A philosophical radical and protégé of the Utilitarian legal and civic reformer Jeremy Bentham, Chadwick believed just as strongly in full employment at high wages. (Low wages meant poverty, and poverty after the Industrial Revolution meant the proliferation of crowded slum tenements and slum areas where diseases due to lack of adequate food and sanitation were endemic.) Yet, as a New Poor Law administrator, Chadwick sought to make residence in the infamous workhouses like "a cold bath—unpleasant in contemplation but invigorating in its effects."[10]

The patronizing appellation of "naïve liberal" was even less suited to the author of one of the books that helped swell the European demands for clean water, safe housing, sanitary sewage disposal, and the high wages that were and remain the first line of defense against environmental decay. His name was Friedrich Engels. The son of a wealthy German-Jewish textile manufacturer in the Rhineland, Engels was so appalled by the conditions he found in Manchester that he embarked upon the systematic series of personal observations which were to culminate in the publication of his *The Condition of the Working Class in England in 1844,* issued originally in German in 1845 and in English translation in 1887. Engels' angry book ranks with Chadwick's Parliamentary *Report on the Sanitary Conditions of the Labouring Classes,* issued in 1842, as a savage dissection of the social biology of poverty in industrializing societies. Yet Engels saw no hope for the improvement of the human condition in the sanitary and health reforms for which Chadwick was to fight for the rest of his life.

Engels, the disciple, friend, and collaborator of Karl Marx, called for total social revolution, rather than for sanitary and public health legislation.[11]

Not every leader of the Health of Towns Association shared the same set of political ideals advanced by Chadwick. The Association's founding committee of 1844 included people of the landed aristocracy; merchants; the physicians John Simon (Galton's instructor in anatomy during Galton's aborted medical education), Joseph Toynbee, and R. D. Grainger; bishops and vicars; Whigs; Tories; and "even radicals like Hawes and Sheil."[12]

The founding committee's Tories included the forty-year-old activist in the Young England wing of British conservatism, Benjamin Disraeli. Two decades later, as the Prime Minister fighting for the passage of the landmark Public Health Act of 1875, Disraeli was to tell the Parliament that "the public health is the foundation on which repose the happiness of the people and the power of a country. The care of the public health is the first duty of a statesman."

OUR FIRST ENVIRONMENTALISTS

Disraeli had learned the lessons of preventive medicine well during his years of association with the large public health movement of Victorian England that was never so much as mentioned in any of Galton's books or major addresses. The barrister-at-law Edward Jenkins was another nonphysician who had learned these lessons during his involvement in the British public health movements. Jenkins' famous report to the British Social Science Association in 1867, *The Legal Aspects of Sanitary Reform,* opened with the words:

"Public health is public wealth. Every person laid aside by ill health is so much subtracted from the power and capacity of the state: and more than this, every person so laid aside is a drain on the resources of the state."[13]

Another well-known but ignored contemporary of Galton's, Alexander P. Stewart, M.D., an active member of the large and vocal group of British physicians and civic reformers who helped achieve the framing and the passage of the long-needed Public Health Act of 1875, was the author of an equally historic report to the Social Science Association in the same year, *The Medical Aspects of Sanitary Reform.* Unlike Galton and the eugenicists, who wrote that the contagious diseases of the poor were the inevitable products of their inferior hereditary endowments, Dr. Stewart and his peers proved again and again that when it came to human biology and human health Galton and his acolytes simply did not know what they were talking about.

Stewart and his collaborators proved time and again that a society whose leaders suffered no eugenic illusions about the causes of the commonest contagious and deficiency diseases of urban industrial towns could—even more than a century ago—cut the death tolls from such diseases in half in less than twelve months. There was, for example, Dr. W. S. Trench, the health officer of the city of Liverpool, whose successes in reducing the deaths from four of the most endemic preventable diseases in a single year were summarized in a table and described in Stewart's 1867 report:

Deaths From	1864	1865
Small Pox	121	37
Scarlatina	81	37
Measles	76	21
Typhus (including Typhoid Fever)	71	63
	349	158

From Stewart and Jenkins, 1867, p. 70.

This reduction in Liverpool deaths from what were eventually classified as two bacterial and two viral diseases was accomplished without vaccines of any sort. Only one vaccine was then available: the smallpox vaccine Jenner had developed in 1796. But the Malthusians who still controlled the governmental purse strings had seen to it that the Vaccination Act of 1853 was turned into a dead letter for want of funding. The germ theory of Pasteur and Koch was, in 1864—the year of this Liverpool episode in the history of social medicine—still more than a decade into the future. The sulfas and antibiotics that act against the microbial agents of scarlet fever (a major cause of deafness in childhood), typhoid fever, and typhus were nearly a century away. The isolation of the measles virus and the creation of the anti-measles vaccine made with this virus were another century into the human future.

The Liverpool "miracle" was accomplished by much less scientifically sophisticated means. As Stewart described the efforts of Dr. Trench:

In [Trench's] report for 1865, he gives the following summary of sanitary operations during the year:—"Pigs were removed from 62 places where the keeping of them was a cause of nuisance to the neighborhood; 124 privies have been converted to water-closets; 21 pits of stagnant water were drained; 2,451 nuisances arising from obstruction of drains, defective traps, &c., were reported by the inspector, and proceedings taken to remedy the same; 4,827 houses, containing 19,263 apartments, were visited with a view to improving their sanitary condition; 1,043 lime-washing notices were served upon occupiers of dirty houses, and attended to by them; 25 cellars, used as dwellings, were vacated; 70 overcrowded houses had their numbers reduced; 37 persons were convicted before the magistrates of offenses against sanitary laws, and penalties amounting in the aggregate to £18, 10s, inflicted."[14]

All of which meant that the public health measures responsible for the successes in Liverpool against four of the then common contagious diseases of urban poverty were confined solely to eliminating the kinds of urban filth attacked by Chadwick in London; by Florence Nightingale in the military hospitals in the Crimea; by Max von Pettenkofer in nineteenth-century Munich; and by public health workers in our own and other industrial nations to this very day.

There have, of course, been many changes in the living environments of industrial nations since Chadwick. One thing that, however, has not changed from the time of the first environmentalists a century and more ago has been the timbre of the blood-chilling cries of self-pity and warnings of impending doom that, historically, have always arisen from the owners of the prime sources of environmental health hazards every time they are asked to mend their ways. Charles Dickens, writing of the factory owners of Manchester— the archetypal new industrial center he called Coketown in his 1854 novel, *Hard Times*—said it very bitterly:

> Surely there never was such fragile china-ware as that of which the millers of Coketown were made. Handle them never so lightly, and they fell to pieces with such ease that you might suspect them of having been flawed before. They were ruined, when they were required to send labouring children to school; they were ruined when inspectors were appointed to look into their works; they were ruined when such inspectors considered it doubtful whether they were quite justified in chopping people up with their machinery; they were utterly undone, when it was hinted that perhaps they need not always make so much smoke. . . .
>
> Whenever a Coketowner felt he was ill used—that is to say, whenever he was not left entirely alone, and it was proposed to hold him accountable for any consequence of his acts—he was sure to come out with the awful menace, that he would "sooner pitch his property into the Atlantic." . . .
>
> However, the Coketowners were so patriotic after all that they never pitched their property into the Atlantic yet, but, on the contrary, had been kind enough to take mighty good care of it. So there it was, in the haze yonder, and it increased and multiplied.[15]

Some 115 years later, as a new American environmental movement —much more modest in scope and purposes than the Health of Towns Association of Disraeli and Chadwick—started to talk about conserving trees and reducing air and water pollution, John Ehrlichman, special counsel to the President of the United States, turned up at a John Muir Society conference in Aspen, Colorado, to voice much the same threats as Dickens' Coketowners. Nothing, Mr. Ehrlichman told the American environmentalists in 1969, was dearer to the heart of the Nixon administration than a lovely "entente with nature." However, he went on to warn them, they had better come up with some "politically feasible solutions" to the problems of industrial pollution, because industrial growth was the key to national growth, and "a nation which does not grow perishes."[16] In short, he told the new environmentalists that they were free to clean up all the air and water pollution they liked as long as they did not interfere with the sacred processes of industrial growth.

Ehrlichman's warning of 1969 has since been often repeated, in terms even more reminiscent of those Dickens satirized in 1854, by the major producers of industrial and transportation pollution, as well as by Mr. Nixon's chosen successor, Gerald Ford. "We must free the business community from regulatory bondage," President Ford declared in an address before a business group in Washington, D.C., on June 17, 1975, and he singled out regulations to protect the environment as the kind of government regulation that "often does more harm than good." Like Mr. Nixon, Mr. Ford said that "the issue is not whether we want to control pollution. We all do. The question is whether the added costs to the public make sense when measured against actual benefits."[17]

Presidents now do the main polluters' wailing for them.

THE CRITICAL CHOICES OF MODERN TIMES

In our own day, more intensive uses of sanitary, medical, and social modalities than those enjoyed by the Victorian Sanitary Reformers have worked even more dramatic "miracles" than they achieved against the major preventable diseases that were, as late as 1900, the major killers of our population. These included pneumonia, influenza, diphtheria, scarlet fever, nephritis, meningitis, measles, and tetanus, to name but a few.

We now have an enormously wider array of sophisticated scientific tools and strategies for the prevention of the costly physiological and mental disorders of malnutrition, of microbial and viral infections, of chemical and radiological and other industrial and environmental causes of killing diseases than Dr. Stewart could ever have dreamed of in 1867, when he bemoaned the hard and ugly medical reality that "the mortality from small-pox is greater in the country of Jenner than in any other country of Europe."[18]

Prominent among these new strategies of human health are longitudinal studies of human development under different sociobiological conditions, and long-range intervention experiments on the effects of providing various otherwise unavailable human needs to developing human beings. Some ongoing examples of both of these encouraging modern scientific techniques are described in many chapters of this book. They range from the already historic studies of the nature and prevention of "the continuum of reproductive casualty" by the psychiatrist Benjamin Pasamanick and his wife, the pediatrician Hilda Knobloch, to the exhilarating Milwaukee Project. In this project the psychologist Rick Heber and his University of Wisconsin collaborators have demonstrated how very, very simple it is, indeed, to enrich the growth environment of the children of randomly selected inner-city poor parents with mean IQ test scores of under 80 with what such deprived infants require—in the way of a few very well-known cultural, emotional, and physiological necessities—in order for these children to achieve mean IQ test scores of 118 at the age of six.

While the new scientific racists peddle the nineteenth-century nostrums of eugenics and Social Darwinism, our twentieth-century life, behavioral, and related scientific research communities—*their work paid for largely out of our taxes*—have presented all of us with infinitely more useful basic knowledge about the nature of our bodies, our minds, and our human potentials in this century than had been provided during the previous millennium of basic clinical, biological, chemical, and physical research.

Because of the work of our real scientists, our entire species, for the first time in the history of science and civilization, now is in possession of viable bases of hope for the entire human condition; for the human futures of our children and their grandchildren.

This scientific knowledge is part of the heritage of all Americans.

Scientific racism is the chief obstacle to the fullest and swiftest use of this vast knowledge of human life for the maximum benefit of the greatest number of our infants, our children, our mothers, ourselves.

It is, therefore, impossible to cope successfully with the multifactorial dangers that scientific racism now inflicts upon all of us without an overview of some of the hard knowledge and directions of the sciences dealing with the genetics, the bodies, the minds of ourselves and our children. Most of the closing four chapters of this book are devoted to just such a review of the real sciences that provide our only satisfactory answers to the quack "science" of the sons of Galton, Spencer, and Malthus.

There is nothing about the work and findings of the biomedical sciences that cannot be presented in terms that every literate human being can readily comprehend—hopefully, with enjoyment, but certainly with ease. For readers who are, like myself, neither physicians, biologists, nor psychologists, there is a glossary of many of the scientific words used in this book starting on page 670.

Finally, before you turn to Part Two, it might be well to fix in your memory two symbols of what the preachments and social policies of scientific racism in this century have meant to our national welfare (see pages 64–65).

The first symbol is the flyer distributed by the American Academy of Pediatrics as part of the publicity campaigns that became the government's substitute, in 1973 and again in 1974, for the meager funds previously provided for vaccinations in the Federal Vaccination Assistance Act, which our Congress in its infinite wisdom allowed to expire on June 30, 1969.

The second mnemonic is a table, medically linked to the information in the Academy of Pediatrics flyer, on the two most sensitive indices of health in all human societies as they now stand in the world's richest nation—and in a dozen other industrial nations; and in two characteristic nonindustrial and very poor nations.

This, then, is what a century of abusing our native-born white poor, slandering the Irish, baiting the Jews and the Italians and the other immigrants, and Jim Crow—all unleashed upon the rest of us in the scientistic names of eugenics, and Social Darwinism, and IQ test scores—has already cost the people of the United States:

> One third of our children not vaccinated against the known destroyers of body and mind for which vaccines already exist.[19]

TWO AMERICAN LEGACIES OF SCIENTIFIC RACISM

One in three preschoolers aren't immunized.

That means more than five million children one to four years old are unprotected against either polio, measles, rubella (German measles), whooping cough, diphtheria or tetanus. Is your child one of them?

Check the table below and ask your family doctor if your child is up to date on his shots. It could save his life.

Measles, rubella, polio, whooping cough, diphtheria and tetanus are not just harmless childhood diseases. All of them can cripple or kill.

All are preventable.

In order to be completely protected against diphtheria, tetanus and whooping cough (pertussis) your child needs a shot of the combination diphtheria-tetanus-pertussis (DTP) vaccine at two, four, six, and eighteen months and a booster when he goes to school. When he gets his DTP shots he should also get a drop of the oral polio vaccine.

After his first birthday your child should have a shot for measles, rubella and mumps. This also can be given in one combination shot. He should be tested for tuberculosis in his first year, and when he's 14 to 16 years make sure he gets his tetanus diphtheria booster shot.

If you don't have a family physician call your local public health department. It usually has supplies of vaccine and may give shots free.

Recommended by The American Academy of Pediatrics	DTP	Polio	Measles	TB Test	Rubella	Mumps	Tetanus-Diphtheria
2 months	▓	▓					
4 months	▓	▓					
6 months	▓	▓					
1 year			▓	▓	▓	▓	
1½ years	▓	▓					
4-6 years	▓	▓					
1-12 years			▓		▓	▓	
14-16 years							▓

Published by the American Academy of Pediatrics

	Infant Deaths per 1,000 Live Births Yearly	Estimated Life Span at Birth	
		Male	*Female*
Sweden	9.2	72.12	77.66
Finland	10.1	65.89	74.21
Netherlands	11.0	71.2	77.2
Japan	11.3	70.49	75.92
Iceland	11.4	70.7	76.3
Norway	11.8	71.24	77.43
France	12.1	68.6	76.4
Denmark	12.2	70.7	76.10
Switzerland	13.2	70.15	76.17
Luxembourg	13.5	NA	NA
Canada	15.5	69.34	76.36
England & Wales	15.9	68.9	75.1
German Democratic Rep. (East Germany)	16.0	68.85	74.19
Byelorussian S.S.R.	16.0	68.0	76.0
New Zealand	16.2	68.19	74.30
United States	16.5	67.4	75.2
Chile	70.9	60.48	66.01
Liberia	159.2	45.8	44.0

NA = Not available
From *United Nations Demographic Yearbook* (1974).

Millions of needlessly dead infants.

Even greater numbers of children and adults who die long before their genetically programmed times, needlessly and for want of social and medical legislation whose principles were known as far back as the first public statement of the Health of Towns Association in 1844.

All this, and the irretrievable loss of the optimal genetic potentials for industrial, scientific, and artistic accomplishment in the life-stifled genes of at least one third of our adult population during fully half of the years of our history as a nation.

This is what scientific racism is all about—and why it should concern us all.

PART TWO

From Malthus to Hitler: The Rise, Conquests, and Twilight of the Old Scientific Racism

4 Natural Law and the Undeserving Poor: The Birth of the Old Scientific Racism in Europe

> Instead of recommending cleanliness to the poor, we should encourage contrary habits. . . . we should . . . crowd more people into the houses, and court the return of the plague. . . . But above all, we should reprobate specific remedies for ravaging diseases; and those benevolent, but mistaken men, who have thought they were doing a service to mankind by projecting schemes for the total extirpation of particular disorders. If by these and similar means the annual mortality were increased from 1 in 36 or 40, to 1 in 18 or 20, we might possibly every one of us marry at the age of puberty, and yet few be absolutely starved.
>
> —PROFESSOR THOMAS MALTHUS, in *An Essay on the Principle of Population,* Book IV, Chapter 5 (second edition, 1803)

Because of the close ideological and collaborative links between the American eugenics movement and the leaders of Nazi Germany—working ties forged long before Hitler took power in 1933—most American scientists believed that American scientific racism had become one of the unmourned casualties of World War II. (See chapters 15 and 16.)

However, by 1966 the clamor in dozens of American state legislatures from California to Mississippi for punitive sterilization laws against welfare mothers and other traditional human targets of the old scientific racism, coupled with the 1954 revival of the IQ testing myths in terms that had not been heard in the land since the presidencies of Warren Gamaliel Harding and Calvin Coolidge, had long since become everyday news in the mass media.[1] It was not by accident that, in 1965, 20 percent of all Americans polled by the Gallup organization recommended that the solution to the problems presented by the fecundity of welfare mothers was to "sterilize the women." Or that, before and since that Gallup poll, swelling proportions of educated Americans had been declaring that, since the IQ test scores *proved* that some races (meaning the nonwhites) were by heredity ineducable, it was a social crime to continue to waste tax dollars in futile attempts to make them as smart as their genetic superiors (the whites).

The organized societies of professional life, behavioral, and social sciences, and finally the entire American Association for the Advancement of Science (AAAS), issued strong statements warning against the further revival of the scientific racism that had come to full flower in the sterilization mills and extermination camps of Nazi Germany.[2] These declarations of principles by America's scientists have had little effect upon the successes of the postwar revival of the scientific racism of Galton, Davenport, and Hitler's "raceologists" and ethologists from Alfred Rosenberg to Konrad Lorenz.

To people old enough to remember how the widespread corporeal applications of Galton's teachings followed the advent of Nazism in Germany, there was a chilling *déjà vu* quality to the post-Nazi era resurrection of eugenics and other branches of scientific racism. Few people have expressed this feeling more powerfully than did the Columbia University professor of anthropology Morton H. Fried, in the paper he presented at the symposium on science and the concept of race held at the 1966 AAAS meetings in response to the initial challenge of the new scientific racists.[3]

Dr. Fried's paper, "On the Need to End the Pseudoscientific Investigation of Race," embraced much the same position taken by the AAAS and other scientific bodies, namely, that there was no basis in science and its data for the revival of the racist myths of eugenics and its Cult of the Nordic Superman. Then, after making a factual reply to the claims of the revivalists of the creed of Galton, Spencer, and Hitler, Professor Fried confessed to something else:

> As I wrote these remarks, two notions preyed on my mind. One was that few persons would find anything to disagree with in the general points I have been making. That is, apart from some of the details, I suspect that educated people are willing and able to discriminate between serious, methodologically well-founded studies and those which I have labeled as pseudoscientific. And with this thought arose another: *that all of this had occurred before, not once but several times,* that Herskovits and Klineberg and other fine social scientists had passed this way before, and often with more cogent and more elegant arguments than these now presented. I do not say that they failed in their intentions; certain ameliorations have occurred. But one thing they did not do is stop the nonsense. The pseudo studies go right on. What is more, some fine scientific journals throw open their pages to serious discussion of this nonsense. Those who oppose the pseudo study of race are called "equalitarians," and this term has been skillfully manipulated to make it appear as if there were two valid camps participating in the normal scientific exchange. But this is not a question of digging the "Mohole" or not, or whether *Homo habilis* is or is not an Australopithecus. *It is more like dividing on the question of whether or not to exterminate the Jews: one side says no and presents its arguments, and the other side says yes and presents its arguments, and this too becomes a debatable scientific question.*
>
> Participation in a debate over racial differences in intelligence, ability, or achievement potential is not a means of asserting and spreading knowledge of the views of professionally concerned scientists. Quite the opposite, it is a means of lifting in the public eye the status of studies that are otherwise disqualified and rejected by science. *There is a need to stop the pseudoscientific investigations of race.* There is even more a need to end practices whereby such studies are treated as serious intellectual endeavors. If we are to recognize them for what they are, *expressions of bias and propaganda tracts favoring certain social arrangements,* it will be easier to confine them to the pages of such magazines as the *Mankind Quarterly.* [Italics added.]

But three years later, Arthur Jensen's famous eugenics propaganda tract, *How Much Can We Boost IQ and Scholastic Achievement?,* was printed not in the *Mankind Quarterly,* a British eugenics journal, but in the *Harvard Educational Review.* And fully five years after Professor Fried warned of the social dangers of taking pseudoscientific race studies seriously as *science,* the senior science adviser to the President of the United States was quoted in *Science,* the weekly magazine of the AAAS, as declaring that the refusal of the National Academy of Sciences to spend tax funds on the studies of racial intelligence demanded by the eugenicist William Shockley was a failure to experiment that would "bring the best part of American society as we know it today to a halt."

As the post-Auschwitz revival of scientific racism in this country soared from peak to peak, it began to take on the cold ambience of a nightmare trip in a time machine, one of those terrifying nightmares in which the infernal machine hurtles us backward in time to the worst rather than the best of times removed. Not to the discovery of the tubercle bacillus but to the passage of the Nuremberg Laws; not to the discovery of the germ theory but to the siege of Paris; not to the invention of the microscope but to the Thirty Years' War; not to the medical treatises of Maimonides but to the Holy Inquisition.

Where did it all begin, this perversion of science for purposes of social regression? Can we truly understand the magnitude of the dangers the revival of scientific racism poses to our children, our society, our future as a nation, without our first knowing at least some of the bare outlines of the roots from whence it springs? I doubt it. Painful as it might prove, I think the time has come to examine these roots, to relive certain episodes of our civilization's not so distant past.

These were unlovely episodes, for the most part, with many fools and clever villains, but with few heroes. Most of this minority of heroes were destined to die knowing that in the main they had failed.

It was an aspect of human history whose every era was marked by the flight of reason. There was, for example, one era in the present century in which patricians and pundits joined forces with the Ku Klux Klan and the American Federation of Labor in common cause against Jews, Catholics, blacks, and foreigners. And in which professors and scientists, senators and Presidents, believed and mouthed the Aryan-supremacy slogans that were later to be used word for word by Hitler, Göring, and Goebbels. An era in which the noble and life-enhancing scientific works of Pasteur and Koch, Binet and Simon, Darwin and Wallace, Mendel and Morgan, were perverted into pseudoscientific arguments for the "eugenic" breeding of races of long-headed, Nordic, and Aryan supermen—and into parallel arguments for genocide as the ultimate instrument for protecting the genes of the supermen from the "dysgenic" threats posed by the "inferior" races.

And in the nineteenth century, in which—before they bought their first lace curtain, owned their first bank, elected their first President—the peasant Irish refugees from the famines of 1848 and the racist abuses of their British masters shared, with the emancipated blacks, the main thrust of the genteel as well as the gutter racism of post-Civil War America. An America whose educated classes, Ivy League professors, and fashionable WASP clergy packed the university auditoriums and lyceums of the land to cheer and applaud as the Oxford professor of history Edward A. Freeman delivered his favorite mot: "The best remedy for whatever is amiss in America would be if every Irishman should kill a negro and be hanged for it."[4]

It was also the century in which William Graham Sumner, Social Darwinist, Episcopal rector, and professor of political and social science at Yale, threw the combined weight of God and Nature at the non-affluent, warning them that while "a man may curse his fate because he is born of an inferior race," God would not pay the slightest heed to such "imprecations."[5] And in which social reformer Henry George, as the frontier lands of the West were nearly all homesteaded by 1883, and as immigrants from many European lands continued to arrive in this nation, asked: "What, in a few years more, are we to do for a dumping-ground? Will it make our difficulty less that our *human garbage* can vote?" (italics added).[6]

Yes, it was a most unlovely history, in which reformers and scientists, clergymen and professors, used the proudest achievements of human culture to justify the most inhuman treatment of the poor, the lowly, the hewers of wood and drawers of water, and even the "middling classes." Because there are a number of excellent histories of the costly political triumphs of the old scientific racism,[7] I need not tell this story outrage for outrage, holocaust for holocaust, each of the tragic acts of Congresses and Presidents and all. In this section of the book, I will therefore confine myself to merely enough of this history, and some of its makers, to present a clear idea of why these relics of our not too distant past should concern us so very much today. Particularly if we are among those who are parents or grandparents, both human responsibilities that cannot be lightly avoided.

What's past is prologue, said the cynical brother of Prospero. But prologue to what? To a replay of the same human and social tragedies? Or to an educated awareness of the dangers inherent in such racial scapegoating for the exacerbations of our times? An awareness that can and must keep us from repeating and compounding the intellectual and moral errors of yesterday—at costs in human health and hope now much too steep for any human society.

What's past need *not* be prologue to a new Inquisition; a new St. Bartholomew's Day Massacre; a new Thousand-Year Nordic and Eugenic Reich. It can also be prologue to a new era of the kind of equality of human opportunity that, by the mid-nineteenth century, had proven to the poor of this nation and the world that illiteracy, lowly occupations, and poverty were not the inexorable products of Divine or Natural Law, but, rather, of repressive and backward societies that denied all but a few "wellborn" children the human and civilized rights to health, education, and entrance into the skilled trades and the noble professions. And an era in which, at a pace unmatched in the history of any earlier civilization, the full benefits of research in the life, behavioral, and social sciences will be applied to the lives of every human being from the moment of conception. Real science is, as we shall see, society's best antidote to the poisons of scientific racism.

STREWER OF THE SEED OF SCIENTIFIC RACISM

Whether it be Sanitary Reform or scientific racism, no major historical development—whether for good or for evil—arises *de novo* from a temporal or cultural vacuum.

Racism—which according to Webster is "the program or practice of racial discrimination, segregation, persecution, and domination, based on *racialism,*" which in turn is defined as "the doctrine or feeling of racial differences or antagonisms, especially with reference to supposed racial superiority, inferiority, or purity"—was not new to the nineteenth century. Nor, for that matter, was it new to our millennium.

What was new, and distinctly a part of the nineteenth century, however, was *scientific racism,* or the creation and employment of a body of legitimately scientific, or patently pseudoscientific, data as rationales for the preservation of poverty, inequality of opportunity for upward mobility, and related regressive social arrangements. In the performance of these functions, scientific racism has often also institutionalized and lent scientific respectability to racist dogma and practices that were all far, far older than science itself.

Three historical developments combined to make England the logical site on which scientific racism was to be born. The first two events—the Agricultural Revolution, and the subsequent drop in national death rates caused by the gains of both the Agricultural and the subsequent Industrial Revolution—were of course not confined to England alone. The third event, the Georgian Enclosures, which ended the era of the small farmer and the cottage or village industries, was uniquely British.

The Agricultural Revolution that, starting around 1700, swept through Europe was, as John W. Osborne wrote, "the greatest move forward in agriculture since neolithic times."[8] It introduced new crops, new methods of food and fiber production such as crop rotation and animal husbandry based upon the introduction of turnips and other root foods for animals, new legumes and grasses that enabled farmers to raise more cattle, and new methods of planting and harvesting. The new legumes, with bacteria in their root nodules that fixed nitrogen in the soil, and the manure of the vastly increased herds of cattle also helped improve the food-producing capacities of the European earth itself.

In England, where the Agricultural Revolution's major advances were not felt until after 1750, "grain output increased 43% during the 18th century and even more rapidly after 1800,"[9] when the mechanical contributions of the young Industrial Revolution contributed to the growth of food and fiber yields per acre. The introduction of turnips and new root fodder crops, which made winter feeding of sheep and cattle feasible, led to significant increases in the size and weight of meat animals. "Contemporary estimates of livestock suggest strongly that the output of wool and mutton increased considerably during the eighteenth century—much faster than the rate of [wheat] production."[10]

During the eighteenth century, when wool cloth was England's chief export, the source of this wealth was produced in the rural areas by the cottage carders, spinners, weavers, and dyers. These familiar village artisans managed to meet their families' biological needs by also keeping a cow or two, raising poultry enough for the family table, and growing potatoes and other vegetables in the village common lands. These commons, essential to the functions of the cottage industries, were not public lands, as a rule, but the legal property of the local manor lords. As the Agricultural Revolution set new horizons of profitability for large-scale agriculture, the legal owners of these traditional common lands started to expel the village artisans and other small farmers from these now potentially profitable lands. Some users of common lands, with quasi-legal claims to their use of them, were bought off by the manor lords. Those villagers who resisted expulsion from the pastures and garden plots of the common lands were legally driven off with the aid of the Enclosure Acts passed by Parliament.

These Enclosure Acts enabled the large landowners to apply to their own vast holdings the discoveries of the Agricultural Revolution. Food production soared. But with the Georgian Enclosures came the end of the cottage artisans, starting around 1760. In order to survive, the village craftsmen and the new landless small farmers migrated in search of jobs in the new industrial towns and in the mines whose coal powered the looms of the new Industrial Revolution.

Ultimately, this rise in British food production, and its wider distribution via the new steam railroads, and the production of such products of the new factories as cheap washable cotton cloth, soap, and the pumps and pipes required to set up new municipal and regional water systems, were to lead to vast improvements in the national health. "In the first four decades of the 18th century excessive indulgence in cheap gin and intermittent periods of famine took a heavy toll of lives; but between 1740 and 1820 the [British] death rate fell almost continuously—from an estimated 35.8 [deaths per 1,000 population] for the ten years ending in 1740 to one of 21.1 for those ending in 1821."[11] This slash in the death rates, in turn, led to a net British population growth rate of about 1½ percent per year between 1801 and 1831.[12] This increase in the population growth rate was, of course, due to the decline in the death rate, and not to any increases in national fecundity.

In human terms, the Georgian Enclosures and the Industrial Revolution, for all of their long-range gains to the nation, became for thousands of English families the instruments of harsh changes, dislocations, forced reliance on private and parish charity, and the making of new and precarious livelihoods as miners, mill hands, and servants in the warrens and the mansions of the new industrial towns and cities. Hard times and poverty had by no means been unique in the village life of the pre-Enclosure eras; hard times and poverty in the pollution-choked, grimy new factory and coal towns, however, caused one in every two children born in the new industrial towns to die before reaching the age of five.[13]

In an era when farm, mine, and mill wages were too low to meet family nutritional and other survival needs, some concerned reformers in and outside Parliament sought the passage of meaningful minimum-wage laws. The minimum-wage concept was savagely attacked by William Pitt (the Younger),

Prime Minister from 1783 through 1801, and by nearly all other Tories as impracticable and conducive of increased unemployment.[14]

The various Poor Laws originally intended to provide relief to keep the displaced rural English families alive through these terrible generations of transition to a viable industrialized society only made life worse. They offered next to nothing to the new urban poor, who were not only underpaid but, in increasing numbers, rendered unemployable by mine, mill, and other industrial accidents suffered on their jobs. They also turned the new class of rural poor into virtual serfs at a time when "well over a third of the occupied population earned their living in agriculture."[15]

Instead of minimum-wage laws, most agricultural parishes settled for copies or variations of the Berkshire Bread Act initiated in 1795 at Speenhamland. Under the notorious Speenhamland system, the local parish tax funds (poor rates) were used to provide outdoor relief—as against the indoor relief of residence in the parish workhouse—in the form of wage supplements, based on the current price of bread, to employed farm workers. This enabled the underpaid farm laborers and their families to stay alive and put in a productive day's work despite their miserably low wages. These outdoor or out-relief bread subsidies "rapidly came to be regarded as the poor man's right, his compensation for the enclosure of his common land."[16]

The same "poor man's right" was also the large landowners' windfall, since it subsidized and in effect froze scandalously low farm wages at the expense of all parish taxpayers. Throughout England, no class of taxpayers paid a higher price for these government subsidies of the rich than did "the small farmers who employed little or no labour and so did not enjoy the rather dubious benefit of parish-subsidized labour, and this rate [tax] burden was perhaps one factor in the small man's decline."[17]

Nobody in England liked the Speenhamland and other Poor Laws— least of all the poor themselves, who but for the same unfair laws would have starved to death.

One man in England, however, opposed these Poor Laws for purely ideological reasons. He attacked them not because they offered the poor far too little in relief and human dignity but because, in his firm opinion, they handed out far too much of both. He held that "we are bound in justice and honour formally to disclaim the *right* of the poor to support." His name was Thomas Robert Malthus, and he made this dogmatic statement in Book IV, Chapter 8, of the second edition of his famous book, *An Essay on the Principle of Population* (1803).

Born in 1766 in the middle of the seventh decade of the Agricultural Revolution that saw his native land doubling its food and fiber production, Thomas Robert Malthus spent most of his adult life as a political economist. Earlier in his career, Malthus had taken holy orders and had served as a curate in Surrey between 1798 and 1805. In that year, he became England's first professor of political economy, holding the chair in the new East India Company College at Haileybury. Malthus retained this academic post until his death, in 1834.

During his lifetime, Malthus had watched British and European food production climb to ever new records as a result of the explosive development of new crops and farming techniques, as well as the endless series of planting, harrowing, plowing, reaping, harvesting, threshing, and transporting machines invented and put into mass production to expand the increased food yields made possible by the Agricultural Revolution. Where the Industrial Revolution added the benefits of mechanization and steam power to the Agricultural Revolution of the previous century, the scientific community expanded its theoretical and empirical foundations. Between 1790 and 1802, two university departments of agriculture were opened, in Edinburgh and Oxford. Many similar institutes of agricultural sciences were established in Germany and France between the opening of the Holstein Institute, in 1803, and 1843, when the great Rothamsted Experimental Station was opened in England. Clearly, the European scientific community did *not* share Malthus' patently uninformed pessimism about the natural limits of agricultural growth.

Malthus was not, as modern mythology would have it, a simple country curate to whom God had imparted a Divine Revelation about population as His humble messenger went about his pastoral chores. Malthus was, rather, a sophisticated dabbler in political economy, who was well aware that history had long since disproven his great "discovery" that man's ability to produce babies will always exceed his capacity to grow enough food to feed them. If this famous Malthusian "Law" had indeed ever been true in England, it was, thanks to the Agricultural Revolution, no longer true at the time Malthus was born.

The myth that mankind was doomed never to be able to grow enough food to keep up with its own birth rates grew less and less true during every year of Malthus' life on this planet. Nor, as a matter of fact, has it become any less untrue since Malthus passed on to the Final Enclosure.

For example, in 1831, three years before Malthus died, the young American Cyrus McCormick gave his first public demonstration of his horse-drawn reaper, for which he was awarded a patent in 1834. Thanks to the mechanical reaper and other aspects of the mechanization of American agriculture, in an America where in the late 1830's "grain had to be imported from Europe to make up for the shortages in the American crop, [and] the shipment of wheat from Chicago amounted to only seventy-eight bushels in 1838, . . . ten years later, Chicago alone was shipping two million."[18] So great was the grain production explosion in America that, when the grain crops failed in Europe for the three years starting in 1861 and the wheat-growing northern states of the Union enjoyed three years of successive bumper crops, Abraham Lincoln was able to turn the American wheat surplus into one of his most effective diplomatic arguments against the French and British desires to send military aid to the rebellious Confederate States.[19]

In our own times, we have seen the United States soybean crop soar from a harvest of 14 million bushels in 1930 to a harvest of more than 1.5 billion bushels in 1973. Not only do soybeans constitute a major source of protein and fat for human consumption, but the soybean residues left when the oils are extracted are then converted into soybean meal and soybean cake used in the feeding of cattle. This makes the soybean crop extremely valuable in the production of meat for human consumption. "As recently as 1950, [U.S.] soybean exports were negligible; last year they had a value exceeding $3 billion, or about 5 percent of all [American] income from exports."[20] Soybeans, in fact, now lead all other agricultural exports of the United States in dollar value.

All these and other equally revolutionary innovations in the world's continuing Agricultural Revolution could have been foreseen—not, of course, in specific detail, but in broad general principles—by the scientific contemporaries of Malthus. They could not have been able to predict that the then largely Chinese soybeans would become, among other things, a major American source of new cattle foods and meat-substitute protein foods for people. However, given the corollary experience of witnessing the rises in British and European meat production following the introduction of turnips as a new cattle food during the eighteenth century, they could have and most probably did predict that in generations and centuries to come other food scientists on various continents would continue to develop equally revolutionary new sources of food for the cattle, poultry, and fish raised for human consumption.

Given the nineteenth-century data on the effects of the McCormick reaper and other engines of the Industrial Revolution on the earlier achievement of quantum leaps in the levels of European food production during the early phases of the Agricultural Revolution, Malthus' educated contemporaries could and probably did predict that by the twentieth century these agricultural advances would have seemed puny. Those of Malthus' peers who understood both the Agricultural and the Industrial revolutions better than he did would not, for example, have been surprised to learn that in the twentieth century American food harvests would continue to grow in geometric leaps while, at the same time, the number of individual farmers and farm workers decreased in even greater measure. In the same America where soybeans were to become the nation's major agricultural export in less than twenty-five years of intensive cultivation starting in 1950, the number of persons supplied with food, fibers, and other agricultural products per farm worker rose from 10.7 in 1940 to 47.1 in 1970. During the same three decades, the U.S. index of farm output per man-hour climbed from 21 in 1940 to 112 in 1970.[21]

Even in Malthus' lifetime, the specialists and teachers of agricultural science and economics knew very well that the ratio between the levels of human food and baby production was in no way fixed by Natural Law, as Malthus stubbornly maintained. Events before and since the publication of the first edition of his *Essay,* in 1798, have provided ample evidence that man's food-production capacities are no more fixed for life than is his intelligence, even if the dogmas and the shibboleths of scientific racism remain immutable. In the real sciences, the most cherished of personal hunches and working hypotheses are constantly altered, and even refuted, by the new findings of hard facts.

Malthus, however, like the modern scientific racists, was not a man to be easily intimidated by mere facts. To his dying day, he was to cling to his central dogma of the naturally ordained and permanent gap between man's dual capacities to father babies and to feed them. First revealed in the original edition of 1798, it was retained in each of the subsequent editions of his book that were published for the rest of Malthus' life. This core myth of the alleged immutable lag between both human capacities was the foundation of the Natural Law on which Malthus based his principles of how to treat the poor and the "middling classes."

An Essay on the Principle of Population was not a book about demography, or population dynamics, or agriculture at all. Malthus' book was, rather, to use Professor Fried's later phrase about the writings of Malthus' successors, a "propaganda tract favoring certain social arrangements." Between 1798 and 1826—the years of the first and sixth editions of the *Essay*— these social arrangements were (1) the abysmally low wage scale and (2) the nearly total absence of such tax-supported social services as clean water systems, health services, and education. Both of these social arrangements combined to maintain and increase the profit rates of the large landowners and the new classes of industrialists and mercantilists.

RAISE THE WAGES AND RUIN THE NATION

The scion of a family of some means, Malthus became one of the most articulate spokesmen of the new classes of primary beneficiaries of the Agricultural and the Industrial revolutions whose vastly increased wealth had endowed the college at Haileybury and Malthus' chair. He looked upon the landless rural poor, artisans and yeomen alike, who swarmed into the new towns in desperate quest of newly created mill, mine, and menial jobs, as important sources of national wealth.

The new urban poor were not only a source of cheap adult labor; they also constituted an ever replenishing reservoir of even more inexpensive and docile child labor. If the new population of wage earners suffered from endemic infectious and deficiency diseases while trying to make a life on the low wages of the new mean towns, this did not indicate that these hardworking people were the victims of poverty. Rather, Malthus wrote, such afflictions were the price of yielding to their own innate low natures. Their chief interests, according to Malthus, were clearly not Christian piety and saving for a rainy day but, instead, vice and debauchery. The famines, plagues, and wars that periodically decimated the ranks of the world's poor and other sinners were, simply, the Judgment of Nature and/or God on the types who ignore or flout the Natural Laws of Population, Prudence, and Probity.

When a poor man persists in marrying and having children, and then hard times deprive him of the opportunity to work and he turns to society for relief, he becomes, in Malthus' view, an enemy of society. The punishment of marrying though poor, Malthus wrote in the second (1803) edition of his *Essay,* is a matter of Natural Law: "the punishment provided for it by the laws of nature falls directly and most severely upon the individual who commits the act, and through him, only more remotely and feebly, on the society. When nature will govern and punish for us, it is a very miserable ambition to wish to snatch the rod from her hand and draw upon ourselves the odium of executioner. To the punishment of nature he should be left, *the punishment of want"* (italics added).

It was, in short, not low wages and cyclical employment and unemployment but the hereditary low nature and irresponsibility of the poor that brought on their woes. The poor were poor because Nature decreed their poverty; to attempt to alleviate in any way the poverty of the poor, and its effects upon their innocent children, was therefore to act counter to the Laws of Nature, which in turn were also the Laws of God.

Poor people, in Malthus' scheme of things, had neither the right to be born nor the right to live. "A man who is born into a world already possessed, if he cannot get substinence from his parents on whom he has a just demand," Malthus wrote in his 1803 edition, "and if the society do not want his labour, has no claim of *right* to the smallest portion of food, and, in fact, has no business to be where he is. At nature's mighty feast there is no vacant cover for him. She tells him to be gone, and will quickly execute her own orders, if he do not work upon the compassion of some of her guests."

Some 161 years later, the vice-president of the American Eugenics Society, Professor Garrett Hardin, was so taken with these words, red in tooth and claw, of Malthus' on Nature that he not only reprinted them in his 1964 college anthology, *Population, Evolution, and Birth Control* (pp. 88–89), but added a poem of his own, called "To Malthus." The poem opened:

> Malthus! Thou shouldst be living in this hour:
> The world hath need of thee: getting and begetting,
> We soil fair Nature's bounty.

Hardin's dithyramb to the father of scientific racism ended with the lines:

> Confound ye those who set unfurled
> Soft flags of good intentions, deaf to obdurate honesty![22]

It was, Malthus wrote in 1803, the misdirected compassion of people deaf to the obdurate honesty of his own words that led to such crimes against Nature's Laws as the agitation for free education for the children of factory workers and miners; small compensatory pensions (outdoor relief) for the people fenced off from some five million acres of village common field by the Georgian Enclosures of 1760–1810, depriving rural laborers and artisans "not merely of the right to grow crops, graze cattle or gather fuel on them, but destroying the last shred of [their] economic freedom";[23] minimum wages and other sentimental excesses that could only spoil the innately undeserving poor. As we saw earlier, Malthus held that since the wages of sin can only be death, he could only "reprobate specific remedies for ravaging diseases," let alone denounce "those benevolent, but mistaken men, who have thought they were doing a service to mankind by projecting schemes for the total extirpation of particular disorders." It was these ravaging diseases, such as tuberculosis, and particular disorders, such as hunger, that God and/or Nature had created for the specific purpose of maintaining a proper balance between human and agricultural fecundities.

The moralist and political economist Malthus also took great pains, in the first (1798) edition of his *Essay,* to deflate the claimed social value of living wages in two passages that revealed his real purpose in writing his book: "Suppose that by a subscription of the rich, the eighteen pence a day which men now earn was made up to five shillings, it might be imagined, perhaps, that they would then be able to live comfortably and have a piece of meat every day for their dinners. But this would be a very false conclusion. The transfer of three shillings and sixpence a day to every labourer *would not increase the quantity of meat in this country"* (italics added).

This was written at the end of the century during which the Agricultural Revolution had doubled the production of meat in England—despite Malthus' pseudoscientific law. Far from putting meat on the laborer's table, then, since the quantity of meat per se is permanently limited by Malthusian Law, all that would be accomplished by raising wages, according to Malthus, would be to raise the cost of meat to levels beyond even the increased purchasing power of the five-shillings-per-day-rate laborers. Worse than that "the spur that these fancied riches [*sic*] would give to population would be increased."

Without pausing to explain why raising wages would trigger a rise in the birth rate, Malthus continued on to declare: "The receipt of five shillings a day, instead of eighteen pence, would make every man fancy himself comparatively rich and able to indulge himself in many hours or days of leisure. This would give a strong and immediate check to productive industry, and in a short time, not only the nation would be poorer, but the lower classes themselves would be much more distressed than when they received only eighteen pence a day."

Adequate wages, then, were positively ruinous to society as a whole and, in particular, to "the labouring poor, [who] to use a vulgar expression, always live from hand to mouth." It was therefore seditious to talk to such low types about *human* rights. Especially since the poor not only lived from hand to mouth but also because "their present wants employ their whole attention, and *they seldom think of the future"* (italics added).

Malthus could never forgive contemporaries such as Robert Owen and Tom Paine for addressing themselves to the human needs and human aspirations of "the lower and middling classes." Malthus and his friends in Parliament beat back every effort of Chadwick and the other Benthamites to have the nation open schools for the education of the laboring classes. Of Paine, Malthus wrote: "The circulation of Paine's *Rights of Man . . .* has done a great mischief among the lower and middling classes of people in this country" (Book IV, Chapter 5). Nothing, Malthus added, would so effectually "counteract the mischiefs occasioned by Mr. Paine's *Rights of Man* as a general knowledge of the real rights of man."

Malthus not only had but shared this general knowledge. He knew that there is "one right which man has generally been thought to possess, which I am confident he neither does nor can possess—a right to subsistence when his labour will not fairly purchase it." To let the poor think otherwise was an "attempt to reverse the laws of Nature." Such attempts to act counter to what Malthus chose to term Nature's Laws would not only fail, but "the poor, who were intended to be benefitted, should suffer most cruelly from the human deceit practiced upon them."

As Malthus reminded his upper-class peers: "In the history of every epidemic it has invariably been observed that the lower classes of people, whose food was poor and insufficient, and who lived crowded together in small and dirty houses, were the principal victims. In what other manner can Nature [*sic*] point out to us that, if we increase too fast for the means of subsistence, so as to render it necessary for a considerable part of society to live in this miserable manner, we have offended against one of her laws?"

To be sure, there were in the land various "benevolent, but mistaken men" who held that the way to cut down on epidemic deaths due to the inadequate diets and housing of the poor was to raise wage scales to levels that could provide adequate food and shelter. But Malthus knew better, for was it not Natural Law that God, in His Boundless Wisdom, had arranged so that man would never be able to grow enough food to feed all of his and/or His children? And was it not obvious to any but godless radicals such as Tom Paine and meddling fools such as Jeremy Bentham that, in the same outpouring of His Infinite Mercy, God had personally arranged for the least worthy and most sinful of these excess progeny (at least in times when there were shortages of mine and mill jobs for little children) to die young?

MALTHUS' "NECESSARY STIMULUS TO INDUSTRY"

When Benthamite social reformers, such as Francis Place, and political radicals, such as Richard Carlile, editor and publisher of *The Republican,* distributed information about contraception, including a simple technique for the prevention of unwanted pregnancies, Malthus was not delighted with such solutions to what he decried as overpopulation. The issue of *The Republican* (May 6, 1825) in which Carlile reprinted the full text of Place's famous *To the Married of Both Sexes of the Working People* (1823) was one of the many issues edited from the Dorchester Gaol, to which the editor had been remanded for the crime of publishing the text of Paine's *Age of Reason.*

Birth control was, as Place pointed out, not intended to make sinning easier but to "destroy vice and put an end to debauchery."

The Place broadside went on to explain why the needs for safe contraception were greatest among the new industrial poor:

> It is a great truth, often told and never denied, that when there are too many working people in any trade or manufacture, they are worse paid than they ought to be paid, and are compelled to work more hours than they ought to work. When the number of working people in any trade or manufacture has for some years been too great, wages are reduced very low, and the working people become little better than slaves. *When wages have been reduced to a very small sum, working people can no longer maintain their children as all good and respectable people wish to maintain their children, but are compelled to neglect them; to send them to different employments; to Mills and Manufactories, at a very*

early age. The misery of these poor children cannot be described, and need not be described to you, who witness and *deplore* them every day of your lives. [Italics added.]

This was a far cry from the story of industrial wages and hours, let alone the morality of child labor, as Malthus told it. And Place endeared himself to Malthus even less by what followed:

Many indeed among you are compelled for a bare substinence to labour incessantly from the moment you rise in the morning to the moment you lie down again at night, *without even the hope of being better off.*

The sickness of yourselves and your children, the privation and pain and *premature death* of those you love but cannot cherish as you wish, need only be alluded to. You know these *evils* all too well.

And what, you will ask, is the remedy? How are we to avoid these miseries? The answer is short and plain: the means are easy. *Do as other people do, to avoid having more children than they wish to have, and can easily maintain.* [Italics added.]

Place, unlike Malthus, described the premature deaths of poverty and the middling classes as societal evils, and not as the judgments or blessings of a Merciful God. After stating the family welfare and broader socio-economic reasons for preventing the birth of unwanted children, Place proceeded to describe the simple contraceptive method Robert Owen had discovered in France for the benefit of the employees of his cotton mills at Lanark. This called for merely "a piece of soft sponge . . . tied by a bobbin or a penny ribbon, and inserted before the sexual intercourse takes place." This was all that was needed, he added, to insure that "both the woman and her husband will be saved from all the miseries which having too many children produces." Place predicted: "By limiting the number of children, *the wages both of children and of grown up persons will rise,* the hours of working will be no more than they ought to be; you will have more time for recreation, some means of enjoying yourselves rationally, some means as well as time for your own and your children's moral and religious instruction" (italics added).[24]

To Malthus the ordained moralist and professional political economist, this was both heresy and sedition. It was not for the poor to aspire to longer lives, fewer sicknesses, higher wages, shorter working hours, the end to child labor, and the education of themselves and their children. Far from hailing simple and inexpensive contraceptive techniques as means of ameliorating the social problems caused by having more children than families of the lower and middling classes could "easily maintain," Malthus declared war on contraception and its advocates. In the fifth (1817) edition of his *Essay,* Malthus declared: "I should always particularly reprobate any artificial and unnatural modes of checking population, both on account of their immorality and their tendency to remove a *necessary stimulus to industry"* (italics added).

The "necessary stimulus to industry" that birth control threatened was, of course, the hopeless miseries of poverty that, as Place had observed, compelled the working poor to send their children out to work in "Mills and

Manufactories at a very early age." This was only one of the reasons why Malthus attacked contraception. He went on to claim: "If it were possible for each married couple to limit by a wish the number of their children, there is certainly reason to fear that the indolence of the human race would be very greatly increased; and that neither the population of individual countries, nor of the whole earth, would ever reach its natural and proper extent."

Much as Malthus ranted against "overpopulation," he feared even more a *low* birth rate that would permit the working poor to demand and get higher wages and shorter working hours, as well as to earn enough money to protect their children from the normal prospect of working the usual dawn-to-dusk shifts in mines and factories. To end low wages and child labor was to deprive industry of valuable stimuli and therefore to trifle with Natural Law.[25]

The Natural Law under which population increased at rates faster than man's Divinely Limited capacity to grow enough food to fill these new hungry mouths was simply a mathematical statement of God's Will. The same *permanent* gap between the rates of baby and food production was, at the same time, the Natural Law which made poverty not only inevitable but even socially desirable as God-Nature's provident mechanism for protecting the Deserving Rich from the excess fecundity of the Undeserving Poor.

All of the Malthusian "Natural Laws" ordaining the permanence of poverty and illiteracy were, of course, the start of scientific racism in Europe.

The poor, in the eyes of Malthus, were a race apart from the nonpoor, as, indeed, were the "middling classes" created lower and apart from the gentry by the workings of the same Natural Laws.

The high death rates of poverty, like the high birth rates of the poor, were God's mechanisms for keeping the poor in the wages-and-hours bargaining position and the proportions reserved for them by Natural Law. Therefore, as long as the Undeserving Poor remained in a state of sin, living only for base sexual gratification, and not setting aside tidy nest eggs of cash for the morrow, any attempts to spoil them further with higher wages and shorter hours—let alone such sinful frills as education and contraception—constituted blasphemous contraventions of Divine Will and unscientific violations of Natural Law.

This was a transitional posture, to be sure, in that it combined Scripture with science, theology with mathematics, and Heavenly Revelation with political economy. It was, nevertheless, the first major theory of the human inferiority of "the lower and middling classes" to be presented in the language of science. From it many less theological and more mathematical systems of scientific racism would flow, but Malthus was the progenitor of them all, from Galton and Gobineau to Lapouge and Davenport.

The scientific racists recognized this, too. In 1925, in the first issue of the *Annals of Eugenics,* edited by Karl Pearson, and issued by the Francis Galton Laboratory for National Eugenics in the University of London, the new organ of "scientific" eugenics (as against what Pearson, on page 1, dismissed as the so-called eugenics produced by "propagandists and dilettanti") announced that each issue would open with the Annals of Eugenics Portrait

Series. Naturally, Malthus was accorded the honor of being celebrated with Portrait Number One, for he was, indeed, as the legend under his portrait read: "Strewer of the Seed which reached its Harvest in the Ideas of Charles Darwin [*sic*] and Francis Galton."

THE LEGACY OF MALTHUS: SICKLIER LIVES AND PREMATURE DEATHS

By "proving" that the lower and middling classes were subhuman creatures who persisted in wallowing in the joys of gin and sex and in squandering all of their "capital" without a thought "of the future," Malthus thereby made societal aggressions against the poor—such as child labor, lower wages, unsafe working conditions, overcrowded and filth-ridden slum housing, and compulsory illiteracy—emerge in the consciousness of the educated and affluent classes as perfectly proper and natural approaches to their own social and human responsibilities to the men, women, and children who dug the coal, manned the factories, and produced the foods and fibers on which their own expanding wealth was based.

To justify the continuation of the inhuman treatment of the working poor, it was first necessary, in Christian England, to deprive them of their right to be considered fully as human as were their employers, their landlords, their rulers. This right to be considered human was a right Malthus had helped tear from their patrimony with his pseudolaws of population, production, and the limits of agricultural growth. The sole function of Malthus' *Essay on Population* was to preserve the status quo of low wages, child labor, and the absence of education and health care for the white, Anglo-Saxon, Protestant families who worked for wages on the land, in the factories, and in the mines.

In this objective, Malthus succeeded beyond his wildest dreams. To this day, Malthus' "Law"—refuted by the *realities* of the Agricultural Revolution before he was born—that "the power of population is indefinitely greater than the power in the earth to produce substinence for men" is accepted by highly educated professors and statesmen as a legitimate *scientific* law.

More than that, Malthus' harsh judgments of the poor and the middling classes as subraces of the population, and his strictures against coddling these subhumans with such spoiling mechanisms as higher wages, free education, healthier housing, the right to vote, remedies for and the prevention of ravaging diseases, are, alas, still shared in many powerful governmental quarters. They have remained the raw materials from which are presently forged the insights and the value judgments of far too many of the educated men and women who in this century have designed our governmental policies concerning public health, mass education, and all general welfare.

The words of few other authors of the nineteenth century have been as awesomely honored in the observance as have been the antiscientific, anticultural, and antihuman homilies of Professor Malthus. More than the thoughts of Tom Paine and Thomas Jefferson, of Louis Pasteur and Abraham Jacobi, of Jeremy Bentham and Florence Nightingale, it has been the philosophy of Malthus—rather than of the creators of what is finest and most life-enhancing in our moral, medical, and cultural heritage—that has most influenced the values and actions of many of those among us who today are directly involved in the major societal decisions dealing with everything that makes us as human, or as inhuman, as we are capable of becoming as adults.

5 Natural Law and the Inferior Races: The Spread of Scientific Racism in Europe and America

> As regards the democratic feelings, its assertion of equality is deserving of the highest admiration so far as it demands equal consideration for the feelings of all, just as in the same way as their rights are equally maintained by the law. But it goes farther than this, for it asserts that men are of equal value as social units, *equally capable of voting,* and the rest. *This feeling is undeniably wrong, and cannot last* [italics added].
>
> —SIR FRANCIS GALTON, in "Hereditary Improvement," *Fraser's Magazine,* January 1873

While Malthus divided the population of his world into the innately Deserving Rich and the naturally Undeserving Lower and Middling Classes, he never felt the need to employ the ancient phobias and bigotries of white Christian England in his lifelong crusade to preserve this natural basis for permanent poverty. It fell to Malthus' nineteenth-century successors in Europe and America to employ variants of the old-fashioned gut racism in their contributions to the same goal. To the mix of Malthusian Natural Law and the rope-and-fagot racism that came before Malthus, the second wave of scientific racists added various elaborate structures based on the counterfeiting of the data and the laws of science and history to bolster their Malthusian claims that poverty, illiteracy, and contagious diseases were human biological fates predetermined by hereditary inferiority.

None of Malthus' moral and intellectual heirs tampered with his central objectives. They merely added systems of racist and scientistic cant to reinforce the Malthusian dream of a permanently frozen social structure: a status quo in which the lower and middling classes accepted their Natural Destiny to live "without even the hope of being better off" (see Place, 1823, pages 80–81), and raised no impertinent and even seditious demands for the education, the medical care, and the upward social mobility of their innately inferior children.

Ironically, just as Malthus ignored the realities of the Agricultural Revolution of 1700 onward in the fabrication of his pseudolaws of food and baby production, his nineteenth-century successors based their dogmas of scientific racism on a hereditary concept of biological determinism that had been exploded to smithereens by the hard lights of legitimate science as early as 1768.

PREFORMATION, EPIGENESIS—AND REALITY

One of the early products of the applied sciences from which the Industrial Revolution developed—the telescope patented by Jan Lippershey of Holland in 1608—was to play a major role in creating the Age of Reason. Once Galileo pointed a telescope at the heavens in 1609, he was able to confirm what Copernicus, in 1543, could only conjecture in his *De revolutionibus orbium coelestium:* man was not at the center of the universe, and our earth was indeed a very minor planet that revolved about the sun.

Four decades later, when the optical principles of the telescope were manipulated by other Hollanders to produce the microscope, the way was technologically cleared for Marcello Malpighi, professor of anatomy at Bologna, to do two things. Malpighi used his microscope to discover the hair-like small veins through which our blood passes from our arteries to our larger visible veins. He named these previously invisible veins *capillaries,* from the Latin word for hairlike. With this discovery, just as Galileo had used the telescope to confirm Copernicus' version of the universe, Malpighi was able, in 1660, to provide the final biological and anatomical proof needed to confirm the theory of the circulation of the blood proposed by William Harvey in 1628. This proof was furnished by Malpighi three years after the death of Harvey himself.

Malpighi also trained his microscopes on plants, worms, and animal and human tissues. Out of this curiosity about the microstructure of living things, hitherto invisible to the naked eye, were to come the sciences of comparative anatomy, embryology, histology (microscopic anatomy), and cellular pathology. But there would also come something far more revolutionary.

Until Malpighi cut open and examined various worms and caterpillars under the lenses of his crude microscopes, all that naturalists had been able to find inside of such lowly creatures was a formless, and somewhat repellent, slime. Under the microscope, however, this apparently formless slime proved to be interacting complexes of digestive, circulatory, and nervous systems that were, quite plainly, the biological analogues or counterparts of our own.

Three years after this discovery, the Curator of Experiments for the British Royal Society, Robert Hooke, used his microscopes to discover and name the cells found in all growing plants. By 1674, Antony van Leeuwenhoek of Delft, draper and handyman, used his homemade single-lens microscopes to discover the protozoa and, a year later, the bacteria. Hooke's discovery of the cells of wood and cork was to lead to the Schwann and Schleiden cell theory of life in 1838. Leeuwenhoek's discoveries of what he described as his *animalcules* (little animals) led inevitably to the creation of the biological sciences of protozoology, bacteriology, microbiology, and virology, let alone to the twentieth-century elucidation of the role of the nucleic acids in heredity. Nevertheless, the most shattering of the first discoveries of microscopy were undoubtedly those made by Marcello Malpighi in the innards of worms.

For if man, because of Copernicus and Galileo, was no longer the center of the universe, then after Malpighi discovered the organs within the worms it was also evident that man was no longer a Special Act of Creation. If man was indeed, as the Bible said, made of the stuff of Divinity, then so too were the lowliest of his compatriot organisms, the beasts of the field, the fishes of the sea, the worms of the earth.

Had the Holy Inquisition understood the biological implications of Malpighi's first studies on the comparative anatomy of man and lowlier creatures a fraction as much as its leaders grasped the cosmological implications of Galileo's astronomy, the talented anatomist of Bologna would very likely have been given the choice of being burned at the stake for heresy or, like Galileo, renouncing the reality of what he had discovered. Because, scientific development being the cumulative process it is, Malpighian comparative anatomy had inevitably to lead to Darwin and Wallace and their theory of variation, natural selection, and evolution.

The theory of evolution did not follow immediately: science does not always work that quickly in the absence of theories that help account for new discoveries in nature. Often, in the historical vacuums that are created between the moment of a major scientific discovery and the time it is understood and applied, the history of ideas and cultures appears to teem with what hindsight reveals to be preventable epidemics of false moves, foolishness, and very costly errors. As, for example, the persistence with which wise and competent men—including Malpighi and Leeuwenhoek—clung to the concept of Preformation.

This was the belief that human beings are created *sui generis* as homunculi, or preformed but microscopically miniature human beings. Once created, these preformed human beings lived in the sperm of the male or the ovum of the female, and simply grew in size in the womb, and continued to grow after birth. Children, therefore, were merely larger homunculi, and adults fully grown homunculi.

Opposed to the concept of Preformation was the theory of Epigenesis. That is, the theory that "development starts from a structureless cell, and consists in the successive formation and addition of new parts which do not pre-exist in the fertile egg."[1] That is, the theory that, in human beings, life develops from the fusion of genetic materials, present in the sperm of the father and the ovum of the mother, into the zygote or fertilized ovum—and that subsequently, changing form from stage to stage of development and growth, the ovum develops into an embryo, which then develops into a fetus, which then develops into an infant. After birth, of course, the infant develops by various stages to maturity.

Both theories were, at the start of the seventeenth century, quite ancient. Aristotle, basing his concepts on his personal observations of the growth and development of embryos to chicks in fertile hen's eggs, and Hippocrates proposed essentially epigenetic views of development. But in their times, as in later millennia, the preformationist concept had considerably more appeal to elitist types who looked to Nature to account for the wealth of their parents and the poverty of their servants.

This theory of preformation was, curiously enough, to dominate the worlds of biology and natural philosophy for well over a century *after* the advent of microscopy. One of the accidents of scientific history that helped keep preformationism alive long beyond its time was a mistaken observation made by Malpighi himself in 1672. In an egg that he erroneously believed to be neither fertile nor incubated (it had been left out in the sun), the great anatomist discovered a well-developed chick embryo. To Malpighi, this proved that the "preformed embryo" was present from the moment the "infertile" egg was formed. Earlier, in 1669, the great Dutch anatomist Jan

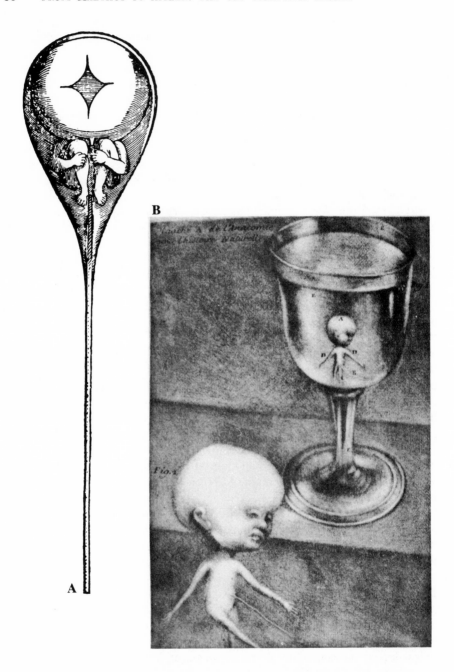

The idea of preformationism—a mechanism of heredity in which human beings were supposed to preexist in the male semen and female ova as fully formed but microscopic
homunculi, growing larger *in utero* and after birth—dominated European scientific thought
for over two centuries. In his 1694 *Essay de dioptrique* Hartsoeker included a drawing
(A) of a homunculus within a human spermatozoon. In 1750 the French naturalist and
illustrator J. Gautier d'Agoty published a color plate (B) featuring two drawings of a
fetus "discovered" in male semen that had been ejaculated into a glass of cold water.

Swammerdam had found a pupal butterfly in a caterpillar larva and, mistaking it for a fully formed butterfly, had joined the ranks of the believing preformationists.

"By 1720," writes Joseph Needham,[2] "the theory of preformation was thoroughly established, not only on the erroneous grounds put forward by Malpighi and Swammerdam, but on the experiments of Andry, Dalenpatius, and Gautier, who all asserted that they had seen exceedingly minute forms of men, with arms, heads, and legs complete, inside the spermatozoa under the microscope. Gautier went so far as to say he had seen a microscopic horse in the semen (he gave a plate of it) and a similar *animalcule* with very large ears in the semen of a donkey; finally, he described minute cocks in the semen of a cock."

One of the barriers to fuller understanding of such "scientific observations" was, as Needham notes, the fact that "biology at this period was still laboring under the disadvantage of being without the cell-theory, and therefore unable to distinguish between an egg and an egg-cell." It was this lack of cell theory that acted to keep preformationism alive and flourishing even after Kaspar Friedrich Wolff, a twenty-six-year-old member of the St. Petersburg Academy in Russia, published his *Theoria generationis* in 1759. "On the practical side, Wolff's work was indeed of the highest importance," writes Needham. "If the embryo pre-exists, he argued, if all the organs are actually present at the very earliest stages and only invisible to us even with the highest power of our microscopes, then we ought to see them fully formed as soon as we see them at all."

In 1768, in his *De formatione intestinorium,* Wolff, according to Needham, "ruined preformationism." Wolff showed that the intestine could not possibly be preformed, and from this he went on to propose "an epigenetic theory which applied the same process to all organs."

Wolff's writings were, however, rejected by most of his scientific contemporaries. By the fourth quarter of the eighteenth century, the epigenesists were still in a minority. Nearly another century was to pass before Rudolf Virchow, the father of cellular pathology, formulated the doctrine that every cell derives from a parent cell (*omnis cellula e cellula*); and before C. P. Reichert, like Virchow a former student under the great biologist Johannes Müller, carefully traced the microscopic stages of development during which the segments of eggs divide into cells, and in which various body structures develop *in stages* from these cells.

The development of the science of embryology did manage, by the middle of the nineteenth century, to discredit preformationism as a biological reality, and to establish the validity of our modern epigenetic concepts of human biological and mental development. But, as it turned out, the essential quarrel between the preformationists and the epigenesists was to be transformed into what was subsequently to become known as the controversy between the eugenically oriented hereditarians on the preformationist side and the genetically oriented developmental biologists (who were to be mislabeled "environmentalists" and "egalitarians" by the latter-day-preformationists).

Post-Malthusian scientific racism became, among other things, the continuation of preformationism by other names.

THE INEQUALITY OF RACES

Not the least of the historical changes caused or accelerated by the contiguous Agricultural and Industrial revolutions was the evolution of political structures more responsive to the needs and thinking of the various essentially new classes of industrial, mercantile, and financial millionaires who could not tolerate the autocratic governments of hereditary princes. As more of the powers of the hereditary nobility had to be yielded to the new masters of the newly industrialized societies, the sons and lackeys of the now obsolete classes often shared fantasies of somehow restoring the social order of the *ancien régimes* toppled by both bloody and bloodless revolutions. An archetype of these foundlings of history was a French count, Joseph Arthur de Gobineau (1816–82), who believed fervently that his mother, Anna-Louise Magdeleine de Gercy, was the daughter of an illegitimate son of King Louis XV. His father, Louis de Gobineau, was a captain in the Royal Guard.

Gobineau began his career as a royalist journalist. After the Revolution of 1848, he entered the French diplomatic service. A dabbler in the literature of science and philosophy, Gobineau set out to justify his abhorrence of the rising tides of democratic sentiment in Europe by fabricating a "scientific" theory of the inborn—that is, the *preformed*—inequalities of all human races.

His major work, *Essai sur l'inégalité des races humaines,* was published in two volumes between 1853 and 1855. By the time it was translated into English, in 1915, as *The Inequality of Races,*[3] Gobineau's scientific explanation of why the lower races could never achieve higher levels of civilization had become the Bible of generations of elitists. Americans quoted Gobineau to defend black slavery in the South; Europeans to rationalize the human costs of imperialism in their Asian and African colonies.

Gobineau wrote that the Teutons were the most superior of all the master Aryan racial strains, and that civilized European history began only after the Teutonic invasions. As the pure Teutons declined in number, so did European civilization at its best decline, and the passing of the Civilization of the Great Race resulted in such lower-class horrors as democracy, the French Revolution, and industrialization.

In his foreword to the second edition of this book Gobineau revealed that his racial theory was "a natural consequence of horror and disgust with democracy."[4] The fall of Louis Napoleon and the "inevitable" decay and degeneration of France after 1870, Gobineau wrote, were caused by the fact that the "Gallo-Roman rabble" and the "bourgeoisie" had taken over the destinies of the nation, and supplanted the superior Teutons.

Gobineau, like most of his class, was a profoundly anti-Semitic author, but his primary thrust was color. To Gobineau, the whiter the race—and in his eyes the Teutons were the whitest of all the white peoples—the nobler its inborn qualities and its destinies. The whites had honor. On the other hand, the very word "honor" did not, Gobineau claimed, even exist in any of the languages of the yellow man and the black man. But while clearly inferior to the white race, the yellow race—Gobineau wrote—was superior to the black race, since the yellows made docile subjects while the blacks were born unruly, stupid, gluttonous, and sex-mad.

Great civilizations, according to Gobineau, decline as their pure Teutonic or Nordic racial stock is thinned by too much intermixing with the blood of inferior races. This was a theme nineteenth- and twentieth-century scientific racists, and their dupes and disciples in government and education, were to embrace and elaborate from that day forward. But long before Vice-President Calvin Coolidge published his variation of Gobineau's dicta on the dangers of non-Nordic race mixing in 1920 (see pages 174–75), the embittered and self-proclaimed bastard great-grandson of Louis XV was discovered and embraced by the anti-Semitic Teuton composer Richard Wagner.

Wagner visited Gobineau and became so enamored of his views that, in 1881, he founded the Gobineau Society, in Bayreuth, to spread the creed of Teutonic or Nordic supremacy to the world. The Gobineau Society published the works of the French Teutonist count in various editions, all of which were studied by Houston Stewart Chamberlain (1855–1927). The son of a British admiral, Chamberlain hated England and all things British as thoroughly as Gobineau hated the France that in his view had turned from its allegedly Teutonic nobility to what Gobineau characterized as the government of the "Gallo-Roman rabble" and the French-speaking but racially mongrelized "bourjeoisie." Chamberlain, who was to become a German citizen and the son-in-law of Richard Wagner, wrote the two-volume *Foundations of the Nineteenth Century* (1899), which added overtly violent Jew baiting to Gobineau's passively anti-Semitic Teutonism.

One of Chamberlain's greatest admirers was Kaiser Wilhelm II, who presented copies of Chamberlain's book to officers of the Imperial Army and members of the German diplomatic corps. The Kaiser also made certain that *Foundations of the Nineteenth Century* was stocked in every German library and used as a text in all German schools.

The defeat of Kaiser Wilhelm's Germany in World War I left Chamberlain in a state of profound despair for civilization. However, before his death, Chamberlain was to have his dreams of Aryan supremacy kindled anew. As Konrad Heiden has written: "Chamberlain, who had fallen sick and silent after the war, privately uttered the hope that from the bands of [German] volunteers on the Russian frontier a young, unknown leader might arise, who would clean out democracy." And, in answer to the old scientific racist's prayer, such a young scourge of democracy did emerge.

"When he actually came," writes Heiden, "he came from the Austrian, not the Russian, frontier; and by meeting him Chamberlain definitely became a key figure of world history. In October, 1923, Hitler went to Bayreuth," where he was received by the Wagner family, and charmed the paralyzed old Chamberlain, who the next day dictated a letter telling Hitler "that this meeting was his strongest experience since 1914 . . . and he found it miraculous that Hitler, the awakener of souls" so restored his faith in Germans that "at one stroke you have transformed the state of my soul."[5]

During the visit that transformed the state of Chamberlain's soul, Hitler's entourage, according to Heiden, had included "Alfred Rosenberg, the Russo-German and prophet of the [Protocols of the] Wise Men of Zion [an anti-Semitic forgery manufactured by the Czarist secret police in 1904]." It was "Rosenberg who had acquainted Hitler with Chamberlain's ideas, and since Hitler was too impatient for steady reading, had underlined the most important and useful sentences in the thick tomes. Rosenberg regarded himself as Chamberlain's disciple and successor, and as an executant of his spiritual legacy. After Chamberlain's death in 1927, the Russo-German wrote a biography of the Anglo-German, published an anthology of his writings, and himself wrote a book the very title of which was intended to indicate that it was a continuation of Chamberlain's chief work: *The Myth of the Twentieth Century.*

Just as the Kaiser, after writing to tell Chamberlain that "it was God who sent your book to the German people and you personally to me," ordered *The Foundations of the Nineteenth Century* to be used in all schools and libraries of Hohenzollern Germany, so too did Hitler order *The Myth of the Twentieth Century* to be utilized in all the schools and libraries of Nazi Germany.

So, it seems, the scientific racism of psychotics and crackpots such as Gobineau and Chamberlain and Rosenberg—for all of its many literary and intellectual similarities with the racist graffiti found on the walls of some American public toilets—has not been without its direct and wholesale life-destroying effects on world history.

THE LONGHEADS, THE ROUNDHEADS, AND THE CHOWDERHEADS

Where nonscientists such as Gobineau and Chamberlain looked to racial inheritance for nonspecific biological mechanisms of inborn human superiority and inferiority, various well-intentioned eighteenth- and nineteenth-century scientists (and some malignant contemporaries on the outer fringes of the legitimate sciences) looked to the structures of the human anatomy to account for what they believed to be inherited neurological "organs" and other factors associated with human mentality and behavior.

Franz Joseph Gall (1758–1828), anatomist of Vienna and Paris, will always be respected as a serious investigator who helped advance the fields of neuroanatomy and neurophysiology. The modern concepts of the functions of the cortex of the brain all derive from Gall's discovery that the cortex—until then thought to be merely an inert covering for the brain—was itself composed of neural cells that obviously were involved in the processes of brain function.

Yet Gall was also, like Malpighi, Darwin, and other mortals in the legitimate sciences, capable of making serious mistakes about the structures and functions of life. Because Gall believed unquestioningly in the inheritance of human moral and behavioral characteristics, he felt that where differences

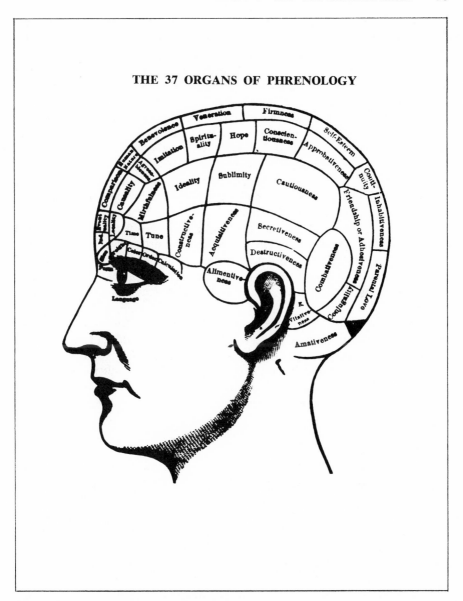

THE 37 ORGANS OF PHRENOLOGY

From O. S. and L. N. Fowler, *New Illustrated Self-Instructor in Phrenology and Physiology* (New York: Fowler and Wells, Publishers, 1859).

could be observed in the performance and nature of the brain and the psyche, these variations of mental ability and moral standards had to be the products of differing inborn anatomical structures on the surface of the human cranium. Gall was also convinced that he had discovered the specific inherited organ or anatomical structures associated with an array of thirty-seven mental, behavioral, moral, and artistic traits. These included such attributes as mechanical aptitudes, cautiousness, verbal memory, self-esteem, conjugality, combativeness, and secretiveness.

Gall's disciple, Johann Kaspar Spurzheim, quickly developed this mess of scientific errors into phrenology, a fashionable and lucrative form of quackery that was immediately accepted as a great scientific advance by August Comte and other European intellectuals, and by all sorts of Americans, including President Quincy of Harvard, the educator Horace Mann, and the writers Ralph Waldo Emerson, Walt Whitman, and Edgar Allan Poe.

Overnight, phrenology replaced physiognomy—the then fashionable "science" of determining human worth and character by measuring and categorizing the angle and other external aspects of the human face—as the character analysis fad of the era. By feeling the head of a customer for the bumps and ridges of the "organs" of each of the thirty-seven inherited traits, the skilled phrenologist could determine in a single examination whether or not the owner of this skull was born musical or moral, talented in languages or mathematics, imaginative, amative, skilled in mathematics, mirthful or morbid.

Phrenology was, ultimately, to be succeeded by another scientific set of head measurements claimed to "reveal" the innate mental and moral worth of all races of man. In 1840, Anders Adolph Retzius of Sweden published a "scientific" system of measuring the contours of human skulls to arrive at the "cephalic indexes" of inferiority, mediocrity, and superiority in individuals and races. Although himself a scientific nonentity, Retzius won the initial agreement of many normally competent scientists. Among their number was to be counted Paul Broca (1824–80), a talented and resourceful French surgeon and anatomist, one of the founders of the Society of Anthropology of Paris, and the inventor of the craniograph, the facial-angle goniometer, the occipital goniometer, and other measuring instruments of anthropometry.

Like Gall, Broca was a first-rate neuroanatomist, and much of his work as an investigator helped advance our understanding of the cortical localization of brain functions. Broca was, however, saddled with many old and basically nonscientific concepts of human worth. He professed to be able to "detect the primitive type in a deformed cranium," and to find the differences in race and character associated with variations of head form.

The Retzius-Broca "cephalic index" was derived from measuring "the breadth of the head above the ears as expressed in percentage of its length from forehead to back . . . assuming that the length is 100, the width is expressed as a fraction of it. As the head grows proportionately broader—that is, the more fully rounded, viewed from the top down—this cephalic index increases. When it rises from 80, the head is called *brachycephalic;* when it falls below 75, the term *dolichocephalic* is applied to it. Indexes between 75 and 80 are characterized as *mesocephalic."*

In plainer language, roundheads (brachycephalics); longheads (dolichocephalics); and in-between-heads (mesocephalics). Many European and American scholars, notably the German anthropologist Otto Ammon, and

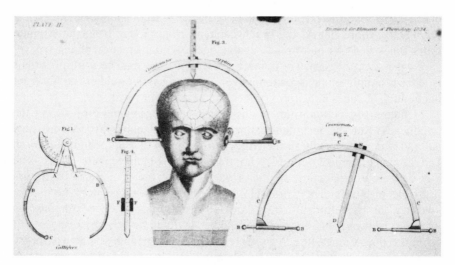

The first measuring instruments of the cephalic index or head-form cult were the crani-ometers and calipers previously employed by the phrenologists.

From George Combe, *Elements of Phrenology* (3d American ed., Boston: Marsh, Capen & Lyon, 1835).

Carlos Closson, instructor of economics at the University of Chicago and "the American apostle of the Social Darwinist school of anthroposociology,"[6] professed to be able to judge the races and qualities of people by their cephalic indexes. Ammon, in fact, based a law on the alleged fact that urban populations tended to be more on the long-headed side because dolichocephalics "showed a stronger inclination to city life and a greater aptitude for success" than did the brachycephalics.

Closson held not only that the cephalic index was "the best single test of race" but also that the "greater taxpaying capacity of the dolichocephalic population" of the world was due to the fact that they were Nordic, Protestant, and enterprising, whereas the roundheads or brachycephalics were Alpine, Catholic, and unimaginative.

Few European authors took cephalic indexes of mankind more seriously than did Count Georges Vacher de Lapouge, a disciple of both Closson and Gobineau. Like Gobineau, Lapouge, who dabbled in craniology and history, was also troubled by what he conceived as catastrophic ethnic shifts in the population of their native France. In this Gallic *Götterdämmerung,* Lapouge envisioned the *endemic* blond, blue-eyed, tall, Protestant, intelligent, dominating, and, naturally, dolichocephalic Nordics being replaced by the *immigrants*—the brunet, dark-eyed, short, Catholic, unintelligent, servile, and, of course, brachycephalic Alpines. In his 1887 essay, "La dépopulation de la France," and his 1899 book, *L'Aryen: son rôle social,* Lapouge spelled out the dangers to civilization inherent in the ascendancy of the Alpine and non-Aryan roundheads over the Nordic and Aryan longheads. For example, on page 563 of his book on the social destiny of the Aryan, he wrote:

"The brachycephalic states, France, Austria, Turkey, even Poland which no longer exists, are a long way from offering the vitality of the United States or England. Nevertheless, the mediocrity of the brachycephalics is in itself a force. Dark, short, heavy, the brachycephalic reigns today from the Atlantic to the Black Sea. As bad money chases good, so the brachycephalic has replaced the superior race. He is inert, he is mediocre, but he multiplies. His patience is unlimited; he is a submissive subject, a passive soldier, an obedient functionary. He does not carry grudges and he does not revolt."

Under normal circumstances, the dolichocephalics would prevail over the roundheads. But the race-denigrating Catholic Alpine brachycephalics now had a very dangerous ally in the struggle for control of European civilization. And Lapouge spelled out just who this ally of brachycephaly was: "The only dangerous competitor of the Aryan is the Jew. . . . He is by nature incapable of productive work. He is a courtier, a speculator; he is not a worker, an agriculturist. . . . A predator, nothing but a predator, he is a bourjeois; he is not, and does not want to be, anything but a bourjeois."[7]

Like the Alpines, the Jews were also stereotyped in the conventional wisdom of the educated classes as being all round-headed, dark, and short. They were so similar that, according to a twentieth-century disciple of Lapouge's, Professor Carl C. Brigham of Princeton, at least the northern Jews were, in fact, "Alpine Slavs." All in all, Lapouge saw nothing but trouble for the world in the proliferating menace of the non-Aryan brachycephalics. "I am convinced," Lapouge wrote in the 1880's, "that in the next century millions will cut each other's throats because of 1 or 2 degrees more or less of cephalic index."[8]

Other equally ill-informed and biased authors in Europe and America—notably the American economist and self-styled "raciologist" William Z. Ripley, in his book *Races of Europe* (1899)—helped make the cephalic index the successor, in many scientific and cultural circles, to phrenology. But whereas Gall and his followers used phrenology to analyze individuals, Lapouge and Europeans such as Karl Pearson and Americans such as Madison Grant utilized the cephalic index as an instrument of scientific racism, and as a rationale for the destruction of Jews, Italians, Slavs, and other non-Aryans. Whereas phrenology was at base the pseudo-anatomy of fools and innocents, craniology—the measurement of skulls for their cephalic indexes—became the pseudo-biology of the prophets of genocide.

Gradually, saner voices began to prevail over the voices of the Ammons and the Lapouges in scientific quarters. In 1889, the British anthropologist A. H. Keane demonstrated that craniology was unable to provide "sufficient, or even altogether trustworthy, materials for distinguishing the main divisions of mankind."[9] Keane's finding that the cephalic index reflects no racial characteristics was repeated, in 1894, by the anatomist Johannes Rank, of Munich. Rudolf Virchow, not only the founder of cellular pathology but also a pioneer of physical anthropology, joined the growing numbers of legitimate investigators who demonstrated that all types of head forms could be found in people of all races.

Broca himself, a serious investigator, was ultimately to agree with Virchow and other scientific critics, and to write that "individual variations [in the cephalic index] are always greater *within* a given race than the distance that separates it not only from neighboring races but sometimes from all other races."[10] In short, the cephalic index could neither describe any anatomic differences among a European, an African, and an Oriental nor reveal the mental traits of any individual of any race.

Unfortunately for mankind, while the legitimate sciences and their serious, trained investigators consigned the cephalic index to the same museum of shattered illusions in which reposed the detritus of phrenology, phlogiston, and preformation, this affirmation of scientific method by the real sciences was without any effects on the zeal and the faith of the scientific racists on the fringes of the sciences. In the firm see-no-science, hear-no-science traditions of Malthus when confronted with the multiplied yields of the Agricultural Revolution, and of Galton when confronted with the life-extending and birth-rate-reducing effects of the national sanitary reforms in England, the true believers, such as Ripley, Grant, Karl Pearson, and Osborn, continued to proclaim the myths of cephalic indexes and the "defect of brachycephaly" as if they were established *scientific* facts. As late as 1961, a retired professor of histology at the University of North Carolina Medical School, in a report commissioned by John Patterson, governor of Alabama, for that state, *The Biology of the Race Problem,* made a stirring defense of the racial myths of the cephalic index in his attack on all modern concepts in physical and cultural anthropology.

SCIENTIFIC RACISM AS A RELIGION

Craniology was one of the essential ingredients in the stew of anthropometric measurements employed in 1846 by the Belgian philosopher and statistician Lambert Quételet to arrive at his "law of deviation from an average." This law, considered to be the basis of what was then called "statistical anthropology," had a profound effect upon Francis Galton, who utilized it and other of Quételet's methods in his own measurements of genius, moneymaking, intelligence, and other human behavioral characteristics and conditions he was convinced were hereditary.

Where cranial measurements and the examinations of the phrenologists —in whose "science" Galton also believed—failed to supply Galton with statistical "data" that confirmed his belief that the Natural Laws of heredity preformed the economic, social, cultural, moral, and health levels of mankind, he placed equally unquestioning reliance on the gossip of anonymous and scientifically illiterate "travellers" and on the racist myths of the Victorian drawing rooms and countinghouses.

It was, for example, from such travelers—none of them able to speak a single African language, or with any evident knowledge of African history, art, and culture—that Galton derived his "scientific" data on the intelligence of black people. He reduced such "data" to elaborate statistics that were taken seriously as "science" for over a century, and which proved to Galton and his followers that there is an inborn or hereditary difference of at least three magnitudes of intelligence between white people and black people.

In the following passage from the chapter entitled "The Comparative

Worth of Different Races" in Galton's masterwork, *Hereditary Genius* (1869), we find this characteristic substitution of the gossip of white lay tourists in Africa for the observations and measurements of trained scientists:

> A native chief has as good an education in the art of ruling men as can be desired; he is continually exercised in personal government, and usually maintains the place by the ascendancy of his character, shown every day over his subjects and rivals. A traveller in *wild* countries also fills, to a certain degree, the position of commander [*sic*], and has to *confront* native chiefs at every inhabited place. It is seldom that we hear of a white traveller meeting with a black chief whom *he* feels to be the better man. I have often discussed this subject with competent persons, and can only recall a few cases of the inferiority of the white man—certainly not more than might be ascribed to an average actual difference of three grades, one of which may be due to the relative demerits of native education, *and the remaining two to a difference in natural* [i.e., hereditary] *gifts.* [Italics added.][11]

Unlike Arthur Jensen and other modern authors, Galton did not need IQ test score statistics to tell him that the causes of the gaps between the average intelligence grades of blacks and whites were at most one third environmental and the rest genetic. Similarly, Sir Francis needed no studies in physical anthropology to tell him that the Irish were, by heredity, the inferiors of the English. Not, that is, so long as he could report: *"Visitors* to Ireland after the potato famine generally remarked that the Irish type of face [*sic*] seemed to have become more prognathous, that is, more like the negro in the protrusion of the lower jaw; the interpretation of which was that the men who survived the starvation and other deadly accidents of that horrible time were more generally of a low and coarse organization" (italics added).[12]

Who these visitors were, or by what scientific methods they made such anthropological deductions, Galton never revealed. On the other hand, when Galton wrote to the Swiss botanist Adolph Candolle in 1884 that "it strikes me that the Jews are specialized for a *parasitical* [Galton's emphasis] existence upon other nations, and that there is need of evidence that they are capable of fulfilling the varied duties of a civilized nation by themselves,"[13] he felt himself perfectly competent to make such racist judgments about the Jews without the eyewitness gossip of anonymous travelers.

If Galton was as anti-nonwhite as Gobineau, and as anti-Semitic as Lapouge and Chamberlain, he nevertheless based his entire system of eugenics—the political and religious movement he organized "to give the more suitable races or strains of blood a better chance of prevailing speedily over the less suitable"—on the Natural Laws of Thomas Robert Malthus. To Galton, democracy itself was a menace, because "it is in the most unqualified manner that I object to the pretentions of natural equality."[14] And because democracy "asserts that men are of equal value as social units, equally capable of voting, and the rest." Along with the most benighted members of his caste, Galton was haunted by the specter of poor English people electing Parliaments that would vote in such offenses against Natural Law as universal free education, minimum wages and hours standards, healthful housing and civic sanitation standards, and meaningful public health laws.

To be sure, as a medical school dropout Galton could not help being aware of the demonstrable relationships between a sound mind and a sound body. In 1873, for example, he conceded: ". . . unless the body be in sound order, we are not likely to get much healthy work or instinct [sic] out of it. A powerful brain is an excellent thing, but it requires for its proper maintenance a good pair of lungs, a vigorous heart, and especially a strong stomach, otherwise its outcome of thoughts is likely to be morbid."[15]

Nevertheless, since most people were by natural inheritance "ineffectives" and "mediocrities," and since "the phrenologists" (sic) had long since delineated the "great variety in the morals of the human race," and since Galton himself had shown that "our moral nature is as unfitted for a high-toned civilization as our intellectual nature is unfitted to deal with a complex one," the founder of eugenics therefore felt it would be a mistake to waste good health and good education on classes of inferior heredity. What Galton therefore advocated as a more suitable way of making the most of the nature-nurture equations was to seek out and increase the birth rates of people of "superior breeding stock," and, at the same time, to reduce or terminate the proliferation of ordinary people. Once this was done by an England awakened to the possibilities of the laws of heredity, then ". . . the present army of ineffectives would disappear, and the deviations of individual gifts toward genius would be no less wide or numerous than they now are; *but by starting from a higher vantage-ground they would reach proportionately further"* (italics added).[16]

What this added up to, of course, was an 1873 preview of Harvard professor Richard Herrnstein's later schema of a meritocracy of the superintelligent based on the introduction of general environmental advances that enriched the genetically superior types even more than they could possibly affect the development of the hordes of genetically less fortunate people who surrounded them. One does not make silk purses from sows' ears, however well nourished, immunized, housed, and educated the sows. Like Herrnstein a century later, Galton held that to improve the national qualities of life for *all* people would only widen the gaps in income and achievement between the preformed geniuses and the equally preselected mediocrities. Worse, by giving the Undeserving Poor the blessing of better health, education, and welfare, they would not only breed more of their dysgenic babies but more of these ineffectives would survive to breed more like themselves.

Galton not only urged seeking out, making studies of, and if need be financially subsidizing the accelerated breeding of hereditary genius types. He also foresaw that such studies would encourage "a sentiment of *caste* among those who are naturally gifted." What was more, publication of the results of these studies of the "naturally gifted" would cause them to form *rural* communities of people bearing purer blood than is found in the veins of the naturally ungifted people who overpopulated the cities.[17] Once such communes graced the countryside, in short order their "selected race will have become a power," and, of course, "a considerable increase will have taken place in the number of families of really good breed."

But what of the eternal Undeserving Poor? How would they fit into the world in which Galton's new caste of state-subsidized hereditary leaders copulated themselves into total supremacy? Galton had thought of this problem, too, in 1873: "I do not see why any insolence of caste should prevent the gifted class, when they had the power, from treating their [lower-caste] compatriots with all kindness, *so long as they maintained celibacy. But if these continued to procreate children,* inferior in moral, intellectual and physical qualities, it is easy to believe that *the time may come when such persons would be considered as enemies to the State,* and to have forfeited all claims to kindness." (Italics added.)[18]

This passage not only reads like a paraphrase of the writings of Konrad Lorenz and other Nazi-era authors on the menace of inferior races to the German state. It also shows quite clearly that, just as Galton anticipated Herrnstein's meritocracy ploy,[19] he was equally alert to the "genetic enslavement" menace Jensen rediscovered in 1969.[20]

Galton firmly believed that, when the "truth of heredity as respects man" was fully known, then "a perfect enthusiasm for improving the race might develop itself among the educated classes." It was, therefore, to the educated classes that Galton directed his cult of eugenics which combined, as the long-time general secretary of the British Eugenics Society, C. P. Blacker, observed in 1952, the elements of science, religion, and social practice. Or, as Galton himself wrote: "There are three stages to be passed through. Firstly, it [eugenics] must be made familiar as an academic question until its exact importance has been understood and accepted as a fact; secondly it must be recognized as a subject whose practical development deserves serious considerations; and thirdly it must be introduced into the national consciousness as a new religion."[21]

That eugenics did emerge as a science, a new religion, or a new theory of social practice—or as two or all three of these—to literally thousands of university-educated formulators of British, American, and German social legislation and practices during the century and more that followed the publication of *Hereditary Genius* (1869) mandates a closer look at the tenets of this cult.

THE WORDS AND DREAMS OF THE PROPHET GALTON

Even by the standards of Victorian-era scholarship, the only remarkable thing about Galton's *Hereditary Genius*—whose thesis was that the rich and the famous are great because of their superior genes rather than their inherited socioeconomic (i.e., environmental) opportunities—was the fact that so many people in his time *and our own* took it seriously as a study of *hereditary human traits.* After all, most literate people by 1869 were well aware of the total lack of significant contributions to the arts, the sciences, and the other cultural activities that make men human by the lineal ancestors and descendants of Shakespeare, Goethe, Beethoven, Aristophanes, Aristotle, Mozart,

Shelley, Rembrandt, Cervantes, Harvey, Malpighi, Newton, Galileo, Lavoisier, Faraday, Priestley, Leeuwenhoek, and other great achievers. Just as most educated people were equally aware of the long lines of incompetents, half-wits, psychotics, and wastrels who had succeeded to the thrones of great monarchs, the ownership of family banks and industries, and other social, military, and economic fiefdoms by virtue of being the biological descendants of exceptionally gifted and/or remarkably lucky parents.

The quintessence of Galton's "scientific" orientation and insights are nicely summed up in the opening page of *Hereditary Genius:* "I propose to show in this book that a man's natural abilities are derived by inheritance, under exactly the same limitations as are the form and physical features of the whole organic world. Consequently, as it is easy, notwithstanding those limitations, to obtain by careful selection a permanent breed of dogs or horses gifted with peculiar powers of running, or of doing anything else, so it would be quite practicable [*sic*] to produce a highly-gifted race of men by judicious marriages during several consecutive generations."

Everything that is wrong—let alone socially and historically malignant—about Galton and the scientistic cult of eugenics that he founded is contained in this paragraph. Out of his simplistic extrapolations of commercial animal-breeding techniques and their products to the culturally and psychologically far more complex human species came his sick dream of producing a "highly gifted race of men" by state-controlled and planned human matings. He clearly did not understand the biological harm done to farm and work animals by turning them into evolutionary cripples when they were bred for those few hereditary traits that made them economically valuable for our species.

The wealthy Victorian dilettante understood even less the nature of human traits and characteristics, and the varying degrees to which such conditions as literacy, or the ability to learn mathematics, or music, or languages, or history, or gourmet cooking, were behavioral traits acquired after birth from the cultural influences of family, community, and school—and that none of these conditions was in any way performed in the human germ cells. Galton had little if any understanding of the enormity of the biological and cultural *differences,* as well as observed biological and behavioral similarities, between men and all other species. Nor, obviously, did Galton understand that correlation is not causation.

It was, perhaps, this great lack of solid biological and psychological knowledge that gave Galton his lifelong addiction to sweeping generalities of the Flat Earth or "It's obvious!" category. For example, five years before *Hereditary Genius,* in an article in *Macmillan's Magazine,* "Hereditary Talent and Character," Galton declared: "We cannot doubt the existence of a great power ready to hand and capable of being directed with vast benefit as soon as we shall have learnt to understand and apply it."

What Galton was talking about here was the power to breed people as we breed pigs. Thirty years later, he was still at it. In "The Part of Religion in Human Evolution," in the August 1894 *National Review,* Galton wrote: "It has now become a serious necessity to better the breed [*sic*] of the human race. *The average citizen is too base for the everyday work of modern civilization.* Civilized man has become too possessed of vaster powers than in old times for good or ill, but has made no corresponding advance in *wits and goodness* to enable him to direct his conduct rightly." (Italics added.) Meaning that intelligence (wits) and moral values (goodness) were in one's genes, and in no way the product of his growth and learning experiences.

Where other Victorians, from Dickens to Disraeli, saw the biological inequities of industrial urban poverty as hazards to all human health, Galton saw only that "the ordinary struggle for existence under the bad sanitary conditions of our towns . . . selects those who are able to withstand zymotic [infectious] diseases and impure and insufficient food, but such are not necessarily foremost in the qualities which make a nation great. On the contrary, it is the classes of coarser organization who seem to be, on the whole, most favoured under this principle of selection, and who survive to become the parents of the next generation."[22]

Even in terms of available nineteenth-century knowledge of natural selection and social medicine, this was an extraordinary exhibition of intellectual and moral bankruptcy. Only a cruel and thoroughly prejudiced ignoramus in British medical history could have, as late as 1873, written that in the new urban slums Nature selects only the least fit to survive the infectious and other deficiency diseases of poverty. As any first-year medical student could have told Galton, between the passage of the Registration (of medical statistics) Act of 1836 and the Public Health Act of 1848—which established England's first General Board of Health—England had long since begun its generations of world leadership in the creation of productive programs for the prevention of contagious and deficiency diseases associated with the new industrialized urban environments.

By the time "Hereditary Improvement" was published in the January 1873 issue of *Fraser's Magazine,* many of the societal hygienic reforms that the forthcoming germ theory would make mandatory by the turn of the century had already been instituted in England as a result of the empirical observations, by civic reformers and doctors, that filth and inferior living conditions caused higher disease rates.[23] Because of the health reforms of the Sanitary Movement that started in England, and then spread to the rest of the industrializing world, by 1873 far too much was commonly known by physicians and educated laymen to allow for Galton's brand of biomedical know-nothingism to be either accepted or forgiven by any intelligent reader.

It was only because Galton's talent for ignoring the hard data of the legitimate biomedical sciences equaled Malthus' capacity to ignore the equally hard data of the Agricultural Revolution that he was able to write of a world in which, by heredity alone, the human race could be divided into seven letter-graded classes according to their innate abilities. Class A people, the lowest Galtonian class of all, had an order of ability of 1 in 4: that is, one in every four people matched their low levels of inborn ability. People in Class F—Galton's "eminent" class—came along only once in every 4,000 births. Finally, the people of very exceptional hereditary endowment who made up the superior or genius humans of Galton's Class G emerged only once in every 79,000 live births.

These are very impressive grades and numbers and ratios. Their only flaw is that none of them have any conceivable basis in valid biological fact or viable sociobiological statistics. Galton's didactic deployment of these pseudo-data survives as a valuable reminder of the fact that while numbers are, indubitably, the language of science, they are only too often also the lexicon of lunatics and con men.

Like the Bourbons, Galton forgot nothing and learned nothing during the three decades of great advances in the scientific knowledge of human heredity and development since the original publication of these pseudo-data in 1869. Thus, in 1901, in his Huxley Lecture to the British Anthropological Institute, "The Possible Improvement of the Human Breed under the Existing Conditions of Law and Sentiment," Galton was still measuring individuals by the size of their incomes. He wrote that "the brains of our nation lie in the higher of our classes," and asserted that if the types of people he categorized as being by heredity Class W and X types—that is, great money earners, even rarer than Galton's Class G genius types—"could be distinguishable as children and procurable by money in order to be reared Englishmen, it would be a cheap bargain for the nation to buy them at the rate of many hundred or some thousands of pounds per head."[24]

It was plain to Galton that "the worth of an X-class baby would be reckoned in thousands of pounds," since such hereditary tycoons "found great industries, establish vast undertakings, increase the wealth of the multitudes and amass large fortunes for themselves." Other high-class types gave the nation its ideals and raised "its tone." A modest man, Galton admitted that he had "not yet succeeded to my satisfaction to make an approximate estimate of the worth of a child according to the class he is *destined* to occupy when adult. It is an eminently important subject for future investigators, for the amount of care and cost that might profitably be expended in improving the race clearly depends on its result" (italics added).

Since breeding up was so vital, Galton concluded that "enthusiasm to improve the race is so noble in its aim that it might well give rise to a sense of religious obligation." Eugenics was and is, after all, a religion. And, Galton hoped, this religious effort "would cause the [Class] X women to bring into the world an average of one adult son and one adult daughter *in addition* [Galton's emphasis] to what they would otherwise have produced." As a mathematician, Galton was able to postulate that "100 [Class] X parentages can be made to produce a net gain of 100 adult sons and 100 adult daughters," who, in turn, would each breed hundreds of superior children and grandchildren, so that "the total value of the prospective produce of the 100 parentages can then be estimated by an actuary, and consequently the sum that is legitimate to spend in favoring an X parentage."[25]

Eugenics was not only religion; it also had a distinctly countinghouse theology.

The most serious attempt to turn eugenics into the religion of the better classes of mankind was to be made by Galton's WASP disciples in the United States, most notably Charles Benedict Davenport. However, even though eugenics became a very pervasive gospel in America, Galton took a dim view of the transplanted Anglo-Saxons who controlled the lost colonies.

Better-class Englishmen, Galton wrote, "prefer to live in the high intellectual and moral atmosphere of the more intelligent circles of English society, to a self-banishment among people of altogether lower grades of mind and interest." As a result, Galton added: *"England has certainly got rid of a great deal of refuse, through means of emigration"* (italics added).

The emigration of the human refuse who, Galton wrote in *Hereditary Genius,* were "not wanted in old civilizations" to lands of lower grades of minds and interest—such as America, Canada, and Australia—saw England become "disembarrassed of a vast number of turbulent radicals and the like, men who are decidedly able but by no means eminent, and whose zeal, self-confidence, and irreverence far outbalance their other qualities."[26]

THE UNION OF EUGENICS AND SOCIAL DARWINISM

Not every American would have quarreled with Galton's characterization of America's Nordic immigrants as refuse. There were, for example, the original inhabitants of this land—the Cherokee, the Sioux, the Arapajo, the Mohawk, the Navajo, and the other natives of this continent—whose lands had been torn from them in three centuries of brute force and even more savage consolidations of the ill-gotten gains of naked aggression that, by modern standards, could only be called genocide. By the third quarter of the nineteenth century, however, those few American Indians who could read the forked tongues were not reading Galton.

The more educated descendants of Galton's human refuse, on the other hand, were reading and enjoying Galton, and with acquired Yankee shrewdness had all managed to interpret Galton's blunt phrase as applying not to the Britons who emigrated to the United States but to their lesser brethren who settled elsewhere.

Until the combination of potato famines in Ireland and the sudden demand for cheap and docile labor to build the networks of new American railroads, cities, sewers, and bridges, saw first the starved Irish and later the Chinese coolies imported by the thousands, the American Indians had been the chief object of the hatred of those people who must have other groups of people to hate. The great American Nativist or Know-Nothing Movement had, as John Higham noted,[27] roots deep in the Protestant hatred of Roman Catholics, as well as in the garden or secular varities of xenophobia found in Catholic, Hindu, Moslem, and Marxist societies in equal measure. But the ethnic or social groups we hate most are, as a rule, those we wrong the most. It was therefore no historical accident that the first Christian minister in the New World to be turned out of his pulpit by the state, Roger Williams, incurred the wrath of the Bay Colony solely for preaching that this land belonged to the Indians, and had to be purchased, not seized, from its rightful owners. Pastor Williams was subsequently tried for high treason and forced to flee for his life into the deep snows of the forests around Salem.

By the time Galton's *Hereditary Genius* was published, the property and the human rights of the American Indians had long since been cut and carved to the point where those few redskins who survived the coming of Western civilization lived in poverty-ridden, germ-infested, famine-cursed concentration camps called reservations. But, the railroads and the sewers having been built by the Chinese coolies and the Irish peasants, history in the form of the Emancipation Proclamation and the defeat of the slavocracy now added to the free labor market a new group of seasoned contributors to the wealth and comfort of America to hate.

Not all affluent people in America were of one mind about the hordes of Chinese, Irish, Poles, Hungarians, Italians, and emancipated blacks who were all ready and willing and anxious to work. To the builders of railroads and water systems, the operators of mines and factories, and the owners of steamships and fallow lands and retail stores, these new non-Protestant, non-white, and even non-English-speaking peoples represented fresh and reliable reservoirs of cheap labor, and *customers* for passage, land, and goods. They were also a form of cheap insurance against the threat of trade unionism.

To the native-born white American poor, however, these European immigrants, and native black southern migrants, seemed to represent growing threats to both their jobs and their wage scales. These fears helped speed the growth of an American trade-union movement that was, traditionally, to be as xenophobic and as racist as the most bigoted of its economic and political adversaries.

Herbert Spencer, the guru of Social Darwinism—with its fierce injunctions against the Undeserving Poor; against free universal education; against free meals for indigent schoolchildren; against clinics, hospitals, and social services for the nonrich; against all laws that either regulated working hours or called for minimum standards of occupational safety and health in mines and factories; against laws establishing minimum standards of health and safety in dwellings built, sold, and rented for human habitation; and, above all else, against trade unions, which Spencer saw as instruments of human tyranny that would destroy civilization—quickly became the favored philosopher of the affluent. He not only proclaimed the moral rights of the Deserving Rich to heaven; Spencer also denounced the immorality and impracticality of health, education, safety, and welfare programs that would have materially increased their taxes here on earth.

Spencer's American disciples, starting with William Sumner of Yale, were quick to add the hereditary postulates of Sir Francis Galton to the laissez-faire nihilism-of-the-rich postures of their master. It made a very neat combination for the tax-hating new billionaires of those pre-income-tax generations, many of whom professed faith in the most pietistic forms of Christianity. For where Spencer offered "evolutionary" rationales for low wages and subhuman working and living conditions, Galton offered the "hereditary" reasons in the Natural Laws of biology for not wasting sympathy, money, education, and, above all else, health care on biologically low-class types who were destined by the Will of God and/or Nature to be nothing but drains on society and a rapidly proliferating population of hereditary paupers, thieves, and parasites.

Let Spencer write that by aiding the children of the poor to stay alive in times of societal or personal adversity, the misguided philanthropists become guilty of serious crimes against society, because when "they aid the offspring of the unworthy, they disadvantage the offspring of the worthy through burdening their parents by increasing local [tax] rates," and the eugenicists would be quick to cite Galton's commandment to prevent the poor from having any offspring. Together, from the era of the Robber Barons forward, the Social Darwinists and the eugenicists—whose various American societies and circles had interchangeable directorates and memberships—started to press for the compulsory sterilization of the mentally ill, the feeble-minded, the epileptics, the "habitual paupers and criminals," and all other "dysgenic" types.

The cruel Spencerian concept of millions of inferior people born worthy only of a quick and unmourned death, and of far lesser numbers of superior people prospering because they were *born fittest to survive,* formed an important element of the conventional wisdom of the *educated* classes of the nineteenth century. So pervasive was this common error that "the survival of the fittest" was what evolution was all about that most educated people also believed that it was Darwin, and not Spencer, who had coined this phrase originally.[28] Although Darwin considered Spencer to be a conceited and ill-informed boor who made sweeping scientific statements on the basis of inadequate evidence and personal observation, this did not stop educated people from regarding Spencer as the man who had applied Darwinian evolution to sociology.

A believer in the Lamarckian illusion of the *hereditary* transmission of postnatally acquired physical and behavioral characteristics, Spencer wrote that by manipulation and selective breeding the inherited and acquired characteristics of the *fittest* could result in races of the super-fit. As he wrote in his *Principles of Psychology* (1870):

> While the modified bodily structure produced by the new habits of life is bequeathed to future generations, the modified nervous tendencies produced by such new habits of life are also bequeathed; and if the new habits of life become permanent, the tendencies become permanent. . . . it needs only to contrast *national* characters to see that *mental peculiarities caused by habit become hereditary. We know* that there are warlike, peaceful, nomadic, maritime, hunting, commercial races[29]— races that are independent or slavish, active or slothful; *we know* that many of these, if not all, have a common origin; and hence it is inferable that these varieties of disposition, which have *evident relations* to the mode of life, have been gradually produced in the course of generations. [Italics added.]

Spencer's solution to the problem of inferior races was an exercise in pragmatic eugenics: he proposed that "science" be used to select the best characters of the various inferior races and then breed them in scientific mixtures planned to salvage whatever rudimentary human worth was present in the pooled hereditary endowments of the lesser breeds of humanity.

Two of Spencer's American contemporaries, David G. Croly and George Wakeman, saw an affinity in evolutionary inferiority between what they and their New World peers held to be the two lowest breeds of humanity—the Irish and the Negroes. Perhaps because they lived in an America that was going through an accelerated phase of its Industrial Revolution as well as a Civil War that was ending black chattel slavery, they had a somewhat greater sensitivity toward the economic values of an ever-replenishing pool of low-paid labor than did the ivory tower Teutonists of Oxford and Harvard. In their 1864 book, *Miscegenation: The Theory of the Blending of the Races, Applied to the American White Man and Negro,* they proposed what they termed the "melaleuketic union" of the Irish and the blacks. This act of guided evolution (as genetic engineering was then called) would, in turn, create a cheap and abundant labor pool whose members would be just a mite smarter than the Negroes and a bit stronger than the Irish.

This idea was to be embraced by Professor Joseph Le Conte, Georgia-born physician, geologist, and racial sociologist, and an original thinker who, in an 1878 essay, "Scientific Relation of Sociology to Biology," revealed to the readers of *The Popular Science Monthly* that "natural law is the mode of Divine activity," and that therefore the social and individual inequities of the times were simply instances of "God working to a given end without the conscious cooperation of individuals." A fervent disciple of Spencer's, Le Conte agonized over the dual menace posed to the greater society by the presence of the Irish and Negro races.

Le Conte recognized that, ideally, the best solution to the problem would have been to cross-breed both of these inferior races with the most superior of the white races, "the fair-haired Teuton," but he knew that this would be politically and morally unacceptable. He therefore settled for what he felt was the second-best solution science had to offer: this was a plan for breeding the Negroes with the lowest of the "marginal varieties" of the white races, which, as everyone Le Conte knew agreed, was the genteel euphemism for the Irish Catholics.[30]

In this proposal, Le Conte was somewhat more humane than was Oxford University's professor of history Edward A. Freeman. That noted scholar, a leader of the Teutonic cult that had burgeoned in Europe and America in the wake of Gobineau's *Inequality of the Races,* had a somewhat less eugenic solution to the problem. During his extremely profitable lecture tours of the United States in 1881 and 1882, Professor Freeman was shocked to find that the Teutonic and Aryan Anglo-Saxon blood lines had been corrupted by the Irish, the Negroes, and the Jews. The Irish were by far the worst offenders. ". . . alas, alas," Freeman wrote home from the citadel of human refuse, "in the oldest of the wooden houses where I went to find the New England Puritans, I found Ould Ireland Papishes-Biddy . . . instead of Hepzibah."[31] As we saw in chapter 4, Professor Freeman proposed a very logical mechanism whereby every Irishman would kill a Negro and be hanged for it.

THE OAK TREES OF THE OLD GERMANY—AND
THE NEW XENOPHOBIA

The Teutonic (later rechristened Nordic) cult in America, which was to dominate so much academic thinking as to make the professor at any of our better universities who was not a Teutonist the exception to the general rule between 1870 and April 1917, was ultimately to produce the 1896 *Atlantic Monthly* article "Restriction of Immigration,"[32] which marked the opening of the political phase of the movement for scientific racism in America. The author of that article was General Francis Amasa Walker, president of the Massachusetts Institute of Technology.

General Walker was one of the many leaders of American thought who stood in awe of Spencer and Galton, Freeman and Lapouge, Gobineau and Ripley. A man of some talent in statistics, he was nevertheless a confirmed Teutonist, a Social Darwinist, and a believer in Galton's concepts of "dysgenesis" due to the influx of breeding stock of less suitable races and strains.

The burning social question of the hour, Walker told his *Atlantic Monthly* readers, was not so much a matter of preventing European human refuse from packing "our alms houses, our insane asylums, and our jails," as everyone knew they did. Rather, he wrote, it was a matter of "protecting the American rate of wages, the American standards of living, and the quality of American citizenship from degradation through the tumultuous access of vast throngs of ignorant and brutalized peasantry from the countries of eastern and southern Europe." Walker, who had earlier served in Washington as chief of the Bureau of Statistics and Superintendent of the Census, observed that "the immigration during the period from 1830 to 1860, instead of constituting a net reinforcement to the population, simply resulted in a replacement of native by foreign elements."

The perils implicit in so simplistic a deduction about the meaning of this variation in fertility rates lay in the fact that, by so limiting his horizons, General Walker could conclude only that "while the population of 1790 was almost wholly native," by 1896 the non-native elements were beginning to outnumber and outbreed the native Americans. By "native Americans" Walker emphatically did not mean the native American Indians; he meant the white European aggressors who had robbed the Indians of their lives, their lands, and even their right to term themselves native Americans.

Between 1830 and 1840, the immigration of "large numbers of degraded peasantry," according to Walker, "created for the first time in this country distinct social classes, and produced an alteration" of conditions which could not fail to powerfully affect population rates. Walker asserted that the appearance of low-class, non-Nordic, grossly inferior types with "repellent habits" caused Americans to shrink "alike from social contact and economic competition thus created. They became increasingly unwilling to bring forth sons and daughters who should be obliged to compete in the market for labor and the walks of life with those whom they did not recognize as those of their own grade and condition."

Walker skirted but failed to come to grips with the advances in baby, child, and maternal care created by the Agricultural and Industrial revolutions. "It has been said by some," he wrote, "that during this time habits of luxury were entering, to reduce both the disposition and the ability to increase our own population. In some small degree, in some restricted localities, this undoubtedly was the case; but prior to 1869 there was no such general growth of luxury in the United States as is competent to account for the effect seen."

Like Sir Francis Galton, General Walker had not grasped the historical reality that the sociobiological by-products of what he called "luxury"—that is, the more hygienic and healthier standards of living being enjoyed by greater numbers of middle-class and wealthier families—had begun to significantly reduce the infant, child, and maternal mortality rates of these social segments in England and the United States. And that lowered infant, child, and maternal death rates were then and still are always followed by declines in the live-birth rates.

Walker conceded that foreign immigration "prior to 1860 was necessary in order to supply the country with a laboring class which should be able and willing to perform the lowest kind of work required in the upbuilding of our industrial and social structure, especially the making of railroads and canals." As a result, "when working on railroads and canals became the sign of a want to education and of a low social condition, our own people gave it up, and left it to those who were able to do that, *and nothing better*" (italics added).

According to Walker, as the Italians (apparently much lower on the evolutionary and racial value scale than even the lowly Irish Catholics) started to arrive, the Irish began to stand aside from the worst and dirtiest of the jobs involved in building America's industrial structure. The General minced no words:

"Does the Italian come because the Irishman refuses to work in the ditches and trenches, in gangs; or has the Irishman taken this position because the Italian has come? The latter is undoubtedly the truth; and if the administrators of Baron Hirsch's estate send us two millions of Russian Jews, we shall soon find the Italians standing on their dignity, and deeming themselves too good to work on streets and sewers and railroads. But meanwhile, what of the republic? What of the American standard of living? What of the American rate of wages?"

Walker, like Galton, was terrified by the dysgenic demographic effects of human refuse. "There is no reason," he wrote, "why every foul and stagnant pool of population in Europe, in which no breath of intellectual or industrial life has stirred for ages," should be admitted as immigrants to the United States. "The problems which so sternly confront us today are serious enough without being complicated and aggravated by the addition of some millions of Hungarians, Bohemians, Poles, south Italians, and Russian Jews." We had, after all, built our railroads, our canals, our sewers, our urban streets by June 1896. Consequently, to continue to admit immigrants of these inferior races was only to degrade our own "native American" race.

And General Walker summed up the case of the educated classes against the Hungarians, Bohemians, Poles, south Italians, and Russian Jews in words that would be cited again and again, on the lecture platforms, in the class-rooms, in the editorial columns of our leading newspapers, magazines, and learned journals, and in the halls of Congress until April 6, 1917, when the Noble, Nordic Teutons became, by an Act of Congress, the Vile, Alpine Huns:

"These people have no history behind them which is of a nature to give encouragement. They have none of the *inherited* instincts and tendencies which made it comparatively easy to deal with the immigration of the olden time. They are beaten men from beaten races; representing the worst failures in the struggle for existence. . . . They have none of the ideas and aptitudes which fit men to take up readily and easily the problem of self-care and self-government, *such as belong to those who are descended from the tribes that met under the oak-trees of old Germany to make laws and choose chieftains."* [Italics added.]

Francis Amasa Walker did not, of course, start the scientific racism movement in America. He did, however, provide one of the essential ingredients for scientific racism in American political life with this historic *Atlantic Monthly* article. Walker's disciples ranged from the economist Jeremiah Jenks, who was to write an extremely important textbook on immigration published in 1912, to the historian Woodrow Wilson, who was to be elected President of the United States in the same year.[33] In a very real sense, as we shall see in succeeding chapters, with this article in a leading journal of American culture, Walker had turned the first shovelful of earth in digging the foundations of the impenetrable barriers that were, starting early in 1933, to prevent millions of Hungarians, Bohemians, Poles, south Italians, Jews and other "beaten men from beaten races" from obtaining in this country even temporary asylum from extermination in the fires kindled with the oak trees of the same old Germany.

6 From Teutonism to Eugenics: The Torch of Scientific Racism Inflames America

> There has been started here a Record Office in *Eugenics;* so you see the seed sown by you is still sprouting in distant countries. . . . We have a Superintendent, a stenographer and two helpers, besides six trained field-workers. . . . though our work is mostly in "negative eugenics," we should put ourselves in a position to give positive advice. We cannot urge all persons with a defect not to marry, for that would imply *most* people, I imagine, but we hope to be able to say, "despite your defect, you can have sound offspring if you will marry thus-and-so."
>
> —CHARLES BENEDICT DAVENPORT, in a letter to Sir Francis Galton, October 26, 1910[1]

The educated people of Boston were outraged when, in 1834, a berserk mob of four thousand righteous Boston nativists, inflamed by anti-Catholic and anti-Irish tracts and rumors, sacked and burned a Charlestown "convent of young women, a seminary for the instruction of young females, turning them out of their beds, half-naked in the hurry of their flight, and half dead with confusion and terror."[2] "The reaction in the press and among the better classes of townspeople was generally one of outraged indignation . . . even the powerful Rev. Beecher delivered three sermons in which he expressed the horror of all decent Protestants at the outrage."[3]

Exactly fifty-eight years later, when the Boston municipal government elected to close the Boston Public Library on St. Patrick's Day, the poet and editor of *The Atlantic Monthly,* Thomas Bailey Aldrich, violently opposed this decision. "Columbus didn't discover America; it was St. Patrick," Aldrich wrote in opposition to this gesture. "He is in full defiant possession now, his colors waving everywhere, 'The Grane above the Red'—white and blue."[4] Far from reacting in cultured distaste to this vulgar echo of the Charlestown ravagers and arsonists of 1834 by a proper Bostonian of 1892, the educated people of the Athens of America hastened to renew their subscriptions to *The Atlantic.* For by 1892 the postulates of scientific racism had made the xenophobia of the Charlestown lynch mob of 1834 the cherished dogma and the moral philosophy of a vast segment of the educated classes.

In the few years that elapsed between the publication of Gobineau's Teutonist *Inequality of the Races* in 1853 and, say, the delivery of a doctoral thesis on Anglo-Saxon land law by the young Harvard graduate student Henry Cabot Lodge, son of a wealthy China-trade merchant, the tenets of the cult of Teutonism had become the conventional wisdom of the higher reaches of American university life. Young Lodge's Harvard mentor, Professor Henry Adams, and Lodge's other teachers, friends, and fellow Harvard faculty members—including such molders of American opinion as the editors of *The Atlantic Monthly,* James Russell Lowell and Thomas Bailey Aldrich, John

111

Fiske, Herbert Baxter Adams of Johns Hopkins University, Barrett Wendell, and A. Lawrence Lowell, the future president of Harvard—were all smitten by the Gobineau bug, and remained Teutonists for most of their active lives. In his 1876 doctoral dissertation, Lodge reflected the shared concepts of these and other American professors, editors, and lecturers, when he wrote: ". . . the purity of the race, the isolated condition of the country, and the very slowness and tenacity of intellect . . . gave a scientific development to the pure German law hardly to be found elsewhere. Free from the injurious influences of the Roman and Celtic peoples, the laws and institutions of the ancient German tribes flourished and waxed strong on the soil of England."[5]

Like Francis Amasa Walker of MIT, Lodge was typical of the new generation of American scholars in his Teutonism—and in his insistence that America close the gates against the "beaten men from beaten races" who threatened to dilute the Anglo-Saxon–Teuton blood that had made America great. This exclusionism, however, was a note new to the songs of Galton and Spencer that had previously enthralled the merchants, steamship operators, manufacturers, moneylenders, and lawyers who had fathered the New Teutonists. The fathers, too busy amassing the fortunes that had enabled their sons to live the lives of full-time scholars and gentlemen and even statesmen, had by no means shared the feelings of their educated sons about the continuation of unrestricted immigration of non-Nordic immigrants.

To the fathers of the new educated Teutonists and declared foes of further European immigration, this same immigration still represented an ever flowing sea of cheap labor; a bulwark against American trade unionism; and solvent customers for steamship passage, for land along the rights of way of the railroads the fathers controlled, and for the goods the fathers manufactured, imported, wholesaled, and retailed. Immigration was good business. Although, for example, the China trade profited enormously by satisfying the narcotics demands of the America in which, by 1869, nearly 4 percent (1,250,000) of the entire population was "caught up in some form of the opium habit,"[6] and in which "the per capita importation of crude opium increased from less than 12 grains annually in the 1840's to more than 52 grains in the 1890's,"[7] possibly even greater profits were earned by the ships of the same China trade for the importation of Chinese coolie labor for the building of railways, canals, sewers, and other lasting bases of American wealth.

Nevertheless, the fathers of the new racist restrictionists looked on in amused wonder as General Walker and other academic immigrant baiters not only blamed the immigrant laboring classes for the low wages paid by the employers of America but also actually managed to convince the native-born laboring classes that it was the foreign immigrants, *and not the native employers,* who established America's scales of wages and rewards for hard work. In the eyes of the fathers, any movement that could so divert the wrath of the American trade-union movement and its members from their employers to their brothers in toil could not be all bad. More than that, if college education could enable their sons and their sons' professors to succeed as famously as they did in selling this unique bit of lunacy to the growing

American trade-union movement, then it certainly proved the *practical* value of higher education.

Therefore, when in 1894 three wealthy young members of the Harvard class of 1889—the merchants' sons Prescott Farnsworth Hall and Robert DeCourcy Ward and the lawyer Charles Warren—founded the Immigration Restriction League in Boston, their families did not disown them. And when, as its anti-immigration weapon of choice, the new IRL settled upon the literacy test for immigrants that former Harvard history professor and now U.S. senator Henry Cabot Lodge introduced into the Senate in 1895, the bankers, merchants, manufacturers, and railroad magnates whose sons financed and staffed the IRL and its full-time lobby in Washington even contributed enough financial aid to keep the cauldron simmering. However, while the cash donations of the fathers who profited from unrestricted immigration helped bring various branches of the Knights of Labor and the American Federation of Labor into the coalition of immigrant-baiting societies headed by the Brahmin IRL, the elders at the same time refrained from exercising any of their own considerable political influence to cause the enactment of any of the anti-immigration legislation their sons demanded.

Within a decade of its founding, the national committee of the IRL was to include the presidents of Harvard, Bowdoin, Stanford, Western Reserve, Georgia Tech, and the Wharton School of Finance (University of Pennsylvania), as well as scores of other university presidents and professors. Many of the nation's leading bankers were to be represented on its board, as were Henry Holt and other major American publishers. The overt anti-Semitic, anti-Catholic, and anti-anything-but-Aryan purposes of the League's propaganda attracted the allegiance of racists and Yahoos of all stripes—from the Teutonists with Ph.D.'s to the night riders of the revived Ku Klux Klan.

The three IRL-sponsored immigration bills calling for literacy tests—of which Hall wrote on page 274 of his 1906 book, *Immigration and Its Effect upon the United States:* "The principal advantage of the educational test is that it is a definite rule of exclusion"—were each passed by Congress and vetoed by the sitting Presidents. The first literacy test bill was vetoed by Grover Cleveland in 1897 on the grounds that while it purported to concern itself with improving the quality of individual immigrants it was nothing but a hypocritical excuse for reducing the quantity of all immigration. The second League-packaged literacy test bill was vetoed by William Howard Taft in 1913 on the advice of his political counselors. Woodrow Wilson vetoed the literacy test bill sent up by Congress in 1915—but, this time, Congress was able to pass the anti-immigration measure over the President's veto.

Until the year before the Congress voted to overturn President Wilson's veto of the IRL reading test bill, the opposition of the heads of the giant corporations who owned and operated the vast industrial and food producing complexes, the rail and transit systems, the mines, real estate holdings, banks and mercantile establishments that produced most of the nation's wealth— and most of the financial support needed by both of our major political parties —had constituted an impregnable barrier to immigration restriction in this country. On the eve of World War I, one of the publications that spoke for this class, New York's *Wall Street Summary,* had attacked the literacy test route to immigration restriction in very blunt terms, noting that "all the leading papers of the Nation oppose the feature [literacy test] of the [1912] immigration bill," and calling upon Congress to back up this demand because: "We need help to harvest our crops, to develop our mines, to do a thousand and one things that cannot now be undertaken because of a dearth of labor. The native supply is non-existent; the immigrant is the solution."[8]

Four new factors, three of them bearing upon these urgent needs for immigrant labor, had, however, entered the scene as of 1915. The first was a technological development, the semi-automated industrial assembly line, which created newer and cheaper techniques of mass production and modified the acute need for highly skilled craftsmen. The second development was the large-scale introduction of a new generation of sophisticated farm machinery into American agriculture that, on the eve of World War I, had (like the eighteenth-century Georgian Enclosures) turned millions of farm workers and technologically displaced small farmers into hungry members of the growing native urban labor pool. The third major event was the outbreak of World War I, which ended the traffic in skilled and unskilled immigrant labor and thereby caused the in-migration of poor southern American whites and blacks from their native farms and hills to the mines, mills, and transportation systems of other areas of the country, where they now took the lowly places formerly filled in prewar America by cheap immigrant labor. The fourth revolutionary innovation was the formation of the powerful American eugenics movement, and the early marriage of love and convenience between this wellspring of scientific racism and the older Immigration Restriction League.

"THE MENACE TO SOCIETY OF INFERIOR BLOOD"

From the formal inception of the American eugenics movement, Charles Benedict Davenport was to reign as its scientific pope. The movement itself was, originally, a committee of the American Breeders' Association (ABA), a society organized by government and university agricultural breeders in 1903 as a direct result of the 1900 rediscovery of Mendel's 1865 paper on the heredity of sweet peas. (Mendel's paper had been rediscovered at least twice before: in 1874, by the Russian botanist I. F. Schmalhausen, who described it in an article in a Russian journal, and by W. O. Focke of Germany, who cited it on page 209 of his 1881 book, *Pfalnzenmischlinge*—and, quite possibly, by other workers in other countries between 1865 and 1900.) Not until 1906, largely at the urging of Davenport, did the ABA set up its Eugenics Section, under his stewardship, "to investigate and report on heredity in the human race" and to "emphasize the value of superior blood and the menace to society of inferior blood."

This sanguinary language could only have been Davenport's. He was formally appointed secretary of the new Eugenics Section and, from the start, he actually ran it. The chairman, David Starr-Jordan, president of Stanford University, biologist, active anti-imperialist, profound and courageous pacifist

(he was nearly lynched for his opposition to World War I in Palo Alto)—
and a true believer in Anglo-Saxon superiority and Mexican inferiority—
leaned heavily on what he felt was Davenport's superior knowledge of
genetics *and genealogy* for three decades.

The other members of America's first formal eugenics committee included the confirmed restrictionist and Teutonist Frederick Adams Woods,
a teacher of biology at MIT and a man described by Henry Fairfield Osborn
as "the American Galton"; Luther Burbank, the empirical plant breeder;
Roswell H. Johnson, then a geologist, later to become one of the first university professors of eugenics and co-author of the widely used college textbook *Applied Eugenics;* the University of Chicago sociologist Charles R.
Henderson; W. E. Castle, Harvard biologist and former student under Davenport; and Alexander Graham Bell, telephone inventor and student of what
he believed to be hereditary deafness in man.

Under Davenport's leadership, the ABA Eugenics Section established ten
research committees, dealing with the Heredity of Feeble-Mindedness; the
Heredity of Insanity; the Heredity of Epilepsy; the Heredity of Criminality;
the Heredity of Deafmutism; the Heredity of Eye Defects; Sterilization and
Other Means of Eliminating Defective Germ-plasm; Genealogy; the Inheritance of Mental Traits; and, of course, Immigration.

Davenport chose the IRL's Hall and Ward to be the leaders of the
Committee on Immigration. Hall saw in eugenics so powerful a propaganda
weapon for his cause that he tried to have the name of the IRL changed,
officially, to the Eugenics Immigration League. Although Hall even went
so far as to have new letterhead samples printed with the name of the Eugenics
Immigration League, he was not able to get the idea approved by the rest
of the IRL board, who felt that their priority gave them seniority in the
nomenclature of American scientific racism.

The appeal of the new eugenics movement, here as in England, was very
broad. Some of its earliest American adherents were overt racists and elitists,
such as Davenport, Woods, Johnson, Osborn, Hall, Ward, and Madison
Grant. Others were people who saw in eugenics a scientific method for possibly improving life for everyone, and for studying and eliminating what
might prove to be the clearly hereditary diseases of mankind. The ranks of
these people of goodwill were broad enough to include the anthropologist
Franz Boas, who was initially attracted to the mathematical theories of
Galton and Pearson; the psychiatrist and pioneer in the treatment and rehabilitation of the mentally retarded, Walter Elmore Fernald, who saw in
eugenics a possible tool for the prevention of what could prove to be hereditary psychiatric diseases; and Fernald's friend and collaborator, the Russian-
born Abraham Myerson, professor of neurology at Tufts University.[9]

To the agnostic Fabians, from Beatrice and Sidney Webb and the
mathematician Karl Pearson to the protean George Bernard Shaw, eugenics
provided the evolutionary evidence of their own inborn superiority over
lesser mortals. To various cliques of minor poets and literary salon keepers
in England and the United States, eugenics presented itself as a scientific
license for wholesale lechery, and was extensively used as such with great

and lusty vigor. Out of this so-called free-love wing of the world's eugenics movement came the popular but, alas, apocryphal story of the dancer Isidora Duncan's proposal to George Bernard Shaw that they mate to produce a child or two with his mind and her body, and Shaw's refusal on the grounds that the eugenic progeny might have her mind and his body. To the American geneticist Hermann Muller, a lifelong Marxist, eugenics was another reason for hastening the Dictatorship of the Proletariat, which he saw as the only kind of a society in which eugenics could be used to produce supermen.[10]

What gave eugenics such appeal to the elitists, faddists, and the democratic and Marxist idealists was its utter simplicity. To the early eugenicists who were also biologists, such as Davenport, each human trait or characteristic was produced by a specific gene for a specific *unit character*. These unitary traits or "unit characters" were discrete particles of hereditary information. In human beings, according to Davenport and the other self-styled "Mendelians," good blood consisted of a hereditary endowment brimming with excellent "unit characters" for physique, morality, intelligence, and specific skills, such as music, mathematics, dancing, statesmanship, and moneymaking. The hereditary endowments of social classes and races, the eugenicists wrote, also included the unit characters that gave a person inborn constitutional immunity to tuberculosis, diphtheria, typhoid fever, measles, and all of the other infectious diseases that were far more prevalent in the overcrowded slums and sparsely populated hovels of rural poverty than among the affluent in their spacious town houses and gracious country estates.

Bad blood, according to the eugenicists, consisted of a hereditary endowment made up primarily of the unit character of *pauperism*—in the eugenic literature the major *genetic* defect of the poor—as well as the unit characters for insanity, epilepsy, criminalism, immorality, low Binet-Simon IQ test scores, graft (at least, wrote Davenport, in the Irish), nomadism, shiftlessness, pellagra, laziness, feeblemindedness, asthenia (general physical weakness), lack of ambition, and general paralysis of the insane. It was also a major postulate of eugenics that inferior heredity was, as well, a human genetic endowment in which the blood of an individual was lacking in any of the unit characters (genes) that were supposed to provide the body with inborn immunity or resistance to tuberculosis, pellagra, infant diarrhea, dysentery, measles, malaria, cholera, pneumonia, influenza, and all of the other deficiency and infectious or parasitic diseases associated with poverty.

This basic concept of genetically determined unit characters was as easy to understand as was the Flat Earth Theory—and as logically foolproof as was the ancient Preformationism from which it had evolved. Now, thanks to the statistics of Pearson and Galton, the new "Mendelian" biology of Davenport, and the new eugenical psychology of Goddard, all that society had to do was to organize the selective mating of people born with varying endowments of superior unit characters in the blood of the superior Nordic and dolichocephalic (long-headed) races—and we would have the *positive eugenics* of race betterment. At the same time, by mandating and enforcing segregative and surgical strategies of preventing the further procreation of "beaten men from beaten races" with inferior unit characters in their bad bloods, society would arrive at the social benefits of *negative eugenics,* or the prevention of racial deterioration (now known as "genetic enslavement").

Such scientistic ideas had socioeconomic implications that were all but irresistible to the budget-conscious statesmen and their more affluent taxpaying and re-election-campaign-supporting constituents. Obviously, if human intelligence, health, and morals were *preformed* in one's hereditary Nature—and *not* in any way the end products of the Nurture provided by schools, vaccinations, clinics, hospitals, better living conditions, safer mines and factories, equality of opportunity, and other aspects of the total growth and developmental environment of a human being—then it was no longer necessary to waste any more tax-raised dollars on health, education, job safety, and employment benefits for races and social classes *doomed by their inferior genes to be ineducable, unhealthy, accident-prone, immoral, and unemployable.*

Cutting back on all private and public support of health, education, welfare, and other such "charity" was therefore not a matter of vulgar race, religious, and class prejudice; now, thanks to the Science of Eugenics, such acts of social regression were merely the prudent applications of the findings of value-free Science.

As individuals, Galton and Pearson both balked at accepting the Mendelian discrete-hereditary-particle mechanisms for this otherwise purely Galtonian version of twentieth-century Preformation. Davenport had originally joined his masters in opposing Mendelism. He had, however, too much exposure to the literature of biology, and to the company of close friends such as the Columbia University geneticist Thomas Hunt Morgan, to fail to realize how the scientific community respected Mendel's lost work. He not only accepted Mendel's findings as realities, but he even took a critical step *backward* from their implications.

From around 1905 onward, Davenport used the lexicon and authority of Mendelian genetics to confirm the predictions and dogmas of Galtonian eugenics. His mechanistic schemes for the application of his "Mendelian" positive and negative eugenics might, possibly, have stemmed from his earliest training, for before he studied biology at Harvard he had earned a degree in civil engineering from the Polytechnic Institute of Brooklyn, New York.

The eighth son and eleventh child of Amzi Benedict Davenport, a teacher and private-school owner turned affluent Brooklyn real estate operator, Charles Benedict Davenport was born on the family's farm near Stamford, Connecticut, in 1866. The elder Davenport, a man of Puritan stock who had authored "an elaborate genealogy of the Davenport family that went back continuously to 1086," was a highly neurotic entrepreneur, plagued by chronic rheumatism or arthritis, who kept young Charles out of school for all but two years of his childhood, tutoring him himself, and exploiting him as an office boy, stableboy, and office janitor.

At fourteen, Charles was permitted to enter Brooklyn Polytech, graduating in 1886 at the head of his class. His independently wealthy mother—her maternal grandfather, Teunis Joralemon, having acquired a farm in Brooklyn Heights on land that was to become a considerable part of downtown municipal Brooklyn—supported him in his desire to study biology at

Harvard. "With his mother's backing, tutoring, and a job in the Division of Water Works of the Massachusetts State Board of Health, he obtained funds for the first frugal student years. Later, the university provided various offices, culminating in an instructorship (1893–1899). He received an A.B. from Harvard in 1889 and a Ph.D. in 1902."

Davenport went to England in 1897 to meet Galton and Pearson and became an early convert to eugenics. At Harvard, he taught zoology and morphology, and in 1899 his *Statistical Methods with Special Reference to Biological Variation* was, according to his friend and fellow biologist Oscar Riddle, "the first book to bring the newer investigations of Karl Pearson to popular attention in the United States."[11]

In 1898, Davenport was appointed director of the Summer School of the Biological Laboratory of the Brooklyn Institute of Arts and Sciences, at Cold Spring Harbor, Long Island. During the academic year, Davenport was at the University of Chicago, first as an assistant professor of biology and, from 1901 to 1904, associate professor.

THE SCIENCE OF EUGENICS VS. THE MADNESS OF CHARITY

When the Carnegie Institution of Washington was organized in 1902, Davenport began a concentrated effort to get the new Carnegie agency to set him up at Cold Spring Harbor in a year-round biological farm and laboratory. His relentless pursuit of the Carnegie endowment succeeded, and in 1904 he was able to quit his job in Chicago and become the founding director of the new Carnegie Institution Station for Experimental Evolution at Cold Spring Harbor.

"Soon after Davenport's appointment," wrote Riddle, the great Columbia University cell biologist Edmund B. Wilson "was appointed a special adviser of the Carnegie Institution on the organization and work of the laboratories then being established at Cold Spring Harbor and Dry Tortugas. In this capacity, Wilson wrote a letter to Davenport suggesting a conference. This letter was curtly dismissed as 'interference' and it seems that Wilson made no further attempt to fulfill his mission with respect to this laboratory."

Davenport, Riddle went on to recall, had not only suffered from dyspepsia and a self-described "nervous temperament" for most of his adult life, but he also "did not seek, nor often accept, the advice of distinguished biologists or his staff concerning policies of high importance." In fact, Riddle concluded, this "lack of balance in Davenport's several abilities markedly limited the magnitude of his total contribution to science."

However meager his real achievements as a scientist, as a promoter Davenport soared from peak to peak. In 1910, he talked Mrs. E. H. Harriman, widow of the railroad tycoon, into putting up the money for the establishment of the Eugenics Record Office (ERO) at Cold Spring Harbor. She was to pour over a half million dollars into the eighty acres of real estate and running costs of the ERO before the Carnegie Institution took title to the Eugenics Record Office in 1918. It remained as part of the

Carnegie Institution Department of Genetics until 1940—at which time it was closed down, and its field worker "eugenic records" turned over to the Dight Institute for Human Genetics at the University of Minnesota.

Until the organization of the American Eugenics Society after World War I, the Eugenics Record Office, along with its subsidiary Eugenics Research Association, was the active center of all organized eugenical research, propaganda, and political activities in the United States. Under the direction of Davenport, and his fanatical second in command, Harry Hamilton Laughlin, the political activities of the Eugenics Record Office—which were concentrated primarily on the achievement of immigration restriction and the compulsory sterilization of human beings of "inferior blood"—constituted the major function of both the Davenport eugenics fronts.

The Eugenics Record Office, modeled on the Eugenics Record Office previously established by Sir Francis Galton in London, was from its inception the institutional vehicle of Davenport's thoughts, dreams, plans, and actions. Thanks to the generosity of Mrs. Harriman, Davenport was able to inform Galton himself, in a letter written on October 26, 1910, that "we have a plot of ground of 80 acres near New York City, and a house with a fireproof addition for our records. We have a Superintendent [H. H. Laughlin], a stenographer and two helpers, besides six trained field-workers. These are all associated with the Station for Experimental Evolution, which supplies experimental evidence of the methods of heredity. We have a satisfactory income . . . and have established very cordial relations with institutions for imbeciles, epileptics, insane and criminals."

In the final paragraph of the same letter, after hailing Galton as "the founder of the Science of Eugenics," Davenport concluded with what were to prove to be ominously accurate predictions: ". . . as the years go by, humanity will more and more appreciate its debt to you. *In this country we have run 'charity' mad.* Now, a revulsion of feeling is coming about, and people are turning to your teaching" (italics added).

What Davenport meant by "charity" were all private and governmental programs funded to improve public health, family health, occupational safety, and mental health—from sewers and water systems to clinics and quality education for all children. This was not only in line with Galton's teaching, but, of course, with the teachings of Galton's inspiration, Malthus. *"Most* people," he had written in the same letter, underlining the word "most," were defective: it was only the few people of "superior blood," such as Galton and Davenport, who were totally free of defects.

Through the Eugenics Record Office and its classes in eugenics for its field workers; its published Reports on specific projects and studies; its monthly *Eugenical News,* a newsletter written in very simple lay language, whose readers included influential Americans in government, publishing, and education; its Eugenics Research Association, whose presidents (hand-picked by Davenport) were to include the chairman of the U.S. House of Representatives Committee on Immigration and Naturalization; its corps of lecturers and visiting experts on eugenics and dysgenics; and its promotion of Aristogenics, or the guided selection and mating of individuals with superior blood to produce a new American Race of Super-Nordics—Davenport had fantasies of making his mark upon history.

Like Galton, Davenport relied heavily upon questionnaires on the genealogy and body measurements of individual families. These questionnaires were sent by the thousands to bankers, brokers, ministers, professors, corporation presidents, political officeholders, newspaper and magazine editors and publishers, and other people Davenport considered to be eminent. Along with the questionnaires went letters explaining that these data were needed in order to make the maximum use of the Science of Eugenics in helping America breed more leaders like the recipients of the questionnaires. The flattered leaders not only supplied their physical measurements and the answers to the questions; they also became the linchpin of the funding edifice Davenport was always building for his projects.

Identical questionnaires were taken into the homes of eminent families of "superior blood" and into the hovels of notorious families of "inferior blood" by the ERO's trained eugenics field workers. These well-motivated ladies not only obtained the anthropometric measurements and the answers to the questions from America's successes and failures. Once they were turned loose, they also roamed the community collecting thousands of old wives' tales, neighborhood myths, and unverified and usually malicious gossip about the sanity, the health, the morals, the intelligence, and even the legitimacy of the nineteenth-century forebears of close living neighbors— let alone of people dead for as long as Abraham Lincoln's grandmother.

The genealogical and eugenical "documents" thus collected by the eugenics field workers constituted by far the larger portion of the ERO "scientific family histories" now in the proud possession of the Dight Institute for Human Genetics in Minnesota. They were painstakingly gathered and put into carefully written reports, complete with pedigree trees for individuals and for generations of entire families, by lady amateurs in genetics, biology, sociology, and psychiatry who truly believed that they were scientifically trained to make such scientific observations and judgments.

SCIENTIFIC DIAGNOSIS "AT A GLANCE"

The eugenics field worker was, in fact, the chosen instrument for conducting the ERO and other Galtonian studies of America's good and bad blood lines. This new type of "scientific investigator" had been fashioned by Galton and Pearson in London, and in this country jointly by the psychologist Henry H. Goddard, in Vineland, New Jersey, and Davenport at Cold Spring Harbor. They were, in the main, young ladies of good families, some but by no means all of them college graduates. Very few of them had had any postgraduate university education or clinical training in psychology, psychiatry, sociology, anthropology, neurology, genetics, or biology.

During their training as field workers, Goddard wrote, these young ladies "spent weeks and months in the institution [i.e., the Vineland, New Jersey, Training School for Feeble-minded Girls and Boys], talking with all grades of defectives."[12] Over and above this on-the-job "training," a few weeks of lectures in eugenics at either Cold Spring Harbor or Vineland were all it took to turn out a eugenics field worker—whose like was never again to be seen.

These Goddard-trained and Davenport-trained eugenics field workers could, *at a glance,* spot and diagnose various hereditary mental conditions ranging from "dementia" and "shiftlessness" and "criminalism" to the most dangerous flaw of all, "feeblemindedness." They could also dig up the health, social, and genetic history of an individual or a family through interviews with neighbors and other contemporaries; through verbal accounts handed down for five to ten generations; through hospital, prison, church, and welfare agencies—and then put all of this "documented" information into concise "scientific" reports.

Since all of the classic studies of familial degeneracy and/or superiority written by leading eugenicists, such as Davenport and Goddard, and all of the new college textbooks in eugenics were based on the exhaustive case records prepared by these eugenics field workers, the question of the evidentiary worth of their reports becomes paramount in any modern consideration of the books and journal articles based on their "data." From the perspective of a half century past the golden hour of the field worker, the ladies trained by Goddard and Davenport come off with brilliant marks in the view of the modern eugenicists Elizabeth and Sheldon Reed, of the Dight Institute for Human Genetics. As they wrote in the preface to their *Mental Retardation: A Family Study,* in 1965, the ninety persons who received training as field workers at Davenport's Eugenics Record Office between 1911 and 1936 "were college graduates with majors in sociology, biology, or other appropriate areas. *They were well trained in the techniques of family history taking, interviewing, and pedigree construction"* (italics added). As we saw earlier, two of these Davenport-trained and ERO-paid field workers, working out of the State School and Colony (now Hospital) for the Feebleminded and Epileptics at Faribault, Minnesota, collected the family-history materials used as the foundation of the Reeds' 1965 study.

Fifty years before the Reeds' evaluation of their scientific worth, a somewhat contradictory assessment of the Goddard and Davenport field workers was offered by a contemporary psychologist, Samuel C. Kohs, who had observed them in action. Dr. Kohs, who was on the staff at the Chicago House of Correction, presented his analysis at the Eugenics Section of the 1915 annual meetings of the American Association for the Advancement of Science. In his paper, Dr. Kohs, according to the summary published in *Science* (September 24, 1915), "warned those doing research in the heredity of human psychical traits that they were in many cases wholly superficial, and that a definition of the traits which they discussed was a prerequisite of intelligent treatment."

The Chicago clinical psychologist, after citing ongoing research in psychiatry (by Freud and various other scientists) which indicated that many so-called hereditary traits "might in reality be due rather to impress on the unconscious mind during the early years of childhood," then added:

. . . most of the work on the inheritance of mental characters in man is of doubtful value . . . because of any one or more of the following reasons: 1. Inaccurate tools with which to measure the ability or capacity. 2. *Amateur field workers.* 3. The use of the *questionnaire method.* 4. Where more than one field worker was necessary for obtaining the data, the differences in the individual standards of the field workers vitiated the results. 5. Being told for what to look, *and possessing the popular conceptions regarding the inheritability of traits,* it is only natural to assume that many of the assistants [field workers] very easily found what was not there. 6. The study of character and personality is still in its infancy. To assume that certain peculiarities are due to the presence or absence of specific determiners [genes] can, in our present state of knowledge, hardly be substantiated by actual fact. [Italics added.]

A somewhat more detailed analysis of the people Kohs described as "amateur field workers" was provided by another clinical contemporary, Abraham Myerson, professor of neurology at Tufts, in his 1925 book *The Inheritance of Mental Disease.*[13]

Dr. Myerson noted that Goddard, in his book on feeblemindedness (*Feeblemindedness: Its Causes and Consequences,* 1914), had decided:

> feeble-mindedness *or the liability to become feeble-minded* is a Mendelian trait. . . . Goddard and Davenport are in full accord as to the traits *they* regard as neuropathic and to the value of the technique they employ in obtaining their data. *The keystone of the arch of their results is the field investigator and her surmises as to the mental and physical state of the dead and the quick;* and the cement is the theory that 30 or 40 different conditions are neuropathic traits and due to the lack of a unit character.
>
> I cite as an example of Goddard's investigations the famous Kallikak family, or rather the famous *account* of the family, quoted in all the lay literature and held up as a spectre of the threatened predominance of the feeble-minded. No royal family has enjoyed quite such a prestige as this group, except their associates in notoriety, the Jukes and the tribe of Ishmael. . . .
>
> Quite considerately Martin Kallikak performed an experiment for the sake of the writer [Goddard] of the book and the rest of society; he united himself first with this "nameless feeble-minded girl" and started this long row of degenerates—feeble-minded syphilitics and alcoholics —and then, reforming, used his germ plasm in orthodox fashion by marrying a nice girl who bore him nice children and started a row of nice people—all nice, no immoral, no syphilitics, no alcoholics, no insane, no criminals—as perfect in its way as its sub-rosa begotten half-relatives were imperfect.
>
> I confess to shame in the presence of the work done by the field worker in this case. I have had charge of a clinic where alleged feeble-minded persons were brought every day, and I see in my practice and hospital work murderers, thieves, sex offenders, failures, etc. Many of these are brought to me by social workers, keen intelligent women, who are in grave doubt as to the moral condition of their charges after

months of daily relationships, after intimate knowledge, and prolonged effort to understand. . . . And I have to say of myself, with due humility, that I have had to reverse my first impressions many and many a time.

Judge how superior the field workers trained by Dr. Goddard were! Not only does their "first glance" tell them that a person is feeble-minded, but they even know, without the faintest misgiving, that a "nameless girl" living over a hundred years before in a primitive community is feeble-minded. They *know* this, and Dr. Goddard, acting on this superior female intuition, founds an important theory of feeble-mindedness, and draws sweeping generalizations, with a fine moral undertone, from their work. Now I am frank to say that the matter is an unexplained miracle to me. How can anyone know anything definite about a nameless girl, living five generations before, whom no one has seen? Granting for the sake of argument that perhaps she had a mole on her left arm above the elbow by which she may have been identified by her contemporaries, how can anyone know that she was feeble-minded?

In his chapter on epilepsy in the same book—in which the Tufts professor of neurology presented some of the then already abundant medical evidence showing that epilepsy was neither an inherited disease nor even a specific disease entity in itself, but rather convulsive symptoms that followed various well-defined injuries or diseases—Myerson made a point of discussing the contribution of the eugenics field workers to Davenport's published reports on epilepsy as a hereditary disease. Myerson wrote:

In much of [Davenport's] work he has had medical collaboration, but except in a few instances he has directed the work of the medical men along his lines of thought. . . . I shall cite his work frequently and must therefore give some attention here to his general method.

Patients are selected, usually from an institution, and then their ancestry and relatives are looked up by a social worker [i.e., eugenics field worker], usually a woman with very limited training in medicine, who goes into the community interviewing relatives, neighbors, friends, social agencies and studying records. This information is *the chief basis for the conclusions reached,* and to a medical man the sang-froid with which the social [field] worker makes diagnoses on people she has never seen, or else met in a casual way, is nothing short of appalling. *Really, it seems utterly unnecessary to have laboratories, blood tests, psychological tests, clinical examinations, and to take four years in a medical school plus hospital experience, etc., when a woman can as a result of a dozen or two of lectures make all kinds of medical, surgical and psychiatric diagnoses in an interview or by reading through a court record.* [Italics added.]

The medical, surgical, and psychiatric diagnoses, and genetic pedigree charts, made by these painfully amateur eugenics field workers became the core "data" in the files of the Eugenics Record Office, the Vineland Training School, and other centers of American eugenical research that, in turn, formed the "scientific evidence" upon which the proliferating new college textbooks in eugenics were based. These included *The Social Direction of Evolution: An Outline of the Science of Eugenics* (1911), by the Goucher College professor of biology William E. Kellicott; *Being Well-Born: An Introduction to Eugenics* (1916), by Michael Guyer, a professor of zoology at the University of Wisconsin; *Applied Eugenics* (1918), by Roswell H. Johnson, professor of eugenics at the University of Pittsburgh since 1912, and Paul Popenoe, editor of the *Journal of Heredity;* and, among many others, *The Trend of the Race* (1921), by Professor Samuel J. Holmes, a zoologist who also gave courses in eugenics at the University of California.

By 1914, Frederick Adams Woods was teaching courses in eugenics at MIT, and, writes Mark Haller, "courses devoted in whole or in large part to eugenics" were being offered at Harvard, Columbia, Cornell, Brown, Wisconsin, Northwestern, Clark, and Utah Agricultural College, while in many other colleges and universities "eugenics entered the curriculum through courses in biology, genetics, sociology, or psychology."[14] The organized and institutionalized penetration of the propaganda of eugenics into the minds of the next generation of leaders in government, education, and publishing was well under way long before America entered World War I, and the eugenics field worker reports were the raw materials of the "scientific" pronouncements and pretensions of the Ph.D.'s who taught eugenics in our universities.

THE GELDING OF AMERICA'S CONSCIENCE

Considerable as the social and cultural effects of such college courses in eugenics were to prove, it is probably also true that few people made historically more devastating use of the eugenics field workers' reports than did Dr. Harry Hamilton Laughlin, superintendent of the Eugenics Record Office. Born in Oskaloosa, Iowa, in 1880, graduated Sc.D. from Princeton, Laughlin was working as a teacher of agriculture at the North Missouri State Normal School in 1910 when Davenport hired him to be second in command at Cold Spring Harbor. Laughlin was to remain at the Eugenics Record Office for the next thirty years, including the years in which he doubled in brass as the official Expert Eugenics Agent of the U.S. House of Representatives' Committee on Immigration and Naturalization.

It was under Laughlin that the two primary political objectives of the Cold Spring Harbor eugenics institutions were advanced. Some of the highlights of the roles of Laughlin, Davenport, and other activists of the eugenics movement in the successful crusade for the restriction of non-Teutonic, non-Nordic immigration may be found in later chapters. The second of the eugenics movement's principal legislative efforts, the campaign to have every state enact laws for the compulsory sterilization of the "socially inadequate," owed more to Laughlin than to any other person.

The movement to sterilize the "hereditary paupers, criminals, feeble-minded, tuberculous, shiftless and ne'er-do-wells" in this country did not originate with Laughlin and Davenport. As early as 1897, a bill to sterilize people of "bad heredity" was introduced in the Michigan state legislature, but was roundly defeated.

Two years later, without the benefit of any enabling legislation, Dr. Harry Sharp, physician at the Indiana State Reformatory at Jeffersonville, began to sterilize by vasectomy those young inmates he deemed to be hereditary criminal or otherwise genetically defective types. Dr. Sharp cited Francis Galton's "scientific discoveries" that all human physical and behavioral traits were purely hereditary as his moral and legal basis for demanding that society cease "permitting idiots, imbeciles, and degenerate criminals to continue the pollution of the race [sic] simply because certain religionists teach that 'marriages are made in heaven' and that the 'function of procreation is divine.' To me [Sharp] these are the most damnable heresies."

Sharp's answer to such heresies was compulsory sterilization of "the unfit." In 1905, the Senate of the Commonwealth of Pennsylvania passed "an act for the prevention of idiocy." This act—which declared: "Whereas, Heredity plays a most important part in the transmission of idiocy and imbecility"—decreed that it should henceforth "be compulsory for each and every institution in the state, entrusted exclusively or especially with the care of idiots and imbecile children," to appoint on its staff one skilled surgeon "whose duty it shall be, in conjunction with the chief physician of the institution, to examine the mental and physical condition of the inmates." If, upon this examination, the surgeon and the house doctor found that the "procreation [of the inmate] is inadvisable, and there is no probability of improvement in the mental and physical condition of the inmate, it shall be lawful for the surgeon to perform such operation for the prevention of procreation as shall be decided safest and most effective."[15]

Governor Samuel W. Pennypacker, fearful that his solons had taken leave of their senses, returned the bill with the observation that "the plainest and safest method of preventing procreation would be to cut the heads off the inmates, and such authority is given by the bill to this staff of scientific experts." Governor Pennypacker, in the same veto message, noted that the forced surgical sterilization bill would "inflict cruelty upon a helpless class in a community which that state has undertaken to protect." He therefore vetoed the bill on the grounds that it not only "violates the principles of ethics" but was "furthermore illogical in its thought. Idiocy will not be prevented by the prevention of procreation among these inmates. This mental condition is due to causes many of which are entirely beyond our knowledge." The governor also reminded his lawmakers that the retarded children had been entrusted to the state homes for the "purpose of training and instruction" and not "to experiment upon them."

Not until 1907, when the Indiana legislature passed, and the governor approved, "an act to prevent procreation of confirmed criminals, idiots, imbeciles, and rapists" held in state institutions and certified as beyond rehabilitation, did a compulsory sterilization law appear on the books of any state. The Indiana act declared that "heredity plays a most important part in the transmission of crime, idiocy, and imbecility," a statement that Galton, Pearson, Hall, Davenport, and other early propagandists for eugenics had made often, and which by 1907 was being taught as an undisputed scientific fact in most colleges.

When the state of Washington, in 1909, passed a bill to prevent the "procreation of feeble-minded, insane, epileptic, habitual criminals, moral degenerates, and sexual perverts, who may be inmates of institutions maintained by the State," it was the first of three compulsory sterilization bills to be passed that year. California and Connecticut passed nearly identical bills.

With these four laws, the cause of forced sterilization marked time until the advent of the Eugenics Record Office, its director, and its superintendent. Once the ERO propagandists and lobbyists entered the fray, the legislatures of a majority of the forty-eight states would fall in line and pass equally Draconian bills for the mandatory sterilization of what Laughlin called the "socially inadequate." In some states, such as New Jersey, where Governor Woodrow Wilson signed into law the brutal sterilization bill passed by the legislature in 1912, the state Supreme Court found such laws to be unconstitutional.

In others, such as Vermont, Nebraska, and Idaho in 1913, the governors vetoed such bills as being "unfair, unjust, unwarranted and inexcusable discrimination which cannot be tolerated"; and a violation of "Section 9, Article 1, of the Bill of Rights, which prohibits cruel and unusual punishment" and other provisions of the Bill of Rights, as well as being "more in keeping with the pagan age than the teachings of Christianity"; and because the gelding bills victimized "persons who by reasons of such confinement [in state institutions] are the least menace to society"; and, finally, because "the scientific premises upon which these laws are based are still too much in the realm of controversy."

In Oregon, the sterilization law passed by the legislature and signed into law by the governor in 1911 was repealed at the polls by a popular referendum in 1913. In 1921, the second sterilization bill to be passed in the twentieth century by a Pennsylvania legislature was vetoed by another governor.

By World War II, thanks to the relentless campaign directed by Laughlin and Davenport out of the Eugenics Record Office, laws for the compulsory sterilization of the poor, the helpless, and the misdiagnosed were on the books of thirty states and Puerto Rico as well. In some states, such as Minnesota—the *Eugenical News* reported in 1925 (p. 71) that "due largely to the efforts of the Minnesota Eugenics Society, under the presidency of Dr. C. F. Dight,[16] the State of Minnesota has enacted a sterilization law"—the burden of the victory was assumed at the local level. In most other states, the direction and impetus had come from Cold Spring Harbor.

A BULWARK AGAINST "THE FECUNDITY OF THE UNWORTHY TYPES"

One of the political triumphs of the ERO sterilization effort was the 1914 report of the Committee to Study and to Report on the Best Practical Means of Cutting Off the Defective Germ-Plasm in the American Population, set up by the American Breeders' Association's Eugenics Section in 1911. It was popularly called the Van Wagenen Committee, after its chairman, Bleecker Van Wagenen. The secretary and active director of the operation was Harry Laughlin.

Among the experts consulted by the committee was former President Theodore Roosevelt, an old adherent of eugenics and racial superiority, who on January 14, 1913, sent the group a letter that read, in part: "As you say, it is obvious that if in the future racial qualities are to be improved, the improving must be wrought mainly by favoring the fecundity of the worthy types and frowning on the fecundity of the unworthy types. At present, we do just the reverse. There is no check to the fecundity of those who are subnormal, both intellectually and morally, while the provident and thrifty tend to develop a cold selfishness, which makes them refuse to breed at all."

The report of this ABA Eugenics Section committee laid down the "scientific" foundations for the persistent eugenical myth that at least 10 percent of the American population is, by heredity, socially inadequate, and should not be permitted to breed. This report was sent to most American legislators, mass-media editors, clergymen, and its leading professors of sociology, political science, and history. Because its conclusions were incorporated, totally, into the conventional wisdom and hence the value systems of the educated classes who then (and now) formulate American social policies, it is essential that the key provisions of this document be exhumed for the modern reader.

The 1914 report also introduced another favorite bit of eugenical dogma. This was the Galtonian theological concept that "society must look upon germ-plasm as belonging to society and not solely to the individual who carries it"—a purple bit of chromosomal metaphysics that persists in some minds to this day, as witnessed by Professor Garrett Hardin's 1970 editorial in *Science,* "Parenthood: Right or Privilege?" Nearly sixty years after the ABA statement of this theme, Hardin's editorial declared: "Biologically, all that I give 'my child' is a set of chromosomes. Are they *my* chromosomes? *Hardly.* . . . 'My' child's germ plasm is not *mine;* it is really only part of the community's store. I was merely the temporary custodian of it."[17]

The committee report also, possibly with a tongue-in-cheek bow to the twentieth century, declared: "It now behooves society in consonance with both humanitarianism and race efficiency to provide more human means for cutting off defectives. . . . Humanitarianism demands that every individual born be given every opportunity for decent and effective life that our civilization can offer. Racial instinct demands that defectives shall not continue their unworthy traits to menace society. There appears to be no incompatibility between the two ideals and demands."

This pious and self-serving bow to the ideal of equality of opportunity for *all* newborn children was, obviously, intended to serve as documentary proof that Laughlin et al. were not opposed to the legitimate aspirations of the Deserving Poor, in whose veins flowed pure Nordic blood chock full of constructive unit characters.

The second portion of the 1914 Van Wagenen Committee report, titled "Classification of the Socially Unfit from Defective Inheritance: The Cacogenic Varieties of the Human Race," survives as one of the century's most spine-chilling blueprints for planned genocide, in a class with *Mein Kampf* and the gamier passages in the works of Vogt and the Paddocks. Like *Mein*

Kampf, this section was presented as a *scientific* rationale for the steriliza-tion of human beings by the state for literally hundreds of physical, mental, social, moral, philosophical, and economic conditions that, even in terms of the scientific knowledge of 1914, were quite plainly not of genetic or, in most instances, of even exogenous biological origins.

This portion listed ten "cacogenic"—that is, dysgenic or causing race deterioration—varieties of the human race: (1) The Feeble-minded Class. (2) The Pauper Class. (3) The Inebriate Class. (4) The Criminalistic Class. (5) The Epileptic Class. (6) The Insane Class. (7) The Asthenic (lacking in strength) Class. (8) The Diathetic (constitutional susceptibility to specific diseases) Class. (9) The Deformed Class. (10) The Cacaesthetic (having defective sense organs, "as the blind or the deaf") Class.

These ten "classes" of what the authors of the report categorized as *hereditary* biological, economic, mental, and moral defects found in at least one tenth of the American population, whom the authors condemned as being born "socially inadequate," were the basis of the central dogma of the Daven-port or eugenical hypothesis of "Mendelian" inheritance. As the report put it on page 17: "This classification of the socially inadequate is obviously partly legal and partly medical, *but it is in the most part biological* [i.e., genetic], although a purely biological classification would be extremely complex, since it must be based upon *unit traits of defective inheritance* and their combina-tions into personalities of the various legal, medical, and social types" (italics added).

In short, the entire "scientific" structure rested upon the belief that *unit characters*—that is, genes for specific physiological and behavioral traits—controlled the different mental, physical, moral, and emotional traits of every human being. The unit-character theory was the foundation on which Galtonian and pseudo-Mendelian concepts of human heredity rested, but, as we shall see, by 1914 the work of such serious biologists and pioneer geneticists as Johannsen, Boveri, Sutton, Hardy (a mathematician), and Weinberg in Europe, and Davenport's intimate friend Thomas Hunt Morgan at Columbia University and his formidably talented young collaborators in this country, had ruined the unit-character hypothesis of heredity as com-pletely as Kaspar Friedrich Wolff destroyed Preformation in 1768.

The discussion of the nature and the social menace of the ten genetically cacogenic (dysgenic) classes, running from pages 16 through 44 of the Van Wagenen report, spelled out in some detail the moral as well as the scientific limitations of its authors.

Of Class 1, the "Feeble-minded Class," the report declared that "the greatest of all eugenical problems in reference to cutting off the lower levels of human society consists in devising a practicable means for eliminating hereditary feeble-mindedness. . . . The chronological age of such individuals is always somewhat and maybe greatly in excess of their mental years." The "mental age" of these people was based on early uses of the Goddard versions of the Binet-Simon IQ tests.

This genetically inferior class, the born feebleminded, had "strong vicious or criminal propensities," but were so innately stupid that "under the selfishly severe stress of a primitive order of social affairs," said the authors of the report, "natural selection would readily cut off these lowest classes." However, the nobly intended but dysgenic "charity" or social welfare programs of the misguided idealists and sentimental do-gooders, let alone naïve liberals, were keeping this cacogenic "class" of the hereditary feebleminded alive and able to perpetuate by their excess sexuality the degeneracy packed into their individual and class genetic endowments of inferior unit characters.

Of the "Pauper Class," that malignant invention of Malthus and Galton, the report (p. 20) found:

> Individuals belonging to this class fall quite naturally into the following three groups: 1. Tramps; 2. Beggars; 3. Ne'er-do-wells.
> Many of these individuals belong properly to the feeble-minded class. Oftentimes their special defect or deficiency takes the form of *shiftlessness or laziness*. . . . Adults of normal traits, who have been socially adequate, but have, through an absolute lack of training and opportunity, become defective and dependent upon charity are not . . . to be included in the pauper class. It is only with the individual of a hereditary, degenerate make-up which manifests itself in an inability to get on, or lack of ambition, or laziness which drives him or her beyond the bounds of self-maintained usefulness in an organized society that this study is concerned. These individuals [the Undeserving Poor] are so strikingly anti-social that society is justified, if the general uselessness can be shown to be hereditary, in cutting off the descent line of this whole group of individuals, *even if their specific traits and defects cannot be catalogued*. [Italics added.]

The moral courage, and the boundless scientific ignorance, of Harry Hamilton Laughlin were displayed in the three paragraphs on page 25 dealing with the "Epileptic Class." The first and third read:

> Among degenerates epilepsy is so common that it deserves a separate classification under the anti-social group. Functionally, this disease is often associated with feeble-mindedness, crime, inebriety, and insanity, but, on the other hand, *sometimes* it is associated with sterling personalities of great social worth. . . .
> *No clearer cases of specific hereditary degeneracy than those of epilepsy have been established.* Even when associated with sterling traits in worthy persons, epilepsy is a deteriorating factor. When associated with other defects, they appear to be inter-accelerating causes of deterioration. [Italics added.]

As it happened, Laughlin was an epileptic himself. For this reason, although married and a true-believing eugenicist and therefore opposed to birth control, which the orthodox eugenicists still described as "race suicide," Laughlin was childless by choice.

As it also happened, by indulging his profound faith in Galton as the world's leading authority on the etiology of epilepsy and ignoring the available hard *scientific* information on epilepsy, this "sterling personality of great social worth" shortchanged himself and possibly posterity. For, by 1914, most *medical* researchers were in agreement with the great nineteenth-century

British neurologist and contemporary of Galton's, Hughlings Jackson, that epilepsy is a *nongenetic* and intermittent disorder of the nervous system due to a sudden and excessive discharge of cerebral neurons (nerve cells). The more responsible medical literature of Laughlin's day—like our current medical textbooks—described epilepsy not as a specific disease but as a family of convulsive disorders caused by an array of environmental insults, from brain injuries at birth or at forty, to infectious thrombosis of the cerebral arteries or veins, malignant tumors, malnutrition-caused disorders such as hypocalcemia and hypoglycemia, alcoholism, drug addiction, and other nongenetic causes.

Of the "Insane Class," the report said nothing that had not previously been said, with equal assurance, by Galton. Including:

> There is no class of anti-social individuals more definitely and sharply marked off from the general social body, so far as their principles of conduct are concerned, than the insane class. With this class heredity plays an important part, and here again the basis of social classification is purely functional, while that of eugenics is hereditary.

Between Galton's heyday and 1914, psychiatry had undergone whole series of significant and massive expansions of scientific knowledge of mental disorders and their multiple and interacting constitutional and postnatal causes. This revolution in scientific thinking, led by such giants as Charcot, Janet, Hughlings Jackson, Franz, Breuer, Freud (who had lectured in the United States in 1910, and whose new psychoanalytical theories were well known by 1914), and many other major workers in the behavioral sciences, had by 1914 long since shattered the hoary simplisms that Galton and other hidebound Victorians put forward as "scientific" explanations for human behavior.

Not a single reference to any of these by 1914 well-recognized advances in the world's psychological knowledge appears in the section of the report that dealt with the "Insane Class." What did emerge, however, both in this section on people with psychiatric disorders and in the section on the "Diathetic Class," were some most ingenious perversions of the scientific findings of the new biological science of medical bacteriology—such as the germ theory, which demolished Galton's moralistic postulates about the hereditary etiology of the contagious diseases of urban poverty—into "scientific proofs" of the medical wisdom of Sir Francis Galton. Thus, on page 27:

> In relation to practical eugenics a specific psychosis may be directly inherited as such, in which case the disease will appear in due ontogenetic [developmental] sequence. Or its diathesis [predisposition] only may be transmitted. In some types, such as chronic alcoholism and paresis, *heredity appears to be the foundation factor,* but the poisons respectively of alcohol and of *treponema pallidum* [the bacterium that causes syphilis] *must conspire with this defective background in order to produce the disease.* So in the group of *so-called functional psychoses* there may be either a weak or a strong diatheses—the one requiring a relatively great stress and the other a relatively little stress to bring on the ailment. To the extent that a given strain possesses a hereditary con-

stitutional make-up liable to display a psychosis under anything less than an extraordinary formidable stress of circumstances, there exists in such strain a cacogenic variety of the human race. [Italics added.]

In plainer English: chemical poisons and pathogenic microbes are more apt to affect people of inferior blood than people better endowed to resist alcohol and germs because of the greater quantities of unit characters (or genes) for disease resistance in their superior blood. This was further developed in the pages of the report that dealt with the differences between "direct heredity . . . the transmission of a trait or a quality that will," in spite of controlled environment, appear at some time in the course of development of an individual, and

> the second type of heredity, which might well be called "indirect heredity," or "hereditary-diathesis," "susceptibility," or "predisposition."
> In this sort of heredity environment plays a much greater part in determining the human trait or condition than it plays in direct heredity, *but even in such cases the exogenous* [environmental] *forces are not all-important.* Heredity is as it were the foundation upon which environment builds the trait. In such cases heredity, although a less powerful factor, is just as definite as with direct inheritance, and the end product is a composite of hereditary and extrinsic factors. Thus, *people do not, biologically speaking, directly inherit tuberculosis and yet they inherit directly a constitutional make-up possibly both functional and chemical, as well as structural, that causes them to fall an easy prey to this disease.* . . . Thus, in reference to their susceptibility and immunity, there appears to be a chemical difference in persons which is directly hereditary, but it requires the presence of an exogenous agent, in addition to the innate [genetic] lack of resistance, to cause the affection. [Italics added.]

All of this was the eugenics movement's way of saying, in answer to the medical scientists—who for decades had published ever increasing volumes of scientific evidence proving that the higher incidences of measles, influenza, tuberculosis, scarlet fever, and other infectious and contagious diseases among the poor than among the nonpoor people of the same Nordic and other ethnic strains were due to the overcrowding, the paucity of running water, baths, inside toilets, sanitary waste-disposal systems, and other environmental realities of nineteenth- and twentieth-century poverty—that such "exogenous forces are not all-important."

Heredity was also (p. 28) responsible for the "Asthenic Class." "Physical weakness, if hereditary, is cacogenic, for a race of weaklings cannot long endure. Physical weakness is not *the menace that feeble-mindedness is,* but it is, nevertheless, great" (italics added).

"The menace of the feebleminded" was a well-used fright phrase in eugenical and popular journalism during the early decades of the twentieth century. It referred, essentially, to the eugenical postulate that feeblemindedness was the cause of most crimes, and ranked second only to "pauperism" as a hereditary cause of racial deterioration.

Nor was there, in this life, any escape from eugenic Predestination:

Hereditary traits do not date from birth, for birth is only a change of environment. The hereditary potentialities of an individual are determined past recall when the two parental gametes [sex cells] meet in fertilization to form the zygote [the fertilized ovum].

This bit of ponderous metaphysics had a direct bearing on all societal expenditures designed to improve the total environment in which every individual is conceived and carried *in utero,* and in which, after birth, people develop and grow through infancy, childhood, adolescence, and maturity. As the committee report, after reviewing all of the proposed solutions to the problems of human betterment, from birth control and polygamy to education and euthanasia, noted:

> *General environmental betterment.* It is held by some schools of social workers that better schools, better churches, better food, better clothing, better living and better social life will remedy almost any social inadequacy in individuals. *The studies of this committee point strongly in the opposite direction.* They *prove* conclusively that much social inadequacy is of a deep-seated biological [i.e., genetic] nature, and *can be remedied only by cutting off the human strains that produce it.* As a rule, a good ancestral germ-plasm will furnish a good environment for the offspring and a bad ancestral germ-plasm will add to the degenerate hereditary gifts of its offspring a poor environment. [Italics added.]

What this meant, therefore, was:

> It is the bolstering up of the defective classes by a *beneficent society that constitutes the real menace to our blood,* because it lowers the basis of parenthood. . . . There must be [eugenic] selection not only for progress, but even for maintaining the present standard. To the degree that we inhibit natural selection, we must substitute rational selection, else our blood will deteriorate. [Italics added.]

Rational selection, of course, came at the cutting edges of the gelders' knives, for in the matter of

> . . . the cutting off of the supply of innate [genetic] social misfits . . . it is the duty of human society to grasp every possible means for its amelioration, and, if it finds in the segregation and sterilization of defectives a means for improving the innate [hereditary] qualities of future generations . . . it is the duty of society even at great cost and effort to bestir itself in applying such remedy. . . . A successful society must under all hazards protect its breeding stock, and since, under modern conditions, a vigorous program of segregation supported by sterilization seems to present the only practicable means for accomplishing such an end, a progressive [*sic*] social order must in sheer self-preservation accept it.

Up to this point in the long report of the eugenics committee to protect America from genetic enslavement, the authors had admittedly concentrated only on negative eugenics. There was also, Aristogenics, or the "positive side" of eugenics—"that of encouraging increased fecundity and fortunate matings among the better classes." But all this increased fecundity of the people of superior blood, the committee report warned, could be offset by the fecundity of the lesser breeds. Salvation was possible, but it would take time and money:

> The program as outlined by the committee calls for a task that will require two generations for the completion of its first [or negative eugenics] stage. No matter to what extent laws may be passed, *unless the eugenics program becomes a part of the American civic religion* the financial support necessary to put it into execution cannot be secured from the several [state and national] legislatures. . . . A quickened *eugenics conscience* is one of the prerequisites necessary to the working out of a successful eugenics program. *Eugenics must be diffused through our religious and moral codes. It must be taught throughout our national educational system.* [Italics added.]

These hopes were to be realized. Eugenics did become a vital part of the "American civic religion"—that is, of the value systems of the civic leaders who established our social programs, and of the voters who elected them to office to administer such programs. And for generations to come, eugenics was to be taught *as a science,* and as a scientific rationale for medically, and culturally, and humanly inadequate social programs dealing with health, education, and family welfare, "throughout our national educational system."

THE MODEL EUGENICAL STERILIZATION LAW FOR THIRTY AMERICAN STATES—AND NAZI GERMANY

To assist the states in enacting compulsory sterilization laws swiftly, Laughlin drew up a Model Eugenical Sterilization Law, which was distributed in huge quantities to governors, legislators, newspaper and magazine editors, clergymen, and teachers by the Eugenics Record Office and by the American Eugenics Society, formed after World War I. Special mailings of Laughlin's Model Eugenical Sterilization Law went out to the leaders and propaganda directors of hundreds of nativist and xenophobic pressure groups.

Certain key clauses of Laughlin's Model Eugenical Sterilization Law merit modern rediscovery. For example:

> *Persons Subject: All* persons in the State who, because of degenerate or defective hereditary qualities are potential parents of socially inadequate offspring, *regardless of whether such persons be in the population at large* or inmates of custodial institutions, regardless also of personality, sex, age, marital condition, race or possessions of such person [italics added].

Laughlin's proposed law was to be enacted as:

> AN ACT to prevent the procreation of persons socially inadequate from defective inheritance, by authorizing and providing for eugenical sterilization of certain potential parents carrying degenerate hereditary qualities.

Section 2 of the Laughlin law defined a socially inadequate person "as one who . . . fails chronically . . . to maintain himself or herself as a useful member of the organized social life of the state. . . ."

The language of Laughlin's law was unambiguous:

> The socially inadequate classes, *regardless of etiology or prognosis* [that is, regardless of the cause of the condition or the chances of eliminating it by means other than gelding], are the following: (1) Feeble-minded; (2) Insane (including the psychopathic); (3) Criminalistic (including the delinquent and wayward); (4) Epileptic; (5) Inebriate (including drug-habitués); (6) Diseased (including the tuberculous, the syphilitic, the leprous, and others with chronic, infectious, and legally segregable diseases); (7) Blind (including those with seriously impaired vision); (8) Deaf (including those with serious impaired hearing); (9) Deformed (including the crippled); and (10) Dependent (including orphans, ne'er-do-wells [*sic*], the homeless, tramps, and *paupers*). [Italics added.][18]

Since, under Laughlin's Eugenical Sterilization Law, these conditions rendered any individual subject to mandatory sterilization *"regardless of the etiology or prognosis"* of his "social inadequacies," this meant, for example, that the law compelled the states to sterilize the orphans of American soldiers killed in the World and Presidential wars of our century. The Eugenical Sterilization Law also demanded the gelding of Americans rendered homeless by floods, earthquakes, and the Great Depression of 1929–41; the people crippled by the viruses of poliomyelitis and rubella infection; and the people blinded and/or deafened by industrial accidents or explosions, if their clearly nongenetic handicaps prevented their being totally self-supporting. The Model Eugenical Sterilization Law also ordered the compulsory sterilization of epileptics, and of victims of microbial and other parasitic infections that were due, in many instances, to the failure of government public health agencies to provide the minimal immunization and civic hygienic services that are routinely available in civilized societies.

Laughlin's Model Eugenical Sterilization Law was aimed as much at the minds of the American taxpayer as it was at the gonads of the poor. It offered the taxpayers a simplistic alternative to costly tax-supported government "charity" programs administered for the healing of the sick and the injured, the care of orphans, and the shelter of the homeless victims of wars, poverty, aging, economic crises, and industrial accidents. To the Yahoos who were also taxpayers, the sterilizer's knife loomed as a cheaper and more equitable instrument of coping with the health, welfare, and geriatric problems that were and are the province of government social agencies.

Most of the compulsory sterilization laws passed by thirty states followed to one degree or another the principles of Laughlin's model law, although as a rule they obliged the states to sterilize only the inmates of prisons, hospitals, geriatric, and mental institutions.

Some state legislators added categories of cacogenics that not even Laughlin and Davenport had discovered. In 1929, for example, Missouri solon G. E. Balew introduced House Bill No. 290 in the 55th General Assembly of the state's legislature. Section 1 of this bill mandated the forced sterilization of people "convicted of murder (not in the heat of passion),

rape, highway robbery, chicken stealing, bombing, or theft of automobile."
The bill was defeated—but not forgotten.[19]

By 1968, 65,000 Americans had been sterilized against their will in the
thirty states that had passed such laws, more than 52 percent of them for
being labeled as "mentally retarded." In most states the diagnosis of "mental
retardation" was made on the IQ test scores of the victims of forced eugenical
sterilization.

The influence of the Eugenics Record Office's Model Eugenical Sterilization Law was not confined to the legislators and governors of thirty of the
United States. In 1924, an early Austrian convert to eugenics (*Rassenhygiene*), and an admirer of the works of Madison Grant and Lothrop Stoddard and their eugenical co-religionists in the United States, wrote a long book
in which, describing the pure-blooded Nordic state of the future, he said:

> The folkish State has to make up for what is today neglected in this
> field [eugenics] in all directions. It has to put the race into the center of
> life in general. It has to care for its preservation in purity. In this matter
> the State must assert itself as the trustee of a millennial future, in the
> face of which the egoistic desires of the individual give way before the
> ruling of the State. In order to fulfill this duty in a practical manner, the
> State will have to avail itself of modern medical discoveries. *It must pro-
> claim as unfit for procreation all those who are afflicted with some
> visible hereditary disease or are carriers of it; and practical measures
> must be adopted to have such people rendered sterile* [italics added].[20]

A decade later, when the author of these lines emerged as Der Führer
of the Third and Nordic Reich, his regime wasted little time in putting these
principles into law. On July 14, 1933, the Nazi government passed an amendment to the 1927 German sterilization law, which had legalized the *voluntary*
sterilization of people for medically certified health reasons. The Nazi addition, taken almost *in toto* from Laughlin's Model Eugenical Sterilization Law,
the *Gesetz zur Verhüting erbkranken Nachwuches* (Act for Averting
Descendants Afflicted with Hereditary Diseases), mandated the compulsory
sterilization of people found by the Nazi Eugenics Courts to be social inadequates. These included the people ruled by the Eugenics Courts to be
genetically mentally deficient, schizophrenic, manic-depressive, epileptic,
blind, deaf, chronic alcoholics, and victims of any grave physical defect ruled
to be inherited.

When Hitler's Thousand-Year Reich fell in 1945, it was revealed by the
German Central Association of Sterilized Persons that at least two million
human beings had been ruled in the Eugenics Courts to be eugenically unfit
(dysgenic) and sterilized against their will during the twelve years of the
Nazi version of Laughlin's Eugenical Sterilization Law.[21] Under the voluntary
sterilization law of the Weimar Republic overthrown by Hitler, between 1927
and 1933 a *total* of less than 500 Germans—about 85 people a year, most of
them women whose health would have been jeopardized by pregnancy—had
been voluntarily sterilized. Under the Nazis, an average of 165,000 Germans
of both sexes were sterilized annually against their will—at the rate of 450
forced sterilizations *per day*.

THE NEW KEEPERS OF THE EUGENICS FLAME

The obsession of the American eugenics movement with the gonads of the inferior races and subraces, and its swift successes in winning the leaders of state and national governments to the cause of the forced sterilization of the poor, the unfortunate, and the victims of industrial accidents and other traumas from wars to preventable infectious diseases, was on the eve of World War I to invest the scientific racists of America with the leadership of the world eugenics movement. The passing of the flame from the Old World to the New was effected as early as 1912, at the First International Congress of Eugenics, held at the University of London a year after Galton's death.

The president of the Congress was Major Leonard Darwin, son of Charles Darwin, and living proof that neither scientific genius nor great intelligence is hereditary, or transmissible in the genes of the parents. The English vice-presidents of the Congress of Eugenics included a future Prime Minister, First Lord of the Admiralty, Winston Churchill; the Bishop of Oxford; the Lord Chief Justice, Lord Alverstone; and the president of the College of Physicians, Sir Thomas Barlow. The German vice-presidents included M. von Gruber, professor of hygiene at Munich, and Dr. Alfred Ploetz, president of the International Society for Race Hygiene. The American vice-presidents included Gifford Pinchot, a future governor of Pennsylvania; Charles W. Eliot, president emeritus of Harvard University; Alexander Graham Bell; David Starr-Jordan, president of Stanford University; and Charles B. Davenport, listed on the program as secretary of the American Breeders' Association.

Davenport presented a paper, "Marriage Laws and Customs"; his collaborator in studies of epilepsy, David Fairchild Weeks, M.D., delivered a paper entitled "The Inheritance of Epilepsy"; and, a year prior to the preparation of its final report, Bleecker Van Wagenen gave "A Preliminary Report of the Committee of the Eugenics Section of the American Breeders' Association to Study and to Report on the Best Practical Means for Cutting Off the Defective Germ-Plasm in the Human Population." As the chairman of this committee, Van Wagenen informed the world's eugenicists that the headquarters of the American Breeders' Association was in Washington, and that its president "is the Hon. James Wilson, Secretary of Agriculture, and member of the President's [William Howard Taft's] Cabinet."

The committee, Van Wagenen reported, "is seeking and receiving assistance from many sources, both *public* and private" (italics added). The purposes of the committee were spelled out quite plainly by the delegate from the Vineland, New Jersey, Training School for Feeble-minded Girls and Boys:

> In recent years society has become aroused to the fact that the number of individuals within its *defective classes* has rapidly increased both absolutely and in proportion to the entire population; *that eleemosynary expenditure is growing yearly;* that some normal strains are becoming contaminated with anti-social and defective traits; and that the shame, the moral retardation, and *the economic burden of the presence of such*

individuals are more keenly felt than ever before. Within the last three years especially there has been a marked development of public interest in this matter. The word "Eugenics" has for the first time become known to thousands of intelligent people who now seek to understand its full significance and application. [Italics added.]

In the cause of reducing or eliminating the economic burden of the national "eleemosynary expenditure"—which in the well-understood code language of eugenics and its true believers meant not only private charity and assistance to the poor and the unfortunate but also *all* public expenditures to improve the health, education, and well-being of every nation's total population—America had taken the lead in the passage of compulsory sterilization laws designed to rid society of the recipients of such private and public assistance. Van Wagenen revealed that, between 1907 and 1912, the states of Indiana, Connecticut, California, Iowa, Nevada, New Jersey, Washington, and New York had passed laws authorizing or requiring the sterilization of "certain classes of defectives and degenerates."

The report was documented with a table which showed that in five of these states the "motive" for the passage of the gelding laws was "Purely Eugenic"; "Purely Punitive" in Washington and Nevada; and "For the physical, mental or moral benefits of inmate" in California. These states, however, were but the harbingers of the eugenics harvest, for:

> The committee has recently received letters from the Governors of Vermont and Kentucky asking for information regarding legislation, and strongly endorsing the proposition that defectives, degenerates, and confirmed criminals should be sterilized. . . . From officials in several states Inquiries have been received regarding legislation and what has been done elsewhere . . . similar laws will soon be enacted in other states.

With the presentation of this documented report on the increasingly successful crusade to enact compulsory eugenical sterilization laws in the United States, the leadership of the world's eugenics movement passed to Davenport, Laughlin, Osborn, Grant, and their collaborators in the Eugenics Record Office, the Immigration Restriction League, the Eugenics Research Association, and other related societies.

The next two International Congresses of Eugenics were to be held in New York, in 1921 and 1932, under the presidencies of Osborn and Davenport. With the exception of the 1933–45 period—the holocaust years when Hitler's Germany, with two million forced sterilizations and nine million exterminations of dysgenic human breeding types in its Race Hygiene camps, led the world in the protection of "the race" against the perils of "genetic enslavement" lurking in the gonads of Germany's native social inadequates and all of Europe's Jews, Italians, Poles, Slavs, Frenchmen, and other non-Nordic, non-Aryan breeding stock—America was to remain the motherland of scientific racism.

7 The Dogmas of the Old Scientific Racism Become the Conventional Wisdom of America's Educated Classes

In a very definite way, the results we obtain by interpreting
the army [intelligence test score] data by means of the race
hypothesis support Mr. Madison Grant's thesis of the superiority
of the Nordic type.

—PROFESSOR CARL C. BRIGHAM, Princeton University, in
A Study of American Intelligence (1923)

. . . the Jukes-Kallikaks "bad heredity" concept may have
been too enthusiastically rejected by perfectionists. . . . Can it
be that our humanitarian welfare programs have already
selectively emphasized high and irresponsible rates of reproduction to produce a socially relatively unadaptable human strain?

—PROFESSOR WILLIAM SHOCKLEY, Stanford University, in
"Possible Transfer of Metallurgical and Astronomical Ap-
proaches to the Problem of Environment versus Ethnic
Heredity," unpublished presentation before the National
Academy of Sciences, October 15, 1966

The dogmas and the concepts of nineteenth-century scientific racism became the warp and the woof of much of the conventional opinion in the nation for most of this century. The efforts of the scientific racists to portray as socially dangerous all medical, hygienic, cultural, and other environmental social programs to improve the physiological, educational, and economic well-being of the *entire* populations of industrialized nations soon bore bitter fruit in the altered value systems of the educated classes of countries such as Germany and the United States, where the propaganda tracts of eugenics and other branches of scientific racism were widely taught as legitimate "science."

Because of the tenets of scientific racism, millions of Germans and other Europeans were, between 1933 and 1945, to lose their dignity, their freedom, and ultimately their lives. Yet the atrocities committed upon millions of people on eugenical grounds during the twelve years of the Nazi holocaust were the least of the depredations of scientific racism. In terms of sheer numbers of people victimized, the human costs of the Nazi holocaust do not begin to equal the price that successive generations of Americans have—for well over half a century—paid, and *still continue to pay,* for the malignant effects of the dogmas of scientific racism on the health, education, environment, and quality of life of the entire nation.

Thanks to the pernicious influences of eugenics and other forms of scientific racism on America's educated classes, at no time in this century of explosive progress in the biomedical and behavioral sciences have our tax-funded health, education, and family welfare programs ever been able to deliver more than a fraction of the known benefits of modern scientific research to the great majority of all Americans born in this century.

If today the United States stands, *as it does,* far behind other industrial nations in the duration and the quality of human life—a tragic reality well documented by all of the world's vital statistics—much of the blame can be charged directly to the enormous successes of the eugenics movement in influencing the main currents of American conventional and institutional wisdom from the turn of the century to today.

Of all the 1900–35 propagandists for scientific racism, four stand out as prime movers. Their writings helped structure the thinking, the scientific concepts, and the human value systems of many of the century's Presidents of the United States and of thousands of the nation's college professors and elementary and high school teachers. And, for over six decades, of millions of well-intentioned, God-fearing, patriotic taxpayers, citizens, and voters.

These four molders of American conventional and institutional wisdom were: (1) Prescott Farnsworth Hall (1868–1921), (2) Charles Benedict Davenport (1866–1944), (3) Henry Herbert Goddard (1866–1957), and (4) Madison Grant (1865–1937).

Just as the transcripts of the White House tapes during the Nixon presidency provided the insights that ultimately enabled this country to remove a cancer from the corpus of American political life, the actual words of these archetypal architects of the conventional wisdom of our educated classes can help liberate our children and grandchildren from the body- and mind-crippling social policies that their malignant perversions of science have handed down as continuing pollutants of our national heritage.

"A FRAIL LITTLE HOTHOUSE PLANT"

"From the moment that Prescott Farnsworth Hall was born," his widow wrote in a privately published memorial volume (1922), "his mother, who was then 45 years of age, was an invalid. Mrs. Hall's only other child, William Farnsworth Kilbourne, had died at the age of two and one half years. This made her all the more insistent concerning the health of the new baby. Consequently, he grew up a frail little hothouse plant, for he was never allowed to romp, to climb and be reckless, as other boys were."[1]

Given this start in life, it is little wonder that Hall was a lifetime hypochondriacal neurotic, who ultimately suffered what was genteelly termed a massive nervous breakdown that incapacitated him for the last years of his obsession-haunted life. He dabbled in the mystic and the occult, via the Rosicrucians and other groups, and while "the spiritual life greatly appealed to him," Hall "could not endure 'tailor-made religion.' " Hall "was very fond of the German language and literature, also German music. The German thinking mind appealed greatly to him," wrote his widow.

He also dabbled in poetry, as evidenced by his unpublished poem "Ye Immigrant," whose ringing opening verse,

Enough, Enough! we want no more
Of Ye immigrant from a foreign shore
Already is our land o'er run
With toiler, beggar, thief and scum. . . .

speaks for his deep spiritual beliefs, as do the closing lines speak for the impact of General Francis Amasa Walker and Sir Francis Galton on this hothouse plant of a man:

> If war and blood we would avoid
> There must be no delay but of one accord
> That our lovely shores you shall no longer use
> As a dumping ground for foreign refuse.

Hall's intellectual life, after graduating from Harvard Law School in 1892, seems to have fallen into two distinct phases: (1) before and (2) after he discovered eugenics. While he did, for a time, practice law—even, between 1904 and 1907, in partnership with a Jewish lawyer, Edward Adler—most of Hall's postgraduate energies were devoted to the Immigration Restriction League and the cause of immigrant baiting.

Before Hall discovered Galton, he had sought far and wide for authentically scientific knowledge about how best to combat "Ye Immigrant," who in Hall's mind turned out to be primarily "Ye Jew." Hall's love of German *Kultur* and his diligence led him to such fonts of anthropological data as the *Amerikanische Antisemitische Association* of Brooklyn, New York, and its 1896 flyer "Protest against Jewish Immigration." Now in the collection of the Immigration Restriction League papers at Harvard's Houghton Library, it still bears Hall's neat personal underlinings concerning the objectionable immigration of this Russo-Jewish people "alien in race and religion."

Prior to the time he discovered Galton, Hall guided the Immigration Restriction League's activities along classical, methodical, Teutonic lines. Essentially, the propaganda message of the League was the nineteenth-century revelation of General Walker's: to wit, if we allowed enough of the non-Teutonic beaten men from beaten races onto our Golden Shores, then as surely as night follows day, the inferior Gaelic, Mediterranean, Slavic, and Hebrew racial stocks would exterminate and supplant the noble Teutonic (Nordic) "native" stock. As Hall's co-founder and co-leader of the IRL, Harvard climatology professor Robert DeCourcy Ward, had stated the problem: "If we have an Italian slum problem, and a Jewish slum problem now [1904], what shall we have when perhaps 3,000,000 Russian Jews come to us, and when 5,000,000 Italians are living here?"

To Ward, Hall, and the rest of the IRL cabal, the key "element is the number of American children who, because of the pressure of foreign immigration, *have never been born. . . . The question is a race question, pure and simple.* Many of our recent immigrants, not discouraged by the problem of maintaining high standards of living with their many children, are replacing native [*sic*] Americans. It is fundamentally a question as to *what kind of babies shall be born;* it is a question as to what races shall dominate in this country. . . . General Walker believed that foreign immigration into this country has, from the time it assumed large proportions, not reinforced *our* [*sic*] population but replaced it."[2] (Italics added.)

This, in essence, was the IRL's "party line." The Lodge literacy test,

American Anti - Semitic Association

A. A. A.

Amerikanische Antisemitische Association.

CENTRAL OFFICE, 1154 MYRTLE AVE.

F. GROSS, PRESIDENT.

Brooklyn, N. Y.,   APRIL 16,   189*6*.

PROTEST AGAINST THE JEWISH IMMIGRATION.

RESOLUTIONS PASSED AT THE REGULAR MONTHLY MEETING OF THE A. A. A.
DELEGATES FROM BRANCH ORGANIZATIONS WERE PRESENT.

WHEREAS, THE PRESENT LAWS FOR EXCLUDING THE OBJECTONABLE
RUSSO-JEWISH IMMIGRATION HAVE PROVEN ENTIRELY INSUFFICIENT, AND
THE MORAL AND SOCIAL DANGER OF A CONTINUOUS INFLUX OF THESE
PEOPLE IS STRONGLY AGITATING THE PUBLIC MIND;

WHEREAS, IN A HISTORY OF OVER 2000 YEARS THE JEWS HAVE
NEVER SHOWN AN EXAMPLE OF ASSIMMILATION WITH OTHER NATIONS,
AND THE PRESENCE OF THIS PEOPLE, ALIEN IN RACE AND RELIGION,
STRIVING AFTER THE FINANCIAL AND POLITICAL SUPREMACY IN THE
CHRISTIAN COUNTRIES OF THE WORLD, IS CREATING SERIOUS DISTURB-
ANCES EVERYWHERE;

RESOLVED, THAT WE PROTEST AGAINST THE PASSAGE OF THE
McCALL EDUCATIONAL TEST BILL AS ENTIRELY INADEQUATE TO DEBAR
THESE PEOPLE, NEARLY ALL OF THEM BEING ABLE TO READ AND WRITE
THE HEBREW JARGON;

RESOLVED, THAT WE FAVOR THE STONE CONSULAR CERTIFICATE
BILL FOR THE COUNTRIES SENDING US THIS UNDESIRABLE IMMIGRATION,
VIZ: RUSSIA, AUSTRO-HUNGARIAN EMPIRE, ITALY, ROUMANIA, SERVIA,
BULGARIA, TURKEY, ASIA, AND AFRICA, WITH THE INSTRUCTIONS TO
THE U. S. CONSULATES ABROAD, THAT ONLY SUCH APPLICANTS BE CON-
SIDERED DESIRABLE IMMIGRANTS, WHO FOR FIVE YEARS PREVIOUS HAVE
BEEN ACTIVELY ENGAGED IN AGRICULTURAL OCCUPATION WITH THEIR OWN
MANUAL LABOR.

E. Finderson
Chairman of Committee.

E. Aug. Schuermann
The Secretary of the A. A. A.

A prototypical nineteenth-century Harvard Teutonist, Prescott F. Hall sought out all sorts of "scientific" evidence of the inborn inferiorities of Jews, Italians, Irish, and other non-Nordic people. The underlinings on this flyer issued by the *Amerkanische Antisemitische Association* of Brooklyn in 1896 are by Hall.

From Personal Scrapbooks of P. F. Hall, Volume 1: 1894–1900, Houghton Library, Harvard University.

and other mental and physical "fitness tests" proposed by the League, were its chosen instruments for keeping America safe for "those who are descended from the tribes that met under the oak-trees of old Germany to make laws and choose chieftains."

However, by the time Hall wrote his book *Immigration and Its Effects upon the United States* (1906), he had already discovered eugenics. While, like Walker and Ward, Hall wrote that "the racial effects of immigration are more far-reaching and potent than all others," he went on to say, in the same paragraph, that "recent discoveries in biology show that in the long run heredity is far more important than environment or education; for though the latter can develop, it cannot create." This, of course, was orthodox Galtonism.

Hall also asserted, in the pseudo-evolutionary jargon of the eugenics cult, that "in the United States, through our power to regulate immigration, we have a unique opportunity to exercise artificial selection on an enormous scale." Because "the question as to the racial effects of immigration is not, as most people assume, a question between us and the immigrants, but between our children and grandchildren and theirs. We are the trustees for the future, and with us is the decision what races and what kind of men shall inherit this country for years after we are gone."

For the most part, however, the "science" in Hall's book on immigration was of the "three races of Europe and their head forms" variety. So that in dealing with the data that the new immigration did not consist of the "superior" Nordics but of the "inferior" Alpines and Mediterraneans, Hall warned that this would not only lead to racial deterioration but, also, "apart from this deterioration, the skull will become more of the brachio-cephalic type, the average stature will be lower and the average complexion will be darker."

The sadistic insensitivity that marked the writings of men like Grant and Davenport in their treatment of Italian and Jewish immigration was equally evident in Hall's book. Italians were menaces to the American workingman because they were content "to live on stale beer and bread" (p. 129) and because "fruit and other food which would be rejected as unfit by most other races furnishes a diet upon which the Italian seems to thrive" (p. 164).

Of the Russian pogroms and other European persecutions that caused Jews to emigrate, Hall wrote: ". . . in the case of the Jews, it is probable that the numbers fleeing from *actual* persecution are relatively small, and that the bulk of the [Jewish] immigration comes from *fear* of persecution and to escape the grinding oppression which, however hard to bear, is not to be confused with the fanatical outbreaks of slaughter and violence" (p. 20) (italics added). Hall much preferred for the Jews to stay home and be raped, lynched, shot, slashed, beaten, pillaged, their homes burned to the ground, their children tortured by drunken Cossacks, first; then, *if* any of them survived such *actual* persecution, Hall conceded that they had a right to emigrate —to Australia.

The Hall book on immigration and race, which was to prove a lifelong influence on the thinking of people as influential as the editor of *The Saturday*

Evening Post, George Horace Lorimer, and the New York University professor of sociology Henry Pratt Fairchild,[3] was also notable for some of the naked truths it revealed about the real nature of the immigration restriction crusade. For example, far from pretending that the purpose of the literacy tests was to select a higher type of immigrant, Hall on page 274 bluntly admitted that "the principal advantage of the educational test is that it is *a definite rule of exclusion*" (italics added).

The League issued its own publications to spread the word. Its Bulletin No. 51 (1908), *Eugenics, Ethics and Immigration,* by Prescott Hall, was one of the first writings on eugenics by that xenophobe. In it, after paying due tribute to Galton and Pearson for introducing the word "eugenics" to cover all attempts to improve breeding stocks in man, Hall declared that the "marvellous [*sic*] work of Luther Burbank [the plant breeder] and others has opened our eyes" to more obvious possibilities. So that "people are now asking why the breeding of the most important animal of all should, alone, be left to chance." In this task of creating "supermen," the "science" of eugenics helped by showing that *"too much emphasis was laid upon the environment and too little upon heredity"* (italics added).

In a posthumously published paper, "Birth Control and World Eugenics" (1922), Hall wrote that the miseries of poverty made the poor "only too ready to relieve the dreary round of a narrow life by means of the most exciting passion known to man." And this sexual passion contributed to the higher birth rates of inferior types. Hall conceded that "although the birth rate has fallen," it was also true that "the death rate has fallen still more."

It was this "catastrophic" drop in the death rate that exacerbated the perils of the American race. For example: "I know an Irish immigrant of 1850 who had fourteen children, only three of whom survived him. Today he might have only ten," but, Hall warned, "most of them would grow to maturity." Thanks to advances in modern sanitation, wages, and medicine, too many Irish and other "dysgenic" races and classes were now surviving the hazards of childhood.

In view of the improving infant and child mortality rates of the Undeserving Poor, Hall seemed to be terrified that "the demands of the world's trade unions not only for the free rights of migration" but also "for wages based upon the cost of adequately supporting a family of five or six children" would, in his opinion, only "put more of the burden upon the more successful portion of the community, which does not undertake to have six children." While Hall and his Brahmin peers made loud noises about the threats that low-paid foreign immigrants posed to the decent living standards of our "native" American workingmen, the well-heeled restrictionists were genuinely alarmed by the awesome possibilities that some alien-infiltrated trade unions might start insisting upon adequate and standard wage scales for *all* workingmen, native-born and immigrant alike.

For most of the years between the formation of the Immigration Restriction League in 1889 and 1922, Prescott Farnsworth Hall, working through the League's paid Washington lobbyist, James H. Patten; through the League's hired secret publicity agent, Jeremiah Jenks; and on his own, peddled these xenophobic, eugenic, anti-medical-care, and anti-labor concepts to every American President. Hall also kept up a steady stream of private letters, reports, bulletins, and flyers to influential senators and congressmen; to leading educators; to the leaders of the IRL-organized coalition of hyperpatriotic, nativist, and restrictionist societies, plus the renascent Ku Klux Klan, plus many unions of the American Federation of Labor who had all joined in the crusade for the end of Jewish, Italian, and other non-German, non-English-speaking immigration; and, above all others, to the editors and publishers of America's leading newspapers and magazines.

Both the *Annals of the American Academy of Political and Social Science* and the executive editor of the Philadelphia *Public Ledger* sought not only writings under his name but also his recommendations about what they should publish about immigration. Typical of the respect with which Hall was held by the mass media was the May 21, 1921, editorial, "Self-Extermination," one of a continuing series of editorials and articles in favor of ending non-Nordic immigration run over a period of years by *The Saturday Evening Post,* then America's most influential magazine. This editorial said, among other things:

"So great is the sterilizing effect of in-coming low-grade aliens that Mr. Prescott Hall, a high authority on the subject, declares that despite the fact that upwards of 33 million foreigners have been admitted to the republic during the last century, most students [*sic*] are agreed that if we had no immigration since 1820 our fine old stock of pioneer days would have so multiplied as to make our present population even greater than it is."

THE BAD AND THE BEAUTIFUL

Of the four giants of the early eugenics literature in America, the best remembered today is not remembered because he was the first to adopt and introduce the Binet IQ tests here, nor because he helped reify "the menace of the feebleminded"—the myth that it is the mentally retarded who are the major cause of crime and corruption.

Henry Herbert Goddard's lasting fame is based almost entirely on his having combined the Greek words *kallos* (beauty) and *kakos* (bad) to arrive at the pseudonym he bestowed upon a native American family who, for what Goddard proclaimed to be hereditary reasons, had during three centuries been born either wholly *kallos* and good, or totally *kakos* and bad.

Goddard named this clan the Kallikaks.

There had been other studies of native American degenerate clans before Goddard, in 1912, published his little book *The Kallikak Family: A Study in the Heredity of Feeble-Mindedness.* The first had been the study of the equally pseudonymous Jukes, *The Jukes: A Study in Crime, Pauperism and Heredity,* by the New York merchant and prison reformer Richard Dugdale, in 1874, of which Galton himself had written so highly in his *Inquiries into Human Faculty* (1883). Between the time Dugdale published his account of how, between 1730 and 1874, the 709 descendants of a colonial frontiersman named Max had cost their taxpaying upstate New York peers $1,308,000 in welfare, health, prison, and other institutional care, and the publication of Goddard's book on the Kallikaks, other eugenics scholars had dug up and written on many equally degenerate families. There were the Nams and the Nats of New York, described by Charles B. Davenport; the Indiana tribe of Ishmael; the Hill Folk and the Happy Hickory family of Ohio; the Dacks, descended from a degenerate Irish family in Pennsylvania; and the Forke-mites of Virginia.

These studies were important to the eugenics myths in that they drama-tized the heredity of the "unit character" they called "pauperism," as well as the feeblemindedness, tuberculosis, syphilis, criminality, epilepsy, and shift-lessness that, the eugenicists declared, *invariably* went with pauperism. At the same time, however, most of these inbred families of hereditary paupers, criminals, and drunks shared an embarrassing set of traits: they were (1) of old, white, "native" American stock; (2) generally of Anglo-Saxon, German, Dutch, or otherwise Nordic Protestant origins; and (3) had lived in America since colonial times. All of which also meant that the moralistic studies of these genetic ne'er-do-wells also posed some considerable peril to the cred-ibility of the Cult of the Nordic.

Worse than that, the most famous study of them all, the Jukes book, had been written by a do-gooder who was not nearly as certain as had been Galton and his acolytes that heredity alone was to blame for the plight of of the Jukes. In fact, Dugdale had written (and Galton had ignored) in his book: "The tendency of heredity is to produce an environment which perpe-tuates that heredity. Thus the licentious parent makes an example which greatly aids in fixing habits of debauchery in the child. . . . *Where environ-ment changes in youth, the characteristics of heredity may be measurably altered*" (italics added).

Far from demanding the segregation and incarceration and/or steriliza-tion of the Jukes and other inheritors of fiscal poverty, Dugdale had proposed that the children of criminal and pauper parents be removed to more whole-some settings, and given vocational education, to enable them to develop into useful noncriminal and self-supporting human beings.

Dugdale was obviously far from convinced that the unsocial behavior of the Jukes was the product of bad heredity alone. Goddard felt that in his book on the Kallikaks he had resolved all such doubts, and proven that pure and impure blood do tell in the end.[4]

Unlike the Reeds, who required 289 institutionalized probands for their 1965 study of familial mental retardation (feeblemindedness) in Minnesota, Goddard based his even more famous study of the good and the bad Kallikaks on only one proband. This historic propositus was a young girl to whom he gave the pseudonym Deborah Kallikak. In November 1897, when she was eight years old, this child had been accepted as an inmate of the Training School for Feeble-minded Girls and Boys at Vineland, New Jersey. Illegiti-mate, illiterate, and, on admission, a poor student, Deborah nevertheless had, by the 1908 standards of Alfred Binet and Théodore Simon—published in English by Goddard's Department of Research at the Vineland school in 1916—become quite able to care for herself.

According to Goddard's book on the Kallikaks, by 1911, when Deborah was twenty-two years old, she had mastered many skills. She could, among other things: read and write English; do arithmetic problems; read music and play the cornet in the school band; make handsome embroidered dresses; wait on table; do excellent carpentry and wood carving; "write a fairly good story"; use a sewing machine; make shirtwaists; use tape measures accurately; and otherwise handle herself as a competent, self-supporting adult. Thus, by the standards of Simon, Binet, and most other psychologists of Goddard's day and our own, *this girl was anything but mentally retarded*—and living proof of Dugdale's belief that "where environment changes in youth, the characteristics of [apparent] heredity may be measurably altered."

However, when given the Goddard-Binet IQ tests in 1911, the girl did consistently badly, and tested as one who "had the mentality of a nine year old child." To Goddard, this did not mean, *as it should have meant,* that in at least this individual the IQ test could not be used to differentiate between a normal and a feeble mind. As with so many other second-rank psychologists then and since, Goddard was never to learn how to differentiate between a living person and that person's IQ test score. All that Goddard could learn from the IQ tests he administered to this obviously competent girl was that she was "feeble-minded."

Goddard assigned his favorite eugenics field worker, Elizabeth Kite, to the task of studying the behavioral, medical, and criminal histories of all the relatives, living and dead, of Deborah Kallikak. This she did in the manner in which Goddard had trained her to do.

In his preface to *The Kallikak Family,* Goddard, addressing himself "to the scientific reader," asserted that "the data here presented are, we believe, accurate to a high degree. *It is true that we have made rather dogmatic statements and have drawn conclusions that do not seem scientifically warranted from the data.* We have done this because it seems necessary to make these statements and conclusions for the benefit of the lay reader, and it was impossible to present in this book all of the data that would substantiate them. We have, as a matter of fact, drawn upon the material which is soon to be presented in a larger book." (Italics added.)

The larger "scientific" book from whose raw materials Goddard drew the data presented to the lay reader in *The Kallikak Family* was published two years later under the title *Feeble-Mindedness: Its Causes and Consequences* (1914). Both books presented exactly the same caliber of data, collected by exactly the same people, and interpreted in exactly the same way by Goddard in accordance with the same eugenical principles. Both books were of identical scientific worth—or worthlessness.

The first chapter of the 1912 book dealt with the girl who had been born in an almshouse near her native New Jersey Pine Barrens in 1889, and raised at the Vineland school after the age of eight. Of this child, Goddard wrote:

The question is "How do we account for this kind of [feebleminded] individual?" The answer is in a word "Heredity"—bad stock. We must recognize that the human family shows varying stocks or strains that are as marked and that breed as true as anything in plant or animal life.

Formerly such a statement would have been a guess, a hypothesis. We submit in the following pages what seems to us *conclusive evidence of its truth* [italics added].

The second chapter, "The Data," described the way the raw materials had been collected, and summed up the main lines of the *kallos* and *kakos* blood lines of the prolific Martin Kallikak, Sr. Goddard's description of how he trained the field workers who gathered these data and made these field diagnoses of the living and the dead Kallikaks reveals that Kohs and Myerson were not exaggerating when they categorized these eugenics field workers as amateurs incapable of performing such clinical and epidemiological functions. By his own testimony in both of his books, Goddard actually did truly believe that, after a few *weeks* of internship at a school for "feebleminded" children, his field workers really could become as expert in the diagnosis of mental and physical disorders in living people *and their dead ancestors* as were physicians, neurologists, psychiatrists, and social work *professionals* with years of college, graduate school, and clinical training and experience.

On page 14 of *The Kallikak Family,* for example, he wrote:

> In determining the mental conditions of people in the earlier generations [that is, as to whether they were feebleminded or not], one proceeds in the same way as one does to determine the character of a Washington or Lincoln or any other man of the past. Recourse is had to original documents *whenever possible.* Oftentimes the absence of these . . . is of itself significant. For instance, the absence of a *record* of a marriage is often quite as significant as its presence. Some record or *memory* is generally obtainable of how the person lived, how he conducted himself, whether he was able to make a living, how he brought up his children, what was his *reputation* in the community; *these facts* [*sic*] *are frequently sufficient to enable one to determine, with a high degree of accuracy, whether the individual was normal or otherwise.* [Italics added.]

This key passage has to be reread to be believed. As, indeed, does the following paragraph on the next page of Goddard's book:

> After some experience, *the field worker becomes expert in inferring the condition of those persons who are not seen,* from the similarity of the language used in describing them to that used in describing persons she has seen [italics added].

It was largely from the reports of his most famous "trained field worker," Elizabeth Kite, that Goddard was able to reveal that from the illegitimate boy born of the fateful fornication of Martin Kallikak, Sr., and the nameless "feebleminded" tavern wench of 1776:

> . . . have come four hundred and eighty descendants. One hundred and forty-three of these, *we have conclusive proof,* were or are feebleminded, *while only forty-six have been found normal* [italics added]. The rest are unknown or doubtful.
>
> Among these 480 descendants, 36 have been illegitimate.
>
> There have been 33 sexually immoral persons, mostly prostitutes.

There have been 24 confirmed alcoholics.
There have been three epileptics.
Eighty-two died in infancy.
Three were criminal.
Eight kept houses of ill fame.
These people have married into other families, generally about the same type, so that we now have on record and charted 1146 individuals.
Of this large group, we have discovered that 262 were feeble-minded. . . . frequently, they are not what we should call good members of society.

In wholesome and even eugenic contrast to the hordes of dysgenic, degenerate and otherwise doomed hereditary paupers who, for six generations, succeeded the bastard son of Martin Kallikak, Sr., and the nameless but addled town doxy, there were the children and their descendants from the famous fornicator's second sexual entanglement. For, as Goddard reminded us, whatever his transient dalliances, genetically he was a lad of "good English stock of the middle class," and:

> . . . on leaving the Revolutionary Army, straightened up and married a *respectable girl of good family,* and through that union has come another line of descendants of radically different character. These now number 496 in direct descent. Three men only have been found among them who were *somewhat degenerate,* but they were *not* defective. Two of these were alcoholic, and the other sexually loose.
>
> All of the legitimate children of Martin Sr. married into the best families in their state, the descendants of colonial governors, signers of the Declaration of Independence, soldiers and even the founders of a great university [Princeton]. Indeed, in this [*kallos* blood line] family and its collateral branches, *we find nothing but good representative citizenship.* [Italics added.]

The seventeen pages of Kallikak pedigree trees Goddard then presented with such pride (pp. 33–49; Chapter 2, in a section called "The Charts") were typical of the "scientific" documents he and Davenport and Pearson in England delighted in preparing and publishing in the leading scientific journals. To the biologically, genetically, neurologically, psychiatrically, and medically untrained readers, such pedigree trees were formidably Scientific. But even in 1912, to any life or behavioral scientist who knew anything at all about the major ongoing developments in the biological and psychological sciences, they were scientific nonsense.

Goddard, however, described them as summing up "a natural experiment of remarkable value to the sociologist and student of heredity. That we are dealing with a problem of true heredity, no one [*sic*] can doubt . . ."

The peerless research efforts of Goddard's field workers had, in their Kallikak studies, removed all doubts about the hereditary effects of bad blood. The *kakos* Kallikaks

> were feeble-minded and *no amount of education or good environment can change a feeble-minded individual into a normal one, any more than it can change red-haired stock into a black-haired one.* The striking fact

of the enormous proportion of feeble-minded individuals in the descendants of Martin Kallikak, Jr. [born of his father's gametes and the hussy's], and the *total absence* of such in the descendants of his half brothers and sisters [on the *kallos* or beautiful side] is conclusive [*sic*] on this point. *Clearly it was not environment that has made that good family. They made their environment; and their own good blood, the good blood in the families into which they were married, told.* [Italics added.]

This, of course, was the central dogma of eugenics. Poverty and its pathologies, like affluence and its comforts, were in the blood—and not in the environments in which human beings were conceived, born, and developed. Therefore, it was not only futile but also wasteful of tax dollars to legislate and administer public and family health, housing, educational, and vocational training programs. *Born* ne'er-do-wells were beyond such help—while the nation's bankers, university founders, corporation presidents, and other leaders were *born* into "good representative citizenship."

THE MENACE OF THE FEEBLEMINDED

Of all the types of feeblemindedness that made for born criminals, the type to which the Kallikak girl at the Vineland school belonged, "the high grade, or moron" type of feeble mind, was the one society had to fear the most. Goddard felt it his duty as a taxpayer and citizen to expose them to the unwary:

All the facts go to show that this type of people makes up a large percentage of our criminals. We may argue *a priori* that such would be the case. Here we have a group who, when children in school, cannot learn the things that are given to them to learn, because through their mental defect, *they are incapable of mastering abstractions. They never learn to read sufficiently well to make reading* pleasureable or *of practical use to them.* The same is true of number work. *Under our compulsory school system and our present courses of study, we compel these children to go to school* and attempt to teach them the three R's and *even higher subjects.* Thus they worry along through a few grades until they are fourteen years old and then leave school, not having learned anything of value or that can help them to make even a meager living in the world. [Italics added.]

In terms of this certitude, here bluntly expressed, that it was a waste of teacher time and tax dollars to provide quality compulsory education to the *hereditarily ineducable* children of the poor, Goddard was clearly a half century and more ahead of Jensen, Banfield, and various other academic thinkers of modern times. Certain of the other ideas that were to be revived in our times by such intellectual heroes of *The Atlantic Monthly* as Banfield were spelled out quite clearly by Goddard in his 1912 *oeuvre*.[5] Banfield's description of the poverty and the manners of the poor, deriving from their lack of a philosophy of the future and their living for the moment in terms of quick gratification of sexual and other coarse appetites, is clearly foreshadowed by Goddard's portrait of the mother of the Kallikak child he called

Deborah. On pages 65 and 66, Goddard wrote of how his field worker, "in visiting with the mother in her present home," discovered that:

> . . . there has been no malice in her life nor voluntary reaction against the social order, but simply a *blind following of impulse* which never rose to objective consciousness. Her life has utterly lacked coordination —there has been *no reasoning* from cause to effect, no learning of any lesson. *She has never known shame.* . . . At times, she works hard in the field or as a farm hand, so that it cannot be wondered that her house is neglected or her children unkempt. *Her philosophy of life is the philosophy of the animal.* . . . There is no rising to the comprehension of the possibilities which life offers or of directing circumstances to a definite, higher end. [Italics added.]

Because the behavioral and venereal histories of the good and the bad blood lines of Kallikaks—living and dead—were so crystal-clear to Goddard and in the all-seeing glances of his field workers, he was convinced beyond all doubt that this "natural experiment" in eugenics and dysgenics made "inevitable" the conclusion that:

> . . . all this degeneracy has come about as a result of the defective mentality and bad blood having been brought into the normal family of good blood, first from the nameless feeble-minded girl and later by additional contaminations from other sources.
>
> The biologist could hardly plan and carry out a more rigid experiment or one from which the conclusions would follow more inevitably.

Elizabeth Kite, the primary field worker of the Kallikak project, was not a biologist. But, according to Goddard, she did have the professional skills needed to derive and verify the "facts, figures and charts" published in *The Kallikak Family* and *Feeble-Mindedness: Its Causes and Consequences,* which all showed "conclusively the difference between good heredity and bad and the result of introducing mental deficiency into the family blood." (A line of inquiry that, as we have already seen, was to be re-explored in modern times by the Reeds in their *Mental Retardation: A Family Study,* in 1965, which their Nobel laureate admirer, William Shockley, said proved "that more than 5,000,000 of our 6,000,000 mentally retarded could have been prevented by eugenics.")

Goddard followed his praise of Miss Kite's competence in medical and psychiatric diagnosis with a passage on the effects of bad blood on good housing (pp. 70–71) that might have been lifted directly out of Banfield's 1970 book, *The Unheavenly City* (pp. 62–63 and 126–27), or vice versa:

> . . . no amount of work in the slums or removing the slums from our cities will ever be successful until we take care of those who make the slums what they are. Unless the two lines of work go on together, either one is bound to be futile in itself. *If all of the slum districts of our cities were removed tomorrow and model tenements built in their places, we would still have slums in a week's time, because we have these mentally defective people who can never be taught to live otherwise than as they have been living.* Not until we take care of this class and see to it that

their lives are guided by intelligent people, shall we remove these sores from our social life.

There are Kallikak families all about us. *They are multiplying at twice the rate of the general population,* and not until we recognize this fact [*sic*], and work on this basis, will we begin to solve these social problems. [Italics added.]

Despite its authorship by Galton and diffusion by Davenport, the myth of the exploding birth rates of the world's inferior types was fiction, not fact. Goddard, like his field workers, kept the faith, however, and did not question the postulates of the Masters. Nowhere is this blind faith in the dogma of Galton and his disciples more evident than in the passages of Goddard's book dealing with Elizabeth Kite.

After more than sixty years have passed, as we read about her adventures in Kallikakland, we can understand, even empathize, as she reacts to life among the *kallos* and *kakos*. What emerges is the touching portrait of a well-motivated and decent young lady sincerely convinced (i.e., brainwashed), by a professional psychologist who should have known better but did not, that her few weeks of "training" under his bumbling direction actually enabled her to make professional psychiatric and medical diagnoses of the quick and the dead, and to recognize *at a glance* such "hereditary traits" as "criminalism," "pauperism," and "shiftlessness." Miss Elizabeth Kite had been as much a victim of this hoax as were the poor South Jersey pineys she set out to help—let alone the readers of both Goddard books based on her field reports, many of them highly educated, who took these "scientific" books seriously as both fact and science.

On pages 76–79, for example, we find Miss Kite in a Pine Barrens town, visiting the family of a boy of the bad type of Kallikak blood who, when younger, had suffered "a severe attack of scarlet fever which deprived him of his hearing." It was a cold February day when Miss Kite knocked on the door of the family's hovel, and when it was opened:

Used as she was to the sights of misery and degradation, she [Kite] was hardly prepared for the spectacle within. The father, a strong, healthy, broad-shouldered man, was sitting helplessly in a corner. The mother, a pretty woman still, with remnants of ragged garments drawn about her, sat in a chair, the picture of despondency. Three children, scantily clad and with shoes that would barely hold together, stood about with drooping jaws and *the unmistakable look of the feeble-minded* [italics added].

The deaf boy was brought out to her. To be sure, Miss Kite knew before she met him that his deafness had been caused by a disease endemic in the biological environment of poverty. She had, however, been trained to classify even the infectious diseases known to be associated with the poor sanitary facilities, and the inadequate housing, heating, nutrition, and health care of poverty, as *hereditary* diseases of hereditary paupers:

A glance sufficed to establish his mentality, which was low. The whole family was a living demonstration of *the futility of trying to make desirable citizens from defective stock through making and enforcing compulsory education laws.* Here were children who seldom went to school *because they seldom had shoes,* but when they went, had neither will nor power to learn anything out of books. The father himself, though strong and vigorous, *showed by his face that he had only a child's mentality.* The mother in filth and rags was also a child. [Italics added.]

Such glances at the faces of the bad-blood Kallikaks were scientifically meaningful because Miss Kite and her sister field workers had been *trained* to make psychiatric diagnoses by this method. However, Goddard warned in his closing chapter, "What Is to Be Done," that it was not that easy for amateurs to make equally accurate diagnoses:

A large proportion of those who are considered feeble-minded in this study are persons who would not be recognized as such by the *untrained observer.* . . . They are the [poor] people who have won the pity rather than the blame [*sic*] of their neighbors, but no one has seemed to suspect the *real cause of their delinquencies,* which careful psychological tests [i.e., the Goddard-Binet IQ tests] have now determined to be feeble-mindedness. [Italics added.]

Most of this feeblemindedness (mental retardation) was due to bad genes—which Galton and Davenport had also exposed as the basic causes of hereditary pauperism, or poverty. In the feebleminded, Goddard wrote, "we have found the hereditary factor in 65% of the cases; while others place it as high as 80%."

The solution, as proposed by Goddard in *The Kallikak Family,* in 1912, was to segregate the feebleminded into all-male and all-female sheltered populations—as in workhouses, mental asylums, fenced-in reservations, and state institutions for the feebleminded and the epileptics—until they were past breeding age. Goddard was aware that his friends Davenport and Laughlin advocated a far more economical Final Solution to the problem of what to do with people with low IQ test scores: sterilization. Goddard, however, was somewhat more sensitive to the moral nuances of a surgical operation that sterilized without taking away the other "sex qualities." He felt that more attention had to be paid to "the social consequences" of such eugenical mutilations: "What will be the effect upon the community in the spread of debauchery and disease through having within it a group of people who are thus free to gratify their instincts without fear of consequences in the form of children?"

The important thing to remember, Goddard emphasized, was that *"feeble-mindedness is hereditary and transmitted as surely as any other [unit] character."* Once and for all, society had to step in and for its own protection halt the further breeding of people of inferior stock in whose blood coursed the unit characters for feeblemindedness, pauperism, and hereditary deafness.

Two years later, in his 599-page treatise, *Feeble-Mindedness: Its Causes and Consequences,* Goddard avowed that about 80 percent of the cases of feeblemindedness he dealt with were either demonstrably or probably hereditary. He also confessed that, although he had originally been reluctant to accept the idea "that the intelligence even acts like a unit character," he now realized that "there seems to be no way to escape the conclusion."

In sum, then, Goddard's two major books presented what seemed to those influential nonscientists who happened to be municipal, county, state, and national legislators, mayors, governors, and even Presidents, to be a highly convincing "scientific" case against squandering any further public funds on the schooling, health care, job training, and healthful housing of hereditary and ineducable paupers, feebleminded losers, and degenerates. There was no way to overcome the genetic destinies of the poor—who were doomed, while still microscopic zygotes in their mothers' wombs, to get low IQ test scores and to turn any houses where they lived into instant slums.

This concept was to be taught to, and accepted unquestioningly by, generations of college-educated Americans, and to be cited by them for many years thereafter in their political words and deeds. It was the ancient Preformationism in all but name, reclaimed from the middens of science's discarded errors, and enthroned by generations of brainwashed college graduates as one of the major segments of American conventional wisdom.

THE KALLIKAK MYTHS SURVIVE THEIR MAKER

During the Great Depression of 1929–41, for example, Theodore Russell Robie, M.D., of the Essex County Mental Hygiene Clinic in New Jersey, presented a paper entitled "Selective Sterilization for Race Culture" at the Third International Congress of Eugenics, held in New York in 1932 under the presidency of Charles Benedict Davenport. In this "scientific paper" Dr. Robie, citing the "detailed case reports of many generations of hereditary defectives found in the extensive researches made by Goddard on the Kallikak Family" and the more recent words of "Dr. Raymond Pearl, of Johns Hopkins, who has recently said, 'The wrong kind of people have too many children and the right kind too few,'" called for the wholesale sterilization of the poor and the unemployed. This doctor of medicine said he assumed that there were few in his audience

> . . . who have not read the descriptions of the trail of crime, murder, pauperism, prostitution, illegitimacy and incest which is found in the history of the famous Jukes and Kallikak families. It was demonstrated that the main factor in these ignoble family histories was *mental deficiency. It would have cost but $150 to have sterilized the original couples,* to cut off the seemingly endless social sores resulting wherever members of these families have settled. Yet the actual cost of relief alone of only one of these families was estimated at over $2,000,000 in 1916, as there were at that time 2,000 members of that socially unworthy clan. . . .
>
> . . . there are those who believe that our population has already attained a greater number than is necessary for efficient functioning of the race as a whole. *Certainly our present picture of millions of unemployed* would point to the belief that this suggestion is a *reasonable* one

> *. . . a major portion of this vast army of unemployed are social inade-*
> *quates, and in many cases mental defectives,* who might have been
> spared the misery they are now facing if they had never been born.
> [Italics added.]

Dr. Robie was not talking about non-Nordic, non-Christian, non-white
groups. He was talking of the white Anglo-Saxon Protestant (WASP) combat veterans of World War I; the WASP factory, office, and farm workers;
and the WASP schoolteachers, nurses, doctors, professors, failed bankers and
brokers, bankrupt corporation presidents, and involuntary college dropouts
who, in 1932, constituted the *majority* of the twenty million Americans
rendered jobless, careerless, bankrupt, homeless, and hopeless by the economic depression brought on, *not by defects in their own genes,* but by the
collapse of the national economy that followed the Wall Street crash of
November 1929. Dr. Robie had, clearly, learned his lesson well from Goddard—and from Goddard's peers and fellow authors in the eugenics cult.

However obsolete and idiotic Dr. Robie's words and ideas might strike
us as being today, we must not forget that, in 1932, they were shared by a
majority of his educated peers. The Kallikak myth was so firmly integrated
into the dogmas and the subconscious of our educated classes that, as late as
1955, the chairman of the Department of Psychology at Columbia University,
Professor Henry E. Garrett, was still presenting it as a body of legitimate
genetics data in his standard college textbook, *General Psychology.*

To the extent that the college students whose education in human
psychology *and genetics* was based on this highly respectable textbook then
went on to become influential formulators of current conventional wisdom,
Garrett's book alone has helped preserve the formative powers of the pseudogenetic Kallikak myth on all modern social legislation concerning health,
education, welfare, housing, job training, and, above all else, child care.

Nor was Garrett the last of our professors to accept Goddard's "genetic"
data at face value. Only five books were among the twenty-nine references
cited by Sheldon and Elizabeth Reed in their 719-page *Mental Retardation:
A Family Study* in 1965; one of these five was Goddard's *Feeble-Mindedness:
Its Causes and Consequences* (1914), the "scientific" version of the earlier
The Kallikak Family.

THE EUGENICS CULT'S TEACHER OF THE TEACHERS

The pseudoscientific concepts about the totally hereditary causes of such common human scourges as "pauperism," "feeblemindedness," tuberculosis,
epilepsy, blindness, deafness, inebriation, and "shiftlessness," previously
encountered in the 1914 final report of the Committee to Study and to Report
on the Best Practical Means of Cutting Off the Defective Germ-Plasm in the
American Population, had all been taken, lock, stock, and barrel, from the
magnum opus of Dr. Charles Benedict Davenport—*Heredity in Relation to
Eugenics* (1911).

While the Davenport book was never the popular success that *The
Kallikak Family* and Madison Grant's *The Passing of the Great Race* (let

MARTIN KALLIKAK

He dallied with
a feeble - minded
tavern girl

He married a
worthy Quakeress

She bore a son
known as "Old Horror"
who had ten children

She bore
seven upright
worthy children

From "Old Horror's"
ten children came
hundreds
of the lowest types of
human beings

From these seven worthy
children came hundreds
of the highest types
of human beings

As late as 1961, this schematic visualization of the Kallikak myth was solemnly presented as a legitimate scientific concept in a widely used college textbook written in 1955 by the then chairman of the Columbia University Department of Psychology and revised six years later with the collaboration of psychology professor Hubert Bonner of Ohio Wesleyan University.

From Henry E. Garrett and Hubert Bonner, *General Psychology* (2d rev. ed., New York: American Book Company, 1961).

alone the best-selling books by Grant's protégé, Lothrop Stoddard, *The Rising Tide of Color* and *The Revolt Against Civilization*) were to become, it is safe to say that, as a textbook, it was used in the systematic indoctrination of more professors and their students in the life, behavioral, and social sciences than any other book produced by Galton's disciples between 1869 and 1920. Not only did generations of American professors of eugenics, sociology, and history accept and teach its concepts, but the even more numerous classroom teachers in our elementary and high school systems were to pass on its simplisms to millions of American children and young people.

It is, therefore, as the eugenics cult's major teacher of the teachers that Davenport has to be seen in history. There was not only his important 1911 book, but also his leadership of the American eugenics movement that translated the pseudoscientific postulates of this book into the conventional wisdom *and the laws* of the nation. There were also the streams of articles that flowed from Davenport to the scientific journals, the middle-brow monthlies, and the lower-brow mass media. There were even, in the 1930's, the series of network radio broadcasts on science and society that Davenport made as a spokesman for the entire American scientific community.

Three of Davenport's guiding articles of faith were stated in *Heredity in Relation to Eugenics*. They were the concept of man as biological breeding stock (p. 1); the concept of biological "unit characters," in the blood of all newborn infants, for all human traits from madness to making millions (pp. 6–10); and, finally, the dogma that the presence of a particular unit character, for the condition the eugenicists called "pauperism," was the genetic or biological cause of what noneugenic social scientists and other observers call poverty.

The first of the Davenport postulates was not original with Davenport. It was, of course, handed down in nearly identical words by the founder of the faith, Sir Francis Galton. As Davenport expressed it: "Man is an organism—an animal; and the laws of improvement of corn and of race horses hold true for him also. Unless people accept this simple truth and let it influence marriage selection human progress will cease."

One does not have to be a geneticist or a physiologist to understand the vast array of damage such simplistic attempts to breed people as farmers breed pigs could and would cause in our species. As Thomas McKeown and C. R. Lowe write in their *Introduction to Social Medicine* (1974): ". . . although cattle are bred with a single object such as increased milk yield or improved quality of beef, no single object would be acceptable in man. *It is also conceivable that improvement of one feature might be accompanied by deterioration of another*" (italics added).[6]

The "laws of heredity" that led Davenport to think there was scientific plausibility in Galton's 1869 fantasies about the extrapolation of barnyard animal breeding tricks to the creation of a new race of super-WASPS were spelled out in Davenport's book. Here the magic phrase was "unit characters." To Davenport, these never-to-be-found genes for every single biological and behavioral trait of man were the real "Mendelian" keys to the mechanism of eugenics, or the selective breeding of human animals.

The "physical, mental, and moral traits" of mankind were all, Davenport wrote, discrete, unitary genetic characteristics. These unit characters ranged from a genius for making music, money, or mathematical equations to the innate sins of graft in politics, "nomadism," and, above all else, "pauperism." As Davenport put it (p. 6): ". . . these characteristics are inheritable, they are independent of each other, and they may be combined in any desirable mosaic."

This was published in 1911, just as Davenport's lifelong friend, Thomas Hunt Morgan, then professor of biology at Columbia University, and his young collaborators were on the eve of demolishing the simplistic "unit character" hypothesis of heredity so completely that Morgan was to become the first geneticist to win the Nobel prize. By 1910, Morgan had already published some of the discoveries on the functions of the chromosomes in heredity that were to lead to modern genetics. By 1916, the hoary "unit character" or "one gene = one trait" hypothesis was as extinct among working biologists as the once equally mighty tyrannosaurus was in the animal kingdom.[7]

Davenport wrote that there were "thousands of unit traits that are possible and that are known [sic] in the [human] species." Many of these were listed in his 1911 book. Others, such as the dysgenic trait of "Nomadism" and the eugenic trait of "Thalassophilia," were described by Davenport in subsequent books. Of "Nomadism," Davenport wrote that all cases of this irresistible impulse to wander "can be ascribed to one fundamental cause . . . the absence of a simple sex-linked gene that 'determines' domesticity," and that "those who show this trait belong to the *nomadic race*" (italics added). "Thalassophilia," a genetic trait discovered by none other than Davenport himself (and by nobody else since), was a unit character for the love of the sea and its associated irresistible compulsions to leave home for life on the bounding main. Since Davenport found this "genetic trait" to be widely prevalent in the cadre of U.S. Navy officers of World War I, it was, obviously, a good or eugenic trait.[8]

The most racially degenerative of all hereditary unit characters was, of course, the one Malthus had singled out as *the* menace even before it was proven by Galton and Davenport to be a unit character: "pauperism." As Davenport described this genetic defect in 1911 (p. 80):

> Pauperism is a result of a complex of causes. On one side, it is mainly environmental in origin as, for instance, in the case when a sudden accident, like the death of the father, leaves a widow and family without means of livelihood or a prolonged disease of the wage earner exhausts savings. *But it is easy to see that in these cases heredity also plays a part; for the effective worker will be able to save enough money to care for his family in case of accident; and the man of strong stock [i.e., the inheritor of good unit characters] will not suffer from prolonged disease.* Barring a few highly exceptional conditions poverty means relative inefficiency and this in turn means mental inferiority. [Italics added.]

Mental inferiority, as all good eugenicists from Galton to Davenport wrote, was *always* genetic. This genetic flaw was, usually, associated with the equally hereditary defect Davenport et al. called "shiftlessness," which, Davenport wrote, was "an important element in poverty." His scientific studies had revealed: "When both parents are shiftless in some degree about 15 percent of the known [sic] offspring are recorded as industrious. When one parent is more or less shiftless while the other is industrious only about 10 percent of the children are 'very shiftless.' It is probable that both shiftlessness and lack of physical energy are due to the absence of something that can be got back into the offspring only by mating with industry."

In addition to "pauperism," Davenport's *Heredity in Relation to Eugenics* listed, among many others, the following physical, mental, and moral traits as being among the *known* unit characters handed down to all human beings in the combined genes of their parents:

Physical traits: eye color; hair color; hair form; skin color; stature; total body weight; final body height; general bodily strength; general bodily resistance to tuberculosis, measles, and other infectious diseases.

Mental traits: intelligence (however defined); musical ability; calculating ability; ability in artistic composition; memory; ability in literary composition; temperament.

Pathological traits: epilepsy; insanity; narcotism; other nervous (*sic*) diseases, including cerebral hemorrhage, cerebral palsy of infancy, multiple sclerosis, and hysteria; rheumatism; speech defects; ear defects; defects of the eye; skin diseases; cancers and tumors; diseases of the blood; diseases of the thyroid gland; diseases of the vascular system (including heart disease and arteriosclerosis); diseases of the respiratory system (including pneumonia, tuberculosis, catarrh, adenoids, tonsillitis, deafness, bronchitis); diseases of excretion (including kidney stones, gout, and blood in the urine); and skeletal disorders.

Needless to say, as was well known in 1911, outside of certain purely genetic characteristics, such as skin and eye color, few of these conditions are hereditary, and very few behavioral characteristics, such as musical and literary ability or mathematical skills, are biological traits at all.

THE EUGENICS CULT'S ASSAULT ON THE MEDICAL SCIENCES

Although not a physician himself, Davenport always wrote and acted as if he knew infinitely more about the physical and mental health of mankind than did all medical and behavioral generalists, specialists, and qualified biomedical and psychiatric researchers. He not only *knew* that good health and the diseases and disorders in all human beings were due to specific unit characters and what he called their "determiners" (genes) in their hereditary endowments; he also knew that "it is a reproach to our intelligence that we as a people . . . should have to support about half a million insane, feebleminded, epileptic, blind and deaf, 80,000 prisoners and 100,000 paupers [sic] at a cost of over 100 million dollars per year" (p. 4).

And Davenport also *knew* the reason why the American people were saddled with the explosively high birth rates of our hereditary paupers, epileptics, and other born menaces to society:

Modern medicine is responsible for the loss of appreciation of the power of heredity. It has had its attention too exclusively focussed [*sic*] on germs and conditions of life. It has neglected the personal [hereditary] element that helps determine the course of every disease. It has begotten a wholly impersonal [i.e., nonracist] hygiene whose teachings [such as germ theory, nutrition, hygiene, etc.] are false in so far as they are laid down as universally applicable. It has forgotten the fundamental fact that all men are created *bound* by their protoplasmic [hereditary] makeup and *unequal* in their powers and responsibilities [p. iv, Davenport's italics].

Therefore, modern medicine, with its discoveries concerning our ability to improve the human condition by improving the human environment, presented a serious threat to racial purity in America. Particularly at a time when nobly intended but eugenically catastrophic private, institutional, and governmental welfare agencies—born of the Applied Christianity, Sanitary Reform, settlement house, and other social movements—had started deliberately or unwittingly to put America at the risk of race suicide by keeping the immigrant and native-born poor alive by means of preventive medicine and social assistance. Thanks to modern medicine and social work, eugenics had, as yet, been kept from solving "the problems of the unsocial classes, of immigration, of population, of effectiveness, of health and vigor."

Davenport's priority list of solutions to the problems of accelerating the applications of eugenics to race betterment in America was, of course, headed by the compulsory sterilization of all Americans found—by his trained field workers and equally competent experts—to be the bearers of inferior unit characters. This would at least contain the perils implicit in the rising tides of live births among the Under Men. But, as he noted in his preface and elsewhere in *Heredity in Relation to Eugenics,* the nation's misguided do-gooders and sentimentalists, led and abetted by the physicians, were—through free or inexpensive clinics, charitable and public hospitals, life-protecting factory and mine safety laws, minimum legal hygienic and safety standards for human habitations, and improved local drinking water and sewage systems—keeping alive precisely those genetically inferior types whom Nature had tagged for extinction via Mr. Herbert Spencer's Law of the Survival of the Fittest.

As Professor Davenport showed on page 263 and elsewhere in his 1911 masterwork, the naïve environmentalists among the legal profession were as guilty as were their medical and social work peers in keeping these low types alive. To Davenport the "evolutionist," it was clear that those ineffectual persons who had the old, inferior genetic traits had no right to live on:

. . . after the new traits became established and constituted the basis for the new society, those persons who had the old traits stood a good chance of being killed off and many a defective line was ended by their death. We are horrified by the 223 capital offenses in England less than a century ago, *but though capital punishment is a crude method of grappling with the difficulty it is infinitely superior to that of training the feeble-minded and criminalistic and then letting them loose upon society and permitting them to perpetuate in their offspring these animal traits.* Our present practices are said to be dictated by emotion untempered by reason; if this is so, then emotion untempered by reason is social suicide. If we are to build up in America a society worthy of the species *man* then we must take such steps as will prevent the increase or even the perpetuation of *animalistic* strains. [Italics added.]

Davenport was too much of a realist to hope that capital punishment for the social crime of being born of inferior blood could ever be enacted in any country in his lifetime.[9] He therefore allowed himself to settle for advocating the universal compulsory sterilization and less Draconian segregation of the Undeserving (albeit usually native-born and Nordic) Poor.

The annihilation of our native and resident hereditary paupers, however, solved only one aspect of our national eugenics problem. There was also the rising menace of unrestricted immigration. As Davenport showed with his "historical" research data in Chapter 5 of his 1911 book, all of this nation's fine, "native," dolichocephalic, and of course prime Nordic germ plasm—rich in the unit traits for scholarship, culture, morality, sharp trading with dull natives, and free-and-clear homeowning—was in danger of being diluted and contaminated to the point of biological extinction (race suicide) by the incoming hordes of foreign and hereditary paupers, criminals, and epileptics, whose germ plasms were totally barren of the unit characters for high culture, free enterprise, and WASP ethics.

In his section "Recent Immigration to America" (pp. 212–20), Davenport spelled out the genetic pluses and minuses of this influx of foreigners. Here Davenport stated, as proven scientific facts, that the blood of the Germans, the Scandinavians, and the other Nordic races were free of all genetic flaws, and filled to the platelets with such "unit traits" as the German "love of art and music, including that of song birds," and the German hereditary capacities to make "useful clerks," as well as the Scandinavians' inherited "love of independence in thought, action, chastity, self-control of other sorts, and a love of agricultural pursuits."

The hereditary traits of the Irish included a few good characteristics, such as "sympathy, chastity, and leadership of men," while on the pathologic side the Irish were the racial blood heirs to "alcoholism, considerable mental defectiveness, and a tendency to tuberculosis." In politics, the Irish genes, although helpful in enabling the Irish to capture control of city governments, also led them to exercise "favoritism and often graft." The Italians were hard-working farm laborers, but these non-Nordics lacked the "self-reliance, initiative, resourcefulness and the self-sufficing individualism that necessarily marks the pioneer farmer." The Italians also suffered an inherited "tendency for crimes of personal violence."

The Jews, Davenport revealed in this chapter, were more literate than all other immigrants, but "on the other hand they show the greatest proportion of offenses against chastity and, in connection with prostitution, the lowest of crimes. There is no question that, taken as a whole, the hordes of Jews that are now coming to us from Russia and the extreme southeast of Europe, with their intense individualism and ideals of gain at the cost of any interest, represent the opposite extreme from the early English and more recent

Scandinavian immigration with their ideals of community life in the open country, advancement by the sweat of their brow, and the uprearing of their families in the fear of God and the love of country."

Davenport spelled out (p. 219) what all these incoming immigrant blood lines would do to the American gene pool:

> . . . the population of the United States will, on account of the great influx of blood from South-eastern Europe, rapidly become darker in pigmentation, smaller in stature, more mercurial, more attached to music and art [sic], more given to crimes of larceny, kidnapping, assault, murder, rape and sex-immorality and less given to burglary, drunkenness, and vagrancy than were the original English settlers. Since of the insane in hospitals there are relatively more foreign-born than native it seems probable that, under present [immigration] conditions, the ratio of insanity in the population will rapidly increase.

What all the eugenic analyses and field worker reports boiled down to, then, was nothing more and nothing less than an old three-letter word starting with the letter J that had been used as the excuse for massacres of ethnic minorities since long before Columbus discovered America. Stripped of its scientistic cant, Davenport's chief motivation was the blind and unreasoning hatred of "the hordes of Jews that are now coming to us from Russia and the extreme southeast of Europe [Hungary, Poland, Greece, Serbia, and Turkey]."

To protect its superior gene pools from genetic enslavement, Davenport warned that there were only two practical courses open to America: (1) to pass immigration restriction laws that would, at long last, keep the hordes of Jews and other inferior breeding stock out of God's Country and (2) to subject every foreign race polluter residing in America to compulsory sterilization (in lieu, of course, of the eugenical capital punishment so widespread in early-nineteenth-century England).

THE PROPHET DAVENPORT BRINGS FORTH
THE EUGENICS CREED

Davenport was not only the scientific spokesman of the American eugenics movement but, also, its spiritual head. It was in 1916, five years after the publication of his *Heredity in Relation to Eugenics,* that Davenport composed his credo of eugenics. It was unveiled in an address—"Eugenics as a Religion" —Davenport delivered at the ceremonies marking the fiftieth anniversary of the Battle Creek (Michigan) Sanitarium. As Davenport told the gathered celebrants, he had written this creed in response to the fact that "Francis Galton, founder of the eugenics movement, once expressed the anticipation that some day, when eugenics had come into its own, it would be accepted as a religion." What Galton anticipated, Davenport delivered.

In setting the stage for the first reading of the Eugenics Creed, Davenport explained:

> Eugenics has to do with racial development. It accepts the fact of differences in people—physical differences, mental differences, differ-

ences in emotional control. *It is based on the principle that nothing can take the place of innate [genetic] qualities.* While it recognizes the value of culture [i.e., nurture] it insists that *culture of a trait is futile, where the germs [genes] of the trait are absent.* [Italics added.]

Eugenics took up where older religions left off. It was a *scientific* religion, as modern as Lapouge's cephalic index, as Galton's racial rating scales of human worth, and as Davenport's unit characters of heredity. "Have you," Davenport asked the assembled participants in religious history in the making, "the instinct of love of the race? If so, then for you, eugenics may be vital and a religion that may determine your behavior." And, since "every religion, it appears, should have a creed," Galton's colporteur to the human refuse of America was ready with a full-fledged "creed for the religion of eugenics." This creed, every sanctified word of it, ran:

I believe in striving to raise the human race and more particularly our nation and community to the highest place of social organization, of cooperative work, and of effective endeavor.

I believe that no merely palliative measures of treatment can ever take the place of good stock with *innate* excellence of physical and mental traits and moral control [Davenport's italics].

I believe that to secure to the next generation the smallest burden of defective development, of physical stigmata, of mental defect, of weak inhibitions, and the largest proportion of physical, mental and moral fitness, it is necessary to make careful marriage selection—not on the ground of the qualities of the individual, merely, but of his or her family traits; and I believe that I can never realize the ideals I have for my children without this basis of appropriate *germinal* factors [Davenport's italics].

I believe that I am the trustee of the germ plasm that I carry, that this has been passed on to me through thousands of generations before me; and that I betray the trust if (that germ plasm being good) I so act as to jeopardize it, with its excellent possibilities, or, from motive of personal convenience, to unduly limit offspring.

I believe that, having made our choice in marriage carefully, we, the married pair, should seek to have 4 to 6 children in order that our carefully selected germ plasms shall be reproduced in adequate degree and that this preferred stock shall not be swamped by that less carefully selected.

I believe that, having children with the determiners of peculiarly good traits, it is the duty of parents to give particularly good training and culture to such children to secure the highest development and effectiveness of hereditary traits in the nation.

I believe in such a selection of immigrants as shall not tend to adulterate our national germ plasm with socially unfit traits.

I believe in such sanitary measures as shall protect, as far as possible, from accidental and unselective mortality, the offspring of carefully selected matings.

I believe in repressing my instincts when to follow them would injure the next generation.

I believe in doing it for the race.

More than a half century after Davenport presented his Eugenics Creed at Battle Creek, Michigan, a professor of educational psychology at Berkeley, the director of an institute for human genetics at the University of Minnesota, the chairman of the Psychology Department at Harvard University, and the Alexander M. Poniatoff professor of engineering science at Stanford (and a Nobel laureate in the physics of Newton and Einstein, to boot), among various other prominent academics at home and abroad, were still vigorously shouting "Amen" to Davenport's theological dogma that intelligence is an overwhelmingly genetic trait, and that "the culture of a trait is futile, where the germs of the trait are absent."

The Apostle Davenport did not, as duly claimed in the secular press, die on February 18, 1944. He lives on in the pages of the *Harvard Educational Review, The Atlantic Monthly,* the prime-time television talk shows, *U.S. News and World Report,* the *Proceedings of the National Academy of Sciences,* the decisions of the United States Supreme Court, and, of course, *The Congressional Record.*

MADISON GRANT: ANTI-SEMITISM AS A SCIENCE

Like many nonscientists before and after his time, Madison Grant suffered the delusion that if a person *writes* about the sciences—however well or badly —this literary act automatically qualifies him as a working *scientist.* In this respect, Grant's fantasies of scientific competence and professional authority were no more unusual than his "scientific" conclusions or his very old-fashioned pathological anti-Semitism. His letters to high government officials, to clergymen, and to real scientists were punctuated by references to himself as "an anthropologist" and as other types of scientist. In a typical letter, one to U.S. Senator F. M. Simmons, dated April 5, 1912, after Franz Boas had demolished the cephalic index myth of scientific racism, Grant wrote:

"Dr. Boas, himself a Jew, in this matter represents a large body of Jewish immigrants, who resent the suggestion that they do not belong to the white race, and his whole effort has been to show that certain physical structures [head forms], which *we scientists* know are profoundly indicative of race, are purely superficial" (italics added).[10]

Educated largely by tutors here and in Germany and other European countries before entering the class of 1887 at Yale, Madison Grant subsequently took a law degree at Columbia University in 1890. By his own description a lifelong hunter, woodsman, explorer, and conservationist, he was one of the founders of the New York Zoological Society–the Bronx Zoo. "Finding it unnecessary to earn a living," *The New York Times* wrote in his obituary in 1937, Grant took up field zoology, but, according to the same article, "much of his early manhood was spent as a typical New York society and club man, and he belonged, among others, to the Union, Knickerbocker, Century, University, Tuxedo, Down Town Association and Turf and Field Clubs of New York, and the Shikar Club of London." His clubs also included the Boone and Crockett Club of New York, whose president was Theodore Roosevelt.

In 1894, when Roosevelt helped lead the reform movement that elected William Strong mayor of New York City, Grant joined in the fight against the Tammany Hall Irish. This led to the decision of the new city regime to participate in the organization of the Bronx Zoo, on whose board Grant was to serve for the rest of his life, the last twelve years as president.

For most of the years after he was admitted to the bar, Grant maintained a Manhattan law office. However, what with running the Bronx Zoo, the Save the Redwoods League, and various legislative and lobbying crusades for the termination of Jewish and other non-Nordic immigration, and striving for many years to have Franz Boas fired from his job as chairman of the Columbia University Department of Anthropology—let alone serving as the very active treasurer of the Second (1921) and Third (1932) International Congresses of Eugenics, co-founder and most active member of the Galton Society, president of the Eugenics Research Association, co-founder and active officer of the American Eugenics Society, and president of the Bronx Parkway Commission—most of the time Madison Grant actually spent in his law office seems to have been devoted to dictating and signing thousands of letters dealing with the causes dearest to his heart. And, as Grant made plain in hundreds of these letters and in his published writings, no cause was dearer to the heart of Madison Grant than the total annihilation of the Jews.

As early as 1911, in a letter to the admissions committee of the Century Association regarding Grant's application for membership, Professor Henry Fairchild Osborn, then serving as president of both the American Museum of Natural History and the New York Zoological Society, described Grant as being "extremely well informed on matters of history, anthropology, and zoology" and said that he had "produced a number of scientific papers of value. He is one of the active men in the Boone and Crocket Club. From all these standpoints he will be a decided addition to the Century Association."

There were also a few viable but, in Osborn's view, benign worms in this apple: "Mr. Grant has a very positive way of expressing himself on occasions, and *his strong views on certain questions, like Catholicism and the Hebrew race,*[11] have made him some enemies. It has always seemed to me, however . . . provided the man is a gentleman and an agreeable companion, his positiveness of opinions on certain subjects should be no bar to his admission." (Italics added.)

Osborn's reference to "Catholicism" was a euphemism for the Irish, whom his agreeable and gentlemanly friend and fellow scholar and clubman hated only slightly less than he did the Jews (Hebrews). These strongly worded anti-Irish and anti-Semitic views *and actions* of Grant's never made him seem any less the gentleman to his lifelong friend Osborn. Perhaps this was because Osborn shared all of Grant's racial views, and for this reason he wrote the prefaces to the first and revised editions of Grant's *The Passing of the Great Race*. Osborn also served with Grant as an active member of the Immigration Restriction League, the International Eugenics Society and the American Eugenics Society, and various other xenophobic and racist organizations. Grant was one of Osborn's successors as president of the

Bronx Zoo, and he in turn was succeeded in that presidency by Henry Fairfield Osborn, Jr., author of *Our Plundered Planet.*

The organizational passions of Grant and other early American scientific racists were legendary, and helped retard both the progress and the application of the findings of the legitimate sciences. Earlier in the century, for example, the American Anthropological Association had been formed as a society of both professional anthropologists, who were by and large poor men, and solvent amateurs whose dues payments and gifts were expected to help keep the AAA alive. By the time Franz Boas served as its president (1907–08) the AAA was evolving into a society of professional, academically qualified anthropologists, in which a doctorate in anthropology or linguistics or anatomy carried far more weight than an individual's bank statement. Anthropology was coming of age rapidly as a serious science.

The increasing dominance of the AAA by professionally qualified scientists bothered certain amateurs, particularly since the verifiable findings of the professionals in physical, cultural, and social anthropology were demolishing some of the most sacred eugenic tenets of scientific racism. In the spring of 1918, Madison Grant and Charles Benedict Davenport had a series of private meetings to devise a remedy for this professionalization of the science of anthropology. They agreed that with the backing of Osborn it might be possible to set up a rival anthropological society. Grant met with Osborn, who asked him to prepare a memo on the plan, which his agreeable and gentlemanly friend did on March 9, 1918. It said, in part:

> . . . the idea of the new anthropological society grew out of a conference between Davenport and myself last Wednesday night in Washington. . . . My proposal is the organization of an anthropological society (or somatological society as you call it) here in New York City with a central governing body, self elected and self perpetuating, and very limited in members, and also *confined to native Americans,* who are anthropologically, socially, and politically sound, no Bolsheviki need apply. . . . my present inclination is toward a central governing body . . . of not more than seven or nine members, the nucleus of which is to be you, Davenport, Dr. Huntington and myself. We might consider as additional members a man like Prescott Hall of Boston.

By April 12, 1918, the trio of organizers had already drawn up and circularized the charter of the new society which co-founder Osborn—who, in 1880, at the age of twenty-three, had collaborated with Sir Francis Galton on a two-page leaflet, *Questions upon the Visualizing and Other Allied Faculties*—suggested be christened the Galton Society for the Study of the Origin and Evolution of Man. The rump anthropological society was formally organized on April 16, 1918, an event celebrated by a dinner party in the Osborn town house at 850 Madison Avenue.

The chairman of the new Galton Society was Davenport; the secretary was Osborn's protégé William K. Gregory, of the Museum and Columbia University. The charter fellows included Davenport, Grant, Osborn, John C. Merriam, president of the Carnegie Institution of Washington, and Edward L.

Thorndike, professor of psychology at Columbia University. The roster of charter fellows was soon expanded to include Dr. Frederick Adams Woods of MIT and Major Robert M. Yerkes, professor of psychology at Harvard, then on leave as chief of the Division of Psychology, Office of the Surgeon General of the Army, in Washington, D.C.

The Galton Society helped retard the development of modern anthropology in America for many years. Its co-founder Madison Grant's chief contribution to scientific racism, however, was his famous book, *The Passing of the Great Race.*

THE BOOK THAT INFLUENCED COOLIDGE AND HITLER

From its opening chapter, "Race and Democracy," to its closing sentence— "If the Melting Pot is allowed to boil without control and we continue to follow our national motto and deliberately blind ourselves to 'all distinctions of race, creed or color,' the type of native American of Colonial descent will become as extinct as the Athenian of the age of Pericles, and the Viking of the days of Rollo"—*The Passing of the Great Race* was one endless outpouring of hatred against democracy and the modern life and behavioral sciences.

> In the democratic forms of government the operation of universal suffrage tends toward the selection of the average man for public office rather than the man qualified by *birth,* education and integrity. How this scheme of administration will ultimately work out remains to be seen but *from a racial point of view it will inevitably increase the preponderance of the lower types* and cause a corresponding loss of efficiency in the community as a whole. [P. 5, italics added.]

The corrosion of race and society started during the French Revolution, when

> . . . the majority, calling itself "the people," deliberately endeavored to destroy the higher type and something of the same sort was in a measure done after the American Revolution by the expulsion of the [Tory] Loyalists and the confiscation of their lands, with resultant loss to the growing nation of good race strains, which were in the next century replaced by immigrants of far lower type.

This submersion of the Great Race—the Nordics—and its replacement by the inferior races from the dregs of Killarney, Sicily, and the Jewish ghettos of Europe had by 1916, according to Grant, pushed America to the brink of racial degeneration.

> In America we have nearly succeeded in destroying the privilege of birth; that is, *the intellectual and moral advantage a man of good stock brings into the world with him.* We are now engaged in destroying the privilege of wealth; that is, the reward of successful intelligence and industry, and in some quarters there is developing a tendency to attack the privilege of intellect and to deprive a man of the advantage gained from an early and thorough classical education. [Italics added.]

Race was the key to all of these problems arising from democracy and its egalitarian evils. The lack of racial foresight on the part of America's Best Families was in large part to blame for our racial plight:

> During the last century the New England manufacturer imported the Irish and French Canadian and the resultant fall in the New England birth rate at once became ominous. The refusal of the native American to work with his hands when he can hire or import serfs to do manual labor for him is the prelude to extinction and the immigrant laborers are now breeding out their masters and killing by filth and crowding as effectively as the sword.
>
> Thus the American sold his birthright in a continent to solve a labor problem. Instead of retaining political control and making citizenship an honorable and valued privilege, he intrusted the government of his country and the maintenance of his ideals to races who have never yet succeeded in governing themselves, much less any one else.
>
> Associated with this advance of democracy and *the transfer of power from the higher to the lower races* [italics added], from the intellectual to the plebeian class, we find the spread of socialism and the recrudescence of obsolete religious forms [i.e., Roman Catholicism].

In the continued absence of immigration restriction laws, the peril to what was left of the Great Race in America was exacerbated:

> The result of unlimited immigration is showing plainly in the rapid decline in the birth rate of native Americans because the poorer classes of Colonial stock, where they still exist, will not bring children into the world to compete in the labor market with the Slovak, the Italian, the Syrian and the Jew. The native American is too proud to mix socially with them and is gradually withdrawing from the scene abandoning to those aliens the land which he conquered and developed. The man of the old stock is being crowded out of many country districts by these foreigners just as he is being literally driven off the streets of New York City by the swarms of Polish Jews.

All that was fine and good in the entire history of world civilization had been the work of the Great Race, the tall, blond, fair-skinned, blue-eyed, dolichocephalic, and Aryan-talking Nordics. The decimation of the defenders of the Great Race, history's noblest and bravest soldiers, during the centuries of the great wars of Christendom—followed by the egalitarian nightmares of the French and American revolutions—had helped speed the Passing of the Great Race. The furious and fiendishly fecund copulations of the short, dark-haired, olive-skinned, yellow-skinned, black-skinned, round-skulled, and Aryan-mimicking Alpines, Mediterraneans, Jews, Orientals, and Africans were what subsequently put the world and American civilizations at their greatest peril. The handwriting was on the wall—even if *all* of the ideas and many of the words in Grant's gents' room graffiti had come from other authors named Malthus, Gobineau, Retzius, Galton, Spencer, Lapouge, Ripley, Hall, Davenport, and General Francis Amasa Walker. The calligraphy was uniquely that of Madison Grant.

From its fulsome preface by Henry Fairfield Osborn to its *Götter-dämmerungisch* ending, Grant's book was a profoundly vulgar and wholly eclectic compendium of pseudohistory, pseudo-anthropology, and pseudo-biology. None of this prevented *The Passing of the Great Race* from being as highly regarded by *Science,* the official organ of the interdisciplinary American Association for the Advancement of Science, and a number of other learned journals, as it was by *The Saturday Evening Post,* the unofficial but crucially influential voice of the American political, social, and economic establishments. To be sure, a few hard-nosed scientists, such as Franz Boas and the editors of the then new *American Journal of Physical Anthropology,* felt that Mr. Grant's book

> . . . is hardly a subject for review in a scientific journal. It is the attempt to justify a prejudice, not with the thoroughness of a Gobineau or the brilliancy of a Chamberlain, but by a superficial skimming-over of a number of commonplace observations, that are given the proper twist to suit the author's fancies. . . . It is unfortunate that a courteous preface by Prof. H. F. Osborn may convey the impression upon the minds of uninformed readers that the book has merit as a work of science.[12]

But for every such deflation, by a working scientist in a serious scientific journal, there were dozens of laudatory reviews by the likes of MIT's lecturer in biology, Dr. Frederick Adams Woods, in *Science,*[13] whose mindless praises of Grant's scholarship and acumen served, also, as revealing measurements of their own worth and abilities as scientists. While Woods admitted to feeling somewhat less pessimistic about "race suicide" than Grant, he found that his good friend had

> . . . written both boldly and attractively, and had produced a work of solid merit. . . .
> Mr. Grant believes in the inborn value of the Nordic race, that tall, fair-haired, longheaded breed which started from the shores of the Baltic some 3000 years ago, formed the ruling classes in Greece, Rome, northern Italy, Spain, northern France, England and parts of the British Isles, and then, in the southern countries, passed away either through its inability to stand the climate in competition with brunette types, or through dilution and pollution of its blood mixture with inferior peoples.
> The present reviewer accepts, in the main, this racial theory of European historical anthropology.

Of the book's historical "evidence"—so eminently acceptable to Dr. Woods, to the editors of *The Saturday Evening Post* and the *Journal of Heredity,* and to literally hundreds of American university professors of history, psychology, and even biology—time and space will not permit us to see how Madison Grant "documented" his didactic assertions that King David, Jesus Christ, Philip of Macedon and his son Alexander the Great, the Emperor Augustus of Rome, and "the chief men of the Cinque Cento and the preceding century . . . Dante, Raphael, Titian, Michael Angelo, Leonardo da Vinci," had all been long-headed, blue-eyed, tall, blond Nordics who always volunteered before their draft numbers were called. My concern is not so much with the pseudohistory as much as with the pseudogenetics and the pseudobiology of Grant's book—because of their curiously lively persistence in the conventional and political wisdom of our times. There was, however, one of Grant's Nordic claims that cannot be ignored so readily, since it was taken up by no less a maker of history than Grant's fellow Teutonist and Defender of the Great Race—Adolf Hitler.

In the revised edition of *The Passing of the Great Race* (1918, pp. 184–86), Grant claimed that because of the staggering losses of the Thirty Years' War,

> which bore, of course, most heavily on the big, blond fighting man, at the end of the war most German states contained a greatly lessened proportion of Nordic blood. In fact, from that time on the purely Teutonic race in Germany has been largely replaced by the Alpine type in the south and by the Wendish and Polish types in the east. This change of race in Germany has gone so far that it has been computed [Grant neglected to say by whom] that out of the 70,000,000 inhabitants of the German Empire, only 9,000,000 are purely Teutonic in coloration, stature, and skull characteristics. . . .
>
> When the Thirty Years' War was over there remained in Germany nothing except the brutalized peasantry, largely of Alpine derivation in the south and east. . . .
>
> Today the ghastly rarity in the German armies of chivalry and generosity toward women and of knightly protection and courtesy toward the prisoners or wounded can be largely attributed to the annihilation of the gentle classes.

In 1925, Hitler, whose closest advisers were avid readers of Madison Grant and Lothrop Stoddard (and whose Nazi-era scientific advisers and leaders were long-time personal and professional friends who had arranged for the German editions of their books years before the Third Reich), wrote in *Mein Kampf,* in his discussions of Race and State:

> Unfortunately, our German nationality is no longer based on a racially uniform nucleus. Also, the process of the blending of the various primal constituents has not yet progressed so far as to permit speaking of a newly formed race. On the contrary: the blood-poisoning which affected our national body, especially since the Thirty Years' War, led not only to a decomposition of our blood but also of our soul. The open frontiers of our fatherland, the dependence upon un-Germanic alien bodies along those frontier districts, but above all strong current influence of foreign blood into the interior of the Reich proper, in consequence of its continued renewal does not leave time for an absolute melding. . . . At the side of the Nordic people there stand the Easterners, at the side of the Easterners the Dinarics, at the side of both stand Westerners, and in between stand mixtures.[14]
>
> With the complete blending of our original racial elements a closed national body would certainly have ensued, but as every racial crossbreeding proves, it would be endowed with an ability to create a culture inferior to that which the highest of the primal components possesses originally. This is the blessing of the failure of complete mixture: that even today we still have in our Germanic national body great stocks of

Nordic-Germanic people who remain inblended, in whom we see the most valuable treasures for the future.[15]

Thanks to Madison Grant and his protégé Lothrop Stoddard, the Harvard law school graduate with a Harvard Ph.D. in history, Hitler was thus presented with a "scientific" and myth-enhancing alibi for the historical fact that the German armies had been crushed by their "racial inferiors" in World War I. The Kaiser's armies had not been as purely Nordic as Corporal Hitler, Captain Göring, and other of their cadres, and therefore the Imperial German Armies could not have been considered to be racially Teutonic. But "great stocks" of that all-redeeming and wonder-working pure Nordic blood still remained in the gene pool of the true German at the end of World War I, and it was around the nucleus of these "racial treasures" that Adolf Hitler now proposed to forge the future Nordic Reich.

Regardless of how much he gave to the likes of Hitler and Calvin Coolidge, Grant himself derived nearly all of his physical, cultural, historical, and paleontological "data" from two men to whom he gave gracious and adequate credit in the opening pages of his own book, William Z. Ripley, of Columbia University, and Henry Fairfield Osborn. But, like many disciples, Grant carried the great discoveries to their next level of consciousness. Whereas Ripley, for example, in his book *The Races of Europe* divided the populations of Europe into three races—the Teutons (Nordics), the Alpines, and the Mediterraneans—Grant offered even more precise indices than Ripley had for their identification:

"These four characteristics, skull shape, eye color, hair color and stature [as postulated by Ripley], are sufficient to enable us to differentiate clearly between the three main subraces of Europe, but if we wish to discuss the minor variations in each race and mixtures between them, we must go much further and take up other proportions of the skull than the cephalic index, as well as the shape and position of the eyes . . . the chin and other features."

All human traits were inherited, according to Grant and Galton, and this meant that all of the races of man "vary intellectually and morally just as they do physically. *Moral, intellectual, and spiritual attributes* are as persistent as physical characters and are *transmitted* substantially unchanged from generation to generation" (italics added).

All human traits were also racial. In the matter of height, for example, not only are the Nordics the tallest of races, but "no one can question the race value of stature who observes on the streets of London the contrast between the Piccadilly gentleman of the Nordic race and the cockney costermonger of the old Neolithic type [*sic*]."

The Mediterranean and Alpine races entered Europe from Africa and Asia Minor, creating an "ethnic pyramid the base of which rests solidly on the round skulled peoples of the great Plateaux of central Asia. . . . both of these races are, therefore, western extensions of Asiatic subspecies and neither of them can be considered as exclusively European.

"With the remaining race, the Nordic, however, the case is different. This is a purely European type, in the sense that it developed its physical characters and its civilization within the confines of that continent. It is, therefore, the *Homo europoeus,* the white man par excellence."

THE STERILIZATION OF THE "WORTHLESS RACE TYPES"

Grant the Biologist was as certain of his postulates as was Grant the Anthropologist. In each case, his guiding principles were, as observed in Osborn's preface to *The Passing of the Great Race,* derived from the teachings of Galton. But if Grant was a true believer, he was also a practical man of affairs, as pragmatic in his essential idealism as were Hall and Davenport: "The ideal in eugenics toward which statesmanship could be directed is, of course, improvement in quality rather than quantity. This, however, is at present a counsel of perfection, and we must face conditions as they are."

In the great counterhumanist tradition first enunciated by Malthus, Grant reminded the world:

> Where altruism, philanthropy or sentimentalism intervene with the noblest purpose and forbid nature to penalize the unfortunate victims of reckless breeding, the multiplication of inferior types is encouraged and fostered. Indiscriminate efforts to preserve babies among the lower classes often results in serious injury to the race . . .
> *Mistaken regard for what are believed to be divine laws and sentimental belief in the sanctity of human life tend to prevent both the elimination of defective infants [infanticide] and the sterilization of such adults as are themselves of no value to the community.* The laws of nature [sic] require the obliteration of the unfit, and human life is valuable only when it is of use to the community or race. [Italics added.]

Such arguments, even Grant conceded, had been advanced before. Now, however, *science* had acquired the biological confirmation of all of these hitherto philosophical postulates:

"Before eugenics were understood much could be said for a Christian and humane viewpoint in favor of indiscriminate charity for the benefit of the individual. The societies for charity, altruism or extension of rights, should have in these days, however, in their management some modicum of brains, otherwise they may continue to do, as they have sometimes done in the past, more injury to the race than black death or smallpox" [italics added].

As an orthodox eugenicist, Grant had little patience with the epidemiologists, physicians, and anthropologists whose measurements and clinical observations of living people were proving, daily, that improvements in the quality of human life led to measurable and *predictable* beneficial changes in the height, weight, health, and mental capacities of people of diverse racial backgrounds, social origins, and birthplaces. As early as page 16 of his book, Grant disposed of these sentimental heresies in what was to emerge as one of the most famous paragraphs of his work:

> There exists today a widespread and fatuous belief in the power of environment, as well as of education and opportunity to alter heredity, which arises from the dogma of the brotherhood of man, derived in its turn from the loose thinkers of the French Revolution and their Ameri-

can mimics [i.e., Washington, Madison, Jefferson, Lincoln, et al.]. Such beliefs have done much damage in the past and if allowed to go uncontradicted may do even more serious damage in the future. Thus the view that the Negro slave was an unfortunate cousin of the white man, deeply tanned by the tropic sun and denied the blessings of Christianity and civilization, played no small part with the sentimentalists of the Civil War period and it has taken us fifty years to learn that speaking English, wearing good clothes and going to school and church does not transform a Negro into a white man. Nor was a Syrian or Egyptian freedman transformed into a Roman by wearing a toga and applauding his favorite gladiator in the amphitheater. Americans will have a similar experience with the Polish Jew, whose dwarf stature, peculiar mentality and ruthless concentration on self-interest are being engrafted upon the stock of the nation.

Grant's answer to the problem of the social inadequates, the "beaten men from beaten races" with inferior blood in their veins, was the answer of all practical spokesmen for applied eugenics: forced surgical sterilization. Like Davenport, Laughlin, and Van Wagenen, Grant demanded compulsory sterilization for "the elimination of those who are weak or unfit—in other words, social failures—as well as [to] enable us to get rid of the undesirables who crowd our jails, hospitals and insane asylums."

The high Nordic honor of vasectomy, tubal ligation, and/or castration (Grant did not concern himself about *how* the world's social failures were sterilized) could, he wrote, "be applied to an ever widening circle of social discards, beginning always with the criminal, the diseased and the insane, and extending gradually to types which may be called *weaklings* rather than defectives, and perhaps *ultimately to worthless race types*" (italics added).

Grant made no secret of which "race types," starting with the Jews, the Irish, and the Negroes, he judged to be worthless. He felt that their doom would be sealed once America embarked upon the large-scale compulsory gelding of what the eugenics movement called the "social inadequates" of the population: "When this unemployed and unemployable human residuum [Galton's human 'refuse'] has been eliminated together with the great mass of crime, poverty, alcoholism and feeblemindedness associated therewith, it would be easy to consider the perpetuation of the then remaining least valuable types. By this method mankind might ultimately become sufficiently intelligent to choose deliberately the most vital *and intellectual strains* to carry on the race" [italics added].

It was not enough to sterilize at least "ten percent of the community." If the race was to be truly protected, society would have to guard against the evils of *unscientific* race mixing. As Grant asserted throughout the book, from the days of the people he identified as the great "Nordics," such as Jesus Christ, Charlemagne, and "the splendid [and, to every serious anthropologist, Spanish] conquistadores of the New World," whenever the noble blue blood of the Nordics became diluted with the inferior Alpine, Mediterranean, American Indian, and—worst of all—Jewish blood, still another priceless aliquot of greatness passed from the Great Race. This was Grant's message to Nordic America, because: "Whether we like to admit it or not, the result of the mixture of two races, in the long run, gives us a race reverting to the more ancient, generalized, and lower type. The cross between a white man and an Indian is an Indian; the cross between a white man and a Negro is a Negro; the cross between a white man and a Hindu is a Hindu; and the cross between any of the three European races [i.e., Nordics, Alpines, and Mediterraneans] and a Jew is a Jew."[16]

"THE NORDICS PROPAGATE THEMSELVES SUCCESSFULLY"

It is nearly impossible to overstate the profound influence this book of Madison Grant's was to have—*and continues to have*—on American history and public policies. It was through the books of Grant, Stoddard, and others of the eugenics mob that the editors and publishers of the nation's leading newspapers and magazines (in the era before radio and television) were infected with the viruses of a century of scientific racism. George Horace Lorimer, the editor of *The Saturday Evening Post,* was only too typical of his professional contemporaries in his slavish devotion to the ideals and solutions set forth in *The Passing of the Great Race.*

Also a disciple of Prescott Hall's, Lorimer, as early as October 31, 1919, assigned a former Army intelligence captain, Kenneth Roberts, to do a series of articles on immigration for the *Post.* They brought national fame to Roberts, and were republished as a best-selling book in 1922, *Why Europe Leaves Home.* The best description of this *Saturday Evening Post* series, whose tone and pitch made Madison Grant sound like a sheltered altar boy, was one I found in the June 26, 1921, issue of the Boston *Sunday Herald* in the Prescott F. Hall Collection at Harvard's Houghton Library. The headline and subhead told the whole story:

Danger That World Scum Will Demoralize America
If We Don't Do Something About Immigration
We Shall Have a Mongrelized America

In an editorial in *The Saturday Evening Post* of May 7, 1921, called "The Great American Myth," Lorimer paid tribute to Gregor Mendel as the man who "taught us to breed the black sheep out of our flocks . . . because he supplied the data that have enabled scientists to study intelligently the beginnings of our racial degeneration." In the next paragraphs, he hailed the new breed of "scientific writers" who had brought Mendel's work on heredity to his attention, men whose "works teem with jolting ideas, solemn warnings and predictions; and yet the clear ring of truth [*sic*] is in them."

Lorimer then singled out:

Two books in particular that every American should read if he wishes to understand the full gravity of our present immigration problem: Mr. Madison Grant's *The Passing of the Great Race* and Dr. Lothrop Stoddard's *The Rising Tide of Color* [with an introduction by Madison Grant]. The former recounts in glowing words the waxing and waning of the unsurpassed Nordic race, from which sprang Lincoln and Washington and the finest of our early stock. The latter attacks other phases of the same subject and shows in the most impressive manner how white supremacy throughout the world is threatened by the yellow, brown and other colored races. These books should do a vast amount of good if they fall into the hands of readers who can face without wincing the impact of new and disturbing ideas. Both these writers base their theses upon recent advances in the study of heredity and other sciences.

The editorial closed with a long quotation of Grant's Wagnerian ending in which he warned that the American immigration policies would render the noble Nordic germ plasm as extinct as "the Viking of the days of Rollo."

A week earlier, in an editorial called "The Burbanks of a People," the *Post,* citing Stoddard on the dangers of non-Nordic immigration, had concluded: "Congress has easily within its power to use the wise restriction of immigration as an effective and beneficent method of *world eugenics"* (italic added).

In its May 31, 1937, obituary on Madison Grant, *The New York Times* —which described him as a "zoologist" (apparently on the theory that the president of a zoo is by definition a zoologist)—wrote, under the subheading "Authority on Anthropology": "More than 16,000 copies of Mr. Grant's book, *The Passing of the Great Race,* have been published in this country. Besides being a recognized book on anthropology, it has often been called to Congressional attention in the passage of restrictive immigration laws. Since 1920 Mr. Grant wrote much about eugenics and had served as one of the eight members of the International Committee of Eugenics. He was also a member of the American Defense Society. As such he helped frame the Johnson Restriction Act of 1924."

The official name of this act was the United States Immigration Act of 1924. Within less than a decade, this immigration act was to deny sanctuary to at least six million of the Jews, Poles, Italians, Hungarians, and other human beings previously designated by Teutonists such as Grant and Hitler as non-Nordic, non-Aryan, and otherwise intellectually, morally, and biologically "worthless racial types." When this bill was passed by the Congress, all that stood between sanctuary and the gas chambers for millions of European men, women, and children was a presidential veto.

The President who had this power of veto at that moment in history was a gentleman from Vermont named Calvin Coolidge, a steady reader of *The Saturday Evening Post* and, like the editor of the *Post,* an admirer of Madison Grant. In the February 1921 issue of *Good Housekeeping Magazine,* the then Vice-President-elect Mr. Coolidge, in an article on immigration called "Whose Country Is This?," had written:

> . . . we should not subject our government to the bitterness and hatred of those who have not been born of our tradition. . . . American liberty is dependent on quality in citizenship. Our obligation is to maintain that citizenship at its best. We must have nothing to do with those who undermine it. The retroactive immigrant is a danger in our midst. . . . There is no room for him here.
>
> We might avoid this danger were we insistent that the immigrant, before he leaves foreign soil, is temperamentally keyed for our national

background. *There are racial considerations too grave to be brushed aside for any sentimental reasons. Biological laws tell us that certain divergent people will not mix or blend. The Nordics propagate themselves successfully. With other races, the outcome shows deterioration on both sides.* Quality of *mind* and body suggests that observance of *ethnic law* is as great a necessity to a nation as immigration law. [Italics added.]

The "biological laws" of heredity invoked in this slick magazine article by Coolidge were *not* the laws of Mendel, Boveri, Sutton, Johannsen, Morgan, Sturtevant, and the other pioneers in the science of genetics between 1865 and 1921, but, rather, the pseudolaws of Galton, Pearson, Davenport, and Laughlin. The "ethnic law" cited by the thirtieth President of the United States in this 1921 magazine article was decidedly not based on the physical and cultural anthropology of legitimate scientists such as Virchow, Wilder, and Boas, but on the propaganda tracts of scientific racists such as Gobineau, Ripley, and Lapouge. Calvin Coolidge had probably never heard of most of these authors, much less read any of their works. It is, however, quite evident that all of Coolidge's (and his scientific advisers') information and insights in genetics, anthropology, and biology had come—lock, stock, and barrel— out of Madison Grant's *The Passing of the Great Race* and Lothrop Stoddard's *The Rising Tide of Color,* two best-selling American books with completely interchangeable parts.

In this, Mr. Coolidge was far from being alone among the Presidents and other Americans, who, for generations, were to fabricate most of the laws, value systems, and racial postures of this nation.

8 Advances in the Legitimate Sciences Challenge and Threaten to Destroy Scientific Racism in America

> The Chinese, it is true, pretend to trace back their history to a period anterior to our own, but this claim is itself sufficient proof of its own worthlessness. *No one will suppose that the individual Chinaman has a larger brain than the individual Caucasian,* and if not, what folly to suppose that the aggregate Chinese mind was capable of doing that which it is impossible to aggregate Caucasian intellect! The truth is, what is supposed to be Chinese history is a mere collection of fables and non-sensical impossibilities. . . . it is perhaps certain that *Confucius and other renowned names known to the modern Chinese were white men,* and what shadowy uncertain historical data they now possess are therefore likely to have originated from these sources. [Italics added.]
>
> —JOHN H. VAN EVRIE, M.D., in *White Supremacy and Negro Subordination* (1868), p. 80

While the Halls, the Goddards, the Davenports, and the Grants were converting the counterfeit biological laws of the Galtons, the Gobineaus, the Lapouges, and the Pearsons into the value systems and the conventional wisdom of the leaders of American cultural and political life, the qualified investigators in the legitimate clinical, biological, and behavioral sciences were, by their reproducible discoveries, shattering some of the most cherished totems and shibboleths of scientific racism. Between 1869, when Galton's *Hereditary Genius* was published, and 1918, when the revised edition of Grant's *The Passing of the Great Race* appeared, four of the most sacred dogmas of world scientific racism were to be wrecked by what Franz Boas called "the cold enthusiasm for truth"[1] that then and now distinguished scientific research from scientific racism.

These four "scientific pillars" of eugenics and its related pseudoscientific cults were: (1) that, anatomically, the brains of white people were larger, heavier, and structurally different from and mentally superior to the brains of Orientals, American Indians, Negroes, and other nonwhite people; (2) that the head forms of individual people, as measured by their cephalic indexes, were not only determined by their racial origins but also *fixed forever by heredity;* (3) that each of the physiological, mental, and behavioral traits of every individual were preformed by "independent unit characters" or genes transmitted in the blood of all parents to their children; (4) that, of all the degenerate subspecies of native American Nordic subpopulations, the least salvable were the hereditary hewers of wood and drawers of water, the ubiquitous white poor of the South.

The first of these tenets, based on anatomical laws counterfeited by scientific racists in the years following Gall's discovery of the functional role of the cortex of the brain, was stated most characteristically in a widely circulated book published by a New York physician, John H. Van Evrie, in 1868—

White Supremacy and Negro Subordination; or, Negroes a Subordinate Race, and (so-called) Slavery Its Normal Condition. Chapter 11, "The Brain," declared:

> (1) . . . the mental capacity [of man] is in exact proportion to the size of the brain relatively with the body. (2) The brain is composed of anterior and posterior portions—the cerebrum and the cerebellum—the first the centre of intelligence, the latter of sensation, or the first the seat of intellect, and the latter of animal instincts. . . . (3) Every day we meet people with small heads and great intelligence, and with large heads and large stupidities, but a closer examination may disclose the truth that the seemingly small head is all brain, all cerebrum, all in front of the ears, while the large one is all behind, and only reveals a largely developed animalism.

Naturally, Van Evrie added, "there being nothing superior to the Caucasian," it followed that "the negro brain in its totality is ten to fifteen per cent less than that of the Caucasian, while in its relations—the relatively large cerebellum and the small cerebrum—the inferiority of the mental organism is still more decided." All of which told Van Evrie that "the negro is vastly inferior to the white man, [and] the relative proportion of the brain of the animal and intellectual natures adds still more to the Caucasian superiority."

A century before Jensen was to claim that white children had what he called Level II ("abstract problem solving, conceptual learning") types of minds, while black children had Level I ("associative" or rote learning) types of brains,[2] Dr. Van Evrie would, in the same book, assert that "the negro brain is incapable of grasping *ideas, or what we call abstract truths,* as absolutely so as the white child" (p. 129) (italics added). And this was "no mere opinion or conjecture of the author," Dr. Van Evrie added. "It is a necessity of the negro being—a consequence of the negro [brain] structure—a fixed and eternally inseparable result of the mental organism" (p. 130).

These concepts of the purported "differences" in the anatomies of white and nonwhite brains and skulls were not exclusive to Van Evrie. They were an essential element in the widespread scientistic attitudes of racial inferiority and superiority held by people as respectable as Thomas Huxley[3] and as raffish as Van Evrie's night-riding allies in the defense of Negro slavery in the South. Between 1868 and World War I, their lexicon had become more biological and less hortatory, but in essence they remained unchanged. In 1915, for example, Robert W. Shufeldt, M.D., a major in the U.S. Army medical corps, trotted out all of Van Evrie's and similar vintage pseudo-anatomical "facts" about the alleged structural inferiorities of the brain of black people in a book called *America's Greatest Problem: The Negro.*

Shufeldt relied primarily on the classic pseudo-anatomical studies of the old scientific racism—from facial angle to cephalic index—to make his case. However, he also included such items as the report "The Negro, Prostitution and Venereal Disease," written for his book in 1912 by Henry Pelouze de Forest, a professor of obstetrics at the Post-Graduate Medical School and Hospital in New York. This report asserted (pp. 251–52), among other things, that "hospital records show that practically all male city Negroes

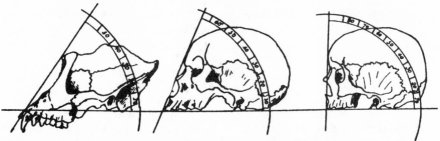

Major Shufeldt not only used the discarded head-form myths of craniology but even resorted to the more ancient facial angle myths of physiognomy to support his claims of the
genetic inferiority of black people. Here he presented drawings of the less than 60-degree
angle of a gorilla (left); the under-70-degree angle of what he categorized as "the negro";
and the nearly 90-degree facial angle of what Shufeldt described as the skull of "the intellectual representative of the white [*sic*] race."

indulge in promiscuous intercourse and carry with them venereal disease."
Since, according to Dr. de Forest, treatment for these ubiquitous venereal
diseases "almost never is carried to the point of cure," the result was "the
almost equally widespread syphilization of the female population."

The man who did the most thorough job of giving the lie to Shufeldt's
book was none other than his old professor of anatomy at Cornell, the now
unjustly forgotten Burt Green Wilder, M.D.[4]

Born in Boston in 1841, the same year as Oliver Wendell Holmes, Jr.,
Wilder—the descendant of *Mayflower* passenger Thomas Wilder—had studied at Harvard under Oliver Wendell Holmes, Sr. The elder Holmes, whose
The Contagiousness of Puerperal Fever (1842–43) had described the hygienic causes of and the simple ways to prevent childbirth fever five years
before Semmelweis made and published the same discoveries in Vienna, was
only one of Wilder's famous teachers at Harvard. Young Wilder was also a
favorite student under Louis Agassiz, the zoologist, geologist, and leading
opponent of Darwinian evolution in America; Asa Gray, the botanist and
foremost spokesman for Darwin's theories of variation, natural selection, and
evolution; and Jeffries Wyman, the great anatomist.

Like the younger Holmes, Wilder interrupted his Harvard studies to
volunteer his services in the Union Army when the American Civil War
broke out. In 1862, upon receiving his B.S. in *anatomia summa cum laude*
from Harvard's Lawrence Scientific School, Wilder postponed his further
medical studies to enter the 55th Massachusetts Infantry (colored) as a
medical cadet. With this active combat unit, Wilder rose in rank to assistant
surgeon and finally surgeon. Wilder also, during his four years of active
service with his regiment, rose to the intellectual ability to see his black
comrades-in-arms for what they were, and not for what prejudiced laymen
and racist scientists such as Louis Agassiz held them to be.[5] For the next
six decades of his life Wilder, as a citizen, was to devote himself to the civil
rights and human dignity of black people.

Wilder returned to Harvard after the war, and received the M.D. degree
in 1866. Two years later, on the recommendations of both Agassiz and Gray,
Wilder was appointed professor of comparative anatomy and zoology at

Cornell University in Ithaca, New York. He was to remain on the Cornell faculty for the next forty-two years, serving as professor of neurology and vertebrate zoology at the time of his retirement in 1910. Among his students at Cornell were great men, such as Theobold Smith (the bacteriologist who, among other things, made such clinical advances as the diphtheria vaccine possible), and lesser men, such as Shufeldt, whose "disgusting" book Wilder, as his old teacher of anatomy as well as America's leading authority on the anatomy of Negro brains, took particular delight in ripping to shreds in a review in *Science*.[6]

As an anatomist and neurologist, Wilder, like his friend and contemporary Franklin P. Mall, the Johns Hopkins Medical School professor of anatomy and embryology, conducted objective scientific studies of large series of the brains of people of all races. These meticulous studies and anatomical measurements helped end the persistent pseudoscientific myths that the brains of black people were anatomically, neurologically, and otherwise biologically different from, let alone inferior to, those of white, Oriental, and other people.

Wilder's 1909 essay, "The Brain of the American Negro," was based on his precise scientific measurements and observations of his vast collection of Negro, Caucasian, and other brains. But he made no secret of the fact that his own experiences in living side by side with black men had added to his insights: ". . . during both my army and university experiences, there have been occasions when I was tempted to exclaim, 'Yes, a white man is as worthy as a colored man—provided he behaves himself as well.' "[7]

The major thrust of Wilder's 1909 paper, however, was anatomical. It concluded, as have all serious scientific studies since then, that "as yet there has been found no constant feature by which the Negro brain may be certainly distinguished from that of a Caucasian."

This was far from an academic argument. By that time the generations of pseudo-anatomical claims by Van Evrie, and such of his many successors as Robert Bean and Shufeldt, of the actual existence of precisely such neuro-anatomical differences had so convinced the editors of *American Medicine* that, in an unsigned editorial on page 197 of their April 1907 issue, they said that Bean, in a "popular article in the *Century Magazine*," had "rightly" stated of the brain of "the negro" that "no amount of training will cause that brain to grow into the Anglo-Saxon form."

Bean, said the editorial, had now given to America "the *anatomical* basis for the complete failure of negro schools to impart the higher studies—the brain cannot comprehend them any more than a horse can understand the rule of three." Had the nation known "these anatomical *facts* when we placed a vote in possession of this brain which cannot comprehend its use," America would have had a very different history. "Leaders in all political parties now acknowledge *the error of human equality,* and the common thought is in the direction of rectifying the matter. . . . it may be practicable to rectify the error and remove a menace to our prosperity—a large electorate without brains." (Italics added.)

The editors of this professional journal were among the many educated

Americans who then accepted the brain weight and brain shape myths of scientific racism as scientific facts.

America's professional anatomists, starting with Wilder and Mall (Bean's mentor), were far less ready to accept as valid Bean's anatomical claims regarded as proven facts by the editors of *American Medicine*. No single qualified anatomist, then or since, has ever been able to find the "racial peculiarities of the negro brain" described by Bean in such lurid detail in the professional *American Journal of Anatomy* and the middle-brow *Century Magazine*.

Like his contemporary Terman, and his post-World War II successors, Bean looked to natural evolution and racial brain differences to account for what they all saw as the inherited inabilities of black people to grasp "abstract" concepts. Thanks to his inferior genetic endowment, Bean wrote in the October 1906 *Century,* "the negro fails to correlate the various ideas when more than one is grasped. . . . The negro is lacking in apperception, faulty in reasoning, and deficient in judgment."

Mall's subsequent study of 106 brains from white and black people produced no findings that even remotely suggested any anatomical differences due to race or sex. The 106 brains Mall examined included 18 brains, 10 from white people and 8 from blacks, in which Bean had previously reported discovering the anatomical variations responsible for black people's failures in reasoning and abstract thinking and learning.

Mall not only studied the same brains that Bean had used—albeit Mall used superior measuring devices—but he was also acutely aware that while science *per se* is free of bias, individual scientists are not. As Pasteur had observed earlier, the easiest thing to discover in a scientific study is what the investigator most hopes to find. Therefore, Mall wrote in his *American Journal of Anatomy* paper, "In order to exclude my own personal equation, which is an item of considerable importance in a study like this, all of the tracings as well as the measurements of all of the areas were made without my knowing the race or sex of any of the individuals from which the brains were taken. The brains were identified from the laboratory records just before the results were tabulated."

In Mall's "color-blind" examination of the brains of white and black people, it developed that "there are great individual variations, but they seem to be of like extent in both the white and the negro brains." This came as no surprise to Professor Mall since, he added, in addition to his brain investigations "I have now had considerable experience in the dissection of the negro [body] and have yet to observe that variations are more common in the negro than in the white."[8]

Mall and Wilder did not stop the pseudoscientific propaganda claims of racial inferiority and superiority in general. They simply put an end to one of the most credible of the pseudoscientific myths about racial intelligence. Unfortunately, they destroyed these costly myths on the eve of the rise of the American IQ-testing movement, which was to create even more formidable racial intelligence myths to replace them.[9]

BOAS DESTROYS THE HEAD-FORM MYTHS

Although the cephalic index or head-form myth so dear to Lapouge and other nineteenth-century scientific racists had been destroyed by Virchow and other great biologists—so that long before the century ended, the legitimate anatomists, anthropologists, and clinicians had abandoned the belief that the cephalic index was Nature's way of labeling the superior (dolichocephalic, or long-skulled) and the inferior (brachycephalic, or round-headed) races of mankind—this advance in the hard data of the legitimate sciences failed to inhibit or intimidate the elitists and confirmed racists who had for a long time been citing scientific and pseudoscientific data as proofs of their most bigoted prejudices. Such a special pleader was the American economist William Z. Ripley, the young protégé of General Francis Amasa Walker.

In 1899, Ripley published his book *The Races of Europe,* in which, as Franz Boas, reviewing the book in *Science,*[10] observed, he agreed "with most authors in recognizing three fundamental types in Europe: the long-headed (*dolichocephalic*) dark Mediterranean; the short-headed (brachycephalic) brunet Alpine; and the long-headed (*dolichocephalic*) blond Teutonic type." Boas noted that while Ripley himself suggested "it would have been desirable to designate the type of Northwestern Europe also by a geographical term— such as Deniker's 'Nordic'—rather than by a national term, such as Teutonic," Ripley used "Teutonic" throughout the book.

Boas, in his review, wrote that "cephalic index alone cannot be considered a primary principle of [race] classification," and also that although Ripley "considers as the most valuable anthropometric characterization the form of the head as expressed by the cephalic index," there already existed considerable scientific evidence to suggest that this was not a valid concept. Even when, as Ripley did in his book, one combined cephalic index and pigmentation, Boas—speaking from facts derived by himself and others— commented that neither head form nor skin color alone provides "a sufficiently broad basis for the characterization of racial types."

By 1908, after Ripley's book had given the racial simplicists their cephalically defined Nordic, Alpine, and Mediterranean "races," Ripley was talking about "race suicide" and delivering impassioned Cassandric speeches about the "hereditary consequences" of mixing inferior and superior European "races" in America.[11] For, according to the MIT and Columbia University economist, the crossing of superior and inferior races could only (as Walker had preached) dilute the good blood of the superior races and result in rising tides of the genetically inferior races.

Objective scientific truths, obviously, lacked the power to penetrate the miasma of Ripley's pathological fear of Jews, Italians, Slavs, and other non-Nordic immigrants.

In 1916, Madison Grant lifted all of Ripley's pseudo-anthropological concepts *in toto* and, substituting the word "Nordic" for the older word "Teuton," incorporated them into the "scientific" message of his influential book, *The Passing of the Great Race.* It was through this book that the cephalic index (head form) myths of the old scientific racism reached the editors of *The Saturday Evening Post* and other important molders of public attitudes.

Scientifically, Franz Boas was no stranger to the nature and meaning of anthropometry (in which his instructor had been Rudolf Virchow himself), the science which deals with the proportions of the human body. As early as 1891, when Boas—then a lecturer in anthropology on the faculty of Clark University—started to study the patterns of physical growth in Worcester, Massachusetts, schoolchildren, he had already quite consciously embarked upon his lifelong study of "the conditions which influence modification of inherited form." Out of this early study of the gene-environment *realities* of human growth and development were to come many of the modern guidelines for dealing with the problems of child health and welfare. The 1891–92 Worcester study demonstrated what hundreds of subsequent scientific and public health studies continue to document to this very day: that the children of this planet's poorer parents grew and developed at rates far slower than those of affluent families.[12]

Boas continued his interest in the processes of growth and development after leaving Clark in 1892, and as part of his new job as curator of anthropology for the World's Columbia Exposition in 1893 he caused similar studies to be undertaken in California and Canada. What Boas and the other involved investigators found in these and concurrent studies was that the growth environment had measurable and predictable influences on the ultimate bodily-form expressions of a child's inborn or hereditary endowment.

With the 1900 rediscovery of Mendel's "lost" paper, Boas' interest in "the phenomena of heredity" was further stimulated. In the spring of 1903, Boas undertook the "study of inheritance of what was for classical anthropology the crucial human trait: the form of the head."[13]

In 1907, President Theodore Roosevelt set up a U.S. Immigration Commission of congressional members and non-governmental experts to study the problems then being raised by the immigration restrictionists. The most influential of these experts, and the intellectual leader of the commission was the economist Jeremiah W. Jenks. Jenks was a disciple of General Walker's, a confirmed Nordicist, and, after 1915, the paid secret propaganda agent—at a fee of $300 per month—of the Immigration Restriction League.[14]

In his direction of the Immigration Commission's scientific research, Professor Jenks relied upon the cephalic index measurements of prospective immigrants as indicators of their innate human worth and potential. In a report to Professor Robert DeCourcy Ward written on the letterhead of the U.S. Senate on February 20, 1909, by the Immigration Restriction League's full-time lobbyist, James H. Patten, it was noted that "Jenks got another $5,000 at the last meeting (the first of December) to take back to Cornell to be used by him and his colleagues in their Dago [Italian] and other skull measurements."[15]

Boas knew that Jenks and the commission would be making various studies to lay the basis for the exclusion of Italian, Jewish, and other non-Nordic immigration, but he also counted on the fervor of Jenks's faith in the prevailing shibboleths of scientific racism. He therefore proposed to

Jenks that he be given a contract to make anthropometric studies of immigrants *and their offspring.*

Jenks, as a true believer in the hereditary principles of Galton and Walker, fully expected that this anthropometric investigation would show that bad stock remains inferior to good stock in all environments. He responded readily to the offer of Columbia University's professor of anthropology to make this study for the Immigration Commission, even though that particular scientist, Franz Boas, was a Jew. Boas was not, after all, a *Russian Jew,* and he was even then not only one of the world's foremost anthropologists, but also a German Ph.D. in physics who had been trained in physical anthropology under the great Virchow. Certainly no one could have accused Jenks of knowing in advance that Franz Boas' anthropometric study for the U.S. Immigration Commission would shatter beyond any repair one of the most sacred of all the reigning tenets of European and American scientific racism.

Boas and his thirteen graduate students and other professional assistants began their intensive study of the bodily measurements of close to 18,000 European immigrants and their foreign-born and American-born children in 1908. Most of these people were Eastern European Jews, Italians, Czechs, and Slovaks, with smaller but statistically significant cohorts of Poles, Hungarians, and Scots. The results of these 18,000 physical measurements—published as one of the reports of the U.S. Immigration Commission under the title *Changes in the Bodily Form of Immigrants* (Senate Document No. 208, 61st Congress) in May 1912—were subsequently reprinted and published in the same year by the Columbia University Press.

What Boas' scientific measurements showed was, simply, that regardless of their national origins "American-born descendants of immigrants differ in type [as defined by height, weight, cephalic index, width of face, color of hair, etc.] from their foreign-born parents." Moreover, these changes were "so definite that, while heretofore we had the right to assume that human types are stable, all the evidence is now in favor of a great plasticity of human types, and permanence of types in new surroundings appears as the exception rather than as the rule." Most significantly, the dramatic differences in the head forms and other physical measurements could, Boas reported, "only be explained as due directly to the influence of the environment."

Boas had proven beyond all doubt not only that the different head forms were not confined to specific races, but also that even in individual families there was nothing genetically fixed in the way of a cephalic index. The head forms—as indeed the heights, weights, and facial proportions—of the children of European immigrants who had been conceived and reared in the old countries turned out to be measurably different from the cephalic indexes and other bodily proportions of their brothers and sisters conceived by the same parents in the United States and raised in the new growth environments of America.

Since all of these findings were documented by his 18,000 anthropometric measurements, Boas recommended to the Immigration Commission that "our fundamental attitudes toward immigration must be decided by" these facts of human biology. The factual findings being as they were, Boas also advised the commission that as a result of what they proved "all fear of an unfavorable influence of South European immigration upon the body of our people should be dismissed."

The findings of scientific investigation had, in short, revealed that it simply was not true that the continued large-scale immigration of Jews, Italians, Poles, Slavs, and other non-Nordic people would cause "racial deterioration" in America—nor that with this "pollution" of the "native American" germ plasm the average American "skull will become more of the brachio-cephalic type, the average stature will be lower, and the average complexion will be darker." (See page 142.)

Boas provided voluminous supporting data for each of his major findings, including the four most famous:

> 1. The influence of American environment makes itself felt with increasing intensity, according to the time elapsed between the arrival of the mother [in this country] and the birth of the child.
> 2. The differences in cephalic index between parents and their own American-born children, born less than ten years after arrival of the mother, and of those born more than ten years after the arrival of the mother, are −.83 and −1.92 respectively.
> 3. The differences between immigrants and *their own* European-born children are always less than those between them and their own American-born children and the differences agree in direction and value with those obtained in the general population. Thus the cephalic-index of American-born children of Hebrew immigrants is by 1.60 units *lower* than that of their European-born children. For Sicilians it is 1.78 units *higher* than that of their European-born children. [The lower the cephalic index, the longer the head; the higher the cephalic index, the rounder the head.]
> 4. The width of face of American-born children is decidedly narrower than that of the foreign-born. Furthermore, there is a decided decline of those born a considerable length of time after the immigration of the mother, so that we get the impression of a cumulative effect of American city environment.[16] [Italics added.]

In a subsequent paper published in the *Proceedings of the National Academy of Sciences* (II, 713) in 1916, "New Evidence in Regard to the Instability of Human Types," Boas wrote:

> The size of the body depends on the condition under which growth takes place. *Growth depends upon nutrition, upon pathological conditions during childhood, and upon many other causes,* all of which have an effect upon the bulk of the body of the adult. When these conditions are favorable, the *physiological form* [the phenotype, in modern parlance] of a certain *genetic type* [genotype] will be large. If there is much retardation during early life, the physiological form of the same genetic type will be small. Retardation and acceleration of growth may also account for varying proportions of the limbs. On the other hand, we have no information whatever that would allow us to determine the cause of the physiological diminution in the size of the face that has been observed in America, nor for the change in the head index that occurs among the descendants of immigrants. Furthermore, *there is nothing to*

indicate that these changes are in any sense genetic changes; that is to say, they influence the hereditary constitution of the germ [the genome or genotype]. *It may very well be that the same people, if carried back to their old environment, would revert to their former physiological types.* [Italics added.]

But by 1916, no serious scientific worker needed this additional Boas paper to set him or her straight. To the community of *professional* life, behavioral, and social scientists, the cephalic index—as a measurement of anything other than the size of a person's hat—was now as viable scientifically as the Flat Earth Theory. Only opinionated and scientifically uninformed amateurs—such as Madison Grant; or his close friend Henry Fairfield Osborn, a once serious paleontologist who had long since put his devotion to his eugenics creed ahead of the scientist's obligation to be governed by objective facts rather than by lifelong prejudices; or the mathematician Karl Pearson, Galton's messianic disciple and successor in the eugenics movement, who

A Comparison of the Round-Headed Savage Prussian Type of Skull (on the Right) and the Gentle Long-Headed Teuton Type (on the Left) Which Professor Osborn Says Represents Now Only Ten Per Cent of Germans.

By Dr. W. H. Ballou,
Member of Academy of Natural Sciences, National Institute of Social Science, Honorary Commissioner United States Department of Agriculture, etc.

THE unparalleled ferocity which the Prussians have shown in the present war has shocked the civilized nations. Scientific examination of the skulls of the Prussians proves, however, that their methods of waging warfare are natural to them, and that, indeed, they could not, being as they are, wage war in

Professor Osborn of the American Museum of Natural History, and Professor Gregory of the Chair of Evolution at Columbia University, Trace the Blood of the German Leaders Back Through the Wild Tartars to the Most Ancient Savages

As late as 1918, the leading scientific racist, Henry Fairfield Osborn, was solemnly employing the head-form myths Boas had shattered to explain that the Kaiser's Germans were not real Nordics but round-headed Tartar impostors. Adolf Hitler was later, in *Mein Kampf*, to use this myth to explain Germany's defeat in World War I, as well as to justify the genocidal Nordic racial policies of the Nazi state. In our times, Arthur Jensen was still including the cephalic index as among the "metrical physical traits" due to "polygenic inheritance" (Jensen, 1969, pp. 32–33).
From *The American Weekly,* Boston Sunday *Advertiser and American,* June 2, 1918.

resorted to cephalic index measurements and statistical analyses *as late as 1925* to "document" the inborn mental inferiority of the children of Russian and Polish Jewish immigrants in London—took the cephalic index seriously as an indicator of human character, intelligence, or even of race after 1912.

To be sure, with characteristic coarseness, Grant, in his 1916 best seller, lashed out at Boas for having had the temerity to subject the dogma of the genetically fixed and trait-revealing cephalic index to the objective scrutiny of scientific test. On page 17 of *The Passing of the Great Race,* in Chapter 2 (entitled "The Physical Basis of Race"), Grant wrote:

> Recent attempts have been made *in the interest of inferior races among our immigrants* [italics added] to show that the shape of the skull does change, not merely in a century, but in a single generation. In 1910, the report of the anthropological expert of the Congressional [*sic*] Immigration Commission gravely declared that a round skull Jew on his way across the Atlantic might and did have a round skull child but a few years later, in response to the subtle elixir of American institutions as exemplified in an East Side tenement, might and did have a child whose skull was appreciably longer; and that a long skull south Italian, breeding freely, would have precisely the same experience in the reverse direction. In other words the Melting Pot was acting instantly under the influence of a changed environment.

Since Grant, who two pages earlier in his book had declared that skull shapes are "to all intents and purposes immutable," and his scientific advisers, such as Charles B. Davenport, wrote this and similarly worded sneers at Boas' 18,000 objective measurements of real people in a real world, some six decades of serious, professional, longitudinal studies of human growth and development the world over by pediatricians, physiologists, anatomists, anthropologists, and nutritionists have more than amply repeated, verified, and expanded upon every finding and conclusion presented by Boas in 1912.[17]

Many scientific racists and xenophobes did as Grant did and simply ignored the scientific truths so explicit in the Boas head-form and other anthropometric measurements. In 1921, for example, the Grant disciple Kenneth Roberts, in one of his series of *Saturday Evening Post* attacks on non-Nordic immigration ("Ports of Embarkation," May 7, 1921), carried on about "long skulls" of the Nordics who made America great, as against "the Alpines, the stocky, slow, dark, round-skulled folk who inhabit most of central Europe" and who "never have been successful at governing themselves or at governing anyone else."

As late as 1937, a professor of anthropology at Harvard University, and member of the Galton Society, Earnest Albert Hooton, was citing the "cephalic index and combinations of hair color and eye color" of criminals—noting that "there is a general tendency for the criminal racial types to be very slightly more brachycephalic than the corresponding civilian types"—as among his chief arguments against the provision of work relief and other social assistance to the unemployed and other victims of the Great Depression. "We must," wrote this holder of both a Harvard professorial chair and the post of curator of somatology at the university's Peabody Museum, "abandon hope of sociological palliatives and face the necessity of dealing with *biological* [i.e., genetic] realities" (italics added).[18]

Since, however, the scientific racists could not refute Boas' hard anthropometric facts, they resorted to slander. The real reason, they claimed, why the head forms of American-born children of European immigrant couples were different from those of their European-born siblings was to be found in the low morals of Sicilian, Jewish, and other immigrant women. Their American-born children, ran this whispered slander, were not the children of the same fathers, but of stray lovers these low creatures had taken in the New World.

This myth proved to be as riddled with contradictions in logic as it was steeped in hatred. For not only did the Jewish families who produced round-skulled children now give birth to long-skulled children in America—but the long-skulled Sicilian immigrant women proceeded to give birth to round-skulled children in the United States. Obviously, even if the Jewish and Sicilian wives had been as wanton as their detractors sniggered, it is improbable that they were so perverse or so skilled in human genetics as to choose only those lovers who caused long-skulled mothers to have round-skulled children and vice versa. Nevertheless, inane and vicious as it was, this slander remained the only consistent answer of the scientific racists to the anthropometric measurements of Boas and his assistants. As late as 1974, John R. Baker, in a book entitled *Race* and dedicated to the long-time general secretary of the British Eugenics Society, saw fit to write of the American-born children of European immigrants whose head forms were measured by Boas: "The possibility must be borne in mind that the true parentage was not in all cases correctly stated."[19]

Baker's and other ugly slurs on Boas and his assistants hastened the end of the uneasy truce that had prevailed between America's leading anthropologist and the hereditarians. As the eugenics creed became an increasingly popular fad of the nonscientific literates, Boas found himself unable to maintain his formerly diplomatic public silence. In a 1916 article entitled "Eugenics," in the *Scientific Monthly* (pp. 471–78), Boas opened by remarking that "the possibility of raising the standards of human physique and mentality . . . has been preached for years by the apostles of eugenics, and has taken hold of the public mind to such an extent that eugenic measures [concerning sterilization of the 'socially inadequate,' new marriage laws, etc.] have even" been passed by a number of states. Boas conceded that "the thought that it may be possible by these means to eliminate suffering and to strive for higher ideals is a beautiful one."

The only flaw in this fancy, Boas observed, was that "there are serious limitations" to the applicability of eugenic measures in human affairs. *"Only those features that are* hereditary can be effected by eugenic selection," Boas wrote. "If an individual possesses a desirable quality [such as education] the development of which is due wholly to environmental causes . . . its selection will have no influence upon the following generations. *It is, therefore, of fundamental importance to know what is hereditary and what not."* (Italics added.)

Boas reminded his readers that "we know that stature depends upon hereditary causes, but that it is also greatly influenced by more or less favorable conditions during the period of growth." A point he had amply proven with his 18,000 anthropometric measurements for the Immigration Commission, which revealed that the American-born children of European immigrant parents were almost invariably taller than their parents, and that the greater the time span between the mothers' arrival and the birth of the American-born children, the greater the gap between the heights of both generations.[20]

Therefore, Boas continued, "the more subject an anatomical or physiological trait is to the influence of the environment, the less definitely can we speak of a controlling influence of heredity, and the less are we justified in claiming that nature, and not nurture, is the deciding element. It would seem, therefore, that the first duty of the eugenist should be to determine empirically and without bias what features are hereditary and what are not. Unfortunately this has not been the method pursued; but the battle cry of the eugenists, 'Nature and not nurture,' has been raised to the rank of a dogma, and the environmental conditions that make and unmake man, physically and mentally, have been relegated to the background."

There followed a paragraph that is possibly more pointed today than it was in 1916, in which Boas defined for all time the differences in scientific acumen between the eugenicists who maintain "that better health depends upon a better hereditary stock" and the life scientists and pediatricians who maintain "that better health may be produced by better bringing up of the existing types of men." The key paragraph read:

> It is easy to see that in many cases environmental causes may convey the erroneous impression of hereditary phenomena. We know that poor people develop slowly and remain short of stature as compared to wealthy people. We may find, therefore, in the poor area, apparently a low hereditary stature, that, however, would change if the economic life of the people were changed. We may find proportions of the body determined by occupations, and apparently transmitted from father to son, provided both father and son follow the same occupation. *It is obvious that the more far-reaching the environmental influences that act upon successive generations, the more readily will a false impression of heredity be given* [italics added].

In view of the known environmental causes, most of them associated with poverty, of so many physical and behavioral disorders, Boas declared that "eugenics alone" could never raise the standards of humanity "by suppression of the progeny of the defective classes." It was clear to Boas, and to a small but growing cohort of younger biologists and geneticists, that

> *no amount of eugenic selection will overcome those social conditions* by means of which we have raised a poverty- and disease-stricken proletariat, which will be reborn from even the best stock, *so long as social conditions persist that remorselessly push human beings into helpless and hopeless misery.* The effect would probably be to push new groups of individuals into the deadly environment where they would take the place of the eliminated defectives. *Eugenics alone can not solve the problem. It requires much more an amelioration of the social conditions of the poor which would also raise many of the apparently defective to higher levels.* [Italics added.]

Boas concluded that "eugenics should, therefore, not be allowed to deceive us into the belief that we should try to raise a race of supermen."

"Eugenics is not a panacea that will cure human ills," Boas warned seventeen years before the *Erbsgesundheitgerichte,* the Eugenics Courts, of Nordic Germany were to prove him only too accurate a prophet; *"it is rather a dangerous sword that may turn its edge against those who rely on its strength."* (Italics added.)

SERIOUS GENETIC RESEARCH WRECKS THE UNIT CHARACTER THEORY

Once it was realized, in 1900, that Mendel's 1865 paper on what he called "the particulates of heredity" in hybrid peas had supplied the missing link in Darwin's theory of evolution—to wit, the crude *mechanisms of heredity—* the world's serious biologists were able to organize the new science of genetics, or the study of the biology of variation, natural selection, and heredity. By 1903, the Danish botanist Wilhelm Ludwig Johannsen (who had earlier proposed the word "genes" for Mendel's hereditary factors) was studying the mechanisms of variation and natural selection in genetically pure beans.[21]

These studies not only helped illuminate the processes of evolution, but in his description of them Johannsen also introduced the modern concepts of the "genotype" and its "phenotype."

The genotype, or genome, is the total assortment of genes that any plant, elephant, or human baby inherits from its parents. These genes, alone and in combination with other genes in the same genetic endowment, then interact with the hurricanes of nutritive, general health, and emotional factors in the total growth and development environment to "express" themselves in very wide arrays of potential end results.

The end results of these ceaseless interactions between the total complement of genes in the genotype and the historical and environmental circumstances under which they are enabled to achieve certain possibilities—or deterred from achieving other possibilities—are called phenotypes.

Johannsen's work with *pure* (self-fertilized) lines of beans demonstrated—in the words of the modern geneticists Cavalli-Sforza and Bodmer— "that variation within such a strain is environmental."[22]

In most instances, as in the European-born and the American-born children of European immigrant families measured by Boas—two groups of siblings with identical genetic endowments (that is, genotypes)—the contributory roles of differing environments, unfavorable and more favorable to human growth and development, are quite evident in the ultimate body forms (phenotypes) of these two genetically identical groups.

People, of course, cannot be bred like Johannsen's beans or Mendel's peas. Every child *usually* represents the mixture of two different genomes, the mother's and the non-blood-related father's. And as Johannsen, and literally thousands of plant, animal, and human geneticists have demonstrated since 1903, the form and nature of most of the phenotypical expressions (or translations into biological flesh and blood) of the inherited genomes of any mature organism are as dependent on their lifelong growth and development environments at least as much as they are dependent on the array of developmental possibilities pre-existent in the genes.

In the human species, moreover, these genetic traits are often *culturally* oriented. Under differing historical circumstances, the acquired phenotypical ability to make fires, for example, can as easily be used to burn dead autumn leaves in the suburbs or live religious heretics at the stake in all seasons.

Certain traits, such as blue eyes or black skin color, are of course developed independently of the growth-development environment of an individual child. But most phenotypical expressions—such as stature, mentality, weight, and temperament—while as genetic as eye or skin color at base, are also at least equally dependent upon vast and interacting complexes of environmental factors ranging from intrauterine and infant nutrition and childhood diseases to prenatal radiation and the housing, health care, and socioeconomic resources of the growing child's family. In our culture, incidentally, most of this prenatal epigenetic and postnatal phenotypic development of the individual's inherited genotype occurs during the six years and nine months prior to the great moment before any given American zygote (fertilized ovum) develops into the child (phenotype) old enough to be given his or her first IQ test.

To be sure, Johannsen, a very human, very susceptible sharer of the illusions of the educated people of his time, felt that his work provided "a complete confirmation of Galton's well-known law of regression." It happened, however, to prove just the *opposite:* Johannsen's classic elucidation of the near-total dependence of the genome on the entire biological and historical continuum of growth and development environment from conception to death drove the final nail into the coffin that Mendel's laws of the particulates of heredity had fashioned for the very mathematical, very impressive, and *totally inaccurate* Galton-Pearson laws of heredity.

Whatever lingering illusions of statistical validity still clung to the ruins of the once mighty scientific edifice of Galton's theories of heredity were swept away in 1908. In that year, the British mathematician G. H. Hardy and the German physician and geneticist Wilhelm Weinberg gave us the biomathematical basis for modern gene-pool or population genetics.[23] This Hardy-Weinberg law, describing the proportions or frequencies of specific genes in given populations, implied, "among other things, that eliminating a [genetic] trait from a population is an extraordinarily long and complex process and thus belies eugenicists' claims that breeding for or against a particular trait is an easy task."[24] A wise discovery that, unfortunately, was to be ignored by the century's pure-race aficionados from Davenport and Laughlin to Hitler and Eichmann—and their living successors.

Most of the serious scientific research in genetics was, from its earliest hours, not in statistics on human phenotypical traits but in the study of what the British geneticist William Bateson termed the "physiology of heredity," or the mechanisms of heredity at the cellular level. In 1902, Theodor Boveri, in Germany, and Walter S. Sutton, in America,[25] made the independent and simultaneous discovery that in all living cells the chromosomes are involved in the ultimate development of all living organisms—and that the absence or even overabundance of any of the normal number of chromosomes will always prevent the normal development of any individual organism of any species.

Davenport and the eugenics true believers, ignoring as was their custom all of the scientific advances that contradicted their simplisms, continued to describe as "Mendelian genetics" their unit character theory of heredity as if it were a viable *fact* of biology. To the eugenicists, the work of such pioneers of the new genetics as Johannsen, Boveri, Sutton, Hardy, and Weinberg simply did not exist. By 1910, however, Davenport could hardly have been unaware of the work that the world's real geneticists were doing in such areas as chromosomal genetics and the unit character hypothesis, since much of this work was being done in the Columbia University laboratories of one of Davenport's oldest and closest friends, Professor Thomas Hunt Morgan. It was in the laboratories of the Columbia professor of biology, and his young and talented collaborators such as Alfred Sturtevant, Calvin Bridges, and Hermann Muller, that the role of the chromosomes in biological heredity became a major area of genetic investigation.

Morgan had found, in the relatively huge chromosomes in the cells of *Drosophila melanogaster,* the common fruit fly, more than the proof that the chromosomes were indeed the cellular sites of Mendel's postulated "particulates of heredity."[26] Shortly after this, "in 1910 and 1911, Morgan came across in rapid succession a series of clearcut mutants whose manner of inheritance demonstrated that the chromosomes contained genes and *interchanged* them with one another."[27]

By 1911, Muller, Sturtevant, and Bridges, all of them working under Morgan's guidance and counsel, were deeply involved in investigations of variations in the different phenotypes of the same genome. As Muller recalled that period a decade later:

> While these investigations were showing that variable characters, apparently obscure in inheritance, were really due to a complex of various genes, each individual one of which was clear-cut and definite, there was a gradual discovery of more and more evidence that even the more definite and easily studied characters also depended on many genes. More and more mutations [that is, atypical phenotypes] were found— chiefly, of course, by Bridges—which affected the same character which we dealt with in the fly really depended upon numerous genes; conversely, other cases kept turning up in which a single gene acted upon several characters. All this evidence gradually brought about the recognition, in the *Drosophila* work, of a point of view that was also developing independently in some other laboratories at that time. . . . This was the general premise that *there was no one-to-one correspondence between visible characters studied and the genes that produced them* [as proposed in Davenport's unit character concordance of Mendelian and Galtonian theories of heredity]; that each character was really the resultant of the balanced action of countless genes, which formed a vast interacting system—reacting with each other *and also with the environment;* that these genes had both convergent effects, many different genes affecting the same character, and also divergent effects, the same gene affecting different characters. Among all these genes, how-

ever, we could only know the ones that had mutated, and then we could not know the effect of the entire gene, but only the *difference* in effect produced by the mutation.

On this theory it was *entirely erroneous to designate the residuum of characters, about which the heredity was unknown, by a single gene,* and as these ideas gradually took root that practice was abandoned. Extensive results leading to similar conclusions had been and were being obtained elsewhere by Nillson-Ehle, Tine Tammes, East, MacDowell, and other investigators, and *all these together have finally spelled the end of the old theory of hard and fast unit characters.*

Along with the death of the old unit-character theory went the death of Bateson's and Pennett's theory of presence and absence—that genes could not change, but only become lost, since it became evident that if characters were not equivalent to genes, the loss of a character, or its non-appearance in a hybrid, could not be proof of the absence of a gene. [Italics added.]

It was now clear, Muller said, "that a gene may change in all sorts of ways, without being destroyed or separated into fragments."

In 1926, still four years shy of forty, Muller succeeded in altering the genes of fruit flies with X rays, causing scores of lethal and bizarre eye, wing, and metabolic mutations to appear in the progeny of X-irradiated flies. For this he was, twenty years later, to be awarded the Nobel prize in medicine.

The unit character—or "one gene equals one trait"—dogma of the eugenics creed had been pronounced dead long before this. As early as January 9, 1914, for example, Thomas Hunt Morgan, in a letter to Henry Fairfield Osborn following a scientific meeting at which their views had clashed, said bluntly:

"The term 'unit character' is little more than the most patent result of an assumed factor, using factor in the sense of differential. It is little more than the ear-mark by means of which we recognize a certain change in difference between two types that are being contrasted. *The term is being largely given up by those working along more modern lines* and belongs to the early days of the [genetic] literature . . . to prevent misunderstanding I should like to repeat once more what I said at our meeting and apparently failed to make clear—that a given factor or determiner [gene] is *only one element in a host of others* which produce a given result [phenotype]."

Thus, in the instance or hornless sheep, Morgan added, "No one who is Mendel-wise would for a moment assume that horns were due solely to a single factor, but this is what I understand you tried to make me say at the meeting. If I had said it, *it would have been an entire perversion of the whole Mendelian point of view.*"[28] (Italics added.)

Morgan's letter to Osborn—as indeed all the basic discoveries in genetics made in Morgan's laboratory by himself and his collaborators—sounded the scientific death knell for Davenport's simplistic unit character theory of human inheritance. But whatever the end of the unit character hypothesis meant to the legitimate sciences and serious scientists, this advance in genetics had little or no effect on the nonscientists, in and out of government office, who continued to believe that such human conditions as poverty, infectious diseases, and illiteracy were caused by the inherited unit characters for these "genetic traits"—and not by low wages and unsanitary environments, as well as the eye, ear, brain, and body damages caused by both of these socioeconomic artifacts.

The new findings of the science of genetics forced the eugenic religion's true believers—like the early Christian martyrs—to place increasing reliance on the kind of faith that moves mountains.

Fortunately for eugenics, if not for the native and foreign-born American people, Osborn and his great friend Davenport did have precisely this caliber of faith in the Gospel of Galton. It would take more than Morgan's hard facts about genetics to change their minds. However much Davenport at times seemed to weaken and backtrack in the matter of unit characters—in 1917, for example, he advised Madison Grant to cut down on if not eliminate the original references to unit characters in the revised edition of *The Passing of the Great Race*[29]—the author of the Eugenics Creed remained steadfast in the Faith. Even at the time of his death, in 1944, Davenport was as firm in the belief that such human traits as morals and intelligence were unit characters controlled by one or more unit genes as he had been in 1910.

THE "GERM OF LAZINESS" UNMASKED

While the Ivy League professors of anatomy, anthropology, and genetics were demolishing the pseudoscientific eugenic tenets about the brain structures, the head forms, and the genes of the "superior" and "inferior" races of mankind, Americans in less prestigious institutions—such as the U.S. Department of Agriculture and the fledgling U.S. Public Health Service—were wrecking one of the hoariest of all the postulates of American scientific racism. This was the dogma that the white Anglo-Saxon Protestant [WASP] poor people of the South—who were, in 1925, to be categorized for the Circuit Court of Amherst County, Virginia, by the supervisor of the Eugenics Record Office as "the shiftless, ignorant, worthless class of anti-social whites of the south"— were genetic or hereditary paupers and misfits.

In 1902, two hitherto unknown Americans, Charles Wardell Stiles and Irving C. Norwood, combined talents to deliver what might have been the final blow to the scientific pretensions of the eugenics cult.

Stiles, a thirty-five-year-old parasitologist, was then with the U.S. Department of Agriculture. Son and grandson of Methodist ministers in Connecticut, Stiles had long had an interest in *Uncinaria vulpis* and *Ancylostoma duodenale*—the hookworms—discovered by Frolich of Germany in 1789 and Dubina of Milan in 1843.

At the age of nineteen, after graduating from Wesleyan University in Connecticut, Stiles set off to Europe to study medical zoology at the Collège de France, the University of Berlin, the University of Leipzig, and the Pasteur Institute. Very early in his postgraduate studies, Stiles had become fascinated by the human problems caused by hookworms. As the child of a family that took seriously the same Christian ethics despised by people like Madison Grant, Stiles was impressed as much by the social as by the medical challenges of hookworm disease.[30]

Biologically, what the hookworm does is to get into the human body and to suck blood for sustenance while it lays its eggs in the human gut. This perpetual consumption of blood by parasitic hookworms causes various problems for their human hosts—all of them associated with the loss of blood and its complements of protein, iron, and other necessities of proper health in human beings. The most common clinical disorders caused by hookworm infections are iron-deficiency anemia and a systemically lowered resistance to infectious and contagious diseases.

The hookworm breeds by sending its eggs into the greater world via the excrement of its mammalian hosts (some hookworms are as partial to cattle, dogs, or sheep as others are to people). There, after growing to threadlike larvae about one hundredth of an inch long, the young hookworms look for mammalian hosts in which to find blood and lay more eggs. One of their favorite portals of entry into human bodies is via the sensitive skin between our toes; they bore into this thin skin with a corkscrew motion that lets the worm into the bloodstream and leaves behind an itch (called, in various cultures, cow-lot itch, foot itch, ground itch, sore feet of coolies, water itch, water pox, and wet-weather itch, the last three all reminders that the larvae of hookworms require a wet, muddy soil and a warm temperature).

Obviously, then, the ideal way to prevent hookworms from causing hookworm disease and any of its cascading medical side effects in human populations is to make certain (a) that human feces are deposited in flush toilets or in more primitive but equally sanitary outdoor privies and (b) that, in areas where people and animals defecate on the ground, people who step in human and animal excrement are protected by wearing shoes too strong to be penetrated by the tiny hookworm larvae. There is also a third way to prevent bodies infested with blood-sucking hookworms from having the hookworm *infestation* turn into the debilitating hookworm *infection* or hookworm disease; that way, of course, is to make sure that the involuntary human hosts of the hookworms eat enough to satisfy the biological needs of both their own bodies and those of their hookworm parasites.

The obvious truth about hookworm disease, as underscored so succinctly by its names in history—miner's disease, brickmaker's disease, coolie's disease, Negro slave's disease, tropical anemia, etc.—is that it was essentially a poverty or social disease. It was not their well-shod overseers, but the barefoot Chinese and African coolies who developed hookworm. Nor, in Europe, was it the well-shod mine or tunnel engineers and the brickwork superintendents, but rather the miners and laborers in torn shoes and shabby clothing who developed hookworm.

Almost the first thing the twenty-four-year-old Stiles did upon returning from his advanced studies in Europe to enter government service in the U.S. Bureau of Animal Industry, in 1891, "was to examine the collection of parasites in the Bureau. . . . In this study I was surprised not to find hookworm specimens from man despite the number of bottles containing hookworms from dogs, cattle, and sheep."

Although there had been few if any reports of American cases of hookworm in the literature, Stiles persisted in the belief that uncinariasis (hookworm disease) was as endemic among the barefoot poor of America's warmer states as it was in the shoeless poor of other lands. As he recalled the events:

> . . . the fact that many of our immigrants had come from parts of Europe, Asia, and Africa in which hookworm disease in man was known to occur, persuaded me that this malady must surely exist here more frequently than reported, but that it had been rather generally overlooked. This view was expressed in the first course on this subject given [by Stiles, of course] in the Johns Hopkins Medical School, and my good friend Dr. William Osler, who was present, took me roundly to task for what I said. He maintained that a disease, as easily recognized as is hookworm disease, could not be generally overlooked in case it was present.

This, of course, was one of the rubs: no cases of hookworm disease originating with parasites native to the United States had been reported. To Stiles, however, this did not mean that native American hookworms were not present—merely that they had yet to be isolated:

> In every lecture (1892–1897) on this malady before the students of Georgetown, Johns Hopkins, and the Army Medical Schools, I made a point to repeat the remark:
> "Gentlemen, if you find cases of anemia in man in the tropics or sub-tropics, the cause of which is not clear to you, consider the possibility of hookworm disease; make a microscopic examination of the feces and look for the eggs."

During the Spanish-American War, an Army doctor, Lieutenant Bailey K. Ashford—who had attended Stiles's lectures on parasitic diseases at both Georgetown and the Army Medical Schools—put Stiles's teachings to the test. Lieutenant Ashford examined twenty Puerto Rican anemia patients, and found the eggs of *Ancylostoma duodenale,* one of the common European hookworms, in the feces of nineteen of them. His microscopic diagnoses, which were started on November 24, 1899, and published the following April, were, Stiles wrote, "the first instance in which hookworm disease, confirmed microscopically as to the etiological [causal] agent, was reported for this island, which by this time had become United States territory."

Stiles knew that it was only a matter of time before hookworm infections originating in mainland soils would be found. Early in 1902, one of the Department of Agriculture field men, Allen J. Smith, sent to Stiles some hookworms from the feces of people with hookworm disease in Texas. Careful study of these and the specimens isolated by Ashford convinced Stiles that a hookworm strain native to America had at last been isolated.

The new type was named *Uncinaria Necator americanus*—Hookworm American Killer. It was under this name and ancestry that Stiles described it at a meeting of the American Gastroenterologic Association in Washington,

D.C., on May 1, 1902. A few days later, when *Necator americanus* was found in the feces of Pygmies from central Africa then on exhibition in an Alexandria, Egypt, music hall, Stiles declared that this new discovery suggested that "the presence of this parasite in America is . . . an inheritance of the slave trade."

In August 1902, Stiles left the Department of Agriculture for a job with the U.S. Public Health Service, "where I immediately invited the attention of Surgeon-General Wyman to the point that while hookworm disease already existed in the United States, the return of our troops from the West Indies and the Philippines might result in the importation of additional infection." And in order to avoid the creation of confusion of the origins of American cases of hookworm, "I requested orders to make a brief survey in the Southern states to establish the existing distribution."

This survey, made in Virginia, North Carolina, and South Carolina in September and October 1902, is today regarded as one of the classic pioneering studies of modern epidemiology. Stiles found, according to his own report, published by the U.S. Public Health Service on October 24, 1902:

> All of the cases [of hookworm disease] are due to *Uncinaria americanus,* demonstrating clearly that this is an endemic infection and totally independent of the cases which have been introduced from Europe, Asia, and northern Africa. . . . There is, in fact, not the slightest room for doubt that uncinariasis is one of the most important and most common diseases of this part of the South, especially on farms and plantations in sandy districts, and indications are not entirely lacking that much of the trouble popularly attributed to "dirt-eating," "resin-chewing," *and even some of the proverbial laziness of the poorer classes of the white population are in reality various manifestations of uncinariasis* [italics added].

According to Stiles, "at a conservative estimate, 80% of the Southern rural schools and churches did not have even a 'sunshine' ('surface open in the back') toilet."

The young parasitologist, with the backing and encouragement of his Public Health Service superiors, took to the hustings to preach the gospel of medical hygiene to the doctors and the laity. An early description of what he had to say and how he said it may be found in an unsigned article on pages 382 and 383 of *The Popular Science Monthly* for February 1903:

> One of the most important symptoms of "hookworm" disease is an extreme lassitude, both mental and physical; this condition is due to the emaciation and to the thin watery character of the blood, which does not properly nourish either the brain or the muscles. Now, curiously enough, it is especially in the sand areas of the south that the poorer whites, known as the "poor white trash," are found, and Dr. Stiles, who has been living among these people for a number of weeks, positively states that it is among these people that hookworm disease is especially common and especially severe. He found entire families and entire neighborhoods affected, and owing to the symptoms which the disease causes, he asserts that this malady is very largely responsible for the present condition of these people. He states in fact that if we were to

place the strongest class of men and women in the country in the conditions of infection under which the poorer whites are living, they would within a generation or two deteriorate to the same poverty of mind, body and worldly goods, which is proverbial for the "poor white trash."

The links between the nature of the environment and hookworm disease were underscored in the next passage, which touched on the fact that the hookworm larvae require the moisture found in sandy or loamy soil in the warmer regions, and not in clay and other drier environments.

It is true that the poorer whites are found on clay soils as well as on sand, but Dr. Stiles maintains that on clay soil these people are healthier, stronger and more intelligent, hence that they are better fitted for the competition in life, from which the hookworm disease practically excludes the poorer whites of the sand farms. *He has further traced families from sand to clay or to the cities and proved their improvement under the new conditions; and conversely he has traced families from clay to sand and proved their deterioration* [italics added].

The same article, written more than a decade before the era of mass IQ testing in this nation, went on to note:

An important point claimed in these investigations is that hookworm disease is especially prevalent among children, and that *it not only interferes with their school attendance, but the children who are afflicted with the malady and who have gone from sandy districts to a city have the reputation among their teachers of being more or less backward and even stupid in their studies.* All this agrees with well-established symptoms of the disease, for it is thoroughly established, not only by Dr. Stiles's investigations, but by observations in Europe and Africa, *that hookworm disease stunts both the physical and the mental development.* Dr. Stiles states in fact that he has found patients of twenty to twenty-three years of age who both mentally and physically were not developed beyond the average boy or girl of eleven to sixteen years old. [Italics added.][31]

As Stiles, many years later, recalled his early efforts to win public support for the eradication of hookworm disease, "I made so many addresses and wrote so many articles on sanitary privies that the noted chemical wit of government service (Dr. Harvey W. Wiley) gave to me the nickname of 'Herr Geheimrath' (namely, 'Privy Councilor')." The long, uphill fight of Stiles for public attention and support was filled with difficulties until the advent of the other key figure in this episode in the history of social medicine: Irving C. Norwood.

In December 1902, Stiles, at the request of Surgeon General Wyman, presented a paper on hookworm disease before the Pan-American Sanitary Conference in Washington, D.C. The reporter who covered this meeting for the New York *Sun* was—Irving C. Norwood. As Stiles recounts the event, Norwood

. . . reported the address with the headline that the *"Germ of Laziness"* had been discovered. This press story was published throughout the world, causing amusement in some circles and indignation in some quarters. My interpretation is that this newspaper reporter contributed an exceedingly valuable piece of work in disseminating knowledge concerning hookworm disease. The "Germ of Laziness" became common information. It would have taken scientific authors years of hard work to direct as much attention to this subject as Mr. Norwood did through his use of the expression "Germ of Laziness." Public health workers and the laity owe him a debt of gratitude.

From that point on, midst cheers and jeers, Stiles carried forward the battle for the eradication of hookworm as a happy warrior certain of victory. Oddly enough, he also had an 80–20 percent formula for his work: in medical meetings throughout the South, writes Greer Williams, Stiles recommended "20% thymol and Epsom salts (treatment) combined with 80% sanitation (prevention)" as the magic formula for the eradication of hookworm disease and its attendant low productivity, low income, and low IQ test scores of the poor-white and -black people of the South. Not the least of his assets in this crusade was the fact that in 1909 John D. Rockefeller, Sr., notified the managers of his public service foundations that, on their advice, he was ready to commit upward of one million dollars to be spent on an "aggressive campaign" against hookworm disease.

By the end of 1914, the Rockefeller commissions had spent $797,888.36, and had cooperated with county health agencies in examining nearly 1.3 million people for hookworm infection and treating 700,000 sufferers with the disease. Some 39 percent of all southern schoolchildren, most of them white, were found to be suffering from hookworm disease, "with a variation [in the incidence of the disease] from one county to another of from 2.5 to 94%."

Many of the highlights of the reports of the Rockefeller agencies found their way into the mass media. Wickliffe Rose, the Tennessee-born educator who served as the staff director of the Rockefeller anti-hookworm crusade, joined the epidemiologists in the field. As Williams describes a few of Rose's discoveries:

> Rose found communities where hookworm disease kept a large proportion of the children out of school. He was particularly impressed with the work of Dr. A. G. Fisher of Emmorton [Virginia]. . . . Dr. Fisher found an average infection rate of 82.6 percent in the school children he examined in Richmond County, the problem being particularly severe among the "Forkemites," people who lived in the vicinity of a wide-spreading fork in a tidewater creek and who for generations were known for their lack of energy, extreme poverty, and low mental and moral state. [Similar, of course, to the Jukes and the tavern wench or dysgenic branch of the Kallikaks.] Of the 40 Forkemite children in the Tortus Key School 38 were infected [with hookworm]. These were a minor problem —there were 45 children with much severer [hookworm] infections who had *never* gone to school. At the school, Rose took pleasure in inspecting two sanitary privies, just built. He also met Willie King, 26 years old, working in the fields. A year before, King had been given up to die, with chronic ulcers eating away his legs. Now, thanks to Dr. Fisher, he was "feeling fine," newly married, and making a corn crop.[32]

To the eugenicists and other American scientific racists, this was a hard pill to swallow. For, true or false, the general public now believed that—far from being a hereditary disease of hereditary paupers—hookworm, and all of its resultant low health, low income, and low Binet test scores, was not in the blood but in the bank balance. It was caused not by the lack or presence of the genes that according to the Davenports caused man to be resistant or susceptible to infection by the "Germ of Laziness," but by the lack of toilets and privies to contain human excrement, and the lack of shoes to wear if one had to step in it. And by the lack of sufficient food to compensate for the nutrients stolen by the parasites. Conversely, it took neither massive injections of superior Nordic blood nor the development of future-orientation to cure the poor people of the South of hookworm or to prevent their being infected with *Necator americanus.*

The most cherished genocidal projects of the eugenics movement were now being subjected to attacks based directly on the work of Stiles and the Rockefeller-funded hookworm eradication crusade. In a review of Eugenics Record Office Bulletins No. 1, 10A and 10B (February 1914), containing the full report of the Laughlin–Van Wagenen committee that recommended the compulsory sterilization of the people Laughlin, Davenport, and their ilk considered to be the bearers of "defective germ-plasm in the American population," by Henry B. Hemmenway in the September 1914 *Journal of Criminal Law and Criminology,* the reviewer declared:

> Just a few years ago the southern states were the home of many people of very low mental development; lazy, shiftless, and to a greater or less degree public charges. Missionary efforts did not seem to help. *In the light of the knowledge of a dozen years ago they were evidently caco-genic, and needed to be suppressed. Now it is known that by killing the blood-destroying hookworm these individuals are made self-supporting, energetic and intelligent members of society.* A fine piece of mechanism may more easily be ruined than one of coarser construction; but the fineness is necessarily for the work to be done, and the destruction is not due to the inherent character of the machine. Just so, though insanity may be based upon the heritable character of the brain, the breakdown is due in probably every case to some extraneous [environmental] cause. *The problem for the Eugenics [Record] Office is to determine how much is due to heredity, and how much is the product of environment.* [Italics added.]

The hookworm findings of Stiles and the Rockefeller commission suggested to this critic of the February 1914 reports of the Committee to Study and to Report on the Best Practical Means of Cutting Off the Defective Germ-Plasm in the American Population:

> . . . it may be seriously questioned if the [Eugenics Record] Office has as yet demonstrated sufficient evidence to warrant the advice which it gives for sterilization. According to the rules of science, all possible factors must be considered. Apparently the [Eugenics Record] Office seems content with simply showing the reappearance of the same trait in successive generations. It is recommending radical operations on *a priori*

reasoning, and on "feeling" generated on hearsay evidence. We well remember when it was universally thought that malarial fever was the product of some miasm; now it has been proven that it is the result of an animal parasite which is developed in a certain genus of mosquito.

The discovery of the role of the environmental hookworm in the behavior and the economics of the poor-white Nordics of the South now threatened to do to the compulsory sterilization crusade of the eugenicists what the discovery of the clinical role of the malaria parasite by Sir Ronald Ross had, in 1898, done to Francis Galton's 1873 claim that the function of malaria was to increase the birth rates of inferior people of the malarious countries, who—according to Galton—were more resistant to malaria than their Anglo-Saxon betters.[33] Something, therefore, had to be done—some great biological *discovery* had to be made by a good Nordic scientist of sound views and eugenic responsibility—that would at one stroke counter the harm done to the Galtonian Creed by Privy Councilor Stiles and the Rockefellers. Of course, the accuracy of the findings of the anti-hookworm crusaders could be challenged. But none of the eugenical stalwarts and bravos dared to attack the hookworm findings of the U.S. Public Health Service and the U.S. Army.

The real tragedy, for the One True Faith, rested in the fact that the cause of eradicating hookworm disease—not only among the poor whites and blacks of the American South but also among those of the nonwhite poor the world over who lived in humid areas warm enough to let them walk around in bare feet—had been taken up by the Rockefeller clan. To attack the "Germ of Laziness" answer to the question of what made the South's 100 percent Nordic "poor-white trash" lazy, shiftless, sickly, "feebleminded," and, above all other sins, *poor,* was now in effect to attack the Rockefeller Foundation and the Rockefellers themselves. And this was a folly that even a quixotic Nordic knight as rich, as malevolent, and as insensitive as Madison Grant— the Aryan *Schreimaul* par excellence—had too much sense to entertain. There are, of course, no inborn instincts in man, but the instinct of self-preservation in *Helden* like Grant and Davenport in the presence of Titans such as the Rockefellers is an awesome sight to behold.

A neutral battleground had to be found on which to test the hereditary mettle of the Nordic and "degenerate" white poor of the South. For Charles Benedict Davenport, this field of honor was soon to emerge in the form of another disease endemic among the poor-white Nordics of the South. It had, however, not a Nordic name but an Italian one: pellagra.

At the very moment when the combined efforts of Charles Wardell Stiles, Irving C. Norwood, and John D. Rockefeller, Sr., had confronted the eugenics movement with its most serious crisis of scientific credibility, Davenport was to extract from pellagra research the raw materials of what was to prove the most successful effort of American scientific racism to control the directions of American social policies prior to our entry into World War I.

9 A Few False Correlations = A Few Million
Real Deaths: Scientific Racism Prevails over Scientific Truth

. . . no pellagra develops in those who consume a mixed, well-balanced, and varied diet, such, for example, as the Navy ration, the Army garrison ration, or the ration prescribed for the Philippine scouts.

—JOSEPH GOLDBERGER, M.D., in "The Cause and Prevention of Pellagra," *Public Health Reports*, XXIX, No. 37 (September 11, 1914)

The constitution of the organism is a racial, that is, hereditary Factor in Pellagra," the last chapter of *Pellagra III*, the final that brown eye color is inheritable. The course of the disease does depend, however, on certain *constitutional, inheritable* traits of the affected individual [italics added].

—CHARLES BENEDICT DAVENPORT, PH.D., in "The Hereditary Factor in Pellagra," the last chapter of *Pellagra III*, the final report of the Robert M. Thompson Pellagra Commission of the New York Post-Graduate Medical School and Hospital, 1917

. . . it may now be stated as established beyond reasonable doubt that pellagra is a vitamin deficiency disease analogous to scurvy and beriberi. . . . there were some 120,000 people in the United States last year who suffered an attack of pellagra . . . why do so many people continue to be stricken with the disease? The answer lies in the fact that *the problem of pellagra is in the main a problem of poverty* [italics added].

—JOSEPH GOLDBERGER, M.D., in an address to the American Dietetic Association, October 31, 1928, printed posthumously in the *Journal of the American Dietetic Association* (March 1929)

Misdiagnosed for centuries as leprosy, scurvy, syphilis, and other diseases marked by rough red skin eruptions, pellagra was described by the Spanish physician Gaspar Casal in 1735 as a distinct disease entity he called *mal de la rosa* (the red disease). Casal categorized the disease as a "kind of leprosy" caused by humid airs, foul winds, and faulty diet. In 1771, Francesco Frapolli, an Italian physician of Milan, described a disease, endemic among the poor Italian peasants, that they called pellagra (angry skin). Frapolli believed that pellagra was caused by the action of the sun on the skin.

Pellagra and *mal de la rosa* turned out to be one and the same disease. It was, in each country, a disease of the very poor, in which the skin eruptions were followed by debilitating sequences of diarrhea, lassitude, dizziness, and various mental disorders ranging from depression to violent lunacy.[1]

Although pellagra had affected many poor people in the United States for many years, it was not until 1906, when George H. Searcy, a southern physician, discovered many cases of pellagra among the inmates of the Mount Vernon insane asylum in Alabama, that doctors and public health officials started to seek out noninstitutionalized cases of the red disease. They soon realized that pellagra was endemic throughout the poorest section of the nation, the southern states. By 1908, various states had established pellagra commissions, and a new National Association for the Study of Pellagra held the first of its annual scientific meetings devoted to reports and discussions on the causes and management of the new plague.

Four years later, when Dr. Edward Jenner Wood, chairman of the Pellagra Commission of the North Carolina State Board of Health, wrote his *Treatise on Pellagra for the General Practitioner,* he revealed that "the cause of pellagra is still unknown, and, indeed, at the present time there is more uncertainty about the whole matter than ever before."

There was, however, no paucity of dogmatically stated hypotheses about the causes of pellagra. Nor was there any shortage of doctors who attributed it to eating too much corn or uncured corn; to submicroscopic agents vectored (carried) by common pest flies; to the parasitic fungus *Aspergillus fumigatus* found in spoiled corn; to the side effects of syphilis; and to unknown infectious agents found in unsanitary environments. These physicians treated pellagra with various remedies aimed at these etiological agents. This meant that the fortunately few pellagrins who did receive any medical care were dosed with Salvarsan and other arsenicals, opiates, alcohol, purges, leeches, and every foul-tasting and evil-smelling useless drug known to clinical and veterinary medicine.

Through all of the medical literature, there was, however, one constant thread of evidence suggesting that pellagra was caused by the inadequate diets of poverty, and that in the presence of enough food to eat the disease never struck. More than this, there was considerable evidence in the literature suggesting that the cure for pellagra was plenty of decent food. The French physician François Thiery, in 1755, wrote that although Casal "regards [pellagra] as incurable . . . he cites the example of a woman of the people who during one of the melancholy deliriums so frequent in this disease, had a great desire to feed herself from cow's butter, for which she spent all of her property, and she was cured."[2] Wood wrote of a French physician, Bouchard, who in 1862 "expressed the opinion that the influence of the sun [as proposed by many experts] was purely secondary and that the real underlying cause was poor nourishment."

The most famous of the authors who pointed to hunger as the cause of pellagra was the nineteenth-century French doctor Théophile Roussel. Of this great French clinical investigator, whose 1866 study showed that European outbreaks of pellagra invariably coincided with periods of shortages of fresh meats and vegetables, and who concluded that pellagra was a disease of malnutrition, Dr. Wood wrote: "Roussel said that the cause of [pellagra] was not bad air nor bad water but bad nourishment. It cannot yet be denied that bad nourishment in the sense of containing something toxic is the cause, but *I can deny that bad nourishment in the sense of insufficient nourishment is a cause*" (italics added).

Five facts about pellagra were, as of 1912, very well proven. The first was that, possibly because of advances in diagnostic techniques, tens of thousands of cases which until recently would have been diagnosed as malaria, syphilis, anemia, or scurvy were now being labeled by their right name— pellagra.

The second was that pellagrins suffered much lower natural levels of resistance to all manner of infectious and parasitic diseases, from common colds, influenza, pneumonia, and tuberculosis to dysentery, hookworm, and ascariasis (or roundworm infection, as prevalent and body-shattering as hookworm itself in the South), let alone malaria and yellow fever.

The third confirmed fact about the ancient scourge was that most of the pellagrins in the nation were poor people in the southern and border states.

The fourth and politically most explosive fact was that pellagra, as of 1913, was playing a major role in impeding the operation and expansion of textile mills and other industries in the South because of the chronic ill health and disease-caused absenteeism of the landless poor-white field hands who migrated from the rural poverty of the country to the new factory towns in search of work. Mill hands chronically deficient in work energy were not worth very much to northern industrialists interested in setting up new mills closer to the sources of cotton and other raw materials and further from trade unions and their rising wage scales.

The fifth bothersome fact about pellagra was "the economic consideration that American corn was having difficulty in securing European markets because of the prevalent belief abroad that corn was responsible for producing the disease."[3]

Stiles's work on the discovery of *Necator americanus,* and the subsequent campaign for the eradication of the equally debilitating disease caused by this native hookworm, gave well-intentioned physicians such as Dr. Wood cause for profound optimism about the early conquest of pellagra. The barefoot and unwashed American pellagrins who were not, at one and the same time, both hosts and victims of hookworms and/or the equally ubiquitous and resistance-lowering *Ascaris lumbricoides* (roundworm, eelworm), were, indeed, medical rarities. It was these synergistic effects of worms and other body parasites combined with those of pellagra that were foremost in the mind of Edward Jenner Wood, M.D., as he closed his 1912 *Treatise's* section on the prognosis of pellagra with these words:

> It is a matter of only a short time, however, before this great resistance lowering agent [hookworm] will be entirely eradicated through the wonderful work of the Rockefeller Commission for the Eradication of Hookworm Disease. *Already throughout the South the poor whites have learned the great benefits of this work and are eagerly availing themselves of the opportunities to cast off lethargy and avail themselves of their pure Anglo-Saxon inheritance.* When this great obstacle is removed there will be nothing left to hold them back physically, and they will be able to cope with the scourge [pellagra] which has so cursed the Italian peasantry. [Italics added.]

Once the "well-known philanthropists" Colonel Robert M. Thompson and J. H. McFadden undertook, in the spring of 1912, to finance the Pellagra

Commission of the New York Post-Graduate Medical School and Hospital, earnest physicians like Wood felt certain that the laboratory and field investigations undertaken by this biomedical commission would quickly be as successful as Stiles and Ashford in discovering the still mysterious infectious or toxic agent that caused pellagra.

The good southern Anglo-Saxon idealists of the Wood type, however, did not see the evils in such types of wholly environmental resolutions of the biological and mental effects of pellagra that were so plain to the super-Nordic of Brooklyn, Charles Benedict Davenport. For let it once be revealed that pellagra, like hookworm, was not in the genes but in the debilitating environment of southern poverty, then the entire eugenic-dysgenic basis for characterizing America's Nordic poor as being *genetically* ineducable, unemployable, and unfit for decent housing went careening down the drain. For now it would be obvious to everyone that the *only differences* between the Nordic cracker and the Nordic chairman of the board stemmed from the money earned and/or inherited by their parents. It would therefore follow that biologically, mentally, and morally the Nordic poor-white trash of the South were, *genetically,* not the inferiors of their fellow Nordics born to parents who could afford to provide adequate measures of the nutrition, plumbing, shoes, shelter, and preschool learning that spelled the difference between the Nams and the Davenports, the Kallikaks and the Laughlins, the Jukes and the Henry Fairfield Osborns.

Dr. Wood probably thought it was a good thing for the South and for medical science when the famous director of the Carnegie Institution's Department of Experimental Evolution and the Eugenics Record Office joined forces with the new pellagra commission. As Elizabeth B. Muncey, M.D., attached to both the ERO and the New York Post-Graduate Medical School and Hospital, revealed, in the *Archives of Internal Medicine* (July 1916), "the desirability of the study of pellagra from the viewpoint of heredity as a causative factor was brought to the attention of the Thompson-McFadden Pellagra Commission by Dr. Charles B. Davenport, Eugenics Record Office, Cold Spring Harbor, N.Y.," and under "the joint patronage of the two offices fieldwork was begun in Spartanburg [South Carolina] in June, 1913" and continued under the direction of Drs. Muncey and Davenport for more than a year.

Up until that moment, the scientific direction of the Pellagra Commission was in the hands of two physicians with considerable experience in the investigation of pellagra, Joseph F. Siler, of the Army Medical Corps, and Philip E. Garrison, a Navy surgeon. Drs. Garrison and Siler were the principal investigators, along with Ward J. MacNeal, Ph.D., M.D., director of laboratories at the New York Post-Graduate Medical School and Hospital. Drs. Davenport and Muncey were listed on the title page of the reports as principal collaborators.

Siler, whose interest in pellagra was sharpened by his earlier finding that pellagra occurs in practically all of the nation's insane asylums, had, with MacNeal, served on an earlier pellagra commission. In 1909, in response to the sudden increase in new cases, or diagnoses, of pellagra in the patient population of the Illinois State Insane Asylum, Illinois governor Deneen had appointed a commission of medical experts, including Siler, MacNeal, and the thirty-eight-year-old Howard Taylor Ricketts, the University of Chicago bacteriologist and pathologist, who within a year was to lose his life in the discovery of the louse-carried microbes of typhus and spotted fever now called the *Rickettsiae*.

During the year after the death of its most famous member, the Illinois pellagra commission, having found over five hundred cases in the state, and having looked into the relationships between many of the theories of the etiology of the disease and the pellagrins of Illinois, agreed on one conclusion. To wit: "According to the weight of evidence, pellagra is a disease due to infection with a living microorganism of unknown nature."[4]

This was not a bad hypothesis; by 1911, all bacteriologists were already aware that submicroscopic viruses—too small to be seen under the light microscope—were the cause of smallpox, measles, rabies, and other killing infectious diseases. Under the auspices of the Thompson Pellagra Commission (McFadden dropped out as a co-sponsor before its third and final report, dated 1917), Siler and Garrison put their infection hypothesis to very rigorous and well-organized scientific tests between 1912 and 1915.

Garrison and Siler were neither medical bacteriologists nor epidemiologists. There was, however, another physician on the federal payroll who was both a first-rate medical bacteriologist and also the very modern version of a working epidemiologist. He was, in fact, the Public Health Service's top field investigator of infectious and parasitic diseases. He had spent five years working as the second in command of Charles Wardell Stiles's parasitology laboratory in Washington. In 1914, because of the rising demand for a quick solution of the nagging social and economic problems caused by pellagra, he was given new orders by the Surgeon General of the United States. His name was Joseph Goldberger, his title was Surgeon, U.S. Public Health Service, and his orders were to organize and direct an investigation designed to discover the cause and the cure of pellagra.

GOLDBERGER DISCOVERS HOW TO CURE, CAUSE, AND PREVENT PELLAGRA

Joseph Goldberger bore the two classic wound stripes of the epidemiologist: a chronic shortage of funds and a history of on-the-job episodes of killing diseases. Goldberger had contracted at least three of the diseases he was investigating. In October 1902, he had contracted yellow fever in Tampico, Mexico. In August 1907, Goldberger had contracted dengue fever ("breakbone fever") in Brownsville, Texas. These infections were caused by exposure to the environmental agents of yellow fever and dengue fever in their natural or environmental settings. But in January 1910, when Goldberger was infected and nearly killed by typhus, it was in a Mexico City laboratory accident: a louse Goldberger had fed on the bodies of typhus patients escaped from a vial and bit him on the hand.

For this work he was paid so poorly that the Goldberger family was chronically hard-pressed to pay its gas bills. And when Goldberger was at death's door from typhus, his wife—six months pregnant with her second child—had to *borrow* the money to pay the train fare from Washington, D.C., to Mexico City, where her husband lay hospitalized.[5]

Goldberger had one other asset in his armamentarium as an epidemiologist: he had been born into poverty. This gave him an insight into the total environment of poverty that enabled him, for example, to write of typhus, which he called one of the "filth diseases," that all studies of typhus had emphasized "the fact . . . that fundamentally sanitation and health are *economic problems*. In proportion as the economic conditions of the masses improved—that is, in proportion as they could afford to keep clean—this notorious filth disease had decreased or disappeared."[6] (Italics added.)

Which explained why, long before Nicolle in France, and Ricketts, Wilder, Anderson, and Goldberger in America, had exposed the louse-carried *Rickettsia prowazekii* to be the microbial cause of typhus fever, there had always been fewer cases of this filth disease among families who lived in spacious homes, bathed daily, and enjoyed freshly laundered clothes and bed linens, than among the less fortunate poor of the urban slums and the rural hovels. Genetically, the poor were no more nor less susceptible to typhus than the rich.

Goldberger also knew, from the histories of his own and other families, why fewer poor children went to high school and college than rich children. He himself was one of eight children—three older brothers, three sisters, and one half brother, Jake, who had gone to America first, become a peddler, then a small grocer, and sent money back to Hungary to pay the fare of the rest of the family for the migration to America. Jake and the three older Goldberger boys never went beyond grammar school, but by the time Joseph was a schoolboy they had earned enough money to allow him to continue on to college.

Any of Davenport's or Goddard's eugenics field workers, making a study of the Goldberger family, could easily "prove," by the actuarial and correlational methods of their cult, that only one in five of the Goldberger sons went to college, and that therefore the children of this immigrant Jewish family on New York's Lower East Side were only one fifth as intelligent as Nordic families of the same size on Park Avenue, where all the sons went to Princeton or Harvard. This, however, proved more about the accuracy of Galtonian correlational analyses of human beings than it did about the inborn intelligence quotients of poor immigrant Jewish families.

When Goldberger was assigned by the Surgeon General to investigate the etiology of pellagra, the disease had been known for far longer than had hookworm disease. One reason why it had never been considered one of the great plagues of modern societies—on a par, say, with malaria, yellow fever, diphtheria, typhoid, and typhus—was possibly because it was so much a disease of the poor alone. The protozoan parasites of malaria, the viruses of yellow fever, the louse-born *Rickettsia* of typhus, the *Salmonella typhosa* of typhoid, the *Shigella* of dysentery—even, from time to time, the hookworms and the roundworms of the very poor—were notoriously no respecters of rank or privilege. The rich, as well as the poor, were known to die of malaria and typhus. The agents of such diseases visited the stateliest of mansions. In 1912, for example, New York State Senator Franklin Delano Roosevelt, although a millionaire residing in the family castle at Hyde Park, and his wife, Eleanor, both came down with typhoid fever (a classic poverty disease) at the very start of his campaign for re-election.[7]

But not one single case of pellagra among the rich, the well-to-do, and even the moderately comfortable (except for individuals malnourished because of alcoholism, anorexia nervosa, or other noneconomic causes of dietary deficiencies) had ever been reported. This basic fact leaped from the pages of the pellagra literature in various languages for all of the 179 years that had elapsed between the publication of Casal's paper on *mal de la rosa* and the day Goldberger was assigned to work on the disease. To Goldberger, who had read it all—in the original languages—this one connecting fact delivered a clear message.

Goldberger had a great respect for both Siler and Garrison. But he had an even greater respect for poverty per se as the underlying cause of the diseases of the poor. Therefore, before setting up a field laboratory and examining the blood, urine, lymph, and stools of the pellagrins, Goldberger made a slow tour of the insane asylums, the orphan asylums, the mill towns and rural slums where pellagra was endemic. His field work started in Spartanburg, South Carolina, where Siler and Garrison were both working out of the new U.S. Public Health Service hospital set up for their study of pellagra as a probably infectious disease. He moved about the town and its rural environs, observing how people lived, particularly the poor mill hands and field hands who provided most of the pellagra cases. Then he moved around other states of the South, as well as to Illinois and other midwestern states, New Jersey, Pennsylvania, Wisconsin.

Wherever there was great poverty, there was pellagra. Wherever people were institutionalized as orphans, as lunatics, as mentally retarded, as prisoners, there was pellagra. And wherever pellagra was widespread, whether in institutions or in cities, Goldberger noticed two things: *Neither the professional nor menial employees of the asylums and prisons and county old-age homes had ever developed a single case of pellagra, and the disease never struck the nonpoor of Spartanburg and other mill towns where it was endemic among the mill hands and their families.* To Goldberger, this meant only one thing, and he had the aftereffects of his own sieges of yellow fever and typhus to support his hypothesis: "This peculiar exemption or immunity was inexplicable on the assumption that pellagra is communicable."[8]

The story—of how Goldberger went on to test this hypothesis, and to prove for all time that pellagra is caused by the lack of meats, poultry, fish, dairy products, fruits, and vegetables containing what he called the Pellagra Preventive (PP) factor, which he realized was a vitamin,[9] and which is present in abundance in foods far costlier than the corn that was the staple of the diet of the poor whites and nonwhites of the South; of how Goldberger showed that pellagra could be prevented and eradicated from orphan asylums and mental hospitals, where it had been endemic, by merely adding meats, vegetables, eggs, and other common foods containing the PP factor to the daily diet; of how Goldberger persuaded a dozen Mississippi white convict volunteers to live on a high-carbohydrate diet with no proteins or fresh green vegetables for six months, and thereby induced pellagra to develop in all of them; and of how he then proceeded to cure pellagra in these same convicts and in thousands of other people who became pellagrous in the prisons without bars called human poverty—is too well known to bear repetition here. Less well known is the fact that Goldberger turned the best economist of the Public Health Service, Edgar Sydenstricker, loose on the problem of "tracking down economic data on labor, family budgets, family dietaries, food prices."[10]

As Robert Parsons, Goldberger's biographer, revealed, by the end of 1915 "sufficient data had been collected to show that, among the poorer classes, the lower the economic status became, the greater was the sacrifice in animal protein foods." And the less animal protein foods, the more pellagra. Goldberger brought Sydenstricker along to various pellagra meetings, such as the Third Triennial Meeting of the National Association for the Study of Pellagra in Columbia, South Carolina, in October 1915. At this meeting, for example, the Public Health Service statistician told the doctors: "The wages statistics show that there has been an increase of not over 25 percent in the wage rates in the period from 1900 to 1913, while in many industries and instances there has been an increase of less than 5 percent since 1907 and 1908 in the South.

"The statistics of retail food prices regularly secured by the same authority show that there has been a 60 percent increase in the average of prices since 1900."[11]

Sydenstricker's data showed that the food-purchasing power of the nation's families was lowest in the South, particularly in the cotton mill towns whose Chambers of Commerce cited the low local wage scales as incentives for relocating northern cotton mills.

Most American physicians and nutritionists are, today, more or less familiar with the most dramatic of Goldberger's experiments on the etiology of pellagra. This was the experiment described in Goldberger's 1916 paper entitled "The Transmissibility of Pellagra: Experimental Attempts at Transmission to the Human Subject."[12] To people of my generation, who were in or near their adolescence during the years when Paul de Kruif's overdramatic books were inspiring more future Nobel laureates in medicine to seek careers in biomedical research than any other single force in our cultural history, this experiment will always be known as the experiment of "Fifteen Men and a Housewife," Goldberger himself being one of the fifteen men and his wife, Mary, the anonymous housewife.

It is the fashion today, among both professional science writers and literary critics, to dismiss de Kruif's purple prose, his later enthusiasm for dubious wonder cures, and his soap-opera dramatics as obsolete and old-hat. But de Kruif wrote about medical research in terms that young people reading him would remember for a lifetime. How, for example, after Goldberger and his assistant, Dr. G. A. Wheeler, on April 25, 1916

. . . drew blood into a clean sterile syringe from the arm of a woman who was broken out and very sick with her first attack of pellagra. Wheeler took off his shirt. Goldberger shot a sixth of an ounce of the blood, still warm from the veins of the sick woman, under the skin of Wheeler's left shoulder. Goldberger took off his shirt. Wheeler shot a *fifth* of an ounce of the sick blood of the woman under the shoulder of Goldberger.

For two days the arms of these adventurers were stiff. . . . That was all.

But Goldberger was a glutton for proofs. The [Thompson-McFadden Pellagra] Commission has said that pellagra spread like typhoid fever, from the bowels of the suffering ones. Well—

On the 26th of April, 1916, alone, he faced it.

He would just be sure the natural acidity of his stomach wouldn't hurt this alleged microbe of pellagra—so he swallowed a dose of baking soda. Now then, ready. . . . Here he stands, alone in this most grotesque of laboratories—the washroom of a Pullman car. Out of his pocket he takes a little vial. Into a pill mass with wheaten flour he makes up the contents of this tube—the intestinal discharge of a woman very sick with a true case of the red disease. He swallows this dose. "And maybe the scales from the skin rash are contagious, too," says Goldberger, who is a thorough man. So for good measure he makes himself a powder from flour and the scaled-off skin from two more people sick with pellagra. He swallows this powder. . . .

And after nothing in the way of pellagra hits the investigator, two weeks later, at the U.S. Pellagra Hospital in Spartanburg, Goldberger, his wife, Sydenstricker, Wheeler, and other doctors working under the epidemiologist get shots of blood from pellagrins, and swallow vials of scrapings from the sores on the skin of pellagrins, and vials bearing little dough balls impregnated with the urine and feces of pellagrins. Four more of these experiments— "filth parties," Goldberger called them in his letters to Washington—were held. None of the human volunteers thus exposed to biological materials from pellagrins came down with pellagra. As de Kruif told it:

> Is adventure dead? All that spring this brown-eyed man, soft-voiced and terribly persuasive, went up and down the South land . . . inciting his cronies, searchers of the Public Health Service, from the Director, George McCoy, down to the cubs of the Service, to join him. He made the experiments better and better and three separate times his good friends tried to infect themselves with the blood and with those unspeakable meals—first recommended to the subjects of Hezekiah by Rabshakeh the Assyrian—from folks dying with pellagra. Always Goldberger was the first to take the dose. Seven times in all did he risk his own skin, and sundry times did he lead fourteen of his mates of the Health Service into the threat of the Valley of the Shadow, and Mary Goldberger, housewife, must not be left out of this reckoning. Bold fools they were, all of them, but now Goldberger *knew* that pellagra was not catching.[13]

So, too, did Siler and Garrison, the principal advocates of the infection hypothesis for pellagra as well as the chief investigating physicians of the Thompson-McFadden Pellagra Commission. After Goldberger and twelve convict volunteers at the Rankin Farm of the Mississippi State Penitentiary, in the spring of 1915, showed how to cause pellagra by living exclusively on the high-carbohydrate, high-fat, no-protein diet of world poverty, Dr. Robert Parsons, friend and co-worker of Goldberger, Siler, and Garrison,

Even before Goldberger's classic Mississippi experiment proved how only the nutritionally inadequate diets of American and world poverty cause pellagra, his earlier discoveries of how to cure and prevent pellagra by adequate diet had won him world fame. In April 1915, Goldberger's previous work on pellagra earned for him the honor of delivering the annual Cutter Lecture in biology at Harvard. Nevertheless, to Davenport and his fellow eugenicists the Hungarian-born Jewish immigrant Goldberger was a nonperson—and so they ignored completely his discoveries dealing with the cause and prevention of pellagra.

recalled, at "the Thompson McFadden commission itself, the news of the prison experiment findings had been something of a blow to it. Garrison was at Spartanburg [the South Carolina mill town where the Commission had set up its main research base] when the report reached him. He packed his suitcase forthwith, and went home. He knew Goldberger. He was through with the Commission, and with problems in pellagra."[14]

A few days after Goldberger's report on his Mississippi experiment was published,[15] the U.S. Public Health Service and the Pellagra Commission held a special pellagra conference at Spartanburg. In a letter to his wife sent from Spartanburg, Goldberger was pleased to note that, at this historic meeting, "Dr. Siler came across very handsomely, stating on the floor that he agreed with every statement in my paper, and later congratulated me personally. Siler is a gentleman."[16]

If Siler and Garrison were glad, Goldberger's long-time Public Health Service co-workers, from Stiles and Milton Rosenau, by then professor of public health at Harvard, to William Welch, the founding dean of the Johns Hopkins Medical School, and Allan J. McLaughlin, the Massachusetts Commissioner of Health, were ecstatic. Rosenau invited Goldberger to deliver Harvard's prestigious Cutter Lecture in Biology for 1915, and in the same year sent him a letter declaring that "your achievement in this disease is equal to any contribution to medical science made in America."[17]

Dr. McLaughlin, in a letter sent two days later, told Goldberger: "Your work in pellagra, coupled with your previous work, . . . stamps you, in my opinion, as the foremost figure in America today in the field of preventive medicine. This is not flattery; it is the careful estimate of a man who is reasonably familar with the advances made in preventive medicine."[18]

"AN INSTANCE OF 'FALSE CORRELATION' "

Not every physician was as ready to accept Goldberger's findings that pellagra was purely and simply a disease of hunger. At the same October 1915 meeting, in Columbia, South Carolina, of the National Association for the Study of Pellagra, when Goldberger and his fellow Public Health Service physician David Willets reported on the prevention and treatment of pellagra by diet, most of the southern doctors present were volubly more impressed by MacNeal's report that the Thompson-McFadden Pellagra Commission had determined that in those Spartanburg homes with inside plumbing and better sewage there was little or no pellagra, while in the neighborhoods where the poor mill hands lacked both plumbing and sewage, pellagra was endemic.

It was at this meeting that Goldberger rose to deliver a comment that, apparently, was lost on so many physicians at the time:

> As for the relation of pellagra to sewerage, as shown by the practical freedom of the sewered sections of Spartanburg, this may be an instance of *"false correlation."* When we recall that pellagra is essentially a disease of poverty, explanation for this seemingly mysterious phenomenon at once suggests itself. In a community having a sewered section and an unsewered section, it is very likely, and in Spartanburg it actually so happens, that the sewered section includes the best residential portion of the town. The sewered section is that part of the town in which the well-to-do people live, in which even the neophyte pellagrologist would not expect to find much pellagra, sewers or no sewers.[19] [Italics added.]

But MacNeal and other physicians at this meeting savaged Goldberger with "false correlations" purporting to prove, statistically, and with hard numbers, that pellagra was anything but a dietary deficiency disease. At one point MacNeal rose to cite even higher authority:

> The problem which confronted us was the analysis of the mass of data. . . . I think the one method commonly used with at least 1,000 comparable observations is that adopted by Davenport and Pearson. . . .
> We must do something with the data besides theorizing with them.

We must try to show some correlation. This is what we have attempted to do by the statistical method. It is probably unnecessary for me to point out that the table Dr. Goldberger cites [on the relationship between diet and pellagra] has very little importance.

It was talk like this that caused Goldberger to lose his temper, and to tell the physicians present at the same meeting that, in reference to their avoidance of even discussing "the very simple or practical device of treating the patient [with food]. . . . There is an old adage to the effect that the proof of the pudding is in the eating thereof, and the only thing I care to say in closing is, suppose we stop talking so much. We do not have to inject salvarsan, etc."

At this point in the history of pellagra, the modern reader may be pardoned for assuming that the prevalence of pellagra, like hookworm disease, was sharply reduced or even eradicated in the South soon after Goldberger isolated the Non-Germ of Pellagra—and that Goldberger's facts triumphed over the unproven myths. This, however, was not at all what happened. The bare pellagra mortality figures tell only the smallest part of what followed. Consider the national totals of deaths reported by state health departments to the Public Health Service before and after Goldberger elucidated the completely nonhereditary mechanisms of the cause, cure, and prevention of pellagra.

Between 1914, the year Goldberger discovered the cause of pellagra, and 1928, the year of his untimely death at the age of fifty-five, the reported pellagra deaths multiplied eightfold, climbing from 847 to 6,523 (see table on page 224).

These data in themselves do not begin to tell the whole story. For one thing, perhaps one in a hundred deaths in which pellagra is the underlying cause is reported as a pellagra death: pellagra mortalities are reported, *and quite accurately,* as being deaths caused by pneumonia, or malaria, or influenza, or measles, or hookworm infection, or ascariasis, or diphtheria, or any combination of scores of other infectious and dietary diseases that are invariably far more lethal in bodies weakened by chronic deficiency diseases such as pellagra than in the bodies of well-fed people.

For another and possibly even more important thing, a diagnosis or an autopsy by definition calls for both a *patient and a physician;* the average person who dies of pellagra or other poverty diseases is in most instances brought into the world and ushered out of it without the presence of a physician, registered nurse, or any other health worker. Therefore, the deaths and the nonfatal cases of pellagra reported to the keepers of health records represent only the tips of vast mountains of mortalities and morbidities from this poverty disease.

There were a number of reasons for the failure of our society to eradicate pellagra from our total population within two or three years after Goldberger showed us how to do just this. Not the least of these pellagra-preserving factors was the con job performed on the Siler- and Garrison-less Thompson-McFadden Pellagra Commission by Charles Benedict Davenport.

PSEUDOSCIENCE PREVAILS OVER
GOLDBERGER'S DISCOVERIES

The sponsorship of Stiles's crusade against the "Germ of Laziness" by John D. Rockefeller, Sr., had, as we saw in the previous chapter, rendered the leaders of the eugenics movement impotent to even attempt to refute the sober scientific discoveries about the side effects of hookworm disease on the economic productivity and the academic achievement and IQ test scores of the poor-white Nordics of the South and their doomed children. The Davenports of this world never attack the deeds of its Rockefellers.

There was nothing the scientific racists could do about the threat posed to their political program by the salutary effects of the hookworm eradication efforts of the Rockefeller commissions on the work and learning patterns of the "poor-white trash." This triumph of epidemiology over eugenics would have to be neutralized by other means.

Goldberger's findings that pellagra, and the mental disease it caused, were neither of them genetic, were targets more in keeping with the courage of the leaders of the eugenics movement. Because of the etiological links between madness and pellagra, Charles Benedict Davenport, the director of the Eugenics Record Office, had as early as 1913 become involved with the Pellagra Commission. When Goldberger made the work of the commission superfluous in 1914, causing Garrison and Siler to leave it, Davenport took control of its work and the preparation of its third and final report, *Pellagra III*.

Two years *after* Goldberger's first report on the real causes of pellagra, Davenport, a zoologist with no clinical training or experience, and Elizabeth Muncey, a physician on the payroll of Davenport's institute, published two articles in the July 1916 issue of the *Archives of Internal Medicine*. During the same month, both articles were reprinted as Bulletin No. 16 of the Eugenics Record Office.

Davenport's article, written by the man who had for years maintained in books and articles and lectures that insanity, mental retardation, imbecility, low family incomes, and low IQ test scores were all hereditary conditions, made much of the fact that reviews of the scientific literature had shown that in Italy "4 to 10 percent of pellagrins are insane," while the Public Health Service study of 1913 had revealed that "in this country about 7 percent of the pellagrins are insane." The fact that pellagra was apparently a disease of poor people who lived in unhygienic rural and urban hovels only reinforced Davenport's belief that pellagra was in large measure a genetic disease, since poverty itself—"pauperism" in the eugenics lexicon—was, according to Galton, Pearson, and all good eugenicists, hereditary. As Davenport observed in this 1916 essay:

> That there should be a correlation between mental insufficiency and pellagra is not strange, since the mentally insufficient are, on the whole, less likely to appreciate the importance of sanitary surroundings and less able to avail themselves of them, and the reports of the [Thompson-

McFadden] pellagra commission *prove* the close relation of pellagra to poor sanitation. No doubt, also, persons who are mentally well developed are, on the whole, more likely to care for their bodies and keep themselves in good condition than are the mentally deficient or unstable. Other things being equal, pellagra is more liable to make headway in "Nam Hollow" than in the cottages on the cliffs at Newport.[20]

Davenport's thesis about the natural history of pellagra was very simple. Pellagra, he wrote, was "the reaction of the individual to the poisons elaborated in the body, probably by a parasitic organism. This accords with the conclusion of Siler, Garrison [*sic*] and MacNeal that pellagra is in all probability a specific infectious disease communicable from person to person."

By March 16, 1916, when Davenport submitted this paper to the *Archives of Internal Medicine,* he *knew* that Garrison and Siler had long since revoked their former infection hypothesis (it was never a "conclusion") in the face of Goldberger's elegant biological proofs that it was incorrect. Davenport also knew very well that Garrison and Siler both agreed, with Goldberger and his Public Health Service colleagues, that pellagra was a simple and predictable sequela of hunger—a classic poverty disease.

Nevertheless, with a Galtonian disdain for the mere facts of human biology, Davenport continued to label pellagra as an infectious disease. More than that, he also insisted that

> in the pellagra reaction [to the mysterious "pellagra germ"] there is a hereditary factor . . . if there is one thing of which experience perfectly assures us it is that different individuals react dissimilarly to the same stimulus. Such dissimilarity of reaction is conditioned both by fundamental dissimilarity in the constitution of the organism and by dissimilarity in antecedent experiences of the organism; but the latter, in turn, is conditioned in part by the former; so that the fundamental dissimilarity of the constitution of the organism must be held to be the principal cause of the diversity which persons show in their reaction to the same disease-inciting factors.
>
> This constitution of the organism is a racial, that is, hereditary factor. And if it appears that certain races or blood lines react in pellagra families in a specific and differential fashion, that will go far to prove the presence of a hereditary factor in pellagra. . . . *colored persons, who differ from most white people in having more or less negro blood, are less subject on the whole to the disease than white persons* [italics added].

Davenport's conclusions were in the true eugenic traditions:

> Pellagra is not an inheritable disease in the sense that brown eye color is inheritable. The course of the disease does depend, however, on certain constitutional, inheritable traits of the affected individual.
>
> Pellagra is probably communicable, but how the communicated "germ of the disease" shall progress in the body depends, in part, upon constitutional factors.
>
> When both parents are susceptible [*sic*] to the disease, at least 40 percent, probably not far from 50 percent, of their children are sus-

ceptible; an enormous rate of incidence in a disease that affects less than 1 percent of the population on the average.

As further "proof" that pellagra was essentially a genetically determined disease, Davenport added the following non sequitur:

Many families never show mental symptoms, while others usually do. . . . In some families the skin eruptions amount to little; other families are characterized by severe ulceration and desquamation of the derma. These *family* differences have all the characteristics of biotypes or blood lines, and afford the best proof that there is, indeed, an hereditary factor in pellagra [italics added].

THE 444-PAGE MEDICAL FRAUD OF THE CENTURY

Both of the articles Muncey and Davenport had previously published on the hereditary causes of pellagra in the *Archives of Internal Medicine,* and reprinted as Bulletin No. 16 of the Eugenics Record Office in 1916, were to be reprinted for a third time—as the closing and summing-up chapters of *Pellagra III,* the third and final report of the Robert M. Thompson [formerly Thompson-McFadden] Pellagra Commission of the New York Post-Graduate Medical School and Hospital, in 1917.

This final report of the Pellagra Commission is now a collector's item, particularly among those fascinated by great American frauds. According to the title page, the principal authors of the report were "J. F. Siler, M.D., Major, Medical Corps, United States Army; P. E. Garrison, M.D., Passed Assistant Surgeon, United States Navy; and W. J. MacNeal, Ph.D., M.D., Director of Laboratories, New York Post-Graduate Medical School; with the collaboration of C. B. Davenport, Ph.D., and Elizabeth B. Muncey, M.D." MacNeal had an extra title-page credit as editor of the report.

In the introduction to this final report, we read that "Dr. J. F. Siler, Major, Med. Corps, U.S. Army, has been detailed to this work continuously . . . from the spring of 1912 to the last of July, 1915. Dr. P. E. Garrison, Passed Asst. Surg., U.S. Navy, was detailed to the [Pellagra Commission] investigation continuously from the spring of 1912 to November, 1913. He was detailed elsewhere from December, 1913, to June, 1914, and after working on pellagra during the summer of 1914, he was compelled to give up further participation in the work [of the Pellagra Commission] by a call to active sea duty."

Since Goldberger's first two reports on the cause and prevention of pellagra had been issued by the U.S. Public Health Service in June and September 1914, the reasons why the U.S. Army and the U.S. Navy—operated by the same government, and paid by the same tax revenues—found other things for Siler and Garrison to do for their salaries after 1914–15 were not military. Their job was done; they were needed elsewhere, now that Goldberger had found what Siler and Garrison had been detailed to look for.

The departure of Siler and Garrison did not, however, bring MacNeal to the front, for as the introduction went on to say: "Dr. W. J. MacNeal, the civilian member of the commission, has continued since the spring of 1912 to devote *part* of his time to the work of this commission, *although by far the major portion of his time each year has of necessity been given to other work in the laboratories of the New York Post-Graduate Medical School"* (italics added).

This, of course, was the academic code language for the fact that MacNeal not only had a job to protect—he was the director of these laboratories in a school quite heavily endowed by Davenport's friends—but he also wanted the American medical community to know that regardless of what they might read in the pages that followed, W. J. MacNeal, like Siler and Garrison, knew that Goldberger was correct, and that the proponents of infectious and genetic etiologies for pellagra were 100 percent wrong.

The final report of the Pellagra Commission was, clearly, the work of the eugenics zealot whose thrice-reprinted 1916 article, "The Hereditary Factor in Pellagra," constituted the eleventh and closing chapter of this so-called scientific report. His name, of course, was Charles Benedict Davenport, and he was not about to let the hard facts of medical science weaken his faith in the religion of eugenics.

Therefore, in this final report of the Pellagra Commission, Goldberger's work was, except for one footnote on pages 226–27, completely ignored. This footnote dismissed, as a failure, Goldberger's careful experiment that disproved the transmissibility of pellagra because Goldberger, according to Davenport, had been ignorant of the non-fact that "the relative insusceptibility to pellagra of young adult men is generally recognized."

The "recently published negative experiment" that Davenport et al. dismissed so cavalierly happened to be the Goldberger experiment described in the *Public Health Service Report* of November 17, 1916—the "Fifteen Men and a Housewife" experiment that wrote finis to the last, lingering chance that pellagra might prove to be an infectious or transmissible disease.

Articles that Garrison and Siler had previously published in various issues of the *Archives of Internal Medicine,* and which both authors had long since agreed had been contradicted by Goldberger's findings, were now reprinted as originally written and/or slightly revised by other hands in the 1917 final report of the Pellagra Commission. These articles by Garrison and Siler constituted the majority of the chapters in the report. The only concession to reality and truth made by Davenport and his co-editors was a footnote to each of the chapters attributed to Garrison and Siler. This bit of fine print informed anyone who could read type that small that "the final copy of the paper itself has been written since Dr. Garrison and Dr. Siler were recalled to active service in the Medical Corps, U.S. Navy, and the Medical Corps, U.S. Army, respectively. They are, therefore, not responsible for the observations of the last two years, for the compilation of the data, or for the deductions drawn from them."

Typical of what Goldberger denounced, in exasperation, as the "false correlations" between the sewage of poor neighborhoods and/or the genealogy of pellagrins, on the one hand, and the real causes of pellagra, on the other, was the table in Chaper 7 of the report. It proved only that poor people are as prone to have the diseases of hunger as were their equally poor neighbors, relatives, and ancestors.

Dr. Muncey's chapter in *Pellagra III* was liberally seasoned with batches of the famous pedigree trees so dear to the hearts of eugenics scholars. This pedigree chart, on page 431 of the report, traced the prevalence of pellagra and other poverty diseases such as "hookworm disease followed by nervous exhaustion and debility" in the family tree and personal history of "Pellagrin #25." This poor woman had died in a madhouse in 1914, after having had hookworm disease initially in 1908 and three subsequent attacks of pellagra-followed-by-madness in 1910, 1911, and 1912. The pedigree chart showed that the woman had three children, two of whom were pellagrins.

Equally impressive, to laymen ignorant of the natural history of pellagra, was Dr. Muncey's Table 5, on page 380 of the report. This table, "Relationship of Pellagrins in Families with Pellagra in the Third Generation," did not, of course, prove that pellagra was in any way a hereditary disease:

Relationship of Pellagrins in Families with Pellagra in the Third Generation

1 grandmother (1911–1913)	————	1 granddaughter (1914)	Direct
2 grandmothers (1863–1911) (1910–1913)	————	1 grandson (1913)	Direct
1 grandfather (1912)	Mother (1913–1914)	2 granddaughters (1913–1914)	Direct
1 grandfather (1909)	[Grandmother] (1911)	2 granddaughters (1913)	Direct
1 grandfather (1912–1913)	Father (1912–1913)	1 grandson (1911–1912)	Direct
1 grandfather (1901)	Mother (1907–1913)	1 granddaughter (1911–1913)	Direct
1 grandmother (1910–1912)	Mother (1909–1913)	2 grandsons (1910–1913)	Direct
1 grandfather (1908)	Mother (1904–1913)	2 grandsons (1912)	Direct
1 grandmother (1910–1912)	Mother (1909–1913)	2 grandsons (1910–1912) (1913)	Direct
1 grandmother (1900)	2 mothers (1905) (1910–1911)	2 grandsons (1911–1912)	Direct
1 grandmother (1910–1914)	4 daughters (1910) (1913) (1914)	2 granddaughters (1914)	Direct
1 grandfather (1913)	Son-in-law (1912–1913)	1 granddaughter (1914)	Direct and indirect
1 grandfather (1910–1912)	Daughter-in-law (1912–1913)	2 grandsons (1913) (1914)	Direct and indirect
1 grandfather (1900)	Daughter and son-in-law (1911–1913) (1912)	1 grandson (1911)	Direct and indirect
1 grandmother (1910)	Son and daughter-in-law (1910–1912) (1910)	3 grandchildren (1912)	Direct and indirect
There are also 1 step-grandfather (1911–1914)	3 step-children (1910) (1910) (1913)	2 step-grandchildren (1910) (1913)	
1 step-grandmother (1912)	1 step-daughter (1910–1914)	3 step-grandchildren (1912)	

Typical of what Goldberger had denounced as the "false correlations" of eugenical perversions of medical and social data was Munsey's Table 5 in *Pellagra III*. Far from establishing, as Munsey and her chief Davenport asserted, that pellagra is a hereditary disease, it proved only that the children and grandchildren of poor people will quite predictably suffer the socially preventable deficiency diseases of poverty.

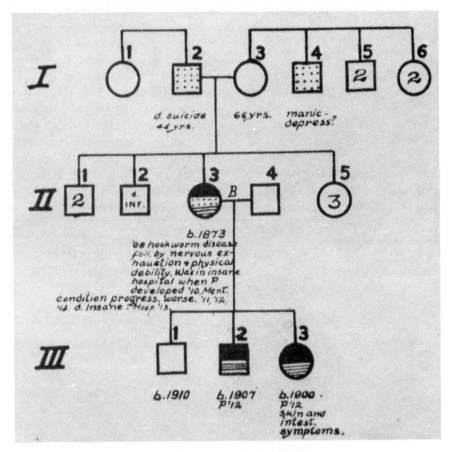

Genealogical or pedigree trees—the famous "black charts" of eugenics propaganda—were employed in 1917 in *Pellagra III* by Davenport and Munsey to "document" the "hereditary" nature of pellagra. Such impressive-looking "scientific" displays were accepted as reliable refutations of Goldberger's findings about the wholly environmental causes of pellagra by a majority of the American legislators responsible for the goals and administration of minimum wage, health, and family welfare laws and programs dealing directly or indirectly with the causes and prevention of pellagra.

This table, however, certainly did prove, as convincingly as Galton's actuarial data in his *Hereditary Genius* in 1869 proved, that children follow in the experiential footsteps of their parents and grandparents. Thus, if the grandparents are bankers, bishops, and brigadier generals, their children and grandchildren are statistically quite prone to become chairmen of corporation boards, college presidents, and cabinet ministers. Similarly, if the parents and grandparents speak only French, *or have pellagra,* the statistical likelihood that the grandchildren will speak French, or have pellagra, is great. But none of these data proved that pellagra was any more genetic than the ability to speak French or inherit banks.

Whether Davenport really believed that he and Muncey had destroyed

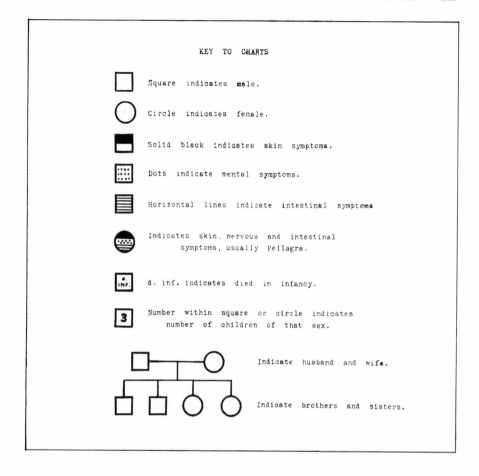

KEY TO CHARTS

Square indicates male.

Circle indicates female.

Solid black indicates skin symptoms.

Dots indicate mental symptoms.

Horizontal lines indicate intestinal symptoms

Indicates skin, nervous and intestinal symptoms, usually Pellagra.

d. inf. indicates died in infancy.

Number within square or circle indicates number of children of that sex.

Indicate husband and wife.

Indicate brothers and sisters.

the scientific validity of Goldberger's findings is, today, of little moment. All that does matter, in history, is that in 1916 and 1917 Davenport and Muncey told the greater society what it would rather hear about this endemic disease of poor people.

If pellagra was indeed, as Davenport claimed, an infectious disease of genetically inferior white Anglo-Saxon Protestant breeding stock, then it could not possibly be either prevented or cured by such social actions as minimum-wage laws that would enable the working poor of the South to earn enough money to provide their families with the meats, eggs, dairy products, and vegetables that Goldberger had demonstrated would prevent and/or cure pellagra. Nor could pellagra be prevented, in the families of America's "definite race of chronic pauper stocks," by the societal provision of food supplements to families that were too poor to buy meat, fish, poultry, and other sources of niacin.

It was cheaper, it seemed, to believe in the pseudogenetic myth of pellagra as an infectious disease of a subrace of inferior hereditary stock. Such self-defeating socioeconomic illusions were, of course, among the primary reasons why this pseudogenetic hypothesis—as spelled out in the third and final report of the Pellagra Commission—was accepted as scientific truth by the lawyers and businessmen who then made and administered governmental priorities and policies concerning health, education, and human welfare.

The state of knowledge of the American medical community in 1917 happened to be not the least of the other reasons why—in state, county, and city health departments and professional medical societies—Davenport's pseudoscientific pellagra hypothesis was able to prevail over Goldberger's testable scientific findings.

As of 1917, a *majority* of American physicians were the graduates of inferior, commercial, scientifically and clinically grossly inadequate medical schools. Most of these "medical schools," even those nominally attached to colleges and universities of high repute, were little more than diploma mills. Outside of Johns Hopkins, and a handful of schools that had joined Johns Hopkins in seeking to emulate the better and more scientifically oriented European medical schools, the average American medical school had yet to catch up with nineteenth-century biomedical advances. As the President of the Carnegie Foundation for the Advancement of Teaching, Henry S. Pritchett, wrote in his introduction to the famous study of medical education in the United States and Canada completed for the foundation by Abraham Flexner in 1910: "For the past 25 years there has been an enormous over-production of uneducated and ill trained medical practitioners . . . due in the main to the existence of a very large number of commercial schools."[21]

These "uneducated and ill trained medical practitioners" were also, in 1917 and at least a dozen years that followed, the medical cohort from which the majority of the nation's municipal, county, state, and federal health officers were recruited.

Thus, the lack of scientifically viable societal policies concerning the standards of medical education, and the clinical training of America's doctors, between 1880 and 1920, was to create a population of American physicians who were intellectually quite capable of accepting the impressive 444-page report of the Robert M. Thompson Pellagra Commission of the New York Post-Graduate Medical School and Hospital as the ultimate scientific conclusions on the nature, causes, and treatment of pellagra.

THE GREAT PELLAGRA COVER-UP OF 1916–33

Davenport, Muncey, and their accomplices in the counterfeiting of fake laws on the natural history of pellagra—and the covering up of Goldberger's discoveries—in the final report of the Pellagra Commission in 1917 were, of course, not the only offenders against the scientific ethic in the matter of this easily preventable deficiency disease of poverty. Some of the nation's well-fed and highly educated true believers in eugenics did not need Davenport's fraudulent *Pellagra III* to persist in the comfortable conviction that pellagra was due to the inborn inferiority of the poor who became pellagrous.

In December 1915, for example—the same year in which Harvard's annual Cutter Lecture in Biology was delivered by Goldberger, the hero of the hour in preventive medicine—the retiring president of the American Association for the Advancement of Science, Charles W. Eliot, for the forty years ending in 1909 the president of Harvard University, delivered a farewell address entitled "The Fruits, Prospects and Lessons of Recent Biological Science."

In his address, which was printed in full in the December 31, 1915, issue of *Science,* the distinguished educator, mathematician, and chemist paid graceful tribute to Jenner and Pasteur; to the Rockefeller Sanitary Commission "through whose well directed efforts . . . hundreds of thousands of persons in the Southern States of the country have been made much more effective laborers, because relieved of the hookworm disease"; to the Rockefeller International Health Commission for taking the fight against hookworm to the barefoot poor abroad; to the work in yellow fever, in syphilis, in improved sanitary food-handling regulations, in bacteriology, in agricultural genetics. Dr. Eliot even found time in his address to predict that the applied biological sciences would soon be employed "in the contest against alcoholism and sexual vice." In regard to alcoholism, he said, "There is every reason to expect that this great field for Christian effort will hereafter be more effectively cultivated than it ever had been."

Dr. Eliot—who four years earlier had served as one of the vice-presidents of the First International Congress of Eugenics, in London—evidently did not think that the eradication of pellagra represented a proper field for Christian effort. Or even one of the sweeter fruits of recent biological science. For the president-emeritus of Harvard had not one solitary word to say about either Goldberger or the conquest of pellagra in his farewell address to the AAAS.

It was not by oversight that Dr. Eliot ignored Goldberger's discovery of the nature and the etiology of this classic disease of poverty, a discovery that had shifted the responsibility of its prevention and eradication from the medical and health professions to the greater society itself.

With the issuance of the Davenport-doctored final report of the Thompson Pellagra Commission, the more backward (if educated) segments of the greater society now had a very official report—over 400 pages long, with mathematical, genetical, and all-so-very-scientific tables and figures, signed by government physicians such as Garrison and Siler as well as by the man billed in the Sunday newspaper supplements as America's greatest biologist, C. B. Davenport—which functioned as a very scientific and authoritative excuse for doing nothing at all about improving the economic opportunities of the poor southern Nordics who developed pellagra for want of adequate foods. As long as *Pellagra III* gave scientific authority to the lie that pellagra was a matter of bad genes and not bad diet, there were no impelling public health reasons for paying the poor southern whites (let alone the poor southern blacks, who, despite the lore of conventional wisdom, were equally prone to suffer from pellagra) enough money to afford the fresh meats, dairy products, eggs, fresh fruits, and vegetables that both prevented and cured pellagra. Or for the greater society to intervene—in the interests of public health—and provide supplemental pellagra-preventing foodstuffs for the tables of those hard-working southern field and mill workers who could not afford them.

At a time when many southern states were boasting of the low native-white wage scales as they wooed northern industries, this meant that Goldberger not only had to face the opposition of old-time eugenical scientific racists such as Davenport and Osborn: he and the findings of his Public Health Service collaborators had also aroused the angry enmity of the southern (and carpetbagger) employers who were—according to Sydenstricker's Public Health Service statistical data—underpaying and overcharging the white southern Nordics of mill and field. If bad genes in the poor were important and sacred to the Davenports, low wages (and high interest rates and food prices) for the white southern mill hand, small farmer, and hired cotton field hands were equally sacred to the large mill owners, landowners, bankers, and shopkeepers of the well-fed (and therefore pellagra-free), well-shod (and therefore hookworm-free), well-connected and/or well-educated (and therefore with far more ready access to the bounties of what Malthus had termed the feast at Nature's table) southern middle and upper classes.

After 1917, while Goldberger devoted the remaining years of his life to the further study of what he termed the Pellagra Preventive, or **PP**, factor in protein and vegetable foods, and to battle for a more mature and moral societal approach to the eradication of pellagra, Davenport and the politicians and publicists who preferred his eugenical blandishments went on writing articles and making speeches about pellagra being one of the genetic defects of the poor, *and therefore inevitable and beyond prevention.* In the September 1920 issue of the *Psychological Bulletin,* published by the American Psychological Association, in an article entitled "Heredity of Constitutional Mental Disorders," for example, Davenport not only described "Feeble-mindedness," "Criminality," and "The Epilepsies" as genetic conditions, but also included the following comments under the section heading "Pellagra": ". . . there are biotypes [genotypes] in pellagra characterized by severity of one or the other but often not all of the principal symptoms of pellagra, namely, local inflammations of the skin, inflammations of the intestinal tract *and nervous and mental disorders"* (italics added).

Goldberger's biographer, Robert Parsons, writing in 1943, noted that "the medical profession still numbers a few members who steadfastly ignore all the proofs Goldberger adduced. As late as 1929, an eminent physician of New Orleans concluded the leading article in a prominent American publication with this statement: 'All the evidence of which I have personal knowledge, to which I am able to attach much weight, favors the opinion that pellagra is due to an infection. I am content to remain with the minority who have not been convinced by *supposed* proof of other causes, and still believe that a specific infection will be found to be the true cause.' "[22]

In 1932, Henry Pelouze de Forest, Ph.B., M.S., M.D., the adjunct professor of obstetrics at the New York Post-Graduate Medical School and Hospital who in 1912 served as Major Shufeldt's authority on syphilis, presented a paper at a medical meeting entitled "Peanut Worms and Pellagra,"[23] in which he suggested that the Indian meal moth *Plodia interpunctella* was the real cause of pellagra. He was aware of Goldberger's work, and in fact even cited a 1928 paper by Goldberger and Wheeler on how they induced a pellagra-like condition, "black tongue," in dogs. However, Dr. de Forest said on page 29:

"All available articles on the subject in French, German, Italian, Spanish and American publications have been consulted, and while these authorities agree as to the characteristic lesions of the disease and while the diagnosis is reasonably certain when the skin manifestations are pronounced, *no one has yet definitely demonstrated the cause of pellagra*" (italics added).[24]

The Davenport-reified pseudo-description of pellagra became so integral an aspect of American conventional wisdom that, as late as 1938, the medical director of the U.S. Public Health Service, A. M. Stimson, had to include a section on pellagra in his history of the Service's bacteriological investigations. "It is known," wrote Dr. Stimson, "that there are still some die-hards who still cling to the infectious theory of pellagra."[25]

For over a generation, our society was not to lack for leaders who, through ignorance of advances in the biomedical sciences—and/or because of their blind adherence to the dogmas of eugenics and other forms of scientific racism—proved unable or unwilling to use the powers of their elective offices to make the benefits of Goldberger's pellagra findings available to every American infant, child, and adult. It is one of the crowning ironies of modern American history that, in the end, pellagra was to be conquered in the United States not by the wise and universal application of Goldberger's discoveries to American social policies, but, rather, by the fallout of the Great Depression of 1929–41.

Once the stock market crash of 1929 triggered the Great Depression that was to last until we entered World War II, and the federal government— with the defeat of Herbert Hoover's bid for re-election—started disbursing food and welfare funds to save the lives of all people rendered hungry and helpless by the economic breakdown, Goldberger's PP factor started to flow to all the hungry, the chronically malnourished of the land. Once the bankers, the manufacturers, the merchants, the office managers, the white-collar hucksters, and the authors of nasty editorials about the hereditary shiftlessness of the South's "poor-white trash" were now themselves—by the hundreds of thousands—bankrupt, penniless, and hungry, nobody raised Malthusian objections that "at Nature's mighty feast there is no vacant cover" for the unemployed and the starving. Suddenly, the born (or undeserving) poor and the Depression-made (or deserving) poor were in the same boat—and it was the same federal relief programs that protected both classes of American poor from pellagra. (See table on page 224.)

Deaths from Pellagra by Color in the United States

Year	Total Deaths	White People	All Other People	Historical Events and Social Conditions During Same Years
1900	2			Most American doctors unaware of pellagra.
1914	847			Goldberger discovers "that pellagra is a vitamin deficiency disease analogous to scurvy and beriberi."
1915	1,058			Goldberger shows how to cause, cure, and
1916	1,807			prevent pellagra. Davenport publishes "The Hereditary Factor in Pellagra" in the *Archives of Internal Medicine*.
1917	2,843			*Pellagra III*, final report of Pellagra Commission, describes pellagra as an infectious disease affecting primarily white people genetically susceptible to its "germ."
1918	3,126			
1919	2,568			
1920	2,122			
1921	2,348			
1922	2,514			
1923	2,245	1,143	1,102	First year U.S. Public Health Service records pellagra deaths of nonwhites. The black 10% of America's population prove to suffer 50% of its pellagra mortalities.
1924	2,206	1,086	1,120	
1925	3,049	1,384	1,665	
1926	3,501	1,724	1,777	
1927	5,091	2,351	2,740	
1928	6,523	2,689	3,834	Death of Joseph Goldberger.
1929	6,623	2,781	3,842	Stock market crash triggers decade of the Great Depression, impoverishing the nation.
1930	6,106	2,722	3,384	
1934	3,602	1,914	1,688	Start of federal work and food relief in 1933 improves diets of poor families. Start of TVA provides many new jobs.
1935	3,543	1,963	1,580	Cheap electric power from new TVA dams
1936	3,740	2,129	1,611	gives South new grain mills, new markets
1937	3,258	1,804	1,454	for local grain, new dairy and poultry in-
1938	3,205	1,707	1,498	dustries, and lowers prices of niacin-bearing foods.
1939	2,419	1,404	1,015	Start of World War II in Europe creates new factories and jobs in the South.
1940	2,123	1,270	853	
1941	1,836	1,137	699	America enters World War II.
1968	15	12	3	During and after World War II, the diets of the southern poor are improved, as continuing industrialization of the South creates more jobs for white and nonwhite poor. Civil Rights Act opens doors to better jobs for southern blacks.[26]

Mortality data from U.S. Public Health Service Center for Disease Control, Atlanta, Georgia

The Great Pellagra Cover-Up of 1916–33, which kept the medical benefits of Goldberger's work on pellagra from the entire nation for two decades, was the greatest triumph of scientific racism since 1826—when Thomas Malthus denounced as reprobates those doctors who devised "specific remedies for ravaging diseases; and those benevolent, but much mistaken men, who have thought they were doing a service to mankind by projecting schemes for the total extirpation of particular disorders."

As every doctor knows, had this society, acting on Goldberger's findings, totally extirpated pellagra by making certain that the wages of the poor allowed them to buy all of the foods needed to prevent pellagra, the same benevolent social action would have set off a series of clinical reactions that, at the same time and for the same investment, would have had equally salubrious effects on scores of other particular disorders of malnutrition. Such as, to name but five groups of the ravaging diseases and disorders of chronic malnutrition: (1) low birth weight; (2) inadequate growth levels during the "period of human development extending from the second trimester of gestation well into the second postnatal year, during which the brain appears to have a once-only opportunity to grow properly"[27]—the period when the human brain achieves 75 percent of its ultimate mature weight; (3) the tragically high rates of often fatal infectious diseases, from measles and diphtheria to pneumonia and tuberculosis, that are always associated with chronic undernourishment the world over; (4) the growth-retarding effects of lifelong malnutrition on the pelvis of the human female, which makes cephalo-pelvic disproportion (CPD) a major cause of injuries during childbirth to the brains and bodies of the children of the world's poor; and (5) the staggeringly high incidences of infection-caused and often fatal dysenteries and diarrheas in the malnourished children of the poor.

It would not be overstating the seriousness of the clinical effects of the Great Pellagra Cover-Up launched by Davenport to say that, in terms of preventable morbidities and mortalities in only these five categories of common diseases and disorders known to be exacerbated by chronic malnutrition, they added up to millions of completely avoidable premature deaths, chronic degenerative diseases, deformations, and otherwise needlessly wasted lives.

The social and historical effects of the Great Pellagra Cover-Up were soon to be eclipsed by two even more crushing victories of scientific racism over scientific truths and moral values. These new triumphs of American scientific racism were to seriously affect the lives and the destinies of millions of human beings, some already born and most of them yet to be born, over the next half century. These two events—the imposition of eugenical perversions of Alfred Binet's benign mental tests on the social decision-making organs of American society, and the U.S. Congress' use of these so-called human intelligence quotient (IQ) test scores in the establishment of America's first exclusionary racial quotas, in the U.S. Immigration Act of 1924—will be examined in the next four chapters.

10 The Ultimate Weapon of the Old Scientific Racism: The IQ Testing That Mislabels 75 Percent of All Adult Americans as Having, by Heredity, the Intelligence of Children Ten to Fourteen Years of Age

> Assuming that these 1,700,000 men are a fair sample of the entire population of approximately 100,000,000, this means that the *average* mental age of Americans is only about 14; that 45 millions, or nearly one-half the whole population, will *never* develop mental capacity beyond the stage represented by a normal 12-year-old child; that only 13 and one-half millions will ever show superior intelligence, and that only 4 and one-half millions can be considered "talented."
>
> —LOTHROP STODDARD, PH.D., on the results of the Army I.Q. tests of World War I, in *The Revolt Against Civilization: The Menace of the Under-Man* (1922)

It is an unfortunate but stubborn fact of American cultural history that for more than a half century following the administration of intelligence tests to 1,726,966 recruits of World War I, the educational policies of this nation were to be based—at their core—on the astonishingly long-lived credibility of the pseudofact that the results of these very "scientific" mental tests *proved* that "the average mental age of all Americans is only about 14."[1] And that, as the chairman of Harvard's Department of Psychology, William McDougall, wrote in his best-selling book of 1921, *Is America Safe for Democracy?*, "the results of the army tests indicate that about 75 percent of the [American] population has not sufficient innate [hereditary] capacity for intellectual development to enable it to complete the usual high-school course. The very extensive [IQ] testing of school-children carried on by Professor [Lewis M.] Terman and his colleagues leads to closely concordant results."[2]

To be sure, the 1,726,966 Army mental test scores were neither as exact, as objective, nor as definitive as Boas' 18,000 measurements of the bodily forms of immigrants. However, they made up in quantity and in universal popular acceptance what they lacked in quality and in real scientific significance when they were stacked against the objective anthropometric measurements of Boas and his associates that had destroyed the last shreds of credibility still clinging to the cephalic index (head-form) myths after Virchow and before Boas.

In the simplistic traditions of scientific racism, they were the logical successors to the cephalic indexes, just as the cephalic index booby trap was the scientistic heir to the mystique of phrenology.

To be sure, the raw data on the Army mental tests, as published by the U.S. Government Printing Office in 1921,[3] even then gave the lie about the purportedly racial and genetic interpretations of the 1917–19 IQ test scores. Few American scholars seemed very interested in the simple truths, originally. But the facts were there for the impartial investigator, such as Otto Klineberg, M.D., Ph.D., and professor of psychology at Columbia University.

By the time Klineberg published his studies of IQ test scores between 1931 and 1935,[4] it was already no secret that, in broad strokes, the Army intelligence test scores showed that: (1) those draftees with the most years of education to their credit scored the highest marks; (2) those draftees, white and black, from those states—chiefly southern states—that spent the least amount per capita on the education of their children invariably scored the lowest marks; (3) among the foreign-born draftees—quite predictably —the longer a foreign-born draftee had lived in this country, and the greater the amount of his pre-draft schooling in American schools, the higher was his IQ test score; and (4) *naturally enough, considering that the tests were written and administered in English,* immigrants from England, Scotland, and other English-speaking countries did, on the whole, somewhat better than draftees who had recently learned English as adults.

In short, what the Army intelligence tests—like the civilian IQ tests written by precisely the same psychologists—measured were not genetic mental levels but, rather, the degrees to which the genotypic intellectual potentialities of 1,726,966 *individual* draftees developed in continuing inter- action with the entire spectrum of widely varying physiological, nutritional, hygienic, emotional, cultural, climatic, socioeconomic, and thousands of other factors that make up the total environment in which every *individual* fetus and child grows and develops. Whatever else it proves to be, *intelli- gence, or whatever is actually measured by the IQ tests, is not a genotype—* like species or eye color—but a phenotype, like height, weight, or health.

Over and above the quite predictable Army mental test scores that demonstrated the phenotypical nature of intellectual development, there were the socioeconomic and sociocultural findings that matched in every significant detail those made by Goldberger, the physician, and Sydenstricker, the Public Health Service economist, as they studied the gaps between wages and family food-purchasing power in their still-continuing pellagra studies. That is, as we saw in chapters 8 and 9, the poorest, most malnourished, and biologically most deprived segments of the population led the nation not only in hookworm and roundworm infections, pellagra, infant mortality and general ill health, but also in the low classroom and IQ test scores that are, like pellagra and hookworm, two of the most ubiquitous products of poverty.

In his book *Race Differences* (1935), which was dedicated to his mentor and friend Franz Boas, Klineberg published two now famous tables based solely on the data in the official report on the World War I mental test scores. The first dealt with "the rather close correspondence between the amount of education and standing in the tests. This was demonstrated clearly by the Army testers, whose results are summarized" in the first table below.

The next table, which dealt with the observed differences between the IQ test scores of Negro and white recruits, noted that while on a nationwide

Alpha Scores and School Grade Completed (Army Results)

Group	0–4	5–8	High School	College	Beyond College
White officers	112.5	107.0	131.1	143.2	143.5
" native-born	22.0	51.1	92.1	117.8	145.9
" foreign-born	21.4	47.2	72.4	91.9	92.5
Colored, north	17.0	37.2	71.2	90.5	
" south	7.2	16.3	45.7	63.8	

average "the Negroes rank below Whites in most intelligence test studies," it was also "well known, for example, that during the war the Army testers found Negro recruits from the North far superior to Negroes from the South, and, in the case of certain northern states, superior also to southern Whites." This is shown in the following table:

Southern Whites and Northern Negroes, by States, Army Recruits

Whites		*Negroes*	
State	Median Score	State	Median Score
Mississippi	41.25	Pennsylvania	42.00
Kentucky	41.50	New York	45.02
Arkansas	41.55	Illinois	47.35
Georgia	42.12	Ohio	49.50

The scientific racists did, of course, have a pseudogenetic explanation for the fact that the scores of northern Negroes were higher, on the average, than those of Negroes who remained in the South. This was the myth of "selective migration"—that Negroes of superior blood (i.e., having more white genes in their veins) tended to migrate to the North more than did their southern kin with fewer white genes in their genomes. The selective migration thesis was, however, *not* used to explain the fact that white southern recruits from the states of the Old Confederacy scored consistently lower in the Army IQ tests than did the equally Nordic, equally Protestant, equally white northern recruits. Nor, of course, did it even begin to explain why the white recruits from the four southern states (Mississippi, Kentucky, Arkansas, and Georgia) *with abysmally low state educational expenditures per pupil* racked up Army IQ test scores significantly lower than those achieved by black recruits from Pennsylvania, New York, Illinois, and Ohio, *which spent considerably more on the education of their children and adolescents* than did the four southern states in Klineberg's table.

That such studies by Klineberg and his predecessors had about as little effect upon the value judgments of the makers of American educational policies as, two decades earlier, the findings of Goldberger, Sydenstricker, and the rest of the Public Health Service team on the causes and prevention of pellagra had upon the conventional wisdom of the makers of America's social and public health policies is, of course, tragic testament to the profound effects of the propaganda of scientific racism on the decision-making organs of our society.

SOME FORGOTTEN EPISODES IN THE PREHISTORY
OF THE ARMY INTELLIGENCE TESTS

The IQ tests of World War I did not spring fully formed into our culture. They evolved, in a general way, from the physiological reaction time tests of Francis Galton, and in a very specific way from the pioneering efforts of one of the founders of experimental psychology, Alfred Binet, to devise "a series of tests to apply to an individual in order to distinguish him from others and to enable us to deduce general conclusions relative to certain of his habits and faculties."[5]

Galton's batteries of "mental" tests were actually tests of physical development. He measured the reaction times of the eyes, ears, and brains of people to different colors, sounds, smells, touches—all of them reactions conditioned by the general health of the individual tested. Healthy people in good health from birth will, of course, have better visual, auditory, and olfactory senses than less healthy people.

As Professor J. McVicker Hunt observed, Galton "saw that if [eugenical] decisions were to be made as to which human beings were to survive and reproduce it would be necessary to have some criteria for survival. So he formed his anthropometric laboratory for the measurement of man, with the hope that by means of tests he could determine those individuals who should survive. Note that he was not deciding who should be selected for jobs in a given industry, but *who should survive to reproduce*" (italics added).[6]

The founder of the eugenics movement never actually called his physiological reaction time tests "mental tests." It was an American graduate student working under the master, James McKeen Cattell—who described Galton as "the greatest man I have known"—who coined the term "mental tests" for what Galton started, and what Cattell was to enlarge into a much broader battery of related measurements. Cattell's "mental tests" covered such major items as "the greatest possible squeeze of the hand" and the height, weight, and memory of the students tested.

Cattell was not the only behavioral scientist to create mental tests based on the concepts of Francis Galton. There were many others, of whom two stand out in history for various reasons. One was R. Meade Bache, an American; the other was K. T. Waugh.

Bache published first. His "Reaction Time with Reference to Race"[7] was based on the use of the most up-to-date "electro-magnetic physiological apparatus." Bache tested three groups of males: 12 Caucasians, 11 American Indians, and 11 American Negroes. The average ages of each group were 19 for the Caucasians, 17½ for the Indians, and 23 for the Negroes (or, as Bache called them, the African Race).

These three groups were tested for the speed with which they reacted to the sight of a pendulum, the sound of something Bache did not identify, and "a slight electric shock given to the wrist of the subject."

The results showed the Indians to have the fastest reaction times, and the Caucasians the slowest. The blacks fell neatly between the two other groups. What this meant to Bache was:

The popular notion that the more highly organized a human being is, the quicker ought to be the response to stimuli, is true only of the sphere of higher thought, not at all that of auditory, visual, or tactile impressions, which invite secondary reflex action. As here stated [in] response to such stimuli . . . the most ordinary intelligence should suffice for its exercise; *in proportion to intellectual advancement, there should be,* through the law of compensation, *a waning in the efficiency of the automatism of the individual* [italics added].

In plainer English, this meant, simply, that the smarter the individual, the more developed his intellect, *the slower his reaction time to ordinary physiological stimuli.* As Bache went on to observe:

Pride of race obscures the view of the white with reference to the relative automatic quickness of the negro. That the negro is, in the truest sense, a race inferior to that of the white can be proved by many facts, and among these by the quickness of his automatic movements as compared with those of the white.

Since Alfred Binet was one of the editors of the journal in which the Bache article appeared, it is not inconceivable that he had Bache, among others, in mind eight years later when he wrote in his *Etude experimentale de l'intelligence* (1903) that "we must always be hospitable to the facts that go counter to our theories."

Binet was not alive to comment when Waugh presented his paper "Comparison of Oriental and American Student Intelligence" at the 1920 meeting of the American Psychological Association in Chicago. However, William McDougall, then professor of psychology at Harvard, was aware of this work, and to him we owe the description of its contents and significance.

What Waugh did was to test students in four colleges of British India, a Chinese college, and some American colleges. The tests, McDougall wrote,[8] "were largely concerned with memory, and were not well suited to test intellectual capacity. They revealed only slight differences, which were slightly in favor of the Indian students—except in one quality, namely the power of concentrating the attention."

Mr. Waugh's results were summarized by McDougall, as follows:

Functions Tested	American	Chinese	Indian
Concentration of attention	75	75	62
Speed of learning	66	62	45
Association-time	46	38	58
Immediate memory	58	38	54
Deferred memory	80	38	88
Range of information	23	15	24

Professor McDougall's comments can best be described only by quoting them accurately. Waugh's results, he wrote:

. . . seem to me extraordinarily interesting and suggestive. For what is this power of concentrating attention? It is essentially will-power. I need only remind you of what William James wrote of this. ("Effort of attention is thus the essential phenomenon of will," *Principles of Psychology,* v. II, p. 562.) Now the more or less orderly and successful government of the three hundred millions of India by a mere handful of British men, during more than a century, is one of the most remarkable facts in the history of the world. . . . Englishmen have marvelled over it. And, when they have sought to explain how it has been possible, they have always come to the same conclusion. They have recognized that the natives of India, or very many of them, have much intellectual capacity; that they are clever, quick, versatile, retentive; that some of them have brilliant intellects. But such observers have frequently expressed the opinion that, *as compared with their British rulers, the natives of India are relatively defective in character or will-power;* and they have found the explanation of British ascendancy in this fact. Now, at the very first attempt [the Waugh study] to apply exact methods in the comparative study of Indians, this opinion finds confirmation. . . .

In this connection let me remind you that the quelling of the Indian Mutiny was, before all things, a triumph of will-power. If even a few of the British leaders . . . had failed, even so little, in the supreme tests of will-power from which they came out triumphant, the British would have been swept from the country, and British rule in India would have been brought to an end about the year 1857 [italics added].

McDougall's interpretation was all properly Pukka Sahib, which was quite fitting for the very proper Englishman that he was. It simply happened to be scientific and historical kitsch. But it was a brand of kitsch that served him well, first as professor of psychology at Oxford, then at Harvard, and finally at Duke University in North Carolina, where he pioneered in studies of the pseudoscience of parapsychology, and continued his lifelong efforts to prove that the Lamarckian hypotheses of the inheritance of acquired characteristics were valid.

The mental tests that succeeded the Galtonian types of reaction tests were originally called, in this country, the Binet tests—which made them a libel against the scientific work and philosophy of Alfred Binet. The mental tests of Binet and his student Théodore Simon, published between 1905 and 1911, were designed to appraise the mental or learning capacities of *individual* students. A superbly trained scientist, with a doctorate in biology, a student under Charcot (along with foreign physicians like Sigmund Freud and French scientists like Richet and Janet), Binet was the father of two daughters, Madeleine and Alice. As have thousands of parents before and since, Binet was made acutely aware of how very different, psychologically, were these children of the same parents, the same environment, the same social history.

A pioneer in differential psychology, Binet studied the different manners in which his daughters learned anything. He was able to find in these sisters two distinct patterns of learning, which he termed the "objective" and the "subjective" types of learning. He also suggested that in people there exists a vague characteristic or complex of traits that, *for want of a better word,* he named "general intelligence." But he was never able to arrive at an exact definition of intelligence that satisfied his own rigid scientific criteria.

In this, Binet, who died at the age of fifty-four in 1911, was more in tune with modern psychologists than were some of his more simplistic contemporaries. In the 1968 edition of the *International Encyclopedia of the Social Sciences,* for example, Professor Robert L. Thorndike (son of E. L. Thorndike) wrote that "in a general way everyone knows what intelligence or intelligent behavior is. But psychologists have had little success in reaching a definition in verbal terms that is much more precise and satisfactory than the common-sense *understanding* of the term held by laymen."

It was not Binet who called the Binet tests "intelligence quotient (IQ) tests." In the Binet scales, the questions were graded in terms of the age level at which the person tested passed all of the tests. The norms for the age levels were based on the average scores made by the selected groups of children of various ages on whom the tests were first tried. In 1912, William Stern, a German psychologist, later driven from Germany because he was a Jew, suggested that when the mental age (MA) was divided by the chronological age (CA) of the person tested, and multiplied by 100, the end result could be considered the "intelligence quotient," or IQ. This meant that the IQ could be stated in mathematical terms as:

$$100 \; \frac{\text{mental age (MA)}}{\text{chronological age (CA)}} = \text{intelligence quotient (IQ).}$$

This was also most unfortunate, because in practical terms it came to mean—before anyone ever knew what intelligence was, or even if it was a single measurable trait, or a combination of factors—that our entire culture was burdened with a very scientific-looking formula *purporting* to represent what the innate intellectual capacity of a human being actually is.

GODDARD AND TERMAN EUGENICIZE BINET'S BENIGN INTELLIGENCE TESTS

One American who had no doubt at all about what intelligence is, and how to measure it, was Henry Herbert Goddard, the psychologist who named and chronicled the Kallikaks of good blood and bad. Goddard translated and adapted two versions of the Binet-Simon tests, in 1908 and 1911, and started using them on the children in the Vineland, New Jersey, Training School for Feeble-minded Girls and Boys. But Goddard agreed with none of Binet's concepts or precautions. As Read Tuddenham observed of Goddard:

> Probably no one else had so much to do with launching the Binet method in the United States. Yet it often happens that the devoted disciple transforms the ideas of the prophet in the very process of transmitting them. So it was in this case.
>
> Accepting Binet's empirical *method,* he substituted for Binet's *idea* of intelligence as a shifting complex of interrelated functions, the concept of a single, underlying function (faculty) of intelligence. Further, he believed that this unitary function was largely determined by heredity, a view much at variance with Binet's optimistic proposals for mental orthopedics.[9]

Goddard was the first of the true-believing eugenicists to perceive what a mighty weapon, indeed, the new Binet intelligence tests could be in the holy crusade to keep America *Judenrein*. By 1912, he was visiting Ellis Island in New York Harbor, then the largest immigration station in the land, to determine whether the new intelligence tests could be used as bulwarks against the despised Jews, Italians, Hungarians, Poles, Russians, and other bearers of the blood of races commonly "known" to be inferior. Goddard's "Binet tests," as *he* interpreted the results, not only "proved" that they could, indeed, spot and stigmatize as ineligible for admission those prospective immigrants who happened to be "mental defectives." Goddard's administration and interpretation of the same tests also laid bare "the inability of these [immigrant] children to handle abstractions."[10] He was invited by the Immigration Commission to come back with some of the young ladies he had trained at Vineland to be professional "Binet testers."

The Binet tests of steerage immigrants at Ellis Island of course "proved" everything that Galton and his true believers had always insisted to be the truth about the inborn human worth of Jews, Catholics, southern Europeans, and other indubitably inferior races and strains of mankind. In no time at all, the Goddard versions of the Binet tests were convincing thousands of educated Americans that 83 percent of all Jews were feebleminded, a "hereditary" defect the Jews shared with 80 percent of all Hungarians, 79 percent of all Italians, and 87 percent of all Russians.

Goddard summed it all up very neatly in a table in his report on the 1913 Ellis Island foray, "Mental Tests and the Immigrant," which was published in the September 1917 issue of the *Journal of Delinquency*.

The Journal of Delinquency

Volume II.	SEPTEMBER, 1917	Number 5

MENTAL TESTS AND THE IMMIGRANT
HENRY H. GODDARD, Ph.D.
Director of Research, Training School, Vineland, N. J.

SUMMARY

1. This is a study not of immigrants in general but of six small highly selected groups, four of "average normals" and two of apparent "defectives," all of them steerage passengers arriving at Ellis Island.

2. The study makes no determination of the actual percentage, even of these groups, who are feeble-minded.

3. It seems evident that mental tests can be successfully used on immigrants, although much study is still necessary before a completely satisfactory scale can be developed.

4. One can hardly escape the conviction that the intelligence of the average "third class" immigrant is low, perhaps of moron grade.

5. Assuming that they are morons, we have two practical questions: first, is it hereditary defect or; second, apparent defect due to deprivation? If the latter, as seems likely, little fear may be felt for the children. Even if the former, we may still question whether we cannot use moron laborers if we are wise enough to train them properly.

Intelligence Classification of Immigrants of Different Nationalities

	Normal		Borderline		Feeble-minded		Moron		Imbecile	
	No.	Percent	No.	Percent	No.	Percent	No.	Percent	No.	Percent
Jews	3	10	2	7−	25	83+	23	76	2	7
Hungarians	0	0	4	20	16	80	16	80	0	0
Italians	3	7−	7	15−	38	79	38	79	0	0
Russians	0	0	4	9	39	87	37	82	2	2.5
Italian F. M.	0	0	1	5+	17	94+	12	63	6	32−
Russian F. M.	0	0	0	0	18	100	14	78−	4	22+

Data and headings from Goddard, *Journal of Delinquency,* 1917.

Whereas Goddard was the first American eugenicist to use the new intelligence tests against the less suitable races, the honor of being the first eugenicist to employ the new IQ tests to certify the children of the less suitable strains of native-born American Nordics as hereditarily "uneducable" went to Lewis M. Terman of Stanford University. The Terman version of the Binet test—better known as the Revised Stanford-Binet—was much more than a mere *revision* of Binet's 1911 scale. According to Terman's own description of his Revised Stanford-Binet IQ test, "after making a few necessary eliminations, 90 tests remained, *or 36 more than the number in the Binet 1911 scale*" (italics added).[11] The basic difference between Terman's and Binet's scales was that Terman's questions were based much more directly on the amount of formal and informal education (that is, school plus family libraries and cultural activities, peer group and community experiences) the children had received prior to taking the new American IQ tests.

Naturally, by 1916 Terman's IQ tests revealed:

It has, in fact, been found wherever comparisons have been made that children of superior social status yield a higher average mental age than children of the laboring classes. . . . In the case of the Stanford investigation, it was found that when the unselected school children were grouped in three classes according to social status (superior, average, and inferior), the average IQ for the superior social group was 107, and that of the inferior social group 93. This is equivalent to a difference of one year in mental age with 7-year-olds, and to a difference of two years with 14-year-olds.[12]

As a true-believing eugenicist, Terman held as an article of faith that this spread between the IQ test scores of the children of lawyers, professors, and bankers and those of the children of the laboring classes was genetically ordained. For in the very next paragraph he hastened to add:

. . . the common opinion that the child from a cultured home does better in tests solely by reason of his superior home advantages is an entirely gratuitous assumption. Practically all of the investigations which have been made of the influence of nature and nurture on mental performance agree in attributing far more to original [i.e., genetic] endowment than to environment. Common observation [*sic*] would itself suggest that *the social class to which the family belongs depends less on chance than on the parents' native [hereditary] qualities of intellect and character* [italics added].

By "character" Terman meant the sum of such things as morality, popularity, grit, chastity, criminality, will power, conscientiousness, and other behavioral patterns classified as both unitary and hereditary "character traits" by Galton and his followers. Terman, also, in his 1919 book, *The Intelligence of School Children,* indulged in the traditional eugenical passion for graphs, charts, equations, and tables to plot the statistical relationships between such "hereditary unit character traits" and IQ test scores.

The social implications of IQ test scores of children in terms of such factors as the educational priorities and policies of tax-supported school systems, and the right of less suitable, *even if Nordic,* strains to have children, were not lost on Professor Terman. As early as 1916, in *The Measurement of Intelligence,* he asked, "What shall we say of cases" in which children "test at high-grade moronity or at the border-line, but are well enough endowed in moral and personal traits [i.e., they are the properly respectful children of the Deserving Poor] to pass as normal in an uncomplicated social environment?"

This was indeed a knotty problem, and Terman did not flinch from facing it soberly:

Among laboring men and servant girls there are thousands like them. They are the world's "hewers of wood and drawers of water." And yet, as far as intelligence is concerned, the [IQ] tests have told the truth. These boys are *uneducable* beyond the merest rudiments of training. *No amount of school instruction will ever make them intelligent voters or capable citizens* in the true sense of the word. Judged psychologically they cannot be considered normal. [Italics added.]

In many instances, as when Terman's collaborators and contemporaries tested Spanish-Indian and Mexican children in the Southwest, and Negro children in California, the IQ test scores revealed:

. . . *Their dullness seems to be racial,* or at least inherent in the family stocks [i.e., strains of given races] from which they come. The fact that one meets this type with such extraordinary frequency among Indians, Mexicans, and negroes suggests quite forcibly that the whole question of *racial differences in mental traits* will have to be taken up anew and by experimental methods. The writer predicts that when this is done, there will be discovered enormously significant racial differences in general intelligence, differences which cannot be wiped out by any scheme of mental culture.

Children of this group should be segregated in special classes and be given instruction which is concrete and practical. They cannot master abstractions, but they can often be made efficient workers, able to look out for themselves. There is no possibility at present of convincing society that *they should not be allowed to reproduce, although from a eugenic point of view they constitute a grave problem* because of their unusually prolific breeding. [Italics added.][13]

In the same book, Terman also proposed (p. 79) an IQ test score scale, based on the Stanford Revision of the Binet-Simon scale, for the *clinical* diagnosis of human mentality.

This IQ test score scale read:

IQ Test Score	Classification
Above 140	"Near" genius or genius.
120–140	Very superior intelligence.
110–120	Superior intelligence.
90–110	Normal, or average intelligence.
80– 90	Dullness, rarely classifiable as feeble-mindedness.
70– 80	Border-line deficiency, sometimes classifiable as dullness, often as feeble-mindedness.
Below 70	Definite feeble-mindedness.

Terman, like Goddard, had taken his doctorate in psychology under G. Stanley Hall at Clark University. Hall, a disciple of Galton and Spencer, was the foremost American exponent of the concept of genetically predetermined behavioral traits.[14]

Scientifically, both Goddard and Terman were Hard-Shell Preformationists. That is, they were each true-believing eugenicists, but all that had changed in Preformationism since 1755 was that the microscopic "homunculi" were now "unit characters," and "intelligence" one of these inherited "unit characters" that were fixed for life at the moment of birth. Therefore the function of *their* variations of "Binet testing" was to measure one's *inherited and fixed* intelligence, rather than to measure the qualities of the cultural and biological life experiences reflected in the answers of children and adults to tests of what they had learned since birth.

Binet himself railed against those who had, as had Terman and Goddard after Binet's death in 1911, "given their moral support to the deplorable verdict that the intelligence of an individual is a fixed quantity."[15]

Such statements caused Binet to "protest" against "this brutal pessimism" about the inborn mental potential of most human beings, since he felt that "a child's mind is like a field for which an expert farmer has advised a change in the method of cultivating, with the result that in place of desert land, we now have a harvest."[16]

TERMAN'S IQ TESTING OF THE DEAD "PROVES" GALTON TO BE 65 POINTS SMARTER THAN DARWIN

Binet was never quite satisfied with the adequacy of his tests. Terman, by contrast, was so certain that his versions of what he called "Binet-Simon tests" had solved the problems of measuring the intelligence of all living children that, after World War I, he and his associate, Catherine Morris Cox, solemnly administered their IQ tests to Beethoven, Darwin, Goethe, Balzac, Washington, Napoleon, Lincoln, and other folks who had died long before the Revised Stanford-Binet IQ test had ever been written.[17]

The methods Terman and Cox used to "test" the intelligence quotients of the illustrious dead were, essentially, identical to those previously devised by Goddard and Davenport for their eugenics field workers. In sum, Terman and Cox dug up whatever written records were available—from school grades and compositions to the memoirs of their subjects and their contemporaries

—and combined such information and gossip with their own highly subjective evaluations of both the work and the personalities of such people as Copernicus, Tennyson, Faraday, Kant, Racine, and Calvin. Since, as a true-believing eugenicist, Terman saw Francis Galton as a far greater scientist than Galton's half cousin Charles Darwin, it was not very difficult by these methods of testing the dead for Terman to confirm his more or less religious belief.

By 1917, Terman had not only managed to measure the IQ of Sir Francis Galton, but had also determined that it came to a figure "unquestionably in the neighborhood of 200, a figure not equalled by more than one child in 50,000 of the generality."[18] In Volume II (1926) of his five-volume *Genetic [sic] Studies of Genius,* Terman and his associates had given Galton's half cousin Charles Darwin an IQ test score of only 135.

Other noted dead who scored a full 65 points less than the great Galton, *when their IQ's were tested by Terman, Cox, et al.,* included: Beethoven, Erasmus, Alexander Hamilton, and Leonardo da Vinci.

Their IQ test scores—which all fell below the IQ score of above 140 that, according to Terman's scale, classified one as being a " 'near' genius or genius"[19]—were still higher than the range of 100 to 110 ("normal, or average intelligence") awarded Copernicus, Cervantes, and Faraday; or the "superior intelligence" range of 110 to 120 achieved by Oliver Cromwell and Rembrandt; and even the IQ test scores in the 120 to 130 ("very superior intelligence") range achieved by Berzelius, William Harvey, Lavoisier, and George Washington. But the IQ test scores awarded Darwin and his intellectual peers by Terman also left them far behind Henry Wadsworth Longfellow, Auguste Comte, Alfred Tennyson, and Alfred Pope, who, *from their tombs,* scored in the 150 to 170 IQ range.

That the author of these sepulchral excursions into storefront spiritualist-medium territory—and his so-called intelligence quotient tests—are both still taken very seriously by so many modern psychologists and educators is, again, living testimony to the lasting power of scientific racism over the minds of educated Americans.

Terman and Goddard—who both regarded the eugenics field workers trained by Davenport and Goddard himself as serious scientific investigators —shared another common illusion about IQ testing. This was the belief that the tests were so foolproof that anyone who could read and write could administer them. As Goddard wrote in 1913 and 1914, the Binet (Goddard version) scale is "wonderfully accurate even down to a variation of only one or two points." It was, he wrote, so efficient that "in most cases the Binet scale itself is ample for the purpose of detecting the feeble-minded."

At Vineland, where nice ladies who wanted to be Binet testers were trained in a special six-week course devised by Goddard, the famous psychologist wrote in the Training School Bulletin of October 1914 that "even novices may use the Binet scale, provided they use it with ordinary good sense, [and] get in many cases all that is needed." During the second decade of this century in America, this meant that novices in psychology, after a few weeks of training, were employed to give people Binet tests to determine whether or not they should be colonized, or involuntarily committed to the state "colonies" for the feebleminded and epileptics.

This use of novices and amateurs in psychiatry to make clinical diagnoses of mental retardation met with surprisingly little resistance among the psychologists. But not all of them thought that Goddard's and other nice ladies who administered Binet tests had either the clinical ability or the moral right to make clinical diagnoses. And one of the most articulate of the professional dissenters was Dr. J. E. Wallace Wallin, a clinical psychologist who, like Goddard and Terman, had trained under Hall at Clark University.

THE SUDDEN DEATH AND QUICK REBIRTH OF THE IQ TESTING MYTHS

Wallin was no stranger to Goddard or Terman. In 1910, as a matter of fact, Goddard had hired him to give a summer course in functional pyschology at the Training School in Vineland while Goddard toured Europe, and in October of the same year Goddard and his Vineland associates recommended Wallin to the directors of the New Jersey Village for Epileptics, at Skillman, where Wallin established the world's first clinical psychology laboratory in any institution for epileptics.

Very early in the course of what was so grievously mislabeled "Binet testing" in America, Wallin started the first of his "long-continued criticisms of amateur diagnosis by Binet testers . . . who had pursued only a six weeks' practicum in Binet testing (the standard practice at that time), who did not possess any fundamental training in psychology or psychiatry." Of the Binet testers who were also classroom teachers, Wallin wrote, most had "pursued only a two years' normal school course."[20]

Somehow, Wallin managed to conduct some "nationwide investigations" of Binet testing in America's school systems in 1913. As he described them:

> I was not fighting a man of straw. In 72% of the 103 school systems out of the 1,350 that supplied the information, individual psychological testing included only the Binet and one or two form boards. Most of the schools that did not reply gave no individual tests at all. Of 115 examiners or testers, 52 were special class teachers, 11 were principals or supervisors of special classes, four were superintendents, five were alienists, 22 were medical inspectors or physicians, eight were psychologists, and 13 were clinical psychologists. On the basis of the data submitted, I felt justified in classifying 74% of the [IQ] examiners as mere "Binet testers" or amateurs.[21]

Needless to say, this investigation did not endear Wallin to the IQ simplicists and/or the eugenicists, who saw in the so-called Binet tests a scientific technique of proving the genetic superiority of some people and the inferiority of most of the others. But Wallin was not out to win popularity contests: he was determined, rather, to protect innocent children and adults from being caged up as menacing feebleminded morons at the hands of unqualified amateurs using only the new Goddard and Terman versions of what *they* called Binet tests to make these clinical psychological diagnoses. And at the 1915 meeting of the American Psychological Association, in Chicago, Wallin drew some real blood from the IQ cult.

A native of Iowa, Wallin had decided to do a little Binet testing in his boyhood haunts during his summer vacation from his job as director of the St. Louis Board of Education's Psycho-Educational Clinic. He described the results of these IQ tests of prosperous farmers, businessmen, and housewives he had known all his life in a paper read before the APA meeting, and subsequently published in the January 1916 *Journal of Criminal Law and Criminology* as "Who Is Feeble-Minded?"

When tested by the 1908 and the 1911 versions of the Goddard-Binet test, *all* of the successful and wealthy Iowans proved to be morons and dangerous feebleminded imbeciles *by the standards of the Goddard tests.* As Wallin told the APA delegates:

> Measured by the standards of one of the best rural communities of the country, socially and industrially considered, and by my own intimate knowledge of the subjects tested during the greater part of my life, not a single one of these persons could be considered feeble-minded. All are law-abiding citizens, eminently successful in their several occupations, all except one (who is unmarried) being parents of intelligent, respectable citizens.

There was, for example, Mr. A, described on pages 708–09:

> Mr. A, 65 years old, faculties well preserved, attended school only about 3 years in the aggregate; successively a successful farmer and business man, now partly retired on a competency of $30,000 (after considerable financial reverses from a fire), for ten years president of the board of education in a town of 700, superintendent or assistant superintendent of a Sunday school for about 30 years; bank director; *raised and educated a family of 9 children, all normal; one of these is engaged in scientific research (Ph.D.); one is assistant professor in a state agricultural school; one is assistant professor in a medical school (now completing thesis for Sc.D.); one is a former music teacher and organist, a graduate of a musical conservatory,* but now an invalid; *one a graduate of the normal department of a college; one is a graduate nurse; two are engaged in a large retail business; one is holding a clerical position; all are high school graduates and all except one have been one-time students in colleges and universities.*
>
> *Mr. A failed on all the new 1911 tests except the six digits and suggestion lines (almost passed the central thought test).* In the 1908 scale he passed all the 10-year [old] tests and the following higher tests; absurdities, 60 words (gave 58 words), abstract definitions, and repetition of sentence. B.-S. age, 109, 10.8; retardation 54 years; intelligence quotient .17. According to the 1911 scale, 10.6 years.
>
> *This man, measured by the automatic standards now in common use, would be hopelessly feeble-minded (an imbecile by the intelligence quotient), and should have been committed to an institution for the feeble-minded long ago.* But is there anyone who has the temerity, in spite of the Binet "proof," to maintain, in view of this man's personal, social and commercial record, and the record of his family, that he has

been a social and mental misfit, *and an undesirable citizen,* and should, therefore, have been *restrained from propagation because of mental deficiency (his wife is still less intelligent than he)? No doubt if a Binet tester had diagnosed this man 45 or 50 years ago he would have had him colonized as a "mental defective." It is a safe guess that there are hundreds of thousands like him throughout the country, no more intelligent and equally successful and prudent in the management of their affairs.* Had he been a criminal when he was tested the Binet testers who implicitly follow these standards would have offered expert testimony under oath that he was feeble-minded and unable to distinguish between right and wrong, or unable to choose the right and avoid the wrong. [Italics added.]

In the same study, as Wallin summarized it for the February 15, 1916, issue of *The Psychological Bulletin,* administration of the revised Goddard-Binet test also revealed: "Of five freshmen in a teacher's college and one high-school junior, at least four of them would be rated as feeble-minded on the most liberal basis of accrediting by the 1911 [Goddard-Binet] Scale."

Wallin told the APA delegates that the purpose of his study was to:

. . . show the unscientific nature of the attempt to differentiate high grade feeble-minded, borderline and backward children and adults purely by a rule-of-thumb procedure based on mental tests and arbitrarily assumed psychological standards. The more we learn of psychological diagnosis the more evident it becomes that psychological tests are just like many of the tests of the physician (temperature, pulse, Wassermann, Noguchi, etc.): they are simply one means for aiding the clinician in arriving at a guarded diagnosis. They do not constitute an automatic diagnosticon, which will enable the examiner to dispense with a thorough clinical examination, or to disregard other clinical findings, nor do they obviate the need of technical training on the part of the examiner. . . . Psychological diagnosis is no easier than medical diagnosis and the *consequences of a blundering psychological diagnosis may be as unfortunate as the consequences of a blundering physical diagnosis* [italics added].

When Wallin carried his campaign for more professional diagnoses of mentality to the American Psychological Association meeting, his attempt to get his peers to join him in "completely rejecting the concept of the high grade moron as determined by the Binet scale from the standpoint of its social and legal implications" was helped enormously by the Chicago press.

But one other research paper on IQ testing, presented at the same APA meeting, attracted even greater attention. As Wallin recalled the incident in his autobiography:

Miss Mary Campbell, formerly connected with the municipal laboratory here, declared that she had given the Binet tests to members of Mayor Harrison's cabinet and to the mayoralty candidates in the last election and almost all of them qualified as morons. The *Herald* published what purported to be the Binet ratings of many of the municipal officials, including Mayor Thompson, who, according to the published report, "could not pass the ten-year old test." The reporters took pains to

EXPERTS ASSAIL MENTALITY TEST RESULT AS JOKE

Hear How Binet-Simon Method Classed Mayor and Other Officials as Morons.

"Mechanical measurement of human intelligence is a fallacy.

"The Binet-Simon test and similar devices (such as used in the city psychopathic laboratory by Dr. W. J. Hickson) are faulty and their results often absurd."

Thus declared speakers before the American Psychological Association in convention at the University of Chicago yesterday.

Among other attacks made against the grading of humans into classes of morons, feeble minded and the like by these arbitrary scales were assertions that Mayor Thompson, present and former city officials and wealthy agriculturists, measured by the Binet scale, had been graded as morons or "children" of 8 to 11 years, mentally, though successful business men.

URGE REFORM MEASURES.

As a result, the convention will be urged by members today to adopt resolutions to provide this action:

Set forth the unreliability of various psychopathic tests, notably the Binet-Simon method, when made by others than experts in psychology, who also have "a bit of common sense."

EXPERTS DISCUSS SUBJECT.

The discussion centered around papers on the subject read by Professors Rudolph Pintner of Ohio State University, Joseph W. Hayes of the University of Chicago, and J. E. Wallace Wallin of the St. Louis Psycho-Educational Clinic.

Among those who joined in the discussion were Professor Robert M. Yerkes of Harvard University, founder of the Yerkes-Bridges point scale of measuring mental deficiency; Miss Mary Campbell, formerly connected with the municipal laboratory here, and Samuel G. Kohs of the Chicago House of Correction.

Mr. Wallin led the attack. He declared that he had examined six prosperous citizens in a Missouri rural community, successful farmers, business men and housewives, and had found all of them mentally under the age of 12, according to the Binet system, and therefore feeble-minded.

TWO WEALTHY "DEFECTIVES."

"Two of these men were worth more than $100,000 each," said Dr. Wallin. "All of them had been successful in life and had been parents of normal children. But if those persons had been accused in a court and sent to the average physician using the Binet-Simon system for examination the experts would have told the court that the persons were irresponsible mentally."

Miss Campbell corroborated the statements of Dr. Wallin. She declared that she had given the Binet tests to members of Mayor Harrison's cabinet and to the mayoralty candidates in the last election and that almost all of them had qualified as morons. Some of her results were these:

Fire Marshal Thomas O'Connor couldn't pass a test proving a mental age of 8 years.

N. L. Pietrowski, former city attorney, now a Herald correspondent in Europe, couldn't pass some of the problems in the 10-year-old test, although he answered the 11 and 12-year-old tests.

Ray Palmer, the electrician in the cabinet, could not pass a 12-year-old test.

Former Chief Gleason refused to let any woman, but his wife "feel his head."

The IQ testing movement was shaken to its roots by the revelation on the front page of the *Chicago Herald* of December 29, 1915, that the mayor and other leading officials of Chicago had been "proven" by the standard American versions of the Binet-Simon intelligence tests to be "morons" with minds too feeble to pass the intelligence test for ten-year-olds. Note how even a hard-core eugenicist, Harvard psychology professor Robert M. Yerkes, joined in the psychologists' attack on the abuses of intelligence testing, warning that such abuses subjected the persons tested to "serious dangers of miscarriage of justice."

advertise the fact that the Binet tests were in use in the municipal psychopathic laboratory.

Wallin—and the intrepid Miss Mary Campbell—had both pointed to the basic weaknesses of the American versions, and interpretations of the test scores, of the Binet tests: the IQ tests measured not that still mysterious quality called "intelligence," but the amount and the quality of classroom learning the people tested had had *recently*. And this was as true among prosperous rural Iowans as among prosperous Chicago mayors and commissioners.

None of the psychologists present at the APA meeting was more vigorous in his support of Wallin than Robert M. Yerkes, then an assistant professor of comparative psychology at Harvard. Earlier in the same year, Yerkes had been the principal author of the Binet-type Point Scale for Measuring Mental Ability, dedicated to "the memory of Alfred Binet and Edmund R. Huey."[22]

"For my part," Yerkes told his professional peers, "I place my reliance upon a psychological examination, not so much upon the method used as on the man who makes up the examination." Like Wallin, Yerkes was equally terrified at the thought of professional clinical diagnoses being made by amateur Binet testers. And his reasons were among the reasons spoken by Wallin: "We are building up a science, but we have not yet devised a mechanism which anyone can operate. There is serious danger of miscarriage of justice," Yerkes concluded, "when courts take the advice of men not familiar with the general [psychological] problems, but who have learned the detail of some systems of tests."

The members of the American Psychological Association were so impressed with the evidence presented by Wallin and Campbell, and the ringing endorsement leaders like Yerkes gave to Wallin's conclusions, that—at this 1915 meeting—they passed a resolution declaring:

> WHEREAS, Psychological diagnosis requires thorough technical training in all phases of mental testing, thorough acquaintance with the facts of mental development and with the various degrees of mental retardation;
> AND WHEREAS, There is evident a tendency to appoint for this work persons whose training in clinical psychology and acquaintance with genetic and educational psychology are inadequate:
> *Be it resolved,* That this Association discourages the use of mental tests for practical psychological diagnosis by individuals psychologically unqualified for this work.[23]

Two years later, the then president of the American Psychological Association, one Robert M. Yerkes, was to take a historic action that ran directly counter to this cautionary resolution of the Association. He presided over the administration of hastily written group intelligence tests to nearly two million military recruits by even more hastily trained testers, who were nearly all of them "individuals psychologically unqualified for this work."

HOW AND WHAT THE ARMY INTELLIGENCE TESTS TESTED

In all the millions of words that have been written by psychologists and lay-men about the intelligence tests of World War I since 1918, and particularly since 1954, very few words indeed have dealt with what these so-called IQ tests tested, let alone with how the tests were administered and evaluated.

The Army tests were written by a committee selected by the chairman of the Eugenics Section of the American Breeders' Association's Committee on the Inheritance of Mental Traits and president of the American Psychological Association, Robert M. Yerkes. The leading members of this seven-man Committee on the Psychological Examination of Recruits were Henry Herbert Goddard and Lewis Terman. "This committee decided on the mental examination of every soldier," Terman wrote in a 1918 report, "The Use of Intelligence Tests in the Army," "and within six weeks had prepared methods adequate for the huge task of testing millions of men."

If it took only six weeks to prepare mental tests "adequate" for this huge undertaking, Terman was also pleased to report that "the intelligence scale devised for this purpose . . . permitted group examining so that *one examiner could test several hundred men in less than an hour. . . .* By the use of scoring stencils the personal equation was entirely eliminated from the grading of papers. . . . *The test papers are in fact scored by enlisted men who know nothing about psychology.*"[24] (Italics added.)

Since these Army intelligence test scores were then used as equivalents of clinical psychological diagnoses of the innate mental acuity of each recruit tested, this in effect gave a few thousand "enlisted men who know nothing about psychology" the professional standing equivalent to that of professional clinical psychologists in the eyes of the government and generations of scholars after World War I.

Terman was not exaggerating the speed at which the Army intelligence testers purported to scientifically measure the mentality of nearly two million human beings. There were two types of Army mental tests—the Alpha, for literates, and the Beta, which was given both to illiterate recruits and to men who took the Alpha examination and failed it. According to the official *Examiner's Guide,* the Alpha tests were given to groups of from 100 to 200 men in 40 to 50 minutes, while the Beta tests were administered to groups of up to 60 inside of 50 to 60 minutes.[25] All who failed Beta were given an individual examination, which would be the Stanford-Binet or the Point Scale for men "who were able to understand English fairly well," and a performance-scale examination to all other testees. The time allowed for the Point Scale and Stanford-Binet examinations was from 15 to 60 minutes.

In content, the questions asked on both the Alpha and the Beta examinations were remarkably similar, in many instances almost identical, to the questions on the Goddard, Stanford (Terman), and rump revisions of the Binet scale, as well as the Binet-scale-like Point Scale developed by Yerkes, Bridges, and Hardwick in Boston in 1915. The differences between the standard IQ tests and the Army tests were: (1) the military mental tests were administered to groups, rather than to individuals, and (2) they were marked by letter grades from A ("very superior intelligence") to E ("very inferior intelligence") rather than by numbers purporting to represent the intelligence quotient (IQ) or the mental age (MA).

While the Alpha questions were for literates, "the test questions were ingeniously arranged so that practically all could be answered without writing, by merely drawing a line, crossing out or checking."[26] As in the Beta test for illiterates, the Examiner in charge of the Alpha tests had to memorize a set of written commands. For example:

> "Attention! The purpose of this examination is to see how well you can remember, think, and carry out what you are told to do. We are not looking for crazy people. . . . You are not expected to make a perfect grade, but do the very best [*sic*] you can."

The Examiner's words and actions were carefully blueprinted. Here are some of the things he had to say and do for Test 1, Form 5:

> 1. "Attention! 'Attention' always means 'Pencils Up.' Look at the circles at 1. When I say 'go,' but not before, make a cross in the first circle and also a figure 1 in the third circle. Go!" (Allow not over 5 seconds.) . . .
> 4. "Attention! Look at 4. When I say 'go' make a figure 1 in the space which is the triangle but not in the square, and also make a figure 1 in the space which is not in the triangle and in the square, and also make a figure 2 in the space which is in the triangle and circle, but not in the square. —Go!" (Allow not over 10 seconds.)
> (N.B. *Examiner.*—in reading 4, don't pause at the word CIRCLE as if ending a sentence.)[27]

To guard against coaching by recruits who had already been tested, five different forms of each scale were used more or less randomly. For example, see Test 8, opposite, which calls for a knowledge of American advertising slogans, trademarks, and professional sports.

The scrambled-sentence and the paired-word tests differed from the analogous scales in the various civilian IQ tests only in their choice of words, not in their form, structure, and function.

Few documents now available to scholars offer more devastating testimony to the fact that all so-called IQ tests yet devised are actually measurements of one's cultural experiences, and not of one's inborn or genetic mental endowment, than the different scales of the Army mental tests. A simple test of the culture-bound nature of the Army mental tests of 1917–19 can be performed in any home or library that has a new or old almanac or encyclopedia listing all of the scientists who won Nobel prizes since the awards were first made in 1903. Go through the list of Nobel laureates in science for the first twenty-three years of the institution—that is, until 1925. What you will find is a population consisting of 20 Germans, 12 Englishmen, 12 Frenchmen, 5 Swedes, 5 Dutch, 5 Americans (by citizenship, but with one exception of French and German birth), 3 Danes, 2 Swiss, 2 Austrians, and one

Test 8

Notice the sample sentence: People *hear* with the *eyes* <u>*ears*</u> *nose* *mouth*
The correct word is *ears,* because it makes the truest sentence.

In each of the sentences below you have four choices for the last word. Only one of them is correct. In each sentence draw a line under the one of these four words which makes the truest sentence. If you can not be sure, guess. The two samples are already marked as they should be.

SAMPLES { People *hear* with the   *eyes* <u>*ears*</u> *nose* *mouth*
{ *France* is in <u>*Europe*</u> *Asia* <u>*Africa*</u> *Australia*

1. The *apple* grows on a <u>*shrub*</u> *vine* *bush* *tree*	1
2. *Five hundred* is played with *rackets* *pins* *cards* *dice*	2
3. The *Percheron* is a kind of *goat* *horse* *cow* *sheep*	3
4. The most prominent industry of *Gloucester* is *fishing* *packing* *brewing* *automobiles*	4
5. *Sapphires* are usually *blue* *red* *green* *yellow*	5
6. The *Rhode Island Red* is a kind of *horse* *granite* *cattle* *fowl*	6
7. *Christie Mathewson* is famous as a *writer* *artist* *baseball player* *comedian*	7
8. *Revolvers are made by* *Swift & Co.* *Smith & Wesson* *W. L. Douglas* *B. T. Babbitt*	8
9. *Carrie Nation* is known as a *singer* *temperance agitator* *suffragist* *nurse*	9
10. *"There's a reason"* is an "ad" for a *drink* *revolver* *flour* *cleanser*	10
11. *Artichoke* is a kind of *hay* *corn* *vegetable* *fodder*	11
12. *Chard* is a *fish* *lizard* *vegetable* *snake*	12
13. *Cornell University* is at *Ithaca* *Cambridge* *Annapolis* *New Haven*	13
14. *Buenos Aires* is a city of *Spain* *Brazil* *Portugal* *Argentina*	14
15. *Ivory* is obtained from *elephants* *mines* *oysters* *reefs*	15
16. *Alfred Noyes* is famous as a *painter* *poet* *musician* *sculptor*	16
17. The *armadillo* is a kind of *ornamental shrub* *animal* *musical instrument* *dagger*	17
18. The *tendon of Achilles* is in the *heel* *head* *shoulder* *abdomen*	18
19. *Crisco* is a *patent medicine* *disinfectant* *tooth-paste* *food product*	19
20. An *aspen* is a *machine* *fabric* *tree* *drink*	20
21. The *sabre* is a kind of *musket* *sword* *cannon* *pistol*	21
22. The *mimeograph* is a kind of *typewriter* *copying machine* *phonograph* *pencil*	22
23. *Maroon* is a *food* *fabric* *drink* *color*	23
24. *The clarionet* is used in *music* *stenography* *book-binding* *lithography*	24
25. *Denim* is a *dance* *food* *fabric* *drink*	25
26. The author of *Huckleberry Finn* is *Poe* *Mark Twain* *Stevenson* *Hawthorne*	26
27. *Faraday* was most famous in *literature* *war* *religion* *science*	27
28. *Air* and *gasoline* are mixed in the *accelerator* *carburetor* *gear case* *differential*	28
29. The *Brooklyn Nationals* are called the *Giants* *Orioles* *Superbas* *Indians*	29
30. *Pasteur* is most famous in *politics* *literature* *war* *science*	30
31. *Becky Sharp* appears in *Vanity Fair* *Romola* *The Christmas Carol* *Henry IV*	31
32. The number of a *Kaffir's* legs is *two* *four* *six* *eight*	32
33. *Habeas corpus* is a term used in *medicine* *law* *theology* *pedagogy*	33
34. *Ensilage* is a term used in *fishing* *athletics* *farming* *hunting*	34
35. The *forward pass* is used in *tennis* *hockey* *football* *golf*	35
36. *General Lee* surrendered at Appomattox in *1812* *1865* *1886* *1832*	36
37. The *watt* is used in measuring *wind power* *rainfall* *water power* *electricity*	37
38. The *Pierce Arrow car* is made in *Buffalo* *Detroit* *Toledo* *Flint*	38
39. *Napoleon* defeated the *Austrians* at *Friedland* *Wagram* *Waterloo* *Leipzig*	39
40. An irregular *four-sided figure* is called a *scholium* *triangle* *trapezium* *pentagon*	40

Nobel laureate each from Russia, Spain, Belgium, Hungary, and Canada. Their number included German Jews, such as Albert Einstein, and French citizens born in Poland, such as Marie Curie.

Most of these foreign Nobel prize-winning scientists were alive and well in 1917 and 1918, when the Army mental tests were given to nearly two million native and foreign-born draftees. Now ask yourself how many of them, with the probable exception of the Canadian discoverer of insulin, Frederick Banting, would have been able to answer questions number 2, 4, 7, 8, 9, 10, 16, 19, 25, 29, 35, and 38 of Alpha Test 8 on page 245.

Yet, because of their quite predictable failure to answer these questions about American baseball, football, trademarks, advertising slogans, and the identity of a fourth-rate British versifier who had a vogue among American middle-brow schoolteachers who lacked the stomach for real poets on the order of Shakespeare and Shelley, these Nobel laureates in physiology, medicine, chemistry, and physics would not have done nearly as well on this "information form" as any American middle-class high school or grammar school student could and did do. This would be particularly true of those non-American Nobel laureates in science who had not bothered to learn English—since the tests had to be taken in English, and not in the native languages of the college-educated foreign-born recruits, or the European winners of the Nobel prizes for science.

For anyone with a decent American grammar school education—and this meant a *minority* of the native-born draftees—the Army Alpha examination was ridiculously easy. For the larger number of graduates and/or dropouts from the very inferior grammar schools in which the *majority* of the literate native-born draftees learned to read and write, the Alpha tests were quite difficult. For those illiterate white and black Americans who had had little or no education in the poorer (chiefly, but not exclusively, southern) states—and the foreign-born draftees of high, low, or no education in their native lands—the Beta tests acted as a guarantee that nearly all the hundreds of thousands of recruits who took them would emerge branded as morons at best.

What the design and administration of the Beta tests might have lacked in sadism in some aspects, they made up for in ineptitude. The Beta materials were shipped to the military training camps in three packages, containing "1. Blackboard frame. 2. Blackboard chart. 3. (a) Cardboard pieces for Test 7; (b) patterns for constructing cubes for Test 2." The blackboard chart on which all the tests were printed was a continuous cloth roll 27 feet long.

The various tests were kept hidden behind a curtain until the Examiner ordered the curtains removed. Instructions to the Examiners in the official testing manuals included such nuggets as:

> With the exception of the brief introductory statements and a few orders, *instructions are to be given throughout by means of samples and demonstrations* and should be animated and emphatic. . . .
>
> Both examiner and demonstrator must be adept in the use of gesture language. In the selection of a demonstrator the Personnel Of-

fice should be consulted. One camp has had great success with a "window seller" as demonstrator. Actors should also be considered for the work. The orderlies [i.e., the demonstrators] should be able to keep the subjects [testees] at work without antagonizing them and to keep them encouraged without actually helping them.

The manual directed that, as soon as the "subjects" were properly seated, pencils and examination blanks were to be distributed. While this was being done,

> . . . E. [Examiner was usually called E] should say, "Here are some papers. You must not open them or turn them over until you are told to." Holding up beta blank, E. continues:
> "In the place where it says name, write your name; print it if you can. (Pause) Fill out the rest of the blank about your age, schooling, etc., as well as you can. If you have any trouble, we will help you." . . .
> After the initial information has been obtained, E. makes the following introductory remarks:
> "*Attention!* Watch *this* man (pointing to demonstrator). He (pointing to demonstrator again) is going to do *here* (tapping black board with pointer), what *you* (pointing to different members of group) are going to do on your *papers* (here E. points to several papers that lie before men in the group, picks up one, holds it next to black board, returns the paper, points to demonstrator and the black board in succession, then to the men and their papers). Ask *no questions. Wait* till I say 'Go ahead!' ". . .
> . . . the orderlies' vocabulary in beta is rigidly restricted to the following words, or their literal equivalents, in Italian, Russian, etc.: *Yes, No, Sure, Good, Quick, How many? Same, Fix it.* Under no circumstances may substitutional explanations or directions be given.

To keep the testees working quickly during the maze test, the orderlies were instructed to yell, "Do it, do it, hurry up, quick," at the native and foreign-born illiterate draftees, as well as at the educated European recruits who were not yet as familiar with English as they would quickly become, and their importuning had the effect—whether intended or not—of harassing and confusing the young men subjected to the Beta tests. Since the orderlies were, themselves, men of fairly limited education, they could be forgiven for merely following the very explicit orders in the official *Examiner's Guide.*

It was, however, not so easy to forgive the college graduates, most of them Ph.D. professors of psychology at Stanford, Harvard, and other leading American universities, who devised the picture-completion test on page 248.

If ever a series of incomplete pictures was designed to be answered correctly only by American, white, middle-class, educated individuals—and *not* by men reared in Mississippi or Wyoming farm and ranch country, in poor Nebraska rural villages, in American and European regions where electric lights were still to come and where phonographs and ladies' vanity tables and powder puffs were as totally unknown as bowling alleys and tennis, or in Orthodox Jewish families where playing cards was forbidden, or in countries where it is not unusual for dancers to put one arm behind the back in the

Test 6

course of a dance—it was this group in Beta Test 6. Klineberg (1935) wrote of it:

"The Army Beta contains a picture-completion test, in which the task is to draw in the missing portions. One of the pictures is of a house lacking a chimney; Boas (private communication) tells of a Sicilian child who drew in its place a crucifix. His particular response had taught him that no house was complete without one."[28]

As Yoakum and Yerkes reported in their book *Army Mental Tests,* published "with the authorization of the War Department" in 1920, the vast majority of the 1,726,966 U.S. Army draftees whose minds were tested by Army Alpha and Army Beta had mentalities ranging downward from the 13–14-year-old to the 10-year-old level. As they summed up the Army mental test findings, only 4 to 5 percent of the draft were A men, of "very superior intelligence." The B rating, for merely "superior intelligence," was obtained "by eight to ten soldiers out of a hundred." From 15 to 18 percent of the draft consisted of C+ men—C+ denoting "high average intelligence." In terms of mental age, A equaled 18–19+ years; B equaled 16–17 years; and C+ a solid 15 years of mental age. These were the mentally competent minority.

For the rest, 25 percent of all soldiers tested were C men—that is, they had a mental age of 13–14 years, and the psychology professors who designed and administered the tests ruled that "C grade is rarely capable of finishing a high school course" (p. 23). Under the 431,742 C men too stupid to finish a high school course were the even more numerous soldiers exposed by the tests to be D men, of "inferior intelligence," and, finally, the D— and the E men, of "very inferior intelligence." The D and E men had mentalities of 11 and 10 years old, respectively.

This all reflected the hereditary endowment of the soldiers, since, according to Yerkes and Yoakum, "the score is little influenced by schooling. Some of the highest records have been made by men who had not completed the eighth grade." Yerkes, a favorite student under Davenport at Harvard, and Yoakum evidently did not consider the privately tutored rich men's sons who never attended grade school or high school to have received at least their equivalent in education at home.

Far from being culturally and linguistically based, the Army IQ tests of World War I, according to Majors Yoakum and Yerkes, were culture-free and got right down to the unit characters for intelligence in the blood of the Naked Adam:

> Examinations alpha and beta are so constructed and administered [they wrote on page 27 of their 1920 book] as to minimize the handicap of men who because of foreign birth or lack of education are little skilled in the use of English. *These group examinations were originally intended, and are now definitely known, to measure native [genetic] intellectual ability.* They are to some extent influenced by educational acquirement, but in the main the soldier's *inborn* [hereditary] intelligence and not the accidents of environment determines his mental rating or grade in the army. [Italics added.]

While, they conceded, the Alpha and Beta tests "do not measure loyalty, bravery, power to command, or the emotional traits that make a man 'carry on,' " Majors Yoakum and Yerkes added that, "however, in the long run these qualities are far more likely to be found in men of superior intelligence than in men who are intellectually inferior [i.e., who get low IQ and other forms of 'intelligence test' scores]."

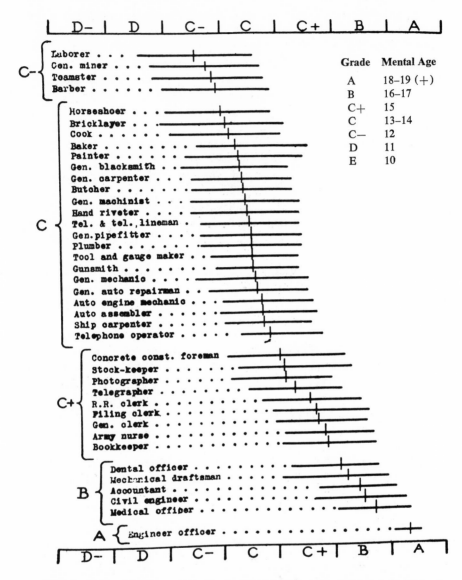

Grade	Mental Age
A	18–19 (+)
B	16–17
C+	15
C	13–14
C–	12
D	11
E	10

Of all the "scientific" displays of the World War I intelligence testing data released by the federal government, few were to have more influence on national and state educational policies than did this official summary of "the intelligence requirements of the occupations" of the men in the armed services. By "proving" that the average serviceman had the mental age of a 13½-year-old, and that even our doctors, dentists, accountants, and civil engineers had only the minds of 16½-year-olds, this table also suggested to government policy makers that it was not at all necessary to increase the tax burden by improving the qualities of our educational systems. (Key to grade scores added.)

From Yerkes, 1921, p. 829.

All things taken into consideration, the surprise is not that Alpha and Beta mental tests—commonly known today as the World War I IQ tests—left most of the 1.7 million young men who took these tests branded as having the minds of 10- to 14-year-old children. The miracle was that, in 1917–18, 95 percent of the recruits tested did not do as badly as did the mayor of Chicago, who in 1915 "could not pass the 10-year-old test."

The Army would have been far better off if, during World War I, they had—like the French, British, and German armies of that war, and the American Army of World War II—not administered any IQ or IQ-like tests to their draftees at all. The recruits could have been spared harassment by the examination teams and the traumatic humiliation of being quite falsely branded as morons and worse.

As a nation, we would have been spared the subsequent generations of massive abuses of our minds, our bodies, our children, and our taxes at the hands of blundering and/or evil authors and public officials who found, in the eugenical interpretations of the Alpha and Beta intelligence test score results, the "scientific" excuses for a half century of catastrophic educational, immigration, and social policies and actions.

11 How the Old Scientific Racists Used the New IQ Test Myths to Manipulate the Minds and Votes of the American People

> In a very definite way, the results which we obtain by interpreting the Army [IQ test score] data by means of the race hypothesis support Mr. Madison Grant's thesis of the superiority of the Nordic type. Our figures would rather tend to disprove the popular belief that the Jew is highly intelligent. . . . Our results showing the marked intellectual inferiority of the negro are corroborated by practically all of the investigators who have used psychological tests on white and negro groups.
> —PROFESSOR CARL C. BRIGHAM, Princeton University, in *A Study of American Intelligence* (1923)

> What are the qualities essential in human beings for the running of a democracy? . . . Is it not possible that the failure of these devices of democracy is due, not to any imperfection inherent in the devices themselves, but to a fundamental [genetic] inferiority in the average intelligence of the voters, which makes them unable to use the method wisely? How can we expect a man with a mental age of less than 10 years to deal intelligently with the complicated questions submitted to the voters in a referendum? Has not the impossible been demanded of a nation, nearly half of whose population is under the mental age of 13?
> —CORNELIA JAMES CANNON, in "American Misgivings," *The Atlantic Monthly,* February 1922

The social policy makers of the nation did not need the test scores of the Army intelligence tests to convince them to preserve the culturally inadequate —but tax-saving—school systems that prevailed before Army Alpha and Army Beta "proved" three out of every four Americans to be ineducable. Nor, for that matter, had they needed Davenport's scientistic "demonstrations" that poverty diseases such as tuberculosis and pellagra were genetic *and therefore not preventable* to put them squarely in opposition to minimum-wage laws that might have given at least the working poor a fair chance to afford enough food to keep their families tuberculosis- and pellagra-free.

The opposition of legislators and Presidents to most tax-supported advances in universal education, clean-water and sewage systems, adequate medical education and training for the nation's doctors, and to such social advances as minimum-wage laws, was a fact of American life long before Goddard "Binet-tested" his first unfortunate Piney in 1908. In each of the four editions of their college textbook, *Applied Eugenics,* Paul Popenoe and Roswell Johnson's two-page segment on "The Minimum Wage" owed nothing to the results of the Army Alpha and Beta tests in arriving at the conclusion: "The minimum wage on the family basis is admittedly [*sic*] not an attempt to pay a man what he is worth. It is an attempt to make it possible for every man, no matter what his economic or social value, to support a family. *Therefore, in so far as it would encourage men of inferior quality to have or increase families, it is unquestionably dysgenic"* (italics added).[1]

This concept stemmed not from the American perversions of Binet's mental tests but from the writings of the founder of scientific racism, Thomas Malthus, who in the first edition (1798) of his *Essay on the Principle of Population,* had warned that raising the wage from eighteen pence to five shillings a day "would give a strong and immediate check to productive industry, and in a short time not only the nation would be poorer, but the lower classes themselves would be much more distressed than when they received only eighteen pence a day."

Thus, while the employment of the Army Alpha and Beta intelligence test scores of 1.7 million World War I recruits by the eugenics movement spokesmen armed the opponents of social advances with new arguments against bettering the status quo in government health, education, and family welfare systems, it did not add anything other than compulsory state sterilization laws to the body of domestic laws and principles. However, when it came to immigration policies, the World War I IQ tests and their civilian progenitors were to forge the last and necessary link for the chain that finally, in 1924, was raised against the further immigration of what Prescott Hall, General Francis Amasa Walker, and the results of the Army Alpha and Army Beta tests proclaimed to be the world's inferior races.[2]

The employment of the prewar and military intelligence tests to ultimately assure the triumph of the immigration restriction crusade launched by Hall, Ward, and other young Boston Brahmins in 1894 did not occur in a historical vacuum. Three important historical changes, each of them related to the increased production demands World War I made of the American automotive, munitions, and other industries starting in 1914, had also changed the labor needs and attitudes of the nation's industrialists. As we shall see, new shifts in the centers of population in the United States had provided new sources of cheap domestic labor. The importation of foreign labor was no longer needed, and the industrialists now left immigration policy to the politicians.

The first of these events was, of course, the complex of military events that cut off the supplies of the cheap labor needed to meet the expanded production demands of American industry. The second was the discovery that the accelerated mechanization of American agriculture had begun to make family farming less feasible, so that the sons of farm families and the small farmers themselves had started to migrate in growing numbers to the jobs becoming available in the expanding industrial centers of the nation. The third was the realization, by personnel recruiters for mines and mills with backlogs of war and war-related orders, that the black people of the South constituted a vast reservoir of candidates for the menial and low-paying jobs that had, until August 1914, been filled by peasant immigrants from Ireland, Russia, Hungary, Italy, and other European nations.

An idea of how these three events caused shifts in the native population can be gleaned from the Census Bureau data on Internal Migration.[3]

In the state of Michigan, for example, which was still largely agricultural at the start of the century, 35,900 of the native white population moved to other states during the decade starting in 1900. During the decade 1910–20, however, as Michigan's young automobile industry grew to giant size, the native white population of the state gained 181,500 by in-migration. Between 1900 and 1910, only 1,900 blacks migrated to Michigan from other states; between 1910 and 1920, 38,700 Negroes moved to Michigan. In absolute numbers, Michigan grew from a total population of 2,420,982 in 1900 to 3,668,-412 in 1920—a gain of 34 percent.

Missouri, to cite a typical contrasting state, lost 228,100 whites and 1,000 blacks to other states between 1900 and 1910, and 178,000 whites and 27,200 of her Negroes by out-migration between 1910 and 1920. As a result, where the total population of newly industrialized Michigan jumped by 34 percent over its 1900 population in twenty years, agricultural Missouri's population remained just about where it was, rising only from 3,106,661 in 1900 to 3,404,055 in 1920.

All three developments—plus the postwar Red Scare phobias about radical ideas that many people believed to be inherent in the blood of inferior European races—combined to remove the bulk of the opposition of the major employers of American industrial and mine labor to restrictive immigration laws. From being the dangerous foreigners, who according to the dogma of Francis Amasa Walker and Professor Robert DeCourcy Ward, were going to lower the scales of American wages, the non-Nordic immigrants from the traditional heartlands of cheap labor had now become the wily undercover agents of the international Jewish Bankers–Jewish Bolshevik conspiracy to organize strikes aimed at forcing good Nordic American employers to pay higher wages than the low-IQ-test-scoring toilers in factories, mines, and fields deserved.

The last barrier to ending the immigration of Italians, Jews, Poles, and the hordes of other polluters of the Nordic germ plasm was now the hard core of the decent, educated, believing Christian people—the proud inheritors of the political and cultural traditions of Jefferson, Lincoln, the Abolitionists, the American sanitary reformers, and the wellborn social reformers who had opened settlement houses for the poor. These were the people who, in their own lifetimes, had contributed money to and taken to the streets in support of such causes as votes for women, birth control for the poor as well as the affluent, living wages, better schools, safer housing and factories for the poor. They were the native-born legislators and editorialists who still had the grace to remember that America was, indeed, a Nation of Immigrants—and that many of them were themselves the descendants of indentured servants (in most cases illiterate), and political prisoners, and religious dissenters, and even deported convicts, who all found in America the opportunities for upward social mobility that had been denied them and their own ancestors from the moment of birth.

These good people had, however, one fatal chink in their moral armor: most of them were better educated than the average American and had been fortunate enough to obtain college and university educations. This very education was to make them particularly susceptible to the very scientific-sounding books and statistical curves and tables of the Ivy League and other prestigious professors of Psychology, Sociology, Zoology, Eugenics, and History who warned America that the Army intelligence test scores had revealed the specter of "the decline of American intelligence."

Whereas the foul mouths and the uglier deeds of the old-time gut racists and convent burners had always made their immigrant baiting even more repellent to the decent educated Americans, the ultra-respectable books of the McDougalls and equally honored professors and the middle-brow monthly pundits could, and in the end did, start to erode the moral objections of the better-educated and well-intentioned people to the anti-Catholic, anti-Semitic,

Plain Remarks on Immigration for Plain Americans
By KENNETH L. ROBERTS

While the xenophobic *Saturday Evening Post* duly trumpeted the wisdom of Lothrop Stoddard, Madison Grant, and other spokesmen for scientific racism on the dangers of non-Nordic immigration, its editors also made equally forceful uses of the less scientific and more overtly political arguments of the era against the continued admission of Jews, Italians, Hungarians, and other non-WASP immigrants. The highly influential magazine's staff immigrant baiter, Kenneth L. Roberts, later made a fortune writing novels claiming that the wrong side won the American Revolution of 1776.

From *The Saturday Evening Post,* cartoon accompanying article by Kenneth L. Roberts, February 12, 1921, p. 21.

and anti-humanitarian immigration policies that would make precisely such immigrant baiting the law of the land.

In the hysterical Red Scare days that followed World War I, the eugenics movement utilized the IQ test scores of 1.7 million draftees to win the hearts and minds of those educated Americans whose support for immigration restriction could not be won by the kind of crude labor baiting indulged in by *The Saturday Evening Post* and other organs of mass opinion.

"IS AMERICA SAFE FOR DEMOCRACY?"

The propaganda line of the eugenicists and other scientific racists was three-pronged:

1. Native white American intelligence was declining. Not only did our native white population have, according to the Army IQ test scores of 1.7 million recruits, an average mental age of 13–14 years, but now this native American ship of fools had added a new ballast of black people who were, according to the same tests, even stupider and less capable of handling abstract thought than were the majority of whites.

2. There was nothing much we could do about the threats of the declining intelligence of the vast majority of the white and black American populations except—as Harvard's Professor McDougall wrote—to prevent "the reproduction of the least fit, especially of those persons who are indisputably feeble-minded."[4]

Exactly how this was to be done was a matter of management, for as far as McDougall was concerned: "It is needless to argue here the relative advantages of sterilization and institutional segregation. Probably both methods will be used."[5] But unless this nation took such Draconian measures to come to grips with the fact that "the birthrate of the inferior half of the population is very considerably higher than that of the superior half," and "if the present state of affairs shall continue, the civilization of America is doomed to rapid decay."[6]

Any other measures, such as better schools, preventive medical care, better housing, minimum-wage laws, and equality of opportunities for employment, were simply a waste of good tax and charity dollars. Because "science" had now shown that you could not make a silk purse out of a sow's ear.

3. As if the decline in native American intelligence were not enough, the nation was now being inundated with hordes of Russian and Polish Jews, southern Italians (the very worst kind), hungry Hungarians, and immigrants of other inferior races proven, by the Army Alpha and Beta test scores, to be genetically even stupider than our nitwitted majority of native Americans. These foreigners not only threatened to add their genes for feeblemindedness to the already three-quarters unintelligent native American gene pool, but were also now exposed by *The Saturday Evening Post* and other authorities to be fiendishly clever strike formenters.

Clearly, then, in order to Keep America Great we not only had to halt the reproduction of that majority of native Americans who scored less than 90 in IQ tests; we also had to slam the gates to America shut to keep out the non-Nordic, non-Aryan, and non-Protestant immigrants "for the good of the race."

The eugenic movement's intelligence test score propaganda for the thinking Americans took the form of serious books, such as (to cite but a few of many examples) *Is America Safe for Democracy?* (1921) by the chairman of the Department of Psychology at Harvard, William McDougall, and Lothrop Stoddard's *The Revolt Against Civilization* (1922); college text-books, such as Popenoe and Johnson's *Applied Eugenics,* whose four editions between 1918 and 1933 were in daily use in American higher education until the start of World War II; *The Trend of the Race* (1921) and *Studies in Evolution and Eugenics* (1923), both by Samuel J. Holmes, Ph.D., professor of zoology at the University of California at Berkeley; and articles in the ultra-respectable middle-brow magazines, such as *The Atlantic Monthly,* by such authors as Holmes, Yerkes, and Cornelia James Cannon.

Professor McDougall's book is best remembered for the doomsday curve with which he mathematically summarized the results of the Army intelligence tests. It was a statistical curve of distribution whose elegance would have warmed the heart of Sir Francis Galton himself. The Harvard professor of psychology's curve was stated in terms of the letter grades used in scoring these tests. As we saw in the preceding chapter, grade A, in this letter scale, was the equivalent of the mental age of 18–19+. C equaled the mental age of 13–14. D equaled the mental age of 10.

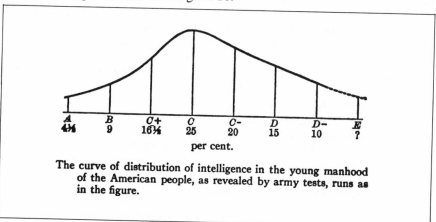

per cent.

The curve of distribution of intelligence in the young manhood of the American people, as revealed by army tests, runs as in the figure.

According to this British-born former Oxford professor, "*A* men are of the grade which 'has the ability to make a superior record in college'; *B* men are 'capable of making an average record in college'; *C* men are 'rarely capable of finishing a high-school course.' And the main bulk of your [*sic*] total population is below the C+ level."[7]

The significant fact that screamed from the dry numbers in the government's official report on the Army intelligence test scores—to wit, that the lower any state's expenditures per pupil per year on elementary and high school education, the lower the average Alpha and Beta intelligence test scores racked up by its young men—presented no more of a problem to the true believers in the racial origins of intelligence and unintelligence than Goldberger's findings on the sole cause of pellagra had to the eugenicists who chose to believe that this disease of hunger was due to bad blood rather than poverty. On page 47 of his 1921 best seller, for example, Professor McDougall, in the tradition of Charles Benedict Davenport, disposed of this obvious sociocultural factor in these pontifical terms:

> All these facts point to only one conclusion, namely, that innate capacity for intellectual growth is the predominant factor in determining the distribution of intelligence in adults, and that *the amount and kind of education is a factor of subordinate importance.*
>
> The superiority of white literates to the white illiterates is due, then, not wholly or mainly to their schooling, but rather to *an inborn greater capacity for intellectual growth.* [Italics added.]

The fact that "the colored men of the Northern States showed distinct superiority to those of the South, in respect to their performance in the army intelligence tests," was, as we have seen, disposed of by McDougall with the racist myth that the more white blood in their veins, the higher the IQ test scores of blacks. And as backing for this "explanation," he quoted a turn-of-the-century racist, Harvard's Professor Nathaniel S. Shaler: "Almost all the Negroes of this country who have shown marked capacity of any kind have had an evident mixture of white blood."[8]

Even for Americans of 100 percent white blood, McDougall warned that "no general raising of the level of prosperity and of the standard of living is likely to have the desired effect" in the matter of making wiser white people.

> Nor would an absolute equality of income for all families meet the case. If that state of affairs could be maintained, it is clear, I think, that *its effects would be positively dysgenic in a high degree.* It is equally clear that the general indiscriminating State endowment of motherhood [i.e., family assistance programs to help meet the nutritional and health needs of growing children], now called for in so many quarters, would have directly dysgenic effects; and it would be disastrous, in that it would go very far to destroy the family as an institution of any nation which should adopt this plan. [Italics added.]

McDougall not only warned against the menace of permitting further non-Nordic immigration. On pages 63–64 and 159 he also quite predictably misinterpreted a study by Ada H. Arlitt of Bryn Mawr of the IQ test scores of native American children of four different social classes, Italian children, and black children. Professor Arlitt's study quite clearly showed the effects of the vastly different social biology of poverty and the social biology of affluence on the IQ test scores of poor versus nonpoor children to be as absolute as were the links between high and low state educational expenditures and high and low Army intelligence test scores. Professor McDougall, however, used her raw data to prove his claim that "the children of Italian immigrants [are] decidedly lower in the scale of intellectual capacity than those of the older white population." (See pages 302–03.)

An even more widely read book than McDougall's *Is America Safe for Democracy?* was Dr. Lothrop Stoddard's *The Revolt Against Civilization: The Menace of the Under-Man.* A long-time protégé and disciple of Madison Grant (who had written the foreword for his earlier book, *The Rising Tide of Color*[9]), Stoddard was proof of the ancient axiom that while it often helps, one does not need to be an illiterate in order to be an opinionated ignoramus in science. One merely has to be arrogant.

The single thesis of Stoddard's book was that the "Under-Man"—whom he defined as "the man who measures *under* the standards of capacity and adaptability imposed by the social order in which he lives"—was taking over civilization. What made the Under-Man such a menace was that sentimental permissivists on the order of Jean Jacques Rousseau, Thomas Jefferson, Abraham Lincoln, Franz Boas, and the rest of *that crowd* had introduced into world thought a mental poison called "Natural Equality." This was the root of all modern evils.

"The idea of 'Natural Equality' is one of the most pernicious delusions that has ever afflicted mankind. It is a figment of the human imagination. *Nature knows no equality.*"[10] (Italics added.)

Small wonder that biased illiterates in biology, such as George Horace Lorimer, editor of *The Saturday Evening Post,* hailed Stoddard and Grant as prime specimens of the new school of 100 percent American "scientific writers" whose works "teem with jolting ideas, solemn warnings and predictions," and "yet the clear ring of truth is in them."[11] Stoddard was speaking for all of their deepest racial hatreds and phobias when he wrote:

"Furthermore, individual inequalities steadily increase as we ascend the biological scale. . . . Thus, we see that evolution means a process of ever-growing *inequality. There is, in fact, no such word as 'equality' in nature's lexicon.* With an increasingly uneven hand she distributes *health,* beauty, vigor, *intelligence,* genius—all the [genetic] qualities which confer on their possessors superiority over their fellows."[12] (Italics added.)

Until the Army mental tests, Stoddard and the other scientific racists had had only the "actuarial" studies of Galton, Dugdale, Goddard, and Davenport to support this primitive gut-racist version of the meaning of evolution and the natural origins of intellectual superiority. Now the Army IQ tests had armed them with millions of mathematically tabulated data, each of them arrived at via the group IQ tests devised for the nation's armed forces by Yerkes, Terman, Goddard, and their associates.

Stoddard would have been somewhat less than human had he resisted loading his chapter "The Iron Law of Inequality" with the Army IQ test score data that "proved" this "law." He reduced them to five tables, three of which are reproduced below as they appeared on pages 68–72 of *The Revolt Against Civilization.* The first, on page 68, showed what the armed forces IQ tests revealed about the intellectual state of the nation:

Grade	Percentage	Mental Age
A	4½	18–19 (+)
B	9	16–17
C +	16½	15
C	25	13–14
C −	20	12
D	15	11
D −	10	10

"This table [above] is assuredly depressing," wrote Stoddard. "Probably never before has the relative scarcity of higher intelligence been so vividly demonstrated." As Stoddard commented, this not only meant that "the *average* mental age of Americans is only about fourteen" but, even worse, "that forty-five million, or nearly half of the whole population, will never develop mental capacity beyond the stage represented by a normal 12-year-old child."

That this alarming state of the American intelligence quotient was not without its racial implications was made abundantly clear—to Stoddard and the rest of the eugenicists—by the next two tables, which are reproduced below exactly as they appeared in his section dealing with "the correlation between intelligence and racial origin." "The [following] table," wrote Stoddard, "needs no comment: it speaks for itself!"

	A	B	C+	C	C−	D	D−	E
White—Draft	2.0	4.8	9.7	20	22	30	8	2
Colored—Draft	.8	1.0	1.9	6	15	37	30	7
Officers	55.0	29.0	12.0	4	0	0	0	0

Percentage of Inferiority

Country of Birth		Country of Birth	
England	8.7	Norway	25.6
Holland	9.2	Austria	37.5
Denmark	13.4	Ireland	39.4
Scotland	13.6	Turkey	42.0
Germany	15.0	Greece	43.6
Sweden	19.4	Russia	60.4
Canada	19.5	Italy	63.4
Belgium	24.0	Poland	69.9

Stoddard saw the Army intelligence test scores as "scientific proofs" of the genetic inferiority of non-Nordic races and, as such, the ultimate argument for ending their immigration to the United States.

The first of these tables also showed that Stoddard was either ignorant of, or aware of but determined to ignore, the equally authenticated *Memoir XV* data showing that the Army mental test scores of Negro recruits from such northern states as Pennsylvania, New York, Illinois, and Ohio were higher than those achieved by white draftees from such southern states as Mississippi, Kentucky, Arkansas, and Georgia.

To Stoddard, the implications of the nativity group's IQ data were crystal-clear: close the gates to the immigration of inferior, non-Nordic, Latin,

Roman Catholic, and Jewish racial stocks. Stoddard's simplistic act of measuring the innate superiority and inferiority of races solely on the basis of their average Army IQ test scores was soon to be emulated by the Expert Eugenics Agent of Congress. (See page 299.)

While McDougall and Stoddard were feeding their eugenical interpretations of the results of the Army IQ tests to the general reader, professors such as Roswell Johnson at the University of Pittsburgh and the zoologist Samuel J. Holmes at the University of California were teaching courses in eugenics to the coming generations of American lawmakers, judges, teachers, and journalists. They were also turning out textbooks in eugenics. Holmes's *Studies in Evolution and Eugenics,* for example, leaned quite heavily on the Army IQ test score results in stating his arguments for immigration restriction. On page 207, he wrote:

> Instead of the English, Scotch, Irish, Germans and Scandinavians who made up the bulk of our immigration before 1880, we have been receiving hordes of Poles, Southern Italians, Greeks, Russians, *especially Russian Jews,* Hungarians, Slovaks, and other southern Europeans—stocks less closely related to us by blood than the northern Europeans [i.e., the Nordics] and less readily imbued with the spirit of our institutions. Our immigrants lodge chiefly in the cities, forming little communities speaking their own language, and preserving, so far as possible, their customs and traditions. *They show a very high percentage of illiteracy* and they furnish a great part of the unskilled labor of our mines, factories, and streets. [Italics added.]

Three pages later, Holmes warmed up to his point:

> Conditions in our cities are bad enough now, with unrestricted immigration they would become almost intolerable.
>
> The greatest permanent danger, however, lies in the likelihood of receiving *stocks of inferior inheritance.* The American is beginning to suspect that some of our racial immigration is of low racial [i.e., genetic] value. *This suspicion has been strengthened by the results of the mental tests applied to the recruits for the United States Army in the late war.* Just as there are families on a low mental level, so there may be peoples [races] on a low mental level. Unquestionably we have been getting much of this kind of human material. [Italics added.]

THE ARMY MENTAL TEST SCORES KINDLE "AMERICAN MISGIVINGS"

Throughout the early postwar years Holmes and other true believers in the eugenical exegesis of the Army IQ test scores also took to writing for the middle-brow magazines, such as *The Independent, Scribner's,* and of course the by then traditional organ of the spokesmen for American scientific racism, *The Atlantic Monthly.* Yerkes joined Holmes in writing for *The Atlantic Monthly* on the scientific and social implications of the Army IQ test score results. But the honor of writing the most famous of *The Atlantic Monthly* articles on the IQ test scores of World War I fell to Cornelia James Cannon, whose lead article, "American Misgivings," appeared in the February 1922 issue.

The release of the Army IQ data gave Mrs. Cannon serious misgivings at nightmare levels about four basic American problems: (1) the caliber of voters required to make democracy work; (2) the danger that we were wasting tax dollars on the education of the genetically ineducable; (3) the danger that our "permissive" immigration laws were flooding the American gene pool with hordes of "eastern Europeans" (meaning, of course, Jews) of low native intelligence; and (4) the existence of black Americans, 89 percent of whom, according to the scientific mental tests of Yerkes, Goddard, Terman, et al., had an average mental age of 13, and "50 per-cent of whom never reached a mental age of 10."

Since the tests showed that, according to the arbitrary "mental age" standards of Yerkes, Goddard, and Terman, which Mrs. Cannon naïvely accepted at face value, "almost half of the white draft, 47.3 per-cent would have been classed as morons," she could have interpreted this as an indictment of, *at the very least,* our entire public education system. And, in view of the dramatic discoveries made by Stiles and Goldberger, between 1902 and 1916, about the roles of hookworm and pellagra in conditioning the biological and mental health of all of the nation's poor—black and white— Mrs. Cannon could not have been faulted for seeing in these Army IQ test scores some pretty strong correlations between poverty and low classroom and IQ test scores. (As the wife of Dr. Walter B. Cannon, the Harvard professor of physiology, she could hardly have been unaware of who delivered the Cutter Lecture at Harvard in 1915.) Instead, Mrs. Cannon came to a number of most orthodox eugenic conclusions:

1. The vast majority of America's voting citizens had the mentality of 13-year-old children and less, and were therefore unfit to vote.

2. The mental ages of every individual, and every ethnic and racial group, were determined by their genes, and not by their education and other developmental conditions.

3. The two lowest-quality intellectual groups were the non-Nordic foreigners, particularly those born in Poland, Italy, and Russia—whence came the bulk of our Jewish and Roman Catholic immigrants—and the native American blacks.

4. Therefore, the only solutions to the perils that confronted all blue-eyed, dolichocephalic, Aryan-speaking, and Nordic Protestant Americans were to: (a) end all further immigration of "the eastern Europeans" (the then-popular code word for Jews), since the Jew "comes to us with a slant toward revolution, a hatred of whatever power there may be"; and (b) abandon the illusion that the Negro was capable of being educated in the abstract thoughts which white students found so easy to grasp.

What made the problem of non-Nordic immigration hopeless, for Mrs. Cannon, was that these people were proven by the Army IQ tests to be beyond redemption by education. "Given a high grade of intelligence . . . the danger [of non-Nordic immigration] is negligible; for education can train in the ideals of democracy." Provided, of course, that people were genetically educable.

> But what chance of this is there with the inferior grade of intelligence? Such individuals form the material of unrest, the stuff of which mobs are made, the tools of demagogues; for they are peculiarly liable to the emotional uncontrol which has been found to characterize so many of the criminals who come before our courts. They are the persons who not only do not think, but are [genetically] unable to think; who cannot help in the solution of our problems, but, instead, become a drag on the progress of civilization.

And this, of course, brought the 1922 *Atlantic Monthly* readers face to face with the menace of race suicide that had been flaunted in their faces daily for over a decade by nativists and restrictionists: "It is not only the individual whom we exclude [from immigration], but that ever-widening circle of his descendants, *whose blood may be destined to mingle with and deteriorate the best we have"* (italics added).

There had also been started, during the war, the internal migration of the southern blacks from rural poverty in the South to urban wartime employment and postwar unemployment and poverty in the North. These migrating blacks created vast educational problems. As Mrs. Cannon reminded her educated readers:

> What light do the intelligence tests throw on our educational problems? The tests here are of a peculiar cogency, for they are tests of [inborn] intelligence, which is a measure of educability. . . .
>
> Educational processes are helpless in the face of native incapacity. Not more than a pint can be poured into a pint receptacle; the rest sinks into the ground and is lost. Professional training is becoming more and not less expensive, and the community has the right to decide to whom this higher education is to be given. *We cannot afford to invest our largest sums in our second-rate men. For our own sakes we must select our best for the types of training that demand a high order of ability.* [Italics added.]

What this meant in terms of the new internal migration of Americans from South to North was, of course, obvious to anyone familiar with the Army's IQ test score data:

> The data from the army tests concerning the negro present the first concrete material, on a large scale, by which we can check up the partisan asseverations of the friends and the critics of the race. Of the entire negro draft 80 per-cent were in the D grade, 89 per-cent under the mental age of 13. . . . in the education of the negro race, we are confronted by an educational problem of a very special kind. Emphasis must necessarily be laid on the development of the primary schools, on the training in activities, habits, *occupations which do not demand the more evolved faculties.* In the South particularly . . . the education of the whites and colored in separate schools may have *justification other than that created by race prejudice. . . . A public-school system, preparing for*

life young people of a race [the blacks], 50 per-cent of whom never reach a mental age of ten, is a system yet to be perfected, if indeed we have so far recognized the urgency of the need for adequate grappling with the problem. [Italics added.]

Clearly, Mrs. Cannon's "scientific" reasons for urging the politically and culturally influential readers of *The Atlantic Monthly* to support both anti-Semitic and anti-Catholic immigration restriction legislation, and the preservation of segregated school systems for black children in the South—and the creation of inferior schools for black children elsewhere—were derived solely from the published results of the Army intelligence test scores of 1917–19.

WAS AMERICAN DEMOCRACY SAFE FOR PEOPLE?

The most influential single piece of writing on the social implications of the Army Alpha and Beta tests was written by a minor participant in the writing and administration of these intelligence tests. His name was Carl C. Brigham, and when his book, *A Study of American Intelligence,* was published in 1923, with a foreword by Robert M. Yerkes, he was, at the age of thirty-three, an assistant professor of psychology at Princeton University.

Although from the way Yerkes wrote about him in his foreword Carl Campbell Brigham seemed to be a Canadian, he was born in Marlboro, Massachusetts, in 1890, and graduated from Marlboro High School in 1908. From Marlboro, Brigham went to Princeton, where he earned a Litt.B. in 1912, an A.M. in 1913, and a Ph.D. in psychology in 1916. His Canadian phase was indeed a short one, for by October 1917 he was back in the States working for Yerkes in the Army.

Early in his years as a Princeton teacher he proved to be just the scholar whom Charles W. Gould—a wealthy septuagenarian lawyer-writer of the Madison Grant type—was looking for to make an analysis of the Army IQ test score results in terms of the Galtonian or race hypothesis. Gould—author of *America, A Family Matter* (1920), a variant of Grant's *Passing of the Great Race* that followed the earlier book so closely that one reads exactly like the other—had an idea that is best described by a paragraph in Yerkes's foreword to the Brigham book:

> It appears that Mr. Charles W. Gould, a clear, vigorous, fearless thinker on problems of race characteristics, amalgamation of peoples and immigration, raised perplexing questions which drove Mr. Brigham to his careful and critical re-examination, analysis, and discussion of army data concerning the relations of intelligence to nativity and length of residence in the United States. In a recently published book, *America, A Family Matter,* to which this little book is a companion volume, Mr. Gould has pointed out the lessons of history for our nation and has argued strongly for pure-bred races.

In his acknowledgments Brigham showed how mightily Gould had influenced him:

Mr. Charles W. Gould suggested this continuation of the army investigations in the first instance, has sponsored [meaning, of course, paid for] the work throughout, has read and re-read all of the manuscript at every stage of its preparation, and is mainly responsible for the whole work. In my treatment of the race hypothesis I have relied on his judgment and on two books, Mr. Madison Grant's *Passing of the Great Race,* and Professor William Z. Ripley's *Races of Europe.*

In the same pages, young Brigham also made "especial acknowledgment to Colonel Robert M. Yerkes, who has read the manuscript several times in its various stages of preparation and has given many helpful suggestions," as well as to Professors E. G. Boring of Harvard, and Edwin A. Conklin for their "invaluable assistance" and wise advice.

Brigham's book opened with a technical description of the Alpha and Beta tests and how they were administered by Examiners and orderlies following instructions to speed up and harry the testees. Brigham had no comment to make about the insensitive methods by which the Army mental tests were administered to native and foreign-born recruits with little or no experience in being tested or in the culture of American middle-class children. However, after stating that "ultimately, the validity of our conclusions from this study rests on the validity of the alpha, beta and the individual [Stanford-Binet] examinations," Professor Brigham added (p. 96):

> It is sometimes stated that the examining methods stressed too much the hurry-up attitude frequently called typically American. The adjustment to test conditions is part of the intelligence test. We have, of course, no other measure of adjustment aside from the total score on the examinations given. If the tests used included some mysterious type of situation that was "typically American," we are indeed fortunate, for this is America, and the purpose of our inquiry is that of obtaining a measure of the character of our immigration. Inability to respond to a "typically American" situation is obviously an undesirable trait.

Wherever possible, Brigham tried to share the judgmental responsibilities with Gould, Grant, Yerkes, Lapouge, and other advocates of gentlemanly genocide. Of Madison Grant's *Passing of the Great Race,* Brigham wrote that "the quotations I have chosen from Mr. Madison Grant's chapter on *Racial Aptitudes* most certainly do not do justice to that author, but they seemed to me to summarize his general position briefly. *The entire book should be read to appreciate the soundness of Mr. Grant's position and the compelling force of his arguments*" (italics added). But even though Brigham kept evoking the authority in the words of others, his name was still on the title page as sole author, and the responsibility for every statement his, including these on pages 182 and 183:

> [1] In a very definite way, the results which we obtain by interpreting the army data by means of the race hypothesis support Mr. Madison Grant's thesis of the superiority of the Nordic type. . . .
>
> Our results based on the army data also support Mr. Grant's estimates of the Alpine race: "The Alpine race is always and everywhere a

race of peasants, an agricultural and never a maritime race. In fact they only extend to salt water at the head of the Adriatic and, like all purely agricultural communities throughout Europe, tend toward democracy, although they are submissive to authority both political and religious, being usually Roman Catholics in western Europe. This race is essentially of the soil, and in towns the type is mediocre and bourgeois" (p. 227 [*Passing of the Great Race*]). . . .

[2] Our results also support de Lapouge [Georges Vacher de Lapouge, *L'Aryen, son rôle social,* Paris, 1899] in his contention that the Nordic type is superior to the Alpine. He says, concerning the Alpine: "The brachycephalic states, France, Austria, Turkey, even Poland, which no longer exists, are a long way from offering the vitality of the United States or England. Nevertheless the mediocrity of the brachycephalics is in itself a force. . . . As bad money chases out good, so the brachycephalic race has replaced the superior [dolichocephalic] race. He is inert, he is mediocre, but he multiplies. His patience is unlimited; he is a submissive subject, a passive soldier, an obedient functionary. He does not carry grudges and he does not revolt."

Brigham quoted the above passage of the Comte de Lapouge's in the original French, rather than in English translation. In French it sounded much more authoritative, particularly to people who could not read French but who would be impressed by Brigham's knowledge of that Alpine-Mediterranean language. But he did not show, *in any language,* how Lapouge's hoary cephalic-index "anthropology" had any relationship at all to the human race, let alone to the Alpha and Beta mental test scores.

Taken all in all, Brigham's analysis of the Army IQ test score data showed that, stupid as the American population had become in recent years, the IQ tests had also revealed the sad fact that the entire Western world was peopled and run by adults whose highest mental age—that of the primarily Nordic cadre of U.S. officers of World War I—was under 19 years of age. And whose genetically most "inferior" races—the peoples of Russia, Italy, Poland, and the colored people of the United States—tested out as having mental ages ranging downward from 11.34 to 10.41 years of age.

This was all summed up in Brigham's famous Figure 36 on page 124. This figure and Brigham's caption (opposite) are reproduced as they appeared in his book.

JEWS, ITALIANS, AND "OTHER GENETICALLY UNINTELLIGENT TYPES"

But what of the *bête noire* race of Madison Grant and Charles W. Gould? That is, the Jews, who had produced such destroyers of racist anthropology as Franz Boas and such defilers of eugenic biology as Joseph Goldberger? Brigham had more than one unkind word for the immigrants who most frightened his sponsor.

Once having proved that the Alpines were the inferiors of the Nordics, Brigham then proceeded to state, on page 190, that "there is no serious objection, from the anthropological standpoint, to classifying the northern Jew

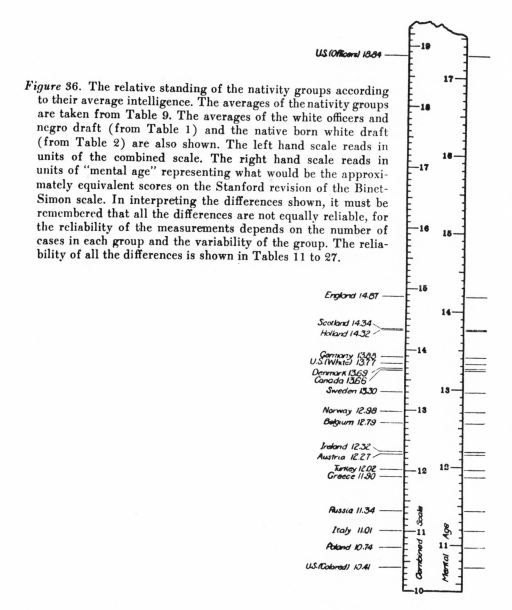

Figure 36. The relative standing of the nativity groups according to their average intelligence. The averages of the nativity groups are taken from Table 9. The averages of the white officers and negro draft (from Table 1) and the native born white draft (from Table 2) are also shown. The left hand scale reads in units of the combined scale. The right hand scale reads in units of "mental age" representing what would be the approximately equivalent scores on the Stanford revision of the Binet-Simon scale. In interpreting the differences shown, it must be remembered that all the differences are not equally reliable, for the reliability of the measurements depends on the number of cases in each group and the variability of the group. The reliability of all the differences is shown in Tables 11 to 27.

Numerically, the chief targets of Brigham's widely cited scale on the average intelligence of "the nativity groups" were the Italians, the native-born American blacks, and the Jews. While the Jews are not identified by name in this scale, Brigham observed on page 189 of his book that "according to the 1910 census, about 50% of the foreign born population reporting Russia as their country of origin spoke Hebrew or Yiddish." The standard code words for Jews among the American eugenicists and less scientific racists were, between 1910 and 1940, "Russians," "Poles," and, most often, "Eastern European immigrants."

as an Alpine, for he has the head form [cephalic index], stature, and color of his Slavic neighbors. He is an Alpine Slav." In the phrase "Alpine Slav" he gave the scientific racists yet another pseudoscientific sobriquet for the Jew. The racial implications of this "discovery" about the real nature of the Jews were quite clear to Brigham:

> From the immigration statistics showing aliens admitted classified according to race or people, we find about 10% (arriving between 1900 and 1920) reported as Hebrew. It is fair to assume that our army sample of immigrants from Russia is at least one-half Jewish, and that the sample we have selected as Alpine is from one fifth to one fourth Jewish.
>
> *Our figures, then, would rather tend to disprove the popular belief that the Jew is highly intelligent.* Immigrants examined in the army, who report their birthplace as Russia, had an average intelligence below those from all other countries except Poland and Italy. It is perhaps significant to note, however, that the sample from Russia has a higher standard deviation (2.83) than that of any other immigrant group sampled, and that the Alpine group has a higher standard deviation than the Nordic or Mediterranean groups (2.30). If we assume that the Jewish immigrants have a low average intelligence, but a higher variability than other nativity groups, this would reconcile our figures with popular beliefs, and, at the same time, with the fact that investigators searching [via forms of the Stanford-Binet IQ tests] for talent in New York City and California schools find a frequent occurrence of talent [i.e., high IQ test scores] among Jewish children. *The able Jew is popularly recognized not only because of his ability, but because he is able and a Jew.* [Italics added.]

After putting the "able Jew"—that is, the Jewish child who racks up a high Stanford-Binet or Otis-Binet test score at the hands of visibly dismayed WASP experimental psychologists—on the same evolutionary plane as a talking dog, Brigham turned his attention to the other major immigrant race that bothered his sponsor and possibly even Brigham himself. That is, the immigrants who came not from Europe but from the rural southern states to the urban North, most of them blacks.

> Our results showing the marked intellectual inferiority of the negro are corroborated by practically all of the investigators who have used psychological tests on white and negro groups. This inferiority holds even when a low intellectual sampling of whites is made by selecting only those who live in the same environment, and who have had the same educational opportunities.

Nevertheless, despite this innate inferiority of blacks to whites, Brigham as a psychologist had to explain the higher IQ test scores of even the black siblings of the same parents who had moved North while their sibs who remained in the South got predictably lower scores. Brigham was equal to this challenge:

> The army tests showed the northern negro superior to the southern negro, and this superiority is attributed to the superior educational op-

portunities in the North. The educational record of the negro sample we are studying shows that more than half of the negroes from the southern states did not go beyond the third grade, and only 7% finished the 8th grade, while about half of the northern negroes finished the 5th grade and a quarter finished the 8th grade. *That the difference between the northern and southern negro is not entirely due to schooling, but partly to intelligence, is shown by the fact* [italics added] that groups of southern and northern negroes of *equal schooling* [Brigham's italics] show striking differences in intelligence.[13]

As a psychologist, particularly one involved with education, it is hard to believe that Brigham was as unaware as he sounded here of the vast differences between the *qualities* of the grammar school education available to black children in the North as against those of the South, particularly the bone-poor counties of the rural South. What Brigham meant by "equal schooling" in the North and South was the number of grades completed, not the quality of the education received in any of these grades. This was merely another of the classic "false correlations" of scientific racism that infuriated honest scientists like Joseph Goldberger and J. E. Wallace Wallin. The fact that Brigham, regardless of what he wrote in the next passage on page 192, was acutely aware of the realities of North-South environmental differences, should be evident in some of his references to precisely these conditioning factors in the excerpt that follows:

The superior intelligence measurements of the northern negro are due to three factors: first, the greater amount of educational opportunity, which does affect, to some extent, scores on our present intelligence tests; second, *the greater admixture of white blood*; and, third, the operation of economic and social forces, such as higher wages, better living conditions, identical school privileges, and a less complete social ostracism, *tending to draw the [genetically] more intelligent negro to the North.* It is impossible to dissect out of this complex of forces the relative weight of each factor.[14] [Italics added.]

The racial implications of his study in terms of future immigration legislation were obvious to Brigham:

The intellectual superiority of our Nordic group over the Alpine, Mediterranean, and negro groups has been demonstrated. If a person is unwilling to accept the race hypothesis as developed here, he may go back to the original nativity groups, and he can not deny the fact that differences exist [italics added].

Since the more white blood Negroes had, and the more Nordic blood mixed white races had, determined their relative superiority in IQ test scores to blacks and whites with lower quotients of white and/or pure Nordic blood (filled with the genes for high IQ test scores), any scientist aware of this fact was also under patriotic obligation to let his fellow Americans in on the Nordic blood titers of the immigrant hordes clamoring to violate our golden shores. This service Brigham performed for his race on page 159, in the form of his Table No. 33:

Tentative estimates of the proportion of Nordic, Alpine and Mediterranean blood in each of the European countries

	Percent Nordic	Percent Alpine	Percent Mediterranean
Austria-Hungary	10	90	0
Belgium	60	40	0
Denmark	85	15	0
France	30	55	15
Germany	40	60	0
Greece	0	15	85
Italy	5	25	70
Netherlands	85	15	0
Norway	90	10	0
Sweden	100	0	0
Russia (including Poland)	5	95	0
Poland	10	90	0
Spain	10	5	85
Portugal	5	0	95
Roumania	0	100	0
Switzerland	35	65	0
Turkey (unclassified)	0	20	80
Turkey (in Europe) (including Serbia, Montenegro and Bulgaria)	0	60	40
Turkey (in Asia)	0	10	90
England	80	0	20
Ireland	30	0	70
Scotland	85	0	15
Wales	40	0	60
British North America	60	40	0

In 1924, these meaningless numbers were accepted at face value as legitimate "scientific data" by the Congress and the President (Coolidge) of the United States. They had and continue to have no factual basis of any kind in biology or physical anthropology.

This table was to enjoy myriad uses in all the xenophobic and genocidal campaigns over the next decade. It was to become a key element in the establishment of the immigration quotas that were to bar American sanctuary to Jews, Italians, Greeks, Spaniards, and other people declared non-Aryan and fit only for extermination by the Nazi regime's *Rassenbiologie* executioners between 1933 and 1945.

Equally well used in the same American efforts to round up the future non-Aryan victims of the eugenics movement for Hitler's hangmen was Brigham's Figure 33, which revealed that the more time immigrants lived in America, the more their IQ test scores climbed.

What made this figure so powerful was the interpretation Brigham gave to its obviously nongenetic data. Far from concluding that these data proved that the longer a foreign-born draftee attended American schools and was exposed to all aspects of American culture, the higher would be his IQ test score, Dr. Brigham, with the aid and money of Charles W. Gould, the writings of Madison Grant and University of Wisconsin professor of sociology E. A. Ross, and his own statistical analyses, came forth with his amazing explanation for the fact that the longer a foreign-born draftee had lived in the

United States prior to the day he took the Army IQ tests, the higher his score always proved to be:

> Instead of considering that our curve (Figure 33) indicates a growth of intelligence with increasing length of residence, we are forced to take the reverse of the picture and accept the hypothesis that *the curve indicates a gradual deterioration in the class of immigrants examined in the army, who came to this country in each succeeding five year period since 1902* [italics added].[15]

At a time when the powerful anti-Semitic, anti-Italian, and anti-Catholic coalition forged by the Immigration Restriction League, the Eugenics Record Office, the American Federation of Labor, the Ku Klux Klan, and literally hundreds of other national and regional pressure groups—with the vigorous backing of the mass and the middle-brow media from *The Saturday Evening Post* and *The Atlantic Monthly* to many of the nation's leading daily newspapers—was vociferously deploring the "increase" in the volume of Jewish and Italian and other non-Nordic immigration, this was a most inflammatory statement to make. Particularly in a book that made the "scientific" statement that the greater the proportion of pure Nordic blood in an American's veins, the higher his IQ test score—and that the higher his *IQ test score,* the higher his *actual intelligence.* For, as Harvard's Professor McDougall had pointed out in his equally "scientific" book, *Is America Safe for Democracy?,* high morals and high intelligence went hand in hand.[16] In the 1920's, such statements as this made by Brigham about the declining human worth of all European immigration since 1902 added great infusions of fuel to fires best left to die.

The 1920's were years tormented by the fears of race and race mixing raised by such racist academics and authors. It was to these fears that Brigham's concluding chapter addressed itself, noting on page 207, for example, that while "our own data from the army tests indicate clearly the intellectual superiority of the Nordic race group," and although "the Alpine race, according to our [Army test] figures, which are supported by historical evidence, seems to be considerably below the Nordic type intellectually," Nordic-Alpine race mixtures were not necessarily cacogenic, or conducive of race degeneration. "The Nordic and Alpine mixture in Switzerland has given a stable people, who have evolved, in spite of linguistic differences, a very advanced form of government."

On the other hand, considering the degeneration of even the lowly Alpine race itself, a Nordic–Alpine Slav cross would equal race suicide:

> The Alpines that our [Army] data sample come for the most part from an area peopled largely by a branch of the Alpine race which appeared late and radiated from the Carpathian Mountains. It is probably a different branch of the Alpine race from that which forms the primitive substratum of the present population of Western Europe. Our data on *the Alpine Slav [that is, the Jew] show that he is intellectually inferior to the Nordic, and every indication would point to a lowering of the average intelligence of the Nordic if crossed with the Alpine Slav.* There can be no objection to the intermixture of races of equal [genetic] ability,

provided the mingling proceeds equally from all sections of the distribution of ability. Our [Army mental test] data, however, indicated that the Alpine Slav we have imported *and to whom we give preference in our present immigration law* is intellectually inferior to the Nordic type. [Italics added.]

This was, of course, perfectly good eugenical orthodoxy: Galtonian Gospel had it that intelligence was primarily genetic; that it was fixed for life; that it was transmissible in the parental genes. Brigham salted his eugenics tract with the usual array of statistics and pseudo-data:

The evidence in regard to the white and negro cross is also indisputable. If we examine the figures showing the proportion of mulattoes to a thousand blacks for each twenty year period from 1850 to 1910, we find that in 1850 there were 126 mulattoes to a thousand blacks, 136 in 1870, and 264 in 1910. This intermixture of white and negro has been a natural result of the emancipation of the negro and the breaking down of social barriers against him, mostly in the North and West. . . .

We must face a possibility of racial admixture here that is infinitely worse than that faced by any European country today, for we are incorporating the negro into our racial stock, while all of Europe is comparatively free from this taint.

"AMERICAN INTELLIGENCE IS DECLINING . . ."

Brigham's concluding two paragraphs sum up the entire purpose of his book, and the role of the IQ test scores in the formulation of American domestic and foreign policy since then:

According to all evidence available, then, *American intelligence is declining, and will proceed with an accelerating rate as the racial admixture becomes more and more extensive. The decline of American intelligence will be more rapid than the decline of the intelligence of European national groups, owing to the presence here of the negro.* These are the plain, if somewhat ugly, facts that our study shows. The deterioration of American intelligence is not inevitable, however, if public action can be aroused to prevent it. *There is no reason why legal steps should not be taken which would insure a continuously progressive upward evolution.*

The steps that should be taken to preserve or increase our present intellectual capacity must of course be *dictated by science and not by political expediency. Immigration should not only be restrictive but highly selective.* And the revision of the immigration and naturalization laws will only afford a slight relief from our present difficulty. *The really important steps are those looking toward the prevention of the continued propagation of defective strains in the present population* [i.e., compulsory sterilization and other "get 'em in the gonads" measures against the Undeserving Poor, black and white]. *If all immigration were stopped now, the decline of American intelligence would still be inevitable.* This is the problem which must be met, and our manner of meeting it will determine the future course of our national life. [17] [Italics added.]

Robert M. Yerkes, at that time (June 1922) chief of the Division of Psychology, Office of the Surgeon General of the Army, recognized the important role Brigham's book could play in legislation aimed at barring the immigration as well as the propagation of Jews, Italians, Negroes, and other "inferior" (that is, non-Nordic) human beings. As he wrote in his foreword to the Brigham book:

> Mr. Brigham has rendered *a notable service* to psychology, to sociology, and *above all to our law-makers* by carefully re-examining and re-presenting with illuminating discussion the data relative to intelligence and nativity first published in the official report of psychological ex-amining in the United States Army. . . . The volume . . . which I now have the responsibility of recommending, is substantial as to fact and *important in its practical implications.* . . . it is better worth re-reading and reflective pondering than any explicit discussion of immigration that I happen to know. The author *presents not theories or opinions but facts.* It behooves us to consider their reliability and their meaning, for no one of us as a citizen can afford to ignore *the menace of race deterioration or the evident relations of immigration to national progress and welfare.* [Italics added.]

The wide uses of Brigham's slender volume—short enough, and simple enough to be understood by the makers of the nation's value systems—were soon to be reflected in local and national governmental attitudes and priorities.

A Study of American Intelligence became a key factor in the establishment or preservation of all local and state orders of priority in education, health, and welfare. Practical school boards and health agencies that made up local or state policies and budgets were not about to waste their taxpayers' money on the education of the genetically ineducable Alpine Slavs, colored people, Italians, and others born to be stupid, to contract infectious and deficiency diseases, to work at menial jobs that "native Americans" were too good to perform, and to die young.

On the national scene, *A Study of American Intelligence* was to become one of *the* key documents used before the Congress by Harry Hamilton Laughlin, the former teacher of agriculture at the North Missouri State Normal School, with the impressive title "Expert Eugenics Agent of the Committee on Immigration and Naturalization, United States House of Representatives." This respected federal official, on loan to his government from the Eugenics Record Office of the Carnegie Institution of Washington biological station at Cold Spring Harbor, New York, where for over a decade he had served as supervisor of the ERO and co-editor of the *Eugenical News,* was to lean heavily on the Brigham book in helping to achieve the greatest legislative triumph ever scored by the world eugenics movement.

12 The Dogmas of the Old Scientific Racism Win the Hearts and Minds of the Congress and the President of the United States of America

Americans, the Philistines are upon us. Rend the fetters with which we have bound ourselves. Nothing but our own folly stands between us and freedom from alien invasion. Speak the word, repeal at once the naturalization laws which corrupt and destroy us. Repeal them, and the thing is done. Yet have we left fifty millions, the greater part of whom can trace descent to Colonial days, a number ample to keep, and which if uncontaminated by foreign blood ever will keep, America for Americans!

—CHARLES W. GOULD, in *America, A Family Affair* (1920)

Within the year Congress will enact a new immigration law. . . . I have undertaken . . . to make a comprehensive study of our immigration problem. . . . It was quite impossible to comprehend our past, present and future without applying to our population all the established truth with respect to anthropology. A failure to take into account family strains, racial characteristics and cultures would have been fatal to correct conclusions. Eugenics entered into it, not here and there but everywhere.

—DR. HIRAM W. EVANS, Imperial Wizard, Knights of the Ku Klux Klan, in "The Menace of Modern Immigration," address delivered on the occasion of Klan Day at the State Fair of Texas at Dallas, October 24, 1923

America must be kept American.

—CALVIN COOLIDGE, President of the United States, on signing the U.S. Immigration Act of 1924, May 26, 1924

Like Lieutenant Colonel Robert M. Yerkes, chief psychologist of the United States Army, Dr. Henry Fairfield Osborn, president of the American Museum of Natural History and a scientific leader of the American eugenics movement, did not propose "to ignore the menace of race deterioration or the evident relation of immigration to national progress and welfare."[1] Osborn did whatever he could to further the openly genocidal twin crusades of the eugenics movement against the native American WASP poor and against the non-Aryan Jewish and non-Protestant Catholic immigrant poor.

As an orator, Osborn took to the hustings on behalf of the crusade to bar further immigration of Jews, Italians, Greeks, and other inferior races. One of his most famous immigration restriction speeches, published as "The Approach to the Immigration Problem Through Science," delivered in December 1923 in New York, dealt with two great advances in modern science that had proven to be

. . . a revelation to our Congress and our Houses of Legislature. *The first was the army intelligence tests.* I believe those tests were worth what the war cost, *even in human life,* if they served to show clearly to our people *the lack of intelligence in our own country, and the degrees of intelligence in different races who are coming to us,* in a way which no one can say is the result of prejudice. It wasn't because one comes from a line of abolitionist ancestors or from a line of men who believed in slavery that he realized that *the negro's intelligence is not to be placed on the same line as that of the white man. . . .* we learned once and for all that the negro is not like us. So in regard to many races and sub-races in Europe we learned that some which we had believed possessed of an order of intelligence perhaps superior to our own were far inferior. . . .

    Intelligence tests were just the opening wedge. Then came the Second International Congress of Eugenics in 1921. . . . The Congress brought together, for the first time, *all the information relating to the laws of heredity;* all the information relating to the advantages and disadvantages of the wonderful law of heredity; all the information relating to *what characteristics are transmitted from parents to offspring* and what are not; *the dangers of the misuse of inheritance in encouraging the multiplication of undesirable stocks* and the advantages of the use of inheritance in encouraging multiplication of desirable stocks. Second, there were assembled unmistakable figures, for the first time, showing that in many sections of the country . . . the best American stock is rapidly dying out. [Italics added.]

Professor Osborn spoke with the assurance that he was not alone in his hopes, beliefs, and fears:

Everyone agrees that it is a crime to bring into the world knowingly a criminal, an idiot, or a hopelessly diseased person, and everyone agrees that it is undesirable to increase in this or any other country, the number of this class, namely, of unintelligent and dependent people. *We don't have to argue about that, because everyone agrees.* This is part, and an important part, of the eugenic program which bears directly upon the immigration program.[2] [Italics added.]

Osborn was, alas, quite right about one thing: "everyone," including most educated Americans from the University of Wisconsin professor of sociology E. A. Ross to the Associate Justice of the Supreme Court Oliver Wendell Holmes, Jr., and seven of his fellow Justices, did indeed believe that paupers and pellagrins, like criminals and imbeciles (*as measured by IQ test scores*), were the products of faulty genes rather than of poverty and other nongenetic factors of human growth and development. But, if they agreed with Osborn and Galton's eugenics, not all of the scientific racists of the era shared Osborn's reverence for intelligence as a quality to be demanded of would-be immigrants.

Ross, for example, in his 1914 book, *The Old World in the New,* devoted his Chapter 7, "The East European Hebrews," to a warning against precisely this quality he believed to be more present in Jews than in "any other recent stream [of immigration], and it may be richer than any large outflow since the colonial era." Ross felt that Jewish "gray matter" was what really accounted for the failure of government statistics to back up anti-Semitic charges: "The fewness of the Hebrews in prison has been used to spread the impression that they are uncommonly law-abiding. The fact is it is harder to catch and convict criminals of cunning than criminals of violence."[3]

And no non-Nordic race was more cunning than the "lower class of Jews of Eastern Europe [who] reach here moral cripples, their souls warped and dwarfed by iron circumstances." These hereditarily inferior types had, Ross wrote, "developed a monstrous and repulsive love of gain," and in America—as Galton himself had warned—the Jews "rapidly push up into a position of prosperous *parasitism*, leaving scorn and curses in their wake."[4]

Since, in the year Ross wrote this book (the same year that an Eastern European lower-class Jewish immigrant named Joseph Goldberger, having pushed himself into the honorable position of Surgeon, U.S. Public Health Service, resolved the mystery of pellagra), "easily one-fifth of the Hebrews in the world are with us, and the freshet shows no signs of subsidence," the super-intelligence of these born parasites on decent society made further Jewish immigration a rising peril to native Americanism. As with Galton and Madison Grant, Ross believed, of the accursed Jews, that "centuries of enforced Ghetto life seem to have bred in them a herding instinct. No other physiques can so well withstand the toxins of urban congestion. Save the Italians, more Jews will crowd upon a given space than any other nationality."[5]

Ross warned that as the Czar's pogroms created new waves of Jewish refugees from terror, the effects in America were disastrous, because "with his clear brain sharpened in the American school, the egoistic, conscienceless young Jew constitutes a menace."

It was all a matter of eugenics, a creed to which Ross subscribed as wholeheartedly as did the Imperial Wizard of the Ku Klux Klan, Dr. Hiram W. Evans. Thus, on page 157, in the section subheaded "Race Traits," Ross wrote: "If the Jews are a race certainly one of their traits is *intellectuality*" (Ross's italics).

And, expanding on this theme, Ross went on to write of the Jews: "Teachers report that their Jewish pupils 'seem to have hungry minds.' They 'grasp information as they do everything else, recognizing it as the requisite for success.' Says a principal: *'Their progress in studies is simply another manifestation of the acquisitiveness of the race.'* . . . The Jewish gift for mathematics and chess is well known." (Italics added.)

Ross's fellow member of the Immigration Restriction League, the banking family heir Joseph Lee, took a similarly dim view of the "hereditary" intellectuality and tropism to hard work of the pushy, parasitic, and born criminalistic Jew. His book, *Constructive and Preventive Philanthropy* (1902, 1910), which on page 8 noted that the "problems of philanthropy [are] largely due to immigration," by page 198 got around to the traits of the Irish, "to whom the social features and athletics make the strongest appeal, and the Jews upon the other hand who never seem to be thoroughly happy unless they are engaged in some kind of hard manual or intellectual work." Far from endearing the Jews to Mr. Lee, these traits caused him to add: "A Jew boy, who has been at school morning and afternoon and worked hard at selling newspapers before and after school hours, would rather spend his evening wrestling over algebraical problems than give his time to frivolous pursuits that make no contribution to his power of achievement."

To Lee, who considered himself to be the father of the American playground movement, the fact that "Jew boys" seemed to prefer hard work and "algebraical problems" to hopscotch, baseball, football, and other children's games was proof of Jewish *inferiority*. The mere *intelligence* of the Jews made them undesirable.

"WE NEED MORE SUPERMEN"

By 1921, when the Second International Congress of Eugenics was convened at New York's American Museum of Natural History under President Henry Fairfield Osborn, time had provided no new scientific or moral justifications for such a congress. Nevertheless, the concepts aired at this world congress of scientific racism were to become the conventional opinions of a *majority* of America's academic, editorial, and political leaders for generations. Since we are still paying vast human and fiscal costs—in such preventable conditions as having one of the industrialized world's highest infant death rates— for the wholesale application of these eugenics ideas to public policies, the modern reader has much to gain by a quick review of what was said, and by whom, at this congress.

Major Leonard Darwin, son of the great Charles Darwin, president of the First Congress of Eugenics in 1912, and living proof that genius or even talent in science is *not* in any way hereditary, was a vice-president and honored guest of the second world convocation. If the future Prime Minister of England, Winston Churchill—a vice-president of the 1912 Congress— was not among the members of the 1921 Congress's General Committee, the soon-to-be President of the United States, Herbert Hoover, was. As were, among many other notables, college and foundation presidents and intellectuals: the future governor of Pennsylvania, Gifford Pinchot; the chief psychologist of the U.S. Army, Robert M. Yerkes; and the chairman of the Psychology Department of Columbia University's Teachers College, Edward L. Thorndike. The principal benefactress of the American eugenics movement, Mrs. E. H. Harriman, contributed generously to help meet the dollar costs of the Congress. She delivered her checks to the chairman of its Finance Committee, Madison Grant.

Grant's protégé, Dr. Lothrop Stoddard, was chairman of the Publicity Committee. Grant's old comrade-in-arms, Harry H. Laughlin, was chairman of the Exhibits Committee. Their old friend (and future president of the universities of Maine and Michigan and, in our own times, the tobacco industry's cancer research expert) C. C. Little was chairman of the Executive Committee.

Some of the men whose names were listed as committee members were not eugenicists and, as biologists, had in fact helped destroy all of its scientific pretensions. These men included Thomas Hunt Morgan, the cell biologist E. B. Wilson, and William H. Welch, then head of the Johns Hopkins University School of Public Health. Each of these men shared a contempt for eugenics as a pseudoscience *and* an all too human reluctance to hurt the feelings of lifelong professional and personal friends, such as Osborn and Davenport, who beseeched their permission to list their names on the literature of the Congress.

Osborn boasted in his address of welcome to the assembled participants at this "international conference on race character and betterment," held at a moment when World War I had cost mankind much of the genetic "heritage of centuries of civilization which can never be regained," that the Second International Congress of Eugenics had indeed brought together the possessors of *all* of the world's knowledge of the mechanics of human heredity.[6]

Osborn, the co-founder of the pathologically anti-democratic Galton Society, described the International Congress of Eugenics as being dedicated to the salvation of the nation's republican institutions by increasing the fecundity of its best breeding stock. "Rampant individualism, not only in art and literature but in all our social institutions" had risen at the end of the nineteenth century to "threaten the deflowering of New England," so that this womb of both branches of the Nordic race, the Teuton and the Anglo-Saxon, "has witnessed the passage of a many-child family to a one-child family. The purest New England stock is not holding its own. The next stage is the no-child marriage and the extinction of the stock which laid the foundations of the republican institutions of this country."

But thanks to the eugenics movement, America had now been alerted to this peril:

> In the United States we are slowly waking to the consciousness that education and environment do not fundamentally alter racial values. *We are engaged in a serious struggle to maintain our historic republican institutions through barring the entrance of those who are unfit to share the duties and responsibilities of our well-founded government* [italics added]. The true spirit of American democracy that *all men are born with equal rights and duties* has been confused with the political sophistry that *all men are born with equal character and ability to govern themselves and others,* and with the educational sophistry that education and environment will offset the handicap of heredity [Osborn's italics].

This, of course, was the theme of most (with some surprising exceptions) of the speakers at the Congress.

The Fourth Section of the Congress was devoted to eugenics and the state. As Osborn said in describing this section:

> The right of the state to safeguard the character and integrity of the race or races on which its future depends is, to my mind, as incontestable as the right of the state to safeguard the health and morals of its people. As science has enlightened government in the prevention and spread of disease, it must also enlighten government in the *prevention of the spread and multiplication of worthless members of society, the spread of feeble-mindedness, of idiocy, and of all moral and intellectual as well as physical diseases* [italics added].

Possibly the most important committee of the Congress was the Exhibits Committee, whose chairman was the supervisor of Davenport's Eugenics Record Office—Harry Hamilton Laughlin, the Mister Fixit of the eugenics movement. Not only did Laughlin supervise the exhibits displayed at the American Museum of Natural History at the time of the Congress, but when the last of the speakers had presented the last bit of "evidence" for the genetic inferiority of the Undeserving Poor, the exhibits were taken down and then remounted in the U.S. Capitol Building in Washington.

There, for the next three years, the representatives and senators pondering the pros and cons of immigration restriction, as well as proposals to make slight improvements in the health, housing, employment, education, and nutrition of the poor, the widows, the orphans, and the disabled, were treated to daily exposure to the maps of Madison Grant on the Passing of the Great Race; exhibits on the heredity of criminality, idiocy, musical talent, epilepsy, and other physical and mental traits; displays on the Jukes, the Nams, the Old Americans, and the Tribe of Ishmael; and various other eugenical arguments against "wasting" tax dollars on social legislation and for ending the further immigration of Jews, Italians, Hungarians, Poles, and other such dysgenic types.

Not all the papers presented at the Congress were incorporated into the exhibit that sat in the U.S. Capitol for the guidance of our solons for the next three years. One of the papers the sponsors tried their best to forget was Paper No. 31, "The Inheritance of Mental Disease," by the professor of neurology at Tufts University, Dr. Abraham Myerson.[7] One of the nonracists attracted to the eugenics movement, Myerson had long since learned what Galton and his followers were really up to, and his paper marked his coming of age.

In this presentation, Professor Myerson showed that contrary to the claims of amateurs in psychiatry such as Galton and Osborn, most mental diseases were demonstrably *not* hereditary; that the mentally sick had *fewer not more children* than other people; that "the transmission of mental diseases is practically nil insofar as the organic disease of the brain goes"; that "it is rare that one can see a direct relationship between a mental disease in a grandparent and one in a child"; that "most of the mental diseases have *no* hereditary relationships" although a few he described do; and, finally, that "the problem of the transmission of mental disease is a clinical medical problem to be studied by laboratory methods as in the rest of experimental medicine" and not by unqualified field workers and other amateurs. Not only did these scientific facts and realistic thoughts constitute heresy to any true-believing eugenicist, but Myerson had gone out of his way to attack, by name, Charles Benedict Davenport as one of the chief sources of dangerous popular myths about the heredity of mental disorders.

The quality of the exhibits of the Congress prepared for the nation's legislators can be gleaned from a quick skimming of the two volumes of its scientific papers. Volume II, for example, opened with an address on the dangers of race mixing by the ancient craniologist, Jew baiter, and Nordicist,

Count Vacher de Lapouge. Very sensibly, on the theory that most readers would not know French, when the proceedings of the Congress were published in 1923 the text of his speech was printed in the original French.

Lapouge, one of the last living links between Gobineau and Osborn—who hailed him as "the leading authority on racial anthropology and earnest exponent of practical eugenic measures by the government"—had during the war been bluntly told by *his* government (in French) to stay away from the Army when he gallantly offered to make cranial measurements of the French soldiers. It appeared that too many French biologists and anthropologists had by then read the scientific reports of Columbia University's Professor Boas on the significance of the cephalic index, and the invaded French Republic had a few more important things to do with its recruits' heads than to bedevil them with craniology or IQ tests.

So that in his talk to the Congress of Eugenics,[8] Lapouge could scarcely be blamed for singling out the man who, he was certain, had given Boas his ideas, and whose pernicious heresies were responsible for the fact that Lapouge's "anthroposociology" and other "branches of science" that told the truth about mongrel races "have been excluded from the curriculums of French universities for political reasons." And then he boldly named names: "Since *Jean Jacques Rousseau* everyone accepts as fact that the differences that arise among individuals emanate from education, and never from heredity." (The press agent of the Congress, Lothrop Stoddard, took an equally paranoid view of Rousseau in five different places in *Revolt Against Civilization* the following year, blaming him for the false and pernicious "doctrine of natural equality," and labeling him a spokesman for revolt, writing that "the tide set flowing by Rousseau and his ilk presently foamed into the French Revolution.")

Now, in the aftermath of World War I, said the Count, "we are witnessing at this time a crisis in which the superior races are at great risk of disappearing, just precisely at the time that we need more supermen."

Along with E. M. East and the modern population extremists, Comte de Lapouge also warned: "The earth itself will soon no longer be able to furnish us with its riches. In a few centuries there will be no more metals, no more coal, no more oil, nor enough food. Soon it will no longer be possible to find capable men."

But the racial and social dangers were far more immediate than coal and oil shortages:

> The time has come for man to decide whether he is to become a God or whether he is to return to the savagery of the days of the Mammoths. *The classes of people which are the least gifted turn on the elite classes, and they are creating a civilization that multiplies their desires far beyond our capacity to fulfill them.* A movement has begun among the inferior classes; it is directed against the whites, the rich, the intellectually superior stock, and against civilization itself. *The war of the classes is in fact a war of the races.*
>
> In my article of February 24, 1887 (*Revue d'Anthropologie,* 1887, p. 549), I said: "The organization of artificial selection is only a ques-

tion of time. It will be possible within a few centuries to totally replace humanity as we now know it with a superior civilization. I am confident that the Anglo-Saxons will be capable of leading this enterprise to its just fruition." At the time I wrote that statement I was all alone in Montpellier. Now I have crossed the ocean and I am surrounded by a large audience. America, the safety of civilization depends on you. [Italics added.]

As this man of 1887 spoke to the assembled scientific racists of 1921, the lunatic asylums of his own and his host nation were filled with men and women whose grip on reality was far more firm than that displayed in his words. At that, Lapouge's "science" was no further off the tracks of the century than most of the other papers, foreign and domestic, presented at the Congress.[9]

Given these scientific standards and guidelines—for what Osborn and Lapouge had had to say was to be repeated again and again by most speakers —it was therefore inevitable that, except for accidents such as the Myerson paper, the so-called scientific papers of this Second International Congress of Eugenics proved to be about as scholarly as Grant, Ripley, Popenoe, and Johnson. A quick glance at the published papers of the Congress pretty well establishes their more or less common ingredients.

"A DEFINITE RACE OF CHRONIC PAUPER STOCKS"

The world's leading authority on the genetics of hereditary pauperism, from which the worst elements of *that "race"* spawned, E. J. Lidbetter of London, England, was invited by his admirer Major Leonard Darwin to present Paper No. 391, "Pedigrees of Pauper Stocks."[10]

Lidbetter was neither a geneticist nor a biologist nor a scientist of any kind. He was a civil service bureaucrat in the London City Council's poor law authority and, since 1898, had been engaged in the administration of welfare services to the poor of London. His hobby, starting in 1910 with an article in Galton's *Eugenics Review,* "Some Examples of Poor Law Eugenics," was the drawing up of the pedigrees of the white Anglo-Saxon Londoners receiving any form of public assistance. These included[11] individuals who were hospital ward patients for any reason; people who were blind, tuberculous, epileptic, alcoholic, and criminal; children who "died within the first year of life"; people certified to be of unsound mind; unemployed people living in workhouses or receiving "outdoor relief" stipends that let them live outside of these miserable shelters; and children maintained in poor law schools or orphan asylums.

Just as Davenport's pedigree trees had, in 1916, "proved" the hereditary transmission of pellagra from poor parents to poor children, Lidbetter's eugenic diagrams proved that the preventable diseases of poverty suffered by all generations of poor people were the products of bad genes rather than of malnutrition, unsanitary slums, and other environmental factors conducive to rickets, tuberculosis, trachoma and other causes of blindness, and the deaths of infants from the deficiency and infectious diseases of childhood.

Needless to say, Lidbetter's paper—which was to be cited endlessly by eugenicists forever afterward—came to predictably Galtonian conclusions. These included:

1. That there is in existence a definite race of chronic pauper stocks. . . .
2. That modern methods of public and private charity tend to encourage the increase of this class by relieving the parents of the normal responsibilities of parenthood. . . .
3. That the reduction of this class may be brought about by a due observance of the laws of heredity.

That is, since paupers are the products of a specific race with inferior genes, society had to keep them from breeding more of their race.

Major Leonard Darwin, in his paper, "Aims and Methods of Eugenical Societies," said that "eugenics aims at increasing the rate of multiplication of [human] stocks above the average inheritable qualities, and at *decreasing* that rate in the case of stocks below the average." Since hereditary paupers were the genetic products of what Lidbetter defined as *"a definite race of chronic pauper stocks* (italics added),", and since the poor were the source of most social inadequates, the son of Charles Darwin and president of the Eugenics Education Society of Great Britain declared there was only one solution. This was for society to face fearlessly "the danger resulting from the unchecked multiplication of inferior types." That is, compulsory sterilization of the hereditary paupers and other socially inadequate "races," as the more advanced one third of the states in America were already doing.

Professor William McDougall presented two papers. The first, "A National Fund for a New Plan of Remuneration as a Eugenic Measure,"[12] was a prime early example of what U.S. Supreme Court Justice William O. Douglas was later to characterize as "socialism-for-the-rich." Briefly, McDougall proposed that society seek out the unfortunately insolvent bearers of super-genes among less affluent teachers, bank clerks, and other members of the "more educated classes" and subsidize their production of children by regular monthly family assistance allotments. McDougall's family assistance plan did not, of course, apply to the Undeserving Poor, because, instead of bearing the genes for high IQ test scores and affluence, *their* blood was contaminated with the genes for low IQ test scores and pauperism.

McDougall's second paper, "The Correlation Between Native Ability and Social Status,"[13] had as its purpose the demonstration of "a positive correlation between social status and civic worth, or, in other words, to show that the economic stratification of society corresponds in some degree with the distribution through the population of the more desirable human qualities, more especially in the quality of intelligence."

The evidence, he reported, was of course derived "from the method of mental testing which Francis Galton was the first to create and apply." Which, in turn, led to the beloved Army intelligence tests of World War I that proved once and for all that "the illiterate recruits attained a considerably lower average mark for intelligence than the literates, and this was true in about the same degree for both white and colored recruits taken separately."

Since "there can be no doubt, I suppose, that the illiterates among the recruits represented socially and economically lower strata of the population than the literates," it was therefore obvious that "the army mental testing has thus provided evidence that *the lower social strata in this country possess less native intelligence than the average citizen*" (italics added).

Harry H. Laughlin also presented two papers. The first, "The Present Status of Eugenical Sterilization in the United States,"[14] proudly revealed that—thanks to the efforts of the American eugenics movement—one in every three states now had compulsory sterilization laws on its books, and that, since 1907, 3,233 human beings had been sterilized by their states, 1,853 of them males, 1,380 of them females—and two thirds of them, 2,558, Californians. All of these victims of eugenical mutilation in California, Connecticut, Indiana, Iowa, Kansas, Michigan, Nebraska, New York, North Dakota, Oregon, Washington, and Wisconsin had been involuntary inmates of state penal and mental institutions. But, said Dr. Laughlin, the success of these compulsory sterilizations of institutionalized Americans to date had shown that "the experimental period is rapidly passing away," and the time had therefore come to face the eugenical truth:

> The extension of the provisions of the sterilization law to *all* cacogenic persons of a given legal standard whether within public or private custodial institutions *or in the population at large,* is both a legal necessity and *a practical requirement for eugenical effectiveness* [italics added].

Laughlin's second paper, "Nativity of Institutional Inmates,"[15] was a statistical study of the "relative inborn social values of recent and older immigrant stocks" in which he showed that:

> . . . the foreign born population of the U.S. is contributing to our custodial institutions [hospitals, insane asylums, and prisons] one and one-third its quota, while our older stock is contributing only about nine-tenths of its percentage allowance. The first generation of our descendants of immigrants show a little higher social value than our most recent immigrants, but here also their quotas of degeneracy are so high as to justify the conclusion that within the last generation the *inborn* physical, mental, and moral qualities of our immigrants have been declining.

By degeneracy, Laughlin meant the victims of tuberculosis and other "contagious segregated diseases," as well as a diagnosis of "feeble-mindedness."

Laughlin's boss and benefactor, Charles B. Davenport, presented a paper called "Research in Eugenics."[16] In this paper, Davenport said that the pedigree analyses which had demonstrated that "feeble-mindedness of the middle and higher grades is inherited as a simple recessive [*sic*]" showed why this nation had to be careful about whom it admitted as immigrants.

In the same paper, Davenport asserted—on no scientific authority higher than his own gut feelings— that "epilepsy, of the ordinary juvenile, dementing [*sic*] type, seems to be due, like feeble-mindedness, to a single developmental defect." Just as another "group of hereditary mental defectives is such because those who belong to it lack a single [hereditary] factor for an adequate developmental impulse."

Davenport fancied himself to be as much of a social philosopher as a biologist, hence:

> The fact that not only our physical but also our mental and temperamental characteristics have a hereditary basis has certain important social bearings. . . . The false doctrines of human equality at birth and of freedom of the will have determined a line of practice in the fields of education and criminology that, it seems to me, is not productive of the best results. In education we must know the child's *native [genetic] capacities* before we can properly train. In dealing with delinquents we must know the *hereditary* mental and emotional make-up before we can get an explanation of the bad conduct and before we can intelligently treat the delinquent. [Italics added.]

The director of the Eugenics Record Office also warned that "a failure to be influenced by the findings of the students of eugenics or a continuance in our present *fatuous* belief in the potency of money to cure *racial* evils will hasten the end" (italics added). By fatuous uses of money Davenport meant appropriations for improved health services, schools, and housing for the native and foreign-born poor who constituted the race known as the Undeserving (and/or Vicious) Poor.

Paper No. 50, "The War from the Eugenic Point of View,"[17] was by Corrado Gini, Mussolini's closest scientific adviser. It proved that, contrary to the "pessimistic views" of the American eugenicist and pacifist David Starr Jordan and the zoologist Vernon Kellogg, war births do not produce frailer children. While "excessive mortality may continue for a time, after war, due to persistence of disease and economic distress," it was also clear to Gini that "such mortality has a favorable selective influence."

THE MENACE OF LITERATE LAPPS

The presentation that would emerge as the ultimate symbol of the entire Congress, however, was Paper No. 7: "Harmonic and Disharmonic Race-crossings,"[18] by Dr. Jon Alfred Mjöen, of the Winderen Laboratorium just outside of Oslo, Norway.

As a scientific presentation, Mjöen's slide lecture was a perfect burlesque of the Nutty Professor routines then so popular in American vaudeville houses. He had graphs, charts, and photographic slides of everything from pure Nordics and Mixed Race (Lapp × Norwegian) types to mixed-breed rabbits and to two types (unfortunate and hereditary) of bare-bosomed prostitutes. He had genetic or pedigree trees by the peck, and his talk was filled with learned citations of the work of other Scandinavian eugenicists.

Like Davenport, who had discovered the genetic traits of "Nomadism" and "thalassophilia" (love of the sea), Mjöen had also made a major eugenical discovery. For it was none other than Jon Alfred Mjöen who had discovered the M.B. Type (M.B. standing for *Manglende Balance* in Norwegian, or Want of Balance in English). As Mjöen reported to the Congress, "the main mental feature of this type was an unbalanced mind." The carriers of the M.B. genes were described by professor Mjöen as "showing as their main symptoms stealing, lying, drinking."

Head form (cephalic index) and other body measurements and observations Mjöen had "taken amongst the Lapps in the north of Norway lead me to formulate the distinction: *Harmonious and disharmonious* crossing." Mixed breeds were disharmonious race crossings, and the nomadic Lapps presented as great a peril of race suicide to the Nordic Norwegians as that which E. A. Ross and other American scientists claimed the Jews presented to Old American (Nordic) stock. His titillating slides on the perils of race crossing lent a dash of added color to his talk.

For the dispassionate but more scientifically oriented eugenicists who believed that one genetic tree was worth a thousand pictures, Mjöen had dozens of pedigree charts such as the one reproduced on page 286.

Clearly, in Oslo, Norway, as in Vineland, New Jersey, and Cold Spring Harbor, New York, the true-believing eugenicists had learned the didactic value of pedigree trees. Mjöen's pedigree trees were every bit as convincing— and as terrifying to faithful eugenicists—as Goddard's pedigree trees of the good and the M.B. type Kallikaks, or Davenport's of the Nams and the white southern pellagrins.

Like all eugenicists since Galton, Mjöen believed that disharmonious race mixing was physically as well as mentally deleterious: ". . . not alone tuberculosis [*sic*] but also many other diseases and social evils, e.g., the increasing criminality (from mentally disturbed race elements), is partly due to disharmonic racecrossing. To find out why is the most urgent work for the racebiologists [i.e., eugenicists]."

But while the "racebiologists" pondered the problems, Mjöen had already come up with the answers, which he provided two paragraphs later: "We have seen today that two individuals from two good stocks can procreate one or more M.B. types *if they belong to different races,* in other words can procreate caricatures of human beings" (italics added).

In his summary, Mjöen not only reiterated that "crossings between widely different races can lower the physical and mental level," but also asserted that "the figures for tuberculosis are the smallest in that part of Norway where the Nordic race is comparatively pure (1.1 to 5) and largest in that part of the country (Finmarken) where the race mixture is the largest (3.6 to 4.0)." And also:

> Prostitutes and the "unwilling to work" are found more frequently among types showing strong race mixtures than among the relatively pure types.
> . . . the immunity which a race or a population has gained towards certain diseases is unfavorably influenced by racecrossings. It seems as if the original disposition for disease which has been eliminated by selection (the weakest dying out) in the original population appears again when the two immune races are being crossed.

This from a man whose curriculum vitae stated that he studied biology and chemistry at the universities of Leipzig, Berlin, and Munich "after several

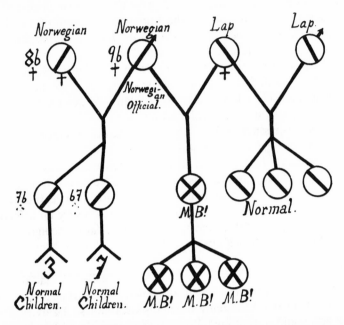

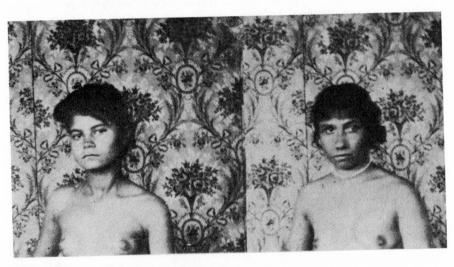

Mjöen's pedigree tree on "Unharmonic Racecrossing" did not, of course, prove the existence of his so-called hereditary "M. B." type. Nor, for that matter, did the essentially pornographic act of photographing two prostitutes with bosoms bared prove that the woman on the left was a pure-blooded Nordic who entered her venerable profession because of "unfortunate surroundings," while the woman on the right became a prostitute "on account of disharmonic race mixture." Mjöen's and other equally pseudoscientific displays from the world eugenics congress were, however, accepted at face value by most members of the Congress of the United States.

years of medical training in American and English hospitals and as assistant to the American physician Professor Squibb, M.D., who is especially known for having introduced the use of ether in the narcosis."

Mjöen's summary included his famous Point 8, which was to bring him lucrative lecture tours of the United States under the auspices of the American Eugenics Society for the next half-dozen years. It ran, in its entirety:

> Until we have more definite knowledge of the effect of racecrossings we shall certainly do best to avoid crossings between widely-different races and nourish and develop a strong and healthy race instinct. By removing the bilinguistic barrier [i.e., enabling the Lapps to learn Norwegian and Swedish] we are building the first bridge, safe and sure, to a blood mixture between the two races which we will deplore and regret when it is too late.

For the balance of the decade, as Mjöen toured the university and other lecture halls of the United States, the inoffensive nomadic Lapps of the Far North of Scandinavia were to stand in as surrogates for the foreign-born Jews and the Italians, the native and indubitably Nordic Jukes and the Kallikaks, the poor-white Nordic trash of the South, the Negroes and the Chinese and the other nonwhites who all combined to imperil the purity of the Nordic-American blood lines—and to waste the taxes of the thrifty and the hard-working by demanding education for their genetically ineducable whelps.

No smart Jewish lawyers like Louis Marshall, no sharp-tongued Italian congressmen like New York's La Guardia, no angry and articulate black veterans of World War I could ever accuse the university, the Legion post, or the women's club that hired Mjöen to present his race-crossing routine of being anti-Semitic, anti-Italian, anti-Negro. There were simply no Lapps on hand to protest that tuberculosis was caused by the bacillus *Mycobacterium tuberculosis* in bodies weakened by the biological sequelae of poverty, and not by Lapp genes, *let alone "disharmonic racecrossings."*

Until the Great Depression made this caricature of a scientist and his humorless burlesque of a scientific presentation economically expendable, Mjöen toured the nation, periodically, speaking to audiences booked by the American Eugenics Society and other like groups. In 1926, Osborn proposed to Davenport that "we should arrange a special dinner and reception for Dr. Jon Alfred Mjöen by the Galton Society," and of course Davenport got in touch immediately with William K. Gregory, secretary of the Society, to bring this reception about. Two years later, Davenport invited Mjöen to join Eugen Fischer (who was already one of Hitler's scientific advisers) on the "committee on race crossing" of the International Federation of Eugenic Organizations, of which Davenport was then president.

THE MENACE OF LITERATE JEWS

Even before the "scientific" papers of the International Congress of Eugenics were published, their 1921 and earlier "data" on the irrevocable (because genetic) ineducability and other physical, behavioral, and moral defects of the inferior races who made up the populations of the Undeserving Poor were lending scholarly respectability to the overtly racist efforts to bar the Jews, Italians, and blacks from colleges, professional schools, and the noble professions—let alone managerial-level jobs or, for that matter, any job above the level of porter or ditch digger. As Professor E. A. Ross had warned in his 1914 book, *The Old World in the New:*

> Twenty years ago under the spoils system the Irish held most of the city jobs in New York. Now under the [civil service merit] *test system* the Jews are driving them out. Among the school teachers of the city Jewesses outnumber the women of any other nationality. Owing to their aversion to "blind-alley" occupations Jewish girls shun housework and crowd into the factories, while those who can get training become stenographers, bookkeepers, accountants and private secretaries.

This proved the racial dangers of letting Jewish girls get white-collar job training. More than that, it was by 1914 clear to Professor Ross:

> One-thirteenth of the students in our seventy-seven leading universities and colleges are of Hebrew parentage. The young Jews take eagerly to medicine and it is said that from seven hundred to nine hundred of the physicians in New York are of their race. More noticeable is the influx into dentistry and especially into pharmacy. Their trend into the legal profession has been pronounced, and of late there is a movement of Jewish students into engineering, agriculture and forestry.[19]

Now, with the Army intelligence tests proving by cold numbers and statistics that these nine hundred Jewish doctors were mental morons if not imbeciles, and with the more than a hundred scientific papers of the world's leading experts on human heredity who addressed the International Congress of Eugenics of 1921 documenting the biological bases for the inferiority of Jews and other low types, the time had come not only to bar them as immigrants but also to keep their American-born children out of our good Nordic colleges, professional schools, and white-collar careers. Already, Ross had written:

> In New York the line is drawn against the Jews in hotels, resorts, clubs, and private schools, and constantly this line hardens and extends. They cry "Bigotry" but bigotry has little or nothing to do with it. What is disliked in the Jews is not their religion but certain ways and manners. Moreover, the Gentile resents being obliged to engage in a humiliating and undignified scramble in order to keep his trade or his clients against the Jewish invader. . . . If the Czar, by keeping up the pressure which has already rid him of two million undesired subjects, should succeed in driving the bulk of his six million Jews to the United States, we shall see the rise of a Jewish question here, perhaps riots and anti-Jewish legislation.[20]

Inspired by Ross and other anti-Semites in the groves of academe, the gentry who sat on the boards of most of our better and less than better private universities began demanding and getting the imposition of admissions quotas whose frank purpose was to keep all Jews, Italians, Poles, Mexicans, blacks,

Orientals, and other non-Aryan students out of their undergraduate and professional schools.[21] By 1920, when the drive to bar the Jews and other "inferior" races from our campuses began, however, it was not conducted in the name of racism but in the name of science.

The board members who demanded their exclusion had, they explained, nothing against the Jews personally; but what with the suddenly revealed *scientific* fact that, as Professor Brigham put it in his great analysis of the Army intelligence tests, *A Study of American Intelligence,* "American intelligence is declining, and will proceed with an accelerating rate as the racial admixture becomes more and more extensive,"[22] the moment of racial decision had come. Now that the Army mental tests had proven, *beyond any scientific doubt,* that, like the American Negroes, the Italians and the Jews were *genetically* ineducable, it would be a most fatuous waste of good money even to try to give these *born* morons and imbeciles a good Anglo-Saxon college education, let alone accept them into our fine medical, law, and engineering graduate schools.

However, keeping the Alpines, the Mediterraneans, and the Jews (or, as Brigham labeled them, the Alpine Slavs) out of medical, law, and engineering schools could not, in itself, protect the Real Americans from what E. A. Ross termed race suicide. The main line of defense against race suicide lay in the formulas advocated by the Immigration Restriction League, of which Ross was a leading member and propagandist, and by noble thinkers such as Charles W. Gould, Madison Grant, Henry Fairfield Osborn, and Hiram Wesley Evans. That was, of course, to end all non-Nordic immigration.

The full story of how the nation's xenophobes, nativists, hand-wringing liberals, and rampaging Yahoos finally put over the genocidal U.S. Immigration Act of 1924 has been well told by many authors, and will not be retold here. Here we will examine, if only briefly, the role of the eugenics movement and other organs of scientific racism in making this victory of the restrictionists a reality.

THE ORIGINS OF THE GENOCIDAL 1924 IMMIGRATION QUOTAS

The primary operator on the scientific racist front in this triumph of the eugenics movement was Harry Hamilton Laughlin. In 1920, he was appointed Expert Eugenics Agent of the House Committee on Immigration and Naturalization by its chairman, Albert Johnson, the high school dropout and semiliterate who owed his job to the superb lobbying effort of the Immigration Restriction League's paid full-time Washington lobbyist, James H. Patten, and the League's many friends in the House, Senate, and governing councils of the Republican Party. As part of the process of manipulating this lumbertown newspaper publisher, the Davenport-Grant-Osborn clique had Albert Johnson installed as the president of the Eugenics Research Association in 1923. This ploy raised many a discreet chuckle from parties to this flattery of the powerful chairman of the House Committee on Immigration.

The opposition had their chuckles too. The irreverent son of a U.S.

Army bandmaster, New York's Congressman Fiorello La Guardia, not only Italian but also Jewish by birth, World War I flyer, and totally lacking in respect for blood-line cults, responded to a magazine article that had "derogated New York City's congressmen as foreigners who represented an alien population" as follows:

"I have no family tree. The only member of my family who has is my dog Yank. He is the son of Doughboy, who was the son of Siegfried, who was the son of Tannhauser, who was the son of Wotan. A distinguished family tree, to be sure—but after all, he's only a son of a bitch."[23]

This was, alas, only gallows humor. And as La Guardia himself knew to his sorrow, on the gallows the *last* laugh always belongs to the hangman. When racist authors such as E. A. Ross found, in the educational quotas of the Czars, the Kaisers, and the Hapsburgs of Old Europe, the magic formula to keep the colleges and professions *Judenrein,* they also ultimately provided the final link in the chains that barred the gates to non-Aryan immigration. This link was the quota system, proposed by the New York lawyer John B. Trevor—Harvard and Columbia Law School classmate of Franklin D. Roosevelt, whose wife, Caroline Wilmerding Trevor, was one of Eleanor Roosevelt's oldest and most cherished friends.

Trevor, an intimate friend and associate of Madison Grant, came up with a quota plan designed to Keep America American. This was to restrict foreign immigration to quotas of 2 percent of the number of foreign-born residents in this country. However, there was a lethal joker in this deck: Trevor's plan called *not* for a quota of 2 percent of the foreign-born residents of the United States as of the latest (1920) U.S. Census, *but as of the Census of 1890.*

The reasons for selecting 1890 as the year on which to base an immigration quota system were quite obvious. As can be seen from the table opposite, the 1890 quotas would have effectively ended nearly all further immigration of Italians, Poles, Greeks, Hungarians, Russians, and Jews.

As can be seen from these Census Bureau data, the Trevor quotas, once enacted, would have established a quota of only 3,652 Italian immigrants per year—as against 55,698 Germans. Countries where most twentieth-century Jewish immigrants and prospective immigrants were born included Russia, Austria, Hungary, Poland, and Lithuania, each of them with comparatively few foreign-born U.S. citizens or residents as of the Census of 1890. The Polish quota, based on 2 percent of the Polish-born population of 1890, came to a minuscule 2,951. The Greek quota was exactly 38.

Few ethnic groups fared worse than the Jews under the proposed Trevor immigration quotas, since as of 1890 most of the Jews resident in the United States were American-born Sephardic Jews who arrived here before the American Revolution and/or native-born children and grandchildren of German Jews who fled Germany after the defeat of the republican Revolution of 1848. Which meant that Trevor's quotas did not recognize any of them as being of other than *American* national origin.

There was, in fact no formal "Jewish quota" at all in the immigration law, any more than there ever was any *official* Jewish quota on admissions in

A GIANT STEP ON THE ROAD TO AUSCHWITZ

National Origins of Past and Prospective Immigrants	Foreign-Born Population in U.S., by Country of Birth, in 1890*	To Be Admitted as Immigrants under Trevor 2 Percent Quotas of 1924
England	909,092	18,182
Ireland (incl. Ulster South counties)	1,871,509	37,430
Sweden	478,041	9,561
France	113,174	2,263
Germany	2,784,894	55,698
Poland	147,577	2,951
Austria	241,377	4,828
Russia (incl. Baltic States and Finland)	182,684	3,654
Greece	1,887	38
Italy	182,580	3,652
Spain	6,185	124
Portugal	15,996	320

* Source: U.S. Bureau of Census.

Although the primary targets of the 2 percent of national origins quotas, based on the foreign-born populations of the U.S. Census of 1890, were the Jews and the Italians, Trevor and the Immigration Restriction League never proposed any Jewish quotas per se. In the case of the Jews, such naked racism was not required; this was because most of the post-World War I Jewish immigrants would have been classified as being under the national origins quotas assigned to Russia, the Baltic States, Austria, Hungary, Poland, and Greece. Most of the Jews residing in the United States at the time of the Census of 1890 were American-born descendants of Spanish, Portuguese, French, and Dutch Jews who arrived in the New World prior to the American Revolution, and of German republicans who fled Europe after the defeat of the German Revolution of 1848. American-born, they were, of course, not included among the foreign-born population in the Census of 1890.

The blatantly Teutonist quotas proposed by Trevor, the eugenics movement, and the IRL action coalition embracing the American Federation of Labor on the left, scores of hyperpatriotic and nativist societies in the center, and the Ku Klux Klan on the right, were obviously intended to end all but token Italian and Jewish immigration into the United States. They succeeded in this objective.

any American university. The Trevor quotas were simply strategies for ending practically all Jewish and Italian immigration completely.

STRAIGHT PSEUDOSCIENCE FROM THE OFFICIAL "EXPERT EUGENICS AGENT" OF CONGRESS

Fully a year before Laughlin had had the exhibits of the Second International Congress of Eugenics set up in the Capitol, he presented an expert report to the U.S. Congress, "Biological Aspects of Immigration," on April 16, 1920. It was to be followed by many others. Laughlin continued periodically to testify before the U.S. Congress as an expert on the biological and eugenical aspects of immigration and deportation until December 1941, when America's entry into World War II rendered both problems academic for the duration.

Laughlin's 1920 appearance, *at the request of the committee,* was as much in the nature of a sales pitch for the employment of Davenport-Laughlin trained field workers abroad to halt Jewish and Italian immigration at their sources as it was for the biological orthodoxies of the eugenics movement.[24] No sooner did he open with "I want to present some of the biological and eugenical aspects of immigration," and continue to explain that "the character of a nation is determined primarily by its racial qualities; that is, by the *hereditary* physical, mental, and moral or temperamental traits of its people," than he was setting forth "a plan which our investigators thought should be enforced in testing the worth of immigrants." The plan turned out to be Davenport's old dream of dispatching an expeditionary force of eugenics field workers to examine would-be Jewish, Italian, Russian, and Hungarian immigrants in their native hovels and certify them as being the carriers of the genes for idiocy, low IQ test scores, tuberculosis, and other undesirable hereditary traits.

Since he was facing a committee of the U.S. Congress, Laughlin treated them like fellow intellectuals: "You have all doubtless heard of the Jukes and the Ishmaels; our field workers went to Indiana to study degenerate families, and found a certain name (now called the Ishmaels) so common that they said there must be something wrong with that family. They began to study it scientifically."

And this scientific study proved that far from being "poor things" who "never had an opportunity" to become chairman of the board, their ancestors "had been deported from England" because even then "it was found that they were the kind who would steal the bishop's silver if they got a chance." Laughlin added that "Dr. Charles B. Davenport, of our association, reported that in 1914 Sydney had an excessively large slum district, populated to a considerable extent by the descendants of the Botany Bayers deported from England. This is one sort of a study our association is making in immigration; we want to prevent any deterioration of the American people due to the immigration of inferior human stock."

> THE CHAIRMAN: The two families you mentioned are who?
> MR. LAUGHLIN: The Jukes and the Ishmaels. Most of the Jukes are in New York State; they are a worthless, mentally backward family or tribe. You have to recognize the fact that although we *give opportunities in this country, everybody is not educable.* There is a third famous degenerate family, the Kallikaks, of New Jersey, and while these three families have been famous in magazines and newspapers, our field workers every month send in case histories that deal with the same human types and conditions. The lesson is that immigrants should be examined, and the family stock should be investigated, lest we admit more degenerate "blood." [Italics added.]

This went on for another twenty pages or so, touching on all the totems and taboos of the eugenics cult, the same tiresome litany about bad stock, and the declining levels of American health, morals, and intelligence.

On page 15 of his statement, Laughlin revealed:

> . . . as in the case of the insane, the feeble-minded and practically all other types of the socially inadequate are recruited more numerously

from recent immigrant stock, in proportion to the total number, than from our older settlers. *Apparently the quality of our immigration is declining. . . .*

Now . . . one notices by the names of the individuals who are found in institutions, that the lower or less progressive races furnish more than their quota. [Italics added.]

Why, in the last reports Laughlin and Davenport had received from their field workers in schools for delinquents in Whittier, California, and Gainesville, Texas, "about half the names were American [*sic*] and the other half Mexican or foreign-sounding." This was alarming because the experts of the Eugenics Record Office had up to April 16, 1920, "failed to find a case in history in which two races have lived side by side for a number of generations and have maintained racial purity." It was a scientific fact, Laughlin told the congressmen, that "wherever two races come in contact, it is found that the women of the lower race are not, as a rule, adverse to intercourse with men of the higher. And that has been true throughout history. It is true now."

On page 18, he presented one of the ERO's famous charts, this one showing that:

. . . of the foreign-born insane in civil hospitals, 32 percent are Irish, whereas of the foreign-born in hospitals for the criminal insane only 25 percent are Irish. On the other hand, the proportions for Italians is 5% and 23% respectively, and for the Poles and the Russians 11.5% and 12.6%. Austrians 4.5% and 5.3%, respectively. This shows that the Italians, Russians, Austrians, largely Jews, constitute a large proportion of the insane.

With these data, Laughlin also presented tables, charts, and fiscal reports, by states, showing that state expenses in 1916 for maintaining institutions for the socially inadequate came to $75,203,239, the sums ranging from the $11,230,856 which put New York in first place, to tiny Delaware, which spent only $91,782. Needless to say, Laughlin continued, the higher the proportion of new immigrants in any state, the higher its percentage of total state expenses for this purpose.

What made this all the more terrifying was what Laughlin called the well-known "fact" that immigrant and native moronic and imbecile women were not only more promiscuous but also more fertile than respectable Nordic-American ladies of good stock.

These were Laughlin's concepts of the biological aspects of immigration in 1920. Two years later, in his report "Analysis of America's Modern Melting Pot,"[25] he arrived at precisely the same conclusions but presented considerably more detailed statistics, analyses, and numbers until they ran out of the ears of his auditors.

During the interregnum between the "Melting Pot" and the next and most historic report of the House committee's Expert Eugenics Agent, some of his close companions-in-arms had their turn before his committee. Starting with Dr. Lothrop Stoddard.

By the time Stoddard testified before Johnson's committee in January 1924, not only was he the author of *The Rising Tide of Color* and *The Revolt*

Against Civilization, but he had been hailed as a prophet of our times by a host of admirers ranging from the recently deceased President of the United States, Warren Gamaliel Harding, to Havelock Ellis, the sexologist who converted Margaret Sanger to the eugenics creed. Dr. Ellis, in fact, had written a highly laudatory three-page review of Stoddard's book *The Rising Tide of Color* for Mrs. Sanger's *Birth Control Review* of October 1920, entitled "The World's Racial Problem." "It is well to remember that [Stoddard's] conclusions are, after all, fundamentally in harmony with those of sober and judicial observers in Europe," wrote Dr. Ellis, no doubt with judicial observers such as Mjöen, Gini, and himself in mind.

Therefore, when Lothrop Stoddard, as an expert witness before the Congress considering America's immigration policies, warned that Jewish money and Jewish guile were inflicting upon this continent a race of low types who had no legal right to be admitted even by the existing legal standards, his words carried great weight. As he explained the "facts":

> . . . the Jews belong to those classes of the community which everywhere are more desirous of coming to America: that is, the lower middle class and peddling classes generally. Then also there is the fact that Jewish emigration has been assisted by various immigrant aid societies by measures of revolving funds of money, etc., and by good team work between those aid societies and various of the steamship organizations, etc. For all these reasons the Jews have succeeded in getting into the quota in numbers ahead of those which they would represent by the mere standards of population.[26]

The quotas Stoddard accused the Jews of cheating were based on the Census of 1910, as set in the then current immigration act up for revision.

Kenneth Roberts, the hired immigrant baiter of *The Saturday Evening Post,* whose series on the menace of Jewish and Italian immigration was published as a best-selling book, *Why Europe Leaves Home* (1922), was yet another of the committee's expert witnesses. Roberts urged all congressmen and other laymen present to read Madison Grant's *The Passing of the Great Race,* and read into the record various of the famous passages in which Grant had held forth on the menace of the dwarf-statured, brachycephalic Polish Jew who spurned the Nordic-American's religion while ravishing his women.

Grant himself did not make a personal appearance at these particular hearings, contenting himself with having the chairman, Albert Johnson, insert into the record a two-page statement, dated January 3, 1924, in which the defender of the cephalic index and other holies of scientific racism pressed for Trevor's 1890 quota, since this would establish modern immigration quotas at 83 percent of the 1920 immigration levels for immigrants from Northern and Western Europe and 11.7 percent of the 1920 immigration totals for Southern and Eastern Europe. These quotas would still give the brachycephalics and other non-Aryan scum a better break than they deserved, "because immigrants from these countries and their descendants certainly do not constitute anything like as much as 11.7% of our total population today. . . . The 1890 census is so fair that there would be no merit in such protests as have been received recently from the Italian Government on the ground of discrimination."[27]

NON-NORDIC BLOOD THREATENS TO WORSEN THE DECLINE OF NATIVE AMERICAN INTELLIGENCE

If he did not deign to materialize in person at the House hearings, Grant had nevertheless been exceedingly active behind the scenes. Much as the whiskey tippling and the coarse manners of the parvenu Albert Johnson possibly offended the customs of the patrician Grant, sheer patriotism (and, perhaps, a soupçon of anti-Semitism) caused Grant and his wellborn friends, such as Charles W. Gould, to receive Johnson in their own homes, and introduce him to their own exclusive clubs, for long briefing and planning sessions out of which came such witnesses as Lothrop Stoddard and Kenneth Roberts. And such statements as that prepared for the committee by the Allied Patriotic Societies of New York, which showed that the change from "superior" to "inferior" immigration had begun around 1882. The Allied Patriots told Congress that the superior "older" immigration from Northern and Western Europe had been supplanted by the vastly inferior newer immigration "composed of peoples widely divergent in their racial qualities" from the races which had made America great.

These alarming demographic and genetic shifts were summarized in their table on the racial nature of our immigration between 1899 and 1902, printed on page 580 of the House hearings:

races or peoples sending the largest numbers during the period 1899 to 1920, and the respective quotas of each, were as follows:

South Italian	2, 898, 499
Hebrew	1, 565, 607
Polish	1, 423, 209
German	1, 100, 058
English (including Canadian)	828, 140
Irish	679, 763
Slovak	484, 110
Magyar	462, 249
Croatian and Slovenian	462, 261
North Italian	551, 348

A glance at this table shows that the bulk of the " newer " immigration is made up of Italians, Hebrews, and Slavs, all of which are races much more widely divergent, biologically speaking, from the basic stocks of the country than the great bulk of the " older " immigration prior to 1882, which was Anglo-Saxon, Germanic, Scandinavian, and Celtic (Irish).

As the statement of the Allied Patriotic Societies of New York declared, the scientific significance of this drastic change in the racial composition of our immigration had been made crystal-clear during the war by the Army intelligence tests:

> The results of these tests . . . have been analyzed by those who were connected with the making of these tests, particularly in the work of Prof. Carl Brigham, of Princeton, entitled "A Study of American Intelligence," published by the Princeton University Press. . . . Prof. Brigham's tables bring out certain startling facts as to the mental capacity of the bulk of our recent immigration. They show that most of the foreign-born groups, *particularly those from southern and eastern*

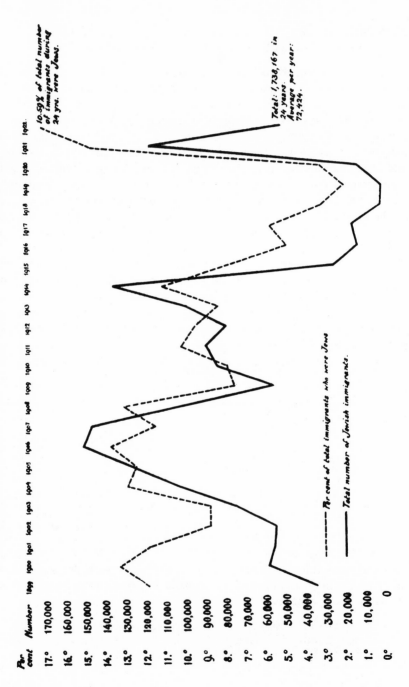

Europe, have contributed people whose average intelligence was far below that of the native-born white draft. One table . . . shows that at the bottom of the list, with respect to intelligence, stand four groups composed of people born respectively in Greece, Italy, Russia and Poland.

Like the Allied Patriots, Harry Laughlin, the committee's Expert Eugenics Agent, let the IQ facts of "science" carry the brunt of his 1924 report, entitled "Europe as an Emigrant-Exporting Continent and the United States as an Emigrant-Receiving Nation."

The full transcript of this hearing, running well over a hundred pages, has been called the proudest moment in the history of the American eugenics movement since the day, nearly fifteen years earlier, when Charles Benedict Davenport first tapped the Harriman legacy to get the Eugenics Record Office established. Fortunately, Laughlin's lifelong passion for illustrating his reports with graphs, tables, pedigree trees, and other scientific and neoscientific symbols makes it possible to sum up his long report in terms of a few of his exhibits, particularly those that appeared on page 1254, and two of the three charts on the Army intelligence tests that appeared between pages 1278 and 1279.

The first exhibit (opposite) showed, quite clearly, precisely where the major danger of further immigration of inferior types was, and to the surprise of nobody at all it turned out to be in the Jews.

The message was plain. Since, as Madison Grant and other leading authors of the day warned, the dwarf-statured Polish Jews liked nothing better than to take unto themselves the willowy fair Nordic maidens, the Jewish Immigration Explosion had to be defused. For not only were the Eastern Jews, or Alpine Slavs as Professor Brigham termed them, bearers of genes that gave them a yen for the blue-eyed dolichocephalic WASP women, but they were also, by scientific test, low-IQ morons.

This was not a matter of biased opinion but of objective IQ test score fact. The superior or old-immigration Nordic and native-born white draftees, as Professor Brigham and Professor Yerkes and Dr. Lothrop Stoddard had shown, had come out much higher in the tests than did the native blacks or the foreign draftees from non-English-speaking countries. But if, as Stoddard acutely observed, the tests showed that "the average mental age of all Americans is only about fourteen," Professor Brigham had also revealed that where the average white draftee had a mental age of 13.77 years, the U.S. colored draftees had a mental age of only 10.41, only slightly lower than the mental ages of 10.74, 11.01, and 11.34 rung up by draftees born in Poland, Italy, and Russia in that order.[28]

In an age of declining national intelligence, then, Laughlin reminded the committee that it was imperative we face the fact that, as Professor Brigham had shown on page 198 of his scientific study of American intelligence, "there is no doubt that the more recent immigrants are intellectually closer to the negro than to the native born white sample." Laughlin summed it all up for Congress in his charts 7 and 8 following page 1278 of the report.

SUPERIOR INTELLIGENCE.

Race, Country of Birth, or Army Rank, and Percent of Men in Each Group, by Army Mental Tests, Showing Superior in, or Very Superior in Intelligence.

1	████████████████████ 84.9	COMMISSIONED OFFICERS
2	██████████████ 46.9	NON-COMMISSIONED OFFICERS
3	███████████ 39.9	NEGRO OFFICERS
4	██████ 28.9	PRIVATES
5	█████ 19.7	ENGLAND
6	████ 13.0	SCOTLAND
7	████ 12.1	WHITE DRAFT
8	███ 10.7	HOLLAND
9	███ 10.5	CANADA
10	███ 8.8	GERMANY
11	██ 5.4	DENMARK
12	██ 4.3	SWEDEN
13	██ 4.1	NORWAY
14	██ 4.1	IRELAND
15	██ 4.0	ALL FOREIGN COUNTRIES
16	██ 3.4	TURKEY
17	██ 3.4	AUSTRIA
18	██ 3.4	NORTHERN AMERICAN NEGRO
19	█ 2.1	RUSSIA
20	█ 2.1	GREECE
21	█ .9	ITALY
22	█ .9	BELGIUM
23	█ .5	POLAND
24	█ .9	SOUTHERN AMERICAN NEGRO

NOTES: Groups recorded in ranks 1,2,3 and 4 are from the Army; all others are from the Draft. (This chart is based upon the records of men psychologically examined by the United States Army, Tables 217, 268, 406, pp 698, 734, 855. Mem. XV, Nat. Acad. Sci.)

It is impossible to overestimate the impact of Laughlin's report, which based itself so solidly on the Yerkes and Brigham studies of the Army IQ tests, on the Congress and the nation. As Madison Grant, that nonstop dictator of letters, wrote to the president of the Carnegie Institution of Washington, Dr. John C. Merriam, on November 26, 1924, about Laughlin's report:

I had already talked over this paper with Albert Johnson, Chairman of the House Immigration and Naturalization Committee, but I was not prepared to find how extensive and thorough the document is. It shows long and detailed study, the results of which have been set forth in graphs and tabulated statistics. As this work is being done under the auspices of the Carnegie Institution of Washington I want to congratulate that organization. . . . I hope you will encourage Dr. Laughlin to pursue these studies as I do not know of any subject more important to the nation at large than the formulation of proper restriction laws *based on accurate and impartial scientific data* [italics added].

INFERIOR INTELLIGENCE

Race, Country of Birth, or Army Rank, and Percent of Men in Each Group, by Army Mental Tests Showing Inferior (D), or Very Inferior (D-,E) Intelligence.

1		COMMISSIONED OFFICERS
2		NON-COMMISSIONED OFFICERS
3		ENGLAND
4		HOLLAND
5		NEGRO OFFICERS
6		DENMARK
7		SCOTLAND
8		PRIVATES
9		GERMANY
10		SWEDEN
11		CANADA
12		BELGIUM
13		WHITE DRAFT
14		NORWAY
15		AUSTRIA
16		IRELAND
17		TURKEY
18		GREECE
19		ALL FOREIGN COUNTRIES
20		NORTHERN AMERICAN NEGRO
21		RUSSIA
22		ITALY
23		POLAND
24		SOUTHERN AMERICAN NEGRO

Notes: Groups recorded in ranks 1,2,5 and 8 are from the Army; all others are from the Draft. (This chart is based upon the records of men psychologically examined by the United States Army. Tables 211,260, 306, pp 690,704,877, Mem. XX, Nat. Acad. Sci.)

Like Lothrop Stoddard and other true believers in eugenics and natural (genetic) intelligence, Laughlin used the intelligence test scores of the recruits of World War I to chart the nations and social classes of hereditarily superior and inferior intelligence. With these charts(above and opposite), from his 1924 report to the Congress as the "expert eugenics agent" of the House Committee on Immigration and Naturalization, Laughlin helped convince the Congress and the President of the United States that continued Jewish, Italian, Polish, and other non-Nordic immigration would irreparably lower the native American levels of what the Laughlin report characterized as natural intelligence.

From "Europe as an Emigrant-Exporting Continent and the United States as an Emigrant Receiving Nation," statement of Harry H. Laughlin, Hearings, Committee on Immigration and Naturalization, March 8, 1924.

Dr. Hiram Wesley Evans, Imperial Wizard of the Ku Klux Klan, an intellectual who cited various scientific books and learned authorities, ranging from *The Immigration Problem* by Jeremiah Jenks and W. Jett Lauck to *The Revolt Against Civilization* by Lothrop Stoddard, in his widely distributed 1923 address, "The Menace of Modern Immigration," reserved his most extravagant praises for Laughlin. He hailed "Dr. Harry H. Laughlin, the noted eugenics expert," as the authority for "the safe estimate that fully a million defectives, dependents, and criminals are now requiring custodial care in local, state and national institutions." And that Dr. Laughlin had "made for the Committee on Immigration a remarkable study of the relative degeneracy of all classes of social inadequates except the venereal." This remarkable Laughlin study, Dr. Evans said, revealed: "Insanity and crime are the worst manifestations of social inadequacy. Insanity among immigrants, Dr. Laughlin shows, is nearly two and two-thirds times more frequent than in our older native stock. Among Russians, Finns and Poles, Bulgarians, Irish and Serbians the proportions are highest of all. The Irish proclivity to insanity is over four times that of our older native stock as a whole."[29]

All of the reports Laughlin made for the House Committee on Immigration and Naturalization over a four-year period helped establish a solid, "scientific" rationale for the final triumph of the immigration restrictionists. The Immigration Act of 1924 (often referred to as the Johnson Act or the Johnson Restriction Act) was so sweeping, and reduced the flow of Italian and Jewish immigration to such a minute trickle, that there was no longer any need to plead for the sending of brigades of Davenport's eugenics field workers to Italy, Russia, Poland, and Hungary to dig up the home-town gossip on every prospective immigrant.

"BUT WE HAVE NO PLACE TO DRIVE THE JEWS TO . . ."

The immediate results of the passage of the Immigration Act of 1924 read like the most rosy dreams of the old Teutonists, from Henry Cabot Lodge and Francis Amasa Walker to Prescott Hall and Robert DeCourcy Ward, come to fruition. In one year, the number of immigrants from Italy had fallen by 89 percent; the number of immigrants from Russia, Poland, and Central and Eastern Europe, whence had come most of America's Jewish immigrants, was down by an average of 83 percent; even Great Britain, home of what Osborn called the Anglo-Saxon branch of the Nordic race, suffered a 54 percent drop in immigrants between 1924 and 1925. Germany, however, the land of "those descended from the tribes that met under the oak-trees of old Germany to make laws and choose chieftains," showed a drop of only 39 percent.

In terms of what was to follow in the near future, the human results of the racial quotas of the Immigration Act of 1924 may be found on page 56, *Historical Statistics of the United States,* published by the Government Printing Office in 1960. Between 1900 and 1924, the combined number of immigrants from Italy, Russia, Poland, and other Central, Eastern, and Southern European countries was 10,870,225, for an average of 434,810 a year. Between 1925 and 1939, when World War II erupted, under the racial quotas of the U.S. Immigration Act of 1924 we were to accept a total of 366,446 immigrants from these countries, for an average of 24,430 per year.

If the immigration from these countries between 1925 and 1939 had been at the pre-1924 quota rates, the total would have been 6,432,150. Deduct from this projection the 366,446 human beings from Southern and Eastern Europe who were admitted as immigrants, and this leaves a grand total of 6,065,704 Italians, Jews, Poles, Hungarians, Balts, Spaniards,

Greeks, and other European people trapped in Europe when war broke out in 1939.

How many of these 6,065,704 immigrants excluded by the racial quotas of 1924 would have emigrated to America between 1924 and 1939 will, of course, never be known. One thing is certain: most of the Jews, Poles, Russians, and other people marked by the Nazi race biologists as dysgenic who were trapped in territories controlled by Germany between 1933 and 1945 were rounded up and thrown into Auschwitz and other Nazi extermination camps. There, over nine million human beings died in the gas chambers or were slowly starved and tortured to death. Untold millions of other people trapped in Europe after 1939 were killed by air raids, artillery fire, ground fire, and war-caused diseases—ranging from tuberculosis and pneumonia to malnutrition and heart failure—during the six years of World War II.

The imposition of the anti-Italian, anti-Semitic, and anti-Slav immigration quotas of 1924 was such a monumental victory for the old scientific racism that, in due time, even Galton's Vicar in America, Charles Benedict Davenport, stopped pressing for his favorite ploy of sending eugenics inspectors to Europe to examine all would-be immigrants for genetic flaws. Not that he was completely satisfied with the results of the 1924 legislative triumph. As Davenport wrote to Madison Grant on April 7, 1925—less than a year after Calvin Coolidge, declaring that "America must be kept American," put his signature to the Johnson Restriction Act—it did not go far enough:

"Our ancestors drove Baptists from Massachusetts Bay into Rhode Island but we have no place to drive the Jews to. Also they burned the witches but it seems to be against the mores to burn any considerable part of our population. Meanwhile we have somewhat diminished the immigration of these people."[30]

Nearly eight more years were to pass before "these people" could be driven by the millions into Nordic institutions where it was not "against the mores to burn any considerable part" of their number.

13 Even as the Legitimate Sciences Demolish the Myths of Hereditary Feeblemindedness, the Discredited Claims of Eugenics Are Upheld, by a Vote of 8 to 1, in the U.S. Supreme Court

> . . . some recent philosophers appear to have given their moral support to the deplorable verdict that the intelligence of a child is a fixed quantity. . . . We must protect and act against this brutal pessimism. . . . A child's mind is like a field for which an expert farmer has advised a change in the method of cultivating, with the result that in place of desert land, we now have a harvest. It is in this particular sense, the one which is significant, that we say that the intelligence of children may be increased. One increases that which constitutes the intelligence of a school child, namely, the capacity to learn, to improve with instruction.
>
> —ALFRED BINET, in *Les Idées modernes sur les enfants* (1909)

Not all psychologists and other educated Americans of the post-World War I years went along with the widely accepted claims that the Army intelligence tests actually measured the amount of one's inborn intelligence. There were, along with the older mavericks such as the iconoclastic clinical psychologist J. E. Wallace Wallin, growing numbers of younger psychologists, such as Dr. Margaret Wooster Curti, of Smith College,[1] and Dr. Ada Hart Arlitt, of Bryn Mawr, who had escaped infection by the virus of eugenics and who now began to look upon the IQ scores with more objective eyes.

It was as early as May 1921 when Professor Arlitt published her significant report, "On the Need for Caution in Establishing Race Norms," in the *Journal of Applied Psychology*. Dr. Arlitt's data derived from a study made of 343 schoolchildren taken from the primary grades in a single urban school district. "Of these 191 were children of native born white parents, 87 were Italians and 71 were negroes. All of the Italians spoke English without difficulty."

When these 343 urban children were tested with the Stanford (Terman) Revision of the Binet tests, the children of the native-born white parents achieved a median IQ test score of 106.5, while that for the Italian children was 85, and for the Negro children 83.4. As Dr. Arlitt, a psychologist with a background in zoology and medical research noted, if (like Yerkes and Brigham) one believed that these IQ test score differences "had been due to race alone [they] would have demonstrated beyond a doubt the superiority of the native white group." However, Dr. Arlitt was no biased simplicist but a responsible behavioral scientist, and she therefore looked at one of the many other variables that were known, through the work of people like Boas and Fernald, to influence human development.

Arlitt broke her study group down into socioeconomic classes, as well as by race.

The native born white children were divided on the basis of social status into five groups with reference to the occupation of the father. The separation into groups by occupation followed Taussig's division into the five non-competing groups, i.e., (1) Professional classes; (2) Semi-professional and higher business; (3) Skilled; (4) Semi-skilled; (5) Unskilled. . . . The last two groups . . . contained too few children to be treated separately . . . they were therefore combined.

When the factor of family income was evaluated, it developed that 37 percent of the native white group came from socioeconomic Classes 1 and 2, while 93 percent of the Negro and 90 percent of the Italian children came from families in the two lowest-earning social classes, at and below the poverty line. When social class was broken down, the IQ test scores could be stated as follows:

Population	Median IQ Test Score
Native whites of Social Class 1	125
Native whites of Social Class 2	118
Native whites of Social Class 3	107
Native whites of Social Classes 4 and 5	92
Italians, all in Social Classes 4 and 5	84
Negroes, all in Social Classes 4 and 5	83

Professor Arlitt warned that when IQ test scores were evaluated, "race norms which do not take the social status factor into account are apt to be to that extent invalid," since her study had shown that:

. . . the difference in median IQ which is due to race alone is in this case at most only 8.6 points whereas the difference between children of the same race but of Inferior [Classes 4 and 5] and Very Superior [Class 1] social status may amount to 33.9 points. It is apparent that *such differences as we have between the negro and Italian children and between these and children of native born white parents are not nearly so striking as the difference between children of the same race but of different social status.* Of the two factors social status seems to play the more important part. [Italics added.]

Dr. Arlitt's 1921 study was not only a break with the eugenic concepts of Goddard, Terman, Yerkes, and their circle; it was also a harbinger of the more objective directions most subsequent scientific studies of human behavior would take for the rest of the century.

Other members of Dr. Arlitt's generation of younger professionals were aware of her work. Such a contemporary was the thirty-three-year-old editor of the *New Republic,* Walter Lippmann. No psychologist himself, he managed, in a memorable series of *New Republic* articles, to do to the reigning IQ test score myths what the guileless young child in the ageless legend did to the professional tailors who made the emperor's new clothes.

On October 25, 1922, Lippmann ran the first article of a six-part series, "The Mental Age of Americans," "a critical inquiry into the claim, now widely made and accepted, that the psychologists have invented a method of measuring the inborn intelligence of all people." The entire series emerges in history as one of the glories of American journalism.

In his opening paragraphs, Lippmann picked up and ridiculed Lothrop Stoddard's assertion that the Army tests had proved that "the average mental age of Americans is only about fourteen." With the apt comment that "the average adult intelligence cannot be less than the average adult intelligence, and to anyone who knows what the words 'mental age' mean, Mr. Stoddard's remark is precisely as silly as if he had written that the average mile was three quarters of a mile long," Lippmann performed a masterly job of demolishing a set of high-blown hypotheses with an array of hard, deflationary facts.

It was a relentless job of deflation in which, because "Mr. Stoddard did not invent this conclusion" but "found it ready made in the writings" of others who had reached that conclusion "by misreading the data collected in the army intelligence tests," Lippmann went on to show that even the Army mental test data "themselves lead to no such conclusion."

Along the route, Lippmann treated his readers to a short but factual review of the history of IQ testing, starting with Binet's establishment of "norms" for his experimental group of "two hundred school children who ranged from three to fifteen years of age": "Whenever [Binet] found a test that about 65% of the children of the same age could pass he called that a Binet test of intelligence for that age. Thus a mental age of seven years was the ability to do all the tests which 65 to 75% of a small group of seven year old Paris school children had shown themselves able to do."

Lippmann then told of how, after Binet's death, Terman found that "the Binet Scale worked badly in California. The same puzzles did not give the same results in California as in Paris." This, of course, was quite obviously due to the fact that the cultural environments of both places were so different, a fact that Terman just as patently did *not* take into consideration as he started to revise the Binet tests.

"Like Binet, [Terman] would guess at a stunt which might indicate intelligence, and then try it out on about 2,300 people of various ages, including 1,700 children 'in a community of average social status.' By editing, rearranging and supplementing the original Binet tests he finally worked out a series of tests for each age which the average child of that age in about one hundred California children could pass."

This, Lippmann showed, was the basis of the Stanford Revision of the Binet-Simon Scale. "The aspect of all this which matters is that 'mental age' is simply the average performance within certain arbitrary problems. *The thing to keep in mind is that all the talk about a 'mental age of fourteen' goes back to the performance of eighty-two California school children in 1913–1914.* Their successes and failures on the days they happened to be tested have become embalmed and concecrated as *the* measure of human intelligence. By means of *that* measure writers like Mr. Stoddard fix the relative values of all the peoples of the earth and of all social classes within the nations." (Italics added.)

"GROSS PERVERSION BY MUDDLE-HEADED AND PREJUDICED MEN"

The intellectual and *moral* qualities of Terman and Yerkes and their supporters led Lippmann to suggest that "the real promise and value of the investigation which Binet started is in danger of *gross perversion by muddle-headed and prejudiced men* (italics added)." It was not Binet and the concept of mental testing that Lippmann attacked: it was the gross perversion of Binet's tests by eugenical true believers such as Terman and Goddard and the overtly racist interpretations of IQ test scores by the likes of Laughlin, Davenport, Gould, and Stoddard that offended Lippmann's brain and soul.

As one reads through these six 1922 articles after more than a half century of discoveries and sophisticated investigations in the behavioral and life sciences, the one consistent quality of Lippmann's words that strikes one is how modern they sound today. For example:

> But intelligence is not an abstraction like length and weight; it is an exceedingly complicated notion which nobody has yet succeeded in defining. . . . The intelligence tester starts with no clear idea of what intelligence means. . . . He proceeds, therefore, to guess at the more abstract mental abilities which come into play again and again. By this rough process the intelligence tester gradually makes up his mind that situations in real life call for memory, definition, ingenuity and so on. ["The Mystery of the 'A' Men," November 1, 1922.]

> . . . the whole claim of the intelligence testers [i.e., Terman, Goddall, et al.] to have found a reliable measure of human capacity rests on an assumption, imported into the argument, that *education is essentially impotent because intelligence is hereditary and unchangeable.* This belief is the ultimate foundation of the claim that the tests are not merely an instrument of classification but a true measure of intelligence. It is this belief that has been seized upon eagerly by writers like Stoddard and McDougall. It is a belief which is, I am convinced, *wholly unproved,* and it is this belief which is obstructing and perverting the practical development of the tests. [Italics added.] ["The Reliability of Intelligence Tests," November 8, 1922.]

> One has only to read . . . the work of popularizers like McDougall and Stoddard to see how easily the intelligence test can be turned into an engine of cruelty, how easily in the hands of blundering or prejudiced men it could turn into a method of stamping a permanent sense of inferiority upon the soul of a child . . . most of the more prominent testers have committed themselves to a dogma [eugenics] which must lead to just such abuse. They claim not only that they are really measuring intelligence, but that intelligence is innate, hereditary, and predetermined. They believe that they are measuring the capacity of a human being for all time and that this capacity is fatally fixed by the child's heredity. Intelligence testing in the hands of men who hold this dogma could not but lead to an intellectual caste system in which the task of education had given way to the doctrine of predestination and infant damnation. ["The Abuse of the Tests," November 15, 1922.]

It is possible, of course, to deny that the early environment has any important influence on the growth of intelligence. Men like Stoddard and McDougall do deny it, and so does Mr. Terman. But on the basis of the mental tests they have no [scientific] right to an opinion. Mr. Terman's observations begin at four years of age. He publishes no data on infancy and he is, therefore, generalizing about the hereditary factor *after* four years of immensely significant development have already taken place. . . . He cannot simply lump together the net result of natural endowment and infantile education and ascribe it to the germ-plasm. In doing just that he is obeying the will to believe, *not the methods of science.* [Italics added.] ["Tests of Hereditary Intelligence," November 22, 1922.]

This, of course, was the crux of the entire triumph of the IQ cult in the hands of willful believers in eugenics from Terman and Davenport to Laughlin and the Honorable Albert Johnson: the IQ cult represented not science but *anti-science.* The blatantly anti-scientific nature of its prophets was nowhere better dramatized than in the answers of Terman and McDougall to Lippmann's *scientific* critiques.

Terman's answer, published as an article, "The Great Conspiracy of the Impulse Imperious of Intelligence Testers, Psychoanalyzed and Exposed by Mr. Lippmann" (*New Republic,* December 27, 1922), compared Lippmann to William Jennings Bryan, then the leading opponent of the Theory of Evolution, and to Wilbur Voliva, head of the Flat Earth Society. Terman said, further:

"Let there be no misapprehension; the principle of democracy [*sic*] is at stake. The essential thing about a democracy is not equality of opportunity, as some foolish persons think, but equality of mental endowment. Where would our American democracy be if it should turn out that people differ in intelligence as they do in height . . . ?"

This boorish sarcasm was dropped when Terman came to the core of Lippmann's exposé: "The validity of intelligence tests is hardly a question the psychologist would care to debate with Mr. Lippmann; nor is there any reason to engage in so profitless a venture."

This was not a scientific answer but a display of petulant name calling, as was the statement that "nearly all the psychologists believe that native ability counts very heavily. Mr. Lippmann doesn't."

Terman's article ended in a citation of the differences between the 80-average IQ test scores of Portuguese children tested in California and those of the average Japanese child, who, in California, "soon develops an IQ not far below that of the average California white child of *Nordic* descent." Lippmann had suggested, in his articles, that "nursery" factors—by which he clearly meant preschool mental development via parent-child games, toys, books, drawing books, etc.—had to be taken into consideration in the measurement of the mental development of children.

Now, citing the differences between Mediterranean Portuguese and true Nordics, Terman, basing his comments solely on race and not on family socioeconomic conditions, wrote: "In this case the nurse girl factor is eliminated; one might also say, the nursery itself. But of course there are the toys, which are more or less different. It is also conceivable that the more liquid Latin tongue exerts a sedative effect on infants' minds as compared with the harsher Japanese language, which may be stimulating by comparison."

McDougall's answer was published as a letter in the May 23, 1923, issue of the magazine, and is interesting for a number of reasons. Like Terman, McDougall compared Lippmann to Mr. Bryan and "other clever journalists" who had built a very plausible propaganda case against the theory of organic evolution. Then he went on to say that "the sociologist who accepts the theory [of evolution] . . . cannot cite any conclusive proof in support of it. He can only point to the fact that almost all the men who have devoted their lives to the study of biology have accepted the theory."

McDougall then asserted that the sociologist "who accepts as highly probable . . . the heredity theory as the partial explanation of the *correlation of intelligence* with *parental status* is in a similar position. He cannot demonstrate the truth of his view; but he can point to the consensus of opinion among such lifelong students of the question as Francis Galton, Havelock Ellis, Karl Pearson and Lewis Terman."

Because Lippmann rejected the opinions of these anti-Mendelian, anti-scientific eugenicists whose authority McDougall chose to revere *after* the genetic and anthropometric scientific *in vivo* studies of Mendel, Johannsen, Morgan, Sturtevant, Bridges, Muller, and Boas had annihilated them, this meant that Lippmann was "denying, also, the theory of organic evolution, and he should come out openly on the side of Mr. Bryan. *For the theory of the heredity of mental qualities is a corollary of the theory of organic evolution. The latter cannot be true if the former is not true.*" (Italics added.)

The only basis for this statement, of course, was McDougall's lack of understanding of the theories of Mendelian and chromosomal genetics as well as the theory of organic evolution. But McDougall had authorities more radical than Galton and Terman to cite: there were also, he wrote, Mr. and Mrs. Sidney Webb, who, like Karl Pearson, were Marxist socialists. And in their book *The Decay of Capitalist Civilization,* the Webbs had written of "the intensification, among the least provident and most casual of all classes, notably in the poorest stratum of irregularly employed laborers of our great cities (who are to a large extent condemned to a perpetual interbreeding), of a reckless propagation that may well be eugenically as adverse in its consequences to the community as the exceptional restriction of the birth-rate among the provident and the prudent."

This quotation from the work of the Webbs did not, in any way, answer any of Lippmann's criticisms of the IQ test mythology, much as it did dramatize the fact that the Webbs, like their equally socialist friend Karl Pearson, were true-believing eugenicists with as great a bias against the Undeserving Poor as that exhibited by the most reactionary Tory. McDougall concluded his non-answer to Lippmann's critique of the IQ tests by striking a pose that could have been the template for that assumed by Professor Garrett Hardin a half-century later: "I should like to come before the public as the champion of the downtrodden and the oppressed, and, clothed in the shining armor of righteousness and charity, to proclaim once more that 'all men are created equal.' But I believe that a man of science who, after long study and reflection, arrives at an opinion on a question of great social importance, should publish his opinion no matter how distasteful it may be to himself or to other persons."[2]

Nearly a decade was to pass before the sheer power of incontrovertible facts was to bring the majority of America's behavioral scientists into agreement with Lippmann and the professional geneticists about both the reliability of IQ testing and the heredity of mentality. In the interim, we entered upon an era in which generations of college and graduate students were to be subjected to texts such as Popenoe and Johnson's *Applied Eugenics* and the various eugenically oriented and widely used textbooks of Samuel J. Holmes, such as *The Trend of the Race* and *Studies in Evolution and Eugenics*.

Many of the ideas of eugenics were incorporated into the texts of other subjects, from political science to elementary biology. Holmes's 1926 *General Biology,* for example, "An Introduction to Life and Evolution" for college undergraduates—that is, a basic textbook in biology—was a cultural Trojan horse in which Lidbetter's construct of a "definite race of chronic pauper stocks" as presented at the Second International Congress of Eugenics in 1921 was reified by Holmes on page 418. There, writing of the Jukes and the Kallikaks in orthodox eugenic terms, Holmes declared, in italics: *"It is a fact of the greatest importance that, sooner or later, bad heredity makes bad environment, and that good heredity makes good environment."*

Textbooks that peddled such eugenics dogmas as scientific *facts* formed essential segments of the core curricula of most American college and university education from, roughly, the end of World War I to the start of World War II. Taken as a whole, they were not without some significant share of responsibility for the outbreak of World War II itself.

PEARSON'S CORRELATIONS "PROVE" THAT NUTRITION IS NOT A FACTOR IN BRAIN GROWTH AND DEVELOPMENT

While America brainwashed its collegians—and the children so many college graduates of the eugenics generation taught in high schools and grade schools—people in England like Pearson, Ellis, Raymond Cattell, Leonard Darwin, Cyril Burt, and the Webbs exhorted their countrymen to do likewise. Of our British cousins, none were more slavish in their imitation of Davenport, Grant, and Laughlin than the statistician Karl Pearson, the socialist half of the team of Galton & Pearson. As we have seen earlier, in October 1925, in the opening issue of the *Annals of Eugenics,* Pearson ran a eugenical study, "The Problem of Alien Immigration into Great Britain, Illustrated by an Examination of Russian and Polish Jewish Children," whose overt purpose was to induce England to follow America's lead and pass its own version of the Johnson Restriction Act to keep out the Jews and other such—to use Galton's charming term—human "refuse."

Not only did Pearson stick to the ancient myth of cephalic indexes and take the head-form measurement of the Jewish children to prove their innate inferiority—finding, of course, that "the defect [*sic*] of brachycephaly in our alien Jewish boys seems noteworthy." The Marxist Pearson also took out after Boas (p. 27) for having had the temerity to show that an improved environment can cause measurable differences between the head forms of European immigrant parents and their American-born children, as well as between European- and American-born sibs of the same parents.

Most revealing of all the Pearson "scientific" findings of this study were those with which he was certain he had demolished the concept, increasingly accepted by life scientists following the work of Stiles and Goldberger on hookworm and pellagra, that few links in human development are firmer or less debatable than those that bound good (or poor) child nutrition to good (or bad) classroom or IQ test scores. Anemia was "judged by the colour of the face" (p. 78). Earlier (p. 31) Pearson had defined the way he and his army of field workers had tested the nutritional status of the children: it was all a matter of the sulci—that is, the grooves—between the ribs. "The well-nourished condition was defined by no sulci showing between ribs; the moderately nourished by well-marked intercostal sulci; and the badly nourished by depression between the ribs showing on front of chest above nipple level."

Scientifically—remember that this classic of eugenics was written a full decade after Goldberger, in his work on pellagra, had also started the new medical discipline of clinical nutrition—this was plain nonsense. What Pearson's field workers, probing between the ribs of the children for layers of fat, were actually measuring was just the opposite of what they reported these fat layers to be.

Actually, the well-nourished child is a child whose diet is so free of the excess fats and starches and sugars of the diets of the poor as to let him develop without the fatty accumulations caused by the high-fat, high-carbohydrate, and low-protein diets of the poor. However, since an insufficient amount of all kinds of food—good as well as bad—also leaves a child too thin, and with hollow grooves instead of fat between the ribs, the literally starving child and the well-nourished child will present the same kind of intercostal sulci. In other words, what Pearson's "scientific" measurements proved was, at best, nothing at all; at worst, they suggested that many of the world's most undernourished children were *demonstrably* well nourished.

Impressive as were the mathematical formulas and tables with which Pearson presented his conclusions about the links between nutrition and mental acuity, they did not in any way prove, as Pearson wrote (p. 78), that in them "we have therefore further confirmation that nutrition and intelligence are not correlated in any sensible degree." All they did prove was that while Pearson's anthropology went back to Gobineau and Galton, his knowledge of clinical nutrition went back to the pre-Goldberger days when, as Terman and Goddard and Davenport claimed, and as some adults actually believed, there were absolutely no correlations between a starved body and a starved brain. Even in 1925 Pearson was at least a decade behind the times in clinical nutrition, let alone the genetics and biology of human growth and development.

The tragedy of all this was not only that a contemporary European author named Adolf Hitler was then saying the same things about the Jews in far less scientistic prose. The greater .tragedy lay in the fact that Britain's intellectuals, from the Webbs and Harold Laski to Havelock Ellis and Leonard Darwin, accepted such intellectual trash as true scientific writing.

A GREAT MAN SPEAKS FOR SCIENCE—TO DEAF EARS

The overt racism of the eugenics movement and its leaders was not, of course, lost on all of their peers. In 1927, for example, when Osborn's protégé, Dr. William K. Gregory, solicited statements of tribute to Osborn, in his capacity as secretary of the Committee for the Celebration of Professor Osborn's Seventieth Birthday, not all of the scientists he approached proved amenable. The American anthropologist and ethnologist Robert H. Lowie, who from 1909 to 1921 had been on the museum's staff—and whose work Osborn had cited most favorably in his opening address at the Second International Congress of Eugenics in that year—declined, adding that "I am not a friend or admirer of President Osborn's. Specifically, his sponsorship of Mr. Grant's book [*The Passing of the Great Race*], in my judgment, puts him beyond the pale as a representative of science."

And when, as Geoffrey Hellman notes in his history of the museum, Gregory took exception to this statement, Dr. Lowie expanded on his original "over-concise" letter, saying:

> To President Osborn I have never imputed Mr. Grant's sadism. However, his active sponsorship of the book, as attested by his two prefaces, proves to my mind that he substantially shares his friend's prejudices and condones even his counterfeiting of biological laws. . . . I admit that his conduct, inexcusable as I consider it, does not automatically exclude constructive work. But I regard the offender as "being beyond the pale" *as a spokesman and interpreter* of the scientific spirit [Lowie's emphasis]. . . . I personally find it impossible to render homage to President Osborn. . . . The issues are important moral ones, I think. . . .[3]

But 1927 was not the year to raise moral issues. Besides, Lowie was one of that generation of American anthropologists who had trained under Franz Boas, and that of course made anything he said about Henry Fairfield Osborn suspect.

Less suspect as an individual, however, was Walter E. Fernald, M.D., twice (1893 and 1924) president of the American Association for the Study of the Feeble-Minded. Dr. Fernald was, by acclamation, the leading authority in the nation on the nature and the care of the feebleminded. In his 1924 presidential address, "Thirty Years' Progress in the Care of the Feeble-Minded," Dr. Fernald paid due tribute to Goddard's "recognition of the vast significance of Binet's theory and technique for measurement of variations of human intelligence," and to Goddard's having "made the Binet test practical," and applying it to "thousands of normal and defective children."

As Fernald put it, "the theory and practice of mental testing and the discovery of the *concept of mental age* did more to explain feeble-mindedness . . . than all the previous study and research from the time of Seguin." But, with these early studies, Fernald noted, the eugenic and dysgenic papers of Davenport and Goddard had made a "profound and widespread impression on public sentiment" by strengthening Galton's original myth about the menace of the feebleminded. Not the least of their faults had been their so-called pedigree or family tree studies.

> The family tree method of eugenic research incidentally revealed a large number of law-breakers and criminals among the feeble-minded and their kinfolk. . . . At this time no paper on mental defect failed to emphasize strongly the criminal and antisocial tendencies of the feeble-minded with little reference to the non-criminal defectives.

The result of this concentration on the "hovel type" of people led to what Fernald named "The Legend of the Feeble-Minded."

> This legend conveyed the impression that the feeble-minded were almost all of the highly hereditary class; they were almost invariably immoral, most of the women bore illegitimate children; nearly all were antisocial, vicious and criminal; they were all idle and shiftless and seldom supported themselves; they were highly dangerous people roaming up and down the earth seeking whom they might destroy.
> *The feeble-minded person became an object of horror* and aversion. He was looked upon as an Ishmaelite, a useless, dangerous person, *who should be ostracized, sterilized, and segregated for his natural life at public expense.* [Italics added.]

As Fernald observed, the legend of the menacing feebleminded was based on studies of defectives who already were in prison for bad behavior when they were studied.

> The legend ignored the defectives from good homes with no troublesome traits of character and behavior. It would be equally logical to describe an iceberg without reference to the 87% of bulk invisible below the surface of the sea. But much water has run over the dam since that period of pessimism, say 1911 or 1912.

What had happened, briefly, was the discovery or continuing discoveries of a whole array of environmental, biological, traumatic, and psychological causes—*all of them nongenetic*—of what pediatricians and psychiatrists in ever increasing numbers were beginning to recognize as the causes of *most* instances of mental retardation. As Fernald put it, "the physical examination may throw much light upon the etiology of a given case." Himself a physician and psychiatrist, Fernald quickly reviewed some of the proven and suspected nongenetic causes of mental retardation, including: "anomalies of growth and development" due to "abnormal functioning of the endocrine or ductless gland system"; obstetrical "focal brain damage with resulting paralysis and mental defect"; the amount of time that passed between "an acute illness, a blow, a fall or a convulsion" and the onset of mental abnormalities.

These and many other episodes in the clinical history [of the individual mental defective] are of great significance. . . .

No one questions the accuracy of the *black* charts of the "hovel" type of hereditary defectives, or their significance, but no extensive studies have been made of the heredity of unselected defectives as they appear in school and out-patient mental clinics. Many of these children seem to come from average American homes, from the homes of the poor and of the middle class, and of the well-to-do, with industrious, well-behaved parents. *The clinical history of many of these pupils suggests that infective, inflammatory or other destructive brain disease in infancy was the cause of the mental defect.* Other types of defect are probably non-hereditary—the Mongolian, the Cretin, the Syphilitic, the Traumatic, etc. [Italics added.]

Dr. Fernald's statements of 1924, particularly in the light of what we know today, spoke well for the acuity of his clinical perceptions and scientific insights. In our modern era of vast knowledge about the role of infectious diseases during pregnancy in causing brain, eye, ear, and other organic damage in the fetus—an age in which our leading research pediatricians report that even among twins in the same womb at the same time rubella virus sometimes infects and deforms one twin and not the other[4]—we can doubly appreciate Fernald's 1924 comment that "the assumption that practically all feeble-mindedness is highly hereditary is most disconcerting to normal brothers and sisters of a feeble-minded child."

It took more than an IQ test or a hastily trained eugenics field worker's judgment to make a meaningful diagnosis of feeblemindedness.

. . . the knowledge of feeble-mindedness is gradually increasing as a result of scientific study and observation in many fields—medical, psychiatrical, psychological, educational, sociological, economic, industrial, moral, legal, eugenic, etc.

We know now that feeble-mindedness is not an entity, to be dealt with in a routine way, but it is an infinitely complex problem. The feeble-minded may be male or female, young or old, idiots, imbeciles or morons; from good homes or bad homes; bad defectives or good defectives; well behaved or vicious; industrious or idle; they may live in the city or in the country; in a good neighborhood or a bad one; they may come from family stock highly hereditary as to feeble-mindedness, from one slightly so, or from good stock. No two defectives are exactly alike.

More than this, Fernald now stated that it was no longer necessary or even moral to permanently segregate all mental defectives in institutions. "We believe that the vast majority will never need such provision but will adjust themselves at home *as they have always done in the past.*" Most mental defectives were perfectly capable of holding simple jobs and being self-supporting.

There could be no question but that, when Fernald delivered this address, he spoke for the better-informed elements of the American psychiatric and psychoanalytic community. Even Goddard, by then professor of abnormal and clinical psychology at Ohio State University, was coming around to more modern concepts of the etiology and nature of mental retardation.

To the high priests of the eugenics creed, such as Davenport and Laughlin, this shift in the thinking of America's community of behavioral scientists had become too obvious; to attack Fernald would be to call down the wrath of the better-trained life and psychological scientists. They therefore did the prudent thing: they ignored Fernald's address and all of the thousands of post-World War I scientific data on which it was based.

THE SUPREME COURT HEEDS LAUGHLIN, NOT FERNALD

In 1924, at the height of the eugenics-dysgenics hysteria in the nation, the state of Virginia passed a compulsory state eugenical sterilization law based on the model law proposed by Davenport's Eugenics Record Office. The first person selected to be sterilized under this eugenical law was an eighteen-year-old white girl, Carrie Buck, an involuntary resident of the State Colony for Epileptics and Feeble-Minded (epilepsy being, according to Francis Galton, as certain an indication of inborn criminality as feeblemindedness). Because some vocal Virginia Christians, who held antiquated pre-eugenics ideas of human rights, objected in force, the legal right of Virginia to mutilate a ward of the state was challenged in the courts. The Eugenics Record Office sent Arthur H. Estabrook, one of its crack field workers, to make a "scientific" study of the case, and Harry Laughlin worked up a "Scientific Analysis" of the Estabrook findings. Laughlin then traveled to Virginia to present his analysis before the Circuit Court of Amherst County.

In his very scientific eugenic-dysgenic analysis,[5] Laughlin noted that the propositus (i.e., the original person whose case serves as the stimulus for a hereditary study) had a "chronological age of 18 years, with a mental age of 9 years, *according to the Stanford Revision of the Binet-Simon Test,*" while the mother of the propositus, Emma Buck, proved to have "mental defectiveness evidenced by the failure of mental development, having a chronological age of 52 years, with a mental age, *according to the Stanford Revision of the Binet-Simon Test,* of seven years and eleven months; and of social and economic inadequacy." (Italics added.)

Propositus and mother of propositus, according to the analysis by the co-editor of the *Eugenical News* and the Expert Eugenics Agent of the U.S. Congress, were both notorious for their "immorality, prostitution, and untruthfulness." Worse, the eighteen-year-old inmate of the Virginia State Colony for Epileptics and Feeble-Minded already "has had one illegitimate child, *supposed* to be mental defective." Her whore mother had also had at least one and *"probably* two" other bastards "inclusive of Carrie Buck, a feeble-minded patient in the State Colony, Va." (Italics added.)

Under "Family History," Dr. Laughlin wrote:

These people belong to the shiftless, ignorant, and worthless class of anti-social whites of the South. . . . She attended school five years, and *attained the sixth grade;* she was fairly helpful in the domestic work of the household [into which she had been adopted at age four] under strict supervision; so far as I understand, *no physical defect or mental trouble attended her early years.* She is well grown, has rather badly formed face; of a sensual emotional reaction, with a mental age of nine years; is incapable of self-support and restraint except under strict supervision. [Italics added.]

Under "Analysis of the Facts," Laughlin reported to the court:

Generally feeble-mindedness is caused by the inheritance of degenerate qualities; but sometimes it might be caused by environmental factors which are not hereditary. In the case given, the evidence points strongly toward the feeble-mindedness and moral delinquency of Carrie Buck being due, primarily, to inheritance and not to environment. . . . From the *evidence* of her mother's feeble-mindedness and moral delinquency, and of her own clearly demonstrated feeble-minded character, the chances of Carrie Buck being a feeble-minded person through environmental and non-hereditary causes are exceptionally remote. [Italics added.]

The fact that she has already borne a child demonstrates her to be a potential parent. . . . She is therefore a "potential parent of socially inadequate offspring."

And, as such, Carrie Buck was clearly subject to mandatory sterilization as a "potential parent of socially inadequate offspring" according to the Virginia eugenical sterilization law.

The case was fought right up to the Supreme Court, where the state of Virginia based its right to mutilate Carrie Buck on the scientific analysis by Harry Laughlin that it presented to the Court. It was a distinguished Supreme Court, whose Chief Justice, William Howard Taft, had previously served— albeit without great distinction—as President of the United States. And whose most noble Associate Justice, the erudite and humane Mr. Oliver Wendell Holmes, Jr., in 1927, wrote the majority decision in *Buck* v. *Bell* (Bell being the superintendent of the State Colony for Epileptics and Feeble-Minded).

Holmes was an intellectual, and a brave man who chose to volunteer in the Union Army instead of hiring a substitute to take his place in the fighting. Twice wounded in the war against slavery, the Great Dissenter on the side of justice and law for all, he brought to his work, as his protégé and fellow Justice Felix Frankfurter wrote, "a sense of history. His traditions were founded not on fear but on knowledge, and his rejections came from knowledge, not from the blindness of prejudice. . . ."[6]

Unfortunately, the knowledge on which Justice Holmes wrote the 8-to-1 majority decision in the case of *Buck* v. *Bell* was not the knowledge of modern medicine, psychology, psychiatry, education, sociology, economics, industry, morals, *and the law* that the president of the American Association for the Feeble-Minded had cited in his *published* address of 1924. Holmes's "knowledge" in this decision was, rather, the eugenics catchbag of field worker reports and pseudoscientific pedigree trees and discredited pronunciamentos of the masterminds of the Eugenics Record Office during the primitive days of American "Binet testing," before Wallin revealed that the Binet tests showed us all to be idiots, regardless of how well we did in our trades and professions and high school and graduate school work.

When it came to the etiology of feeblemindedness, the brilliant and libertarian Justice Holmes was as one with the mediocre President Coolidge and the overtly racist editor of *The Saturday Evening Post* in accepting the pseudoscientific hereditary theories of amateurs in genetics, psychology, and biology like Harry Laughlin, and proponents of the long since disproven unit character theories of heredity such as Charles Benedict Davenport, as against the scientific counter-data of clinically and professionally qualified people like Dr. Fernald and the majority of American life and behavioral scientists for whom he spoke in his 1924 presidential address. If ever a case had to be made for the mandatory inclusion of social biology in the college and graduate school curricula of people training for the nonmedical, nonscientific, and nonpsychological professions, it was made in the text of Holmes's tragic decision in the case of Carrie Buck and the state which claimed the right to mutilate her.

I will cite only small portions of this 8-to-1 decision, which took note of the fact that "Carrie Buck is a feeble-minded white woman who was committed to the State Colony," where her mother was also an inmate, and who had already become "the mother of an illegitimate *feeble-minded* child (italics added)." Justice Holmes noted that the state reserved the right to sterilize the feebleminded, since "experience [*sic*] has shown that heredity plays an important part in the transmission of insanity, imbecility, etc."

Then Holmes rendered his decision in terms of concepts that had long before this been proven false by literally hundreds of published reports by responsible scientists the world over.

> The judgment finds the facts that have been recited and that Carrie Buck "is the probably potential parent of socially inadequate offspring, likewise afflicted, and she may be sexually sterilized without detriment to her general health and that her welfare and that of society will be promoted by her sterilization," and thereupon makes the order. . . . We have seen more than once that the public welfare may call upon the best citizens for their lives. It would be strange if it could not call upon those who already sap the strength of the State for these lesser sacrifices, often not felt to be such by those concerned, in order to *prevent our being swamped with incompetence. It is better for all the world, if instead of waiting to execute degenerate offspring for crime, or let them starve for their imbecility, society can prevent those who are manifestly unfit from breeding their kind.* The principle that sustains compulsory vaccination is broad enough to cover cutting the Fallopian tubes. *Jacobson* v. *Massachusetts,* 197, U.S. 11. *Three generations of imbeciles are enough.* [Italics added.]

This decision, by the son of the great American physician who destroyed one of the hoary tenets of conventional wisdom by publishing the real causes of puerperal fever some five years before Semmelweis did one thing for genetics: it once again disproved the Galtonian myth that scientific talent and wisdom are unit traits transmitted in the parental genes. But this was about the only good thing that could be said about the actual relationship between the science of genetics and the Holmes decision giving states the right to surgically sterilize their wards.

Holmes himself was quite proud of this decision. In the published correspondence between Holmes and the British socialist intellectual Professor Harold Laski, the jurist, on April 25, 1927, initially mentioned a decision he had just written "concerning the constitutionality of an act for sterilizing feeble-minded people—as to which my lad tells me the religious are astir. I have sent just what I think to the printer." Four days later, he again brought up the question of the "Virginia act for the sterilizing of imbeciles, which I believe is a burning theme."[7]

In response, Laski made two comments in his letter of May 7, 1927. Speaking of his own work in London, he wrote that "the problems are less interesting than settling whether a feeble-minded Virginian is to remain virgin, but, as Carlyle said, they make 'bonny fechtin.' " And, in closing, he still seemed to think that the sterilization of Virginians found to be imbeciles by the Stanford Revision of the Binet-Simon Intelligence Quotient Tests was jolly good fun. "My love to you both. Get that stomach better, please. Sterilise *all* the unfit, among whom I include *all* fundamentalists (italics added)."[8]

This contempt for the very poor, as exhibited by Marxists such as Laski, the Webbs, and Pearson, was quite typical of the nineteenth-century traditions from which they sprang. As Professor Charles E. Rosenberg observed in his profoundly informative book *The Cholera Years,* "The dichotomy between the deserving and the vicious poor was well-nigh universal in 19th century thought. Karl Marx's distinction between 'proletariat' and the 'lumpen-proletariat' is simply another reflection of this pervasive assumption."[9] When it came to the Undeserving Poor, Karl Marx and Francis Galton, like Harold Laski (who had begun his career as the personal protégé of Sir Francis Galton)[10] and Oliver Wendell Holmes, were united in their contempt for the humanity and the gonads of the wretched of the earth.

Holmes, at least, did not consider himself to be either a socialist or an egalitarian. In his next letter to Laski, in which he spoke of having written a decision "upholding the constitutionality of a state law for sterilizing imbeciles the other day—and felt that I was getting near the first principle of real reform," he also added that "I don't mean that the surgeon's knife is the ultimate symbol." But he preceded these remarks with the observation that "I have no respect for the passion for equality, which seems to me merely idealizing envy."[11]

There was one final reference to the Carrie Buck decision, this one in a letter from Holmes dated July 23, 1927, in which he reported: "Cranks as usual do not fail. One letter yesterday told me that I was a monster and might expect the judgment of an outraged God for a decision that a law allowing the sterilization of imbeciles was constitutional."[12] At no time did Holmes ever express any regret for this decision, which accepted the Stanford (Terman) Revision of the Binet-Simon IQ test as an adequate scientific gauge of mental age and imbecility, and accepted the hoary and long discarded "one gene—one trait" unit character theory of heredity that categorized the average case of such IQ-test-score-diagnosed imbecility as genetic, and therefore biologically predetermined. To Holmes and the eugenicists imbecility and feeble-mindedness (mental retardation) were a single disorder.

The failure of Holmes, and the seven Justices of the Court who voted with him, to be aware as late as 1927 of the fine work and repeated findings of Jacobi, Fernald, Stiles, Morgan, Goldberger, Wallin, and Myerson—to mention but a few of the Americans who scientifically destroyed the myths that mental retardation and shiftlessness and "laziness" were in the genes and not in the growth and development environment; that IQ tests measured innate intelligence; and that most psychoses and other aberrant behavior were the products of bad genes in people of bad stock—represented a major triumph of the previous quarter century of high-powered eugenical propaganda. Instead of heeding the discoveries and the published writings of the men and women whose findings in genetics, psychology, parasitology, clinical nutrition, and psychiatry were earning for American science the admiration of the world's scientists, Holmes and seven of the eight other Justices, including his fellow liberal, Louis Dembitz Brandeis, accepted the validity of the old scientific racism and its dogmas of eugenics and dysgenics. That is, they accepted as science the eugenic doctrines of human development, human mentality, and human worth as preached by the likes of Galton, Gobineau, Grant, Osborn, Davenport, Goddard, Brigham, and Laughlin.

Unfortunately, Pierce Butler, the one Justice who cast the sole dissenting vote, did not write a dissenting opinion, so we will never know his reasons for rejecting the "Scientific Analysis" made for the Court by Dr. Laughlin. Appointed to the Court by President Warren Gamaliel Harding in 1922, Butler sat on that bench until 1939, and is remembered today, by legal scholars, as one of the most reactionary judges to ever be on the Supreme Court. Like Holmes's majority opinion, Butler's dissent serves to prove, once again, that in terms of the competence—based on scientific knowledge and scientific insights—of politicians and judges to make decisions involving science and humankind, few things on this earth are more meaningless than such political or philosophical labels as liberal, radical, conservative, and reactionary.

With the curious exception of Mr. Pierce Butler, the Supreme Court Justices, like Presidents Theodore Roosevelt, Wilson, Harding, and Coolidge, had based their own concepts of human biology and anthropology on the racist dogmas of the eugenicists who had perverted the behavioral and biological sciences of man rather than on the useful and readily available work of the responsible and talented American scientists who were discovering the truths that could have gone a long way toward ridding our population of many of the scourges of *preventable* body- and brain-wrecking diseases, disorders, and maldevelopments. It was decisions such as the one in the Virginia sterilization case that actually created thousands of needless added tragedies of mental and physical disorders.

Thus, in the presence of bad science or pseudoscience accepted at face value as medical, psychological, and social knowledge, Justice Holmes and so many other equally well-meaning and ethical people became the allies of the century's prophets of genocide, from Madison Grant, Charles Benedict Davenport, and Hiram Wesley Evans in the 1920's—to Adolf Hitler in the 1930's. So costly to humankind is the astonishingly long life of a big lie in science.

GODDARD ADMITS FEEBLEMINDEDNESS IS NOT HEREDITARY

The state surgeon's knife that slashed Carrie Buck's Fallopian tubes was hardly out of her body when Henry H. Goddard, the leading American proponent of "the menace of the feebleminded," published a long scientific paper in which he retracted nearly everything he had been writing about the feebleminded and their hereditary social inadequacy for the previous twenty years. In his article "Feeblemindedness: A Question of Definition,"[13] Goddard presented enough evidence against the reliability of the Terman version of the Binet test as a measurement of whether or not a person was feebleminded, and enough evidence against the myth he himself had done so much to broadcast to the effect that feeblemindedness was *the* cause of most crime and other antisocial behavior, as to make the sterilization of Carrie Buck scientifically and morally an act of unwarranted mayhem.

It was a strange article for Goddard to author, and it took considerable intellectual and moral courage to publish it. For, as he observed, as of 1928 we were not even certain of what feeblemindedness was:

> . . . we are in the unpleasant predicament of having a definition that does not define. Our standard definition calls upon three sciences and finally leaves us in the midst of uncertainty. The phrase "because of mental defect" clearly refers to Psychology. "Existing from birth or an early age" is a matter of Biology, while "competing on equal terms or managing himself or his affairs with ordinary prudence" is clearly a matter of Sociology.

Scientifically, this statement of the sheer complexity of the origins and nature of human behavior was light-years away from the pseudogenetic simplistics of Laughlin that the Supreme Court had less than a year earlier accepted as a scientific justification for the mutilation of wards of the state. This, however, was only by way of introduction. There was also that holy of holies, the IQ test score:

> Stern has given us the concept of the "intelligence quotient," and Terman hoped thereby to help us out of our difficulty by fixing a definite I.Q. for the feebleminded. He accordingly set 70 as the I.Q. below which we all were feebleminded. But even he was afraid to draw so sharp a line, so he states that the group with I.Q.'s 70–80 will generally be found feebleminded.
>
> This sounds simple and easy, but as a matter of fact, it throws us into great confusion because it contradicts the standard definition. According to Terman's I.Q. classification [i.e., the Stanford Revision of the Binet-Simon tests] every adult with a mentality of 11 years or below would be definitely feebleminded. Or if we take his next group, I.Q. 70 to 80, every adult up to 12.8 years would be feebleminded. *According*

to the army tests this would include very nearly half the people of the United States. This is, of course, based on Terman's arbitrary selection of 16 years, chronological age, as the limit of mental development, with the consequent rule that we never use a divisor greater than 16. If we should take 19 or 20 which the army test seems to indicate is the highest development of mentality, then Terman's selection of 70 I.Q. as the upper limit is much too high. *Moreover, the I.Q. would cease to be constant, and this cherished concept would be demolished.* Therefore, the I.Q. concept does not help us out of our difficulty. We are forced back to the original [pre-I.Q. testing] definition with all its difficulties. [Italics added.]

All of which meant that not only did the Stanford Revision of the Binet-Simon tests have no value in the diagnosis of feeblemindedness, but the professional and mental competence of any psychologist who continued to use IQ tests for such diagnoses was highly questionable. It was not, said Goddard, the American Binet-testing pioneer, that the tests themselves were unreliable. It was more a matter of realizing that feeblemindedness was "due to other factors than intelligence. These factors coupled with low intelligence turn the tide one way or another."

The crux of the problem, said this lifelong eugenicist, was not that "intelligence is inherited, while a personality is mostly acquired." It was, rather, and Goddard stated his point in italics for added emphasis: *"that the problem of the moron is a problem of education and training.* Not that education and training will change his moronity or raise his intelligence level, but that it can develop in him the capacity to do certain routine things which will give him happiness and which will enable him to compete in the struggle for existence."

In short, by 1928 Goddard had finally gotten around to accepting the concept Dugdale had originally proposed in his book on the Jukes family in 1874. That is, Dugdale wrote, that "where environment changes in youth, the characteristics of heredity may be measurably altered."

Goddard then proceeded to concede what happened to be true before and after he had helped make Galton's "menace of the feebleminded" a standard American nightmare: to wit, that most people classified as "morons" and "mentally retarded" by the Goddard and the Terman (Stanford) revisions of the Binet tests were capable of earning their own living, of managing "themselves of their affairs with a *high degree* of prudence," and "if carefully and wisely trained are not a burden in the community."

Goddard recognized that many of his readers would "object that this plan neglects the eugenic aspect of the problem. In the community these [self-supporting] morons will marry and have children." The Goddard of 1910 would have joined in raising this alarm. But in 1928, Goddard answered: "And why not? When nine-tenths of the mentally 10 year old people are marrying why should the other tenth be denied? Moreover if moronity is only a problem of education and the right kind of education can make out of them happy and useful hewers of wood and drawers of water, what more do we want?"

If, sociologically, this was a far cry from Spencer, Sumner, E. A. Ross, and Henry Pratt Fairchild, the paragraph that followed revealed that the Goddard of 1928 had also been catching up on developments in *modern* genetics:

"It may still be objected that moron parents are likely to have imbecile or idiot children. *There is not much evidence that this is the case.* The danger is probably negligible. At least it is not likely to occur any oftener than it does in the general population." (Italics added.)

This was not genetics as Laughlin preached, and the U.S. Supreme Court of 1927 accepted, genetics to be. But it was a complete repudiation of everything that the gurus of eugenics, from Galton and Pearson to Davenport and Osborn, had presented as the basis of human heredity. Goddard was acutely aware of his apostasy:

> I assume that most of you, like myself, will find it difficult to admit that the foregoing view may be the true view. We have worked too long under the old concept [i.e., of the heredity of mental traits]; we have known too many morons who certainly could never get along in society; and while we admit that they never had a chance, and that they did not have this ideal education of which we speak, nevertheless the concept [of modern genetics] is so unacceptable that we feel like a certain prominent educator who when confronted with the figures which showed that a statement that he had often made was not true, replied, "Well, I have believed it so long that I think I will keep on. It is easier to do that than to change." As for myself, I think I have gone over to the *enemy* with only one reservation [italics added].

The "enemy," of course, was the community of genetical, pediatric, behavioral, and social scientists who insisted that it was not the genes alone but the entire spectrum of qualities of that environment in which a child grew and matured that determined the ultimate expressions of one's inborn genetic potentials. But in the statement of this "reservation" Goddard revealed how completely he had accepted the scientific conclusions of the "enemy":

> This may surprise you, but frankly when I see what has been made out of the moron by a system of education, which as a rule is only *half right* [Goddard's emphasis], I have no difficulty in concluding that *when we get an education that is entirely right there will be no morons who cannot manage themselves and their affairs and compete in the struggle for existence* [italics added]. If we could hope to add to this a social order that would literally give every man a chance, I should be perfectly sure of the result.

As with most other extremists, when Goddard shattered the shackles of one old absolute dogma he replaced them with an equally all-or-nothing dogma: the ideal education as the solution to the learning and behavioral problems of man. Even by 1928, as any pediatrician could have told him, a sound body was still the *first* prerequisite for a sound mind, and the developmental biology of poverty was as much a factor in education as the quality of its teaching techniques. But he had come so far from the orthodoxies of

Galton and Laughlin as to suggest that the names of the institutions for the feebleminded be changed from Institution for the Feebleminded to simply State School, to keep any of the lasting stigma of "the menace of the feebleminded" from marking their students.

THE *MEA CULPA* OF CARL CAMPBELL BRIGHAM

The most dramatic repudiation of eugenical writings of that tragic decade was yet to come, for in September 1929 Professor Carl Campbell Brigham submitted to *The Psychological Review* an article whose innocuous title, "Intelligence Tests of Immigrant Groups," gave no indication that in this short essay Brigham was to completely repudiate every statement he had made in *A Study of American Intelligence,* published only a half-dozen years earlier. Ostensibly, the article was a commentary on various recent journal articles on the IQ testing of various immigrant groups. Actually, as Brigham said in his opening paragraph, he proposed that this article would "direct itself to the more fundamental problem of the meaning of the test scores."[14]

Brigham, noting that "psychologists have been attacked because of their use of the term 'intelligence,' and have been forced to retreat to the more restricted notion of *test score*," said they were no longer as certain as they until recently had been that it was indeed "intelligence" that the IQ tests really measured. So that the psychologists' "definition of intelligence must now be 'score in a test which we *consider* measures intelligence.' "

To Brigham, the increasing attention psychologists were paying to "the more careful analysis of test scores" had already resulted in several important advances. "The question immediately arises as to whether or not the test score itself represents a single unitary thing. This question is crucial, for if a test score is not a unitary thing it should not be given one name."

After some highly technical discussion of the non-unitary nature of the tests and their results, as revealed by recent scientific studies, Brigham observed:

> A far-reaching result of the recent investigations has been the discovery that test scores may not represent unitary things [such as "sensory discrimination, perception, memory, intelligence and the like"]. It seems apparent that psychologists in adding scores in the sub-tests in some test batteries have been doing something akin to adding apples and oranges. *A case in point is the army alpha tests* [italics added].

What followed was a highly technical dissection of the structure and nature of the Army mental tests he had helped write, administer, and evaluate, which showed them to be so inconsistent as to be worthless. And this, in turn, brought Brigham face to face with his moral and scientific responsibilities as both a scientist and a citizen and the interface at which they clashed with his personal loyalties to Yerkes, Madison Grant, and his long-time patron, Charles W. Gould. The result has been called one of the most agonizing retractions in the history of the behavioral sciences:

If the army alpha test has thus been shown to be internally inconsistent to such a degree, then it is absurd to go beyond this point and combine alpha, beta, the Stanford-Binet and the individual performance tests in the so-called "combined scale," or to regard a combined scale score derived from one test or complex of tests as equivalent to that derived from another test or another complex of tests. *As this method was used by the writer in his earlier analysis of the army test as applied to samples of foreign born in the draft, that study with its entire hypothetical super-structure of racial differences collapses completely* [italics added].

A footnote to this last sentence made it clear that he was referring to the "race hypothesis" on page 210 of *A Study of American Intelligence*. But there was more than this one page to discuss; there was also the rest of the book. And this, after a discussion of the importance of cultural factors, starting with languages and vernaculars, in IQ test performances, Brigham did in his two closing paragraphs:

> For purposes of comparing individuals and groups, it is apparent that *tests in the vernacular must be used only with individuals having equal opportunities to acquire the vernacular of the test*. This requirement precludes the use of such tests in making comparative studies of individuals brought up in homes in which the vernacular of the test is not used, or in which two vernaculars are used. *The last condition is frequently violated here in studies of children born in this country whose parents speak another tongue.* It is important, as the effects of bilingualism are not entirely known.
>
> This review has summarized some of the more recent test findings which show that comparative studies of various national and racial groups may not be made with existing tests, and which show, in particular, that *one of the most pretentious of these comparative racial studies —the writer's own [A Study of American Intelligence]—was without foundation.* [Italics added.]

It was a very courageous confession of total error, and it would have been nice if the anti-Italian, anti-Jewish, anti-Polish, and anti-Greek immigration quotas based on the same scientific fallacies were subsequently abolished as a result of Brigham's public retraction. But all Brigham was to save by his *mea culpa* was his own soul.

By January 1930, when Brigham's retraction appeared, Harry H. Laughlin—who had made such devastating use of Brigham's data and authority as Expert Eugenics Agent of the House Committee on Immigration and Naturalization—was involved nearly full time in the major crusade launched by John B. Trevor, the creator of the quotas, to keep Congress from relaxing the barriers against Italians, Jews, Poles, Greeks, and other then and future victims of the torture and death camps of Mussolini and Hitler. And the intellectuals of the eugenics movement never wavered for a moment in their faith in the scientific validity of IQ tests and test scores as determiners of who should live and who should die.

The year 1930 was not the best year for rational discussion of the validity of IQ testing. It was Year Two of the Great Depression.

14 While the Real Sciences Create New Hope for Humankind, the Great Depression Weakens the Appeal of the Old Scientific Racism

> There are those who believe that our population has already attained a greater number than is necessary for the efficient functioning of the race as a whole. Certainly our present picture of millions of unemployed would point to the belief that this suggestion is a reasonable one. It would undoubtedly be found, if such research was possible, that *a major portion of this vast army of unemployed are social inadequates, and in many cases mental defectives, who might have been spared the misery they are now facing if they had never been born* [italics added].
>
> —THEODORE RUSSELL ROBIE, in "Selective Sterilization for Race Culture," presented at the Third International Congress of Eugenics, New York, August 21–23, 1932

As we saw in chapter 12, the memorandum of the Allied Patriotic Societies of New York to the chairman of the House Committee on Immigration and Naturalization of June 7, 1923, deplored the fact that "the bulk of the 'newer' immigration is made up of Italians, Hebrews, and Slavs, all of which are races much more widely divergent, biologically speaking, from the *basic stocks* of the country than the great bulk of the 'older' immigration prior to 1882, which was Anglo-Saxon, Germanic, Scandinavian, and Celtic *(Irish)*."

This was indeed a strange departure from the xenophobia normal in hyperpatriotic circles. As, indeed, was the statement in the *Dearborn Independent,* the magazine in which Henry Ford serialized large portions of the gamy *Protocols of the Learned Elders of Zion*—that anti-Semitic forgery of the Czar's secret police so beloved to Madison Grant and Adolf Hitler as well as the Detroit motorcar magnate—concerning the same ethnic group. In launching his campaign against the Jews, "who had no civilization . . . no great achievement in any realm but the realm of 'get,' " Henry Ford declared, in this magazine distributed to all Ford dealers, that America's "Founding Fathers were men of the Anglo-Saxon-*Celtic* race" (italics added).[1]

The Celtic (Irish) race was a very new arrival into the charmed circle of the superior races that had been put upon the earth to Make America Great. Even before the potato-blight famines of midcentury sent swarms of Irish immigrants across the ocean to build America's sewers and railways and populate its urban slums, the Irish poor were anathema in the eyes and nostrils of their Nordic and Protestant betters in the New World. For example, after the terrible cholera epidemic of 1832, which was most severe in the cities, and whose most numerous victims were the poorest of the slum dwellers inevitably exposed to the greatest concentrations of cholera germs in polluted water—which, at that time, meant the Irish and the freed Negroes—the members of New York City's Board of Health, in their official report on the outbreak, noted that *"the low Irish suffered the most, being exceedingly dirty in their habits, much addicted to intemperance and crowded together into the worst portions of the city.*"[2]

Of course, between the time this report was written and the moment the low Irish became full-fledged Founding Fathers of Nordic America, Sir John Snow had in 1853 shown that cholera was caused by polluted water and not polluted blood lines. And by 1884 Robert Koch had rediscovered Dr. Pacini's dirty-water agent, the *Vibrio comma,* and shown it to be the anything but Celtic (Irish) germ of cholera. Nevertheless, these and similar discoveries in the life sciences had not, as we saw earlier, stopped Oxford professor Edward A. Freeman from thrilling American lecture audiences with the oft-repeated observation that "the best remedy for whatever is amiss in America would be if every Irishman should kill a negro and be hanged for it." Nor did the discovery of the clearly environmental causes for tuberculosis stop Charles Benedict Davenport from solemnly writing that the Irish *"as a nation* lack resistance to tuberculosis."[3]

No. What had turned the Irish into ex-pariahs was something else again. Davenport had referred to it on page 213 of his *Heredity in Relation to Eugenics* (1911): "The Irish tend to aggregate in cities *and soon control their governments,* frequently exercising favoritism and often graft" (italics added). The control of municipal governments of cities such as New York, Boston, Philadelphia, Chicago, and other urban centers which by the 1920's were clearly becoming the centers of American population densities, meant, as well, control of the municipal purchases of literally billions of dollars' worth of automobiles, fire trucks, police cars, steel, cement, paint, bricks, pipes, copper wires, insurance, land, banking services, and scores of other commodities and services offered by old-line Nordic entrepreneurs and real estate heirs. A change in the status of the formerly *low Irish* who now controlled these purchase orders was therefore quite inevitable.

So that the revision of one of the most firmly believed myths of nineteenth- and twentieth-century American racism, scientific and gutter, had nothing to do with any mutations that had miraculously changed the nature of Irish genes between 1911 and 1920. The causes of the changed racial status of the Celtic (Irish) immigrants and their fairly immediate descendants were not at all scientific or even eugenic. They were merely and exclusively economic. Gone were the days when the want ads of the American press contained the familiar refrain "No Irish need apply." Instead there was the very sudden but widespread acceptance of the fact that the formerly low Irish—those newly elected masters of municipal purchase orders for Nordic goods and services—were indeed fully qualified members of the human race, and their formerly Papist catchbag of pagan superstitions a full-fledged Christian religion.

Even as hidebound a Nordicist as Professor Henry Fairfield Osborn bowed to this new order of things. After the conclusion of the Second International Congress of Eugenics in New York in 1921, Osborn wrote to his faithful adjutant, Davenport, that "I think you will agree with me that Archbishop [later Cardinal] Hayes' official approval of the Eugenics Congress is one of the most important results of our Congress."[4] Which meant that not only the purchase orders of Irish mayors and their commissioners but also the moral approval of the Irish Catholic hierarchy had taken on values to members of the Nordic aristocracy that they had never possessed before the Celtic (Irish) breeders of cholera and tuberculosis took over City Hall.

The loss of the Irish scapegoats was not the only post-World War I change in the demonology of all forms of American racism. Thanks to the educational and white-collar job quotas, the residential "gentlemen's agreements" to neither sell nor rent decent homes to Jews, Italians, and other non-Nordic types that took care of such "scum" on the domestic front—and thanks also to the Immigration Exclusion Act of 1924 that finally kept their kind from entering this land as immigrants—this left America's blacks as the logical successors for the primary aggressions of America's hooded and scientific racists. The military draft of World War I, and the recruiting agents of America's munitions and other war industries, had both set off a new flight from Egypt for the southern blacks.

To be sure, once the refugee southern blacks found gainful employment in northern war industries, white resentment of their presence led to serious race riots in cities such as East St. Louis, Illinois—where five people were killed. After the war, in which 367,000 blacks served in the armed forces, the attempts of black doctors and businessmen to buy decent houses led to much more violent and deadly race riots in Chicago, Detroit, and many other cities ranging from Omaha, Nebraska, to Washington, D.C. There were also, writes Samuel Eliot Morison, "at least seven [race riots] in the South, mostly occasioned by Negro veterans of the war having the 'impudence' to demand their rights as citizens."[5]

However, even though these race riots and the less dramatic resistance to the new migration of poor blacks from Dixieland led to such political reactions as the capture of the state government of Indiana by the Knights of the Ku Klux Klan and other northern political inroads by Dr. Hiram Wesley Evans' night riders, various events tended first to slow and then to table for over a decade the concentration of racist attentions on the blacks. These events were, like those that had transformed the innately low Irish scum into authentic native Americans, economic.

In this instance, it was the series of bursting bubbles and antic market spirals that had made the "boundless" prosperity of the 1920's a time of something less than total inward security. And that, on October 24, 1929, culminated in the great Wall Street stock market crash that was to set off the greatest depression, the most widespread unemployment—proportionately as well as numerically—in the history of the Republic.

Nothing could stand above the Great Depression and its human and socioeconomic effects as a concern of modern man. By the third year of the Depression, 1932, this was clearly reflected by the difficulties the president of the Third International Congress of Eugenics, Charles Benedict Davenport, experienced in getting the world festival of scientific racism held at all. Natural attrition had removed from the scene such financial patrons of the 1921 Congress as Charles W. Gould, whose money also financed Carl C. Brigham's historic *A Study of American Intelligence,* but who died in 1931. Mrs. Harriman was still around, and she contributed generously; the Carnegie Institution of Washington came through with some two thousand dollars; but Mrs. Cleveland H. Dodge, whose late husband had been a major contributor in 1921, sent only fifty dollars, for which she was "terribly apologetic" but, explained the treasurer of the Congress, "this was due, as you can realize, to the extraordinary position of the copper industry at the present time." John D. Rockefeller, Jr., on the other hand, who was quite solvent, but who had inherited his father's excellent advisers in biology and medicine, turned down Davenport's request for money in a letter by his secretary explaining that "in view of the policies which guide Mr. Rockefeller in making his commitments, the project does not seem to be one towards which Mr. Rockefeller, Jr., as an individual could contribute." President Herbert Hoover, who while Secretary of Commerce contributed money to the 1921 Congress and served on its General Committee, turned Davenport down cold in 1932.

In the end the Osborn family had to step in and bail out the show. Henry Fairfield Osborn's nephew, retired financier Frederick Osborn, took over the job of treasurer, advanced the needed seed money, and arranged to come in at the end with whatever funds would still be needed in the closing weeks before the Congress assembled at the American Museum of Natural History, the family museum.

Henry Fairfield Osborn, president of the 1921 Congress and honorary president in 1932, lent considerable support and guidance. In a letter dated May 9, 1932, Osborn reported to Davenport that on his recent trip to Europe "I endeavored to interview Premier Mussolini while in Rome so as to insure a very representative delegate from Italy; I presume Corrado Gini is such a one."[6]

ERADICATING THE GENES FOR UNEMPLOYMENT

The makeup of the Program Committee, chaired by Davenport himself, pretty much forecast what the individual papers would say. Some of the members, and their committee responsibility, were: Paul Popenoe, sterilization, fecundity; his collaborator on the textbook *Applied Eugenics,* Roswell H. Johnson, sociology; Robert DeCourcy Ward, immigration; H. H. Goddard, mental defect; S. J. Holmes, bibliography; W. K. Gregory, phylogeny; E. A. Hooton, anthropology; Lewis Terman, psychology; and Albert Edward Wiggam, applied eugenics.

Not even this hereditarian Program Committee could put together a scientific meeting on the problems of mankind in which any problem would be termed more immediate, more critical than the Great Depression. But, where most people blamed the collapse of the economy on its principal managers, to the organizers of the Third International Congress of Eugenics the blame lay in other directions. To the assembled eugenics zealots, the chief authors of the Great Depression were the millions of suddenly unemployed wage and salary workers; the hundreds of thousands of suddenly ruined tradesmen and professionals and owners of small businesses and industries; and the scores of suddenly wiped-out super-rich who took to leaping from high windows in

Wall Street and from the roofs of their huge shuttered mills and assembly plants, or to weeping their way into prisons for various acts of defalcation and other larceny. And the thousands of ruined employers around the nation who took to joining their former employees on the breadlines and in the soup kitchens that had become ubiquitous fixtures in all American cities.

To the true-believing eugenicists it was obvious that the real underlying cause of the American financial breakdown was the delayed expression of the ruined Americans' inherited genes for pauperism, bankruptcy, and commercial failure that had manifested themselves later in life because of generations of uncontrolled, unscientific, and genetically suicidal race mixing. Grave crises required drastic remedies. As the collected papers of the Third International Congress of Eugenics show, the average participant in the program knew exactly what had to be done. Once and for all, we as a race had to make certain that the bearers of bad stock (inferior genes) be prevented, by segregation, by sterilization, by compulsory abortion, and by infanticide, from multiplying and further diluting the American gene pool. This was the major thrust of the scientific thought at the Congress.

There were a few papers on other themes, including even one on craniology, and one by Roswell Johnson, "The Inheritance of Mental Test Abilities," proving, of course, that "a stimulating home environment" was a minor factor, but that *Blut ist alles.* Basically, however, the primary message of the entire Congress was summed up in the opening paragraph of Paper No. 27, "Is the Abnormal to Become Normal?" presented by Lena K. Sadler, M.D., F.A.C.S.:

> The various dysgenic classes which are so rapidly increasing in the United States constitute our vast "aristocracy of the unfit." They are an undesirable group of citizens which the more thrifty, intelligent, and superior stocks willingly tax themselves to support and perpetuate. This increasing horde will ultimately overrun and destroy the diminishing prosperity of the better classes unless a practical program of restrictive eugenics is adopted and effectively executed.[7]

Dr. Sadler's practical program of applied eugenics was based on infanticide—the end to trying "to save every weak child that is born into the world," and particularly to the free medical treatment of "the unfit baby in our welfare stations, dispensaries and clinics," since every such "coddled, protected weakling grows to adolescence and . . . cannot get out of the fourth grade at school." For every such "manifestly defective and degenerate" child who survived such planned Malthusian neglect, Dr. Sadler demanded forced sterilization:

> There is no question that a sterilization law, enforced throughout the United States, would result, in less than 100 years, in eliminating at least 90% of crime, insanity, feeblemindedness, moronism, and abnormal sexuality, not to mention many other forms of defectiveness and degeneracy. Thus, within a century, our asylums, prisons, and state hospitals would be largely emptied of their present victims of human woe and misery. The indigent and aged paupers, and the unfortunate degen-

erates of various types would disappear as a troublesome factor in civilized society.

Galton's dream of selective breeding was advanced in the next paper, "Selective Sterilization for Race Culture,"[8] by Theodore R. Robie, M.D., of the Essex County Mental Hygiene Clinic of Cedar Grove, New Jersey. Dr. Robie, using the Army Alpha and Beta mental test scores of World War I, as well as the pre-World War I studies of the "Kallikaks" by Goddard, as his scientific authorities, claimed that fourteen million Americans were "below the accepted standard for average intelligence." Robie had no doubt at all that, when it came to the millions of Americans rendered unemployed by the lack of jobs in the Depression, "a major portion of this vast army of unemployed are social inadequates, and in many cases mental defectives, who might have been spared the misery they are now facing if they had never been born."

Naturally, since "the greatest single cause of mental deficiency (50 to 65 per cent) is poor heredity," Dr. Robie's proposals for the selective compulsory sterilization of the bearers of the genes for mental retardation, as well as the genes for chronic unemployment and/or pauperism, would indeed have made Dr. Sadler's dream of race purification via the sterilization of the dysgenic classes come true in possibly even less than her hundred years. Scientific documentation for such compulsory selective sterilization was provided by Sir Bernard Mallet, K.C.B., president of the Eugenics Society, of London, in Paper No. 48, "The Reduction of the Fecundity of the Socially Inadequate."[9]

Sir Bernard came to exactly the same conclusions about compulsory sterilization of the unemployed as those advanced by Drs. Sadler and Robie. To Bernard Mallet, it was obvious that "birth control has so far acted dysgenically, by reducing the fertility of the better-endowed strains, while leaving relatively unchanged the birth rate of those who are less fit for parentage."

To cope with this "racially harmful" state of affairs, the London Eugenics Society had established a Mental Deficiency Committee, whose members, by 1929, confirmed the existence of a human subspecies they termed the "Social Problem Group." Now, in 1932, Mallet told the International Congress of Eugenics:

> The existence of such a group has long been realized by eugenists and sociologists, but *the existence among us of a definite race of chronic paupers,* a race parasitic upon the community, breeding in and through successive generations, and only to a small extent recruited either from the ranks of unskilled labourers, or by the sufferers from the fluctuation of employment, was, perhaps, first noted and investigated by Mr. E. J. Lidbetter, who for many years has conducted a study of this group in a Poor Law area in East London [italics added].

The London Eugenics Society's Mental Deficiency Committee had described this special race of the chronic poor as including "a large proportion of the insane persons, epileptics, paupers, criminals (especially recidivists), unemployables, habitual slum-dwellers, prostitutes, inebriates and other social inefficients" found in all societies. The committee estimated that "this group comprises approximately the lowest 10 percent in the social scale of most communities."

It was bad enough to have discovered the existence of a distinct *race* of hereditary paupers, but this was the least of it. For "of this group, the report further states that its anti-social characteristics are the result, mainly, of inferior heredity, and that *its fertility is higher than that of any other social element*" (italics added). This, of course, was the specter that had long haunted Francis Galton and Lothrop Stoddard: the chimera of the undue fertility of the Under-Man.

Birth control had been tried, in England, but far from making things better it had only made them worse, said Dr. Mallet. For now that improvements in the total environment, in medical modalities, and in social conditions had reduced the carnage of infants and children dying before their time, Britain's educated families were addicted to birth control as a safe method of protecting their women from the burden of having too many children in order to make certain of having a few survive to become heirs and parents themselves. But, Sir Bernard declared, "the classes in which the use of contraceptive methods would be eminently desirable are incapable of applying any methods yet devised."

Therefore, pending the development of new birth control devices that even the inferior brains of poor people could manage, Sir Bernard Mallet and the Eugenics Society concluded that "for the time being apart from segregation, which is clearly the best method, *it is only to sterilization that we can look to limit the fertility of mental defectives and of those classes composing the Social Problem Group*" (italics added).

The Social Problem Group was, of course, a flowery pseudoscientific label for the venerable Undeserving Poor. As Eugene S. Gosney, president of the Human Betterment Foundation of Pasadena, California, who followed Sir Bernard to the podium, observed, this Social Problem Group was a product of misguided altruism, and not nature itself.[10] For, as all true-believing eugenicists knew, "it is the unfit that is dangerous to civilization. The real problem is to prevent their inferior posterity from deteriorating the race."

Mr. Gosney, like Dr. Sadler, proclaimed that the greatest enemy of "the race" was civilization itself. "Nature's law of the survival of the fittest took care of that problem in past ages. In those stages of race development it was only the physically strong and mentally alert that could survive." However, said lawyer and banker Gosney, no scientist himself but a confirmed and believing eugenicist: *"With the dawn of the spirit of charity and human sympathy . . . nature's hard but effective law was nullified.* The weak and unfit are nursed to maturity and allowed to reproduce their kind. Our charity organizations have not completed their job until they have made some provision for the prevention of reproduction in the recognized cases of the *hereditary inadequate."*

Segregation of these Special Problem Group people was impractical, said Mr. Gosney, "because no country has sufficient buildings to house more than

a small per cent of the *increasing multitudes of that class.*" Forced imprison-
ment would also, said the sentimental Californian, make them unhappy. There
was no escaping the inevitable:

"Birth control by contraceptives cannot be used by the unfit. They have
not the necessary intelligence, stability, or will power. *Sterilization,* as used in
California continuously for 23 years, *offers the only adequate method of
materially checking this approaching shadow of race degeneracy.*" (Italics
added.)

Although the compulsory sterilization of America's mounting millions of
unemployed human beings, and their children, was the main theme of the
Third International Congress of Eugenics, the Congress itself was not without
its dramatic surprises. Where, in 1921, the Tufts professor of neurology
Abraham Myerson had risen to refute the hereditarian myth that most mental
diseases and neurological disorders such as epilepsy were genetic in origin, in
1932 Professor Hermann J. Muller, the geneticist who helped destroy the unit
character (one gene = one trait) myths on which *all* of the "scientific" hereditarian theories of eugenics were based, arose to explode a bomb entitled "The
Dominance of Economics over Eugenics."[11] For Muller, it now developed,
was one of those rare and atypical left-wing eugenicists who dreamed of synthesizing the two great scientistic religions of the modern intelligentsia, the
Gospels of Galton and Marx.

Whereas the other speakers at the Congress were all rightists who saw
the world's poor, unemployed, and suffering people as the dysgenic barriers to
the breeding of a eugenic race of Nordic supermen, Muller had other adverse
factors in mind:

"We might as well admit that the forces at work are quite beyond the
control of us as eugenists, in the society in which we live. For they are fundamental economic forces. Galton lived too early to appreciate the principle
brought out by Marx that the practices of mankind, in any age, are an expression of the economic system and material techniques existing in that age."

Because of these economic imperatives, Muller concluded, "eugenics
under our social system cannot work. . . . Only the impending revolution in
our economic system will bring us into a position where we can properly judge,
from a truly social point of view, what characters are most worthy of man, and
what will best serve to carry the species onward to greater power and happiness in a united struggle against nature, and for the mutual betterment of all
its members."

A new world was dawning, the world of socialism: "in our day the
writing on the wall is manifest, and they are fools who blind themselves to it."
In this just new world, logic and reason would prevail, and the "present dis-
putes of eugenists about the fates of races will soon appear in vain and beside
the point, when the economic and social reasons for the existence of the
differential fertility of races, as well as for race prejudices, will have disappeared with the general abolition of exploitation. True eugenics will then first
come into its own and our science will no longer stand as a mockery."

So that Muller could end by saying that "it is up to us, if we want
eugenics that functions, to work for it in the only way now practicable, by first

turning our hand to help throw over the incubus of the old, outworn society."

In terms of its effect on the members of the Congress, Muller's paper was far less shattering than its postulates were soon to prove to their author.[12] Muller's Marxist Galtonism was an odd and discordant note sounded in an orchestrated war cry. It was on the seminal issue of the compulsory sterilization of America's unemployed and their children that the Third International Congress of Eugenics based its 1932 message to humanity.

Historically, of course, this crusade was doomed, for one simple reason. In 1921—when the Second Congress of Eugenics declared war on foreigners, particularly low-IQ-test-scoring foreigners—millions of employed, faintly paranoid, and more than mildly hyperpatriotic Americans were ready to do as Laughlin and Davenport and Osborn and Jon Alfred Mjöen and Lapouge bade them do unto would-be non-Nordic immigrants. But by 1932 between three and five million members of the Ku Klux Klan,[13] and the even greater numbers of members of the American Legion, the Sons of the American Revolution, the American Federation of Labor, and other societies that had joined in the crusade of the eugenics movement to get the Immigration Act of 1924 passed, were—by the millions—losing their jobs, their businesses, their homes, their farms, their illusions. Now *they* were the starving, the homeless, the angry veterans of World War I—wearing their old Army and Navy uniforms—being brutally driven out of Washington by federal troops when they arrived to petition for relief from the mounting miseries of the Great Depression.

It was not the gonads of the unsuitable foreigners that the eugenicists were after this time. For all of their former zeal for race purity, the millions of now unemployed Nordics, Celts, and other native Americans had very intimate reasons for suddenly becoming conservatives in the matter of compulsory sterilization. Somehow, they all felt that they themselves were newly labeled as the hereditary "aristocracy of the unfit" and the genetically "social inadequates" Drs. Sadler and Robie had denounced; the American members of the Social Problem Group, the "definite race of chronic paupers," whose forced sexual sterilization had been demanded by Sir Bernard Mallet. They were quite right. And they were not about to yield up their fecundity to the sterilizers' knives so that the likes of Sadler, Robie, Mallet, and Henry Fairfield Osborn could sleep easier nights.

The Third International Congress of Eugenics crusade to sterilize the people the eugenicists declared to be social inadequates collapsed almost as soon as it was launched in 1932. The Galtonian dream of gelding the poor was far from dead, but the golden era of the forced sterilization of the helpless was yet to come.

Years of bitter history and surprising changes would have to pass before the involuntary surgical sterilization of the poor would be resumed by the state *and federal* governments on a gargantuan scale. But, as of 1932, the sterilization crusade of the old scientific racism was rendered moribund by the inane speeches of people such as Sadler, Robie, Gosney, and Mallet at the Third International Congress of Eugenics.

The Great Depression itself had taken care of the demands of the eugenicists that all government agencies should cease appropriating tax dollars for health, education, and family welfare programs. All that remained for the eugenicists and their political allies to do for their cause was to make certain that the anti-Italian, anti-Semitic, and anti-Catholic immigration quotas they helped write into federal law in 1924—their proudest achievement—were not tampered with or relaxed in the fatuous name of humanity, compassion, brotherhood, decency, and other catchwords of the ancient and modern enemies of eugenics. Here they were to taste far sweeter fruit. (See chapter 15.)

A SECOND MESSAGE FROM IOWA

The Great Depression that was, at least for the duration, to deprive millions of fellow travelers of the eugenics cult of their erstwhile enthusiasm for the forced sterilization of the unemployed, also caused sudden and catastrophic cuts in the money available for investigation and experimentation in the life, behavioral, and social sciences. Trained scientists who found work driving cabs or raking leaves were the lucky ones; qualified investigators who wound up on the welfare rolls were the rule rather than the exception.

Despite the disappearance of funds for research, however, and possibly even because of the social effects of the Depression itself, a series of discoveries about the children of parents with low IQ test scores made during the Depression years was to destroy forever whatever rudimentary worth some people still professed to find in IQ test scores as either measurements of transmissible mental traits or as predictors of future development. History being always more prone to dramatic coincidences than fiction, this final blow to the genetic and/or productive worth of IQ tests was delivered in the same prairie state of Iowa where, some two decades earlier, Wallin's work had first demonstrated that the Goddard-Binet tests had mislabeled successful and highly competent farmers, landowners, merchants, laborers, housewives, and even high school and college students as feebleminded morons.

The times being what they were, the now famous "Iowa studies" were not designed as research projects at all; research was the last thing the state of Iowa had in mind, in 1934, when it called upon the clinical psychologists on the faculty of the State University of Iowa for help in the administration of the placement of unwanted children in foster homes. As Dr. Harold M. Skeels, a psychologist on the staff of the Iowa Child Welfare Research Station at the State University, reported at the 1937 annual meetings of the American Association for the Advancement of Science in Indianapolis:

". . . this study was not set up with any preconceived ideas as to the effect of the environment on the growth of intelligence. It grew, instead, out of the regular clinical program. The Bureau of Child Welfare requires that a child remain in a foster home at least 12 months before adoption is completed. On February 1, 1934, a policy was established whereby no child could be adopted until a psychological examination had been made."

Because it lacked the money to hire psychologists to visit foster homes and conduct such examinations, the state of Iowa was forced to turn to the

Department of Psychology of the State University for such people as were already on its payroll. This led to the collaboration of two remarkable men, Harold M. Skeels, of the Iowa Child Welfare Research Station and the State Board of Control, and Dr. George D. Stoddard, then professor of child psychology and director of the Iowa Child Welfare Research Station.

As a state psychologist Skeels was "faced with practical problems concerning the placement and adoption of illegitimate children. These become wards of the state and are available to families who want children. At first, more or less as a routine, the mothers were given Stanford-Binets."[14]

The illegitimate children who became wards of the state were all white, and the offspring of native-born parents. The histories of their true parents were known. Since so many of the mothers had IQ test scores at or below the score of 80 that *supposedly* denoted feeblemindedness, the psychologists fully expected "to find frequent cases of retarded children," but, said Skeels, "when these examinations repeatedly failed to show such retardation, it seemed important that a more extended study on a research basis be made."[15]

This study the Iowa group proceeded to make, with a total of 147 children, all placed for adoption before they were six months old. Three of the tables Skeels showed at the 1937 AAAS meetings summed up the first three years of the study. The first table (p. 35) dealt with the education of the natural and the adoptive parents of these 147 Iowa children:

Educational Level*	True Parents				Foster Parents			
	Fathers		Mothers		Fathers		Mothers	
	Number	Percent	Number	Percent	Number	Percent	Number	Percent
Graduate work (one or more years)	1	1.1			14	9.8	2	1.4
College graduate	3	3.3			20	14.1	22	15.6
Some college work (one or more years)	11	12.2	9	6.6	24	16.9	30	21.3
High school graduate	29	32.2	34	24.8	26	18.3	42	29.8
Some high school work (one or more years)	19	21.1	44	32.1	22	15.5	22	15.6
Grade school graduate	19	21.1	30	21.9	36	25.4	21	14.9
Seventh grade or below	8	8.8	20	14.6			2	1.4

* School grade completed.

Note the far higher educational levels of the foster parents when compared with those of the biological parents of the Iowa children placed for adoption.

In the table comparing the occupational levels of the real and the foster fathers (p. 35), Skeels added a column to show how both groups of Iowans compared with the national population as a whole.

Occupational Groups	True Fathers		Foster Fathers		General Population	
	Number	Percent	Number	Percent	Number	Percent
I	2	1.8	20	13.6		2.5
II	2	1.8	15	10.2		4.7
III	11	10.3	47	32.0		14.4
IV	5	4.7	33	22.4		18.7
V	26	24.3	25	17.0		27.4
VI	21	19.0	5	3.4		13.2
VII	40	37.4	2	1.4		10.0

Group I: Professional   Group II: Semi-professional and managerial

Group III: Clerical, skilled trade, and retail business   Group IV: Farmers

Group V: Semi-skilled occupations, minor clerical jobs, and minor business

Group VI: Slightly skilled trades and occupations requiring little training or ability

Group VII: Day laborers, all classes, rural and urban

As can be seen above, the Iowa children were placed for adoption with foster fathers of far more comfortable means than their natural fathers.

The third table (p. 39) summed up the measured differences between the IQ test scores of the natural mothers and those of their children three years after they had been placed for adoption in families more emotionally secure, better educated, and with greater economic resources than their natural parents:

IQ Level of Mothers	Number	Mean IQ of Mothers	Mean IQ of Children
100 and above	14	105.3	117.6
90 to 99	23	94.3	116.2
80 to 89	19	84.9	116.8
70 to 79	13	75.2	119.7
Below 70	9	61.6	112.0
All IQ's	78	87.0	116.7
Below 80	22	69.6	116.5

Note that the 22 children whose natural mothers had IQ test scores of below 80 achieved a mean IQ test score of 116.5.

As the table shows, in the 78 mother-child pairs for whom comparative IQ tests were available, the mean IQ test score of the real mothers was 87.0 (with 22 of the 78 graded below 80), while the mean IQ test score of their children placed for adoption was 116.7. The differences between the mean IQ test scores of the children whose mothers had low test scores and those of the children whose mothers had high test scores were insignificant.

The Iowa studies were expanded and extended. The new studies simply reinforced the point the pilot study had made: feebleminded mothers—or, rather, mothers *labeled* as being morons and worse by all the American versions of the Binet tests—did *not* transmit anything resembling hereditary mental retardation and other intellectual traits to their children. Moreover, given a better growth and development environment, the children of parents with very low IQ test scores were able, within three years of such access to the emotional, biological, and cultural advantages of families in more fortunate socioeconomic circumstances, to rack up IQ test scores averaging 116.7— which, as Stoddard observed, "equals that of the children of university professors and is about that of the highest occupational groupings."[16]

What this did to the IQ-test-score-reinforced myths of the eugenicists and other scientific racists was devastating. For as George Stoddard wrote:

> . . . as we look into the case histories of the [real] mothers and fathers we discover a picture of economic and social inadequacy, of delinquent and criminal records, of frequent institutional care. There is nothing about the true parents in any way inconsistent with their low mental ratings. They are what is usually designated "poor stock." Nevertheless, they have produced bright children, the only significant factor being that their children were taken from their parents at a very early age and placed in what we believe to be good homes.[17]

Wherever the Iowa psychologists altered the growth and development environment of the infant and child wards of the state, the results paralleled those observed in the pilot group of 147 infants placed for adoption. To me the most significant of these Iowa demonstrations of the response of all children to improved environments was that described by Skeels in his 1966 monograph, *Adult Status of Children with Contrasting Early Life Experiences.*[18]

TWO PITIFUL LITTLE CREATURES

During the Great Depression, the populations of orphaned and abandoned children who became wards of the state increased in quantum jumps all over the country. In Iowa this sudden increase in helpless children caused all of the state's orphanages and other institutional facilities for the care of healthy children to become overcrowded and understaffed. The orphanage in which the 25 children of the Skeels study had been placed in the 1930's "occupied, with few exceptions, buildings that had first served as a hospital and barracks during the Civil War." Infants up to the age of 2 were housed in the infant nursery of the hospital and kept in standard cribs "that often had protective sheeting on the sides, thus effectively limiting visual stimulation; no toys or other objects were hung in the infants' line of vision. Human interactions were limited to baby nurses who, with the speed born of practice and necessity, changed diapers or bedding, bathed and medicated the infants, and fed them effectively *with propped bottles*" (italics added).[19]

These newborn human beings kept in mentally sterile, shielded cribs and fed with propped bottles, with all the human contact experienced by chicks in a tin feeder, were moved into small, toyless dormitories containing from two to five cribs between the ages of 6 to 24 months, and then, "at 2 years of age those children were graduated to the cottages, which had been built around 1860." Not only were 30 to 35 children between 2 and 6 years of age packed into cottages under the care of one matron "and three or four entirely untrained and often reluctant girls 13 to 15 years of age," where their waking and sleeping hours were spent in an average-size room approximately 15 feet square, but "no child had any property which belonged exclusively to him except, perhaps, his toothbrush. Even his clothing, including shoes, was selected and put on him according to size" (p. 4).

After they reached their sixth birthday, they started to attend the grade school on the orphanage grounds. Although the curriculum was the same as that of the local public school attended by children from unbroken homes who lived with their parents, none of the *mentally normal* children from the orphanage cottages did well enough in their school to be "able to make the transition to the public junior high school." They were, in short, doomed to become the greater society's hewers of wood and drawers of water.

Psychological services had not been introduced into this orphanage until 1932 when Skeels, already on the state payroll at the university, became "the first psychologist to be employed by the Iowa Board of Control of State Institutions." In this capacity Skeels and his university colleagues were able to have some otherwise doomed children transferred to other institutions where they at least would have a reasonably better chance of developing their inherited capacities to somewhat happier levels. Similarly, children who tested out as truly mentally retarded were transferred to institutions for the feeble-minded.

Shortly after Skeels was assigned to this work, "two baby girls, neglected by their feebleminded mothers, ignored by their inadequate relatives, malnourished and frail, were legally committed to the orphanage. The youngsters were pitiful little creatures. They were tearful, had runny noses, and sparse, stringy, and colorless hair; they were emaciated, undersized, and lacked muscle tonus or responsiveness. Sad and inactive, the two spent their days rocking and whining" (p. 5).

They were carefully examined by pediatricians, who found "no evidence of physiological or organic defect, or of birth injury or glandular dysfunction." No correctable biological cause or causes for their serious delay in mental growth was overlooked before they were examined by Skeels and other psychologists, who found that although the chronological age of the children was 13 and 16 months, the mental age was 6 and 7 months.

The two unwanted little girls "were considered unplaceable" for adoption, "and transfer to a school for the mentally retarded was recommended with a high degree of confidence. Accordingly, they were transferred to an institution for the mentally retarded . . . when they were aged 15 and 18 months."

Six months later, when his professional duties took him to that institution, Skeels "noticed two outstanding little girls. They were alert, smiling, running about, responding to the playful attention of adults, and generally behaving and looking like any other toddlers. [Skeels] scarcely recognized them as the two little girls with the *hopeless prognosis,* and thereupon tested them again."

Although the test results showed that the children were not mentally retarded, Skeels "was skeptical of the validity of the permanence of improvement and no change was instituted in the lives of the children. Twelve months later they were re-examined, and then again when they were 40 and 43 months old. Each examination gave unmistakable evidence of mental development well within the normal range for age" (p. 6).

Skeels had no reason to doubt the accuracy of the base-line test results at the orphanage as well as all the subsequent mental tests the children had been given at the institution for the mentally retarded. As he concluded in 1966—more than thirty years later—"there was no question that the initial [mental] evaluations gave a true picture of the children's functioning level *at the time they were tested.* It appeared equally evident that *later appraisals* showed *normal* mental growth accompanied by parallel changes in social growth, emotional maturity, communication skills, and general behavior" (p. 6). (Italics added.)

As behavioral scientists, Skeels and the other university psychologists were profoundly interested in discovering logical explanations of these changes in mental, bodily, and emotional development of these two little girls. Skeels summed up his findings in one short paragraph that had much to tell psychologists with open minds:

> The two girls had been placed on one of the wards of older, brighter girls and women, ranging in [chronological] age from 18 to 50 years and in mental age from 5 to 9 years, where they were the only children of preschool age, except for a few hopeless bed patients with gross physical defects. An older girl on the ward had 'adopted' each of the two girls, and other older girls served as adoring aunts. Attendants and nurses *also showed affection* to the two, spending time with them, taking them along on their days off for automobile rides and shopping excursions, and purchasing toys, picture books, and play materials for them *in great abundance.* The setting seemed to be a homelike one, abundant in affection, rich in wholesome and interesting experiences, and geared to a preschool level of development. [Italics added.]

Similar experiences in the development of other Iowa orphan children—who blossomed after being placed for adoption with families where they, too, had access to the stimuli of abundant affection, toys, picture books, trips, and plenty of toys and other play materials of their own—told Skeels all he really had to know about the probable causes of the dramatic change in the mental status of the two children abandoned to the state by their feebleminded mothers. But Skeels was also enough of a child psychologist to realize that since the two formerly mentally retarded girls "were now normal . . . the need for care in such an institution [for the mentally retarded] no longer existed." It could, in fact, even become harmful. "Consequently, they were transferred back to the orphanage and shortly thereafter they were placed in adoptive homes" (p. 6).

Observations of children in institutions and in foster families had shown the Iowa group that "older children who came from inadequate non-nurturant homes were mentally retarded or borderline but that their younger siblings were generally of normal ability, which suggested that *longer residence in such homes has a cumulatively depressing effect on intelligence* (italics added)." That is, just as moving into a nurturant environment could help mentally retarded youngsters develop mentalities, moving into intellectually sterile homes could turn normal children into mental retardates. "Children also became retarded if they remained for long periods in an institution supposedly designed for normal children," Skeels had discovered, but "if children were placed in adoptive homes as infants, or even as young pre-schoolers, their development surpassed expectations, and improvement continued for long periods following placement" (pp. 6–7).

Given these experimental findings, Skeels and his Iowa State University colleagues were now confronted with a dilemma. As psychologists and as human beings, they understood that the longer the children were kept in nonnurturant institutional settings prior to being placed for adoption, the greater were the chances that children born mentally normal would react by becoming mentally retarded. But "since study homes or temporary homes were not available to the state agency at that time, the choice for children who were not suitable for immediate placement in adoptive homes was between, on the one hand, an unstimulating, large nursery with *predictable* mental retardation, or, on the other hand, a radical, iconoclastic solution."

This radical alternative was the placement of other young orphans as "house guests" in other institutions "for the mentally retarded in a bold experiment to see whether retardation in infancy was reversible." To those state officials who feared that placing infants in homes for grown feebleminded women would be harmful to their wards, the Iowa psychologists pointed out that the alternatives to this were *proved* to cause mental retardation. Finally Skeels was allowed to put his hunch to the test with twenty-five children, twenty of them illegitimate, the others taken from homes where they had been subjected to severe neglect and/or abuse. "All the children were white and of North-European [i.e, Nordic] background." All of these twenty-five children were considered "unsuitable" for adoption because of evident mental retardation (pp. 7–8).

Thirteen of these children who tested out as being mentally retarded were selected as the experimental group and, while still under three years of age, "transferred from an orphanage for mentally retarded children to an institution for mentally retarded adult females as 'house guests.'" The other twelve white Nordic children, the contrast group, "initially higher in intelligence than the experimental group, were exposed to a relatively non-stimulating orphanage environment over a prolonged period of time."

Inside of two years, the children in the experimental group, who were sent to live with feebleminded women of all ages who gave them love and tender attention, registered an average gain of 28.5 in IQ test scores. In contrast, the twelve children of matched ages of the contrast group who lived in the orphan asylum for normal children showed an average loss of 26.2 points in their IQ test scores (p. 56).

Skeels was to devote various periods of the next three decades of his life studying these two groups of Nordic Iowa children, first as a state psychologist, and ultimately as a staff member of the NIH National Institute of Child Health and Human Development in Bethesda, Maryland. By the time he finished his study in 1966, eleven of the thirteen children in the experimental group had married and had a total of twenty-eight children; only two of the eleven surviving children in the contrast group had married, and between them had five children. The children of the adults who had formed the group of thirteen mentally retarded children sent to live in a home for mentally retarded adults had a mean IQ test score of 103.9, with a range of 86 to 125, which meant that no child tested below the dull-normal level.

Of the two members of the contrast group who had married and had children, one—a boy whose original mental test scores might have been handicapped by his hearing difficulties—had four children, three with IQ test scores on the Stanford Form L of between 107 and 119, and a nine-month-old infant with an IQ of 103 on the Cattell Infant Test. The father owned a comfortable home in a good middle-class neighborhood, had completed high school and attended college for a short while, and held a good job as a compositor and typesetter. The other parent, a girl, had one child, a mentally retarded son who had been severely beaten and otherwise abused often by her first husband, the boy's father, and who on medical examination showed definite signs of brain damage.

The thirteen children of the experimental group were all self-supporting, and had completed an average of 11.68 years of schooling; some had gone to college and one had completed postgraduate work; and, in the twenty-one years since Skeels had last observed them, they worked at various occupations, ranging from nursing instructor and vocational counselor to Army staff sergeant, real estate salesman, and housewife. All but two of these children, following their life as house guests in state institutions for the mentally retarded, had been placed for adoption in loving and nurturant families; the two children who had not been adopted by foster parents now worked as domestic servants.

Of the contrast group, one was dead, and four of the eleven survivors were still residents of state institutions for the retarded. Two had jobs as dishwashers; one was a cafeteria worker; one an institutional gardener's assistant; one a floater; and one, as we have seen, had graduated from high school and worked as a typesetter and compositor. His income "easily equalled that of all the other employed contrast-group members combined."

In terms of what they contributed and took from society, the cost of institutionalizing the children of the experimental group had come to a total of $30,716—as against the $138,571 the state of Iowa had, as of 1963, spent on the twelve contrast group children who had become its wards in 1936. During that base-line year of 1963, the 100 percent self-supporting young adults of the former experimental group had a combined income of $62,498 and paid $5,485 in income taxes. The seven self-supporting survivors of the contrast group paid a total of $2,238 in income taxes on a combined income of $19,826.[20]

COMPENSATORY INTERVENTION: NEW HOPE FOR HITHERTO DOOMED CHILDREN

The implications of this study were quite clear to Skeels, particularly since at the start of the experiment eleven of the thirteen children in the experimental group had evidenced fairly pronounced mental retardation. As Skeels summed it up in 1966, "The developmental trend was reversed through *planned intervention* during the experimental period. The program of nurturance and cognitive stimulation was followed by placement in adoptive homes that provided love and affection and normal life experiences. The normal, average intellectual level attained by the subjects in early or middle childhood was maintained into adulthood." (Italics added.)

To Skeels, there was little doubt but that if the children in the contrast group "had been placed in suitable adoptive homes" or given some other equally personal equivalent in early infancy, most if not all of the twelve institutionalized children would also have raised their IQ and classroom achievement test scores, and grown up to be as self-supporting and productive as the thirteen equally disadvantaged children of the experimental group.

In this and other studies of the Iowa group, Skeels and his associates were not only in the great American scientific tradition of Franz Boas—whose studies of the changes in the bodily forms of immigrants showed the physiological role of improved environment in human beings—but also in the considerably older tradition of the founder of pediatrics, Professor Abraham Jacobi.

A refugee from his native Germany, where he had been jailed for treason after the defeat of the republican Revolution of 1848, Dr. Jacobi, professor of infantile pathology and pediatrics at New York Medical College, later clinical professor of diseases of infancy and childhood at New York's College of Physicians and Surgeons, had made as many enemies as admirers for himself in 1871, in his first address as president of the New York County Medical Society. Jacobi—who for ten years had served without pay as the physician of the Nursery and Child's Hospital—declared that "in spite of the efforts of the medical staff, and the painstaking and kind-hearted ladies, the probability of the lives of the children entrusted to a public institution is very slim indeed."[21]

More than sixty years before the Iowa group rediscovered the mind- and body-destroying effects of institutional life on infants and children, Jacobi was warning: "The younger the children and the larger the institution, the surer is death. Modern civilization, planning for the best but mistaken about the means, has succeeded in out-Heroding Herod. The poor tenements of our working classes yield better results in their raising of infants."[22]

Since the kindhearted ladies whose charity balls raised the money for the Nursery and Child's Hospital represented the old white Protestant aristocracy of New York, and the lowly Irish Catholic immigrants formed so large a proportion of the 1870 New York working-class tenement population, it is easy to understand why Jacobi's blunt voicing of medical truths caused him to be fired from his nonpaying job at the favorite charity hospital of the city's social leaders. Not until Jacobi's nephew by marriage, whom he talked into emigrating to the United States from a Fatherland where Jewish scientists had very limited opportunities for academic posts, arrived to play the same pioneering role in anthropology that his uncle had played in children's medicine, did the town's elite hate any other scientific figure quite as much for revealing equally blunt truths. The nephew's name was Franz Boas.

Skeels and the Iowa psychologists played a greater role in our scientific history than merely validating on the psychological level the findings of Jacobi and Boas on the life expectancy and the bodily forms of children under various institutional, familial, anad socioeconomic conditions. Their great contribution was the demonstration of the human and societal benefits of early nutritional, psychological, environmental, and educational *intervention* in the growth and development of emotionally, biologically, and culturally deprived children.

What forms of intervention, be they psychological or biological, economic or social, were most important were designations Skeels was too modest to choose. His twenty-five children were, after all, a small study population, and he agreed that "it would be presumptuous to attempt to identify the specific influences that produced the changes observed." But to all but the most unreasonable of scientific racists and other hereditarians, "it seems obvious that under present-day conditions there are still countless infants born with sound biological constitutions and potentialities for development well within the normal range who *will* become mentally retarded and noncontributing members of society *unless appropriate intervention occurs*. It is suggested by the findings of this study *and others published in the past 20 years* that sufficient knowledge is available to design programs of intervention to counteract the devastating effects of poverty, sociocultural deprivation, and maternal deprivation."[23] (Italics added.)

In more normal times the publication of the findings of Skeels and his Iowa associates might have quickly resulted in serious state and federal programs of compensatory intervention in the physiological and mental development of millions of the nation's socioeconomically deprived children. Unfortunately, neither scientific research nor life-enhancing scientific discoveries occur in historical vacuums.

During the Depression years between World Wars I and II, while Skeels, Dye, Skodak, Welman, et al. were publishing the results of their experiments, events in Europe and Asia of more dramatic immediacy were obscuring the domestic dangers inherent in the unscientific use and interpretation of IQ tests. These events—violent, sanguinary, and terrifying—arose from a new historical wave called fascism. These events were also giving those better-educated Americans who would normally be most influenced by the implications of the Iowa discoveries a growing (and well-founded) fear that fascism would drag the nation into another world war.

Americans such as Skeels and Boas were on the wrong end of this wave. Others, such as Madison Grant, Henry Fairfield Osborn, Charles Benedict Davenport, and Lothrop Stoddard, rode its crest.

15 The Old Scientific Racism's Last Hurrah: Six Million One-Way Tickets to the Nazi Death Camps

> One sure service of the able and the good is to beget and rear
> offspring. One sure service (about the only one) which the
> inferior and vicious can perform is to prevent their genes from survival.
> —PROFESSOR EDWARD L. THORNDIKE, in *Human Nature and
> the Social Order* (1940)

Not since the great Kishinev pogrom of 1903 had any event in European history thrilled the true believers of the world's eugenics movement quite as much as the March on Rome of October 8, 1922, which brought Benito Mussolini to power as Italy's fascist Duce and established mass terror as a primary instrument of public policy.

The unanimity of adulation on both Nordic sides of the Atlantic for Il Duce was, on the surface, something of an anomaly. Benito Mussolini was not a Nordic but a Mediterranean; not a fair-haired dolichocephalic but a former dark-haired and now bald brachycephalic; not an Anglican or even a Protestant but a somewhat lapsed Catholic; not a descendant of the highborn Teutons but a lowborn Italian who for most of his life had been a firebrand socialist agitator. Nevertheless, the Osborns, the Grants, the Mjöens, the Eugen Fischers, and the Lapouges hailed Mussolini as if he were Sir Francis Galton risen again and walking upon this hallowed Nordic earth.

The reason for Mussolini's instant election as a major living saint of the eugenics creed was not hard to see. All the learned scientistic verbiage of the Galtons and the Davenports about race and superior or inferior mental, physical, and moral traits had little to do—in reality—with race, nationality, language, or religion. In the final analysis it mattered not whether a man was Scot or Celt, Saxon or African, Christian or Jew, Protestant or Catholic. These shibboleths and totems of scientific racism were—like the myths of the infallibility of IQ test scores so dear to eugenicists past and present—merely rationales of convenience. All that really mattered to the eugenicists were two races—(1) the world's minority of the nonpoor and (2) the Social Problem Group which England's E. J. Lidbetter had described before the Second International Congress of Eugenics in 1921 as "a definite race of chronic pauper stocks," and their evolutionarily close relatives, the working poor.

When Mussolini was called to power by King and Establishment, it required no profound knowledge of Italian internal politics to know, at once, that Il Duce's prime function was to keep this race of Under-Men—Galton's human "refuse" and Terman's IQ-test-score-certified *hereditary* hewers of wood and drawers of water—in its rightful and lowly place. Mussolini's bombast about the Corporate State and his promises to restore Italy to the military glories that were Rome's—like the race hygiene and mental test patter of the eugenicists—were merely layers of velvet cant that covered the iron fist. The historic function of fascism and that of scientific racism were one and the same: to keep what Malthus had contemptuously termed the "lower and middling classes of people" from ever aspiring to rise above their stations at birth.

Since Mussolini understood this quite as well as did the British, American, German, Norwegian, French, and other eugenics movement leaders, it became a matter of love at first sight between them. No sooner did Mussolini's Black Shirts start rounding up, torturing, and killing his non-fascist domestic opponents than he became the darling of the Osborns, the Davenports, and the Grants. No visit to Europe was complete without a visit to Il Duce, who went out of his way to discuss eugenics and other race problems with Osborn, Mjöen, Davenport, and all other eugenics leaders whom his scientific adviser, Corrado Gini, could steer his way. In return, these foreign eugenicists made speeches and wrote articles in praise of the new Caesar.

When Madison Grant's *Conquest of a Continent,* his "racial history of the United States," was published in 1933, he caused quite a problem for R. V. Coleman at Scribner's, who did not know how to address the fascist dictator when Grant ordered the publishers to send him a copy, but was certain that "Boss Benito Mussolini" was not good protocol. Grant also requested that copies be sent to Professor Dr. Eugen Fischer at the Kaiser Wilhelm Institute for the Study of Anthropology, Human Heredity and Eugenics in Berlin, and, of course, to Hitler's number-one scientific adviser, Dr. Alfred Rosenberg, at Schellingstrasse 39 in Munich, as well as to the noted Nazi race hygienist Professor Dr. Fritz Lenz at the University of Munich.[1]

When the German edition of Grant's book was published in Berlin in 1937, it bore not only a foreword by Henry Fairfield Osborn but also a special foreword to the German edition by Eugen Fischer. In this foreword by Hitler's adviser on *Rassenhygiene,* the veteran German eugenicist wrote, among other fine things, that since the publication of the German translation of Grant's *Passing of the Great Race* in 1925 the American racist had been "no stranger to German readers of writings on race and eugenics." In addition:

> No one has as much reason to note the work of this man [Grant] with the keenest of attention as does a German of today—in a time when the racial idea has become one of the chief foundations of the National Socialist State's population policies.

These Nazi state population policies included national compulsory sterilization laws based quite literally on the Model Eugenical Sterilization Law drawn up by the supervisor of the Eugenics Record Office of Cold Spring Harbor, New York, Harry H. Laughlin. A worldly Nazi, Fischer therefore could note in this foreword to the 1937 German translation of Grant's last book:

> No one will be surprised that this work met with the most vehement opposition in the land of its origin, where politicians and scholars, *led above all by the Jewish anthropologist and ethnologist, Franz Boas,* dominated all public opinion with the notion that racial differences were determined by environment and changeable with it, that they were without significance, and that since the founding of the United States, or

at least since the emancipation of slaves in free America, no differences based on race, creed, or color should be recognized. [Italics added.]

As every good German knew, Grant and the Nazi state were both agreed, Fischer wrote, that the racial characteristics inherited with the blood were "ultimately the sole determining basis of history and of all cultural development." Therefore, no good German could "forget what courage, what intellectual independence were required of this scholar [sic], now a man of 72, to proclaim the racial idea as the basis of his country's history, in a world where public opinion, *which in this context means largely the Jewish press,* brands as a National Socialist in disguise anyone who dares even to speak seriously of race, let alone of the Nordic race and its creative qualities" (italics added).

Fischer ended his foreword to Madison Grant's book with the pious and *völkisch* hope: "may this book by an American, which is enjoying immediate success on the other side despite the efforts of the Jewish press to kill it with silence, also find the distribution it deserves in German garb."

Prior to the ascent of Eugen Fischer, his countryman Hans Günther had not only arranged for the German edition of Lothrop Stoddard's *Revolt Against Civilization,* but also paid Madison Grant the greater compliment of borrowing quite heavily from Grant's *Passing of the Great Race* in his own racist volume, issued in English as *The Racial Elements of European History* (1927).[2] Günther's book sounded very much like Grant's, saying, for example, that "Athens sank in the same measure that the blood of her Nordic upper class ran out"; and, as Madison Grant pointed out in the review he wrote of Günther's book in the *Eugenical News* (1928, 118–20), it dealt with "this decline of the Nordic Race, not only in Germany but in all Europe." It was, in fact, simply a rehash of the racial "history," and the vulgar pseudobiology and pseudoanthropology, that Grant had originally borrowed from Ripley, Galton, and Gobineau for his 1916 classic of scientific racism.

Like the Anglo-Saxon Nordics, Günther, Fischer, and the other Teutonic Nordics joined in the worship of and collaboration with Mussolini and Italian fascism with all the fervor they were later to have for Adolf Hitler. In the November 1929 issue of Davenport's *Eugenical News,* which by then was the official organ not only of Davenport's Eugenics Research Association but also of the Galton Society and the International Federation of Eugenic Organizations, we find many testimonials to this adoration of the non-Nordic Mussolini by the Nordics of the world eugenics movement. On the first page, for example, we find the lead item devoted to the Rome meeting of the International Federation on September 27, 1929, in the library of the Central Institute of Government Statistics "as a guest of Professor Corrado Gini."

According to the *Eugenical News,* the meeting had been called to order by its president, Charles B. Davenport, in the presence of eight council members of whom only two, Gini and Pestalozzi, were non-Nordics, in contrast to their fellow council members Fischer, Frets, Mjöen, Nilsson-Ehle, Reichel, and Van Herwerden. No sooner were Gini, Mjöen, and Fischer appointed as a committee to formulate a plan for a permanent secretariat, and Bernard Mallet and Professor Ruzicka elected as vice-presidents for 1929–30, than the council got down to its most pressing task. This was the presentation of a memorandum, drawn up by Dr. Eugen Fischer for the Federation, to His Excellency Benito Mussolini, "the great statesman who, in the Eternal City, shows more than any other leader today, both in deed and word, how much he has the eugenic problems of his people at heart."

This memorandum, which closed with a fervent "Videat Konsul!", was characteristic of the veneration the entire world eugenics movement had for Mussolini and his cruel police state. But Mussolini was, for all the lavish praises of the Fischers and the Osborns, no substitute for a real Nordic dictator dedicated to the advancement of applied eugenics. Davenport, for example, groveled along with other members of the Federation when, in the Piazza Venezia, Mussolini met with their council and exchanged banalities with them after Fischer read their statement of admiration to Il Duce. And Davenport even attended the second Italian Congress of Genetics and Eugenics, held a few days later "under the honorary presidency of S. E. Benito Mussolini," according to the *Eugenical News,* at which he presented a paper— "Is Race Crossing Useful?"—"in which certain dangers, based on the results of study of negro-white crosses, were pointed out."

On the occasion of the Assembly of the International Federation of Eugenic Organizations in Rome, it seems natural and desirable, when considering eugenic problems, that some expression of our hopes and wishes should be addressed to the great statesman who, in the Eternal City, shows more than any other leader to-day, both in deed and word, how much he has the eugenic problems of his people at heart.

Cultures and peoples of the past have, without exception, however high the glamour of their zenith, passed irrevocably. In the opinion of those who have made a scientific study of these catastrophies, one cause—if not the main cause—of this decline has been a fall in the birth-rate of the upper classes, resulting in a lack of leaders.

Europe in each of her separate cultures, shows to-day not merely a decline in birth-rates, but a veritable catastrophic plunge. Means of arrest are everywhere being sought, and amongst the most determined and energetic are those taken in Italy through the personal influence of her great Duce, Mussolini. We who, as men of science, are by profession forced to the study of these population problems, and who, in virtue of our knowledge are called upon to bring before governments and representatives of our peoples the dangers we perceive, beg herewith to give voice to our hopes and ardent desire that all these measures to stem population decrease and to preserve and build up a healthy and numerous posterity, may be improved and increased.

Far more, however, than the generally acknowledged quantitative problem, we are impressed with the urgency of the qualitative side of these issues. It is not only masses that make up a people. 'Men make history' is a saying of the great Italian leader which has been repeatedly quoted.

The gravest concern of all eugenists to-day is the preservation of human quality. It is a possibility! And in view of the tremendous importance for the future of every nation of this objective, no economic sacrifice can be too great. The sacrifices, however, would not be so very considerable. Here it is only possible to suggest how suitable measures in the sphere of property and income-tax, and yet more certainly the inheritance tax can be brought to bear on maintaining families of talent in every social stratum. Such measures, however, should be fitted to the social position of the family, and favor those who have arrived at high position, and require to be so graded to the social rank attained that the best receive the greatest acknowledgment. Such suggestions may seem to sound an anti-social and anti-democratic note. It must, therefore, be borne in mind that each stratum in turn supplies its quota of those favored individuals who have attained social distinction, and the protection and advantages have to do with the family rather than with the individual—the family giving to the State children from amongst whom future leaders can be chosen. Thus every such attempt is in the truest sense of the word one which concerns 'res publica'—in the highest sense democratic. Such administrative and legislative means are without doubt to hand, and can for each country be formulated by those forces in eugenics, in such a way that the legislators can make use of them, for the individual national eugenic organizations stand ready to hand.

Here to-day—in the oldest capital of the world, we beg to express with the utmost solemnity our hope that those great men to whom the destinies of the highly gifted Italian nation are entrusted, will be first in setting a model to the world by showing that energetic administration can make good the damage which has already been done to our culture, by arresting the fall in population and by preserving the best endowed.

We pray that what was denied to earlier cultures may here be achieved in grasping fortune's wheel and controlling and turning it! Quality as well as quantity! The urgency brooks no delay; the danger is imminent.

Videat Konsul!

Nevertheless, by 1929 even Davenport could sense that the eugenics movement still needed a Nordic leader to establish a true eugenical state. This yearning for a blood Nordic orientation was reflected in the masthead of the *Eugenical News,* which, subsequent to Davenport's election to the presidency of the International Federation of Eugenics Organizations in 1927, added a line beneath its original masthead: the publication was now, also, the *Current Record of Race Hygiene,* a recognition of the increasingly Teutonic complexion of the world eugenics movement.

The Teutons in the eugenics movement, moreover, were most appreciative of the words and deeds of their American comrades. In a letter to Madison Grant (published in the *Eugenical News,* 1928, pp. 132–33), the incredible Count Vacher de Lapouge, describing the German reaction to Grant's book, *The Passing of the Great Race,* said that it had "exercised, from the publication of the German edition [in 1925], a powerful effect upon the Nordic movement and everybody knows it. . . . Günther, Holler, Konopseki and Konepath have a great veneration for you; also Lothrop Stoddard. I think that the change of orientation of which they are the protagonists is connected above all with a Nordic movement developed in Scandinavia by Mjöen and Lundberg and ought to be regarded as a contre-coup of the American [eugenics] movement."

Günther's veneration of Stoddard went back at least four years earlier than this communication from Lapouge, as evidenced by Stoddard's letter of

November 4, 1924, to his editor at Scribner's, Maxwell Perkins. In this letter Stoddard noted that it was Günther who first suggested the idea of a German edition of *Revolt Against Civilization* and even recommended a German translator.

Hans Günther, the expatriate German ethnologist and race biologist, was even then a particular favorite of Hitler's. When the Nazi Party took control of the provincial government of Thuringia, Hitler personally saw to it that Günther was given a professorial chair at the University of Jena. It was a new chair in "raceology" and it was given to Günther in 1930, three years before Hitler himself came to national power. Günther's *Kleine Rassenkunde des deutschen Volkes* (*Short Raceology of the German People*), published in 1929, was one of the most frequently printed books in Hitler's Germany, and over 272,000 copies were sold between 1929 and 1943. Along with Eugen Fischer, Günther maintained the closest of ties with the American leaders of the eugenics movement, and it was quite possibly at their recommendation that the University of Frankfurt (the Johann Wolfgang von Goethe University) offered Henry Fairfield Osborn an honorary doctorate of science, which led to "an enthusiastic trip" to pick up the Nazi honor in 1934.[3]

The fact that 1934 was also the year of the great blood purge of June, in which the nakedly brutal treatment of Christian and Jewish adults and children demonstrated to the world the true nature of Hitler's branch of the Nordic movement, in no way diminished the enthusiasm or the collaboration of American eugenicists such as Osborn and Stoddard. The treatment of Germany's Jews and other non-Aryan citizens, while horrifying to most decent Americans, only served to prove to the hard-core eugenicists that the Nazis were the fearless and heroic race biologists they had long awaited.[4] This was made plain in 1940, when Stoddard paid a long visit to his admired Nordic state on the eve of its plunging into war against his native America. The climax of this pilgrimage, described in Chapter 17 of the book Stoddard wrote about it, *Into the Darkness,* was his visit with Adolf Hitler himself.

AN AMERICAN ADMIRER VISITS A NAZI EUGENICS COURT

Although this chapter was entitled "I See Hitler," it was fairly dull, since, as Stoddard pointed out (pp. 201–02), this meeting was not an *interview* but an *audience* (Stoddard's italics), at which the lucky visitor was allowed to see and to listen to the great one—but not to write or even tell about what was said. But the preceding chapter, "In a Eugenics Court," was and remains of great historical and scientific interest. As Stoddard explained, Nazi Germany's ideas about race were its most distinctive feature. However, he added, unfortunately it was easier to talk about than to understand the racial concepts of Hitler, Goering, and Goebbels because the truth about the Nazi race ideas had been terribly "obscured by passion and propaganda."[5]

The "passion and propaganda" were, of course, those of the enemies of the Third and Thousand-Year Reich. Not until Dr. Lothrop Stoddard arrived on the scene could an intelligent evaluation of the whole problem be made. For as he wrote (p. 187): "I have long been interested in the practical applications of biology and eugenics—the science of race-betterment—and have studied much along these lines." During his visit to Nazi Germany he had met with many Nazi raceologists, such as his old admirer Hans Günther, Eugen Fischer, Fritz Lenz, and Paul Schultze-Maumburg, as well as with official spokesmen such as Reichsministers Wilhelm Frick and Walther Darré. Through the kind assistance of these ranking Nazis, Stoddard was invited to join the judges on the bench of the Eugenics High Court of Appeals.[6]

Before sharing with his readers his experiences on the bench of the Eugenics High Court of Appeals, Dr. Stoddard discussed race problems and eugenics in Nazi Germany and its Axis partner, Fascist Italy, in terms that are of considerably more than passing interest in modern times that are experiencing, as they are, the revival of various IQ testing and other ploys of the eugenics movement. Stoddard said, for example (p. 189), that "the purity of the racial strains must be preserved." And, Stoddard explained, "this is the Nazi doctrine best described as *racialism*" (Stoddard's italics). So that, once the Jews and other inferior stocks were annihilated, the Nazi state would be able to concern itself with "improvements *within* the racial stock, that are recognized everywhere as constituting the modern science of *eugenics,* or race-betterment" (Stoddard's italics).

There followed two paragraphs that showed how completely Dr. Stoddard, as a representative eugenicist of good education—he not only had a Harvard Law School diploma but also an earned Harvard doctorate in history —identified with and accepted the Nazi race doctrines and the racial writings of the Nazi state's leading scientific spokesmen:

> The relative emphasis which Hitler gave racialism and eugenics many years ago foreshadows the respective interest toward the two subjects in Germany today. Outside Germany, the reverse is true, due chiefly to Nazi treatment of its Jewish minority. Inside Germany, the Jewish problem[7] is regarded as a passing phenomenon, already settled in principle and soon to be settled in fact by the physical elimination of the Jews themselves from the Third Reich. It is the regeneration of the Germanic stock with which public opinion is most concerned and which it seeks to further in various ways.
>
> There are one or two German ideas about race which, it seems to me, are widely misunderstood abroad. The first concerns the German attitude toward Nordic blood. Although this tall, blond strain and the qualities assumed to go with it constitute an *ideal type* in Nazi eyes, their scientists do not claim that Germany is today an overwhelmingly Nordic land. They admit that the present German people is a mixture of several European stocks. Their attitude is voiced by Professor [Hans F. K.] Günther when he writes: *"The Nordic ideal becomes for us an ideal of unity. That which is common to all the divisions of the German people is the Nordic strain. The question is not so much whether we men now living are more or less Nordic; the question put to us is whether we have the courage to make ready for future generations a world cleansing itself racially and eugenically."* [Italics added.][8]

As Günther explained in his *Kleine Rassenkunde des deutschen Volkes,* ideally "the Nordic race is tall, long-legged, slim, with an average height, among males, of above 1.74 meters . . . narrow faced, with a cephalic index of around 75 and a facial index above 90. . . . The skin of the Nordic race is roseate-bright and the blood shines through. . . . The hair color is blond; among most of the existing types it can extend from a pink undertone of light blond to golden blond up to dark blond."[9]

The genes for this basic ideal type of Nordic, as Hitler had observed in *Mein Kampf,* were still Germany's "most valuable treasure for the future." As Madison Grant and other great Nordic raceologists had revealed, the Thirty Years' War and subsequent cacogenic historical events had somewhat diluted this ideal Nordic type with unwelcome infusions of Alpine, Mediterranean, and other inferior types of blood, but people of genuine racial insight, such as Hitler himself and Rosenberg and Günther and Eugen Fischer, were, as Lorenz wrote later of graylag geese and such, "able to separate the essentials of type from the background of little accidental imperfections."[10]

The basic mechanisms of the National Socialist German Workers' State for separating out these little accidental imperfections from the pure Nordic bloodlines of the *nordische Volksdeutschen* were spelled out quite explicitly by Konrad Lorenz in the German *Journal of Applied Psychology and Characterology:*

> There is a close analogy between a human body invaded by a cancer and a nation afflicted with subpopulations whose inborn defects cause them to become social liabilities. *Just as in cancer the best treatment is to eradicate the parasitic growth as quickly as possible, the eugenic defense against the dysgenic social effects of afflicted subpopulations is of necessity limited to equally drastic measures.* . . . When these inferior elements are not effectively eliminated from a [healthy] population, then—just as when the cells of a malignant tumor are allowed to proliferate throughout a human body—they destroy the host body as well as themselves. [Italics added.]

Along with the mass extermination camps from Auschwitz to Treblinka, one of the most familiar Nazi instruments for "eradicating," as Lorenz put it, the national "cancers" of "subpopulations whose inborn defects cause them to become social liabilities" was, as Stoddard revealed, the gift of the American eugenics movement to the Nordic world: the compulsory state sterilization of the "genetically unfit."

As soon as Hitler took power, one of the first things he did was to replace the mild Weimar Republic law permitting voluntary sterilization for medically certified reasons with a national law styled directly on the Eugenics Record Office's Model Eugenical Sterilization Law published by H. H. Laughlin in 1922. Under this Nazi Act for Averting Descendants Afflicted with Hereditary Diseases, the famous Eugenics Law (*Erbgesundheitsrecht*), the Eugenics Courts of Nazi Germany, each local court consisting of two doctors to discuss the medical aspects and one judge appointed by the local *Gauleiter* to lay down the law, were set up to rule on who should be sterilized in the interest of the ideal Nordic blood type. The American origins of this Nazi sterilization law, as spelled out by Stoddard (p. 191), are obvious:

The grounds for sterilization [via the Eugenics Courts of Germany] are specifically enumerated. They are: (1) Congenital Mental Deficiency; (2) Schizophrenia, or split personality; (3) Manic-Depressive Insanity; (4) Inherited Epilepsy; (5) Inherited Chorea; (6) Inherited Blindness; (7) Inherited Deafness; (8) Any grave physical defect that has been inherited; (9) Chronic alcoholism, when this has been scientifically determined to be symptomatic of psychological abnormality [sic]. [See above, pages 128–29, 134–36.]

The forced sterilization of Germans became so common a fact of life in Nazi Germany that in the popular vernacular a sterilization operation became known as a *Hitlerschnitt*—literally "Hitler cut," but actually a play on words with the German term for a Caesarean operation, *Kaiserschnitt*.[11]

Up to the beginning of World War II, wrote Wallace R. Duell, the *Chicago Daily News* correspondent in Germany, the Nazi "regime had sterilized approximately 375,000 persons."[12] Duell summed up the official rationales for these forced sterilizations as follows:

Congenital feeble-mindedness	203,250
Schizophrenia	73,125
Epilepsy	57,750
Acute alcoholism	28,500
Manic-depressive insanity	6,000
Hereditary deafness	2,625
Severe hereditary physical deformity	1,875
Hereditary blindness	1,125
St. Vitus' dance	750
Total	375,000

"In examining supposedly feeble-minded persons to decide whether or not they are subject to compulsory sterilization, the nazis gave them an intelligence test devised by the Reich government for this purpose," Duell wrote.

These tests, Duell added, had to be changed often, since supposedly feebleminded persons who took them memorized the questions and passed them on to their friends, who "in many cases passed the intelligence tests with flying colors." This did them little good, though, for as an official report on the enforcement of the sterilization law, quoted by Duell, observed: "Among the feeble-minded there is a large number who have a certain mental agility and who answer the usual easy questions quickly and apparently with assurance, and who only after a more searching examination betray the utter superficiality of their thinking, and their inability to reason and their lack of moral judgment."[13]

In short, people tagged for compulsory sterilization by the Nazis, and who passed the intelligence tests, were still found to be feebleminded by the Eugenics Court for moral reasons.

Of the 171,750 Germans sterilized before World War II as having diseases other than congenital feeblemindedness that were also classified as *hereditary*—such as epilepsy, deafness, blindness, and even St. Vitus's dance (Sydenham's chorea, a complication of rheumatic fever), let alone alcoholism—most had conditions acquired not in the genes but in prenatal and post-natal infections, deprivations, traumas, and emotional stresses. "Hereditary" blindness, deafness, and physical deformities in children born so afflicted usually prove to be congenital, and caused by infections of pregnant women by rubella virus, cytomegalovirus, and other environmental pathogens.

By 1940, one did not have to be a biologist or a psychiatrist to realize that the essentially pseudogenetic provisions of the Nazi compulsory sterilization law gave the Nazi judges the authority to order the sterilization of any political opponent of the Nazi regime, or any personal enemy of any of the local *Erbgesundheitsgericht*. To a visiting true believer in eugenics, however, these laws were God's answer to Galton's prayer for racial purity.

Of Stoddard's session as a visiting American eugenicist on the bench of Germany's Eugenics Supreme Court in Charlottenburg, before which appeals from the sterilization decisions of the local Eugenics Courts were heard, little need be said other than what his chapter on it proved about Stoddard himself. He shared the bench with a regular Nazi judge, in solemn court robes and cap, with a well-known psychopathologist, Professor Zutt, and with a younger man—a criminal psychologist—all of whom would frequently explain to Stoddard finer points of law and degenerate psychology.[14]

In a Germany where, under the German version of Harry Laughlin's Model Eugenical Sterilization Law, the Nazi Eugenics Courts had between 1933 and 1939 caused the forced sterilization of 375,000 human beings—and where, during the entire twelve years of Hitler's Thousand-Year Third Reich, according to the Central Association of Sterilized Persons organized in Germany in 1945, the "total number of sterilizations amounted to two million"[15] —Stoddard sat in on what was, possibly, the most extraordinary session in the entire history of this supreme court of German eugenics.

As Stoddard watched and listened, there paraded before the highest eugenics bench: (1) an "ape-like" man with a receding forehead and flaring nostrils with a history of homosexuality, a marriage "to a Jewess" with whom he had had three ne'er-do-well children, who now that "that marriage had been dissolved under the Nuremberg Laws" sought permission to marry a "woman who had already been sterilized as a moron"; (2) an obvious manic-depressive of whom Stoddard wrote that "there was no doubt that he should be sterilized"; (3) an eighteen-year-old deaf-mute girl; and (4) a seventeen-year-old feebleminded girl who "was employed as a helper in a cheap restaurant. . . . The members of the High Court examined this poor waif carefully and with kindly patience."

Far from sustaining the friendly neighborhood Eugenics Courts' decisions against the gonads of these four suspects, the supreme court of eugenics ordered further clinical and psychological studies in the first three cases—and found that the poor little seventeen-year-old waif "was not a moron within the meaning of the law and therefore should not be sterilized."

There were other cases that day, not described by Stoddard but enough to tell him—as a member of the American bar—that the law was administered with sense and in Stoddard's judgment a bit too much mercy. There was no question in Dr. Stoddard's mind but that the "Sterilization Law is weeding out the worst strains in the Germanic stock in a scientific and truly humanitarian way."[16]

It never occurred to Dr. Stoddard that, on this surrealistic day in Charlottenburg, he had witnessed a dumb show staged expressly for his American readers, a charade so patently crude that it would have rated only a sneer of contempt from any two-bit con man in his native Boston.

LAUGHLIN'S LAST CAPER

The fervent support the world's eugenicists gave to Mussolini, Hitler, and fascism in general widened the gap between the science of genetics and the religion of eugenics. By 1934, the Columbia University geneticist L. C. Dunn had read and seen enough of the abuses of eugenics to question the scientific wisdom as well as the morality of compulsory sterilization as an instrument of human betterment.

Dunn visited Germany and Italy, where he had a good chance to observe applied eugenics in action. On July 3, 1935, in his capacity as an adviser to the Carnegie Institution of Washington, Dunn sent its president, the eugenicist John C. Merriam, a letter suggesting that "the techniques of measurement, description and diagnosis have now reached a stage where the material that goes into the records can be much more accurate and significant than was possible in the past," when Davenport's eugenics field workers collected the Eugenics Record Office data on human beings. To Dunn this all meant that, the ideas of its director (Davenport) and supervisor (Laughlin) having been proven, by advances in biology in general and genetics in particular, to be without scientific merit, "the only peculiar or unique part of the eugenics laboratory at Cold Spring Harbor seems to me to be the archives."[17]

Although it was not until 1939 that the Seventh International Genetics Congress, meeting at Edinburgh, "made an official condemnation of eugenics, racism, and Nazi doctrines," by then it was only institutionalizing the world genetics community's feelings. "Not surprisingly, [the] geneticists' renunciation of the eugenics movement at this time contributed to the movement's ultimate downfall."[18]

The Carnegie Institution, by then anxious to rid itself of what had become the incubus of the Eugenics Record Office, induced Davenport to retire in 1934, and by 1939 changed its name to the Genetic Record Office. The following year "the institution hastened Laughlin's retirement and thus severed its connections with eugenics."[19] However, while this made the Carnegie Institution's skirts somewhat more presentable, it did not end the eugenics career of Harry Hamilton Laughlin, the erstwhile Expert Eugenics Agent of the House Committee on Immigration and Naturalization.

Laughlin had one last caper left as a eugenical hatchet man. Now, under the banner of the American Coalition of Patriotic Societies—a catchbag of various groups that included the Sons and Daughters of the American Revolution, the Veterans of Foreign Wars, and the American Legion Auxiliary—led by the inventor of the anti-Jewish and anti-Italian quotas of the U.S. Immigration Act of 1924, John B. Trevor, Laughlin lent his scientific expertise to their alas only too successful campaign to keep what they called "international sentimentality" from lowering the 1924 bars against those few European Jews and anti-fascist Italians lucky enough to escape from the lands of applied eugenics.

The full and terrible story of how Trevor and Laughlin, playing all the racist, xenophobic, and eugenical tunes in their repertoire, succeeded in making certain that even those few Jews and Italians who escaped to non-fascist nations on the European continent were ultimately rounded up and destroyed in the Nazi death camps has been told so well by Arthur Morse in his *While Six Million Died*[20] that repetition is needless. And it is from Morse's book that the following variation of Laughlin's on Ku Klux Klan leader Hiram Evans' classic question: "Would you have your daughter marry a Jew?" is quoted. I should point out that the Klan's Imperial Wizard went on to say that even if we Nordics were willing to let this happen, "the Jew would stand against it."[21]

By 1934, according to Laughlin, "the Jew" had withdrawn his opposition to mixing bad Jew blood with fine Nordic blood. For now, Laughlin warned the Congress of the United States: "If they who control immigration would look upon the incoming immigrants, not essentially as in offering asylum nor in securing cheap labor, but primarily as 'sons-in-law to marry their own daughters,' they would be looking at it in the light of the long-time truth. Immigrants are essentially breeding stock."[22]

It was not merely the Congress of the United States which accepted this naked racism as scientific truth; it was also the President of the United States, Franklin Delano Roosevelt, the Harvard and Columbia Law School classmate of John B. Trevor. In an America where men as brilliant as Justice Oliver Wendell Holmes could be brainwashed by Laughlin and other peddlers of the pseudoscientific dogmas of eugenics, it was not surprising that somewhat lesser men, such as President Roosevelt, went along with anti-Semites like Trevor and Laughlin, rather than with those concerned members of his own cabinet and his own official family who saw the human costs of Nazism in more intelligent terms.

For Roosevelt was a creature of his times, and the years of his youth, his education, his formative social contacts had been years in which his class had, as a whole, accepted without question the legitimacy of eugenics as wholeheartedly as did his blood relation President Theodore Roosevelt. Politicians then and now took their science and their scientific ideas from the same universities where they had taken the rest of their education. And far too many of the major scientific reputations of the eras of both Roosevelt Presidents belonged to academic figures who, if anything, had been even more eager to accept the elitist reassurance of eugenics than had the century's politicians.

A PROFESSOR FLAYS THE HELPLESS EPILEPTICS

An archetypal example of an influential American academic so thoroughly the product of the antiscientific and pseudogenetic dogmas of eugenics was Edward Lee Thorndike (1874–1949). Professor of educational psychology at Teachers College, Columbia University, since 1904, Thorndike was one of the prominent collaborators in the development of the Army intelligence tests of World War I, and was the author of many of the standard textbooks on mental development, educational psychology, and child psychology used in the professional training of generations of teachers in the twentieth century. He was one of the leading "established" symbols of knowledge and scholarship on the national horizon, and few people in the sciences were held in higher regard by Presidents, preachers, editors, fellow academics, Supreme Court Justices, and other mentors and monitors of societal value systems. In 1934, Thorndike was one of the various prominent adherents of the eugenics creed to serve as president of the American Association for the Advancement of Science.[23]

In 1940, the year he retired from his Columbia Teachers College chair, Thorndike gave his nation the sum total of his thoughts after his four decades of active participation and, indeed, leadership in the nation's scientific and intellectual life. It was in the form of a huge book, 963 pages long, *Human Nature and the Social Order*. What it proved was how little Thorndike, a lifelong hereditarian, had actually learned of human heredity, human development, the biology of brain growth and development, and above all else the great advances in the hard knowledge of child development and genetics since, say, 1913. By the eve of World War I, Thorndike had long since confused eugenics for genetics, and the dogmas of the nongeneticist Galton for the scientific findings of working geneticists starting with Mendel, Johannsen, Sutton, Boveri, and Weinberg—and ending with his Columbia University colleagues Thomas Hunt Morgan and Alfred Sturtevant.[24]

Thorndike, by 1940, had still not found in any of the historic advances in biological and genetical knowledge *since 1913* any evidence that caused him to change his profound eugenic convictions. So that, at the age of sixty-six, he was still peddling the long discredited myths about epilepsy that Galton had revived when Thorndike was a boy of nine.

On page 455, Thorndike made the point that the possible disadvantages of the compulsory sterilization laws of the majority of our states should not blind us to what he saw as the fact that their "operations seem more beneficent than those of an equal amount of time and skill spent in 'social education.' Indeed the first lesson in social education for an habitual criminal or a moral degenerate might well be to teach him to submit voluntarily [*sic*] to an operation which would leave his sex life unaltered but eliminate his genes from the world. *The same would hold for dull or vicious epileptics* and for certain sorts of dull and vicious sex perverts. The genes of a few of these persons might be up to the ordinary human level but on the average they would be exceedingly low." (Italics added.)

In terms of the mountains of biomedical knowledge available to Thorndike during the three decades prior to his writing these totally inaccurate words about the hereditary, behavioral, and moral nature of epilepsy, there was literally no excuse for such views being either written or published. To be sure, Galton had written that "the criminal classes contain a considerable portion of epileptics and other persons of instable, emotional temperament, subject to nervous explosions that burst out at intervals and relieve the system. . . .

The highest form of emotional instability exists in confirmed epilepsy. . . . Madness is often associated with epilepsy; in all cases it is a frightful and *hereditary* disfigurement of humanity, which appears, from the upshot of various conflicting accounts, to be on the increase."[25] But since Galton and Davenport and Goddard, the overwhelming weight of medical evidence proved the nongenetic and nonvicious nature of epilepsy.

The sad truth was that, as far as epilepsy and other human disorders, traits, and talents were concerned, Thorndike turned not to science but to the religion of eugenics. It was the Third Law of Galton and Pearson in flower: "Why bother with the facts of biology when you already have the dogmas of eugenics?"

Thus, this twentieth-century professor of educational psychology could write, on the next page:

> The training their parents would give them [children who because of traumas or illnesses develop epilepsy] might in a few cases of reformed criminals or marriage to superior persons be up to the ordinary human level but on the average it would be extremely bad. *Indeed, the principle of eliminating bad genes is so thoroughly sound that almost any practice based on it is likely to do more good than harm* [italics added].

Time had stood so still for Thorndike since 1913 that he went on to say:

> Add (1) the facts [*sic*] of correlation whereby defects and delinquencies imply one another so that moral degenerates tend to be dull, imbeciles to be degraded, *epileptics to be dull and degraded, etc.,* (2) the facts [*sic*] of homogamy, that like tends to mate with like, and (3) the fact [*sic*] that *genes which make able and good people* also tend to make competent and helpful homes, *and the argument for sterilizing anybody near the low end of the scale in intellect and morals whenever it can be done legally is very strong* [italics added].

Despite Thorndike's use of such twentieth-century scientific words as "genes," and his advocacy of the then current Nazi Eugenics Courts practice of sterilizing people who got low marks on intelligence tests and for "inferior" morals, this was, essentially, the 1869 Gospel of Galton, the eugenical orthodoxy that all mental disorders and diseases were at least 80 percent genetic and at most 20 percent environmental. But, between 1883, when Galton revived the superstitions about epilepsy and other mental disorders that Hippocrates had attacked in 400 B.C., and 1940, the scientific community, as Fernald noted, had shown "that feeble-mindedness is not entity" caused by bad genes. And that epilepsy—as so many of the world's physicians and psychiatrists, including one of Thorndike's contemporaries, Tufts professor of neurology Abraham Myerson, had shown—was positively not a genetic curse of the gods but, as Hippocrates had described it, an ordinary acquired disease (or, rather, one of a number of acquired diseases characterized by fits or seizures).

It was not merely books such as Myerson's 1925 *Inheritance of Mental Disease* that Thorndike ignored or rejected in his *Human Nature and the Social*

Order. He also ignored or rejected the total scientific literature, by 1940 comprising literally hundreds of scientific books and monographs and thousands of scientific journal articles by life and behavioral scientists the world over, on the various *environmental* conditions—from maternal malnutrition prior to and during the nine months of gestation to the known effects of obstetrical injury on the brains and nervous systems of newborn children, to the postnatal infectious and inflammatory diseases such as measles, scarlet fever, tuberculosis, and polio—that by 1940 were well known to cause mental retardation.

There was, to cite but one typical example, a widely known review of the scientific literature, "The Fate and Development of the Immature and of the Premature Child," by Aaron Capper, M.D., published in 1928 in the *American Journal of Diseases of Children,* which dealt with 303 books and articles published since 1877. As all properly educated pediatricians, and most serious psychologists working in the areas of child psychology and the mental development of infants and children, well knew by 1928, the immature—that is, the underweight—and the premature children were as a rule far more likely to be slower in learning how to sit, to walk, to talk, to read, and to learn than children born of weights in the normal ranges and after full-term pregnancies.

Dr. Capper's paper, which also described clinical studies made while he was attached to the Children's Clinic of the University of Vienna, confirmed the fact that Austrian children, like French children and English children and Bantu children and American children, when born too small and/or too soon, were subject to the same sets of postnatal developmental risks. As Dr. Capper put it, "the immature infant becomes the backward school child, and is a potential psychopathic or neuropathic patient and even a potential candidate for the home for imbeciles and idiots."[26]

To Dr. Capper, this was cause for neither infanticide nor the sterilization of the parents of children born too small and too soon. By 1928, most pediatricians and obstetricians knew that immaturity and prematurity in newborn infants had far more to do with the nutrition and environmental well-being of the parents than with their genes. Therefore, Capper—speaking the language of what even by 1928 was common scientific knowledge—concluded:

"Efforts should therefore be directed not only to the care and preservation of the immature child, but more especially *to the prenatal care and hygiene of the mother, prevention of maternal infection and correction of anomalies of the birth passages before pregnancy"* (italics added).[27]

If Thorndike did read any portions of the vast scientific literature on human genetics, growth, and development, he evidently dismissed it as cavalierly as, in his 1940 testament to posterity, he dismissed the solid scientific findings of Boas, Ballantyne, Arlitt, Klineberg, Fernald, Skeels, and other life and behavioral scientists whose work created such viable bases of hope for human betterment. In his final chapter, starting on page 957, Thorndike summed up the thinking of a much honored lifetime in twenty "principles of action," ranging from No. 1 (*"Better genes"*) to No. 19 (*"Quality is better than numbers"*) and the more modest No. 20 (*"Reasonable expectations"*). Thorndike's principle No. 1 read in full:

Better genes. A Man's intelligence and virtue can work for welfare only for a life-span, but his genes can live forever. By selective breeding supported by a suitable environment we can have a world in which all men will equal the top ten percent of present men. One sure service of the able and good is to beget and rear offspring. *One sure service (about the only one) which the inferior and vicious [sic] can perform is to prevent their genes from survival.* Any forces which increase the relative birth-rate of superior men should be treasured, and the effect of any alleged reform or benevolence upon the selective birth-rate should be considered. [Italics added.]

The remaining nineteen points were merely developments of this basic one on "better genes" and "inferior and vicious" genes. Taken all in all, Thorndike's twenty "sound principles of action" described quite clearly the philosophy and practices of the "quality is better than quantity" eugenics program of Nazi Germany which, as Stoddard wrote in that same year, was "weeding out the worst strains in the Germanic stock in a scientific and truly humanitarian way."

Thorndike was far from being, as Jensen wrote in 1970, "probably America's greatest psychologist."[28] But he was certainly, during the three decades ending in 1940, one of its most influential ones, as well as one of the most respected spokesmen for the pseudobiology of eugenics. What Thorndike thought and wrote carried far more weight with our Presidents and our opinion molders than did the thoughts and scientific findings of Boas and Klineberg, Skeels and Dunn—for Thorndike gave the blessings of science to the most baseless fears, the meanest prejudices, the most niggardly impulses of the men who made our laws, our values, our history.

"YOU CANNOT CHANGE THE LEOPARD'S SPOTS"

The literary epitaph for the decade in American history which began with the two great triumphs of the eugenics movements—the anti-Jewish and anti-Italian quotas that barred unwanted ethnic stocks from American colleges and from immigration to America—was in a sense the Last Hurrah of the old scientific racism. It was a volume that not only celebrated a total victory for scientific racism, but was also distributed in significant numbers to legislators and mass-media editors by Trevor's Coalition of Patriotic Societies, founded in 1927 to defend the racial immigration quotas of 1924.

The idea for the book itself, *The Alien in Our Midst or "Selling Our Birthright for a Mess of Pottage"*—an anthology of great Americans of past and present on the subject of "Immigration and its results"—came from the brain of Madison Grant. Since Scribner's had published his previous books, he took it to them. They turned it down as a Scribner's book, but made a deal with Grant and his principal collaborator, Charles Stewart Davison, to print the book for them at their own expense. Accordingly, the book came out, in 1930, just as the Great Depression was getting under way, as a publication of the Galton Publishing Company, Inc.

The authors, willing and unwilling, ranged from Madison Grant, Lothrop Stoddard, Charles W. Gould, Harry H. Laughlin, and other living race warriors to Thomas Jefferson, Patrick Henry, James Madison, and George Washington, who were powerless to protest. Since many Establishments in this life have to change a little—in order to prevent major changes—the opening essay in the book showed how far even Madison Grant would bend to save the race.

As the burgeoning Depression gave millions of suddenly unemployed and bankrupt Americans the ghastly feeling that the bottom had dropped out of everything, not only did Grant solicit an essay from the president of the American Federation of Labor, William Green, but in honor of the new national mood about the wisdom of Wall Street and the captains of American industry, Green's contribution, "Immigration Should Be Regulated," was the opening essay in the book.[29]

Green, a companion-in-arms of '24, had little to say about immigration that had not previously been said by other contributors to the book, such as Albert Johnson, Charles B. Davenport, Henry Pratt Fairchild, Paul Popenoe, Kenneth Roberts, and Henry Fairfield Osborn. He opened by reminding everyone that "during the past quarter of a century the American Federation of Labor has advocated legislation to control immigration" not only for economic reasons but also because "to guard our gates we feel is necessary to the preservation of our national characteristics and to our physical and our mental health."

When it came to keeping the Italians and the Jews out of this country, Green could be as scientific as Madison Grant. Our "republican form of government is" not only "the result of centuries of growth and of training in the exercise of self-restraint in political affairs. Its maintenance depends on a *high degree of intelligence.*" Mr. Green did not have to remind the readers of this book what the Army intelligence tests had told us about which pure races produced the A men and which mongrel races produced the D and E men.

The nation's ranking labor leader concluded on a note derived from Madison Grant's *Passing of the Great Race:* "Our republican institutions are the outgrowth of ten centuries of the same people in England and America. They can only be preserved if the Country contains at all times a great preponderance of those of British descent."

Here was a fitting and proper Nordic keynote for this collection of pieces by the architects of the barriers to non-Aryan immigration erected in 1924, and the field marshals of the subsequent crusade to keep these walls intact against the refugees from Mussolini, Hitler, and Franco until December 7, 1941. In this fight to keep the non-Aryan brachycephalics out of this land, the restrictionists remained as devoted to pseudoscience as they had been at the opening of this decade of racial victories.

Harvard's Professor E. M. East, for example, in his essay "Population Pressure and Immigration,"[30] not only cited his own studies proving that the American food production rate could not continue to keep up with the native American birth rate—and that therefore, of course, we had perfectly scientific Malthusian reasons for keeping foreign snouts out of the, alas, finite American trough. He also asserted: "It has been *proved conclusively,* by many similar modes of research, that individuals vary in inborn mental capacities just as they do in physical traits. . . . How could one assume, then, that all people were constitutionally the same in brain power? Apparently it was because we had adopted the *fiction* that all men were created equal . . ."

He admitted, of course, that "naturally, opportunity plays an important role. A may be a better man than B because he has had training that B was unable to undergo, but *neither can progress beyond a limit set by the mental constitution he has inherited."* (Italics added.)

What this meant, of course, was that all the education and environmental advantages in creation could not raise the IQ test score of the born dullard more than a few points. Therefore—as Herrnstein was to reassert in his *Atlantic Monthly* article forty years later—to improve environmental growth and development conditions for any population would only increase the wide gaps between the born losers and the born-to-succeed. To East, who shared the ancient eugenic nightmare about the decline in American intelligence, there were "only two sensible and practical ways" to end this decline, both of them eugenic. *"The present differential birth-rate of the unintelligent over the intelligent can be eliminated, and immigration of the unintelligent can be forbidden"* (italics added).

Lothrop Stoddard, the hard-core racist, anti-Semite, and ardent admirer of the Nazi regime, borrowed an anti-Semitic canard from the literature of the Russian Black Hundreds—the superpatriotic pogrom corps of the Czars—to grace his essay, "The Permanent Menace from Europe."[31] This was the "anthropological" discovery that the Jews of Eastern Europe were not the Jews of the Bible, whence had sprung Jesus of Nazareth and his first Apostles.

On the contrary, Dr. Stoddard wrote, the East European Jews *"are not Semitic 'Hebrews,'* but are descended from West-Asiatic stocks akin to Armenians, and from a Central Asiatic (Mongoloid) folk, the Khazars. Eastern Europe thus presents a bewildering complexity of races, creeds, and cultures, which reaches its climax in the Balkans—that unhappy abode of jarring, and to an extent, half-barbarian peoples." (Italics added.) Nothing, Stoddard concluded, should induce us to ever relax our vigilance against the immigration of such scum.

If the opening essay, by William Green, was a strategic bow to changed conditions, the selection of the contribution of Professor Robert DeCourcy Ward of Harvard University to close the anthology was a gracious nod to the past. For it was Ward who, with the departed Prescott Hall, had in 1894 founded the Immigration Restriction League. The title of the aging climatology professor's piece was nearly as long as the essay itself: "Fallacies of the Melting-Pot Idea and America's Traditional Immigration Policy."[32]

Its opening lines summed up the entire chapter: " 'Never shall ye make the crab walk straight. Never shall ye make the sea-urchin smooth.' Thus, many centuries ago, Aristophanes disposed of the fallacy of the Melting-Pot. Up to recent times we have ignored the principle of selection in our immigration legislation."

Once the Johnson Act was passed, however, we were able to keep out the undesirable and Jewish D and E immigrants, and admit only the desirable and Nordic A and B (high IQ-test-scoring) future Americans. And Ward quoted

Osborn's address of welcome to the 1921 International Congress of Eugenics, in which his fellow contributor had declared that "we are slowly awakening to the consciousness that education and environment do not fundamentally alter racial values."

For as Ward now reiterated, "if the material fed into the Melting-Pot is a polyglot assortment of nationalities, physically and mentally below par, then there can be no hope of producing anything but an inferior race."

Ward cited a maxim of Lothrop Stoddard's to underscore this point: "The admission of aliens should, indeed, be regarded just as solemnly as the begetting of our own children, for the racial effect is essentially the same."

Nor was it really true that America was ever meant to be the "asylum and a haven of refuge for the poor and the oppressed of every land." Because of such sentimental rot, Professor Ward said, "it has become obvious to all thinking Americans that our 'asylum' had become crowded with alien insane and feeble-minded; that our 'refuge' was a penitentiary and was becoming filled with alien paupers."

Like his Harvard faculty mate Edward M. East, Ward could not have been unaware, by 1930, that modern science had refuted all of the basic myths about the hereditary lunatics, mental retardates, criminals, moral lepers, and paupers who, according to the sacred tenets of eugenics, made up the sub-species called the Undeserving Poor. Ward had to be cognizant of the work in modern genetics that had exposed, for all time, the reality that eugenics was not a science but a pseudoscience.

What he did in the face of the mountains of scientific data that refuted everything he had ever written and said about racial traits and human worth was to turn to the eugenical faith, whose high priests had long ago formulated two rules for what to do when confronted with the type of ugly facts that shatter cherished hypotheses. Rule I was to quote Galton. This Ward did, with the following Galton line: "A democracy cannot endure unless it is composed of able citizens: therefore it must in self-defence withstand the free introduction of degenerate stock."

Rule II, of course, was to quote Pearson. Which Ward did on the very next page, where, describing the British socialist, mathematician, and self-proclaimed authority on nutrition and intercostal sulci as "one of the best-known of modern writers on heredity," he quoted Pearson as writing: "You cannot change the leopard's spots, and you cannot change bad stock to good. You may dilute it; spread it over a wide area, spoiling good stock; but until it ceases to multiply it will not cease to be."

Ward's faith was rewarded. The Johnson Act quotas of 1924 were kept inviolate until Pearl Harbor day, 1941. At least nine million human beings of what Galton and Pearson called degenerate stock, two thirds of them the Jews Dr. Stoddard had malignantly mislabeled as Central Asiatics posing as Semitic Hebrews, continued to be denied sanctuary at our gates. They were all ultimately herded into Nordic *Rassenhygiene* camps, where the race biologists in charge made certain that they ceased to multiply. And ceased to be.

PART THREE

The Phoenix:
The Rise of the New
Scientific Racism

16 "Overpopulation" and Pollution Become the Opening Gambits of the New Scientific Racism— But Its Human Targets Remain Unchanged

> . . . experts agree unanimously that it would be a good thing if we could reduce the statistical differential that now runs so heavily in favor of the unfit. If it is maintained indefinitely, there will be a wholesale degeneration of the American stock, and the average of the sense and competence of the whole nation will sink to what it is now in the forlorn valleys of Appalachia.
> There are, plainly enough, only two ways to get rid of this differential. One is for the people of the upper IQ brackets to develop a birth rate higher, or at least as high, as that prevailing among the undernourished; the other is for the undernourished to reduce their birth rate to something approximating that of the smart and the swell.
>
> —H. L. MENCKEN, in "Utopia by Sterilization," *The American Mercury,* August 1937

Scientific racism went into deep hibernation in America and England during World War II. The Nazi employment of eugenics as an instrument of national and foreign policy was one reason for the seeming disappearance of American scientific racism after December 7, 1941. There was also the historical reality that, between 1941 and 1945, the Nordic Nazis and their Japanese allies were engaged in the wholesale killing of American, British, and other Allied soldiers, sailors, and civilians. This hardly sat well with the eugenics movement's hallowed Cult of the Nordic. Nor, for that matter, did the deaths of the millions of our Russian allies in the common war for survival against the Nazi race purifiers and defenders of *die nordische Seele.*

There were many other reasons for educated Americans to hope, after V-J Day, 1945, that the American people had suffered the last of the aggressions against their health and their dignity that the scientific racists had been unleashing upon successive generations here since the rise of Spencer's Social Darwinism immediately after the end of the American Civil War. Most of these reasons had to do with the great advances in the behavioral and biomedical and biological sciences between the passage of the U.S. Immigration Act of 1924 and the surrender of the German and Japanese enemies in 1945.

In the behavioral sciences, the demolition by Columbia University's Otto Klineberg of the eugenics myth of "selective migration," along with the work of Skeels and the other Iowa psychologists, had pretty well ended the myths of racial and genetically fixed intelligence. Klineberg, who had a doctorate in both medicine and psychology, published, between 1931 and 1935, a series of studies he made in France, Germany, Italy, and the United States which showed that—in all racial groups, whether European Nordics, Alpines, and Mediterraneans, or American blacks—the children who lived in the cities got higher intelligence and classroom test marks than did

362

children of the same racial and ethnic groups who lived in the country. The results were the same in four countries and among children of eight different racial groups and subgroups. In the cities of each country, where family incomes were on the average considerably higher than in the rural areas, and where the urban schools were invariably far superior to the farm area schools, the city children all got consistently higher marks for intelligence and academic work than their rural peers.[1]

The American studies were particularly dramatic, for they showed that even as short a family move as from a southern rural community to the nearest southern city—segregated schools and all—resulted in a predictable jump in the intelligence test scores of the family's children. Klineberg showed that the further black children moved from the South, and the longer their families lived in northern cities—such as Baltimore, Philadelphia, and New York—the higher their IQ test scores climbed. In two of his 1935 publications[2] Klineberg ran two tables that became well known to all professional educational and child psychologists. One of them showed what happened to southern- and northern-born twelve-year-old black girls when they took the then standardized National Intelligence Tests (NIT):

NIT Score and Length of New York Residence

Residence Years	Number of Cases	Average Score
One and two years	150	72
Three and four years	125	76
Five and six years	136	84
Seven and eight years	112	90
Nine and more years	157	94
Northern-born	1,017	92

In his *Negro Intelligence and Selective Migration,* Klineberg ran a table that paralleled the study he had made in 1931, in which he compared the effects of urban and rural residence on the IQ test scores of white children in Germany, Italy, and France:

Migrants from City and Country (Yates)

Years of residence in New York City	Average Stanford-Binet IQ test scores	
	Southern city-born black children	*Southern country-born black children*
1–2	76	49.6
3–4	81.1	67.4
5–6	94.34	84
7–8	99.4	104
9–10 and 11	103.33	101.5

What the table showed, of course, was that it was not the allegedly greater proportions of "superior" white genes in the "blood" of the migrant black children but the relatively superior growth and development environment of New York City living that produced the IQ test score differences he found in the children studied.

In the biomedical fields, there was an endless series of advances in clinical nutrition, hematology, obstetrics, and surgery that had made many brain and body birth defects preventable—and had turned diseases such as diabetes and pernicious anemia (which, as late as 1921 and 1926, respectively, were untreatable and invariably fatal diseases) into medically manageable disorders that allowed people to continue in their jobs or careers.[3] On the eve of and during World War II, the new era of chemical and biological antibiotic drugs was begun. These lifesaving drugs now made possible the successful treatment of formerly killing infectious diseases, such as pneumonia, for which the only previous medical treatment had consisted of a reassuring bedside manner.

Finally, in 1944, work at the Rockefeller Institute in New York by various investigators—including the biochemist and cell biologist Alfred E. Mirsky and the physicians Oswald T. Avery, Colin M. Macleod, and Maclyn McCarty—led to the identification of deoxyribonucleic acid (DNA) as the primary material of heredity. This discovery, of course, was the start of the modern era of biochemical genetics and molecular biology.[4]

In addition to the scientific reasons for believing that scientific racism had lost its former hold in America, there were also some very compelling moral reasons.

During the immediate postwar months, when the heartrending eyewitness accounts by thousands of American soldiers, doctors, nurses, and journalists who entered the Nazi camps of applied eugenics at Auschwitz, Buchenwald, Dachau, Maidenek, Treblinka, and other corners of hell with the liberating Allied troops, brought home in full measure the human costs of the holocaust of Eugenics Race Biology Triumphant, American scientists were filled with a sense of revulsion—and guilt. Many of them had believed in and taught the eugenic concepts of the *Herrenvolk* to young people. Others had not quite believed the reports of competent journalists and European refugees prior to the war—when they might have raised their voices against the Greens, the Grants, the Davenports, the Laughlins, the Stoddards, and the Easts who fought to maintain the 1924 Immigration Act quota barriers against Hitler's targeted victims. The word "eugenics" was now no longer even whispered on America's campuses, in America's laboratories.

During those immediate postwar seasons, as United Nations and American volunteers labored to restore some sparks of life and human hope to the bodies and minds of the "inferior race types" who still breathed on the day their concentration camps had been liberated, and as the testimony from the masters of German eugenics at the Nuremberg trials drove home the truths, the American scientists and intellectuals who had lived through the rise and triumphs of the American eugenics movement between 1906 and 1941 shook their heads in shame and in sorrow and thought: "Never again."

Never was to last less than a year.

The New Scientific Racism was soon to rise, like the Phoenix, from the flames which consumed the Old Scientific Racism that had lasted from Malthus to Hitler. Ironically, it was not by some new magic touchstone that the new scientific racism found the secret of eternal life, but in the basic myth that had formed the trunk on which Gobineau and Galton, Retzius and Van Evrie, Spencer and Davenport, Yerkes and East had added deadly new limbs after 1798.

The mechanism of regeneration was, of course, the original Malthus myth, the pseudonatural "Law" that man's ability to produce babies would always and forever be greater than his "finite" capacity to grow food. Therefore, unless the exploding human birth rates were slashed, our species faced famine and extinction. This "Law of Population" was purely a figment of Malthus' imagination. Some seven decades before Malthus was born in 1766, the European Agricultural Revolution, "the greatest move forward in agriculture since neolithic times," had proven—and continues to prove, abundantly, in our own times—that Malthus' famous "Law" was a totally false description of the realities of food production and human reproduction on this planet.

The old scientific racists were able to obscure this truth, and to hide even the existence of the Agricultural Revolution of 1700 onward (and of which our modern Green Revolution is simply a modern phase), for many years. However, in an era in which, for example, the national government issues an annual *Statistical Abstract of the United States*—which each year includes a very factual section on agriculture—the new scientific racism was forced to put the old specter of too-many-mouths-for-too-little-food on the back shelf and to raise a scare about a specter that was much more viable.

This new specter was based on the reality that the pollution of our living environment by the criminal abuses of technology—that is, by the fossil fuel and chemical waste products of big industry; mismanaged water resources; unsanitary sewage and garbage-disposal systems; criminally negligent urban planning; the post-World War II substitution of the present one-man–one-car system of daily transportation for the prewar less pollutive, less dangerous, and less expensive electrified inner-urban and inter-urban trolley cars and commuter trains—was a major health problem in our society. The pollution of the environment by mismanaged technology and urbanization is, indeed, one of our most serious national health problems.

What was not true, however, was the anti-human myth the new scientific racists were to graft on to the basic problems of environmental pollution— and that is the lie that it is people, and not motor vehicles, that pollute; that it is people, and not toxic industrial chemicals, that pollute; that it is people, and not the combination of fossil fuel pollutants and cigarette smoke, that are responsible for our growing epidemic of lung cancer. By 1954, the scientific racists were to embellish this new myth with the revival of the old eugenics myth of "the decline of American intelligence"[5] as a scientific rationale for segregating, undereducating, and sterilizing people.

While the concept of People Pollute was to be added to the equally mythic American Population Explosion—an event that the U.S. Bureau of the Census, the agency that measures the realities of American demography, shows to have never existed—the goals of scientific racism per se remained unchanged. All that had changed was the scientific rationale for doing nothing about making life safer and longer for all people. Now the new myth of the new scientific racism had it that, until we had far fewer people, the healthier and longer we made life for the greatest number of people, the more the Menace of Pollution, caused by the Law that People Pollute, would hasten us to death and extinction.

As in the 1920's, when the nation's decent people were betrayed by their own education into accepting, as a scientific truth, the crude eugenic myth—"proven" by the civilian and Army IQ test scores—of "the decline in American intelligence," and therefore decided to throw their support behind the anti-Italian, anti-Semitic, anti-Catholic immigration restriction demands of the old scientific racists, the contemporary effects were tragic. Some of the best-educated and best-intentioned people in our society began to wear People Pollute buttons on their lapels, and to become true believers and vigorous fellow travelers in the pseudo-environmental crusades of the new scientific racism.

In future generations, educated people will be astounded, and even highly amused, at how the well-educated people of our times swallowed this crudely baited hook—just as we, in turn, chuckle at the naïveté of the eminent Victorian savants and nineteenth-century American college presidents who seriously accepted such a ridiculous fraud as phrenology as a "biological science." There was only one difference between the neo-Malthusian People Pollute crusade of our times and phrenology: Professor Gall's phrenology hurt nobody, and peddled nothing but harmless illusions.

PLUS ÇA CHANGE . . .

Late in 1945, while the crematoria ashes of Auschwitz and other centers of applied eugenics in *Festung Europa* were still warm, the Population Reference Bureau of Washington, D.C., issued a slim monograph, *Population Roads to Peace or War,* by Guy Irving Burch and Elmer Pendell, which was revised and reissued in 1947 as a commercial soft-cover book retitled *Human Breeding and Survival,* with a foreword and postscript by Professor Walter B. Pitkin, of Columbia University.

Burch was the director of the Population Reference Bureau and editor of its *Population Bulletin.* Pendell, a Ph.D. economist, then teaching at Baldwin-Wallace College in Ohio, was a member of the board of Birthright, Inc., an offshoot of the old eugenics movement that worked for surgical sterilization.

The book's blurb described Burch as a charter member of the Population Association of America, a consultant in the framing of the draft act of 1940, and "the author of numerous scientific papers." While the publishers did mention the fact that Burch was a director of the American Eugenics Society, they said nothing about his having been part of the last great legislative triumph of the old scientific racism—the preservation of the anti-Semitic, anti-Italian, and anti-Slav immigration quotas of 1924.

Burch, who shared the racial views of John B. Trevor, became one of the active leaders in the Coalition of Patriotic Societies, which Trevor had formed in 1927 to keep these barriers to non-Nordic immigration inviolate. A onetime lobbyist for the National Committee on Federal Legislation for Birth Control, Burch explained to the chairman of the Southern Baptist Convention in 1934 that he was active in Mrs. Sanger's birth control movement because "my family on both sides were early colonial and pioneer stock, and I have long worked with the American Coalition of Patriotic Societies to prevent the American people from being replaced by *alien or Negro stock,* whether it be by immigration *or by overly high birth rates among others* [*sic*] *in this country"* (italics added).[6]

Burch hated the poor and the Roman Catholics of his native land as much as he despised people of "alien or Negro stock." As he wrote in a review of Harry Laughlin's 1924 report to Congress on the menace of Jewish, Italian, and other non-Nordic immigration, in Margaret Sanger's *Birth Control Review* of November 1926: "America has traded the high birth rate of her pioneer stock . . . for steerage immigration." The only way to keep America American was to make certain that "Scientific Birth Control must be practiced by ignorant, diseased and poverty stricken families." But, Burch concluded, "the most uncompromising organized opponent of Scientific Birth Control, the Roman Catholic Church, has increased its numbers in the United States from one-hundredth part of the total population in 1790, to one-fortieth part in 1820, and to one-sixth part in 1920."

Burch remained active in Trevor's racial exclusion coalition until July 23, 1942, when the Coalition of Patriotic Societies, along with the Ku Klux Klan, the Silver Shirts, the German-American Bund (the American branch of the Nazi Party), and the Black Legion, was indicted for sedition in the District of Columbia Federal Court.

While their book was an essay in orthodox eugenical scientific racism, complete even unto citations of General Francis Amasa Walker's Greshamite population law (under which an inferior non-Teuton population replaced the descendants of the tribesmen who had "met under the oak-trees of old Germany to make laws and choose chieftains" in the "Native American" bloodstream), neither Burch nor Pendell were standpatters. Unlike Charles Benedict Davenport, who in 1932, at the Third International Congress of Eugenics, was still insisting that "we have come a long way from the standpoint of the medical man who said, in effect, tuberculosis is due to the *bacillus tuberculosis,"*[7] Burch and Pendell knew the importance of changing the key, and even adding a variation or two, to preserve the ancient myths of scientific racism.

Observe, for example, how carefully Burch and Pendell utilized the words of a great liberal, Oliver Wendell Holmes, as the epigraph of their Chapter 8, "Can Sterilization Help?"—not only as an excuse for the compulsory sterilization of Americans drawing relief checks, "persons who are in bad health," and "habitual criminals" and other "defective or socially inadequate" types, but also as the benediction for their brand-new postwar plan for American-sponsored genocide on a world scale. The chapter opened as follows:

CHAPTER EIGHT
Can Sterilization Help?

"We have seen more than once that the public welfare may call upon the best citizens for their lives. It would be strange if it could not call upon those who already sap the strength of the state for these lesser sacrifices, often not felt to be such by those concerned, in order to prevent our being swamped with incompetence."

—Justice Oliver Wendell Holmes, upholding the constitutionality of the Virginia sterilization law in the decision of the Supreme Court case, Buck *vs.* Bell 274 U.S. 200.

Previous chapters have shown overwhelmingly that population is inevitably limited. If it is not limited through conscious controls, then it is limited by malnutrition, famine, disease, and war.

Population is a dynamic force that we cannot ignore without catastrophe. As a problem that we are bound to face, it takes the form: *What methods of limitation are appropriate?* . . .

We have seen that control has been attempted in various lands by means of abortion, infanticide, migration, contraception, and sterilization. Abortion and infanticide may be less horrible than malnutrition, famine, disease, and war, but among the control possibilities, abortion and infanticide rank too far down the scale to claim discussion in a book so brief as this.

And it progressed inexorably to the logical conclusion with which it closed:

Since blind population forces are the most persistent influences barring the way to the world-wide attainment of freedom from want and from war, and the attainment of government by the people, then to sponsor those goals is sanctimonious twaddle or pious fraud, *except as one is realistically ready to control the population forces.*

In connection with sterilization, it appears that what the United Nations needs to do is to recommend to all nations the adoption of laws which will

(a) actually lead to the sterilization of all persons who are inadequate, either biologically or socially, and

(b) encourage the voluntary sterilization of normal persons who have had their share of children.

It was, of course, the *Weltanschauung* of Malthus and Gobineau, Galton and Davenport, modernized to project the fledgling United Nations as the instrument of a United States in pursuit of the eugenical dream of the forced sterilization of the "aristocracy of the unfit."

By 1945, when the human costs of the war and the Hitler regime had stripped anti-Semitism—and racism itself—of the respectability it had still enjoyed in many educated quarters, Burch and the other unreconstructed eugenicists needed a new package wrapper for their old bill of goods. They found it—in peace.

Hitler's *Lebensraum* rationale for his acts of aggression against other nations was accepted at face value by the authors of *Human Breeding and Survival*. Rewriting history to suit their long disproved thesis, Burch and Pendell presented as hard fact a dream world in which "if the birth rate is not kept high to produce a bumper crop of oncoming youth, in time the nation not only decreases its potential cannon fodder but also has an abnormally large proportion of oldsters to take care of. On the other hand, if the birth rate is kept high the total population eventually reaches the 'must expand or explode' stage. . . . The way is now prepared for the despot." And the despot leads the "overpopulated" nation into war.[8]

To this fairy tale, which among other things offered excuses for Hitler's role in World War II, the authors added a revived dose of Malthusian population mythology. The world was running out of fertile soil, they warned, and would soon be unable to produce enough food to meet its present population's biological needs, let alone to feed the millions of gaping new mouths produced by the reckless breeding of the Under-Men. "The differences in birth rates are, for the most part," they claimed, "a phase of the displacement principle of population: some people have few babies *because* other people have many" (p. 97).

From this statement, which has *no* basis in biology or demography, it was an easy step to the main point of the book: the sterilization of the poor in the name, not of eugenics, but of *peace*. "If we are willing to keep the focus on *undesirable parentage* . . . then sterilization can play a rather large part in the attainment of the peace goals" (p. 97; italics added).

Then, in words that sound like a replay of the 1920 writings of Davenport, Goddard, and Grant as well as a preview of the 1970 writings to come from the new scientific racists, Burch and Pendell made themselves perfectly clear. "What are the social bases on which sterilization might be indicated in the program to attain the peace goals?" Their answer:

> Looking toward a possibly *economic* test, are persons who are on relief to be encouraged to reproduce while they are on relief, as they have been? . . . Are their children more likely to be social burdens than are the children of those who are in better control of their own environment? . . . Is it reasonable to ask other citizens to pay more taxes in order that relief recipients may reproduce? Is it reasonable to impose the heavier tax burden when that additional pressure on many taxpayers will be just enough to prevent their own reproduction [p. 97]?

This, of course, was merely the old eugenical nightmare of the inferior stocks outbreeding their superiors, dressed up in terms of tax relief for the solvent.

The welfare clients whose right to have children and whose basic human worth Burch and Pendell denied in 1945 were, of course, not the welfare mothers and the "relief chiselers" who gave the economic racists of the 1960's and the 1970's such running nightmares. They were, rather, the more than 20 million American victims of the Great Depression of 1929–41, most of them native-born white Nordic Protestants, for whom welfare in the form of relief checks and work-relief jobs had been all that stood between them and either a life of crime or death by starvation.

An even more modern twist was given to the old fear of dysgenesis via the breeding of unhealthy types. Some states, they noted, have premarital health tests, but "these are only designed to patch up ailments, not to discourage reproduction by persons whose health is poor." Then, in tongue-in-cheek words that would have brought a chuckle to the heart of Madison Grant, the two humanitarians asked, "Is that *all* we should do . . . for the cause of prosperity, peace and democracy? Or is some sort of a *health test* called for as a *pre-requisite to conception?*" (Italics added.) Few of the prewar scientific racists had had to face what Burch and Pendell recognized as the political necessity of invoking the word "democracy" to justify their proposed and accomplished acts of aggression against the poor of alien and Negro stock.

In a clever updating of the old eugenic bogey of the genetic monster in human form whose bad genes program him to be only a "habitual criminal," Burch and Pendell asked, "Do their [the habitual criminals'] children stand as good a chance as the children of non-criminals of being social assets?" Since most people educated in America between 1900 and 1935 or even later were taught the Galtonian gospel that criminals are born and not made, the authors did not bother to answer their own question, but moved on to a point of social philosophy. "Whether or not there is social justification for permitting a criminal to die of old age, a separate question is here in point. Are we to choose the criminals to constitute the most influential phase of the environment for a part of the citizenry of tomorrow?" (pp. 97–98).

In former years, they would have buttressed their plea for sterilization of the poor, the sick, and the unfortunate with quotes from Galton, Davenport, Lothrop Stoddard, Grant, Henry Fairfield Osborn, and Goddard. But these were different times, and besides the racist gospels of the old scientific racism had been a little too similar to the racist gospels of Hitler and Mussolini and Goering to sit well with American readers in 1945 and 1947. An authority not associated with the old and conservative voices of the WASP Establishment was needed. And who could be better for such purposes than the irreverent baiter of all conservative types, the hero of many a fight for freedom to write literature for adults, the scourge of the Establishment— H. L. Mencken?

Now, in their population tract, Burch and Pendell played their trump card:

> In basing sterilization on social criteria such as criminality, low earnings, poor health, and lack of education, H. L. Mencken has gone probably farther than anyone before him, in suggesting a large-scale use of the economic test [of the right to procreate]. In the *American Mercury* for August, 1937, he observed that in general the sterilization laws apply only to persons who are defective in some gross and melodramatic way. Said he: "Let a resolute attack be made upon the fecundity of *all* the males on the lowest rungs of the social ladder, and there will be a gradual and permanent improvement." There was inventiveness in the

method that Mencken presented: A sum of money was to be offered to the prospective sterilees—$100, $50, $25—some amount not large enough to result in a stampede to the sterilizing physicians or for the new money to start an inflation. The Federal government or state government or private philanthropy might finance the plan [p. 99; italics added].

Writing immediately after World War II, Burch and Pendell commented that sterilization, "this substitute for contraception involving only one instance of inconvenience, and surer than mechanical contraception itself, would have special appropriateness in China, and India, and Puerto Rico, where domestic facilities for the use of contraceptives are few" (pp. 99–100).

All this was, of course, the old eugenics canon warmed over in the coals of Pearl Harbor and Buchenwald, seasoned with a dash of peace and democracy, and even a smidgeon of "autonomy" *in re* the nonwhite peoples' then demanding not autonomy in the white nations' empires but independence. It was a vigorous pioneer try, and Professor Pitkin, author of *Life Begins at Forty,* was inspired to write a postscript filled to the brim with Galtonian eugenical orthodoxies:

Reckless breeding has become strangely like a social cancer. . . . Unless men see the problem and work on it, America soon after the year 2000 will be a nation of high-grade morons ruled by a few surviving clever people. It will be no more of a democracy than any other monkey house. But of course the mass, sitting on bleachers and watching free prize fights, will shout all the ancient slogans of democracy [pp. 130–31].

HOW MEDICAL RESEARCH "THREATENS HUMANITY EUGENICALLY"

Two years later, Professor Garrett Hardin, of the University of California at Santa Barbara, was to cite Burch and Pendell's *Human Breeding and Survival* as the first of five "selected references" for further information on the themes of the closing chapter of the college biology textbook he published in 1949—*Biology: Its Human Implications.*[9]

In this chapter, entitled "Man: Evolution in the Future," Hardin charged not only that the world population was controlled by the industrial and agricultural practices of any given time in history, but also that "in the world as a whole, the population seems to be very close to its momentary limit, as is suggested by the fact that the deaths due to starvation and crowd diseases run into the millions every year." He also went on to say that the sheer *qualities* of this needlessly large world population were declining, because the people with the most education were having the fewest children, and the people with the highest IQ test scores had declining rates of reproduction, while the lower a group's aggregate IQ test score, the more its rate of reproduction increased.

To help document these assertions, Hardin ran two very impressive-looking tables, one from a 1943 Milbank Foundation study on differential fertility, and the other from a 1934 book written by the treasurer of the Third International Congress of Eugenics, Frederick Osborn, and Frank Lorimer.

The key to understanding the meaning of these tables, according to Hardin (pp. 610–12), was the fact that intelligence, which "is measured in terms of a statistic called 'Intelligence Quotient,' or 'I.Q.' for short," is hereditary. "Ideally, the I.Q. is independent of the amount of schooling, and within a given society, *among those who have had some schooling,* it approaches the ideal closely enough" (italics added). Hardin cited the fact that "studies made of the IQ of parents and children show that parents with higher-than-normal IQ's have children with higher-than-normal IQ's on the average. Similarly, parents with low IQ's produce children with low IQ's." Hardin did not identify these studies any further, but it was clear that he was *not* referring to any of the many studies of Skeels, Dye, Wellman, and the other psychologists of the Iowa group whose data proved that these assertions were not merely simplistic but also dead wrong.

Hardin then followed this with a paragraph whose theological roots were fairly obvious:

> In passing, it should be pointed out that there are many people who, fearful of the possible consequences of admitting that intelligence is partly determined by heredity, would deny the role of heredity entirely, implying that "all men are created equal." There is a dull kind of safety in adopting this position because, among humans, it is difficult to disentangle environmental and hereditary factors. However, in other animals, where experimentation is possible, it has been clearly shown that there are inheritable factors that determine the limits of intellectual ability. To assert, either explicitly or implicitly, that the case is otherwise with humans is to espouse a doctrine of exceptionalism that is repugnant to scientists.

Equally repugnant to scientists and scientific discourse is the setting up of otherwise unidentified straw men, such as the "many people" who shy from "admitting that intelligence is *partly* determined by heredity"—and the blind references to unidentified purportedly serious scientific studies which clearly showed that "there are inheritable factors that *determine* the limits of intellectual ability." I can think of no serious modern biologist or psychologist who ever denied that intelligence—whatever intelligence proves to be—is not *partly* determined by heredity any more than human height, human weight, human health, and human nervous systems can escape being partly determined by heredity.

By 1949, however, the differences between a human genotype and its phenotype were quite well understood even by nonbiologists. By the same year—exactly forty years after Boas' *Changes in the Bodily Form of Immigrants,* and ten years after Skeels and Dye and Marie Skodak published their studies on the effects of differential stimulation on mentally retarded children and on the mental development of children in foster homes— scientists understood that the full *range* of phenotypic potentials (the norm of reaction) already present in the genes we inherit at birth are not only genetically but also environmentally finite. Theoretically, that is.

Each gene-environmental interaction can produce variations in the characteristics of an individual, such as height, weight, strength, and general health. The quantities of interactions between an individual's genes and his environment throughout the course of an average lifetime are, in reality, so numerous as to be for all practical purposes infinite. In a 1946 essay, "The Interaction of Nature and Nurture,"[10] the noted British geneticist J. B. S. Haldane calculated from his equations that, working with only ten genotypes and ten environmental factors, there are 7×10^{144} types of possible interactions between genotype and environment—that is, seven times a number made up of 1 followed by 144 zeros.[11]

In other words, human genetics is not quite the simple slot-machine mechanism that Hardin, in the 1951 edition of his biology textbook, made it out to be. More than the right gene or combination of genes alone is needed to produce the finished human taker of IQ tests.

To be sure, Hardin was writing as a professor of biology, and not as a psychologist. But just as Edward Lee Thorndike in 1940 had owed it to his readers to catch up on the extensive literature of the biology of heredity and development before writing about these subjects, Hardin could also have benefited from a quick survey of the scientific studies of IQ testing more recent than those of Lorimer and Osborn in 1934. There was, for example, the very well-known report in the *American Psychologist* in 1948 by Hardin's University of California colleague Professor Read H. Tuddenham, "Soldier Intelligence in World Wars I and II."

After working out a concordance between the mental tests of both world wars, Tuddenham not only revealed that the mental test scores of the average World War II draftees were higher than the scores of the World War I draftees—but also demonstrated that in both eras the more years of education the average individual draftee completed prior to being tested, the higher was his intelligence test score. Since the mean education of the World War I draft population was only eight years of school, while the World War II draftees had completed an average of ten years of schooling, this was reflected in the proportionately higher intelligence test scores of the second war's recruits.

But Tuddenham was concerned with more than simply the raw intelligence test scores. Or the obvious correlations between the amount of schooling completed and the ultimate mental test scores. In education, as in other matters, quality had to be considered along with quantity. "If additional allowance is made for the progressive increase in the length of the school year and for improvements in school facilities and in the professional preparation of teachers, it is evident that the superior test performance of the World War II group *can* be accounted for largely in terms of education."

Largely, but not completely. For even though Tuddenham was not a biologist, he was enough of a scientist to know that man is, indeed, from his genes to his fingertips, a biological entity who happens to be born with certain mental capacities that are also biologically controlled. Eyes, ears, bodies weakened by chronic hunger, and deformed by chronic if preventable infectious and inflammatory diseases, can wreak havoc on a brain with the genetic potential of an Einstein. He therefore also took due notice of the fact that, as of 1948, "numerous investigators have reported that as a nation we are increasing in height, in weight and in longevity. The indirect influence of improvements in public health and nutrition may have operated to increase test performance, though to an unknown degree."

Had Hardin read (or, having read them, digested) both the Haldane and the Tuddenham article, he might—certainly as a biologist—have seen the nature and the implications of the two tables on what he presented as the decline in American intelligence in a somewhat more modern scientific light. Instead, citing studies *other* than those of Lorimer and Osborn (but not further identified), Hardin wrote: "In all cases, the studies indicate *that as long as our present social organization [democracy] continues, there will be a slow but continuous downward trend in the average intelligence*" (Hardin's italics, pp. 611–12).

So here we were, back to Brigham in 1923 and Yerkes in 1921, with a revival of their hoary myth about the declining American intelligence, in a post-World War II college textbook. And a revival of the solutions proposed by the cult whence the myth derived: negative and positive eugenics.

Negative eugenics was described by Hardin as

> . . . measures aimed at discouraging the breeding of the less-desired types of humans. The principal measure used to date has been that of sterilization of the extremely unintelligent. . . . Observation [*by whom not specified*] has shown that, almost without exception, two feeble-minded parents can produce only feeble-minded children. There seems to be little danger of society's being deprived of something valuable by the sterilization of all feeble-minded individuals. The legal right of society to do this has been upheld in the courts [i.e., by Holmes's 8-to-1 Supreme Court decision in the tragedy of *Buck* v. *Bell* in 1927].

This would not only protect society from the "menace of the feeble-minded" but would also save us the costs of locking up these extremely unintelligent Americans until they were past breeding age. So that, taken all in all, as Hardin took it on page 613, "it is difficult, on rational grounds, to object to the sterilization of the feeble-minded." As long as we also realized that "more spectacular results could be obtained by *preventing the breeding of numerous members of the subnormal classes higher than the feeble-minded*" (italics added).

But while concerning ourselves with the gonads of low IQ test score achievers, we should not at the same time lose sight of the benefits of what Hardin called *positive* eugenics. For example:

> It should also be possible to raise the average IQ by positive eugenics, that is, by encouraging the reproduction of those individuals whose IQ is higher than the average of the general population. If such people can be persuaded or enabled to produce *more children* on the average than are produced by other members of society, the average IQ may be expected to rise.

There was little time to waste. For by page 618 Hardin was warning America's biology students:

Sooner or later . . . human population will reach a limit. . . . Sooner or later, not all the children that humans are willing to procreate can survive. Either there must be a relatively painless weeding out before birth or a more painful and wasteful elimination of individuals after birth. . . . If we neglect to choose a program of eugenics, will the production of children be nonselective? . . . People with low I.Q. are reproducing at a faster rate than those with high I.Q.

It was not only the gonads of the people with low IQ test scores that threatened Hardin's world. There were also the dollars of those misguided millionaires, such as the old John D. Rockefeller, who paid for the historic pre-World War I crusade against hookworm in the South. For these millionaires, unlike Hardin, were not biologists, and did not realize that *"every time a philanthropist sets up a foundation to look for a cure for a certain disease, he thereby threatens humanity eugenically"* (italics added). Nor was it only by funding research in the nature and prevention and cures of diseases that our wealthier people committed dysgenesis.

> Again, consider the matter of charity. When one saves a starving man, one may thereby help him to breed more children. This may be a good or a bad thing, depending on the facts. Some people maintain that very poor people are, on the average, less able and intelligent than the rich, and that their deficiencies are, in part, due to hereditary factors. Others maintain that pauperism [sic] is exclusively a matter of bad luck; or that paupers are better genetic material than millionaires. [Alas, none of these people were identified by name.] There is a need here for indisputable facts; *but whatever the facts, aid to paupers undoubtedly has genetic consequences.* A more subtle form of aid to low-income groups [sic] is the graduated income tax which taxes the rich proportionally more heavily than the poor. The precise genetic effect of this may be hard to predict, but it undoubtedly has effects. [Italics added.]
>
> It is obvious that we have been dealing with highly controversial matters, matters which one ordinarily thinks have no place in a textbook of biology. We do not propose to discuss them further, except to stress once more the point that it is not possible to avoid eugenic action; that every time we support a charity, endow a research institute, or promulgate a new taxation scheme, our actions, whether good or bad, have eugenic consequences, however unconscious we may be of them. We cannot live in the world without acting on it.[12]

From here it was only a short downhill run to the Population Bomb hysteria of the 1960's, and the pseudo-environmental movement which blamed the proliferation of the poor, who owned neither cars nor factories, for pollutants delivered into the atmosphere by the suicidally high (and economically wasteful) ratio of motor vehicles to people[13] and by the chimneys of the nation's industrial plants. From here, the revival of the IQ test score mystique, and compulsory sterilization on more massive scales, and *de jure* as well as *de facto* infanticide and geronticide were only hours away.

What should have been obvious with the publication and mass reception of harbingers such as *Human Breeding and Survival* in 1945–47, and Vogt's *Road to Survival* (with a glowing foreword by the adviser of many Presidents, Bernard M. Baruch) in 1948 (see page 378), was neatly underscored by the 1949 release of Hardin's *Biology*. Scientific racism had not died at all. It had merely gone into hibernation for the duration of World War II—the war with the fascist nations that had made scientific racism integral to their philosophy and genocidal practices.

AIR POLLUTION AND MIND POLLUTION

The biological dangers presented to man and his environment by industrial and automotive pollution were legitimate concerns of serious biologists that were converted into powerful propaganda weapons of the new scientific racism of the post-Nazi world. Few people took the problems of natural conservation and preventable environmental pollution more seriously than did the president of the Bronx Zoo, the son of Henry Fairfield Osborn, who wrote under the name of Fairfield Osborn. Unlike his father, Fairfield Osborn was neither a reactionary nor a racist.

In 1948, when his subsequently popular little book, *Our Plundered Planet,* was published, Fairfield Osborn made it plain that he had long since rejected the old scientific racism of his father. Whereas his father, and the older Osborn's great friends Madison Grant and Charles B. Davenport, had written reams of pseudoscientific rubbish about biological impossibilities such as "pure" inferior and superior races of mankind, the 1948 president of the Bronx Zoo wrote: "The saying 'We are all brothers under the skin' has a basis in scientific fact. . . . *The antipathies of nations and races, the cults of 'superior' and 'inferior' races, cannot be founded on biology"* (italics added).[14]

Fairfield Osborn was sixty-one when he wrote these lines, and the book in which they appeared was offered as the statement of a lifetime that had witnessed vast and dramatic changes in the world. The author had lived through two world wars, the second one a war in which his own nation had been pitted against the protagonists of the racial and eugenic views of his own father, and in which these racist dicta played an important role.

Because acts that disrupt the ecosystem affect all people, and because "countless thousands of landowners have in this very way" brought bankruptcy to themselves and environmental disaster to the nation, Osborn asked what to his father would have been a forbidden question: "How equitable are our present moral codes?"

His answer followed in the next sentence: "There is nothing revolutionary in the concept that renewable resources are the property of all the people, and, therefore, that land use must be coordinated into an over-all plan." Such planning, Osborn continued, had already begun in the unified program of the nine southern states of the Tennessee Valley Authority. Of the still controversial TVA program, Osborn wrote that "ably administered, it has, within the span of little more than a decade, justified itself not only as a social experiment but as an effort to harmonize human needs with the processes of nature."[15]

These were the thoughtful words of a man of goodwill. Unfortunately, they were also the words of a man who was too modest not to seek the advice of "experts" he felt had more hard knowledge of things than he did—and this modest son of an immodest father had the hierarchical and economic resources to obtain the advice and assistance of many people he considered to be experts.

One of these people happened to be Guy Irving Burch, co-author of *Human Breeding and Survival.* Another was William Vogt, chief of the Conservation Section of the Pan American Union. The result, in terms of what such advice did to Fairfield Osborn's book on conservation and the need for systematic planning programs to preserve our natural resources, can be compared only to the success won by clever corporation lawyers in their efforts to convert the Sherman Anti-Trust Act of 1890 into a legal weapon against American trade unions after 1900.

The gentle and socially concerned Osborn II was gulled into repeating Burch's (and Hitler's) *Lebensraum* theory of modern war being a product of "population pressures." And into the writing of Burch-like lines such as: "the problem of the pressure of increasing populations—perhaps the greatest problem facing humanity today—*cannot be solved in a way that is consistent with the ideals of humanity*" (italics added).[16]

His father went back to Galton, but Fairfield Osborn was propelled back even further into history. "Shades of Dr. Malthus!" wrote Fairfield Osborn in 1948. "He was not so far wrong when he postulated that the increase in population tends to exceed the ability of the earth to support it."[17] The still continuing Agricultural Revolution that started long before Malthus was born had destroyed all realistic grounds for this Malthusian pseudolaw, but Fairfield Osborn swallowed it whole.

He went on to note that the "gloomy doctor did not foresee" several developments in 1798, among them the Industrial Revolution when it spread to agriculture, the opening of the vast New World agricultural resources known as the United States and Canada, and, Osborn wrote, "above all [Malthus] did not envision *the invention of the internal combustion engine* which has so incredibly accelerated the capacity to exploit the earth's resources of forests and croplands. This invention has brought its innumerable benefits and *wreaked its irreparable damage.*" (Italics added.) But, despite what Malthus missed knowing in 1798, Osborn found himself sharing Malthus' gloom. And in the closing paragraph of *Our Plundered Planet,* he wound up declaring that "the tide of the earth's population is rising, the reservoir of the earth's living resources is falling. . . . Man must recognize the necessity of cooperating with nature. . . . The time for defiance is at an end."[18]

The advice and the "factual" data proffered him by advisers such as Burch and Vogt turned Fairfield Osborn's well-meant book into the first major vehicle of the new or postwar scientific racism. In linking the socially constructive causes of conservation of natural resources and the maintenance of a viable environment to the birth rates of the poor of the world and the nation, Burch and Vogt constructed a high-powered delivery system for their socially regressive cause of the obliteration, by sterilization and starvation, of the world's "alien or Negro stock" who happened to be poor.

Over three million copies of *Our Plundered Planet* were read by Americans as well-meaning as its author. When they finished reading this book, there were to be newer books by Vogt and others to take them further along the road to generally unwitting advocacy of that act of mass massacre which had not even had a proper name before the war. That ancient act now had a new name. The name was—genocide.

THE SCOURGE OF THE UNTRAMMELED COPULATORS

Vogt's *Road to Survival,*[19] published almost immediately after *Our Plundered Planet,* was an instant best seller, the selection of major book clubs, and was hailed as a masterpiece by literary critics from coast to coast. Like the Osborn book, Vogt's book offered as scientific fact the Malthusian myth of a planet that lacked the natural resources required to feed its human inhabitants. Like the Osborn book, the Vogt book also paid tribute to Guy Irving Burch, who had been so "extraordinarily helpful with advice, bibliographic suggestions, and critical discussion."

The author's love of nature was particularly evident in those passages that dealt with the birds, wildflowers, and skunk cabbages. When it came to people, Vogt sang a different tune. "It is certain that," he wrote (p. 47), "for all practical purposes, large areas of the earth now occupied by backward populations will have to be written off the credit side of the ledger." Vogt explained, in vivid prose, just how this could be done. The remedy was simple. It was called death.

Of Chile, for example, Vogt wrote (p. 186) that "one of the greatest national assets of Chile, perhaps the greatest asset, is its high death rate." Of China, an ocean away from Chile, Vogt said (pp. 214–15) that "the greatest tragedy that China could suffer, at the present time, would be a reduction in her death rate." Therefore the United Nations "should not ship food to keep alive ten million Indians and Chinese this year, so that fifty million may die five years hence" (pp. 281–82).

Since "overpopulation," in Vogt's view, constituted the chief menace to human survival, and the leading cause of all wars and pogroms, it was the bounden duty of the world's greatest nation to show the rest of the world how to eliminate this peril. Here his native land let Vogt down, because "we have not been willing to seek this remedy in our own continental slum areas such as South Boston [the bastion of the Irish Catholic poor], nor in Puerto Rico, nor have we been willing to advocate it in international organizations . . ." (p. 218).

In a famous passage titled "The Dangerous Doctor," Vogt minced few words in putting the blame for all human misery squarely on the shoulders of the physicians and the sanitary reformers and other serious environmentalists:

> The modern medical profession, still framing its ethics on the dubious statements of an ignorant man [Hippocrates] who lived more than two thousand years ago continues to believe it has a duty to keep alive as many people as possible. In many parts of the world doctors apply

their intelligence to one aspect of man's welfare—survival—and deny their moral right to apply it to the problem as a whole. Through medical care and improved sanitation they are responsible for more millions living more years in increasing misery. Their refusal to consider their responsibility in these matters does not seem to them to compromise their intellectual integrity. . . . They set the stage for disaster; then, like Pilate, they wash their hands of the consequences.[20]

Nothing in this world seemed to bother Vogt more than childbirth among the poor and the lowly. "Gresham's Law applies to labor as it does to money: cheap labor tends to thrive at the expense of, and drive out, higher-priced labor. Why the United States . . . should subsidize the unchecked spawning of India, China and other countries by purchasing their goods is difficult to see" (p. 77).

When an Indian scientist made the suggestion that the world still contained in North and South America, Australia, and other nations millions of square miles of wholly or partially unpeopled vast open spaces where hard-working people of so-called overpopulated lands could immigrate and grow food and fibers for the rest of the world, Vogt's response (p. 228) was a livid one: "In other words, Australia, Brazil, the United States and Canada should open their doors to Moslems, Sikhs, Hindus (and their sacred cows) to reduce the pressure caused by *untrammeled copulation*. Our living standard must be dragged down, to raise that of the backward billion of Asia." (Italics added.)

Like Burch, Vogt kept harping on population pressure as a major cause of modern war. And of all the overpopulated nations that menaced the world, few were as dangerous to world peace as the nation he called the Outsize Bear. Russia was not only "certainly overpopulated," but its "mounting population pressure" made Russia into "the major threat in Asia." A threat made more acute because Russia "has deliberately embarked upon a planned program of population expansion." This military threat from Russia was matched by another specter that haunted William Vogt: "A heavily industrialized India, backed up by such population pressure, would be a danger to the entire world."[21]

For industrialization was another of his phobias. To Vogt, the mechanization of agriculture—which had, early in the contiguous histories of the Agricultural and Industrial revolutions, helped sweep the Malthusian predictions into the dustbin of history while the gloomy dominie manqué still lived—was at best a dubious boon in that it caused more problems than it solved, and led to Too Many Cities. Vogt sneered at "Winston Churchill, that magnificent anachronism," who, in "complete ignorance of the findings of Empire scientists, subscribed to the fatuous American statement: 'There is enough for all. The earth is a generous mother; she will provide a plentiful abundance for all her children if they will but cultivate her soil in justice and peace.' "

Vogt was not a man to suffer in silence any such heresies about the earth's natural resources and the value of industrialization. He was well aware, of course, that by 1930 the world's population rates had been stabilized into their present patterns, in which the birth rates of the nonpoor and industrialized minority of the world's nations are exactly half as large as those of the majority of the world's nations, which are both nonindustrialized and poor. "As a class," notes Kingsley Davis, director of the University of California Center for International Population and Urban Research, "the nonindustrial nations since 1930 have been growing in population twice as fast as the industrial ones."[22] Vogt, however, was not a man who could be easily dissuaded by such historical and biological realities.

The Road to Survival predicted (p. 72) that many famines would take place between 1948 and 1978, notably in Great Britain. For there, "the Socialist government, counting on 'economic' and 'political' prestidigitation that hung in the air without any base on the land, promises to lift the United Kingdom by its own bootstraps, without recognizing that the bootstraps had been worn to the breaking point. Unless we are willing to place fifty million British feet beneath our dining-room table, we may well see the famine once more stalking the streets of London. And hand in hand with famine will walk the shade of that clear-sighted English clergyman, Thomas Robert Malthus."

Other nations in which Vogt predicted imminent famine included Japan and Germany, both of which "outbred the carrying capacity of their own land," and hence were on the road to extinction. Japan was a particular offender, for "with the introduction of modern industrial methods and modern sanitation, the Japanese population tripled in about 75 years, and it is clear that the Japanese cannot possibly feed themselves at a decent living standard."

In a memorable chapter called "The Kallikaks of the Land," Vogt proved himself to be a well-educated product of the old scientific racism of Davenport and Goddard (p. 145). "The question of how to solve our forest problems opens up a wide, grim vista of ecological incompetence. The Jukes and the Kallikaks—at least those who are obtrusively incompetent—we support as public charges. We do the same with the senile, the incurables, the insane, the paupers, and those who might be called the ecological incompetents, such as the subsidized stockmen and the sheepherders. . . ."

Nor was Vogt blind to what had to be done about the Jukes and the Kallikaks who were not sheepherders, but who did pose dysgenic dangers to America. "There is more than a little merit in the suggestion . . . made . . . by H. L. Mencken, of 'sterilization bonuses' " for the genetically inferior. As Vogt observed in the closing pages (pp. 282–83) of *The Road to Survival,* "Since such a bonus would appeal primarily to the world's shiftless, it would probably have a favorable selective influence." Like Madison Grant, Vogt was a keen student of evolution and natural selection. "From the point of view of society, it would certainly be preferable to pay permanently indigent individuals, many of whom would be physically and psychologically marginal, $50 or $100 rather than support their hordes of offspring that, by both genetic and social inheritance, would tend to perpetuate the fecklessness."

The chief message of the book was that we were faced with what Vogt described as a Hobson's choice. We were being blackmailed by the countries hardest hit by the war, and "we have literally no choice but to accede to this blackmail. . . . Unless we pay it, we shall leave a vacuum that would suck in the police state from the east; and there would be no more of this self-determination nonsense." We would be "stupid, indeed, not to draw the fuse" of another European war if we could.

But, he warned, "we shall be even more stupid if we do not recognize that the overpopulation that has contributed so much to past European disorders is a continuing and growing threat." Our medical and food and financial aid were exacerbating this biological time bomb by reducing European death rates. "Birth rates are falling, but not fast enough to be much help. . . . Anything we do to *fortify the stench*—to increase the population —is a disservice to both Europe and to ourselves" (italics added).

Vogt then proceeded (pp. 210–11) to tell his countrymen how to cope with this mess. Blackmail had to be countered with super-blackmail. "Any aid we give should be made contingent on national programs leading toward population stabilization through *voluntary* [Vogt's italics] action of the people. We should insist on freedom of contraception as we insist on freedom of the press; it is just as important. And as we pour in hundreds of millions of American taxpayers' dollars we should make certain that substantial proportions make available educational and functional contraceptive material."

If I have dealt at such length with *The Road to Survival,* and quoted so lavishly from its most typical passages, it was not for amusement in terms of today's hindsights, let alone the non-happening of the British and German and Chinese and Japanese famines Vogt so didactically predicted would mark our era. It is, rather, because for the next three decades, every argument, every concept, every recommendation made in *The Road to Survival* would become integral to the conventional wisdom of the post-Hiroshima generation of educated Americans—as the eugenical concepts of the old scientific racism of Galton, Davenport, Goddard, Grant, Terman, and Thorndike were integral to the concepts, dogmas, and value systems of most educated Americans prior to Pearl Harbor. The neo-Malthusianisms of Vogt, from his updating of Malthus' denunciations of doctors who healed the sick and discovered how to achieve "the total extirpation of particular disorders," to his twentieth-century version of the Malthusian food and population growth hoax, would for decades to come be repeated, and restated, and incorporated again and again into streams of books, articles, television commentaries, speeches, propaganda tracts, posters, and even lapel buttons.

Out of *The Road to Survival* and its many literary and intellectual descendants were to come the Zero Population Growth and other popular movements, as well as some of the most pervasive slogans of modern America.

A whole generation of impressionable young people were to come under the influence of Vogt and his book during their most formative years. One of them was a freshman at the University of Pennsylvania, Paul R. Ehrlich, who now traces his interest in what he helped make known as the American Population Explosion to his reading of Vogt's book during his freshman year.[23]

THE CAMPAIGN TO CHECK THE POPULATION EXPLOSION

One of the many influential older Americans inspired by the Malthusian teachings of Vogt was the Dixie Cup king, Hugh Moore. As *The New York Times* observed in its obituary for Mr. Moore in 1972, "the latter part of Mr. Moore's life . . . was dominated by his concern with the danger of overpopulation. His attention was drawn to it by a book, *The Road to Survival,* by Dr. William Vogt."[24]

Lawrence Lader, in his highly authorized biography of Moore, *Breeding Ourselves to Death,* echoed this appraisal: " 'Moore was the first business-man willing to stand up and be counted on this issue, the first to stick his neck out,' commented Dr. William Vogt, former national director of the Planned Parenthood Federation of America and author of the influential book *The Road to Survival,* which first stirred Moore's interest in popu-lation."[25]

Moore, whom Lader described as the "showman-salesman for popula-tion control," put his immense organizing and fund-raising talents—as well as considerable amounts of money—into selling the Malthusian tenets of Vogt's book to the people who make American foreign and domestic policies, the voters who elect them, and the teachers who teach their children.

Under such organizational banners as the Hugh Moore Fund and the Campaign to Check the Population Explosion, the Moore crusade for some years took one- and two-page advertisements in *The New York Times,* the *Washington Post,* the *Washington Star, Fortune,* the *Wall Street Journal, Harper's, Saturday Review,* and *Time.* These ads warned literate Americans that all of the problems that beset modern man—from war and crime to hunger and environmental pollution—were caused by the explosively high birth rates (of soldiers, or perhaps of foreign-policy makers?), of criminals, hungry people, and babies.

In the classic tradition of big-business tycoons who take over faltering companies and turn them into profit makers, and who merge their own new companies with successful older ones competing for the same market, Moore also went to work on the organizational bases of the nation's population and sterilization movements. Thus, in 1954, when Guy Irving Burch, whom Lader describes as "a prominent demographer," told Moore that the Popu-lation Reference Bureau he had headed since 1929 had gone broke, Moore reorganized it. He advanced seed money "through which it could embark upon public fund-raising campaigns," and "by 1966, Moore had helped raise the Bureau's annual budget to $400,000."[26]

A true disciple of Vogt's, Moore looked to sexual sterilization as the ultimate solution to population problems that could not be resolved by less traumatic methods. When Moore took over the presidency of the nation's leading sterilization society in 1964, Lader writes, the salesman-showman of population control insisted that it change its name from the prissy Human Betterment Association (née Birthright, Inc.) to the more meaningful Association for Voluntary Sterilization, Inc. Things began to happen in a big way. Moore "raised money to move the office to a midtown New York suite just off Fifth Avenue, and employed an experienced executive director and staff. Since the principal obstacle to sterilization was the lack of public understanding of its legality and medical acceptance, the Association appointed Dr. H. Curtis Wood, Jr., a Philadelphia obstetrician, as medical director. Crisscrossing the country on speaking tours, Wood made as many as 110 lectures and radio and television appearances a year."[27] (See pages 406–07.)

In 1961, Moore brought about the merger of Margaret Sanger's venerable Planned Parenthood Federation with his own World Population Emergency Campaign. Margaret Sanger's last public appearance was at the Waldorf-Astoria dinner held by the new Planned Parenthood–World Population society in May 1961.

Seven decades into the century in which the automobile, the major single source of American environmental pollution, had proliferated from 8,000 registered motor vehicles in 1900 to over 111 million by the year Moore died—motor vehicle registrations having climbed at the cataclysmic rate of 1,500 percent per decade—the costly full-page advertisements of the Moore population crusade blamed environmental pollution on the American babies whose live births, during the same years, had climbed only 13 percent, and whose live birth rates had declined by 53 percent since 1900. The eminently preventable environmental pollution caused by industry and by automotive transportation was not the fault of industry and of the government's failure to build alternative mass transit systems, the Moore committee advertisements claimed, but of the human victims of this needless pollution themselves—starting with every newly conceived child.

Among the millions of educated Americans who took such patently absurd propaganda seriously as legitimate demographic and sociological facts were the bevy of Nobel laureates in science, the former scientific adviser to President Kennedy, the Secretary of Defense in the Kennedy administration, and the prominent clergymen, corporation chiefs, and elected public officials who signed one or more of the campaign advertisements of the Moore committees. They included individuals of such diverse sociopolitical convictions as Bruce Barton, Dr. Detlev W. Bronk, Van Wyck Brooks, Ambassador Ellsworth Bunker, General William H. Draper, Rabbi Maurice Eisendrath, Rev. Harry Emerson Fosdick, Senator Ernest Gruening, Mrs. Clare Boothe Luce, Archibald MacLeish, Ashley Montagu, Hermann J. Muller, Reinhold Niebuhr, Fairfield Osborn, Linus Pauling, Eddie Rickenbacker, M. Lincoln Schuster, Eleanor Roosevelt, and Governor William H. Vanderbilt. Philosophically, the only thing most of the hundreds of distinguished signers of the Moore crusade manifestos probably had in common was a shared misconception of the controlling mechanisms—well known to the world's professional demographers and epidemiologists—of high and low human birth rates. (See page 404.)

The advertisements and other propaganda of the Moore crusade did

Dear President-Elect Nixon:
The underlying problem facing your administration will not be war, riots or crime, but the population bomb

In the four year term of office to which you have been elected there will be *ten million more* Americans—most of them living in our already overcrowded cities.

And there will be *three hundred million more* people in the world at the present rate of increase—most of them without enough to eat.

Fourteen million people will die of starvation during your term of office unless the present death rate of ten thousand a day is reduced. (America cannot feed the world, as we have found after shipping $15 billion worth of food abroad in recent years.)

There were 2½ billion people in the world in 1953. Today only 15 years later there are *one billion more!* This basic problem, Mr. President, bears directly or indirectly on most of the problems you will have to deal with during your Administration.

To check the population explosion —both in this country and abroad— you must seek appropriations many times larger than those heretofore allocated for birth control by our Government. Surely a nation which approves $15½ billion to develop a new military airplane can afford to devote a comparable sum to the solution of the most urgent problem of the human race?

For there is little doubt that unless population is brought under control *at an early date* the resulting social tensions and misery will inevitably lead to chaos and strife—to revolutions and wars which may make our present experience in Vietnam minor by comparison.

Nothing less than the survival of civilization is in the balance.

CAMPAIGN TO CHECK THE POPULATION EXPLOSION
EMERSON FOOTE, CHAIRMAN

This advertisement appeared in full page space in The New York Times and The Washington Post.

Dear President Nixon:
We can't lick the environment problem without considering this little fellow.

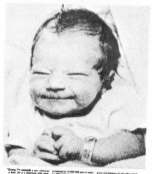

"Every 1% expends a new American or born. He is a deserving little thing, but he appears to scream loudly in a voice that can be heard for seventy years. He is producing by 24,000,000 tons of water... 17,000 gallons of gasoline, 10,150 pounds of meat, 28,000 pounds of milk and wheat... and 9,000 pounds of wheat, and great storehouses of all other lands, stocks, and tobacco. These are the facts, some demands of all the American child economy." Robert and Leona Train Rienow

We applaud your message to Congress, Mr. President, proposing a vast program to improve our environment. You said quite correctly that "While our population grows each one of us keeps using more of the earth's resources." You foresaw the need of spending more than 10 billion dollars to purify our lakes and rivers, and additional billions to cleanse the air in this great country of ours.

But, Mr. President, this will be a losing battle unless we check our rapidly growing population, which is an underlying cause of the pollution of our environment.

Let's take a look at the growth of population in the United States:

1. We had 100 million people as recently as 1920 and never worried about the pollution of our environment.
2. However, we have added another 100 million in the brief period since 1920.

3. And we shall add *another 100 million* at the rate thirty years or so at the present rate, bringing the total number up to 300 million people.

Today, with only 200 million, the water we drink may be contaminated and the air not fit to breathe. Noxious deaths us. Our cities are packed with youngsters—thousands of them idle and victims of drug addiction. And millions more will pour into our streets in the years immediately ahead.

Americans will be jammed together in an anthill society," according to your Secretary of Commerce, Mr. Maurice Stans, "unless government and business join in a coherent national growth policy."

Population Control Is Essential

Last year, Mr. President, you made a ringing pledge to provide leadership in the population field. You said, "When future generations evaluate our time... one of the most important factors... will be the way we respond to population growth. Let us act in such a way that those who come after us... can do so with pride in the planet... with gratitude to those who lived on it in the past and with confidence in its future."

Notwithstanding your strong words Congress has taken the best part of a year to authorize the Commission on Population Growth which you recommended. And we understand that the Family Assistance Program now under consideration would subsidize children at the rate of $100 for each child, up to ten in many cases!

We submit that a government which plans to spend untold billions to improve the environment for our present large population should not encourage large families. Indeed, there should be made available at least $1 billion and make known practical methods of humanity to control its numbers.

Proposed Measures To Control Population

These measures should include, among others, (1) improved methods of contraception (2) basic research in the physiology of reproduction, and (3) the utilization of communication techniques—radio, television and mass media—to break the barriers of illiteracy, ignorance and inertia.

Population control must be a part of a world-looking program to improve the environment. It should be among the very first to take, since the surging growth of population is basically responsible for the pollution of our environment.

THE POPULATION BOMB THREATENS THE PEACE OF THE WORLD

SO WHAT ARE WE DOING ABOUT IT?

Fifteen years ago there were 2.5 billion people on earth. Today there are 3.5 billion—and newcomers are arriving on the scene at the net rate of more than one million a week! In another fifteen short years there will be at least 4.5 billion people on this small planet of ours. Most of them hungry. And make no mistake about it, America cannot long remain an island of prosperity in a sea of poverty and hunger.

If corrective measures to check this human flood are not taken right here and now the resulting world-wide misery, strife, revolutions and wars will make our experience in Viet Nam appear minor by comparison.

The population crisis is the greatest problem humanity faces. And the National Academy of Sciences has said that the Population Bomb can be successfully attacked by developing new methods of family regulation and implementing programs of family planning widely and rapidly throughout the world. The fast-accompanying chart reflects the scant amount of attention the population problem is currently receiving from our Government.

This is your problem and you can do something about it. Tear out this ad and send it to anyone in Washington you think might be helpful. Urge the Government to initiate a crash program for population stabilization. And write to do two things: (1) Measures the Government can take to implement such a program. (2) Aid channel things you can do to help.

We can't afford to wait very much longer. Every day lost will only compound the population problem.

The time to act is *now*.

CAMPAIGN TO CHECK THE POPULATION EXPLOSION
60 EAST 42nd STREET, NEW YORK, NEW YORK 10017
EMERSON FOOTE, CHAIRMAN

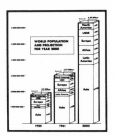

Malthus Lives Again, or the Brainwashing of America's Literates

These examples are typical of the many full-page scare advertisements the Campaign to Check the Population Explosion and the Hugh Moore Fund ran in *The New York Times,* the *Washington Post,* the *Wall Street Journal, Time, Fortune,* and other publications for many years prior to the death of Vogt disciple Hugh Moore in 1972. These advertisements successfully marketed the population myths of Malthus and his followers from Galton to Vogt. Some of the dogmas of scientific racism broadcast in these advertisements included the myths that:

1. Population is exploding independently of such well-defined demographic determinants as infant, child, maternal, and general death rates; family income levels; malnutrition due to both substandard wages and the maldistribution of food.
2. The population bomb is the single most important problem facing the world today.
3. The Roman Catholic Church is a major menace to civilization-as-we-know-it. Here the Pope is blamed for the fact that the world's poor people do not earn enough money to pay for the planet's abundantly available food.
4. Muggings and other violent crimes are among the inevitable end products of the population bomb.
5. Environmental degradation is not caused by the abuses of technology by large industries and the greater society, but rather by the most helpless and totally innocent victims of technological pollution: America's newborn babies.

not tell educated Americans anything they had not learned long ago in college and graduate school—and in high schools and elementary schools from teachers trained in America's colleges. Malthusianism and related systems of scientific racism have been integral elements of American higher education for all of the years of this century; the Moore crusade merely reinforced the Malthusian preconceptions of its educated endorsers and activists.

As in the worlds of science, business, government, and religion, the Moore (out of Vogt, out of Malthus) dogmas and simplisms swept all before them on the American campuses. By 1970, thousands of earnest and idealistic Americans of all ages were swapping their "End the Killing in Vietnam" buttons for more modish buttons bearing the words "People Pollute."

The greatest peace movement ever created in America, a coalition of young people powerful enough to have driven Lyndon Johnson out of public life in 1968, now committed hara-kiri on national network television in the name of the Moore revival of Malthusianism. In this, they did not lack the enthusiastic and expensive assistance of Moore and his hired propagandists. As Lader, an erstwhile worker for Moore's committee, tells us:

> The Nationwide Environmental Teach-In [Earth Day], planned for April 22, 1970, seemed to offer possibilities. But, as Moore read the early plans, the Teach-In appeared to lack any strong population component. Population control was mentioned, but seldom emphasized, and almost never given stature as *the basic factor necessary to reverse the decline of the environment* [italics added]. He set out to help change this.
>
> First, a third of a million leaflets, folders, and pamphlets (including a new pictorial edition of the venerable *Population Bomb*) were produced for campus and community distribution. Next, three efforts stressed the intimate relation between overpopulation and a degraded environment. One was the free distribution to 300-odd college radio stations of a taped radio program featuring Paul Ehrlich and David Brower. The second was provision, for reproduction free by college newspapers, of a score of editorial cartoons highlighting the population crisis. The third was a contest, conducted on over 200 campuses, that awarded prizes for slogans relating environmental problems to "popullution."[28]

With such helpful friends, the student social protest movement had little need for enemies. What the might of the Johnson and Nixon administrations had been unable to do with all of their propaganda agencies and secret police powers, the new Malthusian population-bombers did with taped radio programs, syndicated editorial cartoons, brainwashing prize contests, and some 333,000 folders, leaflets, and pamphlets.

In a holy fervor of opportunism and righteousness, let alone profound relief to find the angry campus generation mesmerizing itself into abandoning the cause of peace in Indochina, President Nixon and various of his high officials joined in the orgies of well-televised oratory and *Kraft durch Freude* fiestas that marked Earth Day, 1970: the day the young people switched goals from the quest for peace to the crusade for Zero Population Growth and Clean Cars. From that point forward, Nixon embraced the new Malthusian "environmentalism" with public passion.

Future historians will also note that, four days after Earth Day, President Nixon—whose air and ground forces had secretly been grinding Cambodia to shambles for two years—now ordered the overt invasion of Cambodia. Within a few days, the same President was to denounce as "bums" those unappreciative young people who, in diminishing numbers, persisted in reviving the student opposition to the war in Indochina. A day or so after the President so characterized the anti-war students, the Ohio National Guard killed four young people on the campus at Kent State University during a demonstration protesting the widening of the war to Cambodia and Laos.

HOW TO PRESERVE ENVIRONMENTAL DEGRADATION AND ENRICH THE MAJOR POLLUTERS

The point at which the classic goals and functions of the environmental movement founded by English sanitary and social reformers early in the Industrial Revolution parted company with the new pseudo-environmentalist People Pollute movement of the new scientific racism was underscored by Vogt's previously quoted attack on "The Dangerous Doctor": "Through medical care and improved sanitation [the world's doctors] are responsible for more millions living more years in increasing misery."

The older environmental movement, which was not dead but merely overshadowed by the fashionable People Pollute movement, clung to the principles spelled out in the founding statement of the Health of Towns Association of England in 1844. These principles lived in the statement "The Effects on Man of Deterioration of the Environment," issued by the World Health Organization in 1973:

> The health and well-being of a population depend both on its degree of socioeconomic development and on the complex of physical, chemical, biological, and social factors that make up its environment.
>
> It has become clear in recent years that environmental degradation, if allowed to proceed unchecked, could result in serious and even irreversible damage to life on this planet. Poor sanitary conditions and the accompanying communicable diseases are the most important causes of morbidity and mortality in the developing [poor] countries, where the majority of the world's population live.[29]

To the new Malthusian "environmentalists," the causes of environmental degradation were much simpler: there were too many people on this planet, and it was people, and nothing else, that caused pollution. Such social factors as poverty, unsanitary conditions, and the abuse of technology, which doctors for two centuries have recognized as major causes of environmental pollution, were not even recognized by the new People Pollute movement. The only non-animate *secondary* causes of environmental degradation that they recognized—the damages caused to trees, flowers, wildlife, and trout streams by the fossil fuel combustion emissions of our over 130 million motor vehicles, our factories, and our electric power plants and by our chemical pesticides—were subject to equally simplified correction. All that was needed, they said, was to pass laws requiring the installation of anti-pollution devices that would make every car an environmentally Clean Car, as well as the passage of a few laws regulating the safe and unsafe amounts of pollution factory chimneys should be allowed to release, and banning pesticides that killed songbirds.

Like the original Law of Population of Malthus, the Gobineau Cult of the Nordic, and the eugenic myth of the Decline of American Intelligence, the simplistic dogmas of the new People Pollute movement addressed themselves to chimeras rather than realities. They also helped hide the real causes and biosocial effects of environmental degradation from many educated but scientifically naïve Americans. Finally, in the classic traditions of scientific racism, the snappy slogans of Zero Population Growth and other wings of the People Pollute movement succeeded in pinning the blame for environmental degradation on the backs of its primary victims—the poor, the near poor, and the lower middle classes, who are condemned, for socioeconomic reasons, to be born and to live out their lives in that urbanized 2 percent of our national territory where well over three quarters of all Americans now live.

It is in the cities and their sprawl of satellites where most of the causes of environmental pollution are both generated and suffered. However, as one reads the writings and listens to the speeches of the spokesmen and spokeswomen of the People Pollute environmentalism of the new scientific racism, it quickly becomes apparent that their concerns are not to make the environment biologically safe, let alone healthier, for the 80 to 90 percent of our population who are forced to live in our urbanized enclaves. They are, instead, passionately devoted to keeping the trees and the game birds, the primeval forests, the wildflowers and the mountain trout, and the *Lebensraum* of our lusher suburbs as green and as lovely as they ever were for that 10 to 20 percent of our population fortunate enough to be born and to live "out where the sidewalk ends." There is nothing wrong with such goals; they merely happen to be totally irrelevant to the environmental health problems and needs of our entire population.

Despite the simplisms of the People Pollute propaganda, there are at least five major causes of environmental degradation in this country. Two existed in the previous century; the other three are new to the twentieth century. The old causes of environmental degradation are: (1) poverty and (2) factory, refinery, and other industrial fossil fuel combustion emissions, noise, and olfactory pollutants. The three twentieth-century sources of environmental degradation are: (1) the fossil-fuel-burning motor vehicle; (2) the cigarette; and (3) the 30,000-plus new toxic chemicals that contaminate the air we breathe, the water we drink, the food we eat.[30]

Poverty and near poverty have been greatly reduced during the twentieth century. However, between one quarter and one third of our population still lives at, slightly above, or considerably below the official poverty level.

Preventable communicable and deficiency diseases—among the very poor and among many employed blue-collar and white-collar wage workers who cannot afford adequate medical care—are still major causes of crippling and even fatal environmental degradation.

The fossil fuel emissions of modern industry were originally proved to cause human cancer in 1775 by Sir Percivall Pott, surgeon at St. Bartholomew's Hospital in London, shortly after England was converted from being a wood-burning to a coal-burning nation.[31] The world's doctors have been increasingly aware of the carcinogenic and noncancerous but equally fatal cardiopulmonary effects of coal and other fossil fuel combustion ever since. Over the past two hundred years, many of them have striven, with generally limited and usually no success, to convince big industry of the social wisdom and personal morality of devising, and/or utilizing already devised, technological systems for making the use of fossil fuels for energy as safe for the human environment as it has been profitable for its users.

The dense clouds of coal and later oil smoke that after the American Civil War started to turn the air of our industrializing cities into cesspools of lethal pollutants were, long before 1900, directly responsible for the first waves of upper-middle-class flights from the older cities to the residences and "gentleman's farms" they established in or near agricultural villages located within an hour or two by steam railway from their profitable offices, department stores, banks, and pollution-generating factories in the cities. These old rural villages became the new suburban towns. Later, starting before World War I, as inexpensive electric commuter trains and interurban trolleys were developed, these quiet and nonpolluting electric trains helped make considerably more modest suburban living feasible for thousands of lower-middle-class and salaried employees equally anxious to spare their children from the malignant slings and the toxic arrows of the degraded urban environment.[32]

The cold statistical data on two of the most common twentieth-century causes of environmental degradation tell their own stories:

Motor vehicles: In 1900, when we had a human population of 76,094,-000, we had exactly 8,000 registered motor vehicles—or a ratio of one motor vehicle to every 9,500 people. In 1974, when we had 211.9 million people, we had 131 million motor vehicles—or a ratio of one motor vehicle to every 1.6 people.

Even if by some miracle, as yet quite beyond the competence of our chemical, physical, and biological sciences, the expensive catalytic converters and other "anti-pollution" tail-pipe devices could ever give us a really "clean" car—that is, a car whose gasoline combustion fumes were truly harmless to our hearts and our lungs and to the green plants of earth and oceans that are the sources of the oxygen needed to sustain human life—this technological breakthrough would still not reduce by one iota the scores of other factors unleashed by this motor vehicle explosion that also degrade the environment.

"More than 1.8 million people have been killed in the United States in motor vehicle accidents during the last 70 years," the New York City Environmental Administration observed in 1970.[33] By 1972, our total of 24,-

850,000 accidents for the year had resulted in 56,600 deaths and 5,190,000 injuries, causing economic losses from traffic accidents to soar to $19,066,-000,000. According to the Insurance Information Institute, given the present density of motor vehicles, by 1980 automotive traffic accidents would kill 86,000 people and injure 6,460,000 annually—at economic costs far exceeding the $19 billion of 1972.[34]

Five years later, the U.S. Department of Transportation reported that "because vehicle crashes kill and injure the young as well as the old, they are equalled only by heart diseases as the major single factor in lost man-years of productivity. And among persons under 35, highway crashes are the major single cause of death."[35]

Neither better driver education, nor stronger car bumpers, nor stricter traffic laws would significantly reduce these automotive death and accident totals. The sole underlying cause of this *needless* carnage is, of course, the sheer density of motor vehicles in America today.

The maximization of profits has had more than a little to do with the maximization of fossil fuel combustion pollutants produced by American motor vehicles. Where in 1925 the average American automobile engine produced 55 horsepower, by 1946 it was up to 100 horsepower, and it reached an average of 250 horsepower by 1968.[36] To see how this more than fourfold increase in *needless* horsepower added to the net amount of the nation's equally needless fossil fuel pollution, consider the facts that motor vehicle registration climbed 153 percent from 49,300,000 in 1950 to 125,-157,000 in 1973, and the consumption of motor vehicle fuel climbed from 40,280,000 gallons in 1950 to 105,061,000 gallons in 1972—a revealing jump of 160 percent.[37]

Polluted air and drinking water, and motor vehicle accidents, are far from being the only serious fallouts of our present biologically intolerable density of one motor vehicle to every 1.6 people. One of the cruelest effects of the present forced reliance upon individually owned and operated cars to travel to and from work, shopping, medical centers, and other essential locations is what it does to the purchasing power of the average American family, whose median income as of 1974 was $12,836 for whites and $8,265 for Negroes and other non-whites.[38] These figures mean, simply, that the average family that owns and drives a car daily can "afford" the excruciatingly high costs of buying, insuring, fueling, repairing, and operating this costly item of technology only at the expense of commodities and services that happen to be vitally necessary for family health and welfare. These family necessities range from proper food and housing and adequate preventive and therapeutic medical and dental care to proper clothing, books and other educational materials, as well as life insurance and savings for emergencies and old age.

The sudden abandonment of cheap, electrified, and infinitely safer trolleys, trolley-buses, and commuter trains between 1932—when General Motors Corporation, during the Great Depression, "became involved in the operation of bus and rail passenger services"—and 1950 forced millions of Americans previously served by mass transit systems to buy their own cars in order to get to and from work daily. By 1949, more than 100 electric surface rail systems had been abandoned "in 45 cities, including New York, Philadelphia, Baltimore, St. Louis, Oakland, Salt Lake City, and Los Angeles."[39]

As Bradford C. Snell, special counsel to the U.S. Senate Subcommittee on Antitrust and Monopoly, observed in a 1974 report for that committee: "The economics are obvious: one bus can eliminate 35 automobiles; one street-car, subway or rail transit vehicle can supplant 50 passenger cars; one train can displace 1000 cars or a fleet of 150 cargo-laden trucks."[40]

The Senate document further noted:

> By 1949, a Chicago Federal jury convicted General Motors of having criminally conspired with Standard Oil of California, Firestone Tire and others to replace electric transportation with gas- or diesel-powered buses and to monopolize the sale of buses and related products to local transportation companies throughout the country. The courts imposed a sanction of $5,000 on General Motors. In addition, the jury convicted H. C. Grossman, the man who was then treasurer of General Motors. Grossman had played a key role in the motorization campaigns and had served as a director of Pacific Electric Railway when that company undertook the dismantlement of the $100 million Pacific Electric system. The court fined Grossman the magnanimous sum of $1.[41]

Of the buses that replaced the electric trains and trolleys, the same Senate report stated: "Due to their high cost of operation and slow speed on congested streets, however, these buses ultimately contributed to the collapse of several hundred public transit systems and to the diversion of hundreds of thousands of patrons to automobiles."[42]

By misleading millions of well-meaning people about the vast array of environmental damages caused by motor vehicles—and by concentrating the hopes of a pollution-sickened people on the illusion of a Clean Car as a cure-all, rather than on the restoration of cheaper, quieter, safer, and less-polluting systems of mass rapid transportation—the People Pollute movement acts to protect the profits of the automobile manufacturers, and not the quality of the environment in which we live.

Cigarettes: In 1900, the annual consumption of cigarettes in the United States was two cigarettes—one ounce of tobacco—per person. By 1940, when heart disease, cancer, and strokes had become the three leading causes of death in this country, our annual cigarette consumption had climbed to 1,976 cigarettes—5.35 pounds of tobacco—per person.[43]

Starting in 1964, the U.S. Surgeon General's Office began issuing the historic studies called *The Health Consequences of Smoking.* These periodic studies have proven that, particularly in urban environments also contaminated with automobile and industrial fossil fuel pollution, cigarette smoking is a major cause of cancer and heart disease in all people and of reduced birth weight and other adverse conditions in the babies of mothers who smoked during pregnancy.[44]

Smoking is, of course, a world problem, and in its continuing efforts to help people cope with it the World Health Organization (WHO) appointed a WHO Expert Committee on Smoking and Its Effects on Health. The report[45] issued by this panel of American, European, and Asian physicians and scientists in 1975 opens with the statement:

> Epidemiological evidence from many countries implicates tobacco smoking as an important causative factor in lung cancer, chronic bronchitis and emphysema, ischaemic heart disease, and obstructive peripheral vascular disease. It also shows that smoking plays a part in the causation of cancer of the tongue, larynx, oesophagus, pancreas, and bladder; abortion, stillbirth, and neonatal death; and gastroduodenal ulcer.

The People Pollute movement treated the hard data on this widespread environmental agent of lung pollution as the old scientific racists had earlier treated the pre-World War I life-enhancing studies of hookworm and pellagra by two earlier U.S. Public Health Service scientists, Stiles and Goldberger. That is, they ignored the Surgeon General's Advisory Committee on Smoking and Health, as earlier the final report of the Pellagra Commission had, in 1917, ignored Goldberger's historic 1914 discovery concerning the cause of pellagra.

By 1973, the United States led the world in the consumption of these proven environmental lung pollutants:

Nations Where Cigarette Smoking Is Heavy
(1973; per capita consumption for all persons 15 years of age and older)

United States	3,812
Japan	3,270
United Kingdom	3,190
Italy	2,774
West Germany	2,624
Denmark	1,972
Sweden	1,680

From *The New York Times,* June 8, 1975, whose sources were the U.S. Department of Agriculture and the World Health Organization.

In 1975, Sweden, where the infant-mortality rates are lower and the life spans of all people are longer than those in the United States (which consumes almost two and a half times as many cigarettes per person as Sweden), felt that the risks of continued cigarette smoking were biologically too great. That country therefore organized a program to tax the cigarette trade out of Sweden, while at the same time launching serious efforts to stop smoking among parents, teachers, doctors, dentists, nurses, and other adults in contact with children. The Swedish government also prepared legislation that would bar smoking in all public places.

In America, the People Pollute zealots still do not acknowledge the existence of this major source of environmental lung pollution. Their attitudes toward the dangers of chemical pollution, and the pollution caused by the government's inexplicable failure to follow the path opened in 1956 by England with the passage of its Clean Air Act—under which British industry was forced to spend around a billion dollars in then available technological systems for trapping many of the worst pollutants in coal and oil combustion fumes[46]—have been equally cavalier. And when the Nixon administration set up elaborate new environmental protection bureaucracies to issue reams of self-serving publicity releases, measure environmental pollution, establish standards of safety for air and water pollutants—but *not* to seriously enforce the observance of any of these standards—the People Pollute crowd accepted Richard M. Nixon as a true environmentalist after their own hearts.

In 1968, the new People Pollute school of environmentalism received its Magna Carta in the form of a since endlessly anthologized essay, "The Tragedy of the Commons," by Garrett Hardin,[47] the quintessential eighteenth-century Malthusian who teaches biology in twentieth-century California. The subtitle of the Hardin essay read: "The population problem has no technical solution: it requires a fundamental extension in morality."

In this essay, Hardin compared modern society to a finite and geographically limited commons in which lived a race of recklessly breeding human beings. Hardin, citing Malthus as his ultimate authority, wrote that not only is our world finite but also that "a finite world can support only a finite population; therefore, population growth must eventually equal zero." The tragedy of freedom in a commons of herdsmen sharing one pasture is that "each man is locked into a system that compels him to increase his herd without limit—in a world that is limited. Ruin is the destination toward which all men rush, each pursuing his own best interest in a society that believes in the freedom of the commons. *Freedom in a commons brings ruin to all"* (italics added).

Then, extrapolating from the simple sheep-grazing commons to the infinitely complex structures of modern industrial societies, Hardin announced that "the tragedy of the commons reappears in the problems of pollution." For, to Hardin, as to Burch and Vogt, "the pollution problem is a consequence of population . . . as population becomes denser, the natural chemical and biological recycling processes become overloaded, calling for a redefinition of property rights."

However, the property rights Hardin had in mind were definitely not those of the automobile companies, the electric power plants, the steel mills, and other industrial producers of environmental pollution: "Analysis of the pollution problem as a function of population density uncovers a not generally recognized principle of morality, namely: *the morality of an act is a function of the state of the system at the time it is performed"* (Hardin's italics).

Since, in Hardin's eyes, it is not the density of motor vehicles and standing industrial producers of toxic chemical wastes and fossil fuel smoke particles, but the human "population density" that pollutes the environment, the very freedom to procreate human beings becomes socially and morally intolerable: ". . . to couple the concept of freedom to breed with the belief that everyone has a born equal right to the commons is to lock the world into a tragic course of action."

Garrett Hardin, father of four, had a better idea: "The only way we can preserve and nurture other and more precise freedoms is by relinquishing the freedom to breed, and that very soon. . . . Only so can we put an end to this aspect of the tragedy of the commons."

Thus spoke the new People Pollute Malthusian "environmentalism."

Less than a year later, in his presidential address, "Social Health," to the American Academy of Pediatrics, Dr. Hugh C. Thompson dealt with the problems of the human environment in the less metaphysical lexicon and philosophy of the classic environmental movement—the one that sprung from the etiological discovery of Sir Percivall Pott and the human concerns of the early-nineteenth-century sanitary, medical, educational, and social reformers. Dr. Thompson said, among other things:

> The ecology in which the manifestations of social ill health occur includes sub-standard living conditions and poverty, which are interrelated and singly or together play varying roles in each of the symptoms of social ill health. . . . Poverty itself is often self-perpetuating and operates in a cycle of social ill health and environmental deprivation of children which leads to impaired development, school failure, and behavioral abnormalities, which in turn predispose to more poverty and yet another generation of social ill health.

Hugh Thompson's address was duly published in *Pediatrics* in December 1969. It has not been reprinted nearly as widely as Hardin's December 1968 People Pollute manifesto quoted above. Nor, alas, has it caused any changes, however minimal, in the perceptions or insights of the zealous followers of Vogt and Hardin.

COMMIT GENOCIDE AND SAVE WESTERN CIVILIZATION: TRIAGE

The effects of the Vogt-Moore-Burch-Pendell-Hardin-Nixon vision of the population-pollution dilemma found equal acceptance in pop and academic cultures. Robert Ardrey, America's foremost pop prophet of the people-are-lousy-animals philosophy of Dr. Konrad Lorenz, announced in his best-selling *The Social Contract* (1970) that we are doomed. According to Ardrey, nothing—not automobile accidents, drug addiction, "cardiac and other stress diseases," not even "nuclear entertainments" (since modern "wars simply do not kill enough people")—would save us from the demographic and ecological consequences of our untrammeled copulations. The fact was: "The tragedy and magnificence of *Homo sapiens* together rise from the same smoky truth that we alone among animal species refuse to listen to reason."

Less apocalyptic a phrase slinger than Ardrey, but philosophically his equal, is Sheldon C. Reed, Ph.D., director of the Dight Institute for Human Genetics at the University of Minnesota. His article, "Toward a New Eugenics," in the British *Eugenics Review* for June 1965, effected a working synthesis between the shibboleths of the old scientific racism and the tribal slogans of the new:

"The need for eugenic concern is *greater* today than ever before because of the population 'explosion' and the automation 'explosion.' It is not realistic to encourage the more intelligent to increase their birth rate greatly because of the menace of overpopulation. It is imperative that the less intelligent be discouraged from reproducing as much as at present *because machines are rapidly taking over the jobs previously held by the least able of our fellow men.*" (Italics added.) (See chapter 2.)

A 1971 book, *The Case for Compulsory Birth Control,* by Dr. Edgar R. Chasteen, father of three, associate professor of sociology at William Jewell College in Liberty, Missouri, and member of the national board of Zero Population Growth, Inc., proposed legislation limiting the number of children per family to two. Portions of this book were run in *Mademoiselle,* a mass circulation magazine.

The very influential book *Famine—1975! America's Decision: Who Will Survive?* by Paul and William Paddock—one a diplomat, the other an agricultural expert—said that since the world could never produce enough food to keep up with the birth rates, and since America had vast food surpluses, the time had come for a Pax Americana enforced by America's food allocations.

While the Paddocks had nothing but contempt for well-meaning people who raised what they termed the false hopes of birth control and sterilization, they reserved their major ire for those demographers who raised what they termed the false hope of industrialization: "Although improved living conditions, such as those brought on by industrialization, are often given as the reason for decreased birth rate, *actually the evidence seems to be the opposite*" (italics added).[48]

Since every U.S. government, U.N. special agency, and nongovernmental compilation of the world's vital statistics abounds in evidence that industrialization is—invariably, and on all continents—associated with lower death and birth rates than are found in the nonindustrial nations, the Paddocks' claim that "the evidence seems to be the opposite" is meaningless in the absence of the "evidence" they have in mind.

Because famines were inevitable, the Paddocks wrote, the only thing that remained to be done was to seize this God-given opportunity to Make America Great.

> Let history's tribunal record that, although the United States could not prevent the Times of Famines, it nevertheless accepted this period as a challenge to its ingenuity and power. Let history record that because the American people met this challenge . . . out of the . . . Times of Famines came the foundation on which man built an era of greatness . . . not for the United States alone but also for the hungry nations.[49]

This greatness was to be achieved by the application of scientific method. The word for this method, alas, was not American but French: *triage.* The Paddocks did not try to change the word itself: they merely broadened its application.

Triage, from the French for "sorting," is a term used in military carnage and civilian disasters for the medical screening of the human casualties to determine the priority of their treatment. Under this system, patients are classified as being among (1) those who cannot be expected to survive even with medical care; (2) those who would probably recover even without medical help; and (3) those whose sheer survival depends upon prompt and effective medical care.

The Paddocks translated these priorities of military medicine into the priorities of American greatness in the face of what they saw as a world situation in which only the United States had, and would have, the surplus grains required to sustain life in those hungry countries whose populations *deserved* saving. Therefore, all U.S. food aid to the hungry nations of the nonindustrial world would have to follow the new triage system suggested by the Paddocks.

This one, which might be called the Pontius Pilate or triage approach to humanity, the Paddocks described as follows:

> (1) Nations in which the population growth trend has already passed the agricultural potential. . . . Those nations form the "can't-be-saved" group. To send them food is to throw sand in the ocean. (2) Nations which have the necessary agricultural resources and/or foreign exchange for the purchase of food from abroad and which therefore will be only moderately affected by the shortage of food. They are the "walking wounded" and do not require *food* aid in order to survive. (3) Nations in which the imbalance between food and population is great but the *degree* of the imbalance is manageable. . . . These countries will have a chance to come through their crises provided careful medical treatment is given, that is, receipt of enough American food and also of other types of assistance.

Then, for reasons they explain in their book, the Paddocks give their proposed triage classifications of seven representative nations, each of them with a nonwhite population. India, Egypt, and Haiti were abandoned as being in the lowest (Triage I) category—the "can't-be-saved" nations. The Paddocks put Gambia and oil-rich Libya in the "walking wounded," or Triage II, group. Finally, the "should receive food," or Triage III, nations were listed as Tunisia and Pakistan.[50]

The Paddocks, both of them very patriotic, were willing to waive some of their triage proscriptions for reasons of *Realpolitik*. Panama, for example, to protect the Canal. Bolivia, "to keep the flow of tin open." The Philippines, because the "Filipinos do retain a pro-American attitude" of a quality rarely shown by other have-not nations. They were therefore gracious enough to include the Philippine Republic on their "should receive food" list until her population exploded once too often—or until the surplus food reserves in America's vast granaries were reduced to the levels where we were forced to make an agonizing choice between the friendly but fertile Filipinos and other deserving poor countries with more Draconian population control programs.[51]

Like William Vogt, the Paddocks had nothing but overt scorn (and ill-concealed bad will) for China, which was then in the eighteenth year of its revolutionary transformation from a land of endemic floods, famine, and preventable infectious diseases to a reforested land which, by the Paddocks'

Doomsday Year of 1975, would be found by visiting American experts in agriculture to be "by far the largest producer of rice in the world today."[52] Of this then rapidly changing China, the Paddocks in their book could only say:

"Now [1967] China is on the direct road to famine. The collision between population explosion and static [sic] agriculture seems truly inevitable. . . . All power rests in Peking, and the government there does appear to the outside world as a confused, unstable company."[53]

China, the Paddocks wrote in their Jeremiad of 1967, was one of those countries where "mass starvation seems likely to hit . . . within five to ten years, or even sooner." I reread this prediction on page 160 of the Paddocks' book when the May 1975 issue of the *WHO Chronicle* arrived in the mail with an eyewitness report, "Health Care in the People's Republic of China," by Drs. Victor W. Sidel and Ruth Sidel. The Sidels found that "there is no evidence of the malnutrition, infectious diseases, or other manifestations of ill health that accompany this level of [technological] development."[54]

Writing in the same year for which the Paddocks had predicted only famine for China, the Sidels concluded that "the greatest lesson that China teaches is that it can be done—that a nation can within one generation move from a starving, sickness-riddled, illiterate, elitist, semifeudal society to a vigorous, healthy, productive, highly literate, mass participation society."[55]

All of the Paddocks' evaluations of the food-producing and baby-producing capacities of China, and all of this world's other nations, and of their scientifically realistic chances of even surviving the twin perils of producing too little food and too many new babies by the magic year 1975,[56] were accepted, unquestioningly, as agricultural, demographic, and sociological *facts* by Vogt's freshman disciple, Paul R. Ehrlich. Now grown to man's estate, and a full professor of biology at Stanford University, Dr. Ehrlich was soon to write what was to become the popular handbook of the population explosion movement. Dr. Ehrlich, whose scientific work has concerned insects, emerged as the most charismatic of the spokesmen for this wing of the new Malthusians, although in other aspects he considers himself an opponent of racism and has clashed with his faculty colleague, the true believer William Shockley, on the question of race and intelligence.

A STAR IS BORN

Much of what the young Paul Ehrlich was subsequently to tell his fellow Americans in his astonishingly successful paperback tract, *The Population Bomb,* was included in the paper he delivered at the University of Texas on November 16, 1967, during a symposium entitled "Limitations of the Earth —A Compelling Focus for Geology."

His talk opened with a short sentence: "The facts of human population growth are simple." There followed a few paragraphs of the gospel according to Burch, Vogt, and the Paddocks, ending with a most extraordinary declaration for a working biologist to make. The human population had to stop growing, he said, and "this halt must come through a decrease in the birth rate, or an increase in the death rate, or a combination of the two. A corollary of this is that *anyone or any organization opposing reduction in the birth rate is automatically an agent for increasing the death rate"* (italics added). Whatever applications this law might have in the insect species with which Professor Ehrlich is professionally familiar, it happens to constitute almost a perfect non sequitur in our own species. For if the health statistics of the past century have shown anything at all, they have shown that while increases in birth rates are not followed by increases in a nation's death rates, it is well known that decreases in human death rates are *followed*—but rarely if ever preceded—by decreases in national birth rates. "There has never been a people on the face of the earth that, having acquired literacy, education and a fair level of living, did not reduce its birth rate," observed Professor Philip Hauser, director of the University of Chicago Population Research Center.[57]

Dr. Ehrlich averred that, in connection with the timing of the world famines, "my guess is that the Paddocks are more likely correct, but in the long run it makes no difference. A great many people are going to starve to death, and soon. There is nothing that can be done to prevent it." This was followed by the Stanford biologist's exposition and endorsement of the Paddocks' triage concept. Speaking for himself, in a comment on the triage idea, Ehrlich said that "India, where population growth is colossal, agriculture hopelessly antiquated, and the government incompetent, will be one of those we must allow to slip down the drain."

Whereupon he unveiled the program he has been pushing, with a few modifications, from time to time, ever since:

> I think our first move must be to convince all those that we can that the planet Earth must be viewed as a space ship of limited carrying capacity. It must be made crystal clear that population growth must stop, and we must arrive at a consensus as to what the ideal size of the human crew of the Earth should be. When we have determined the size of the crew, then we can attempt to design an environment in which that crew will be maintained in some sort of an optimum state. . . . I think that 150 million people would be an optimum number to live comfortably in the United States.

His first three steps dealt with setting up a Federal Population Commission; a change in tax laws "to discourage rather than encourage reproduction"; and "federal laws which make instruction in birth control methods mandatory in all public schools . . . and . . . also forbid state laws" against abortions approved by a woman's physician.

His fourth point sounded the same "kill the healers" note previously struck by Vogt, Malthus, Hardin, and Burch. ". . . we should change the pattern of federal support of biomedical research so that a *majority* of it goes into the broad areas of population regulation, environmental sciences, behavioral sciences and related areas, *rather than into short-sighted programs on death control*. It is absurd to be preoccupied with the *medical quality of life* until and unless the problem of the *quantity* of life is solved." (Italics added.)

All of these four steps, Dr. Ehrlich said, "might produce the desired result of a reversal of today's population growth trend. If they should fail, however, we would be faced with some form of compulsory birth regulation. We might, for instance, institute a system which would make *positive* action necessary before reproduction is possible. This might be the addition of a temporary sterilant to staple food, or to the water supply. An antidote would have to be taken to permit reproduction." And of course, if need be, the antidote would be doled out by Big Brother in ratios small enough to "produce the desired constance of population size."

The entomologist from the Golden West ended his talk with a program for immediate action. It was a five-point program, starting with the announcement "that we will no longer ship food to countries where dispassionate analysis indicates that the food-population unbalance is hopeless." That is, triage as prescribed by the Paddocks.

The second point was for the United States to "announce that we will no longer give aid to any country with an increasing population until that country convinces us that it is doing everything within its power to limit its population." The third point provided for massive aid "to all interested countries in the technology of birth control." The fourth point called for massive aid to help poor countries increase their yields of food production per acre to levels approaching those of the nonpoor and nonhungry countries.

The fifth point called for the United States to bring "extreme political and economic pressure" to bear on "any country impeding a solution to the world's most pressing problem. A good place to start would be breaking off diplomatic relations with the Vatican until that organization [*sic*] brings its policies into line with the desires of the majority of American Catholics."

Between the delivery of this talk at the University of Texas learned symposium and the publication of *The Population Bomb* in May 1968, Dr. Ehrlich evidently decided to give the Vatican a second chance, for this plank of the Ehrlich plan to save humanity was not included in the book. Aside from this one diplomatic revision, the plan of the Texas University speech was the solid core of all the recommendations made in the book.

The opening chapter of the book was exactly one paragraph long. Titled "The Problem," it described how Dr. Ehrlich, his wife, and his young daughter, "one stinking hot night in Delhi a couple of years ago," had the wits frightened out of them as their flea-ridden taxi crawled through a crowded slum area on the way back to their hotel. Through a haze of dust and smoke, the cab plowed through seas of people. "People eating, people washing, people sleeping. People visiting, arguing, screaming. People thrusting their hands through the taxi window, begging. People defecating and urinating. People clinging to buses. People herding animals. People, people, people, people. As we moved slowly through the mob, hand horn squawking, the dust, noise, heat, and cooking fires gave the scene a hellish aspect. Would we ever get to our hotel? All three of us were, frankly, frightened." Somehow, they did manage to escape alive, and, wrote the ardent population-bomber, "since that night I've known the *feel* of overpopulation."[58]

The scenes that gave Dr. Ehrlich the *"feel* of overpopulation" happened to match the scenes of human misery and degradation that Hogarth found and drew on Gin Lane and the other London slum streets where the dis-possessed of early-eighteenth-century England—two centuries before the "population explosion"—slept and defecated and snatched purses and knocked heads and drank themselves into oblivion. To Hogarth, however, these crowded slum streets were not the product of overpopulation but of poverty.

The degradation that confronted the Ehrlichs that night in Delhi was simply a function of poverty, and of nothing else. And this poverty was not the product of India's birth rates but of her history for at least the previous century of exploitation as a colony of a non-Indian European power.

Many of the book's other nonbiological, nonentomological judgments, made with equal assurance, were of the same order of accuracy and insight. There was, for example, the sentence on page 66 to the effect that "too many cars, too many factories, too much detergent, too much pesticide, multiplying contrails, inadequate sewage treatment plants, too little water, too much carbon dioxide—*all* can be traced easily to too many people."

There is nothing in biology or sociology that requires that all human populations copy the United States in riding to and from work in private cars as against cheaper and safer trolleys, trains, and buses, let alone bicycles. Or that even a single factory has to be permitted to pollute the environment when the technology to prevent such pollution already exists. Or that people should stop using biodegradable soap and start, as Americans have done, to use environmentally dangerous detergents. Or that people should use pesticides that menace other forms of life instead of pesticides safe for the ecosystems. Or that governments must not provide adequate and safe instead of unhygienic sewage treatment plants to the taxpayers and voters who elect them to office.

To blame the growth in any population for the stupidity of governments, and the cupidity of manufacturers of needless and even socially harmful products, is to unwittingly cast serious aspersions on the schools which gradu-ated an adult who writes such books telling mankind how to live.

In chapter 4, "What Needs to Be Done," after hailing the Paddocks' triage idea as "the only realistic suggestion in this area," Ehrlich urged the adoption of triage policies under which we would send food aid to "the Pakistani government, under the tough-minded leadership of President Ayub Khan," but not to India. "West Pakistan might receive aid, but not East Pakistan." And "perhaps we should support secessionist movements" in some poor nations, and rearrange the boundaries of others (pp. 159–60).

Our government, charged Ehrlich, made a big mistake in not supporting an Indian politician who suggested sterilizing all males who had fathered three or more children. "We should have applied pressure on the Indian government to go ahead with the plan. We should have volunteered logistic support in the form of helicopters, vehicles, and surgical instruments. We should have sent doctors to aid the program by setting up centers for train-ing para-medical personnel to do vasectomies."[59]

Finally, touching on a theme dear to the hearts of his forerunners, Ehrlich noted that from times ancient and less ancient, population pressures caused European wars. "In more recent history we have the stunning ex-

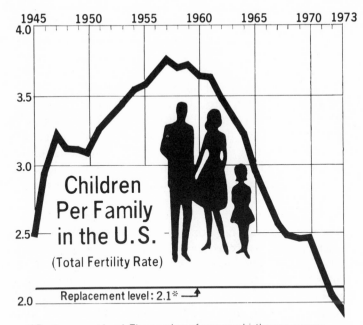

Children
Per Family
in the U.S.
(Total Fertility Rate)

Replacement level: 2.1*

*Replacement level: The number of average births per woman
over her lifetime necessary for the population eventually to
reach zero population growth. This would take about 70 years.

Sources: National Center for Health Statistics, Census Bureau

For more than a decade before the 1968 publication of the People Pollute movement's
handbook, Paul Ehrlich's *The Population Bomb,* the actual U.S. birth and fertility rates
were clearly dropping to and below the so-called zero population growth level.

ample of Nazi Germany's drive for 'Lebensraum,' and Japan's attempts to
relieve the crowded condition of her small islands. Whether or not things were
really all that difficult for the Germans is a point for debate." But "it is
certainly clear that if population growth proceeds much further the proba-
bilities of wars will be immensely increased."[60]

MALTHUS, EHRLICH, AND THE DEMOGRAPHIC TRANSITION

The book was an instant hit. From that moment forward, as edition after
edition hit the stands, Paul Ehrlich became a familiar figure on the TV talk
shows, the lucrative lecture circuits, and the Sunday newspapers. He was
elected president of Zero Population Growth, Inc., and was cited frequently
as our greatest living authority on population and pollution problems.

Ehrlich's stature rose to that of a national institution. Let him write, as
he did in his Malthusian tract, that the psychiatric problems of the population
explosion included "finding substitutes for the sexual satisfaction which

many women derive from childbearing and finding substitutes for the ego satisfaction that often accompanies excessive fatherhood,"[61] and in no time at all such vulgar parodies of Freudian psychoanalysis were being used as "scientific" arguments for the sterilization of welfare mothers who bore children out of wedlock. Question his words or concepts in any way, and the answering torrents of literate abuse let the heretics know they were in Coventry—as the genial and often incisive general columnist on the *San Francisco Chronicle,* Charles McCabe, learned early in 1970.

In January of that year, McCabe wrote two columns about Ehrlich and his advocacy of Malthusian social policies for the United States. McCabe, a divorced man, made it clear that although reared as a Catholic, "I think, as do many other Catholics, that the Pope is bang-on wrong about birth control. Further, I do not believe making love is a matter of faith or morals, in which things I am supposed to accept papal infallibility." He did, however, seriously question Ehrlich's infallibility.

"This guy Ehrlich has to be read to be believed," McCabe wrote. He saw Ehrlich as a social menace and quoted, verbatim, from some of the Ehrlich canon, previously cited in this chapter, on denying aid to nations with population policies unapproved by us; the absurdity of improving the medical quality of life before solving the problem of its quantity; the wisdom of dropping sterilants in staple foods or water supplies; and the making of birth control instruction in all schools mandatory by federal law. "These," commented McCabe, "are the views of a zealot, not of a rational man."

The McCabe columns drew an angry letter of protest, signed by the chairman and fifteen other members of the Stanford Biology Department, that the *Chronicle* printed on February 5, 1970. The letter, which deplored the fact that McCabe and other skeptics had "raised doubt about Dr. Ehrlich's own professional qualifications" to make judgments about demography and world food policies, insisted that, on the contrary, the insect biologist Ehrlich had "excellent credentials for speaking authoritatively on questions of human population growth and environmental quality."

This had to be the most extraordinary letter ever written by one group of academic scientists in defense of a colleague's misuse of science in the furtherance of *political* goals since 1923, when Terman, McDougall, et al. protested the *New Republic* series in which young Walter Lippmann had subjected the equally sacrosanct IQ test score mystique to the same deflationary scrutiny. As it happened, in 1970, as in 1923, the lay critic of scientific racism did not deserve this kind of an attack from the professors.

There is nothing in biology in general, or in genetics and population genetics in particular, that gives a young, healthy, well-paid American in Palo Alto, California, the moral, scientific, or political right to blandly and not too blandly urge the termination of food, medicine, and all other aid to India and various other countries whose birth policies offend him. Nor is there any justification in biology for him to advocate the diminution of medical assistance to the world's poor, abroad and at home, or the reordering of the nation's biological research priorities in terms of less support for investigations of the major cause of death and chronic diseases and more support for research into new techniques of birth control.

There is nothing biological or scientific about such suggestions. They are based on subjective moral judgments about the intrinsic value of the lives of human beings other and poorer than oneself—and *not* on the biological imperatives of society. In putting them forward Dr. Ehrlich is not offering a body of *scientific* recommendations but, rather, a program of *political* action.

As a moral philosopher, and as an open and blunt advocate of genocidal political policies such as the triage ploy developed by the Paddocks, Dr. Ehrlich has neither the intellectual and professional right, nor the moral authority, to speak for biology in particular and for the scientific community in general. Genocide remains genocide, whether advocated in a Munich beer hall in 1920 or in a Texas college auditorium in 1967—and neither the brown shirts of its earlier German advocates nor the graduate degrees and academic posts of its latter-day American proponents make it any less a political rather than a scholarly proposal.

Nor were the targeted victims of the genocidal policies proposed by Ehrlich confined to merely the nonwhite citizens of East Pakistan (now Bangladesh), India, Nigeria, and other nations judged by the Paddocks and himself to be in the "can't-be-saved" triage category. Consider the implications of his demands that the direction of most federally funded biomedical research be diverted from "short-sighted programs on death control" and the "medical quality of life" to "broad areas of population regulation" and what Ehrlich calls "environmental sciences, behavioral sciences and related areas," when these proposals are examined not in the context of the overheated prose of Ehrlich's propaganda tract but in the context of the hard, statistical realities of life and death in this nation in this century.

The message of these data has long been clear. In 1927, for example, before Ehrlich was born, Franz Boas dismissed the dysgenic population explosion scares that East and other old scientific racists were already beginning to raise, with the observation that "the well-to-do have, ordinarily, a low birth rate and a low mortality. Among the poor, the reverse is true."[62]

This, of course, was before our modern *professionally qualified* demographers coined the useful phrase "the Demographic Transition" to account for the well-known demographic reality that as a nation's family income and living standards rise, and its death rates fall, its birth rates start to decline.

In the United States during this century, the levels of our population growth have been determined not by our declining live-birth rates but by the improvements in general living standards that have enabled us to live healthier, longer lives than we lived before 1900. By 1972 our live-birth rates had, for this reason, dropped below the 2.1 replacement level required for zero population growth. The replacement level has continued to drop since then, having fallen below 1.9 by 1974; for the same socioeconomic reasons, the live-birth rates of all industrial and industrializing nations, from France and England to Russia and China, have also continued to fall.

The Demographic Transition

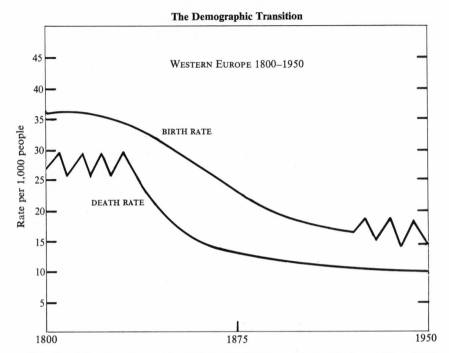

From Population Council report for 1952–64 (July 1965), reprinted in *The Lancet* (December 3, 1966), p. 1239.

The demographic transition, writes Princeton demographer Ansley J. Coale, is "the central event in the recent history of the human population. It begins with a decline in the death rate, precipitated by advances in medicine (particularly in public health), nutrition or both. Some years later the birth rate also declines, primarily because of changes in the perceived value of having children." Before the demographic transition, "the birth rate is constant but the death rate varies; afterward, the death rate is constant but the birth rate fluctuates."

Historically, the demographic transition, as in Western Europe and the United States, has always followed such life-enhancing developments as the Agricultural Revolution; the Industrial Revolution; the contiguous births of germ theory and immunology; the nineteenth- and twentieth-century enactments of social laws providing free education, vaccinations, and medical care; and enforced statutes mandating minimum wages, maximum working hours, and healthier standards of human housing.

Nearly half a century before Professor Coale wrote the above definition of the demographic transition (in the September 1974 *Scientific American*), the anthropologist Franz Boas, writing in the February 1927 *Current History,* observed that ". . . the well-to-do have, ordinarily, a low birth rate and a low mortality. Among the poor, the reverse is true." As of now, it is the poorest states of this republic and the most impoverished nations of this planet that have the highest death rates and the highest birth rates. Among the more affluent states and nations, "the reverse is true." See also Sir Dugald Baird's 1946 table, "Mortality and Live Birth Rates in England and Wales, 1841–1939," on page 429.

U.S. Demographic Statistics Since 1900 (or closest year)				
	1900	1973	% Up	% Down
Live-birth rate per 1,000 population	32.3	15.0	—	53
Expectation of life at birth	47.3 years	70.4 years (1969)	49	—
Infant mortality per 1,000 live births	99.9 (1915)	19.2 (1971)	—	81
Death rate per 1,000 population	17.2	9.3 (1971)	—	46
Median household-family size	4.23 people	3.53 people	—	17
Total number of births per year	2,718,000 (1909)	3,141,000	16	—
Total U.S. population	76,094,000	203,184,772	167	—

From *Statistical Abstract of the United States*, 1974.

As these birth-rate and death-rate and life-span data clearly demonstrate, the major reasons for the 167 percent increase in the total living U.S. population in this century have not been the *declining* birth rate but the declining death rate and the 49 percent increase in the longevity of the average American. Increases in family incomes which in turn resulted in rising standards of nutrition, housing, medical care, and education, were responsible for the 81 percent drop in infant-mortality rates and the 46 percent drop in the total death rates—as well as for the 53 percent decline in the live-birth rates to a new all-time record low by 1972. This *predictable* decline in live-birth rates as living standards rise, and infant and general death rates fall, is known to the world's professional demographers as "the Demographic Transition."

The rise in the net living U.S. population since 1900, then, has been due, essentially, to the declines in infant, maternal, child, and general mortality—particularly from environmental diseases eliminated or contained by classic sanitary reforms, such as clean-water and hygienic waste-disposal systems. Other major factors in this over-all decline in the U.S. death rates are equally well known. They include: (1) the elimination or reduction of deficiency diseases, brought about by wage increases that expanded the food-purchasing power of the average family; (2) the role of new vaccines and antibiotics in preventing and/or curing once invariably fatal environmental infectious diseases, such as tuberculosis, pneumonia, diphtheria, streptococcal nephritis, and infant infections.

What all this means, according to classic Malthusian dogmas on the dangerous doctors and the even more dangerous menace of living wages, is that the only feasible way to achieve Paul R. Ehrlich's dream of attaining a permanent U.S. population of 150,000,000 is probably by wiping out the real causes of our falling death rates.[63]

Establishing substandard wage scales, and ending biomedical research into the "total extirpation of particular disorders" that would enable us to further reduce the death rates, was of course what Thomas Malthus, the founding father of scientific racism, proposed as far back as 1826. This does not make such social goals more acceptable; it merely makes them more explicable.

17 People Pollute, so Kill the People: The New Scientific Racism Scores Its First Major Legislative Victory

PRESIDENT SIGNS BIRTH CURB BILL
Programs Are Expanded—
Nixon Kills Measure on More Family Doctors

President Nixon has signed into law a $382-million, three-year
expansion of the family planning service and creation of a
Federal office to coordinate ways to control population growth.
The White House also announced that Mr. Nixon had vetoed
a bill that would have set up a three-year $225-million program
to train family doctors.
—*The New York Times,* December 27, 1970

The postwar population explosion hysteria initiated by Guy Irving Burch and Elmer Pendell in 1945, injected by Burch and Vogt into the body of Fairfield Osborn's benignly intentioned book on natural conservation, and carried to full intellectual fruition by the Paddocks, Ehrlich, and Hardin, succeeded far beyond the wildest hopes of the old-time eugenicists who started it all. Out of it came not only mass movements, such as Zero Population Growth, Inc., with chapters of active members in many American cities, but also new causes for older conservationist societies, such as the venerable Sierra Club. In the conventional wisdom of the era, the conservation of green plants, wildlife, and virgin forests became as one with the crusade to lower birth rates: it was people, and not machines, that polluted the environment. Population Control and Pollution Control were seen by most Americans as one and the same problem, and to be against compulsory birth control was to be for pollution. Worst of all, from the standpoint of history and common morality, the efforts of the medical and allied public health professions to prevent and/or cure the widespread diseases of the poor of this and other nations were being increasingly characterized in Vogt-Ehrlich terms as misguided and harmful efforts that only exacerbated the population problem.

The cry of millions of educated Americans became "People Pollute." A junior college curriculum for a course unit, "Man and Environment," disseminated by the U.S. Office of Education for courses taken by over 1,400,-000 students between 1971 and the fall of 1972, described the first objective of this course on population dynamics as being: "to make the student aware that overpopulation is the underlying cause of our environmental problems."[1]

In 1973, Dr. H. Curtis Wood, Jr., for fifteen years the president of the Association for Voluntary Sterilization, Inc., and now its medical consultant, declared in an article in a medical publication: "People pollute, and too many people crowded too close together cause many of our social and economic problems. These, in turn, are aggravated by involuntary and irresponsible parenthood. As physicians we have obligations to our individual

406

patients, but we also have obligations to the society of which we are a part. The welfare mess, as it has been called, cries out for solutions, one of which is fertility control [sterilization]."[2]

This argument became so integral an element of the conventional wisdom of America's teachers and of America's literate families that, in 1973, the assistant director of the Population Council, Stephen Viederman, reprinted in some dismay the last stanza of a poem called "Overpopulation" that had recently been written by two sixth-graders in Kensington, Maryland, a suburb of Washington, D.C.:

> If we didn't have people
> We wouldn't have pollution,
> Get rid of the people
> That's the only solution.[3]

The results of such widely held views were not long in making themselves felt in the reordering of national priorities and the shaping of societal programs involving health, research, and public welfare. By 1970, in an America where at least one third of the pregnant women gave birth without ever having had one moment of professional prenatal medical care, and in which the U.S. Public Health Service estimated that this nation had at least 50,000 physicians fewer than our population needed, Congress passed a bill appropriating $225 million dollars to help meet the costs of training some of the new family doctors the federal government's health agency said the nation needed. This bill was vetoed by President Nixon—who that same day also signed another bill authorizing the expenditure of $382 million for a program of distributing birth control advice and contraceptive devices to clients of public and private organizations involved in efforts to "control population growth."[4]

Thus, only three years after Paul Ehrlich first demanded, in November 1967, a reversal of federal research funding priorities so that a majority of such support would go "into the broad areas of population regulation . . . rather than into short-sighted programs on death control," since he felt it "is absurd to be preoccupied with the medical *quality* of life until and unless the problem of the *quantity* of life is solved," his prayers were answered by a President who had to veto the Congress's physician training program to do it. And this was only one of hundreds of similar triumphs for the Malthusian and/or eugenical cause of Burch, Vogt, Moore, Ehrlich, and Hardin.

HOW TO REDUCE BIRTH RATES

In a nation where professors of biology could solemnly write and probably even believe the malevolent Malthusian propaganda about the healing arts being the nemesis of humanity, the fact that a U.S. President could veto a bill providing for the training of more family physicians, while at the same time signing into law an act spending even more money for population control, without thereby stirring up a storm of horrified protests from men and women of goodwill stands as a monument to the societal effects of the population explosion hysteria. The public health effects of the population-bombers go much further than these two executive decisions, however.

The soaring dollar costs of the endless Indochina wars claimed, as their earliest victims, all of the once great federal programs of research into the major killing and crippling diseases, and support for the training of physicians, nurses, and other health personnel. Long-standing programs of research in heart disease, stroke, and cancer were allowed to terminate, or were drastically cut back to skeletons of what they were before the wars. Federal aid to medical schools, community health agencies and institutions, and demonstration clinical centers was sharply slashed, year after year, as the war dragged on. Young investigators with doctorates in the biomedical sciences found themselves unemployed as their federal grants ran out and their federally supported laboratories and hospitals were left without the funds to continue their work.[5]

Throughout this long era of attrition and retrenchment in biomedical research and education, there were only two areas of clinical investigation which suffered no shortage of federal funding support, and even received increasingly large grants of national tax dollars. One was drug abuse—and the other was and is birth control (or human fertility and/or population research). But here all of the human fertility studies were and continue to be tragically misdirected because of the universal acceptance of the myth that population problems were human behavioral problems entirely independent of all other sociobiological factors. That is, that birth rates were high because some people were stupid, superstitious, sex-mad, and slothful—and low because their betters were smart, educated, and sexually prudent.

The fact remains, as any student of demography or even history knows, that in Europe, America, and Japan birth rates started to fall and kept right on falling after industrialization brought about those improvements in the duration and quality of life that, historically, always are followed by declining birth rates. So that by 1930, and up until the present, the birth rates of the industrial (and nonpoor) nations of Europe, Asia, America, and Oceania had stabilized themselves at half the birth rates of the world's non-industrial (and poor) nations.

Given these kinds of hard factual data, any meaningful and constructive birth studies would direct themselves toward determining just what single quality-of-life factor or combination of factors was most responsible for the lower birth rates of the industrially advanced lands.

Whether these factors had to do with low infant and child death rates, or high protein-calorie intakes, or improved water supply, sewage disposal, and other aspects of environmental hygiene, or the combination of all of these factors, is something nobody has yet tried to study. But one thing is certain, and that is that whatever the birth control techniques and devices used in the industrial nations during the two centuries of declining birth rates that started in the 1830's, they were more than adequate to the task of cutting their birth rates in half.

The reasons for these manifestly *voluntary* reductions in the live-birth rates of the industrial nations are well known to serious students of human behavior and history. They are no mystery to those scholars who look upon people as members of the human race. Unfortunately, the number of literate Americans who see people as human beings, with uniquely human characteristics, seems to be declining.

When the post-World War II flight from reason got under way, it became the fashion in the pop culture to write and think of people not as people but as aggressive aardvarks, territorial termites, overcrowded rodents, and, in fact, anything but human beings. Yet the very cultural and verbal attributes that in truth do make the behavioral patterns of our species unique in all of nature seem to have been overlooked in the flow of pseudo-intellectual nonsense about naked apes and other simplisms. For all the biological and behavioral traits that we do, indeed, share with many of our compatriot creatures of the land, the sea, the sky, there also happen to be traits that are uniquely human, just as the bearing of their undeveloped offspring in pouches is uniquely marsupial to kangaroos and opossums.[6]

There is, for example, the small matter of verbal language, which not only is an exclusively human possession but is also, in many human cultures, used in silent and permanent forms called written languages. There is another uniquely human trait, the related wonder called verbalized hope.

Many creatures other than man probably have and exercise the capacity to hope. Only man can put his hope into language, spoken or written or both. And it is only in the human species that this verbal construct, this generally unwritten abstraction, can inspire vast numbers to move mountains, to cross oceans—and to have fewer babies.

This capacity to be motivated by hope used to be one of the best-known characteristics of our species. As far back as 1911, for example, when Dr. Scott Nearing was a young instructor in economics at the University of Pennsylvania, he cited this human trait as the answer of history to those Americans—Nearing mentioned President Theodore Roosevelt as one of them—who feared that the legalization of birth control devices would lead to what Roosevelt and the eugenicists called "race suicide." In an article in the January 1911 issue of *The Popular Science Monthly* entitled " 'Race Suicide' vs. Overpopulation," Dr. Nearing offered a graphic demonstration of the role of even the hope of upward social mobility on voluntary restrictions of family size. And he did it in only two paragraphs:

The western world, at the opening of the nineteenth century, presented this significant picture—a high birth rate, a low and decreasing death rate; a phenomenal increase in population made possible by the wealth-producing power of the factory system; and big families treading on the heels of subsistence. Here was ample justification for the pessimistic gloom of Malthus. *Catastrophe seemed inevitable, when democracy entered the field, telling the men at the margin whose families were either unregulated in size or else regulated only by subsistence, that they were free and equal to every other man and had a like right to "rise."* The thought was new. "How can I rise?" asked the laborer. "Stop having children," replied the economist. The advice was followed. The family of eight is replaced by the family of two and thus unencumbered of an enormous burden, the laborer is enabled to raise his standard of life.

Until 1750 any great increase in population was prevented by a high death rate. In the succeeding century, *as a result of science and sanitation, the death rate was gradually reduced,* and an overwhelming increase in population was prevented in only one way—by decreasing the birth rate. The decline in the birth rate therefore saved the modern civilized world from overpopulation and economic disaster. [Italics added.]

This was written before the eugenicists stopped calling birth control "race suicide" and started exhorting about population explosions and demanding the compulsory sterilization of the 25 million-plus Americans at and slightly above the official poverty level.

ABORTION, INFANTICIDE, AND POPULATION CONTROL

Neither contraceptive techniques nor abortion were particularly new or novel by the start of the twentieth century. For the affluent and the educated elements of all Western societies, including our own, contraception and abortion were extensively employed to prevent unwanted births. Contraception had always been safe, but by 1930 over a century of trial and error had made contraceptives much more effective. At the start of the nineteenth century—before the germ theory and the antiseptic surgical techniques that arose from the work of Pasteur and Koch—abortions were as likely to kill the mother as her unborn fetus. After Lister's work in the 1860's abortion became a safe surgical procedure. The advent of antibiotics in 1945 reduced the risks of abortion septicemia further.

In most states, while both contraception and abortion were illegal by state law, they were practiced very extensively by families and single women who could afford them. Voluntary abortions might have been illegal, but a curettage performed to save the life of a pregnant woman who started to bleed was perfectly legal. People of means, with family physicians and surgeons, resorted to such elegantly mis-labeled abortions.

For the poor, who had less ready access to contraceptives than the non-poor, abortion was a more frequent experience in this country between 1900 and 1970 than for the affluent. But, lacking family physicians and surgeons ready to falsify a record to perform a sanitary surgical abortion in a clean hospital, the poor had to rely upon rogue abortionists—with or without medical training—to perform illegal operations in generally filthy settings. The needless infections and injuries caused by such illegal abortions killed thousands of women yearly.[7]

Thanks to the fear of an ephemeral population explosion, let alone the fear of more inner-city poor (and often black) people proliferating and swelling the already swollen welfare rolls, state after state started to repeal their laws against the dissemination of birth control information and the sale of birth control devices. After the wave of inner-city violence that exploded in the Watts district of Los Angeles and spread to Washington, Newark, Detroit, and many other cities during the long hot summers of the

1960's, the movements to repeal the laws against abortion grew in strength. Although originally organized by people determined to make surgically safe abortions as accessible to the poor as they had been to the rich for nearly a century, the earliest successes of these efforts to accomplish the repeal of state anti-abortion laws owed far more to the fears inspired by the angry blacks of Watts and Detroit than to the humanitarian pleas of the reformers.

In New York State, where abortions became legal on July 1, 1970, some 278,122 legal abortions were performed during the first eighteen months of the new dispensation. With the arrival of safe, legal, and, for many indigent New Yorkers, free hospital abortions, the New York maternal mortality rate dropped from 5.3 deaths per 10,000 live births in 1969 and 4.6 in 1970 to an all-time low of 2.9 in 1971. What this meant, of course, was not that more families resorted to abortion, but that abortions had now become better maternal health risks.[8]

The figures for the first eighteen months revealed a few other facts worth noting. About 65 percent of all the women who had legal abortions in New York City hospitals were out-of-state nonresidents who, according to the records, "tended to be younger, white, and terminating a first pregnancy, indicating that more nonresidents were unmarried. It is also probable that nonresident women came from higher income groups, since they had to add the cost of a trip to the cost of their abortion."

"By contrast," observed the Health Services Administration report, "only 42.2% of New York City residents were ending a first pregnancy, while 22.6% had had three or more births or pregnancies. These data seem to indicate that New York City women tended to get abortions to limit family size, while nonresidents tended to get abortions for a first, out-of-wedlock pregnancy . . . 46.2% of the residents were white, 43.4% were non-white, and 10.4% were Puerto Rican, compared to 89.1% white, 10.4% non-white, and 0.5% Puerto Rican for nonresidents."[9]

By 1973, after the Supreme Court upheld the right of women to have abortions, many states passed abortion laws similar to the New York law. Two years later, a National Academy of Sciences study of the health effects of the nation's new liberalized state abortion laws revealed that, in New York, passage of such legislation was quickly followed by a precipitate decline in maternal deaths and injuries caused by abortions. In New York City alone, according to this study, hospital admissions for both septic abortion and incomplete abortion—both classic side effects of illegal abortions—fell from 6,524 in 1969 to 3,253 in 1973. The National Academy of Sciences investigation also disclosed that 25 percent of the women who had abortions were married, and that 68 percent of them were white.

What these figures also show is, of course, that as a crash procedure to slash the birth rates of the inner-city and disproportionately black poor, abortion will probably not become the hoped-for shortcut to lower birth rates. Safe abortions will, in fact, keep more poor and black women alive.

When this fact of demographic life becomes clear, therefore, we can safely expect the population-bombers to start agitating for voluntary and mandatory infanticide. That is, the murder of children born to people on welfare, to the chronic poor, to people with IQ test scores below 75, and to other unfortunate elements. Such suggestions have not been without many supporters in this century, and the eugenic and dysgenic arguments advanced for infanticide by acts of commission or omission are familiar to all students of pediatrics, history, and scientific racism. (See page 327.)

When such demands become as strident as the present demands for an end to medical research and medical care for the proliferating poor, it might be well to recall the comments of L. Emmett Holt, Sr., M.D., L.L.D., professor of diseases of children, Columbia University College of Physicians and Surgeons. In his presidential address before the American Association for the Study and Prevention of Infant Mortality, Dr. Holt said:

> It is not true, as has sometimes been assumed, that a nation as a whole is improved physically by a high infant death rate. Visitors to the marasmus wards of a modern infants' hospital often will remark on the uselessness or futility of saving these infants. They look upon the effort as misguided philanthropy and almost as a perversion of medical science, arguing that however praiseworthy from a humanitarian point of view to save such infant lives, it is false economy and does not improve the race; that it interferes with the law of natural selection, which is the survival of the fittest; that by efforts to keep the feeble alive, the degeneration of the race rather than improvement in it is favored.
>
> The argument is not a strong one and is based upon erroneous premises. In the first place, most of those who in infancy are regarded as physically unfit were healthy at birth and are merely the victims of a bad environment, improper feeding and neglect—conditions which it is quite possible to remove. When these obstacles are overcome these infants not only have the same chance to survive, but to grow into healthy, even robust, children as have others. How many of the world's brightest geniuses would have been lost had this law been rigidly applied, and who can say. It is hard to tell who are the unfit. *A high infant mortality results in a sacrifice of the* UNFORTUNATE, *not the* UNFIT.

Dr. Holt delivered this talk on the 14th of November—1913.[10]

NOT A POPULATION EXPLOSION—BUT A POPULATION SHIFT

While the population explosion crusades were attaining ever higher plateaus of hysteria—with their gurus keening doleful prophecies about doublings and triplings of national and planetary populations—what was happening in reality on the various human fertility fronts here and abroad?

In real life, in terms of actual biological rather than postulated births, what had happened in this country, starting as early as 1955, and continuing to the present moment, was that the U.S. birth rate had entered into a steady decline toward what by 1972 the U.S. Bureau of the Census would characterize as the lowest U.S. live-birth rate ever recorded in this nation.

Where, in 1955, the U.S. live-birth rate had stood at 24.9 per 1,000 population, by 1974 it had dived to 15.0 per 1,000. For the first time since 1940, the nation's rate of population growth, in 1968, dropped below 1 percent. By 1971, the national population growth rate stood at 0.98 percent. By 1972 it had fallen below the level of 2.1 children per completed family, which the demographers define as the point of exact replacement—or zero population growth.

Far from having been subjected to the population explosion of the uninformed if phenomenally articulate extremists, this nation had in reality been experiencing what the Washington Center for Metropolitan Studies called "the Baby Bust."[11] In terms of the general fertility rate—that is, the rate of annual births per 1,000 females between 15 and 44 years of age—which most professional demographers consider to be a better indicator of population dynamics than the live-birth rate per 1,000 population, the U.S. Baby Bust of 1955–19?? has been even more destructive of the myths of the population-bombers.

The general fertility rate, which in the United States had risen from 118.7 in 1955 to 123.0 in 1957, then proceeded to sag to 97.6 by 1965 and to an all-time low of 68.5 in 1974.

These rates of decline in live-birth rates and general fertility were not, however, uniform in all fifty states of the Union, or in all ethnic, racial, and socioeconomic groups. Among poor whites, as among poor nonwhites, both the death rates and the birth rates continued to be higher than among the nonpoor whites and nonwhites.

The University of Chicago demographer and director of that institu ion's Community and Family Study Center, Professor Donald J. Bogue, in a paper delivered at the 1971 annual meeting of the American College of Obstetricians and Gynecologists in San Francisco, told the assembled physicians that "in fact, the WASP and the Jewish population of this country are already reproducing very close to the replacement level, if not below, and most of the growth we are experiencing today is contributed" by what Dr. Bogue described as the "high-fertility remnants," in which he included "the Negro population, the Spanish-speaking population, the rural population and a few segments of the Roman Catholic population."[12]

Even in these "high-fertility remnants," however, many families were, in increasing proportions, "experiencing especially rapid declines in fertility." Far from endorsing the amateur extremist projections of a U.S. population of over 300 million by the year 2001, Dr. Bogue declared that "a reasonable demographic prediction for the year 2001 for the United States is that there will be a population of between 240 and 260 million persons with a crude birth rate of about 14 or 15 per thousand." The demographer, assuming a population of 250 million and a birth rate of 14.5—the middle of these ranges—therefore obtained a prediction of 3,625,000 births per year.

> Those who are familiar with vital statistics will grasp immediately that the figure of 3.6 million births is almost identical with the number of children to be born in the United States in 1971. In other words, we are predicting that the annual number of births in the year 2001 will be the same as during the current year, or a zero increase in the number of

births. If this is true, the maternity and infant health profession has entered a stage of "zero customer growth."

When he turned his attention to the dream of zero population growth, Dr. Bogue was equally non-incendiary:

> At the present moment, the crude birth rate of the U.S. is about 17.8 per thousand. This implies an average of about 2.5 children per completed family. Demographers have placed 2.1 children per completed family as the point of exact replacement (zero population growth). Already, large segments of our population are reproducing at the replacement level *or below,* and this 0.4 child is contributed by a few special groups whose fertility is also declining. . . .
> There is no guarantee that the birth rate will level off at the replacement level when it reaches this point. *There is plenty of precedent, both in this nation and in Europe, for birth rates to sink below replacement levels* and to remain there for prolonged periods of time. . . .
> Many people who were sensitized in the 1950's to the dramatic announcements about the "baby boom" are unaware that since 1957 the U.S. has been in the grip of a sharp "fertility recession" and that in most years since that date there have been fewer births than in the preceding year. [Italics added.]

In demography, the only explosion that occurred in the United States in our times was the taking and analysis of the 1970 national census. Publication of these census results exploded to tiny fragments the myths and the half-truths of the population-bombers.

But since the same census data dealt with the realities rather than the myths of American population dynamics, they also described the mass migrations and racial shifts that helped fan the flames of mass fear so closely associated with the population explosion hysteria that had been ignited by the extremists. And these mass population shifts—*not increases*—were far from illusory or ended.

How great these population shifts have been in the America of modern times was summed up in a perceptive article in *World Health,* the magazine of the World Health Organization (WHO), in May 1972 by Dr. Martin Kaplan, director of the WHO Office of Science and Technology. In his lead article, entitled "Environmental Hazards for Human Health" (the entire issue of the magazine was devoted to this problem), Dr. Kaplan, reviewing the problems of the *entire world,* noted:

> In some highly industrialized countries large portions of the population change residence frequently (*in the 10 years between 1950 and 1960, for example, 52 million persons in the U.S.A., one in four of the population, moved from one state to another, and about the same number moved from one part of a state to another*). This is likely to cause stress in families, particularly for young children and the aged, who may have considerable difficulty in making new friends and generally adapting to new environments [italics added].
> Uprooting and resettlement are more likely to be traumatic where migration has been involuntary, and such phenomena have affected

large populations during this century. Particularly high rates of mental disorder have been found among migrants who have survived mass ill-treatment.

As any American who has had a year or two of even high school modern history knows, mass ill-treatment in our century on this planet—and in this country—has ranged from artillery and aerial bombardment and war-time shortages of food, medicine, and housing, to sudden loss of employment or markets for a population's goods, to racism and its twentieth-century manifestations, from the gas chambers and crematoria of Nordic Germany to the segregated patterns of economic opportunity, education, and housing in all of the southern and many of the northern states of our own country. In some populations, as among the nonwhites and Spanish-speaking people of our own country, the immediate causes of family migration were and are related as directly to the desire to escape the psychological trauma of overt racism as to get away from its economic and educational and biological consequences. However, the majority of the families who migrated to other sections of the nation moved for economic reasons.

THE NEW ENCLOSURES

Interestingly enough, just as Malthus' 1798 *Essay on the Principle of Population* had been preceded by the Agricultural Revolution that more than doubled Europe's food production in less than a century, the present population explosion hysteria came on the heels of the greatest increases in agricultural production in the nation's history. Between 1946 and 1970, for example, in such crops as corn, wheat, and the protein-rich soybeans and peanuts, the federal government data reveal that both total production and yields per acre soared to new record high levels.[13]

THE CONTINUING AGRICULTURAL REVOLUTION: U.S.A., 1946–1970		
	1946–1950	1969–1970
CORN FOR GRAIN		
Production	2,808 million bushels	4,110 million bushels
Yield per acre	37.0 bushels	89.9 bushels
WHEAT		
Production	1,185 million bushels	1,460 million bushels
Yield per acre	16.9 bushels	31.1 bushels
SOYBEANS		
Production	230 million bushels	1,136 million bushels
Yield per acre	20.4 bushels	27.5 bushels
PEANUTS		
Production	2,091 million bushels	2,987 million bushels
Yield per acre	742 pounds	2,053 pounds

From *Statistical Abstract of the United States,* 1971.

Livestock rose from 60,675,000 head of all cattle, worth $2.5 billion in 1940, to 114,568,000 head worth over $21 billion in 1970.

The key to these increases lay in the productivity of the average farm worker. In 1940, each American farm worker supplied 10.7 persons with

THE NEW ENCLOSURES

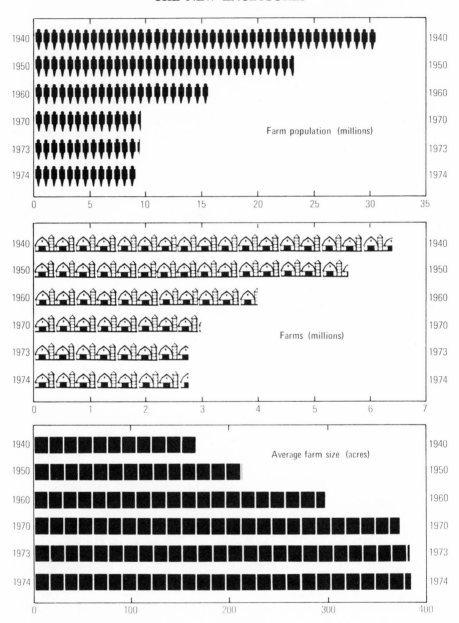

From *Statistical Abstract of the United States,* 1975.

food and other farm products. By 1970, every American farm worker supplied enough foods and fibers to meet the needs of 47.1 people. What made each farm worker that much more productive in so short a period was, of course, the new generations of farm machinery and chemical fertilizers.

Machines, however, were so expensive that the average small family farm became economically obsolete. Where in 1940 the 30,547,000 people in farm families constituted 23.2 percent of our total population, by 1970 our farm population was down to 9,712,000, and made up only 4.8 percent of our total population. Just as in eighteenth-century England the food production gains of the Agricultural Revolution were followed by the Georgian Enclosures that eliminated the small village farmers, in our own times this vast industrialization of agriculture drove the small family farmers off their lands in record numbers.

These New Enclosures also brought about certain basic changes in crops that added to our pollution problems as well as to the number of uprooted former farmers. In 1950 cotton and wool accounted for 35 pounds of the 44 pounds of fibers used per capita, and synthetic fibers such as nylon for only 1 pound; by 1968, as Barry Commoner wrote, while our total fiber consumption had climbed to 49 pounds per capita, "cotton and wool accounted for 22 pounds per capita, modified cellulose fibers for 9 pounds per capita, and synthetic fibers for 18 pounds per capita."[14] Not only did this reduce the need for workers in the production of natural fibers, but the increased manufacture of synthetic fibers added to the total load of environmental pollution, already worsened by the new chemical fertilizers.

As the New Enclosures drove the family farmers to the job markets of the urban and industrial centers, once thriving towns and villages became abandoned ghost towns in the midwestern states of the nation. In 1971, *The New York Times* ran a revealing map demarking the states most affected by these vast migrations (see next page).

Most of the millions of people involved in these migrations were of white Nordic ancestry. But the New Enclosures in the South also altered the racial composition of the states of the Old Confederacy. So by 1974 the descendants of the black slaves who had helped create the region's wealth were no longer concentrated in the South. Where, as recently as 1940, 70 percent of the nation's black people were living in the southern states, by 1970 more than half of our black Americans were living outside the South.

The blacks, to be sure, constituted a *minority* of the rural population displaced by the New Enclosures. But if this fact was known to the U.S. Bureau of the Census, which did not hide it, it was also well ignored by the demagogues of the hustings, the media, and the new scientific racism, who have succeeded in creating just the opposite impression in the minds of millions of Americans.

Since the white rural Nordics displaced by the New Enclosures of 1940–75 were as ill trained for jobs in industry and commerce as were the black agricultural workers, equal proportions of white and black American migrants were forced to seek public assistance in the absence of jobs in their new states. This meant that, between 1950 and 1970 for example, the total

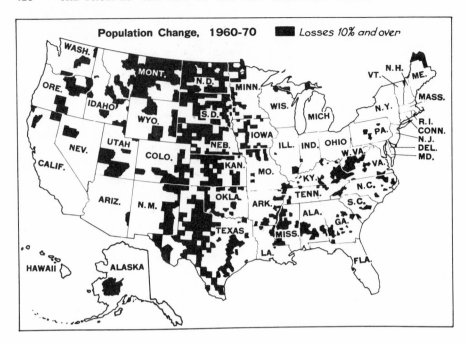

Population Change, 1960-70   ■ Losses 10% and over

federal, state, and local tax dollar costs of public assistance (welfare) climbed from $3 billion to over $16 billion per year, 57.9 percent of this welfare money coming from the federal treasury, and 42.1 percent from state and local tax revenues.

Contrary to popular myths disseminated by population and racial extremists, the vast majority of Americans on welfare are *white,* not black. According to the Bureau of the Census study of poverty-level U.S. populations released in November 1971, some 60 percent of the 25.5 million Americans living at or below the official federal poverty level were white English-speaking people; another 9 percent were Spanish-speaking Americans; and 30 percent were English-speaking American Negroes.[15]

The proportions of poor to nonpoor in the various ethnic groups that make up the American population varied greatly. They ran from 34 percent of all Negroes to 4.5 percent of all Americans classified as of Russian descent but who are known to be primarily Jews descended from Russian Jewish immigrants. Nearly 25 percent of all Americans of Spanish-speaking origins were classified as living at the poverty level. Thus the proportions of black and Spanish-speaking (that is, primarily Chicano and Puerto Rican) poor are much greater than the proportions of people of other American ethnic groups who live in poverty. The Americans of Irish origin, for example, are 10.5 percent poor. More than 8.5 percent of the English, Scottish, Welsh, and German Americans are poor, as against 6.1 percent of all Italian Americans and 5.3 percent of all Polish Americans (including a large segment of Jews of Polish immigrant origins).

Another interesting section of the Bureau of the Census report dealt with the fact that about half of the nation's poor are either too young or too old to work. Some 34 percent of the poor are under fourteen years of age, and another 19 percent are over sixty-five. About 30 percent of the nation's poor now live in the cities, while another 21 percent live in the metropolitan areas surrounding the cities.

These 25 million Americans living at or below the official poverty level, plus an equal number of Americans living on incomes fractionally above the official poverty level, provide the vast majority of the millions of white and nonwhite displaced Americans. As is well known, a large proportion of these displaced elements have to seek public assistance during their periods of adjustments to their new environments. These new groups of welfare clients, however, do not represent a population explosion but, rather, a historic migration caused by the quantum jump in the industrialization of agriculture in America since World War II.

Migrations from and to various states are not the same thing as an increase in the net rate of population growth. It is true that these population shifts have increased materially the tax and social burdens of the established and solvent populations of the urban centers to which the white and black rural-to-urban migrants gravitated. It is also obvious that these fiscal and related societal burdens have made millions of Americans only too ready to believe that the increased numbers of poor white and black migrant families on public welfare in their communities represent a population explosion rather than a population migration.

These false and self-defeating notions are a testament to the broad penetration of the myths of the population-bombers. They have no more basis in fact than does the well-exploded claim that this country has been living through decades of rising birth and fertility rates.

The United States does not have, *and never did have,* the post-World War II population explosion described in such harrowing terms by the neo-Malthusian extremists.

THE DEMOGRAPHIC TRANSITION HAS NOT BEEN REPEALED

It is not possible to make an equally accurate statement about the existence or nonexistence of the well-ballyhooed population explosion in the rest of the world, particularly in those poor, nonindustrial, nonwhite nations in Asia, Africa, and Latin America where, according to Ehrlich, the population "doubling times . . . range around 20 to 35 years."

There are, on this planet, two types of population data. The first are those collected and prepared by the census agencies of the minority of nations that, like the United States, are industrialized, nonpoor, and—with the exception of Japan, China, Australia, and New Zealand—European or North American. These nations have not only the statisticians, demographers, and clerical and supervisory workers with the professional and quasi-professional skills required for scientific population studies, but also the electronic data-handling systems, the transportation and communications networks, and, above all, the economic resources to make and utilize modern population census studies. The population data gathered by the U.S. and other census

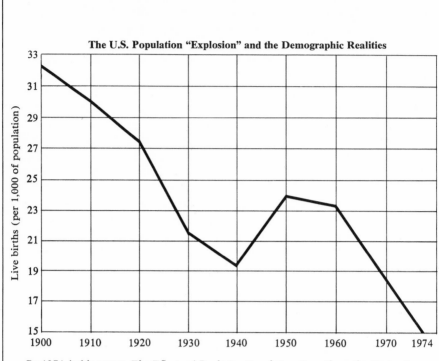

By 1974, in his report, *The Effects of Declining Population Growth on the Demand for Housing,* the Department of Agriculture economist Thomas C. Marcin noted: "The current fertility rate is . . . the lowest in American demographic history."
From *Statistical Abstract of the United States,* 1975; U.S. Bureau of the Census, 1976.

bureaus are regarded as fairly representative of population realities, give or take margins of error of from 5 to 15 percent.

The population data of the majority of the world's nations, the non-industrial, poor, and for the major part nonwhite nations, are at best what Vogt characterized as "approximations" and at worst what might well ultimately turn out to be cynical lies designed to frighten wealthy Western nations into pouring cash earmarked for "family planning" into the coffers of various hungry governments and/or the numbered Swiss bank accounts of tinhorn dictators. Even Ehrlich, by 1972, agrees that in this world "demographic statistics are often incomplete or unreliable." The reasons are as multiple as they are obvious: the poor nations lack the human, mechanical, and cash resources to make anything resembling accurate national census counts. As a result, the numbers bandied about by population extremists and their intellectual dupes in discussing world, regional, and national population trends bear what has to be, at best, an accidental and coincidental relation-ship to the demographic facts as they are.

The European population data, being the products of scientific and costly census taking and processing, are quite reliable. They also show that in Europe, as in the United States, the population explosion has existed only in the writings of the irresponsible extremists, and not in the maternity wards of the average European nation. A 1971 study of the European population data made by Jean Bourgeois-Pichat of the French National Institute of Demographic Studies revealed that birth rates have been falling sharply throughout Europe. In seven countries—Sweden, Denmark, Finland, Portugal (a Catholic nation), West Germany, Czechoslovakia, and Hungary—the birth rates have already dropped below the replacement (or zero growth) level. Switzerland, Austria, and East Germany are now growing at very close to the replacement level. In France, whose demographers consider the nation to be seriously *underpopulated,* the live-birth rate has dropped from its all-time high of 20.9 per 1,000 population in 1950 to the 16.7 rate of the years since 1968. The French nation, which averaged 2.6 children per family in 1967—when General de Gaulle established the present national population goal of at least three children per family—is still short of this goal.[16]

French demographer Georges Mauco, the head of the national com-mittee on population that formulated de Gaulle's three-children-per-family policy, is highly critical of the Zero Population Growth hysteria that has gripped many publicists and other demographic amateurs in the United States for the better part of a decade. According to Rodney Angove, who interviewed Mauco for the Associated Press late in 1971, the French demographer "thinks ZPG is a mistake for the United States. And he suspects that its sponsors are thinking more about reducing the black birth rate than the white."[17]

The AP news story also quoted Mauco as declaring that "the United States is not overpopulated. Its resources are enormous. It is technically advanced. . . . By reducing its birth rate, America would be reducing its vitality." And, he added, if the American birth rate does drop to ZPG level, the United States would be forced to import manual and other laborers, just as France has been doing from Poland, Italy, Algeria, and Portugal for many years.

There is little hard data on the new African nations, but some interest-ing reports from Asia suggest that the Paddock-Vogt-Ehrlich predictions of imminent famine in India and China are not quite about to come true. What is happening in India is particularly interesting, since it has seemed to be the one nation about which the population extremists were in closest agreement. Whatever happened anywhere else, India was doomed by the Malthusian equations to widespread famines and early doom; small wonder, then, that in his first major speech on the world population crisis in 1967, Erhlich declared that "India, where population growth is colossal, agriculture hopelessly antiquated, and the government incompetent, will be one of those we must allow to slip down the drain."

Not every American scholar who studied the population dynamics of India was as cocksurely genocidal as Professor Ehrlich. In the August 1970 issue of *Nutrition Reviews,* Alan Berg, senior fellow of the Brookings Institution of Washington, published an article based on a paper he had previously presented at an international nutrition conference at the Massachusetts Institute of Technology. In this article, Dr. Berg made a number of interesting observations.

> With current estimated infant and adult death rates in India, a couple must bear 6.3 children to be 95 percent certain that one son will be alive at the father's 65th birthday. The average number of births in India per couple is 6.5 which tends to support the increasing body of opinion that parents will continue to bear children until reasonably sure of the survival of at least one son.
>
> It is the combined reality of desire for adult sons and high child mortality that poses the crux of the population dilemma. *Paradoxically, the best way to lower the population growth rate may be to keep children alive* [italics added].
>
> . . . In some countries an effective family planning program may be difficult if not impossible without better nutrition. A policy of pursuing either alone may eventually prove untenable.

Dr. Berg's suggestion that "the best way to lower the population growth rate may be to keep children alive" was, of course, based on all modern findings in the history and mechanisms of demography. But what gives his second suggestion—improved nutrition as the essential precondition of an effective family planning program—such immediacy now is that in April 1972 India made what its officials described as a "breakthrough" in its food output, becoming a surplus nation in both wheat and rice.

In a long news dispatch that appeared in *The New York Times,* it was revealed by the Indian Minister of State for Food, Annasaheb Shinde, that the year's total grain harvest was 5 million tons higher than the 1971 record-breaking level of 108 million tons. Wheat production alone reached 26 million tons, and the Food Minister said that at the rate the Indian wheat production per acre was increasing, the American wheat output of 40 million tons would be surpassed inside of "seven or eight years." It was also revealed that, despite the fears of the Malthusian extremists, India was now self-sufficient enough to cease importing wheat from the United States under the Food-for-Peace Program. India, in fact, instead started her own Food-for-Peace Program under which she provided 700,000 tons of wheat to neighboring Bangladesh, the former East Bengal province of Pakistan which the Paddocks and Ehrlich suggested was not entitled to food aid from us.[18]

The Indian officials told the *Times* that, thanks in large measure to its own efforts in agricultural research and crop planning, India was also developing the capability of coping with droughts and other natural disasters. Under the direction of Indian agricultural scientists, for example, farmers in the chronically rice-short West Bengal had experimented successfully, said the *Times,* "with the raising of wheat during the summer season when their rice fields normally are left fallow. The state no longer depends on monsoon rains but on the vast reservoir of underground water from the Ganges River."

There would still be crop failures and even hunger in India after 1971. But after 1974, when India detonated the first of the nuclear bombs designed by Indian scientists, and built by Indian industry, it was clear that India had the scientific and technological resources required to modernize Indian agriculture. All that was now needed was a national priority system that put food and the family incomes to buy it above nuclear bombs.

The fact that the Indian government had not yet instituted the land reforms, the abolition of village usury, and food distribution policies that stand between the Indian people and chronic and/or cyclical hunger does not mean that India lacks the natural and scientific resources to end famines as effectively as they were ended in neighboring China. It can also well mean that—in our times—the Indian people might decide to end malnutrition and infant mortality the way their Chinese neighbors ended both scourges. Should this come to pass, it will not be because the hungry Indians love freedom less than their well-fed sister nations in North America. It will be because, in this world, in families haunted by chronic hunger and seared by the chronic trauma of seeing their babies die, the two freedoms they value above the freedom of speech, the freedom of religious belief, and the freedom to own and operate banks, steel mills, and oil refineries, are: (1) the freedom from hunger and (2) the freedom from the repeated tragedy of needless infant deaths. In India, where some 125 million people suffer from trachoma, an easily preventable and curable infectious disease that causes blindness, the freedom from preventable blindness, deafness, and other handicaps is yet another freedom of more urgency than, say, the freedom of the press.

In any review of the world population realities versus the neo-Malthusian myths, China is of prime interest for many reasons, not the least of which is the fact that, in 1972, two very competent and well-funded census organizations—the United States Bureau of the Census and the United Nations Population Division—both made studied estimates of the population of mainland China. Both of these are very careful, very professional, very objective institutions, with no axes to grind, no population hobbyhorses to ride. The objective estimates of these highly professional census bureaus happened to have been *120 million people apart.* The U.S. census bureau estimated the mainland Chinese population to be about 871 million people; the U.N. census bureau put the Chinese population estimate at 753 million.

This was quite a variance. The sum of 120 million people just about equals the combined populations of Australia, Canada, France, the Netherlands, Norway, and Spain. And these 1972 estimates were the informed guesses of census professionals, not woolly amateurs whose sole qualifications for the task of estimating the world's population consist of a typewriter and a publisher.

As it happened, a few months after these informed U.S. and U.N. guesses were made, the Chinese government in Peking released the results of its own census of its national population. This was not an informed guess but an actual head count made by methods as modern, and by census workers and professional supervisors as competent, as any in the world, by one of the two Asian nations (industrialized Japan being the other) with the economic and professional resources required to take a reliable census. The Chinese census authorities, by counting and not *estimating* (that is, *guessing*), came up with a census count of 697,260,000 Chinese.

This actual head count, then, was some 56 million lower than the 753 million population guessed by the U.N. Population Division, and 174 million less than the estimate of the U.S. Bureau of the Census. Putting these variances another way, the differences between the U.S. *estimate* of the Chinese population and the actual count came to the equivalent of the combined populations of Algeria, Argentina, Ethiopia, Germany, Greece, Libya, Hungary, Syria, Ireland, Lebanon, Israel, and New Zealand.

The official Chinese head count is of interest for even more immediate reasons.

It goes without saying that China has *not* suffered the endless succession of famines, epidemics, and insufferable population explosions predicted for it in prose of deepening shades of purple by Vogt and other amateur demographers. Nor did the new rulers of China heed the advice of Vogt, who wrote that the high Chinese death rate was its greatest national asset, and that medical care for the young and the sick would only make bad things worse in China. As the world knows, the Chinese not only went on graduating doctors but also trained a generation of "barefoot doctors"—the equivalent, roughly, of Western public health nurses or college-trained physician's assistants—to bring the rudiments of modern health care to the millions of peasants who traditionally never enjoyed professional medical care from cradle to grave.

These "barefoot doctors" also teach family planning and distribute pills, loops, and other contraceptive devices. In a recent article in *The New York Times*,[19] Dr. Jaime Zipper, professor of physiology at the University of Chile Medical School, one of a group of Chilean doctors who visited China in 1971, reported that the Chinese birth rates were falling much faster in the urban than in the rural areas of China. In Peking, the Chilean doctors were informed by their Chinese medical hosts, "the birth rate was down to 6 and 7 per thousand, and . . . the rate of population increase was registered at 0.1 percent annually, the lowest in the world. For Peking province as a whole, Hopeh, the birth rate was given as 17 per thousand." (The U.S. birth rate then stood at 15.0 live births per 1,000 population; Sweden's birth rate was then 14.1.)

If these Chinese birth rates have little in common with the Chinese population explosion described in such mathematical detail by the Malthusian extremists, it is equally apparent that the low-birth-rate China of 1972 had little in common with the high-birth-rate China of 1942. As Tillman Durdin, the old China hand, discovered in 1970, when the Ping-Pong thaw opened that nation to Western journalists, "the present life style in China does not encourage large families, as did the old."[20] For one thing, infant and child mortality have, according to Western physicians who visit China, been significantly reduced from their former tragic levels.

For another, as Mrs. Hu Fang-tsu, the thirty-five-year-old leader of an agricultural production team in the Machiao commune, eighteen miles west of Shanghai, told Mr. Durdin, "Children are a lot of trouble. Nobody wants very many of them any more."

Mrs. Hu and her husband live as no Chinese peasants ever lived before —in a two-room home in an apartment building near the fields in which she works. "The old idea that parents should have lots of children to honor and support them is finished," she told Mr. Durdin. "The largest family [in her commune] has only four." As Durdin reported, the Hus, like most of their friends, "work in the fields or at other jobs on the collective farm. Two very young children can be left at nurseries during the day while the parents are at work, but they need home care and feeding and this is a burden." When peasant children spend their days in nurseries, and child labor is neither needed nor used, birth rates generally drop.

To be sure, the government encourages family planning. But whether contraceptive techniques and voluntary sterilizations are as responsible for the low Chinese birth rates of today as are the other changes the Chinese revolution has brought to Chinese life—starting with family nutrition and infant and child health—is a question prudent observers should hesitate to answer before studying the situation in China, and at first hand. My own guess—and it is only a guess—is that in China, as in all the European, American, and Asian nations that have reduced their birth rates to half the size of those in poorer nations, the sociobiological improvements in the total quality of Chinese life came first, and were followed in turn by the subsequently observed reductions in the live-birth and general fertility rates.

This is not by any means an original idea; it is simply the application of the lessons of demographic history. As *Science*[21] reported from a scientific meeting in Colorado in 1970, Dr. Roger Revelle, head of the Harvard Center for Population Studies, had to warn scientists newly bitten by the population explosion bug "that the only examples of sustained population decline up to now have followed, not preceded, sustained economic growth."

These reports from India and China make it plain that the population facts and problems of the nonindustrial world—where babies and children die at rates of from ten to twenty times those suffered in the industrial nations—are not quite as simple or as clear as the extremists have made them out to be. Why, then, the almost universal acceptance of the dogma of a world population explosion, with world population doubling in less time than it takes to raise one good line of racehorses? Why do the sober and deflationary but factual reports of the U.S. Bureau of the Census about the U.S. birth and growth rates impress intelligent laymen far less than the People Pollute nonsense of the faddists?

WHAT CONSTITUTES A POPULATION PROBLEM?

Since there exists no actual modern census data on any of the poor nations of Latin America, Asia, and Africa where the populations are alleged to be doubling every five, ten, or twenty years—depending upon which population extremist you read last—it must be realized that all the population figures we have are essentially guesses. In the absence of any hard data and tested statistics on what the world's rates of population growth or decline actually are today, the interested reader could do worse than to seek models upon which one can base more accurate guesses than those used by the Doomsday extremists to panic a generation of American book-club subscribers and TV-talk-show fans.

One such model happens to be the United States, whose health, income, and vital statistics are available in dozens of standard almanacs and statistical abstracts—most of them issued by the government—in any decent library.

The United States had, as of 1974, a live-birth rate of 15.0 per 1,000 population. This 15.0 birth rate, however, was only the average birth rate of the entire population.

When the same national birth rate of 15.0 was broken down to its component parts, it could be seen that the U.S. live-birth rate actually ranges from a low of 12.2 in Connecticut, which has the third highest per capita income of all the states, to a high of 24.2 in Utah, which ranks forty-second of the fifty states in per capita income. For reasons equally related to the Demographic Transition, the U.S. infant-mortality rate of 17.7 deaths per 1,000 live births is a statistical umbrella under which huddle the 13.9 deaths per 1,000 live births experienced by white families who live in Hawaii and the 35.6 deaths per 1,000 live births suffered each year by nonwhite Americans in Mississippi. And where the annual deaths of children under five years of age, which average 482.6 per 100,000 population throughout the United States, range from 349.2 in North Dakota to 773.4 in Mississippi.

When the same data are broken down by race, they reveal that the white birth rate stands at 14.0 per 1,000 population, while the nonwhite Americans have a birth rate of 21.4. All of the other indices of health and longevity— and the economic levels associated with them—show similar differentials between the white and the considerably less affluent nonwhite populations of this country. White infant deaths occur at the rate of 15.8 per 1,000 live births, as against the nonwhite infant-mortality rate of 26.2 deaths per 1,000 births.

For every 1,000 live births, there were 10.8 fetal deaths in the wombs of white mothers, versus 18.6 fetal deaths per 1,000 live births among nonwhite mothers. Neonatal deaths—that is, deaths of babies between the moment of birth and the twenty-eighth day of life—stand at the rates of 11.8 per 1,000 live births for whites, and 17.9 for nonwhites.

The color of the mothers—and, of course, the considerable gaps between the average incomes of white and nonwhite families—is even more pronounced in the annual totals of maternal deaths. Such deaths are, by definition, caused by the complications of pregnancy and childbirth, which in turn are largely associated with lifelong and during-pregnancy malnutrition, unhygienic living environments, and the absence of proper medical care. The complications of pregnancy kill 10.7 white mothers per 1,000 live births each year as opposed to a scandalous (because so many maternal deaths are preventable) 34.6 nonwhite mothers per year.

To the old-time eugenicists, such differentials in the health indices of whites and nonwhites were proof positive that the white races were more fit to survive than the nonwhites. But the same data contain a few surprises for people who would like to believe this old myth. In most states, from Connecticut and New York to Alabama and Mississippi, the nonwhites turn out to be Negroes. In other states—such as Utah and Wyoming, or Arizona and New Mexico—the nonwhites turn out to be largely American Indians or Mexican Americans. In California, however, where nonwhites turn out to be Japanese and Chinese, as well as Negro, the hoary Galtonian hypotheses about the inborn biological superiorities of the white races are blown sky-high by all available public health facts.

Dr. Lester Breslow, the former State Health Commissioner, published an interesting study, "Health and Race in California," in the *American Journal of Public Health* in 1971. The data presented by Professor Breslow wreaked havoc on some of the strongest-held illusions of many readers.

While, as in New York and Mississippi, California's black citizens suffer the highest rates of fetal, infant, child, and maternal mortality in the state, the native and *arriviste* whites of California stand in only *third* place in these and other health indices, and in fourth place in others. In nearly every aspect of health, the two Mongoloid or Oriental subpopulations of California are healthier than their WASP and non-WASP white fellow Californians.

In infant mortality—that most sensitive and revealing of the world's health indices—the Japanese Californians suffer only 13.2 deaths per 1,000 live births. This compares quite favorably with the Swedish and Netherlands rates of 12.9 and 13.6, respectively. Chinese Californians have a rate of 15.8. The white Californians have a rate of 18.8 per 1,000 live births, or just under the national *white* infant-mortality rate of 19.2. The black Californians suffer a rate of 30.1, much higher than any other group in the state, but at the same time considerably lower than the aggregate U.S. infant-mortality rate for all nonwhites, which is 34.5, let alone the nonwhite rate of 48.1 experienced in Mississippi.

The yellow-skinned Mongoloid Californians also enjoy considerably longer life expectancies than their white peers. These life-expectancy rates, estimated by the actuarial and demographic techniques used by California public health statisticians show that, at birth, Japanese Californians have male life expectancies of 75 years, and female rates of 81 years. Chinese Californians have life expectancies at birth of 68 and 77 years, for boys in the first instance and girls in the second. While white California boys share the 68-year life expectancy of Chinese male infants, their sisters have life expectancies of 75 years. Among the Negro Californians, the life expectancies at birth now stand at 64 years for males and 70 for females.

In terms of years of life left at age 35, the Japanese Californian figures are 42 remaining years for the men and 49 for the women. But white Californians, at age 35, have an average of 37 years left to live, and the white females of the state have 43 years to go.

Many, but not all, of the reasons for the superior health indices of the Oriental Californians are of clearly socioeconomic origin. Between 1950 and 1960, for example, the percentage of Japanese and Chinese Californians employed in the census bureau's socioeconomic Class I (that is, "professional, technical, and kindred workers") rose from 4.4 to 15.0 for the Japanese, and from 6.3 to 16.9 percent for the Chinese. The white percentage rose from 10.3 to 14.0 during the same decade, while the black Class I percentage climbed from 2.6 to 4.4. Yet, in 1959, the average white male over 14 years of age earned $5,109, while the median income of the Japanese males of the same age group stood at $4,388 and that of the Chinese at $3,803. The Negro males over 14 earned $3,553.

So that rapid upward mobility, regardless of total income, emerges as a probably crucial factor in family health (and family planning). If this is the case, it must be remembered that the formerly impenetrable racial barriers that prevented any significant upward job and career mobility for Oriental Californians are as high and as rigid as they ever were for black Americans, in and out of California.

Outside of California, where the category "nonwhite" means black, Amerindian, Mexican, and Puerto Rican, the socioeconomic barriers to better jobs and upward mobility still make a dark skin fairly synonymous with poverty. And the biology of poverty operates the same in blacks as it does in whites.

Another interesting highway marker for anyone making his own maps of the present world's high and low birth rate populations was provided by the great Scottish professor of obstetrics Sir Dugald Baird in a 1946 *Lancet* article called "Variations in Reproductive Pattern According to Social Class." Professor Baird presented a table showing the relationship between the infant-mortality and the general (or total for the entire population) death rates and the live-birth rates of England and Wales between 1841 and 1939. Dr. Baird included this table because "it shows a very striking fall in the infant mortality [rate] from 150 to 50 in forty years, and a much smaller fall in the general mortality-rate."

I am including it here, however, for an additional reason. The rich, poor, and middle-class populations its data describe were overwhelmingly white, Anglo-Saxon, and Protestant—the original WASPs. Color and/or race had literally nothing to do with the changing mortality and birth rates. With this caveat, then, Sir Dugald Baird's table (opposite) is presented.

The underlying demographic lesson of this table—namely, that in every race, every culture, every continent, and every century, when a nation's total infant and general death rates fall, the birth rates go down—must be applied to the American model for the world's population guesses with some caution.

Even here the comparisons between nations must be relative, not literal: most of the world's poorer nations, where the great majority of mankind struggles to live, have socioeconomic and health conditions that are *inferior* even to those of our poorest states. As of this writing, for example, the infant-mortality rates of India (which has a live-birth rate of 41.7 per 1,000 live births) is 146 deaths per 1,000 live births. This means that the Indian infant-mortality rate is four times as great as the death rate for non-

THE DEMOGRAPHIC TRANSITION IN ENGLAND AND WALES

Mortality and Live-Birth Rates in England and Wales, 1841–1939

Year	Deaths of infants under 1 year per 1,000 live births	Standardized death-rate per 1,000 living	Live births per 1,000 population of all ages
1841–1845	148	20.6	32.3
1846–1850	157	22.4	32.8
1851–1855	156	21.7	33.9
1856–1860	152	20.7	34.4
1861–1865	151	21.4	35.1
1866–1870	157	21.2	35.3
1871–1875	153	20.9	35.5
1876–1880	145	19.8	35.3
1881–1885	139	18.7	33.5
1886–1890	145	18.5	31.4
1891–1895	151	18.5	30.5
1896–1900	156	17.6	29.3
1901–1905	138	16.0	28.2
1906–1910	117	14.4	26.3
1911–1915	110	13.7	23.6
1916–1920	90	13.4	20.1
1921–1925	76	10.9	19.9
1926–1930	68	10.3	16.7
1931–1935	62	9.6	15.0
1936	59	9.2	14.8
1937	58	9.2	14.9
1938	53	8.5	15.1
1939	50	8.5	14.9

The demographic transition in action during the Industrial Revolution, the great sanitary reforms, and the century of social legislation that raised family incomes, reduced working hours, and made jobs and housing safer for most people in England and Wales. Note, particularly, the dramatic decreases in infant mortality that preceded the drop in the live birth rates to record low levels.

white infants in Mississippi and more than eight times as great as the death rate for white infants in America's poorest state. It also means that India's infant-mortality rate is nearly sixteen times as high as affluent Sweden's rate of 9.2 infant deaths per 1,000 live births.

It is a fairly well known fact that none of the infants who make up what is described as the world population explosion can breed new environmental polluters until they themselves have attained sexual maturity. That is, until they are about fourteen or fifteen years old. Therefore, the readers who use the American and world demographic data as models for their own world population trend guesses can also come up with an interesting hunch or two about the probable validity of the Doomsday extremists' estimates about the world population doubling every decade. Particularly when, as is well known, between 40 and 50 percent of all the children born alive in the world's poor and so-called "developing" nations of Asia, Africa, and Latin America die before they reach the age of five years.[22] To be sure, as a 1971 World Health Organization survey of African nations showed, in the poor nations "the 10 principal causes of death in infancy and childhood are diseases that could be prevented." The fact is that they are the infectious and deficiency diseases of poverty, and as long as these nations remain poor their death rates will prevail unchanged.

In making your own estimates of population trends in those nations that cannot afford to take meaningful censuses, all that you need, along with the *Statistical Abstract of the United States,* is a copy of any standard data book on world health. The best of these, the United Nations' annual *Statistical Year Book* and the World Health Organization's annual *Demographic Yearbook,* are available in most good libraries. Their data on infant and child and maternal mortalities, collected by the public health agencies of all the U.N. member nations, most probably err on the conservative side, since most nations are understandably ashamed to reveal the full extent of the rates of medically avoidable deaths in their populations.

The U.S. and U.N. data are probably accurate enough to give the intelligent reader some insight into the realities as against the commonest myths of the "world population explosion." Starting with the basic reality of all demographic history: there never was, and is not now, such a thing as a "population problem" in this or any other nation. There is only a *socio-biological* problem called poverty, of which high death rates and their related high birth rates are and always have been two of the most universal and predictable effects. There is also an equally well understood socio-biological condition called affluence, which is characterized by low death rates and the parallel low birth rates that invariably follow the reduction of family or national death rates.

DEMOGRAPHIC TRUTH AND CONSEQUENCES

As anyone who studies the available data can discover for himself or herself, the death rates of the world's poor nations—and the poor subpopulations of the world's affluent countries—are today high enough to negate any theoretical danger of a population explosion that would make this world or this nation too crowded for *Lebensraum.* Children biologically and statistically doomed to die shortly after birth, and long before they reach breeding age themselves, crowd only the cemeteries, and menace none of us while they live.

The Demographic Transition is not very difficult for the unbiased mind to grasp. A dramatic case in point was provided at the 1974 United Nations world conference on population problems by John D. Rockefeller 3d, the former chairman of the U.S. Population Commission and for forty years an advocate of family planning approaches to the world's population problems. "I have now changed my mind," he said in Bucharest, at the start of the world conference. "The evidence has been mounting, particularly in the past decade, to indicate that family planning alone is not adequate."[23]

In words that sound like a modern echo of those of Scott Nearing's in 1911, Mr. Rockefeller added: "This approach recognizes that rapid population growth is only one among many problems facing most countries, that it is a multiplier and intensifier of other problems rather than the cause of them. And it recognizes that *motivation for family planning is best stimulated by hope that living conditions and opportunties in general will improve"* (italics added).

Facing the hard facts of poverty and its sociobiological consequences in family life, Mr. Rockefeller had the intelligence and the moral courage to call for a "reappraisal of all that has been done in the population field."[24]

Hard facts, however, never bothered the peddlers of demographic and racial myths. In our times, and in such matters, one fake scare seems to nourish and supplement the other. Thanks to the population extremists, conventional wisdom now has it that this nation is presently in the throes of a population explosion; that it is people, and not machines, that pollute; that the undeserving American poor, who are largely nonwhite and Spanish-speaking in many states, but certainly not in all, live only to collect welfare dollars; that these subhuman subpopulations keep breeding babies to collect more welfare dollars; that by nature and heredity these American poor are too dedicated to the pleasures of untrammeled sexuality to do anything to help themselves. Therefore, say the Malthusian zealots, until and unless they can prove their human worthiness by reducing their birth rates to those of the affluent American families, society is justified in writing them off as a hopeless and parasitic drain on the tax dollars of all hard-working, decent, and Silent Americans.

This classic posture of scientific racism is not always, perhaps, conducive of gut racism as well. But only six years after the High Road Osborn and the Low Road Vogt led us by the noses into the morass of the population explosion mania, the United States Supreme Court in 1954 precipitated the continuing racial crisis that in turn led to the full-scale revival of the good old black-baiting, Jew-baiting, poor-baiting, IQ-shouting, eugenically slanted scientific racism of the generation that gave us Auschwitz and Pearl Harbor. The population explosion became an integral and important segment of the new synthesis of the old gut racism and the new scientific racism. It was a combination as natural as the teaming of Nixon and Agnew, as self-evident as Nordic Supremacy.

18 The New Scientific Racism Makes Lavish Use of the New Military Mental Tests— Which in Turn Reveal Either Two Evolutionary Miracles or One Gigantic Hoax

> There is no evidence that feeding people makes them smart,
> but it is indisputable that hunger makes them dull.
> —CHARLES UPTON LOWE, M.D., Chairman, Committee on
> Nutrition, American Academy of Pediatrics, in a statement
> to the U.S. House of Representatives, April 24, 1969

Between the administration of the Binet-type Army Alpha and Beta intelligence tests of World War I to nearly two million military recruits, and the unanimous decision of the U.S. Supreme Court in the case of *Brown* v. *Board of Education* in 1954, many things had happened to the once untouchable IQ testing mystique. There were, among other events, the careful analytical studies of Klineberg that showed the one-to-one relationship between each state's educational expenditure per pupil and the Army Alpha and Beta test scores of its young recruits in World War I. (See page 228.) There was even, by 1947, the belated realization by Lewis M. Terman himself that both the family income and the number of books in a child's home played significant roles in the student's IQ test scores, and that:

"No race or nationality has any monopoly on brains. The non-Caucasian representation in our gifted group [i.e., the California children with high IQ test scores] would certainly have been larger than it was *but for handicaps of language, environment, and educational opportunities*" (italics added).[1]

By World War II, the needs of the armed forces for competent officers and noncoms were too great to permit intelligence tests of the World War I modality to be inflicted upon our draftees. It was far more important to the nation to select the best potential combat officers, squad leaders, and military technicians from the millions of civilian recruits than to judge military potentials on a draftee's ability to identify the best right-handed pitcher of the New York Giants, or correctly differentiate between the advertising slogans of brands of breath deodorants and the Kama Sutra.

In place of the Binet-type tests devised in six weeks by Terman, Goddard, Yerkes, and other early IQ test authors, the armed forces relied on more modern *aptitude* tests to help screen most likely candidates for officer training in the special skills of modern warfare. After the war, the military services evolved the Armed Forces Qualification Test (AFQT) as their "basic military mental test." Dr. Bernard D. Karpinos, the Army's Special Assistant for Manpower Studies, made it clear, however, that as with the older Alpha and Beta tests, and the civilian IQ tests:

The examinee's score on the tests depends on several factors: on the level of his educational attainment; on the *quality* of his education (quality of the school facilities); and on the knowledge he gained from his educational training or otherwise, in and outside of school. These are interrelated factors, *which obviously vary with the youth's socio-economic and cultural environment,* in addition to his innate ability to learn—commonly understood as I.Q. Hence, *the results of these mental tests are not to be considered as measures of I.Q. . . . nor are they to be translated in terms of I.Q.* [Italics added.][2]

When the Supreme Court ruled the segregation of children on a racial basis to be unconstitutional in 1954, however, the enemies of educational equality did precisely what Dr. Karpinos warned could not be done. The AFQT test scores were cited, by the new scientific racists, as proof of all the ancient and discredited claims of the Brighams, the Lothrop Stoddards, and the Cornelia Cannons to the effect that American intelligence was declining, that non-Caucasian children with black skins were *genetically* the mental inferiors of children with white skins, and that the societal provision of equality of educational opportunity to black children was a total waste of taxpayers' dollars.

Since it was Shuey, Jensen, Shockley, and their followers who made AFQT test scores integral to their post-1954 analyses of what they pictured as black intellectual inferiority, I felt it necessary to examine not what special pleaders wrote about the AFQT data but the raw data themselves. By 1971, when I went to Washington to collect the official reports on the AFQT tests, I was of course aware of the findings of many of the studies of human body, brain, and intellectual development made in this century. What I did not anticipate was that, as I encountered the voluminous data released by the Office of the Surgeon General on the physical and mental health of the young Americans who took these AFQT examinations, I would be left with the feeling that I was either (a) being confronted with two major evolutionary miracles that for some reason had been kept a top secret from all civilians but myself, or (b) that I was examining the raw materials of one of the most expensive and massive sociobiological hoaxes ever committed with the taxpayers' money.

If the AFQT test scores could be taken at face value, they meant that not one but *two* new races of young American males had been evolved in our lifetime. The first was a race of white American males with strong minds and weak bodies. The second evolutionary miracle was the simultaneously evolved race of black American males with strong bodies but weak minds. Since both of these new races had apparently evolved in the space of less than a generation, rather than in the normal span of thousands of years Nature has hitherto required to select and evolve new species or new races, a quick review of the scientific consensus on the nature of human physiological and mental development was and remains in order.

THE SOCIAL BIOLOGY OF MENTAL TESTING

Before and after 1969, when Jensen cited the AFQT scores as documentation for his thesis that black children are innately inferior in mental capacity to white children, what Terman described as the "handicaps of language, environment, and educational opportunities" that even he now recognized as contributing causes of the lower IQ and classroom test records of the average black child had been well defined and exhaustively studied.

There was, for example, the matter of the home environment of growing children. In Terman's study population of California children with high IQ test scores, "The number of books in the home library . . . ranged from none to 6,000 *with a mean of 328.* Only 6.4 percent of the homes had 25 books or fewer, as against 16.5 percent reporting 500 or more. Comparative figures for the general population of California are not available, but the mean for the generality would certainly be far less than 328" (italics added).[3]

Poor families that cannot afford to buy their children the minimum amount of food, clothing, shelter, and medical care required for proper mental and physical growth are certainly unable to purchase any books at all. Therefore, it should come as no surprise to read, in the very next paragraph of Terman's 1947 report, *The Gifted Child Grows Up* (Volume IV of *Genetic Studies in Genius*):

> Information on family income (as of 1921) was requested from only one hundred seventy families representing a fairly random sampling of all families [of children with high IQ test scores]. These reported a median of $3,333 and an average of $4,705. Only 4.4 percent of the families reported $1,500 or less, and only 35.3 percent reported $2,500 or less. At the other extreme were 14.1 percent reporting $8,500 or more, and 4.1 percent reporting $12,500 or more. *Although the median for the one hundred seventy families was more than twice as high as for the generality of families in California,* only one family in twelve had an income of above $10,000 [italics added].[4]

These income data meant that a family did not have to be millionaires to be able to afford the books, food, medical care, shelter, and other environmental requirements for proper mental development; they merely had to have incomes only "twice as high as for the generality of families in California" and all other states of the Union. These data also meant that the average California family did not earn enough money to adequately meet the mental growth and development needs of any of their children.

If the various continuing surveys of health and income being conducted by the National Health Survey and other agencies of our government since the close of World War II are accurate, we as a nation have not progressed very much in this respect since 1921, when Terman's group established their California baselines.[5]

The relationships between low income, malnutrition, iron-deficiency anemia, eye and ear defects, low birth weight, the preventable but endemic infectious diseases of poverty, and retarded brain and mental development, were all very well established by 1969. Few people have described this tragic and preventable complex of causes and effects better than the pediatrician and scientific director of the National Institute of Child Health and Human Development, Charles Upton Lowe. In 1969, testifying before the

House of Representatives in his other capacity as chairman of the Committee on Nutrition of the American Academy of Pediatrics, Dr. Lowe, basing his statement on millions of government data, told the Congress that "of the 58 million children in our country under 15 years of age . . . almost 11 million, or 24%, live in poverty." And that "one-fourth of the nation's children, and one out of every three children under 6 years of age, are living in homes in which incomes are insufficient to meet the costs of procuring many of the essentials of life, and particularly food."

Poverty, he reminded the Congress, "is more a white problem than a black problem," since government estimates now showed that "a total of 21 million white people and only 10 million black people live at or below the poverty level."[6]

Dr. Lowe noted that in the United States, as in all other affluent nations, the prematurity rate of the poor—that is, the premature and immature babies born too soon or full-term but with abnormally low birth weights—"is from two to four times as high as it is in middle-class communities. In suburban American counties, the prematurity rate amounts to 5%. In the inner city of certain urban slums, it may amount to 25%. *This figure is of great importance when we realize that up to 75% of all infant mortality in the first month of life results from immaturity [low birth weight]*" (italics added).

Although he was testifying on behalf of efforts to obtain more food for poor children of America, Dr. Lowe was too much of a pediatric scientist to suggest that food alone would solve everything.

> This waste of human lives reflects a constellation of environmental and social conditions and cannot be ascribed to a single variable. What is most critical for the future of our country is the fact that children in the culture of poverty, particularly those prematurely born, contribute the major numbers of those of our population who grow up to be mentally retarded. Depending upon the degree of prematurity or immaturity, up to 50% of low-birth-weight infants may, upon survival, have intellectual quotients [IQ test scores] below 70%. In short, they are too dull to fend for themselves. It has been estimated that *approximately 75–80% of all mentally defective children are born in a poverty environment* [italics added].

The revival of the IQ mystique of the 1920's was making headlines when Dr. Lowe testified, and he had this in mind a few pages later in his statement:

> Although an increasing body of scientific data suggests a relation between ultimate intellectual accomplishment and educability of the child with the nutritive condition of the pregnant woman, the infant and the preschool child, the argument concerning the relative importance of genetic endowment, as opposed to environmental influence and nutrition, is hardly germane. Inheritance may establish a ceiling above which each child cannot achieve. However, when malnutrition is coupled with the constellation of adverse environmental factors that are characteristic of life in poverty, it is clear that intellectual growth will be jeopardized and in this way *preclude attainment of the maximum*

genetic endowment with which each child is born. There is no evidence that feeding people makes them smart, but it is indisputable that hunger makes them dull [italics added].

Dr. Lowe's testimony was given to Congress in 1969, or Year Five of the massive escalation of American aerial bombing and ground fighting in Vietnam. The poverty he described was, of course, then and in subsequent years exacerbated by the predictable inflation caused by this multibillion-dollar-per-year military adventure. By January 1974, for example, the U.S. Senate's Select Committee on Nutrition and Human Needs, reported:

> The prevalence of low birth weight and premature babies among low income families is staggering. For every 10 low birth weight infants born into families with less than $3000 income, 6.2 low birth weight infants are born into families with income above $10,000. The incidence of low birth weight in the United States is increasing. Among white infants, the incidence has hovered about 7 percent. The incidence among nonwhite infants increased from 10.4 percent in 1950 to 13.8 percent in 1964. To a large extent, these small babies are the result of poor fetal nutrition. . . . *All the statistical data available reaffirms the connection between poverty and malnutrition, between malnutrition and disease.* Malnutrition appears to be the common denominator of each of the problems—low birth weight, infant mortality, mental retardation, and intellectual malfunction [italics added].[7]

Since malnutrition and all of the other "adverse environmental factors that are characteristic of life in poverty" are functions of low family incomes, the standard income data published in the *Statistical Abstract of the United States* must always be looked at in conjunction with health and mental and school examination data. For example, if we were to average the per capita personal incomes of the ten most affluent states (Connecticut, New York, Alaska, Nevada, New Jersey, Hawaii, Illinois, California, Massachusetts, and Ohio) as of 1969, and compare it with the average per capita income of our ten poorest states in 1969 (Louisiana, Kentucky, Tennessee, New Mexico, North Dakota, West Virginia, South Carolina, Alabama, Arkansas, and Mississippi), we would emerge with two figures.

The ten highest-ranking states had an average per capita income of $4,236.1, while the ten poorest states had an average per capita income of $2,748.9. These numbers should predict the relative proportions of premature or immature infants born each year, the infant- and child-mortality rates, the mental retardation rates, and other standard parameters of human physiological and mental health and development. They do. In World War I, the comparative state personal income data were equally reflective of state educational expenditures per pupil and of the Army Alpha and Beta test scores of their recruits. That, however, was before the seeming miracle revealed in the AFQT scores, which will be discussed shortly.

But before we examine the anatomy of these miracles, let us examine a few more data derived by pediatricians and other biomedical investigators in this and other countries.

A 1971 study reported in the *Indian Journal of Nutrition and Dietetics,* "Protein Intake and Mental Abilities of Preschool Children," found that in a group of sixty children in a village near Coimbatore City, India, the nutritional status was good for 8 children, fair for 40, and poor for 12.

"The mean height, weight, and head circumference and the hemoglobin levels were highest for the children with good nutrition. The mean IQ's of children in the good, fair, and poor groups were 122, 90, and 76 respectively. Protein intake and the clinical scores of the children correlated significantly with their IQ test scores, and the educational level of the parents had a positive correlation with the IQ's of the children."[8]

The continuing studies of child health and development being conducted in Scotland by Dr. Cecil Mary Drillien of the University of Edinburgh are well known to all the world's better-trained pediatricians.[9] As her table below reveals, while low birth weight is a reliable indicator of low IQ test scores, low family income is an equally reliable indicator of low birth weight.

Mean IQ Scores at 11–13 years by Birth Weight and Social Grades

Social Grade*	596 Edinburgh Children, Birth Weight (in pounds)					
	under 4.4 lbs.		*4.4 to 5.5 lbs.*		*over 5.5 lbs.*	
	N.†	Mean IQ	N.	Mean IQ	N.	Mean IQ
1 and 2	43	100.0	61	105.7	76	109.9
3	35	90.7	78	97.6	54	100.4
4	21	83.7	29	82.6	27	90.9

* Social Grades—1 and 2: middle-class and superior working-class homes; 3: average working-class homes; 4: poorest homes.
† N. = Number of children.

Source: "School Disposal and Performance for Children of Different Birthweight Born 1953–1960," *Archives of Diseases of Children,* 1969, 44:562.

As this well-known study by the pediatrician Cecil Mary Drillien of the University of Edinburgh demonstrates, IQ test scores, as well as academic achievement test scores, are functions of both birth weight and family income. Dr. Drillien found that "the incidence of moderate or severe handicaps increases with decreasing birth weight, particularly at weights of 2000 grams [4.4 pounds] and under."

The 596 children in Dr. Drillien's study population are all white and Nordic. Yet what she found in these Scottish white children was matched in every significant particular by what Dr. Mark Golden, an assistant professor of psychiatry at New York's Albert Einstein College of Medicine, and his collaborators found in their longitudinal study of 89 black children from different social classes. And what they found was that in this group of American black urban children, "there was a highly significant 23-point mean IQ difference in the Stanford-Binet at 3 years of age between children from welfare and middle class black families."[10]

In Dr. Golden's 89 black children in 1968–71, as in the 831 white children Terman and Merrill studied in 1937, family income was a most reliable predictor of children's IQ test scores. The results of the Golden study were summed up in the following table:

Comparisons of Stanford-Binet IQ Scores of Black Children in Longitudinal Study Classified by Hollingshead's Modified System (1968–1971), and White Children from Terman and Merrill's Standardization Sample (1937)

Social Class	IQ
Black Children in Longitudinal Study Classified by Modified Hollingshead System	
1 (middle class)	116
2 (working class)	107
3 (lower class/nonwelfare)	100
4 (lower class/welfare)	93
Terman and Merrill's White Children Classified by Father's Occupation	
I (professional)	116
II (semiprofessional managerial)	112
III (clerical, skilled trades, retail)	108
IV (rural owners)	99
V (semiskilled, minor clerical, small business)	104
VI (semiskilled laborers)	95
VII (unskilled laborers)	94

Adapted from Golden et al., 1971.

What these family income-IQ test score data showed was that now, as in 1937, the more a family had to spend on the health, nourishment, and cultural development of their offspring, the higher the IQ test scores achieved by their children—and vice-versa.

All of the modern investigations of human development, from the massive National Health Survey of millions of American children and parents to the 89-child Golden et al. study, share more than a more or less similar set of results and conclusions. The findings of all of these studies have nothing in common with the results of the more recent Armed Forces Qualification Test (AFQT).

THROUGH THE LOOKING GLASS

Unlike the historic *Memoir XV* of the National Academy of Sciences (1921), in which Robert Yerkes and his collaborators evaluated the Army intelligence test scores of World War I primarily in terms of race, regional and national origins, and previous education of the military recruits tested between 1917 and 1919, the modern reports by the Office of the Surgeon General on the AFQT tests evaluated the results of the mental tests in conjunction with the medical examination findings of the same study population. Thus it was recognized that the human worth and capacities of the draftees could not be measured by IQ or AFQT or any other mental test alone. Like the scientific director of the federal government's National Institute of Child Health and Human Development (see pages 436–36 and *passim*), and the U.S. Senate Committee on Nutrition and Human Needs (see page 436), the medical and behavioral professionals in the Army's medical branch understood quite clearly that, in our species, a sound mind is nearly always attached to a sound body.

Because the medical examination results were available, I decided to see how the AFQT findings[11] dealing with clinical conditions known to have effects upon physical and mental functioning compared with the data available in the hundreds of reports of the National Health Survey and other governmental agencies. These conditions are: (1) eye diseases and defects, (2) ear and mastoid process diseases, (3) psychiatric disorders, (4) asthma, (5) neurological diseases, and (6) rheumatic fever and chronic rheumatic heart disease. All of these groups of disorders are known to have a continuum of adverse effects on the learning processes and the development of the potentials of innate human mental acuities.

Asthma and rheumatic heart disease, while not normally thought of as diseases involved in IQ and classroom test achievements, happen to be, among other things, two major causes of classroom inattention and prolonged absences from school. Asthma is a stress-related disorder. Rheumatic fever, the aftereffect of a "strep throat" and the precursor of chronic rheumatic heart disease, has been, since the advent of penicillin, a medically preventable disease of poverty.

The proportion of the poor to nonpoor being significantly higher among American Negroes than among American whites, it should therefore follow that the preinduction examining physicians would find considerably more rheumatic fever and chronic rheumatic heart disease among black draftees than among their white peers.

This, however, is not what the AFQT data reveal. What the preinduction examination records actually do show is summarized in the following table:

Rejection Rates for All U.S. Draftees for All Medical Defects, Per 1,000 Draftees Examined for First Time, August 1969–January 1970.
(Including Specific Health Disorders Known to Affect Intelligence, School Attendance, Learning Capacities, etc.)

	United States		Middle Atlantic		South Atlantic		East So. Central	
	White	Negro	White	Negro	White	Negro	White	Negro
All medical defects	328	233	368	352	305	180	329	197
Eye diseases and defects	19	21	21	29	18	20	18	21
Ear and mastoid process diseases	22	17	15	15	22	9	15	11
Psychiatric disorders	20	9	37	28	14	5	8	3
Asthma	15	8	18	13	13	5	10	6
Neurological diseases	5	2	6	3	5	2	5	2
Rheumatic fever and chronic rheumatic heart disease	3	2	3	5	2	1	2	1

PERSONAL INCOME PER CAPITA—1969
UNITED STATES—$3,698   *South Atlantic States*—$3,304   *East South Central States*—$2,724
Middle Atlantic States—$4,193   [Del.; Md.; D.C.; W.Va.;   [Ky.; Tenn.; Ala.; Miss.]
[N.Y.; N.J.; Pa.]   N.C.; S.C.; Ga.; Fla.]
INCOME OF FAMILIES BY RACE, U.S., 1969
WHITE—$9,794   NEGRO—$5,999

From AFQT data available in Karpinos, 1969 and 1970, and Bellas and Vinyard, 1971.

The table shows that only one of these six disorders is more prevalent among America's black draftees than among whites, and that this disorder is not rheumatic fever and its sequelae, but eye disorders.

Obviously, eye disorders and defects are also learning handicaps. Reading is, after all, a visual skill. And since most vision-limiting eye defects found in eighteen-year-old males either are environmental in origin, or, even if wholly genetic, can be detected and corrected long before a child enters school, it should therefore be expected that children of school age with undetected and uncorrected eye diseases and defects will get lower IQ and class work test scores than children with naturally good or medically corrected vision. This would therefore offer not only a biological explanation for many of the lower mental test scores rung up by poor white and black draftees: the same biological explanation would also be very much socio-economic, that is—environmental—in origin. Poverty—*even inherited poverty*—is not in a child's genes but in the intrauterine and postnatal environments. It is certainly not a child's genes that determine whether or not his or her parents will have enough money to afford cradle-to-kindergarten pediatric care, eye examinations; and eyeglasses.

Since family income, as Lewis Terman and other psychologists have long reported, is so integral to the mental and academic test score achievements of all children, it might be well to examine some of the standard family income data before looking at the mental test data of the AFQT programs.

According to the *Statistical Abstract of the United States,* the mean annual income of the average American white family in 1969 was estimated to be $9,974. The annual income of the average black family was estimated to be $5,999, or 60 percent of what the average white family earned.

Higher incomes, in this and all other nations, and among people of all races, mean better nutrition, better living conditions, and better health. While within each racial group the mental test score results do vary enormously, and for pretty much identical sociobiological reasons, in the United States the proportion of black poor to black nonpoor are far higher than are the proportions of white poor to white nonpoor. In his 1969 testimony before the Congress, Dr. Charles Upton Lowe estimated on the basis of government data available to him that "40% of the black people in this country live in poverty as compared to 12% of the white population." This being the situation, it would naturally be expected that the white draftees would prove to be physically—that is, medically—and mentally in at least 40 percent better shape than the less nourished, less medically cared for, less affluent black eighteen-year-olds, since the income of the average white family is 40 percent higher than that of the average black family.

As we saw in the table on page 439, the medical findings of the AFQT examination centers ran counter to those of other federal studies of health, disease, and family income. The findings on mental acuity in the AFQT centers were equally anomalous.

In Alabama, for example, forty-seventh in rank order of all states in terms of personal income per capita, and fiftieth in terms of the amount of money—in this instance, $503—spent each year on the education of each elementary and high school pupil, the preinduction mental test scores of white eighteen-year-olds were *higher* than the test scores of white draftees in twenty-one states that spent more on education. Eleven of these states were southern states; the non-southern states included New York, Connecticut, New Jersey, California, Rhode Island, and Arizona, all of which spent between two and three times as much per year per student in their public elementary and high schools.

Nor were the preinduction mental test scores of the white Alabama youths only *fractionally* higher than those of the white draftees in the twenty-one states that spent more than Alabama did on public elementary and high school education. They were *much* higher. In Alabama, only 2.1 percent of all white draftees tested for the first time failed their preinduction mental tests. In other states, the percentages of white draftees who failed their preinduction mental tests included: New Jersey, 7.9 percent; New York, 4.8 percent; Connecticut, 3.9 percent; Hawaii, 5.5 percent; Illinois, 3.5 percent; North Carolina, 6.6 percent; and, among others in the South, Georgia, 6.3 percent; South Carolina, 4.5 percent; and Mississippi, 4.1 percent.

In 1967 nearly five times more blacks than whites were rejected for failing the AFQT mental requirements. When it came to medical disqualifications, however, the rates were 26.9 percent for the whites and only 14.7 percent for the blacks.[12] Nor did family income seem to have any bearing on the health of America's young men in both ethnic groups.

Medically, according to the same armed forces preinduction test data, Mississippi—the poorest state in the Union—was also, at least for white and black eighteen-year-olds, one of the healthiest states in the nation. For in this impoverished state, with infant-mortality rates of 22.9 per 1,000 live births for white babies and 48.1 for black babies—and with the nation's highest incidence of death in children of all races who were between one and five years old—the health of all the state's draftees was unbelievably excellent.

At their first preinduction medical examinations, 33.4 percent of all the Mississippi white youth, *and only 10.7 percent* of all black youths in the state were medically disqualified. These medical examination data compare more than favorably with more affluent states, such as Nebraska, where 35.4 percent of all white draftees and 28.1 percent of the blacks were medically disqualified.

Mississippi was not alone in this medically amazing bracket. In most southern states, while black youths *invariably* showed up as stronger and healthier in their preinduction medical examinations than did the whites of the same states, their rates of medical disqualifications were also, with equal predictability, far lower than those of black youths examined by physicians in most states located outside of the Old Confederacy.

No matter where one looks in these AFQT results, whether in mental or physical conditions, the same glaring disparities between these and all other civilian and military data published by all other agencies of the same federal government become apparent. Even the pathologically biased Yerkes and his similarly oriented collaborators of 1921 made it very plain in *Memoir XV* that, among whites and blacks, the draftees from the largely southern states with the lowest average expenditures per year per elementary and high school pupil achieved the nation's lowest Alpha and Beta Army intelligence test scores.

Consider the matter of hypertension, which is now known to be the number-one killer of black people in this country. Hypertension, or high blood pressure, is a controllable chronic disease on which the public health and medical research agencies of federal and state governments have spent millions of dollars in epidemiological and therapeutic studies. The National Center for Health Statistics[13] has estimated that hypertension is more than twice as prevalent among the black poor as among the white poor—appearing in 37.1 percent of the black males with incomes under $2,000 and in only 16.7 percent of the equally poor and more numerous white males. Even in the income bracket of $10,000 or over, 26.6 percent of the black males as against only 11.6 percent of the affluent white males suffer hypertension. Oddly enough, in the $7,000 to $9,999 annual income bracket, the black males who suffer hypertension add up to only 5.4 percent, while 10.6 percent of the white males in this annual income bracket have hypertension.

Yet, in the populations of young white and black males examined in the AFQT centers, only 21 black draftees per 1,000 examined were rejected on medical grounds for high blood pressure—as against an unbelievable 32 in every 1,000 white draftees.

The more I contemplated these and all the other numbers in the table, the more I began to feel as if I had parachuted into Cloud Cuckoo Land. To make sure that the entire official volume had not been, after all, an antic printer's jape, I checked out all of the preinduction mental and medical examination data in the U.S. armed forces for the previous fifteen years, as well as for the period after 1969. The actual numbers for each year varied, of course, but the essential variations between the examination scores of white draftees and black held more or less constant. According to these armed forces examination data, for over a decade American white youths aged eighteen were physically inferior but mentally *superior* to black youths of the same age in the same states and regions of the Union. Somewhere there had to be a logical explanation for the apparent anomalies—or we were in the presence of an entirely new human biology.

I therefore tried to reach Dr. Karpinos at the Medical Statistics Agency, in the Army Surgeon General's Office, only to learn that he had just retired. From his successors and their assistants, I discovered that I was not alone in my inability to accept the validity of the test score data of the past decade at face value.

On page 17 of the June 1969 *Supplement to the Health of the Army,* I found that these glaring disparities between biological logic and the examination test results of mind and body had also proved to be of deep concern to the armed forces' biometricians and their medical superiors. *"Strange as it may appear,"* wrote Dr. Karpinos in the foreword to this volume, *"low disqualification rates for mental reasons—suggesting relatively better socioeconomic status—coincide with relatively high disqualification rates for medical reasons, and vice versa"* (italics added). The military health statisticians, I was thus relieved to read, were also aware of twentieth-century biomedical realities.

After discussing some suggested explanations for these anomalies, Dr. Karpinos declared that "actually, there is no fully reliable answer at present to the inverse relationship. It is for this important reason that a special medical disqualification study has been undertaken by the Department of Defense."

This report, on draftees disqualified for medical reasons at their first preinduction examinations between 1958 and 1970, was prepared by Colonel Joseph J. Bellas, a physician and chief of the Physical Standards Division of the Army's Health and Environment Directorate, and Mr. John Vinyard, Jr., statistician, Patient Administration and Biostatistics Branch. Through the kindness of the Office of the Surgeon General, Department of the Army, I obtained a copy of their report, dated September 1971.[14]

WHAT THE SURGEON GENERAL'S PROBE REVEALED

The Bellas-Vinyard report offers *no* explanations for the disparities evident in the preinduction examination results. But it does offer some hard new data on health and educational levels that raise even newer questions, as well as some candid insights which suggest that the Army, like any good university faculty, is not bereft of acutely intelligent people.

For openers, the report avoided completely the ultrasensitive area of the anomalous mental test results. This showed rare good judgment, since the Army is notoriously allergic to anything resembling public controversies over any of its policies and operations.

While ignoring the mental disqualifications, the authors of the report, in examining the raw data, did have the mature good sense to recognize that no eighteen-year-old male on this earth lives in a sociobiological or historical vacuum. Thus it comes as a refreshing surprise to read, in this Army study, the recognition that history might possibly play a significant part in these examination results.

On page 23 of the Bellas-Vinyard report, for example, the authors wrote:

> The sharp increase in medical rejection rates noted in 1966 cannot be ignored. It is not believed to be coincidental that this increase occurred when military activity intensified in Southeast Asia. *The general unpopularity of the war coupled with the general unrest of young people and the blossoming of the drug culture must certainly have had an effect on the rejection rate.* Exactly what effect cannot be measured by the draft data in this report, but it is generally conceded that *the war has become more unpopular and youths more restless, and the medical rejection rate has certainly continued to increase.* [Italics added.]

As reported above, most of the increase in medical disqualifications of draftees of all American racial groups had occurred within the previous four years. "The 1966 rate was only 11 percent greater than the 1958 rate, but the 1970 rate was 43 percent greater than the 1966 rate." This was for all races. For white draftees, "there was an 8 percent increase from 1958 to 1966, and a 42 percent increase from 1966 to 1970. Comparable figures for Negroes were 18 percent increase 1958 to 1966, and 52 percent increase 1966 to 1970." Obviously, the youth of America could hardly have become so unhealthy in such great numbers and in so short a time as four years.

The previously unpublished data on the education of the decade's draftees, which the authors extracted from their Selective Service records, were as anomalous as all the other numbers. By definition, a sound mind is more often than not a function of a sound body. But in the U.S. preinduction data, the more educated the draftee, the more *unsound* his body, and the greater his chances of being medically disqualified. In the following table, taken from the Bellas-Vinyard report, we can observe that here another solid axiom of modern biology seems to bite the dust:

The Relationship Between Educational Levels of Examinees and Medical Rejection Rate, August 1969–January 1970

Educational Level	Medical Rejection Rate per 1,000 Examined	
	White	*Negro*
All levels	328	233
Grades 0–8	271	189
High school graduates	301	230
College, 1 year	375	312
College, 4 or more years	410	350

The authors of the report noted that the major factor contributing to the disqualification of the better-educated draftees appeared to be "the simple passage of time. When the older ages are reached some diseases and conditions will be frankly manifest which were latent or nonexistent at an earlier age." But the college graduate, at the ripe old age of twenty-three, can scarcely be classified as being senile or in his physical dotage by virtue of advanced age.

Dr. Bellas and Mr. Vinyard therefore took a closer look at the whole picture of educational levels in relation to the proportions of medically disqualified draftees, and what they came up with bears close scrutiny:

The reasons for the increasing medical disqualification rates associated with increasing educational levels, except for that part due to aging, would seem to be more obscure and complex. Some factors affecting the rate may be:
(1) greater facility for articulation concerning medical conditions
(2) more knowledge about medical conditions and induction standards
(3) poorer motivation for service in the Armed Forces
(4) more likelihood of seeking draft counseling, and perhaps more convenient availability of such counseling.
Regardless of the reasons the results are clear: Caucasians who have completed some or all college work and are over twenty-three years of age are rejected for medical reasons almost 50 percent of the time, and 38 percent of the Negroes in the same [education-age] class are dis-

qualified. This is an area in which *no explanation can be derived from these data* as to why there are differential rates within age groups when sub-classified into different levels of education [italics added].

As we saw earlier on page 439, of all the body- and brain-diminishing visual, hearing, stress, and chronic conditions universally associated with poverty, according to the AFQT raw data only one of them—eye defects—proved to be more prevalent among the much poorer blacks than among the white draftees.

But why do the sociobiological imperatives of human health seem to manifest themselves only when the eyes of eighteen-year-old white and black draftees are examined? Why are other biological organs and systems, from the brains to the bones, apparently immune to the same sociobiological factors in white and black American males of eighteen? The answer, possibly, lies not in biology but in sociology.

In many armed forces examination and induction stations, the majority of preinduction examinations were conducted by local private physicians, who were paid examination fees by the armed services to examine draftees when the regular Army physicians at the stations needed their assistance. These private, community-based physicians had to live and practice in the same towns after peace came and the draft ended. They were subject to various obvious pressures from the people in the communities in which they had their practices. Some hawk families might have been terribly eager to have their sons drafted into the military services. Some dove families might have dreaded the thought of their sons being sent to fight in ongoing and forthcoming presidential wars.

At the same time, the local civilian physicians who helped speed up the examination processes at the induction centers were immediately responsible to the military employers who paid their fees. The primary requirement that the armed forces demand of a soldier deemed healthy enough to shoulder a rifle is that he be able to see well enough to shoot straight. A combat soldier who cannot properly see the enemy cannot kill an enemy—or drive a military vehicle without endangering his own troops.

There is some empirical and quantitative—if not qualitative—support for this hypothesis. (A hypothesis is, by definition, the five-dollar scientific euphemism for a guess.) This hypothetical "explanation" of the higher proportions of black than white disqualifications for eye defects might be logical, but it requires more than the logic born of a simple (and quite possibly biased) hunch to be accepted as fact. In science and medicine, the logical hunches of fallible human beings should never be confused with *proven* mechanisms or structures.

There is no logical explanation in the Bellas-Vinyard report of why the sociobiological laws that are validated in the matter of the expected variations between whites and blacks examined for eye diseases and defects should not, apparently, prevail in the examination data for medical defects equally subject to the same social and biological imperatives. Yet, with the exception of the Middle Atlantic and the East North Central states—*where local and federal and state medical programs provide somewhat more free medical care for the poor than in other regions*—the whites seem to get more chronic rheumatic heart disease than the blacks.

Instead of the expected validation of all known laws of modern biomedical science these Army preinduction medical examination data seem to suggest that a biological organ as vital to human functions as the brain can grow and develop *independently* of the rest of the human body of which it is an integral biological and neurological part.

According to the Army preinduction data, it is the more affluent white draftees who are the weakest, the least healthy, and the most plagued with medical problems—while the far poorer and less nourished and less medically cared for black youths are healthier than the whites in every physical category except vision. Yet, according to these same data, it is the weaker, less efficient bodies of the whites that grow the strongest and most effective brains.

SHADOW AND SUBSTANCE IN THE CULTURE OF TODAY

Although to the serious biomedical, behavioral, and social scientists aware of the scientific realities of human growth and development, the Armed Forces Qualification Test scores might be scientific garbage, the AFQT scores, which appear to absolve poverty and accuse race for the scandalously higher rate of black failures in the AFQT mental tests, are taken at face value as legitimate scientific measurements by too many people of great intellectual and political influence in the greater society.

That the Bellas-Vinyard study of the AFQT scores for the Office of the Surgeon General has by now thoroughly exposed their data to be worth less than nothing does not carry nearly as much weight with the editors of the mass media, the readers of the middle-brow magazines, and the elected makers of our health, education, and welfare policies as does the enthusiastic acceptance of these tainted numbers by spokesmen for the pop culture.

Of all the authors who have commented on the AFQT scores as they interact with the making of social policies, only one, to my knowledge, the economist Ronald W. Conley, seemed aware, as he wrote in his magnificent book *The Economics of Mental Retardation,* that "curiously, there was an inverse relationship between the percentage of inductees medically unqualified and the percentage mentally unqualified. . . . There is no satisfactory explanation of this inverse relationship."[15]

Men of equal and greater prominence, from the presidential counselor and ambassador Professor Daniel P. Moynihan and the Nobel laureate in physics William Shockley to the Berkeley professor of educational psychology Arthur R. Jensen, have all rushed to print and/or the lecture platform to hail the AFQT mental scores as if they were the most profound set of validated numbers in science since Einstein gave us $e = mc^2$. Before them, of course, there were the intelligentsia of the white Citizens Councils, such as the former chairman of the Department of Psychology at Columbia University, Henry E. Garrett. As early as 1967, in a propaganda tract distributed by the Citizens Councils, *The Relative Intelligence of Whites and Negroes: The Armed Forces Tests,* Professor Garrett found in the AFQT scores all the proof he required to conclude: "The persistent and regular gap between Negroes and Whites in mental test performance strongly indicates significant differences in native ability. In short, the case for the genetic basis for White-Negro differences in intellectual capacity is as good as a scientific case can be."

The reception of the AFQT scores as legitimate scientific data by the special pleaders for given social arrangements served to underscore the well-known fact, evident since the days of Galton's so-called mental tests, that the people who write learned books and articles which judge people solely on the bases of their race skin color and mental test scores always reveal considerably more about themselves than about the people tested.

More significantly, the utilization of the discredited AFQT examination scores helped set the stage for the "miracle" that, after 1954, was to make the miraculous and sudden evolution of two new races of "strong body–weak mind" black and "strong mind–weak body" white young males, as revealed by the AFQT scores, appear to be quite ordinary and routine by comparison.

19 The New Scientific Racism
Revives the Moribund IQ Testing Myths of the Old Scientific Racism

> I am convinced that a policy of voluntary [*sic*] segregation of the colored people of the United States is the only sound one. Whether the American nation should provide a territory for them within its own borders, or should seek to secure a suitable territory in Africa or elsewhere, is a question of which I have no decided opinion. . . . In any case, the American nation owes a great act of justice, and reparation to its African population, and that debt can, in my opinion, only be discharged by the expenditures of large sums of money and philanthropic effort in the endeavor to carry through a wisely planned program of segregation.
>
> —WILLIAM MCDOUGALL, Professor of Psychology, Harvard University, 1923

When, on May 17, 1954, the U.S. Supreme Court, in the case of *Brown* v. *Board of Education,* unanimously ruled that the racial segregation of school-children did indeed "deprive the children of the minority group of equal educational opportunities," and that "separate educational facilities are in-herently unequal," the segregation of black children into all-black schools was rendered unconstitutional in the seventeen southern and border states, as well as the District of Columbia, where it had prevailed until then.

This court decision had two significant consequences. The first was the miracle whereby nine out of every ten of the 75 percent of the American people who up to 1954 had been certified by the Army intelligence tests of World War I and subsequent civilian IQ tests to have the average intelligence of a thirteen-year-old child were suddenly "discovered" to possess high-IQ-test-scoring minds. These formerly low-IQ-test-scoring people were the white people; the remaining 10 percent were not only found to be as stupid as they ever were, but the scores of the new Armed Forces Qualification Tests (AFQT) now showed that intellectually the blacks were falling even further behind the suddenly intelligent white majority.

The decision ending school segregation also created the new issue of "forced busing" that would be exploited to great political advantage by the demagogues and gut racists of both political parties for the next generation. Forced busing was, of course, far from new in the South. As Mary E. Mebane, a teacher in the South Carolina State College, observed in an article in *The New York Times* in 1972:

"The south has always bused children great distances to achieve a racial pattern. The pattern was total segregation. Its purpose was to insure that from one-third to three-fourths of its population, that part which was black, would never get the education which would make them economically competitive. It worked."[1]

After 1954, the use of busing as one means of desegregating education threatened to alter the old social arrangements based on cheap and economically noncompetitive black labor.

The public reactions to the 1954 Supreme Court decision took many forms—from quiet thanksgiving to murder of women, children, and clergymen—which have been described in various contemporary news and magazine articles and books. These included the historic confrontation forced by the governor of Arkansas in 1957—in which President Eisenhower, a lukewarm supporter of the *Brown* v. *Board of Education* decision, whose first statement on the state's defiance of the Constitution was "You cannot change people's hearts merely by law," as well as a sympathetic comment that the people of the South "see a mongrelization of the race, as they call it" in this decision, was finally forced to send in regular U.S. Army troops to protect the black children of Little Rock. This protection was not only from the local racist mobs but also from the Arkansas National Guard, which the governor had been using to prevent enforcement of the nation's law.

Of equal prominence was the Negro boycott of segregated public buses which started in Montgomery, Alabama, in 1955 and which was also to launch the short but effective civil rights leadership career of Martin Luther King, Jr. Out of this crusade was to come the voter registration drives and the resulting civil rights legislation of the 1960's—and the tragic morning of September 15, 1963, when a bomb planted by anonymous foes exploded in a Negro church in Birmingham, Alabama, and killed four Sunday-school children.

Such acts of wanton murder, effective as they had proven in the days of the original Ku Klux Klan and its early-twentieth-century commercial revival, were now no longer effective in delaying social changes. Times, high courts, customs, old patterns of racial servility *were* changing. In the midst of these changes, people in many parts of the nation, but particularly in the South, surged toward old values. Many felt impelled to rally round the flag: not the Stars and Stripes but the flag of the Old Confederacy, which suddenly began appearing on automobiles, public buildings, and suit lapels all over the South. New societies, new publications, new national and state lobbies, new crusades to stay change and preserve the values of the prewar, pre-New Deal, pre-twentieth-century days arose. Typical of these new groups was the Citizens Council of Jackson, Mississippi.

Organized on July 14, 1954, at Indianola, Mississippi, "by fourteen men who were concerned about the federal surrender to minority group pressures, and who determined to do something about it,"[2] the Citizens Council developed into a powerful regional movement of autonomous Councils in many states. Commonly referred to as the White Citizens Councils in the mass media, the Citizens Councils are a well-financed political pressure group whose members include many elected officials, business leaders, civil leaders, and professional people in the Deep South.

The five-point "action program" of the Citizens Councils speaks for itself: "(1) Oppose Race-Mixing. (2) Avoid Violence. (3) Maintain and Restore Legal Segregation. (4) Defend States' Rights. (5) Correct the Court and the Congress." Amplifying on Point 5, a recent Council pamphlet declares: "Both the Supreme Court's 'Black Monday' decision [*Brown* v. *Board of Education*] and the Congressional 'Civil Rights' Acts are obviously un-Constitutional, *based on false 'science'* in mockery of the law. If they stand, government of the people will perish from the earth. Such a prospect is intolerable! The 'Black Monday' decision must be reversed, the 'Civil Rights' Acts repealed." (Italics added.)[3]

As an organization, the Citizens Council has not repeated the anti-Catholic postures of predecessor organizations, such as the Ku Klux Klan of the 1920's. It is far more modern in its outlook. If it recognizes that the Supreme Court decision of 1954 and the civil rights acts of the early 1960's are "based on false 'science,' " it is also prepared to disseminate what the Council's leaders consider to be the scientific answers to such scientific errors. The Council Literature List includes dozens of scientific pamphlets and books on race and racial problems, most of them for moderate prices under a dollar, as well as full-scale scientific treatises, such as *The Testing of Negro Intelligence,* by Dr. Audrey M. Shuey ($6.50), a former student of Dr. Henry Garrett, whose monumental volume, *Psychology and Life,* is also sold by the Council at $12.50 per copy.

Other writings by Dr. Garrett, all distributed by the Council at under a dollar each, include: *Race and Psychology; How Classroom Desegregation Will Work; Heredity: The Cause of Racial Differences in Intelligence;* and *The Relative Intelligence of Whites and Negroes,* as revealed by the AFQT mental test scores.[4] Dr. Garrett's contemporary Dr. Robert Kuttner, who like Dr. Garrett served as one of the contributing editors listed on the masthead of the short-lived hate sheet *Western Destiny,* is represented on the Citizens Council Literature List by the pamphlet *A Brief Account of Negro History,* as well as by the anthology he edited, *Race and Modern Science,* with essays by various writers, including Professor Frank McGurk, of Alabama College; C. D. Darlington, of Oxford; and Corrado Gini, of Italy. The Council also distributes various books by Carleton Putnam, such as *Race and Reason* and *Race and Reality.* And, of course, the other major classic of the new scientific racism, *The Biology of the Race Problem,* by Wesley Critz George, Ph.D.

It was largely through the efforts of the white Citizens Councils and their favorite authors that the IQ testing mystique, nearly moribund since World War II, was revived to the powers it had enjoyed between 1913 and the eve of World War II.

THE WHITER THE BLOOD, THE HIGHER THE IQ TEST SCORE

After May 17, 1954, no section of the country put greater stress on the importance of IQ and academic test scores than did the South. There was no absence of irony, in this instance, since it was precisely in the South where state expenditures per pupil for education were the nation's lowest, and where intellectuals and scholars were widely derided as pointy-heads and eggheads by the politicians who consistently won most of the South's elections. It was also very well known, by 1954, not only that the IQ and academic test scores of *southern whites* were invariably lower than those of non-southern whites, but also that the average southern white's mental test scores were often found to be considerably lower than those of many healthier and better-educated non-southern blacks.

This potential embarrassment was bypassed by the new scientific racists of the South, who cleverly started to group all mental and academic test scores on their *national* averages, so that the IQ test scores of the high-achieving Connecticut and California whites *and blacks* were averaged in with those of the low-achieving whites and blacks of Alabama and Mississippi. In 1955, for example, this meant that the test scores of whites and blacks in the Middle Atlantic states (New York, New Jersey, and Pennsylvania), whose personal per capita income averaged $4,461, were averaged in with the mental test scores of the children of the white and black families of the East South Central states (Kentucky, Tennessee, Alabama, and Mississippi), whose average per capita income was only $2,908.

In the same base-line year of 1955, in the U.S. population as a whole, while the infant-mortality rate was 26.4 deaths per 1,000 live births, the rate for all whites came to 23.6—and the "Negro and other" rate was 44.5. The differentials in these rates were caused by poverty, not race.

In 1955, as in 1917, the civilian mental test scores were higher for the whites and nonwhites of the affluent and healthier states than they were for the white and Negro testees from the poorer states with higher infant-mortality rates—and with enormously lower state expenditures per pupil for education.

The authors who did the most active jobs of interpreting the national mental test score data to make a case for the genetic mental inferiority of the blacks included Professor Audrey M. Shuey, of the Randolph-Macon College of Virginia, and Professor Henry E. Garrett, under whom Shuey had taken her doctorate in psychology at Columbia University.

Professor Shuey's major contribution, *The Testing of Negro Intelligence,* published in 1958 by the J. B. Bell Co., in Lynchburg, Virginia, and in a second edition in 1966 by the Social Science Press in New York, was a review of "380 *original investigations* of Negro intelligence, included in 48 published monographs, books or sections of books, 203 published articles, 90 unpublished Master's theses, 35 unpublished Doctor's dissertations, and four other unpublished monographs; as well as 62 *reviews, interpretations or research* pertaining to the topic, and 122 books, articles and monographs dealing with *material related to the tests* used, their interpretation and standardization" (Dr. Shuey's italics).

These studies were of tests given during the previous five decades to both Negroes and whites and to Negroes alone. "Where only Negro groups were examined," she explained, "there are references to norms that have been derived mainly from standardization on white groups."

This research protocol had a fatal flaw, which Dr. Shuey described on page 2 of her introduction: "Specific studies made on whites alone have not been included, as they were when Negroes were tested." Since many of these specific studies made on whites alone were available to Dr. Shuey between

1958 and 1966, *and because so many of them showed that socioeconomic family conditions had infinitely more to do with the IQ test scores of white children than did their genes or skin color,*[5] this omission of a vast number of modern studies also acted to nullify all of the uses of whites as reference populations in the southern psychology professor's book.

What Shuey did, in brief, was to eliminate the socioeconomic and sociobiological factors known to be involved in the IQ test scores of all ethnic groups, and—taking national rather than regional or state or socioeconomic class averages as her base data—she evaluated the inborn mental capacity of human beings on only two measurements: skin color and IQ test score. The 380 studies she reviewed in her book ranged from Daniel P. Moynihan's use of the AFQT mental test scores as a measure of black innate intelligence to such fine studies as those of Otto Klineberg, and even included Carl C. Brigham's 1923 *A Study of American Intelligence.*

It did not seem to matter to Shuey that by 1930 Brigham had completely retracted every racial intelligence claim he had made in 1923.

Without quoting from Brigham's repudiation of *A Study of American Intelligence* and its *"hypothetical structure of racial differences" in intelligence,* Dr. Shuey reduced Brigham's 1930 retraction to the level of a non sequitur. In her oversimplification of the eight-page Brigham article of 1930, she wrote: "Brigham later rejected completely his own and others' findings in the field of natio-racial differences in intelligence on the grounds that the subjects were handicapped by not having been brought up in homes where the vernacular of the test was used (or used exclusively) and that *intelligence tests do not measure a unitary trait.* As regards the latter point, *Garrett believes that Brigham attached too much importance to test purity."* (Italics added.)[6]

In short, Dr. Shuey was saying here, since *Garrett* said that *Brigham's* total retraction was based on what *Garrett* considered to be a trifling objection, it was therefore perfectly all right for Shuey to take at their former face value all of the statements in the 1923 book that Brigham himself had repudiated in 1930 as being entirely "without foundation."

If Shuey had blatantly ignored Brigham's total repudiation of the mental testing and "racial intelligence" concepts and conclusions of his 1923 book, she also just as obviously indulged in a "selective migration" (see page 228) from the data and conclusions offered by other authors she cited, such as Anne Anastasi, Ada H. Arlitt, and Otto Klineberg. Shuey listed no less than six publications by Klineberg in her references, including his 1935 book, *Negro Intelligence and Selective Migration.*

In this instance, Shuey cited only page 66—which indicated that she was aware of, and decided to ignore, all of Chapter 8, "Schools North and South," starting on page 56. On that page, Klineberg had written: "The opinion expressed in the earlier days of the testing movement in America, that the [IQ] tests measure native [genetic] endowment altogether apart from the influence of training and background, is now held by few if any of the psychologists who have concerned themselves with the tests. The question is no longer whether training has an effect, but rather how great that effect can be. Better schooling is not the only environmental factor which influences test scores, but it is probably the most important one."

On the following page, Klineberg cited various studies showing that, for example, "the per capita expenditure for the education of White and Negro children in the . . . Southern states was found to be $10.32 for each White child, and only $2.89 for each Colored child" between 1910 and 1920. Klineberg cited a 1931 study which revealed that in eight southern states the annual "average expenditures of $44.31 per capita for Whites and only $12.50 for Negroes" were both entirely inadequate "when one compares them with the average expenditure throughout the United States as a whole, which is $87.22 per school child." Still a third study quoted by Klineberg on the same page 57 showed that, as of 1930, "South Carolina spends $4.48 per Negro child and $45.45 per White child; the figures for Alabama are $5.45 and $37.63; for Georgia, $7.44 and $35.24; for Louisiana, $8.02 and $46.67; for Mississippi, $9.34 and $42.17; for Virginia, $14.86 and $54.21." "In the light of these figures," Klineberg had observed on the same page, "the educational handicap under which the average Southern Negro child suffers hardly requires further comment."

Shuey not only failed to comment on these preprogrammed educational handicaps that made inevitable and wholly predictable the lower average IQ and classroom test scores of the black children of the South—where most of the nation's black children lived. She also ignored the very existence of these significant data on state educational expenditures per capita. In this posture, of course, she acted in the classic traditions of American scientific racism, from the days when C. B. Davenport ignored Goldberger's 1914 discovery of the real and wholly environmental cause of pellagra to fabricate his own hereditarian pseudo-etiology of this nongenetic plague of poverty in 1916—to that tragic moment in 1927, when Davenport's adjutant, Harry Laughlin, ignoring the findings of the American life and behavioral scientists who had proved otherwise, convinced the U.S. Supreme Court that the "feeblemindedness" proved by IQ test scores in what he described to the Court as "the shiftless, ignorant, and worthless class of anti-social whites of the South" was almost invariably genetic in origin.

The entire Shuey book had only one purpose, and that was to so interpret the Army intelligence test scores of World War I, the civilian intelligence test scores of the twentieth century, and the contemporary AFQT mental test scores, to reach her conclusion (p. 521) that these mental test scores, "all taken together, inevitably point to the presence of native [genetic] differences between Negroes and whites as determined by intelligence tests."[7]

Only the New York City telephone book contains more numbers than Professor Shuey's 521-page *The Testing of Negro Intelligence*. But in terms of their bearing on the genetics and biology of human intelligence, the numbers found in the phone book and in Professor Shuey's study are of equal relevance. Nevertheless, and possibly because too many people believe that numbers per se are the language of science, Shuey's compendium of irrelevant mental test score numbers was to become a standard "scientific" reference book to those scholars, in and out of academic life, who were in search of statistical rationales for their own social and racial biases.

Shuey's study of irrelevant numbers, while a standard work in some quarters, never quite took on the authority of, say, such older standard source works as the venerable *Encyclopædia Britannica.* Individual authors can have opinions; the *Britannica,* most educated people believe, prints only proven facts. Such as, for example, the facts about differential psychology.

In 1929, the fourteenth edition of the *Britannica* came off the presses with an article entitled "Differential Psychology," which ran on page 367–68 of Volume VII and was to be reprinted in every subsequent edition of this highly regarded encyclopedia for the next thirty years. The *Britannica* article had a fairly low opinion of most of mankind, noting as early as paragraph two that in most human traits "mediocrity is the status most commonly encountered, marked superiority or inferiority being relatively and equally infrequent." Some nations, however, produced people superior to others.

"The most extensive comparative data on the intellectual differences as among national groups is that obtained from the tests given in the Army during the World War. These tests indicated a superiority of those foreign-born men from northern countries of Europe over those from central and southern Europe."

The conclusions stated in the rest of the article were in line with the race hypothesis repudiated by Brigham. For example:

"The greater the admixture of white blood, the closer does the negro approach the white in performance" (Ferguson). *"Intelligence tests given to large groups of whites and negroes in the American army place the negro below the white both in tests of language and non-language variety."* Several investigations "have shown *the negro to be more overtly emotional and less inhibited in his reactions than the white"* (Crane). (Italics added.)

If the author of the *Britannica* essay knew *what* "the negro" was less inhibited in reacting *to,* the encyclopedia never revealed. However, the faith of the *Britannica* expert in the intelligence quotient of "white blood" was plain: "The American Indian ranks consistently below the white man on tests of mental capacity; *the greater the admixture of white blood, the smaller the deviation from the performance of the white"* (italics added).

The name of the author of the *Britannica* article on differential psychology was—Henry E. Garrett. When the fourteenth edition of the *Britannica* appeared, Garrett was an assistant professor of psychology at Columbia University; during the next twenty-seven years he rose in rank, becoming professor and chairman of the Psychology Department in 1942, and retaining both posts until he retired, in 1956. He had served as a member of the Division of Psychology and Anthropology of the National Research Council between 1937 and 1940, and as consultant to the Secretary of War from 1940 to 1944. Between 1943 and 1946, Garrett had served consecutive annual terms as president of the Psychometric Society, the Eastern Psychological Association and the American Psychological Association.

On May 17, 1954—the day the Citizens Council has named Black Monday—the Virginia-born Garrett plunged into the battle for racial segregation with a vigor that would have been astounding in much younger men.

Garrett carried the Word to varied forums, from the *American Psychologist,* journal of the American Psychological Association, and *The Citizen* to the Patrick Henry Press, Inc., of various post-office boxes in Virginia, which was operated by a veteran paid lobbyist for white supremacy and against civil rights named John J. Synon. For the Patrick Henry Press, which at some point or other was apparently absorbed by the Citizens Council—which now distributes its publications—Garrett authored a number of widely circulated propaganda tracts (see pages 446–47). These pamphlets were printed in enormous editions, such as the first printing of 200,000 for the 26-page tract on classroom desegregation, followed by a second printing of 85,000. They were distributed by various white-supremacy groups to newspaper editors and columnists all over the nation, as well as to teachers, preachers, politicians, and influential citizens in all fifty states.

The curious nineteenth-century flavor of Garrett's thinking is summarized in an article, "Some Words People Play Games With," which he wrote for the February 1972 issue of *The Citizen,* the official organ of the white Citizens Council. Here is what he had to say about such things as segregation and democracy:

> Segregation was a unique institution, and in many ways worked quite well. . . . It was outlawed in the 1954 decision of the Supreme Court. *Segregation has been steadily lambasted by those who believe the environmental theory, namely, that there are no race differences.* They also believe in the one-man-one-vote principle, no matter if one man is highly educated and another illiterate. One-man-one-vote is incredibly wrong; it is negative, not positive, democracy. But it is popular with dedicated Liberals [italics added].

Thanks to the vast distribution of the works of such authors as Shuey and Garrett by the white Citizens Council and many less respectable anti–civil rights groups, their writings have had a great influence on the thinking and value systems of thousands of people who teach the young and the graduate school students; who select, edit, write, and televise the news; who, as political leaders and officeholders, play major roles in shaping public policies concerning the budgets and the functions of educational, health, housing, welfare, nursery, and child care programs.

THE PSYCHOPATHOLOGY OF THE NIGRA PROBLEM

Shortly after the white Citizens Council discovered the political value of scientistic race studies, the Honorable John Patterson, governor of Alabama, commissioned a study of the Race Problem by Wesley Critz George, Ph.D., a native of Yadkin County, North Carolina, and the chairman of the Department of Anatomy of the University of North Carolina Medical School until his retirement in 1959. Professor George's report, *The Biology of the Race Problem,*[8] issued as an official state of Alabama document in 1961, and since reprinted by the Patrick Henry Press in massively distributed editions—it is still a best seller on the Citizens Council book list—is about as modern as a Confederate flag.

The most frequently cited authority in George's 1961 report was Sir Francis Galton, who died the year George graduated from the University of North Carolina in 1911. The other authorities cited ran from the scientifically discredited anatomist Robert Bennett Bean (see pages 179–80) and Henry E. Garrett to the true-believing eugenicists Jon Alfred Mjöen, Earnest A. Hooton, and Robert M. Yerkes. While Professor George accused the Supreme Court of basing its decision in *Brown* v. *Board of Education* upon allegedly scientific opinions of others "that should have been subjected to critical examination and was not," his own exposition of the biology and genetics of human development was a plain throwback to the "genetics" of the unit character, or one gene = one trait, versions of Davenport, Madison Grant, and Jon Alfred Mjöen, the discoverer of the M.B. type. On pages 9 and 10, for example, George—a half century after the unit character or trait transmission theory of genetics was destroyed by Morgan, Sturtevant, Muller, and Bridges at Columbia University—wrote:

> The unit characters, or the substance that transmits them from generation to generation, exist in the cells as genes, which are arranged in a linear manner in or on chromosomes, like beads on a string. Except in eggs and sperms, all of the cells of our bodies have chromosomes present in pairs. Consequently, *the genes for unit characters are present in pairs.* . . . A man consists of a multitude of characters synthesized in the individual. *Each character [sic] is transmitted from generation to generation through the influence of its pair, or assemblage of pairs, of genes.* Both members of the pair exercise an influence on the resulting character. [Italics added.]

Not even Davenport, who had led in the promotion of this unit character transmission genetics, felt it wise to promote it by 1918. But here was the state of Alabama's hired official authority on human biology solemnly putting forth this scientific offal as "genetics" in 1961.

The George comments on IQ test scores and racial intelligence, crime and race, and "the historical record of the Negro race" were all as modern, and as accurate, as his 1961 promotion as modern genetics of the long-since disproven unit character concepts of 1910. After "proving," in his first nine chapters, that the world's blacks have always produced more criminals and fewer generals, more low IQ test scores than high scores, more sharecroppers and fewer bankers than the whites, Professor George was then ready for his concluding chapter—"The Influence of Franz Boas."

To Wesley Critz George, professor emeritus of anatomy at the University of North Carolina, whose significant contributions to the biological and other sciences can be described with a nice, neat cipher, the corruption of the Supreme Court and society as a whole had come about for two reasons: "(1) The scientific evidence has failed to reach the public mind. (2) *Error, presented as scientific truth and intermingled with scientific truth, has flooded the public mind*" (italics added). The state of Alabama's trained investigator concluded his report by exposing the Machiavellian foreign-born non-Aryan responsible for seven decades of the propagation of equalitarian and environmentalist errors about race biology.

"The story of the origin of the prevailing situation illustrates the influence that flows from a clever and forceful man when supported by other men trained by him. If we disregard the question of motives, which were probably complex, the facts make a fairly straightforward story."

There was no power on earth that could keep Wesley Critz George from telling this straightforward story. He was no dupe like Earl Warren. He could boldly write: "The principal character in this story is Franz Boas, born of *Jewish* parents in Minden, *Germany,* in 1858" (italics added).[9]

Like Madison Grant and Eugen Fischer, the Nazi scientists who had arranged for the German edition of Grant's *Passing of the Great Race,* Wesley Critz George was alert to the menace this Jew Boas presented to all good dolichocephalic Nordics with high IQ test scores. Once having established clearly that Boas was a European-born Jew, George warmed to his task of character assassination:

> Boas seems to have been a man passionately devoted to certain social and political beliefs which he upheld with whatever resources at his command. He was said to have been a *pacifist* at the time of the First World War. This need not concern us here. [*Sic!*] *Later he had various communist-front affiliations and was reputed to be a communist.* This might concern us somewhat more but, since it is difficult to verify this, I do not wish to go into the matter further than to quote Herskovits: "In his *political* sympathies he leaned toward a variety of *socialism* common among Nineteenth Century *Liberals.*" [Italics added.]

A slanderer in the Madison Grant tradition, Dr. George could not resist another cheap shot at his long dead target, this one in the form of a reference to another "learned" expert who had sailed into Boas.

> Boas and his followers have been activists as well as theorists . . . in the field of immigration policy. Boas prepared a report for the Federal Immigration Commission which he called "Changes in the Bodily Form of Descendants of Immigrants" which *purported* to prove that head forms changed with the transfer of southern [Italian] and eastern [Jewish] European stocks to American soil. *This obvious effort to stretch the doctrines of environmentalism to the utmost extreme in the interest of the equalitarian dogma has been sufficiently unmasked* by Professor Henry Pratt Fairchild, past president of the American Sociological Society, whose chapter on Boas in the book *Race and Nationality* makes further comment unnecessary. Suffice it to say that *no other study has supported Boas before or since.*[10] [Italics added.]

It is impossible to believe that a full professor of biology and chairman of the Department of Anatomy at so modern and progressive an institution as the University of North Carolina was not aware of the scores of studies on European and Asian and other immigrants and their children which, between 1910 and 1961, more than amply replicated and validated Boas' findings in his report to the Federal Immigration Commission.

Professor Fairchild, George's secret weapon against Boas' anthropometric findings, was a confirmed and open racist; a disciple, from his undergraduate years at Yale, of the guiding genius of the Immigration Restriction League, the pathological Jew baiter Prescott F. Hall; and a lifelong crusader against the foreign-born and the nonwhites.[11] Aside from this, Fairchild had no professional training or competence in anatomy or general biology, let alone anthropometry, and his attack on Boas consisted of citing what two "careful scholars, G. M. Morant and Otto Samson"—not otherwise identified—had alleged about "the Jewish material" in the Immigration Commission study being inadequately proven.

Professor George closed his report *The Biology of the Race Problem* with a blacklist of "some of the first-generation people who came under Boas' influence, either as students or colleagues," and who were responsible for corrupting American education and culture and, in some instances, for influencing the Supreme Court in its Black Monday decision. These evildoers included Professors Ruth Benedict, Isidor Chein, Kenneth B. Clark, Theodosius Dobzhansky, Leslie C. Dunn, Melville Herskovits, Otto Klineberg, Margaret Mead, Ashley Montagu, Howard Odum, and Gene Weltfish. The Georgia-born Howard Odum had been a colleague of George's for over three decades at the University of North Carolina, where he not only served as professor of sociology but also, as George noted, "developed" that university's "department of sociology and anthropology."

This is what George had to say, in his closing pages, about his own university, where he had studied and taught for the major part of his life:

> At the University of North Carolina there is a course called Modern Civilization. This course is required of all freshmen and is prerequisite to other courses in History. Upon investigation, I found that one of the first required readings in the course is the integration tract by Otto Klineberg in *Columbia University Readings in Race, Personality, and Culture.* I carefully read the article by Klineberg and judged it to be *without scholarly merit . . . literary charm or virtue* [italics added].

Professor George's report to the governor of Alabama ended with a short paragraph worthy of Luther: "I can do little more than present the facts. Study and action by the American people are necessary to correct the situation."

Since that time, a generation of young people, most but far from all of them in the South, have been fed this 87-page catchbag of half-truths, quarter-truths, outright lies, and racist myths as a helping of objective science. This state of Alabama report not only has enjoyed wide use in various state and local school systems, but has been pumped into the pliable and willing-to-believe brains of older people directly involved in the formulation of public value systems and governmental policies.

In 1961, George joined Garrett and other like-minded proponents of "race genetics" in writing the foreword to *Race and Reason,* by Carleton Putnam,[12] a retired businessman who felt that a serious scientific study of the corruption of anthropology since 1930 "might result in a reversal" of *Brown* v. *Board of Education.* By profession a lawyer, Putnam claimed that the reason "many leaders of the modern church support the integration movement" was simply "because they have accepted uncritically the anthropological doctrines of Boas."

A man of firm opinions, Putnam wrote (p. 108): "In my opinion, the Supreme Court has been badly advised, both by the Attorney General and by counsel for the South. *The Boas equalitarian anthropology* has never been properly examined, the rotten core of this rosy apple, which is *the apple upon which integration feeds,* has never been laid bare to the judicial eye" (italics added).

As Putnam noted later about the *Brown,* or Black Monday, case, "the use of Myrdal's [*American*] *Dilemma* and the Boas anthropology on which it is based, was apparently entirely apart from the record and without notice to the parties." And whenever the southerner questions the environmental nature of Negro deficiencies (such as low IQ test scores) "he is met with Boas, and there the debate ends with the North and the court understandably feeling themselves to be the intellectual victors" (p. 110).

After Black Monday seemed to mark the end of the inferior education of black children, Putnam looked into the reasons for this tragic event. All he could find throughout the South was a feeling of betrayal and defeatism: "Among the few Northern defenders of the South the same defeatism existed. Feverish talks about the validity of the 14th Amendment went on, up North, down South, *while no one challenged the assumption at the root of the whole trouble—the validity of Boas*" (p. 20).

Putnam quoted from a long letter in which he had called the Attorney General's attention to the fact that the Warren Court had cited Myrdal's *An American Dilemma* in its opinion, and that (p. 23) "I need hardly dwell upon the highly socialistic bias of its *foreign* author," who "on page 90–91 introduces the doctrines of Franz Boas, a *foreign-born* Columbia University professor who arrived in the United States in 1886, who was *himself a member of a racial minority group,* and who may be called the father of equalitarian anthropology in America. From these pages forward, Myrdal's *Dilemma* is founded upon the philosophy of Boas. . . ." (Italics added.)

Putnam writes (p. 18) that he deliberately began studying Boas' books before learning, from people who had known him over many years, "the facts about Franz Boas himself—*his minority group background,* his arrival from Germany in 1886, his association with Columbia . . . the names of his students, *Herskovits, Klineberg,* Ashley Montagu—the nature of his department at Columbia, the influence in it of an instructor named *Weltfish* who later publicly announced that she had evidence to prove that the U.S. had used germ warfare in Korea. . . ." (Italics added.)

Unlike Wesley Critz George, who came right out with it and wrote that "Franz Boas was born of *Jewish* parents in Minden, Germany," nowhere in his account of Boas' Great Equalitarian Conspiracy does Carleton Putnam ever spell out the name of the "minority group"[13] he repeatedly accuses Boas of belonging to. The explanation for this delicacy of expression probably lies in Putnam's training. Putnam had spent most of his adult life in a competitive retail trade called airline transportation, and ever since Henry

Ford had learned the sales-data way in the 1920's that open and flagrant anti-Semitism causes Jews to avoid one's retail wares, American businessmen learned to avoid the word "Jew" in any public utterance, even when the word was prefaced by the phrase "some of my best friends are."

Putnam's book ran the entire gamut of the literature and dogmas of the new scientific racism, from the history and low IQ scores of the blacks to their low income and high crime rates. He also delved into theology, finding in his subsection called "Christian Ethics" that "Christ was a Man of infinite compassion, but He was not a Man of maudlin or undiscriminating sentimentality. *Christ's life* among other things, *might well be called a study in firm discrimination*" (italics added).[14]

The book closed with a passage from Garrett Hardin's *Science* article, "The Competitive Exclusion Principle" (April 1960):

> As a result of recent findings in the fields of physiological genetics and population genetics, particularly as concerns blood groups, the applicability of both the inequality axiom and the exclusion principle is rapidly becoming accepted. . . . The emotional restrictions of rational discussion in this field are immense. . . . It is not sadism or masochism that makes us urge that the denial be brought to an end. Rather, it is a love of the reality principle, and recognition that only those truths that are admitted to the conscious mind are available for use in making sense of the world.

The Comte de Gobineau, author of *The Inequality of Human Races* (1853–55), might have said it first, but Hardin said it with population genetics and blood groups.

After the Putnam book was published, a group of influential southern gentlemen, including the then lieutenant governor of Alabama, a former governor, and a former lieutenant governor, organized the Putnam Letters Committee to pay for the distribution of the writings and the open letters of Mr. Putnam on the race problem.

The State Board of Education purchased 5,000 copies of *Race and Reason* for use in the schools of Louisiana, thereby helping to indoctrinate an untold number of teachers and students with Putnam's feelings about certain prominent members of what he terms a certain racial minority (called— if not by Putnam—Jews), and another racial minority called Negroes, and a certain venerable American institution called the United States Supreme Court. They have not been the only teachers and students, publishers and commentators, legislators and jurists to be flooded with this and other writings by Putnam. There is more than enough evidence, in terms of changes in public and congressional attitudes since 1961, that Putnam's writings have not been without influence on governmental priorities and legislation.

THE CITIZENS COUNCIL MYTHS BECOME THE NEW CONVENTIONAL WISDOM

The wholehearted acceptance of the Citizens Council's pseudogenetic and scientistic myths by the molders of American conventional wisdom sent shock waves of fear and disbelief through the communities of American life, behavioral, and social scientists. One after another, the professional societies of anthropologists, psychologists, and pediatricians issued statements and passed resolutions seeking to counter the eugenical claims of the Shueys, Garretts, Georges, and Putnams with the hard data of biology, psychology, and the interfaces where both interact with human history and human environments. At the annual national meetings of the multidisciplinary American Association for the Advancement of Science (AAAS) in 1966, two full-dress symposia were held, as Dr. Ethel Tobach wrote in her introduction to the volume which contained the papers of these symposia, as a scientific response to "the barrage of pseudoscientific statements" found most prominently in the post-1954 writings of George, Putnam, and Garrett.

The American Psychological Association (APA), through its Society for the Psychological Study of Social Issues (SPSSI), not only attacked the revival of scientific racism in a 1961 resolution but also sponsored a number of books by APA members on the problems of race, education, and society. One of these SPSSI-sponsored volumes, *Social Class, Race, and Psychological Development,* edited by Professors Martin Deutsch, Irwin Katz, and Arthur R. Jensen, infuriated the primitives. Some of its chapters, such as Jensen's on "Social Class and Verbal Learning," offered basic scientific data on the socio-economic or environmental causes of the differences between the learning achievements of the poor as against the nonpoor. Two years earlier, in an article in *Educational Research,* a British journal, Professor Jensen had flatly stated:

> . . . the fact that Negroes and Mexicans are disproportionately represented in the lower end of the socio-economic-status scale cannot be interpreted as evidence of poor genetic potential. For we know that there have been, and are still, powerful racial barriers to social mobility. Innate potential should be much more highly correlated with socio-economic-status among whites than among Negroes or other easily distinguishable minorities, who are discriminated against on the basis of intellectually irrelevant characteristics. . . .
>
> Since we know that the Negro population for the most part has suffered socio-economic and cultural disadvantages for generations past, it seems a reasonable hypothesis that their low-average IQ [test score] is due to environmental rather than to genetic factors.

However, neither such sobering books and articles, nor the essays by the psychologist and geneticist Jerry Hirsch, the psychologist and pediatrician Herbert Birch, and the anthropologist Morton Fried at the AAAS symposia of 1966, had had any cautionary effects on the mass-media editors, the television and radio commentators, the legislators, and the Presidents to whom the words of the Citizens Council pundits had more meaning than those of the nation's infinitely more distinguished life and behavioral scientists.

In 1966, this tendency on the part of the true believers in Nordic supremacy to find confirmation of their dogmas in the objective data of educational and social achievement came to a serious climax with the almost universal perversion of the raw data and the implications of the 838-page monograph *Equality of Educational Opportunity,* which was then and now better known as the Coleman Report.

This was a study, ordered by Congress under Section 402 of the Civil Rights Act of 1964, "concerning the lack of availability of equal educational opportunities for individuals by reason of race, color, religion or national origin in public educational institutions at all levels," and conducted by a task force headed by the Johns Hopkins University sociologist James S. Coleman. It was essentially a sociological study of the status of educational processes in some 650,000 pupils in over 4,000 public schools in all fifty states. What the Coleman Report showed, in brief, was that the mental millennium had not arrived as yet; that better schooling by itself could not make up for *at least* three generations of sociobiological deprivations known to cause biological and mental underdevelopment in humans; and that, a decade after *Brown* v. *Board of Education,* the raw classroom achievement test scores, *when not weighted for family income, occupations, and value systems,* showed that the *average white student* got better school grades than the *average nonwhite* student. (See pages 472–505.)

In terms of equality of educational opportunity, the Coleman Report found that 80 percent of all white pupils in the first and twelfth grades attended public schools that were between 90 and 100 percent white; that some sixty-five percent of all black children in the first grade attended schools that were 90 to 100 percent black; and that 48 percent of the black twelfth-graders (that is, high school seniors) went to high schools that were more than 50 percent black. This meant, according to the Coleman Report, that in terms of access to facilities related to academic achievement, "nationally, Negro pupils have . . . less access to physics, chemistry, and language laboratories; there are fewer books per pupil in their libraries; their textbooks are less often in sufficient supply."[15]

Perhaps the most striking of the findings of the Coleman Report had to do with the socioeconomic mix of given schools and both white *and black* academic achievement levels. It turned out that the higher the socioeconomic class average of the entire student body—the lower social grade blacks averaged in with the higher social grade whites—the higher the classroom test scores of *all* racial and ethnic strains. Conversely, the poorer the dominant white pupil population of an integrated school, the lower the academic test scores of the entire school population. In short, peer models and peer pressures seemed to have more to do with raising or lowering the academic achievement levels of all racial and ethnic groups—including blacks and whites—than the qualities of teaching in their schools.

It was not exactly hot news, to *professionals* in the field, that classroom education, albeit an important factor, was nevertheless *only one environmental factor* essential to child mental development. And it was not only the biologists and pediatricians and neurologists engaged in highly sophisticated investigations into the biology and chemistry of brain and mental development who knew this. Even some of the plain, old-fashioned eugenically oriented psychologists such as Terman had come to understand the relation of the total environment to mental development fully a generation before the Coleman Report appeared.

If they had learned anything at all, since the *Memoir XV* days of Yerkes, Goddard, Thorndike, and the 1921-model Terman, the professionals in the American life, behavioral, and social sciences had learned to approach serious reports with judicial calm, and above all else to avoid making snap judgments until *all the evidence* was in on any investigation. But in 1966, as before and since, amateurs never hesitated to rush in where the better-qualified feared to tread.

Once the amateurs got their hands on the Coleman Report, it seemed that the less they knew about education and genetics and the gene-environment complexes involved in the processes of human growth and development, the more dogmatic they became in their analyses of its findings. The archetype of them all was, possibly, Robert Ardrey, the playwright and Hollywood screen writer and foremost lay prophet of "the celebrated Konrad Lorenz," who in his 1970 book, *The Social Contract,* let fly with barrages of Putnamisms, including: "In the small black race—which I much suspect, from its numbers, to be the youngest of races—we have such evidence of superiority of anatomical endowment and neurological coordination that it must be regarded as a distinct subdivision of *Homo sapiens.* If racial distinction in the playing field is to be accepted [by whom?], then can there exist theoretical grounds for banishing distinction in the classroom? In the United States the *evidence for inferior learning capacity is as inarguable as superior performance on the baseball diamond;* yet the question of intelligence remains distinctly unsettled" (italics added).[16]

And what makes this alleged "evidence" so "inarguable"? "In 1966, an enormous Federally financed study, *Equality of Educational Opportunity,* reported on the educational achievements of some 600,000 students in American schools."

This document, "referred to usually as the Coleman Report," was statistical proof that: *"The Negro had failed in American schools—failed catastrophically, beyond statistical doubt or sentimental apology, beyond all explanation.* It was not a document to be freely circulated in Congressional areas wherein the Negro commanded the swing vote." (Italics added.)[17]

What such glib reactions to the Coleman Report proved, of course, was that meaning is in the eyes of the beholder. But the opinion of Coleman himself is not without significance, and when I asked him what he thought about its reception he answered that he "was distressed" to learn "that Ardrey had used my report as evidence that there are racial genetic differences. Certainly nothing of the kind can be inferred from the report. The report examined only the relative influence of various *environmental* [Coleman's italics] differences on achievement, principally those in the school. It showed that those influences which come from peers and those which come from the intellectual environment provided by the family, were more important for achievement than those supplied by the schools themselves."[18]

Coleman's distress was mild compared to that of thousands of decent, nonracist liberals who—their minds unfettered by any hard knowledge of the biology of human growth and development—were left in a state of shattering disillusion by the classroom test score data of the Coleman Report. Until 1966 they had, in their innocence of any knowledge of human biology, fervently believed that one or two years of desegregated classroom schooling could undo the predictable biological, psychological, and cultural pathologies suffered by *all* races during at least three generations of body-crippling, mind-wracking poverty.

One of the reasons for the distress of the well-meaning liberals was, of course, the relative prominence given to the writings of instant authorities like Ardrey on the subjects of education and human development, as compared to that given the reports and other published writings of people far better trained to make judgment about such processes.

For example, by the end of 1969, some two million American preschool children in America had participated in the two- to eight-week summer sessions of the Head Start program—that underfinanced, little understood, and widely abused attempt by the guns-and-butter regime of Lyndon Johnson to help the preschool-age children of America's poor prepare to cope with the challenges of American public school education. Since the program included medical examinations of the poor preschool children enrolled in the Head Start projects, the U.S. Department of Health, Education, and Welfare was able to issue a series of reports on this aspect of the program. By 1969, these reports revealed that, of the first two million children enrolled:

        180,000 failed a vision test
         60,000 needed eyeglasses
        180,000 had anemia
         60,000 had bad skin diseases
         40,000 had mental retardation or a learning problem requiring
                 evaluation by a specialist
         20,000 had a bone or joint problem
      1,300,000 had a dental disease
      1,200,000 had not been vaccinated against measles.[19]

The government reports containing these and other data on the socio-biological reasons for the classroom test scores of the Coleman Report were not, however, also treated as major news—possibly because they challenged the widespread if rarely verbalized beliefs of our educated classes that it was the absence of Nordic blood, rather than conception-to-maturity good health, that was responsible for the lower IQ and classroom test scores of the poor. The managers of our mass media and the other makers of American conventional wisdom paid no heed at all to the pediatricians and biologists who protested that the biological prerequisite for a sound mind remained a sound body—in the children of rich and poor, white and black, Aryan and non-Aryan families.

Because our educated opinion shapers were themselves so deficient in knowledge and insights into the social biology of human development, the wise and sobering words of people such as the president-elect of the American Academy of Pediatrics, Hugh C. Thompson, M.D., made no impression upon them. While the well-meaning liberal amateurs in child health and development rested secure in the cozy illusion that a few million dollars' worth of superior preschool teaching techniques could undo, in toddlers, all of the damages inherent in what Jensen in 1967 had accurately described as the "powerful racial barriers to social mobility" that had caused the "Negro population" to suffer "socio-economic and cultural disadvantages for generations past," Dr. Thompson—like most intelligent pediatricians—knew that the solutions to the learning problems of the children of the poor, *the majority of them white,* were far from being that simple. In accepting election to the presidency of the American Academy of Pediatrics in 1968, Dr. Thompson said:

> The fact that many problems [concerning the welfare of children] will take years or decades to solve makes no less important the necessity to address ourselves to them now, before the welfare of a generation of children suffers irrevocably, or before alternative and less desirable solutions are imposed by government or other agencies.
>
> The American people are impatient. Witness only Head Start. The Administration was impatient to start the operation before it was pre-tested so that one year's crop of 5 year olds would not be deprived of help. Many Fellows of the Academy, by the same token, have been impatient because the health aspects are still not functioning smoothly. If Head Start works well across the country in 10 years, it will have been a notable achievement.[20]

Unfortunately, before the Academy and other professional and scientific societies could manage to get through to the public with the message that a sound body was still the biological imperative for a sound mind, the sudden conversion of Arthur Jensen to the eugenics creed of Galton, Pearson, and Jensen's hero, Sir Cyril Burt, was to concentrate the American public's attention on the skin color, rather than on the bodily and mental health, of all children. Jensen's conversion to eugenics became a mass-media "happening" of colossal proportions.

The white Citizens Councils now had themselves a new Nordic Knight, and the chances of helping the primarily white children of America's poor and disadvantaged families help themselves grew swiftly smaller in the acrid and bitter controversies over "racial intelligence" that now swirled to center stage. As this ancient pseudo-question was revived by the new scientific racism, it was to help mislead the American public into seeing the learning deficits of the children of all of our millions of poor and disadvantaged families only in terms of that *minority* of their tragic number whose skins were black.

20 "Genetic Enslavement": The War Cry of the New Scientific Racism

> The differential birthrate, as a function of socioeconomic status, is greater in the Negro than in the white population. . . . Much more thought and research should be given to the educational and social implications of these trends for the future. *Is there a danger that current welfare policies, unaided by eugenic foresight, could lead to the genetic enslavement of a substantial segment of our total population?* [Italics added.]
>
> —ARTHUR R. JENSEN, *Harvard Educational Review,* XXXIX (Winter 1969), 1

> Our nobly intended welfare programs may be encouraging dysgenics—retrogressive evolution through disproportionate reproduction of the genetically disadvantaged. . . . We fear that "fatuous beliefs" in the power of welfare money, unaided by eugenic foresight, may contribute to the decline of human quality for all segments of society.
>
> —WILLIAM B. SHOCKLEY, President, Foundation for Research and Education in Eugenics and Dysgenics, 1969

The year 1969 was the year of the one-day sensation of Windsor Hills, and the birth of the phenomenon called Jensenism, which is still very much with us. Since both events were intimately linked in content and meaning, let us open this chapter with a brief review of the events that followed the publication of a banner headline across the October 12, 1969, front page of the *Los Angeles Times* reading: *Black School Highest in IQ—Is Affluence the Reason?*

What triggered this headline was a survey of all sixth-graders in the Los Angeles public school system which had revealed that "the IQ score of 115 averaged by the sixth-graders of the Windsor Hills Elementary School was the highest in the city. As it happens, 90 percent of the children in this school are black. . . . The black children with the highest IQ scores in the Los Angeles area are, mostly, the children of black doctors, lawyers, teachers, professionals, and business men, and live in one-family West Los Angeles homes valued at from $35,000 to $100,000 each."[1] The same survey also showed that, while the entire student body of the Windsor Hills School had an average reading grade placement score of 6.8, the city school system's lowest reading score, 3.6, was achieved by the equally black students of the Weigand Avenue School, in the Watts district—where 43 percent of the families had incomes below the poverty level; the median family income was below $4,000; and 21 percent of the housing units were classified as "deteriorating" by the U.S. Bureau of the Census.[2]

COMPENSATORY INTERVENTION: ITS REWARDS IN THE CHILDREN OF ITS BENEFICIARIES

When Olive Walker, the editor of *The Integrator,* a Los Angeles magazine, discovered that "selective migration" of hybrid blacks with blood rich in the white genes for intelligence had had literally nothing to do with the affluence of the Windsor Hills homeowners, or the learning abilities of their children, and published the real reason for their upward mobility in the May–June 1970 issue of *Integrated Education: Race and Society,* her factual explanation of the high IQ and reading test scores of the black children of Windsor Hills was not considered the stuff of which banner headlines are forged. The secret lay not in their genes but in recent history:

"The affluence of the black parent at Windsor Hills is not for the most part derived from a middle-class background; rather, these are first-generation middle-class, *often from poor homes.* They made it into the middle class because of World War II and the *G.I. Bill of Rights that enabled them to get a college education."* (Italics added.)

Although born to parents just as black, and to grandparents just as poor, as the Watts and Harlem and Bedford-Stuyvesant Negro families, the children of Windsor Hills had been conceived by black parents who—before their conception—had, thanks to the government subsidy of college and technical educations for military veterans of World War II, managed to escape from the poverty into which they had been born. Thanks to this supportive societal intervention into the welfare of their parents, the black children of Windsor Hills had, from the moment of conception, suffered neither lack of adequate nourishment nor proper medical care. They had been, from the moment of birth, adequately fed, adequately clothed, adequately housed, and had benefited from all of the immunizations and other preventive modalities of modern pediatric practice. They had been raised in families where the median income, *as in Terman's cohort of high-IQ-test-scoring white children of 1921,* was "more than twice as high as for the generality of families in California"[3] —and where books, paintings, educational toys, and other cultural advantages of the nonpoor families of America were in ample supply.

These socioeconomic and sociobiological advantages meant that black children of Windsor Hills were enabled to achieve the optimum levels of the ultimate—that is, phenotypical—development of their genetic, or genotypical, capacities for physiological and mental growth. As the University of Illinois psychologist and *geneticist* Jerry Hirsch has reminded us, "early in this century Woltereck called to our attention the *norm-of-reaction* concept: the same genotype can give rise to a wide array of phenotypes *depending upon the environment in which it develops"* (italics added).[4]

In the affluent and therefore biologically healthy environment of Windsor Hills, the black children developed into phenotypes with the city's highest IQ and reading achievement test scores. In the impoverished and therefore biologically adverse environment of Watts, the equally black children developed into phenotypes with the same city's lowest IQ and reading test scores. The norms-of-reaction of the genes children inherit are never independent of the qualities of the total environment with which they interact. "We are," writes the Nobel laureate in medicine George Wald, "the products of editing, rather than of authorship."

Ms. Walker's article on the historical and socioeconomic factors involved in the norms-of-reaction responsible for the high IQ test scores of the black children of Windsor Hills was, however, ignored by the mass media, and remains little known to professional scholars. A different fate awaited the miracle of Jensenism, born in the same year as the transient sensation of Windsor Hills.

RACIAL DETERIORATION BECOMES GENETIC ENSLAVEMENT: JENSENISM

Not the least of the reasons why the events triggered by the publication, in the Winter 1969 *Harvard Educational Review,* of a 123-page eugenics tract, "How Much Can We Boost IQ and Scholastic Achievement?" by a recent convert to the creed of Galton, Davenport, and Sir Cyril Burt, Professor Arthur R. Jensen, have to be described as a miracle is clear to any reader with a working knowledge of the history of both eugenics and mental testing. That this compendium of the myths and shibboleths of the old scientific racism, from the pseudoneurological anatomy of Dr. John H. Van Evrie to the "race deterioration" bugaboos of Galton, Grant, and Cornelia Cannon, could be taken seriously as "science" by so many educated authors here and abroad can only be ascribed to a miracle of either advanced amnesia or total ignorance of the subject or a combination of both.

The reasons for the wide public acceptance of the Jensen version of the old scientific racism are less mysterious. The Jensen manifesto was not published in a historical or political vacuum: it appeared just at the time when the mass migration of white and black victims of the New Enclosures from the farms to the cities had caused huge increases in the city, state, and federal budgets for health, education, and welfare, and housing.

While the majority of this rural-to-urban migration was white, the increased proportions of black skins among the new urban families were more obvious to many people. At a time when the rising expectations of the civil rights revolution had created such tax-consuming educational projects as the Head Start programs for preschool children, any rationale for abandoning Head Start and related programs of compensatory education for deprived children would have made its author a hero to the solvent taxpayers already suffering from the monetary inflation of the Vietnam War, then in its fifth year of massive multibillion-dollar-a-year escalation.

Into this historical opportunity, Arthur R. Jensen—the co-editor of a 1968 book attacking the concepts of "racial intelligence" (see page 461)— leaped with his 123-page essay, "How Much Can We Boost IQ and Scholastic Achievement?" in the Winter 1969 issue of the *Harvard Educational Review.* Prior to this, as a professor of Educational Psychology sharing the view of most life, behavioral, and social scientists about a sound body being the biological prerequisite of a sound mind, and as a scholarly journal devoted to experimental research in education, neither Jensen nor the *Harvard Educational Review* had ever attracted the attention of the nation's mass media. However, as a sudden convert from the thinking of most modern sciences dealing with human development to the exotic and long-since discredited dogmas of the eugenics "civic religion"—and as the vessel in which the confessions of Saul-called-Paul had been borne to the world—both Jensen and the *Harvard Educational Review* became instant publicity saints.

In terms of the attention it had received in the mass media, the Jensen statement of his total conversion to the hoary hereditarianism of Galton, Pearson, and Cyril Burt was treated as the greatest scientific development since the first nuclear device was exploded at Los Alamos in the closing days of World War II. It rated lead stories in *Newsweek* (circulation 2,150,000), *Life* (7,400,000), and *Time* (3,800,000). On March 10, 1969, it was presented to the 1,625,000 readers of *U.S. News and World Report* in an article entitled "Can Negroes Learn the Way Whites Do?" Various other magazines, newspapers, radio, and television news and discussion shows treated Jensen's revival of the Goddard-Terman-Yerkes-McDougall approach to the "racial genetics" of high or low IQ test scores with far more time and space than they had given to such modern advances in genetics and the human condition as the discovery of the role of DNA in human heredity by Avery and his collaborators in 1944, or the discoveries in basic virology that led to such advances as vaccines against the polio, measles, and rubella viruses and won for Drs. John Enders, Thomas H. Weller, and Frederick C. Robbins the Nobel prize for medicine in 1954.

The title of the *Newsweek* article on the Jensen manifesto—"Born Dumb?"—quite accurately reduced Jensen's basic claim to plain English. And, in equally plain English, the *Newsweek* article (March 31, 1969, p. 84) summed up what the article in the *Harvard Educational Review* was all about:

> Dr. Jensen's view, put simply, is that most blacks are born with less "intelligence" than most whites. The existing statistical evidence, he says, shows that blacks score some 15 points lower than most whites. The reason, he argues, is that intelligence is an inherited capacity and that since a prime characteristic of races is that they are "inbred," blacks are likely to remain lower in intelligence.

The *Newsweek* report on Jensen's conversion to eugenics neatly summed up Jensen's conclusions—at least insofar as they touched on educational policy—in an equally accurate paragraph:

> Jensen's theoretical views lead him in his article to develop some quite practical policy recommendations. *Since intelligence is fixed at birth anyway, he claims, it is senseless to waste vast sums of money and resources on such remedial programs as Head Start which assume that a child's intellect is malleable and can be improved* ("Compensatory education has been tried and it apparently failed," he writes). Instead, programs should concentrate on skills which require a low level of abstract intelligence. [Italics added.]

I have quoted at length from the *Newsweek* report primarily because it was in this form—as well as in the equally professional lay English summaries in the other mass news and comment media—that Jensen's "scientific" conclusions reached millions of American taxpayers, voters, and local, state, and federal government officials who determine the functions, structures, and goals of the entire American educational system.

Newsweek did not pick up the most evangelical Come-to-Galton paragraph of the Jensen testament of the One True Faith. But *Life* and the *Indianapolis Star* (circulation 429,063) and various other newspapers I have read, and television and radio broadcasts I have heard since 1969, have indeed quoted in full the sentences eugenicists will be telling their children about for years to come, the immortal words with which Arthur R. Jensen shook off the Boasian coils:

> Is there a danger that *current welfare policies, unaided by eugenic fore-sight, could lead to the genetic enslavement of a substantial segment of our population?* The possible consequences of our failure seriously to study these questions may well be viewed by future generations as our society's greatest injustice to Negro Americans. [Italics added.][5]

This was the basic point of the Jensen testament—not education. Nor intelligence. Nor justice for Negro Americans. Jensen's article was nothing more, and nothing less, than a propaganda tract for eugenics.

As Professors Mark Golden and Wagner Bridger noted in the October 1969 issue of *Mental Hygiene,* "Jensen does not offer any startling *new evidence* to support his point of view, *but instead dredges up old data and the same sterile arguments from the past"* (italics added). This was the kindest thing that could be said for Jensen's article by most of the front-rank geneticists, such as the Nobel laureate geneticist Joshua Lederberg, and the Harvard psychiatrist Leon Eisenberg, and the nation's ranking authority on intelligence and experience, the University of Illinois psychologist J. Mc-Vicker Hunt.[6] When Drs. Golden and Bridger went on to declare that "politicians may use Jensen's arguments as an excuse to reduce the amount of money made available for compensatory education programs," they expressed the basic reason why so many concerned life, behavioral, and social scientists were aghast at the political dangers inherent in Jensenism.

Intellectually, Jensenism was a throwback to the Terman and Goddard IQ testing concepts of the pre-World War I era, and the biology and genetics of even earlier periods—but it was now presented with mathematical pyrotechnics far more esoteric than the simple pedigree charts, tables, and curves of the Davenports, the Shueys, and the Garretts, all of which could be comprehended by anyone with enough education in mathematics to read a baseball box score in the sports pages of a daily paper.

Separated only by the dimensions of time and lexicon, Jensen's conclusions, in his final paragraph, that society has to "develop the techniques by which school learning can be most effectively achieved in accordance with different patterns of ability" that "schools must also be able to find ways of utilizing other strengths in children whose major strength is not of the cognitive [sic] variety," are, of course, modern hand-wringing liberal variations of Terman's 1916 lament that the poor and non-Nordic schoolchildren of California had inherited so few genes for intelligence that, while they did have the mentality to make "an excellent laborer" or a good servant girl, they were "uneducable beyond the merest rudiments of training."

Where Jensen's closing paragraph goes on to present the separate-but-equal "reasonable conclusion that schools and society must provide a range and diversity of educational methods, programs and goals, and occupational opportunities, just as wide as the [genetically predetermined] range of human abilities," it is clear that Jensen is proposing nothing new to the company of IQ true believers. Although couched in the liberal clichés of 1969, what Jensen was here suggesting was merely Terman's much less pretentious "special classes" into which the world's future "hewers of wood and drawers of water" would be taught their places and, hopefully, encouraged to breed far less prolifically than their poor parents. It is only the catch phrases and the rhetoric, and not the concepts or the conclusions, that have changed between Terman and Cannon in 1916 and 1922 and Jensen in 1969.

Jensen's *Harvard Educational Review* essay, and his books that followed and expanded on the same themes, were packed to the gills with learned and scholarly citations: the trouble was that, when one looked closely at these source references, all too many of them turned out to be data as tainted as the Armed Forces Qualification Test results and the writings of Audrey Shuey.

As a very recent convert to the nineteenth-century creed of eugenics, Jensen found it impossible to ignore all of the socioeconomic and socio-biological data that had accumulated since Galton to suggest that the norm-of-reaction involved in the interaction of genes and developmental environments was, indeed, firmly based on such facts of life as poverty, health, and ambient culture. He got around this ugly fact of genetics and human development by indulging in a "selective migration" from the facts of developmental biology and socioeconomic realities and to soothsayers who provided what to him was personal assurance that they no longer mattered. Thus, in the face of mountains of scientific evidence that malnutrition does predictably affect the IQ test scores of undernourished children, Jensen could write with the same eugenic sang-froid which enabled Davenport to ignore Goldberger's evidence that undernutrition is the *sole* cause of pellagra: "These studies were done in countries where extreme undernutrition is not uncommon. *Such gross nutritional deprivation is rare in the United States*" (italics added).[7]

This was published during the same year that Charles Upton Lowe, M.D., the scientific director of the federal government's National Institute of Child Health and Human Development, testified before the Congress that *"one-fourth* of the nation's children . . . are living in homes in which incomes are insufficient to meet the costs of procuring many of the essentials of life, *particularly food"* (italics added). (See pages 436–37.)

Not content with wiping out poverty and hunger in the United States as a whole, Jensen then took out his magic wand and ended racial discrimination, poverty, and lack of economic opportunity for half of all Negro families in America. This he did in one of the most extraordinarily unscientific passages in his entire essay, on pages 87 and 88, under the heading "Magnitude of Adult Negro-White Differences":

The largest sampling of Negro and white intelligence test scores resulted from the administration of the Armed Forces Qualification Test (AFQT) to a national sample of over 10 million men between the ages of 18 and 26. As of 1966, the overall failure rate for Negroes was 68 percent as compared with 19 percent for whites (*U.S. News and World Report,* 1966).[8] (The failure cut-off score that yields these percentages is roughly equivalent to a Stanford-Binet IQ of 86.) *Moynihan, 1965, has estimated that during the same period in which the AFQT was administered to these large representative samples of Negro and white male youths, approximately one-half of Negro families could be considered as middle-class or above by the usual socioeconomic criteria.* So even if we assumed that all of the lower 50 percent of Negroes on the SES [socioeconomic scale] failed the AFQT, it would still mean that at least 36 percent of the middle SES Negroes failed the test, a failure rate almost twice as high as that of the white population for all levels of SES.

Do such findings raise any questions as to the plausibility of theories that postulate exclusively environmental factors as sufficient causes for the observed differences? [Italics added.]

Such blind reliance on a lay magazine's account of the highly dubious AFQT examination results, as well as upon a man named Moynihan's statement about the socioeconomic status of American Negroes c. 1965, is in no way acceptable as statements of reality; not, that is, from a full professor at a major university making such statements in an Ivy League learned journal. The tainted AFQT examination results—medical as well as mental—and Moynihan's *undocumented* statement about an allegedly huge Negro middle class embracing over 50 percent of all black families, are not, as they stand and as Jensen presents them to us, anything but raw, unverified, unexamined, and certainly untested data. And, at that, even the *U.S. News and World Report* article cited so smugly by Jensen contained one short sentence that gave the hoary hereditarian caper away: "Southern whites are behind whites [in mental test performance] in all other regions of the country; Southern Negroes are behind Negroes in all other regions of the country." Not even the most fervent of White Supremacists had ever claimed that there were racial or genetic mental differences between southern and northern *whites.*

Had Jensen consulted the raw AFQT data or the *Statistical Abstract of the United States,* he could have found at a glance that there was no basis whatsoever for such perversions of the facts on hand. It was true that, according to the raw data, the median income of black families in America had climbed from $1,614 per year in 1947 to $5,141 in 1967—while, at the same time, the income of white families rose from $3,157 to $8,274. This meant that in twenty years, the Negro family income had risen from only 51 percent of white family income to 62 percent of average white family income. Viewed in a historical vacuum, these data were far more impressive than they were when viewed against what $5,141 actually could buy in the way of a year's food, clothing, shelter, medical care, and cultural amenities for an average family of four people in the United States in 1967, Year

Three of the massive escalation of our military involvement in Indochina and the ruinous spirals of fiscal inflation it triggered.

As Dr. Charles E. Greene observed in the *New England Journal of Medicine:* "The infant mortality rate of non-whites of 1940 was 70 percent greater than that of whites. In 1962, 22 years later, it was 90 percent greater; *according to the United States Children's Bureau, infant mortality rises as family income decreases*" (italics added).[9]

Of all the demonstrably income-related health indices known to have a major bearing on mental development, the infant, maternal, fetal, and neonatal death rates have proved—along with abnormally low birth weights—to be the most accurate clinical indicators of future high or low IQ and classroom test scores. In 1969, the year Jensen's eugenics tract was published in the *Harvard Educational Review,* the federal government's annual *Statistical Abstract of the United States* ran the following table on its page 55:

THE DEMOGRAPHY OF BENIGN NEGLECT

Infant, Maternal, Fetal, and Neonatal Death Rates, by Color, 1940 to 1968

Item	1940	1945	1950	1955	1960	1965	1966	1967	1968 (prel.)
Infant deaths[1]	47.0	38.3	29.2	26.4	26.0	24.7	23.7	22.4	21.7
White	43.2	35.6	26.8	23.6	22.9	21.5	20.6	19.7	(NA)
Nonwhite	73.8	57.0	44.5	42.8	43.2	40.3	38.8	35.9	(NA)
Maternal deaths[2]	376.0	207.2	83.3	47.0	37.1	31.6	29.1	28.0	(NA)
White	319.8	172.1	61.1	32.8	26.0	21.0	20.2	19.5	(NA)
Nonwhite	773.5	454.8	221.6	130.3	97.9	83.7	72.4	69.5	(NA)
Fetal deaths[3]	(NA)	23.9	19.2	17.1	16.1	16.2	15.7	15.6	(NA)
White	(NA)	21.4	17.1	15.2	14.1	13.9	13.6	13.5	(NA)
Nonwhite	(NA)	42.0	32.5	28.4	26.8	27.2	26.1	25.8	(NA)
Neonatal deaths[4]	28.8	24.3	20.5	19.1	18.7	17.7	17.2	16.5	15.8
White	27.2	23.3	19.4	17.7	17.7	17.7	17.2	16.5	(NA)
Nonwhite	39.7	32.0	27.5	27.2	26.9	25.4	24.8	23.8	(NA)

Deaths per 1,000 live births, except as noted. Prior to 1960, excludes Alaska and Hawaii.
NA Not available.
[1] Represents deaths of infants under 1 year old, exclusive of fetal deaths.
[2] Per 100,000 live births from deliveries and complications of pregnancy, childbirth, and the puerperium. Beginning 1960, deaths are classified according to seventh revision of *International Lists of Diseases and Causes of Death;* see text, p. 45.
[3] Includes only fetal deaths (stillbirths) for which period of gestation was 20 weeks (or 5 months) or more, or was not stated.
[4] Represents deaths of infants under 28 days old, exclusive of fetal deaths.
From Department of Health, Education, and Welfare, Public Health Service annual report, *Vital Statistics of the United States.*

According to Jensen, the source of his statement that "approximately one-half of Negro families could be considered middle-class and above" was presidential counselor Daniel P. Moynihan. It was Dr. Moynihan who, in a famous report to Nixon, suggested that now that the blacks had risen so high and so quickly, the time had come for the "benign neglect" of America's black population in terms of federal programs to create greater equality of opportunities.

Based on all available U.S. government data, such as the vital statistics summarized above, the nonwhite infant-mortality rate was 35.9 per 100,000 live births, nearly twice the white infant mortality of 19.7. The vast gaps between the white and nonwhite populations' maternal and fetal death rates were even more dramatic. Throughout the world the true socioeconomic status of any family or ethnic group can be determined quite accurately by its rates of such socially preventable deaths as those tabulated in this Public Health Service table.[10]

Had Jensen really been concerned about the biological status of whites and blacks in relation to such factors as low birth weights, he could have asked any of his medical or epidemiological colleagues for some standard available data. Such as, for example, National Health Survey, U.S. National Center for Health Statistics publication, Series 3, Number 6, *1967* (International Comparison of Perinatal and Infant Mortality: The United States and Six Western European Countries), which would have told him what they told Dr. Richard L. Naeye and his colleagues:

"Low birth weights are more than twice as common in nonwhites, but the white infant does not become significantly larger than the nonwhite until the last month of gestational life. Since the greatest effect of undernutrition on fetal growth may be in late gestation [the months of fetal growth most crucial in brain development], one might suspect that nutritional factors are responsible for the reported racial differences in fetal growth and neo-natal mortality. Data in the present study support this hypothesis." (Italics added.)[11]

TWINS, TUBERCULOSIS, AND HERITABILITY

Low birth weight and prematurity are invariably associated not only with low IQ and academic test scores but also with significantly larger numbers of physical deficiencies in childhood and maturity. Many of the modern investigations of infant and child development include studies of the intrauterine and postnatal growth and welfare of identical (monozygotic, or the products of a single zygote) and fraternal or nonidentical (dizygotic) twins. As readers of the lay press were informed, the *pièce de résistance* of Jensen's *HER* article was his use of twin studies.

What most lay readers, however, do not know is that these studies were not conducted by Jensen personally but by four other investigators, including Jensen's new idol, Sir Cyril Burt, a personal protégé of Sir Francis Galton, who had trained under William McDougall at Oxford and to whose memory Jensen was to dedicate his book *Educability and Group Differences* in 1973. These were studies of 122 pairs of white female middle-class identical twins in England, the United States, and Denmark, separated at various times after birth and reared in different foster homes. When given IQ tests, the results quite predictably showed that monozygotic twins reared in presumably similar environments (the foster-child placement services of England, the United States, and Denmark being fairly rigid about such details) wind up having IQ test scores in more or less identical ranges.

Whatever else these studies might or might not prove about these 122 pairs of white female identical twins of middle-class parents, to extrapolate these findings to the allegedly measurable heritability of intelligence of the world's white, black, brown, yellow, red, middle-class, upper-class, and poverty-class twin and single children is—in the 1930 words of the repentant Carl Brigham—"something akin to adding apples and oranges."

Jensen was well aware at the time he wrote this essay that various other twin studies (including the Willerman-Churchill 1967 study cited in his references) did and continue to show that there is more to the intelligence of identical twins than simplistic correlational analyses would seem to indicate. As many studies now show, when the weights of identical twins are not identical at birth—for reasons of environmental inequalities or random events during the nine months of intrauterine development—the mental and physical development of the lighter of the two is nearly always slower and less complete than that of the heavier identical twin.

Three examples, taken at random from many—(1) the 1967 Willerman-Churchill study cited by Jensen; (2) the Sandra Scarr 1969 study, "Effects of Birth Weight on Later Intelligence"; and (3) the 1964 Oregon study by Babson et al., "Growth and Development of Twins of Dissimilar Size at Birth"—all found that the runt of each pair showed "a detrimental relationship between intrauterine runting and intellectual development."

In the Oregon study it was shown that even after eight and a half years, such pairs showed significant differences in intellectual growth. The 52 pairs of white and Negro, male and female twins in the Willerman-Churchill and Scarr studies showed very similar results, the first using the Wisconsin Performance IQ test, the second using the Goodenough-Harris Draw-a-Man Test.

In the Willerman-Churchill study of 27 pairs, the higher-birth-weight twins averaged IQ scores of 94; their lighter-weighted twins averaged 89. The heavier of the twins in the Scarr all-female cohort of identical twins averaged 105 in their IQ tests, as against 96 for their runts.

As Dr. Scarr noted in her paper, "the advantages of the heavier twins in later intellectual performances result from better prenatal conditions and higher birth weight, not from genetic differences or length of gestation, which are identical in monozygotic twin pairs." It should also be noted that, once born, the identical and nonidentical twins in all three of these studies were raised together, in the same environment. The only environmental differences they experienced had been during their intrauterine lives.[12]

While eugenically oriented geneticists such as C. D. Darlington feel, with Jensen, that twin studies have "proved to be one of the foundations of genetics," geneticists J. V. Neel and W. J. Schull, in their textbook *Human Heredity,* spoke for probably a majority of our era's working geneticists when they wrote that in genetic research "the twin method has not vindicated the time spent in the collection of such data." In modern pediatric research, twin studies are seldom designed to test genetic hypotheses but, rather, are conducted as part of the continuing research on the mental and biological sequelae of twinning, low birth weight and prematurity.

Jensen's "selective migration" from the hard facts of twentieth-century human biology stemmed not only from his new faith in Galton's Victorian dogmas but also from his apparent conviction that: (1) At least 80 percent of the intellectual development of human beings—or, to put this in genetic terms, the ultimate phenotypical expression of their genotypical potentials for mental acuity—is controlled by the genes they inherit from their parents, and only 20 percent is attributable to such factors as family income, medical care, nutrition, housing, and family cultural resources, such as books and other educational materials. (2) Human intelligence is a *unitary* trait, like height or weight or eye color, and can be measured as accurately with IQ tests as height and weight are measured with rulers and scales. (3) The degree to which this unitary trait of intelligence is heritable can be measured and quantified with mathematical exactness. (4) Not only is intelligence at least 80 percent genetic, but races differ in their racial intelligence quotients. In fact, as Jensen wrote in his 1973 book, *Educability and Group Differences:*

"The possibility of a biochemical connection between skin pigmentation and intelligence is not totally unlikely in view of the biochemical relation between melanins, which are responsible for pigmentation, and some of the neural transmitter substances in the brain. The skin and the cerebral cortex both arise from the ectoderm in the development of the embryo and share some of the same biochemical processes."

Whatever this passage might sound like to TV commentators and national legislators with no background in biology or neurology, to professional neurobiologists, such as the professor of biology and head of the Brain Research Group at England's Open University, Dr. Steven Rose, this is simply a total perversion of the biology and chemistry of brain function in human beings. "Despite the cautionary tone in which this is worded," Dr. Rose declared during a scientific symposium in London in 1974 attended by Jensen, "these sentences are either quite devoid of meaning or they are making the extraordinary claim that can be paraphrased as 'black skins may cause black brains.' "

Equally meaningless, to neurobiologists, is Jensen's odd passage on brain anatomy in his 1969 *Harvard Educational Review* essay. On page 114, he writes of "certain neural structures" that Jensen feels "must be available" for what he terms Level II mental level abilities to develop. According to Jensen, Level II brains are, by heredity, capable of abstract reasoning and problem solving, while Level I brains correspond to the simple brains of Terman's hereditary "hewers of wood and drawers of water" which would, for genetic reasons, forever lack the capability of being trained to grasp abstract concepts such as eugenics and racial superiority.

As Jensen describes these still to be discovered biological structures, they are *"conceived* of as being different from the neural structures underlying Level I. The genetic factors involved in each of these types of ability are *presumed* to have become differentially distributed in the population as a function of social class, since Level II has been most important for scholastic performance under the traditional methods of instruction." (Italics added.)

My younger friends among the geneticists and the pediatricians who scoff at this conceit as utter if unique rubbish, based as it is on the existence of neural structures that Jensen only *presumes* must exist *but have yet to be found,* are (as I have been trying to tell them for years) sadly lacking in a sense of history. Jensen's proposed discrete, measurable, corporeal structures for high- and low-IQ-scoring brains were, of course, foreseen in some detail in Chapter 6, "The Brain," of Dr. John H. Van Evrie's *White Supremacy and Negro Subordination* in 1868, and in the 1906 and 1915 writings of Robert Bean and Robert Shufeldt, all of which we have already encountered in chapter 8. However, in the interim, the same concepts of brain anatomy had been quite thoroughly shattered by the scientific findings of such American brain anatomists as Professors Burt G. Wilder of Cornell and Franklin P. Mall, of Johns Hopkins. So that where Van Evrie, Bean, and Shufeldt claimed to have actually seen anatomical differences in the brains of white and nonwhite individuals, Jensen could only somewhat wistfully suggest that the wholly imaginary "neural structures" *"must also be available for Level II abilities to develop"* (italics added). This is the equivalent of saying that "my aunt must also have an as yet undiscovered mustache, for she is really my uncle."

If Jensen had to go back to Van Evrie and Bean for his ideas of brain structure, and to Lapouge and Madison Grant for his view of the genetic fixity of the cephalic index (see pages 32–33), his ideas on the etiology of tuberculosis stem from Davenport. In making his case for what he feels is a unit or polygenic trait called intelligence, Jensen offered this analogy on page 45:

> At one time tuberculosis had a very high heritability, the reason [*sic*] being that the tuberculosis bacilli were extremely widespread throughout the population, so that the main factor determining whether an individual contracted tuberculosis was not the probability of exposure but the *individual's inherited physical constitution.* Now that tuberculosis bacilli are relatively rare [*sic!*], difference in exposure rather than in physical predisposition is a more important determinant of who contracts tuberculosis. In the absence of exposure, individual differences in predisposition are of no consequence.

Even in the days when Davenport, Popenoe, and Johnson were using such pseudo-data to prove that the Irish as a nation had a greater predisposition to contracting tuberculosis than the non-Irish, such arguments had long since been discarded by physicians, physiologists, and microbiologists as arrant nonsense. The fact is that tuberculosis bacilli are no rarer now than they ever were. What has altered the pathogenicity of our ubiquitous tubercle bacilli in the affluent nations is also very well known.

As Professors McKeown and Lowe, the authors of the English-speaking world's standard textbook *An Introduction to Social Medicine,* note (p. 14), the most important reason for the decline of tuberculosis *and other infectious disease* mortalities in the second half of the nineteenth century—and since 1900—was *"a rising standard of living,* of which the most significant feature was probably improved diet (responsible mainly for the decline of tuberculosis . . .)" (italics added). Improved nutrition, thanks to higher wages, and better hygiene, thanks to the efforts of the Sanitary Reformers, gave a majority of the people of the industrialized nations a better phenotypical opportunity to develop their inherited, or genotypical, potentials for the production of antibodies and other inborn body defenses against the pathogens of tuberculosis and other infectious diseases. This in turn led, as McKeown and Lowe put it, to "a favorable trend in the relationship between infectious agent and human host."

That, as a professor of educational psychology at Berkeley, Jensen in 1969 chose to selectively ignore these well-known facts of modern medical microbiology and social medicine speaks volumes for his sudden nostalgia, as a new convert to the eugenics mystique, for the biomedical simplisms of Galton and Davenport. It does not, however, constitute even a whisper of biological evidence for the heritability-of-tuberculosis tenets of Davenport and Popenoe, let alone the existence of one type of "neural structure" for the brains of people who get high IQ test scores and a different type for the brains of people who get low IQ test scores. It constitutes, simply, nostalgia.

Scientifically, there is far less here than meets the eye. Historically—there is far more.

MEANWHILE, IN THE REAL WORLD OF FLESH AND BLOOD

In dealing with the biology of intelligence, it is obvious why Jensen prefers hypotheses about nonexistent neural structures and secondhand pseudobiological AFQT data to the hard facts at issue. But by the time he wrote his *Harvard Educational Review* essay on the heritability versus the social biology of intelligence, a very basic study had become an integral part of the literature of American education. In a book published in 1961— *Education and Income: Inequalities in Our Public Schools*—Dr. Patricia Cayo Sexton, associate professor of educational sociology at New York University, had made what was to become *the* basic modern analysis of the relationships between urban American family incomes and the biological imperatives of classroom performance, attendance, and test scores.

The data and their implications are so clear that, in order to make his *HER* thesis stick, Jensen would have had to refute the Sexton findings and prove them wrong or irrelevant or, as the eugenicists have been wont to put it for over half a century, fatuous. Perhaps this was why Jensen ignored them. But as Joe Louis said of the speedy Billy Conn before their second fight, "Conn can run but he can't hide." Here, then, are some of the Sexton data on the social biology of classroom and IQ test scores, as derived from her study of the elementary school population of Detroit.

At the outset of her section "Sickness and Medical Problems," Dr. Sexton writes:

> Lower-income children often have trouble in school because of sickness and health problems. They are much more likely than upper-income children to be sick or diseased *or to suffer from some chronic ailment that goes undetected or untreated.* This is probably a result of the fact that lower-income children are not as well cared for as upper-income children. They are not so well fed, their housing conditions are more conducive to epidemic disease, and they do not as often get proper medical attention and care [such as early immunization against diphtheria, tetanus, polio, measles, and other completely preventable diseases]. [Italics added.][13]

For rheumatic fever, for example, the rate of incidence was *three* times higher among children whose parents earn $3,000 a year or less than among children whose parents earn $9,000 a year. Which means that in Detroit the rheumatic fever rate (per 10,000 students) is 2.6 in the $9,000 income group; 4.8 in the $7,000 income group; 6.4 in the $5,000 income group; and 7.9 in the $3,000 and less group. (See page 439.) These income figures were, of course, in 1961 dollars, which had a purchasing power 65 percent greater than 1974 dollars.

Diphtheria, *a totally preventable disease,* turned out to be exactly that— in the $9,000 income families. Zero cases. Among the slightly less affluent $7,000 income families, the diphtheria incidence was 0.9 per 10,000 students. At the $5,000 level, there were 1.4 cases—while at the $3,000 or poverty level there were a whopping 15.1 cases. The tuberculosis rate was also zero in the $9,000 income families' children, and, on a climbing scale, 0.6 percent among the moderately poor; 3.4 percent among the very poor; and 6.8 percent among the children of the poorest Detroit families. The tubercle bacilli of America's fifth-largest city apparently ignore the Davenport-Jensen laws of tuberculosis etiology.

In terms of the biological traits most intimately related to learning— vision and hearing—this is the way Dr. Sexton summed up her data:

Nursing Divisions (in order of increasing income)	Children with Diagnosed Defects Who Received No Treatment		
	Total Defects	*Vision*	*Hearing*
1 (lowest income)	32.5%	31.1%	53.2%
2	31.7	37.6	41.0
3	31.8	29.8	42.5
4	29.8	29.1	39.8
5 (highest income)	15.5	13.5	34.6

All of this meant that not only were there fewer eye and ear defects in the Detroit children of the highest income groups, but also that "the percentage of untreated defects in group 1 [the poorest children] is more than *twice* what it is in group 5 [the most affluent]" (italics added).

These health data made the reading, arithmetic and the achievement test scores, as well as the IQ test scores, of the children of all social class groups—*regardless of race*—entirely predictable. As Dr. Sexton observed (p. 102): "Obviously, if a child is ill or if he suffers from *undiagnosed* sight, hearing, and speech defect, his chances of doing well in school are greatly reduced."

The children in the schools of Detroit did not enjoy the same equality of opportunity to receive preventive and corrective medical care. While, as Dr. Sexton learned, only 7 percent of the children whose parents had an annual income of $9,000 or more had not been examined by doctors before entering kindergarten, 49.3 percent of the children whose families had annual incomes at or below the $3,000 level had started their school lives without medical examinations to determine whether their eyes, ears, speech, and general health were up to the requirements of modern education.

When it came to IQ test scores, the children of "all income groups below $7,000 have [IQ] scores of less than 4.00, or less than 'C' rating. All income groups over $7,000 have ratings of better than 'C.' *The [IQ] scores tend to go up as income goes up"* (p. 38; italics added). Dr. Sexton summarized her data in the following table:

Major Income Group	IQ Test Score
I ($3,000–)	2.79
II ($5,000–)	3.31
III ($7,000–)	4.55
IV ($9,000–)	5.09
Key to IQ scores: 2.00 = D; 3.00 = C−; 4.00 = C; 5.00 = C+; 6.00 = B	

Given the cognitive effects of the undiagnosed and medically neglected pathologies of poverty on the abilities to see and hear, as well as the vastly greater number of days absent from school because of the *preventable* infectious and nutritional diseases associated with poverty in America, it becomes clear to anyone but a true-believing eugenicist that the differences in the family incomes of the poor and the nonpoor account for considerably more than 20 percent of the differences between the IQ test scores of the poorer southern whites and the scores of the more affluent but equally Caucasian northern whites—and for the differences between the IQ test scores of the *average* white children and the scores of the considerably poorer *average* black children.

However, to base our governmental social policies on guaranteeing the minimum amounts of food, clothing, shelter, medical care, and cultural amenities all growing children require in order to have eyes, ears, bodies, and brains as healthy and as fully developed as they are genetically capable of developing, this nation would have to go out of the business of massive military adventures in places like Southeast Asia and start spending some of the money formerly spent for such activities on the proper care of all of our children—regardless of their parents' social status or race. To base our social policies, on the other hand, on the Davenport-Galton postulate that it is a waste of time and money to try to educate the genetically ineducable, and to prevent infectious, deficiency, ear, eye, and other medically and socially preventable diseases and disorders in the bodies of the children "genetically doomed" by their "inferior genetic endowments" to be blind, deaf, diseased, and stupid, calls for infinitely more modest social programs paid for out of far fewer tax dollars.

THE DISCIPLES OF JENSEN

Jensen's two most prominent disciples, the electrical engineer William Shockley, co-winner of the Nobel prize for physics in 1956 for the development of the transistor, and Richard B. Herrnstein, professor and chairman of the Psychology Department at Harvard, have helped advance his eugenics concepts.

Shockley entered the fray even earlier than Jensen, as attested by a four-page interview in the November 22, 1965, issue of *U.S. News and World Report*—which has been to Shockley, Jensen, and Herrnstein in our times what the old *Saturday Evening Post* was to Madison Grant, Lothrop Stoddard, and Prescott Hall. In this interview, Shockley averred that the quality of the human race is declining in this country because society was not doing enough research into the genetic factors that make people what they are. "Many of the large improvident families with social problems simply have constitutional deficiencies in those parts of the brain which enable a person to plan and carry out plans." Worse, in view of the Vogt-Hardin-Ehrlich Population Explosion, such hereditary menaces tended to produce more children than persons of average or superior ability.

After Jensen's *HER* article appeared in 1969, Shockley mailed reprints of the eugenics tract to each of his fellow members of the National Academy of Sciences. Shockley, a professor of engineering science at Stanford, also set up the Foundation for Research and Education on Eugenics and Dysgenics, with the acronym of FREED, whose original board included the late Sir Cyril Burt, to push his crusade against "genetic enslavement."

Stanford University turned down Shockley's 1972 proposal to teach a course called "Dysgenic Question: New Research Methodology on Human Behavior Genetics and Racial Differences" on the grounds that, as biology professor Colin S. Pittendrigh of the faculty committee that made this recommendation wrote, "no geneticist of stature—indeed, none that I know of—regards Professor Shockley as adequately familiar with the complexities of the genetic problems to which he addresses himself." Three other members of the same faculty committee, psychology professor Robert Sears, statistics professor Bradley Effron, and communications professor Nathan Maccoby, also noted that "the essentially genocidal policies [Shockley] has seemed to propose are not only painful for black people to hear but are abhorrent to all decent people whatever their skin color."

The same fate was meted out to Shockley's annual demands, of the National Academy of Sciences, that it launch a national investigation of the intelligence of Negroes to serve as the basis of what Shockley describes as sound national "eugenic practices." When he abused the National Academy and accused its members of "Lysenkoism" in their rejection of his demands in 1970, seven members of the Academy, headed by the Harvard professor of biochemistry John T. Edsall, and including a prominent geneticist, Professor I. Michael Lerner of Berkeley, sent a letter to the Academy saying:

> . . . Dr. Shockley's proposals are based upon *such simplistic notions of race, intelligence, and "human quality" as to be unworthy of serious consideration by a body of scientists.* A point-by-point refutation seems necessary.
>
> The most detestable aspect of Shockley's proposal is a categorizing of the population into "whites" and "Negroes." Each individual is genetically unique; there is not a single important trait for which there is not a wide overlap between different human populations. *It is basically vicious to evaluate individuals on the basis of the group to which they belong.* [Italics added.]

Despite such peer verdicts on Shockley's simplisms, he became a lasting fixture on television talk shows and on the college lecture hall circuit, where he kept repeating that he was not promoting racism but "raceology" and, on at least one syndicated television talk show I taped in April 1972, deriding the "followers of Boas" (whose name Shockley mispronounced). Whatever his standing among geneticists, pediatricians, psychologists, and other scientists dealing with the qualities of human life, his fervent supporters could be found in the White House itself.

In the mass media and on the lecture platforms, Shockley often cites the surrealistic AFQT scores as statistical evidence of the inferiority of the Negro mind, and works these worthless data into his propaganda for his own "Sterilization Bonus Plan."

Under the Shockley Plan, everyone would benefit, because "every baby born—and the population explosion demands limits—deserves good genes and environment. A sterilization bonus plan should reduce the number of babies who don't get a fair shake from their parental dice cup." What Shockley proposed, therefore, was to "offer bonuses for sterilization. Income-tax payers would get nothing. Bonuses for all others, regardless of sex, race, or welfare status, would depend on best scientific estimates of hereditary factors in disadvantages such as diabetes, epilepsy, heroin addiction, etc. If a bonus rate of $1,000 for each point below 100 IQ, $30,000 put in trust for a 70 IQ moron of 20-child potential it might return $250,000 to taxpayers in reduced costs of mental retardation care."[14]

Shockley's sterilization plan even took care of both the unemployed and the fully employed people whose wages were too low for the proper care of a family. It was something worthy even of Malthus: "A motivation boost might be to permit those sterilized to be employed at below minimum standard wages without any loss of a welfare floor income. Opportunity for those now unemployed?" This was, of course, nothing less than a gonadal revision of the notorious eighteenth-century Speenhamland System, under which tax welfare funds were used to subsidize low agricultural wages in England.

After Shockley announced this sterilization plan he learned that he had been pre-empted in 1937 by H. L. Mencken, the hard-core Babbitt in iconoclast's clothing, a priority he gracefully acknowledged on February 14, 1971. (See epigraph, page 362.) Mencken, though, had himself been preempted by another pop philosopher of his era, Mrs. Margaret Sanger, who had presented exactly the same sterilization plan in 1926.[15] Shockley also observed, in presenting the Sanger-Mencken-Shockley plan for the extinction of people who get IQ scores lower than 100, that Vice-President Spiro Agnew saw things his way about "the unmentionable threat concealed in our nobly intentioned welfare programs—dysgenics: retrogressive evolution through the disproportionate reproduction of the genetically disadvantaged with the consequent explosion of the relief burden." Like the Vice-President, Shockley felt the time was ripe for the kind of " 'hard social judgments no one . . . is willing to even think about.' "

In 1969, Shockley introduced his "white gene" hypothesis, derived from reading a long article by the geneticist T. E. Reed, in *Science,* in which

Reed dealt with a blood fraction called Duffy's Fy^a that is normally found in 43 percent of all Caucasian blood samples, and rarely found in the blood of those various native Africans tested for the presence of this material.

Reed suggested that the presence and frequency of this substance in the blood of American Negroes might help to reveal the presence of white genes added to the genetic heritage of individual black family lines since their ancestors were brought here as slaves. His article in *Science* made no mention whatsoever of intelligence in connection with these Duffy's Fy^a genetic markers. He did, however, cite data by other investigators suggesting that two mental disorders known or believed to be of genetic origin—phenyl-ketonuria and Huntington's chorea—were considerably more prevalent among white than among nonwhite peoples.

Reed's article also reported that in those few sample populations tested, the presence of these Duffy factors was higher among northern than among southern Negroes. He stressed, however, then (1969) and in later papers, that these samples were *too fragmentary* to allow for firm estimates of the degrees of admixtures in any population tested.

From these preliminary data, Shockley charged in with a hot slide rule and came up with a very neat formula: for every one percent of "Caucasian genes" in their blood, the IQ of Negroes climbed one IQ test point. Shockley saw in the "Surgeon General's [AFQT] reports on mental failures on the preinduction tests for 1967 and 1968" and "the Coleman Report results for ninth grade verbal and nonverbal for Northern and Southern Negroes" valid proof that "for low IQ Negro populations, each one percent of Caucasian genes raised the expected IQ by about one point." Previously, on the basis of the AFQT preinduction examinations mentioned above, Shockley had "estimated a drop of about 5 IQ points for Negroes compared to whites from 1918 to 1966," an act of faith in the grievously anomalous preinduction examination data even the Surgeon General's investigators find difficult to explain or accept.

More than this, Shockley also figured out that among blacks "the lower IQ mothers have significantly more children, probably about twelve. Profes-sional Negroes have families of about 1.9 children according to a 1965 article by Moynihan. These figures suggest that the Caucasian component is being reduced in each generation with a resulting lower average IQ for the next generation." And this, Shockley warned, spelled trouble ahead. Because "if those blacks with the least amount of Caucasian genes are in fact the most prolific and also the least intelligent," Shockley observed in an interview in *The Stanford Daily* on December 2, 1970, "then genetic enslavement will be the destiny of their next generation."

However, since the Duffy Fy^a factors, which are *not* genes themselves, are more than twice as prevalent in Oriental blood as they are in Caucasian, with a frequency of 93 percent in the blood of all Oriental people as against only 43 percent in the blood of all Caucasians (personal communication, T. E. Reed), then if they must be classified as genes and given a racial label they have to be called *Oriental* genes. This would then mean, according to Shockleyan logic, that the more *Oriental* genes there are in the blood of white Americans, the higher their IQ test scores—*and the purer Caucasian and less Oriental the blood of a white man, the lower his IQ test score.* Unless, of course, Shockley's famous predecessor in raceology, Dr. John H. Van Evrie, was right in 1868, when he wrote that all the great men of the Oriental races—Van Evrie listed Attila and Genghis Khan, among others— were actually Caucasians, and that it was quite probable that "Confucius and other renowned names known to the modern Chinese were white men."[16]

On the other hand, it is also quite possible that the distinguished Dr. Van Evrie was dead wrong, and that Confucius was not a closet Caucasian.

Despite what most hematologists have long known about the Oriental nature of Shockley's "Caucasian genes" for intelligence, his equation of one IQ test point for every one percent of these Duffy's Fy^a factors in the blood of black people has been accepted by many influential Americans at face value. Even more credible, to newspaper and magazine editors and university professors, has been Shockley's claim that for three decades a eugenic sterilization program has existed in Denmark.

In 1967, in a paper read before the National Academy of Sciences, Shockley told his fellow Academicians that "the lesson to be learned from Nazi history is the value of free speech, not that eugenics is intolerable. A form of eugenics has been in effect in Denmark for 30 years, but I have found no one in this country who has studied it really seriously."

In December 1969, in his chapter of a multi-authored book, *New Concepts and Directions in Education,* after presenting "data of the sort that has aroused my concern for the welfare of the American Negro minority," Shockley added that his study of population figures told him "that 70 or more illegitimate slum Negro babies are born per day with a genetic potential for IQ below 75, approximately the cutoff point for sterilization in Denmark's 30-year-old eugenics program." Shockley cited, as the factual source of his information, a long article, "The 'Unfit': Denmark's Solution," that had appeared in *U.S. News and World Report.*[17]

The opening sentences of this *U.S. News* article, which appeared to be an on-site report, read: "Denmark is turning up as the world's first nation to commit itself to sterilization of the unfit as an important means of solving its social problems. Here, as in the U.S. and other countries, authorities are worried by a tendency of problem individuals to breed unlimited numbers of children likely to populate the jails, mental institutions, and welfare rolls of the future." According to the same article, copies of which Professor Shockley includes in his "kit" for the unconverted, "criminals are offered release from prison as an inducement" to be sterilized.

Shockley and the *U.S. News* writer were not the only authors to make such statements. For example, Jameson G. Campaigne, editor of the *Indianapolis Star,* in a 1969 Sunday column castigating the National Academy of Sciences for rejecting Shockley's demand for a large-scale study of inborn racial inferiority and superiority, wrote: "The problem is not race. It is genetics. For instance, Denmark, since 1935, has had a eugenic program based on long study and experience, which includes sterilization of people who have an IQ of less than 75. Incorrigible prisoners are not released unless they agree to sterilization."[18]

In 1971, in an article called "Dysgenics—The Unthinkable, Unmentionable Threat," in *Experimental Mechanics,* Shockley wrote: "Do our nobly intended welfare programs promote dysgenics—retrogressive evolution through the disproportionate reproduction of the genetically disadvantaged? One incident that led me to express my worries publicly was a news story of an acid-throwing teen-ager, one of 17 children of a mother with an I.Q. of 55. Later I learned of Denmark's sterilization programs with their eugenic implications. The rising per capita homicide rate of Washington, D.C., is fifty times Denmark's falling one. Dysgenics?"

There is only one trouble with all of these statements about Denmark. And that is that Denmark never had, and does not now have, anything resembling a eugenic sterilization program. Denmark did and does have abortion and sterilization laws roughly analogous to the similar legislation in most civilized nations, including the codes of England, New York State, and Switzerland. These laws, as in New York, are designed primarily to protect mothers, children, and families from the biological, economic, and social complications of unwanted and/or medically inadvisable pregnancies.

After the *U.S. News* report of March 1966 was brought to his attention, Erik Olaf-Hansen, medical editor of the Copenhagen *Politikken,* writing that this American news magazine article "compares the situation in Denmark to that of Nazi Germany," noted that the same article created "the impression that the Danish government had adopted a sterilization program aimed at denying parenthood to all citizens with low IQ test scores or handicapped persons." He interviewed Mrs. Aase Willumsen, director of the Danish Health Ministry Mothers' Aid Centers—which administer over 95 percent of all sterilizations in Denmark. Mrs. Willumsen charged that the *U.S. News* report had "completely misunderstood the Danish program," and that "the American [that is, *U.S. News,* Shockley, and Campaigne] published descriptions of the sterilization situation in Denmark are quite distorted."[19]

The entire sterilization program of Denmark is administered jointly by the Mothers' Aid and the national University Institute of Medical Genetics. When I could find no reference of any sort to IQ test scores in the full text of the Danish Law on Sterilization and Castration of June 3, 1967, I asked Professor Jan Mohr, of the Institute, about this. He replied that in Denmark, as in other civilized countries, "oligophrenia [defective mental development] is of course *not* defined primarily by an IQ test." What horrifies Danish physicians and geneticists is the suggestion that the Danish sterilization and abortion programs are eugenic in origin, purpose, or administration.

Since few nations resisted the Nazis and their eugenics laws during World War II more courageously than Denmark, it should come as no surprise that today's generation of health workers in Denmark are so horrified by the widespread American concept that their programs, organized for the protection of the innate rights and dignity of all human beings, are in any way derived from the racist and elitist eugenics of Galton and Hitler. This

bewildered resentment is apparent in the answers received from Magna
Norgaard, head of the Mothers' Aid Centers Department of Information and
Statistics, in reply to six questions that I mailed to her concerning the
repeated American statements about the sterilization program administered
largely by her agency. Answers with which, incidentally, Professor Mohr, who
has read them, concurs.

QUESTION: Is it true that the function of the Danish sterilization program is, as the *U.S. News and World Report* article states, to ameliorate the
nation's social problems by the sterilization of criminals, people on welfare
rolls, and people in mental institutions? ANSWER: No. This entire concept is
at variance with Danish philosophy and ideas of human worth. I can add
that if a person who is in a mental hospital against his or her will, or in a
penal institution, or under outpatient care for mental retardation, applies for
sterilization, this application has to be studied and approved by a socio-
medical council that includes a judge, a psychiatrist, and a social worker.

QUESTION: Is it true that as an inducement for release from Danish
prisons criminals are given their freedom if they agree to be sterilized?
ANSWER: No.

QUESTION: Is it true that "chronic joblessness" and "alcoholism" are
both considered grounds for mandatory sterilization in Denmark? ANSWER:
No.

QUESTION: Is it true that the mothers of illegitimate children who also
happen to have low IQ test scores "are not allowed to mingle with ordinary
people" unless they first agree to being sterilized? ANSWER: No. The socio-
medical council in charge of hearing sterilization applications is extremely
reluctant to permit the sterilization of unmarried women. Of the 1,134
women granted sterilization in Denmark in 1970–71, only 56 were
unmarried. By far the greatest percentage—83 percent—of all women
who apply for sterilization in Denmark are married, and two thirds of those
women are over thirty. I do not know how many of the 56 unmarried Danish
women granted sterilization had children. The question is rather irrelevant,
as unmarried mothers are more respected and accepted in the Danish community than in most other countries in the world. This fact of Danish life is
generally well known in Europe and the United States.

QUESTION: Is it true that a special provision of the Danish law on sterilization "provides for the sterilization of feebleminded persons with an IQ
test score of less than 75"? ANSWER: No. There is no special law for sterilization of feebleminded persons. Mental diseases are one of the grounds on
which persons can apply for sterilization, but there are no special IQ test
score cutoff points in the laws that justify sterilization of people with mental
problems. We have no special data for feebleminded persons being sterilized,
but the numbers are certainly small.

QUESTION: Is it true or false that the sterilization and abortion laws of
Denmark are eugenic in purpose and function? Or are they in effect for the
protection of the health of mothers, children, and families who would suffer
from a spectrum of genetic and/or birth associated but nongenetic disorders
(such as, for example, Down's syndrome in over-age parents; damage to

mother and child due to cephalo-pelvic disproportion and other skeletal or physiological defects caused by malnutrition; rubella virus, cytomegalovirus, and toxoplasmosis infection in the mother during pregnancy, etc.)? ANSWER: The sterilization and abortion laws of Denmark are *not* eugenic in purpose and function, but have the purpose of protecting women, mothers, children and families.

Where Shockley found few Americans ready to reject his widely believed fantasies of Nazi-style eugenic sterilization laws in Denmark, he had a little more trouble selling the credibility of the AFQT mental test scores to psychologists familiar with their worth. In 1971, for example, Shockley presented a paper at the national meetings of the American Psychological Association in Washington, D.C., called "Dysgenics—A Social-Problem Reality Evaded by the Illusion of Infinite Plasticity of Human Intelligence?" Behind this long title was the familiar mix of Jensen's secondhand twin data; the Sanger-Mencken-Shockley sterilization bonus plan for the extinction of people with what Shockley called *hereditary* disadvantages such as diabetes, epilepsy, heroin addiction, arthritis, and color blindness; Shockley's tiresome misreading of the meaning of Reed's Duffy blood factor data, which the Stanford professor of engineering science combined with the contemporary "Army [AFQT] preinduction test data to estimate that for low IQ Negro populations each 1% of Caucasian ancestry raises average IQ by 1 point"; and, of course, the familiar Shockley homily that "the lesson of Nazi history is not that eugenics is intolerable" and therefore "only the most anti-Teutonic racist can believe that the German people are such an evil breed of man that they would have tolerated the concentration camps and gas chambers if a working First Amendment had permitted exposure and discussion on Hitler's 'final solution'—the extermination of the Jews."

Shockley had added only two new ingredients to his familiar eugenic hodgepodge: he now cited "Richard Herrnstein's article in the current *Atlantic Monthly*," along with Jensen's post-1968 writings, as added proof of his own scientific accuracy, and he announced that "the word 'raceology' has been proposed for studies like mine. They are not racism."

To reporters on the scene, such as the *Washington Post*'s Stuart Auerbach, Shockley declared that the primary sources of scientific data on which his "raceologist" claims were based were the 1966 Coleman Report and the decade's Armed Forces Qualification Test results. Of these military mental test scores—as sacred to Jensen as they were to Shockley—visiting APA member Dr. Edward A. Scanlon, a clinical psychologist at the Schuylkill County Mental Health Center in Pennsylvania and a former Army psychologist, told the press that Shockley "ought to quit using them unless he wants to tell a lie." Even Shockley had to admit that Scanlon's comments on the total unreliability of the AFQT mental test results constituted "a very valid area of criticism; it bears on the validity and significance of one of the types of data I am using."

Although shaken by the growing rejection of his AFQT mental test score data by American scientists who knew their worth, Shockley has managed to find, in his own professional training and experience, what he truly believes to be added scientific evidence of the correctness of his views of what the eugenicists and other scientific racists call "racial intelligence."

Most of Shockley's professional life was spent as a staff electrical engineering researcher for the American Telephone and Telegraph Company. This professional experience was reflected in the statement he made in the *Phi Delta Kappan* of December 1972, in which God-Nature was visualized as a Master Telephone Cable Spinner in the Sky:

"Nature has color-coded groups of individuals so that statistically reliable predictions of their adaptability to intellectually rewarding and effective lives can easily be made and profitably be used by the pragmatic man in the street."

THE SQUARE ROOT OF THE MERITOCRACY

The pragmatic professor of psychology at Harvard, Richard Herrnstein, includes himself among Shockley's admirers, as Dr. Herrnstein made clear during a three-page interview in *Science* in 1973, in which, wrote the interviewer: "Herrnstein believes in the validity of Shockley's central thesis: that the way people are breeding now, there is a danger of a large portion of the population becoming 'genetically enslaved.' "[20] In an interview published in the Harvard *Crimson,* Herrnstein also shared Shockley's views about inherited capacities for high or low IQ test scores; he advocates that IQ test scores be recorded by the U.S. Census takers in order to enable our lawmakers to "observe dysgenic or eugenic trends in American society." In this way, Herrnstein explained, "if at some time in the future we decide that our population is getting too large, and we need to limit it, we could use census information on IQ to decide how and when to limit it."[21]

However, most of Herrnstein's genocidal and pseudogenetic ideas on IQ and human worth are derived, as he quite graciously revealed in his famous *Atlantic Monthly* article of September 1971, from Jensen's 1969 *HER* article. Herrnstein's article, "IQ," which was preceded by a fulsome paean of self-praise by the editors of *The Atlantic Monthly* for their courage in publishing it, added little to the basic Jensen thesis. The editors of the venerable fountainhead of nineteenth- and twentieth-century American conventional wisdom —delivered by such *Atlantic* authors as Henry W. Holland, John Fiske, Thomas Bailey Aldrich, Nathaniel S. Shaler, William Z. Ripley, Francis Amasa Walker, Samuel J. Holmes, Robert M. Yerkes, and Cornelia J. Cannon—said in their foreword that they were defying the wishful-thinking liberals and printing Herrnstein because they believed that "it is not only possible but necessary to have a public discussion of important, albeit painful, social issues. The subject of intelligence is such an issue—important because *social legislation must come to terms with actual human potentialities,* painful because the actualities are sometimes not what we vainly hope" (italics added).

At the outset of his article (p. 45), Professor Herrnstein made a flat statement that raised more than a few eyebrows among his professional peers. "The measurement of intelligence," he wrote, "is psychology's most telling accomplishment to date." This declaration revealed infinitely more about the actual dimensions of Herrnstein's knowledge of the amply documented history of IQ testing, and his regard for the equally documented history of psychology, than it did about the nature, function, and scientific worth of IQ testing.

Herrnstein's 1971 article showed, of course, that according to their IQ test scores, the "actual human potentialities" of the American people had not been expanded one iota since February 1922, when in *The Atlantic Monthly*'s lead article, "American Misgivings," Cornelia James Cannon bemoaned the fact that the World War I intelligence tests had revealed that "nearly half [of the U.S.] population is under the mental age of thirteen," and that "almost half of the white draft, 47.3 percent, would have been classed as morons" had they been given IQ tests in civilian life. What Herrnstein added to this ancient eugenic concept was the warning that, since both inferiority and superiority were 80 percent genetic and 20 percent environmental, improving such things as health, education, and equality of opportunities for the total population would only make the genetically high IQ types even smarter than they are now, and thus widen the intelligence gap between them and the vast multitudes of born clods so much that a "meritocracy" of these enriched high IQ minds would replace our present democracy.

Herrnstein's *Atlantic* article (and the book into which he expanded it) also included an alleged "history" of IQ testing far more notable for what it omitted (excluding, as it did, among other highlights, Goddard, Wallin, Klineberg, and the role of the Army intelligence test scores in ending Jewish and Italian immigration in 1924) than for what it included, and for the breathtakingly amazing judgment that Sir Francis Galton was "perhaps smarter" than his half cousin Charles Darwin.[22]

Within hours of the appearance of the September 1971 *Atlantic* on the Washington newsstands, the CBS–TV national prime-time news commentator Eric Sevareid was on the air remarking that now that it had been shown that some people were revealed to be genetically less educable than others, the time had come for an agonizing reappraisal of our governmental educational priorities. Sevareid had gotten the message of the editors of *The Atlantic* that "social legislation must come to terms with actual human potentialities."

When Herrnstein, like Jensen, expanded his article into a book—*I.Q. in the Meritocracy* (1973)—the highly respectful reviews included a politically very significant one by the New York University urbanologist Irving Kristol in the May 1973 issue of *Fortune*. In this review, Kristol quoted Herrnstein as writing that "the heritability of IQ is not a *new* finding, on which the burden of proof suddenly falls, but the standard, virtually uncontested [*sic*] . . . finding since the first decade of mental testing." Then, both paraphrasing and endorsing Herrnstein's comment "Jensen concluded (as have most of the other experts in the field) that the genetic factor is worth about 80 percent and that only 20 percent is left to everything else," Kristol added: "A search of the scientific literature from one end to the other reveals no serious challenge to the view that differences in IQ are largely hereditary [*sic*]." And, Professor Kristol added: "We are dealing with a *scientific finding that,* though challengeable (as is every such finding), *has the weight of evidence clearly on its side"* (italics added).

THE MAKING OF A PSEUDOFACT, 1971–1973

Jensen concluded (as have most of the other experts in the field) that the genetic factor is worth about 80 percent and that only 20 percent is left to everything else.

Richard Herrnstein, *The Atlantic Monthly,* September 1971, p. 46.

A search of the scientific literature from one end to the other reveals no serious challenge to the view that differences in I.Q. are largely hereditary.

Irving Kristol, *Fortune,* May 1973, p. 293.

Both *The Atlantic Monthly,* our leading middlebrow monthly, and *Fortune,* the magazine of big business, ran display heads on the theme that most scientists agree that the differences in IQ test scores of different people are "largely hereditary."

In the only sampling of the opinions of American psychologists, made in 1970 by the sociologist R. W. Friedrichs of Williams College, over two-thirds (68 percent) "either disagreed or tended to disagree" with the Jensen thesis that "genetic factors are strongly implicated in the average Negro-white intelligence difference." (See pages 43–45.)

Kristol cited "the distinguished Harvard psychologist" Herrnstein's book as a valuable antidote to the claims of "our orthodox egalitarians" who were "committed to the proposition that there are no important and ineradicable differences in innate mental ability" that would not be reduced by improving the human growth and developmental environments. Since achieving such improvements in health, education, hygiene, and housing would raise both wages and taxes, Kristol's hearty endorsement of Herrnstein's nineteenth-century concepts of human heredity and development made hard common sense to *Fortune*'s readers—whose ranks just happen to include possibly a majority of the major financial contributors to the coffers of both of our major political parties.

This mindless review of Herrnstein's book was possibly as much of the literature on the ongoing heredity-environment-human worth controversy as most of these very busy bankers, corporation executives, and lawyers would ever read. To these enormously influential Americans, the judgments of Herrnstein and Kristol probably represented the consensus of the thinking of American and European science. For these people are not illiterates: they are the graduates of some of our finest colleges and graduate schools, with the utmost respect for mathematics, the language of science. And they have been made aware that the articles and books of Shockley, Jensen, and Herrnstein abound in dazzling and advanced mathematical equations, correlations, curves, and other figures.

What these influential, college-educated Americans were *not* taught in college and elsewhere was that essentially the identical mathematical documentations were offered by the old scientific racists before this era, on behalf of the same eugenics creed that Jensen and his disciples now promote—and that these earlier mathematical displays of half-truths, and quarter-truths, and outright lies presented as truths, have in this century been responsible for literally millions of needless deaths and preventable morbidities in the United States and other advanced industrial nations.

As we have seen thus far in this book, these earlier abuses of the language and data of the legitimate sciences have included the statistics and pedigree charts with which Davenport—*after* Goldberger proved pellagra to be a preventable disease caused in people of all races by the lack of proper foods—convinced the makers of American health, minimum-wage, and other social policies that pellagra was a hereditary and there-fore *unavoidable* disease of *white* people of inferior breeding stock. And the mathematical displays of the meaningless Army intelligence test scores of World War I, and the accurate enough but grievously perverted ethnic patient and inmate data of America's mental and physical hospitals and penal institutions prepared by the official Expert Eugenics Agent of the U.S. Congress, H. H. Laughlin, which helped close the gates of America to Jewish, Italian, Greek, Spanish, and other non-Aryan immigrants seven years before the start of the Nazi holocaust. And the maniacally mathematical 1925 study of Jewish immigrant children in London by Galton's chosen disciple, the socialist Karl Pearson, in which worthless pseudofacts were summarized in statistical fireworks which Pearson claimed proved that nutrition has no effect at all upon intelligence, and that the Jews are by heredity an inferior race of parasites upon worthy nations.

Much as I admire a well-turned curve of distribution, let alone an elegant differential equation, I am also only too well aware of the well-known mathematical axiom that the square root of 9.5 trillion times zero is still zero. The factual and scientific value of the "data" from the AFQT scores, and the "genetic" studies of mental retardation by Davenport and the Reeds, and the "white genes equal higher IQ test scores" equations of Garrett, Shuey, McDougall (and their nineteenth-century forebears, Galton, Van Evrie, and Gobineau), on which Jensen, Herrnstein, and Shockley have based their mathematical pyrotechnics, are all so totally devoid of anything re-sembling the elementary particles of scientific and historical truth that, in all due caution, logic, and probity, I am afraid that I have to agree with my friend David Layzer, the Harvard astrophysicist, who writes that "although Jensen's sum is good arithmetic, it is bad science."

Professor Layzer, whose skills in mathematics and scientific logic and whose background and training in biology and genetics are certainly at least equal to those of Jensen and Herrnstein, concluded after a careful examina-tion of the Jensen canon that, when it came to the interactions between gene and environment from which the phenotypes are developed, Jensen's "re-marks clearly demonstrate that he understands neither the mathematical nor the practical problems involved in the estimation of interaction effects." A conclusion with which Dr. Layzer's colleague Richard Lewontin, the Harvard population geneticist, concurs. As, indeed, do many working geneticists agree with Layzer's other conclusion that Jensen's hypothesis "of genetic differences in intelligence between ethnic groups is shown to be untestable by existing or foreseeable methods. Hence it should not be regarded as a scientific hypothesis but as a metaphysical speculation."[23]

Mathematical calculations might have enabled our kind to split the atom and reach the moon—but they still lack the capabilities to turn such infectious and deficiency diseases of poverty as tuberculosis and pellagra into hereditary diseases of "inferior" races. Or to prove that intelligence—*however* we define "intelligence"—is 80 percent genetic and 20 percent experiential. Or to convert excrement into emeralds.

21 | The Emperor's New Clothes Revisited: The Legitimate Sciences Launch Factual Counterattacks Against the New Scientific Racism

> Even when the severe effects of intrauterine and neo-natal deprivation are avoided, malnutrition may still increase the incidence and severity of illness, retard the physical and mental development of the young, and, in adults, reduce work efficiency and mental stability. *When all infants are given equal conditions both within and outside the womb, it is likely that many so-called racial characteristics will disappear.* [Italics added.]
>
> —B. S. PLATT and R. J. C. STEWART, in "Reversible and Irreversible Effects of Protein-Calorie Deficiency on the Central Nervous System of Animals and Men," *World Review of Nutrition and Dietetics,* 1971

Despite the wishful thinking of the editors of *The Atlantic Monthly, U.S. News and World Report,* and such nongeneticists as Professor Irving Kristol, the eugenic concepts of Jensen and his modern disciples did not in any way represent anything even remotely resembling the uncontested thinking of the vast majority of the world's genetic, biological, behavioral, and social scientists. To most scientists working in the fields of human heredity and development, a sound body remains the basic biological prerequisite for a sound mind. Similarly, most of the nation's scientists agree that reasonable doses of such social artifacts as the aspiration for upward socioeconomic mobility, and exposure to adequate experiences with spoken and written English, do indeed make vast differences in the reading, writing, arithmetic, and IQ test scores of all American children.

Whatever impressions prominent authors of the minority of true believers in eugenics in the ranks of the sciences dealing with human development might have conveyed to nonscientists, the replicable findings of the main currents of American research into the realities of learning and cognition in America's sixty million children did not support the statisticized assertions of Jensen, Shockley, and Herrnstein. For example, while American social policies and priorities were being based on the tenets of Jensenism, two major papers delivered at the 1971 meetings of the American Psychological Association were to shatter to smithereens all of the choicest shibboleths of the new scientific racism.

The two papers, one by a social scientist, the other by a behavioral scientist, stemmed from two ongoing but unrelated investigations.

The social scientist, Dr. Jane R. Mercer, associate professor of sociology at the University of California's Riverside campus, was concerned with the growing national tragedy of the mislabeling of intellectually competent young people as being "mentally retarded." This false labeling was going on all over the country because of the reckless administration and interpretation of standard IQ tests to the children of poor minority group parents. What concerned the scientific community most about these widespread abuses of IQ test scores was that the mentally normal Chicano, Puerto Rican, and Negro children so mislabeled, and subsequently placed in special classes, schools, and custodial institutions for the mentally retarded, were (and still are) subjected to the educational and other life experiences calculated to turn their false diagnoses of mental retardation into avoidable realities.

The behavioral scientist, Dr. George W. Mayeske, research psychologist of the U.S. Office of Education, was concerned with the over-all quality of U.S. education. Whereas Dr. Mercer based her paper on local investigations of some 644 persons selected from a population of 6,907 people in 3,198 housing units in Riverside, California, Dr. Mayeske's research was based on the socioeconomic and sociocultural data on some 650,000 students and their families in 4,000 schools in all fifty states gathered by the teams of social, life, and behavioral scientists headed by Dr. James S. Coleman.

Dr. Mercer's study, made in collaboration with behavioral and educational scientists attached to the Riverside school system, involved the testing and retesting of people previously classified as mentally retarded because their IQ scores fell below 70. Dr. Mayeske's study, in which he and his group of USOE staff scientists had the close working collaboration of Dr. Coleman, involved putting the massive Coleman data through computers programmed to analyze whether or not, and to what extent, various familial, community, and social factors were involved in classroom achievement scores.

HOW TODAY'S IQ TESTS MISLABEL NORMAL PEOPLE AS BEING MENTALLY RETARDED

Dr. Mercer's study was the product of an eight-year-long study of the epidemiology of mental retardation conducted in the city of Riverside, California, in the Los Angeles metropolitan district. Her investigation showed that while IQ test scores of below 70 were the standard cutoff points under which people were automatically classified as "mentally retarded," a majority of the Riverside adults her group tested and who achieved IQ test scores of under 70 "were, in fact, filling the usual complement of social roles for persons of their age and sex: 83.6% had completed 8 grades or more in school; 82.6% had held a job, 64.9% had a semi-skilled or high occupation, 80.2% were financially independent or a housewife, almost 100% were able to do their own shopping and to travel alone and so forth." In short, Dr. Mercer and her group had, over an eight-year period, discovered in a southern California city pretty much what J. E. Wallace Wallin had demonstrated during one summer vacation in rural Iowa in 1913.

Of course, during the nearly sixty years that had passed between the Wallin and Mercer studies, the social and behavioral sciences had pinpointed the reasons why the standard IQ tests often showed perfectly normal people to be ineducable morons. In Riverside it was primarily the Chicanos and the blacks who were being mislabeled as mentally retarded by the IQ tests, and the Mercer investigation found that "evaluation of adaptive behavior was important in evaluating persons from ethnic minorities and lower socioeconomic levels—persons from backgrounds that do not conform to the modal social and cultural pattern of the community. *Many of them fail intelligence tests mainly because they have not had the opportunity to learn the cognitive skills and to acquire the knowledge needed to pass such tests.* They demonstrate by their ability to cope with problems in other areas of life that they are not comprehensively incompetent." (Italics added.)

The Mercer group concluded that "the IQ tests now being used by psychologists are, to a large extent, *Anglocentric.* They tend to measure the extent to which an individual's background is similar to that of the modal configuration of American society. Because a significant amount of the variance in IQ test scores is related to sociocultural characteristics, we concluded that sociocultural factors *must* be taken into account in interpreting the meaning of any individual score." (Italics added.)

During the period of the Mercer group's studies, the California State Department of Education conducted an investigation of a "sample of Mexican-American (Chicano) pupils enrolled in classes for the educable mentally retarded (EMR) in selected school districts in California." Since these non-Anglo children had been classified as mentally retarded on the basis of their low scores in the Wechsler Intelligence Scale for Children (WISC)—one of the nation's standard IQ tests—the state's investigation was directed to the question whether the children were really mentally retarded or whether, at the time they took the WISC IQ test, "a language barrier prevented them from being assessed properly as to their native [hereditary] abilities to perform cognitive tasks."

The children were therefore retested with the Escala de Inteligencia Wechsler para Niños, "which is the Spanish version of the Wechsler Intelligence Scale for Children." The textbook Spanish in the standard Spanish version of the WISC test was, however, modified to make its Spanish more in keeping with that heard and spoken by Mexican American children in California. For example, *concreto,* which is good textbook Spanish for concrete, was changed to the Chicano vernacular *cemento.*

Test results from the urban and rural school districts, in which the Chicano children labeled as retarded were retested in their own language, showed that where the median score for Chicano children given the WISC test in English had been 70, the median score for the Chicano children previously labeled as mentally retarded climbed to 83, for an increase of 13 points. The scores of some of the "retarded" children climbed over 25 points when they were given the same IQ test in their own language.

The Department of Education investigators concluded that "many Mexican-American pupils may have been placed in EMR [educable mentally retarded] classes *solely on the basis of performance on an invalid IQ test*" (italics added).[1] Professor Mercer's study of IQ testing among Anglo and non-Anglo populations in Riverside, California, dealt with a wider spectrum of sociocultural factors than the test language alone.

When these sociocultural factors were taken into account in IQ tests of one study population consisting of 100 Chicanos, 47 blacks, and 556 Anglos from seven months through fifty years of age, and a second test population consisting of 1,513 elementary school children whose quotient of 598

Chicanos and 339 black children comprised the entire student cohort of the three then segregated Riverside public schools, and whose 576 Anglo children were randomly selected from eleven predominantly Anglo schools in the district, the results were alike. From 25 to 32 percent of the variances in the IQ test scores of both groups could be accounted for by sociocultural differences. "Not only did sociocultural characteristics account for a large amount of the variance in IQ test scores in the large samples which combined all three ethnic groups," Dr. Mercer reported, "but they also accounted for a large amount of the variance in the IQ *within* each ethnic group" (italics added).

Among the Chicanos tested, the four sociocultural characteristics most significant in the IQ test score variances were, in order of their weight: (1) living in a household headed by an employed white-collar worker; (2) living in a family with five or fewer members; (3) having a head of a household with a skilled or higher occupation; and (4) living in a family in which the head of the household was reared in an urban environment in the United States.

Among the blacks tested, the five most significant sociocultural factors influencing IQ test scores proved to be: (1) having a mother reared in the North; (2) having a family headed by a person with a white-collar job; (3) having a male head of household; (4) living in an intact family; and (5) living in a family which is buying its own home.

To the surprise of no social or behavioral scientists, let alone pediatricians, experienced classroom teachers, and school nurses, the Chicano and black children with the highest IQ test scores came from families with sociocultural characteristics "similar to the modal configuration of the community." What this meant to Mercer and her group was, clearly, that "sociocultural factors cannot be ignored in interpreting the meaning of a standardized intelligence test when evaluating the child from a non-Anglo background. *The [IQ] tests are measuring, to a significant extent, sociocultural characteristics"* (italics added).

The Mercer studies showed that when the sociocultural factors were equalized—that is, when the standards of living in the families of black or Chicano children tested happened to have been the same as those of the average whites or Anglos in the test populations—"there were no differences in measured intelligence." *Within* each minority group, as the table below shows, the children tested in the Mercer study proved the importance of these quality-of-family-living factors:

Family Living Standards and IQ Test Scores

	High				Low	
Socioeconomic status (SES)	*5*	*4*	*3*	*2*	*1–0*	*Total*
Chicano children						
Number of children	25	174	126	146	127	598
Mean IQ	104.4	95.5	89.0	88.1	84.5	90.4
Black children						
Number of children	17	68	106	101	47	339
Mean IQ	99.5	95.5	92.8	87.1	82.7	90.5

From Mercer, 1967, p. 26.

What this meant in terms of IQ testing of children of socioculturally deprived ethnic groups, such as the Chicanos and the blacks, was that the educational program of any minority child should be planned "on the assumption that he is a person with normal learning ability who may need special help in learning the ways of a dominant society." This also meant, wrote Professor Mercer, that in the evaluation of supposedly mentally retarded human beings, "information about adaptive behavior, [and] an individual's ability to cope with problems in the family, neighborhood, and community, should be considered as well as his score on an intelligence test in making clinical assessments. Only persons who are subnormal *both* on the intelligence test and in adaptive behavior should be regarded as clinically retarded." Therefore, "the meaning of a particular IQ test score or adaptive behavior score should be assessed not only within the framework of the standardized norms for the general population but should also be evaluated in relation to the sociocultural group from which the person comes."

The conclusion of the Mercer report ran directly counter to the most cherished dogmas of the new scientific racism, or raceology, or whatever else the old-time eugenics religion is called by its various modern communicants:

> When we re-analyzed the survey data from the field survey of the clinical epidemiology using these pluralistic diagnostic procedures, differences between rates for mental retardation between ethnic groups disappeared. Approximately the same percentage of persons were being identified as clinically retarded from each ethnic group [i.e., Chicanos, blacks, Anglos]. *When we re-diagnosed 268 children who were in classes for the educable mentally retarded in two school districts in Southern California using pluralistic diagnostic procedures, we found that approximately 75% of the children in those classes would not have been placed in special education if their adaptive behavior and sociocultural backgrounds had been systematically taken into account at the time of assessment.* When they were taken into account, the proportion of children diagnosed as mentally retarded from each ethnic group was approximately the same as the proportion of children from that ethnic group in the entire public school population. [Italics added.]

The Mercer findings created shock waves that were to be felt in local and state courts the nation over, where lawyers for the National Association for the Advancement of Colored People (NAACP) and other minority group organizations were and still are battling to correct the mislabeling (via the flagrant abuse of IQ test scores) as "mentally retarded" inflicted upon thousands of poor children from coast to coast. What made the Mercer findings so particularly significant in these court cases—some of which are beyond doubt headed for the Supreme Court of the United States—were the Mayeske-Coleman supporting data and concepts presented to the same 1971 national meetings of the nation's professional psychologists.

THE COLEMAN REPORT DATA VERSUS EUGENICS DOGMAS

The information the Coleman investigation collected in 1965 happened to have been, as Dr. Mayeske observed, "the most comprehensive body of data

ever collected on public schools and their students in the United States." Far
from confining itself to the two variables the scientific racists and the editorial
simplicists had picked out of the Coleman Report when it had been issued in
1966—that is, the achievement scores and the races of their achievers—the
Coleman study was an exhaustive and serious inquiry into the sociocultural
and socioeconomic characteristics of the students, their families, their areas
of residence, their schools, their teachers, and their school principals. The
initial report, *Equality of Educational Opportunity,* barely skimmed the surface of its vast reservoirs of hard data.

The Mayeske paper of 1971, *On the Explanation of Racial-Ethnic
Group Differences in Achievement Test Scores,* represented a more intensive
utilization of the same data bank. One of the working papers of the Mayeske
study (*A Study of Our Nation's Schools, 1969*) was 884 pages long, 335 of
them text and the rest tabular and computer displays of the source materials.
The final report, which many psychologists and educators who heard the
APA meetings summary feel is one of the most important scientific studies
of our era, ran to over 400 text pages alone.

It is, of course, impossible to convey much more than some of the
conclusions of the Mayeske report at this writing. Before presenting them,
however, it might be instructive to glance at the meticulously detailed nature
of some of the raw factual materials. The Coleman study's small army of
trained investigators gathered and collated hard information on more than
400 variables. These included detailed information on family structure,
socioeconomic status, attitudes toward life, expectations from education, and
home educational atmosphere; on teacher training and attitudes; on school
standards and resources; on racial and ethnic values, patterns, and interactions with society and history; on the sociocultural differences between
urban and rural residence, as well as between the northern, southern,
eastern, and western regions of the nation.

As did other serious studies of education, the Coleman inquiry investigated and tabulated the relationships between fathers' occupations and composite student achievement scores. But they did this in considerably more
detail than it has usually been done before. So that, for example, the Mayeske
working paper of 1969 could make the presentation from which the following
five lines of data have been extracted:

Average Composite Achievement Scores of Students by Father's Occupation

Father's occupation	Mean Scores for the Different School Grade Levels				
	1st	*3rd*	*6th*	*9th*	*12th*
Professional	55.466	56.833	55.299	56.597	56.012
Manager	53.665	54.586	53.570	53.541	52.771
Farm or ranch owner or manager	53.533	52.897	50.166	50.387	50.707
Workman or laborer	46.487	45.923	49.572	48.657	47.221
Farm worker	46.430	45.684	45.532	43.316	42.478

Behind the obvious sociocultural factors inherent in these socioeconomic comparisons, however, there are various other achievement-influencing factors that are to an amazing and predictable degree associated with the structure and value systems of any given families. There is, for example (to select a characteristically American factor), the amount of time the average student spends watching television at home. The raw data on the relationships between these figures and school marks, particularly at the twelfth-grade level, as exerpted from the 1969 working paper, seems to tell a very straight story:

Average Composite Achievement Scores of Students Indicating Number of Hours Spent Watching TV at Home

Number of TV hours per day	Mean Scores for Different School Grade Levels		
	6th	*9th*	*12th*
None or almost none	47.087	48.262	50.670
About ½	45.796	49.291	52.001
About 3	52.234	51.348	48.403
About 4 or more	49.051	47.280	45.839

However, there is slightly more here than meets the eye. For example, the usual reason for children not watching *any* home television is because their parents are too poor to afford a set. It is also well known that, at the next-higher level of poverty where poor people manage to own television sets, many families make undue use of these sets as surrogate baby-sitters. But this form of child abuse is also quite prevalent in nonpoor but very ignorant families. So that unless, as the Mayeske group did, you relate such isolated facets of the mores and customs of American families to hundreds of other sociocultural variables, you can wind up with a bit of impressive but in itself worthless information.

The Mayeske group, employing sophisticated statistical techniques, broke down all of the 400 investigated variables involved in learning into primary and secondary groups, and then refined and rearranged them further into sets of variables they rated by their relative weight or influence on school achievement scores. Even here the analyses, and the ratings, were complex and relative.

A characteristic example of the kind of sensitive weighting the sheer enormity of the raw data made possible can be seen in one of the elements in the index called Family Structure and Stability. This particular element, Mother's Work, deals with the effects on student school achievement of mothers working full time. On the relative weighting scale worked up by the Mayeske group and their computers, this single element of family life has a weight of 70 for six-year-olds in first grade, who at that age are highly dependent upon their parents. The older the same child grows, however, the greater his or her independence, and the lesser the learning impact of not having a full-time mother—so that by the third grade this factor rated a weight of 57; by sixth grade a weight of 28; and by twelfth grade a weight of 21.

When the Coleman Report of 1966 reported that various environmental factors were even more important to achievement scores than the quality of the school itself, the racists and the simplicists on campus, microphone, and typewriter perverted this general conclusion into the totally unfounded claim that Coleman had found that the quality of education is not nearly as important to society as the hereditary quality of the individual student. To prevent such perversions of their uses of the same Coleman data, the Mayeske group included a review of the relevant scientific literature in which they showed that schools (like calories) certainly do count.

Among other studies included in this section of the Mayeske report was the 1964 investigation conducted by Green et al. on the effects of the shutting down of all the public schools in Prince Edward County, Virginia, by local scofflaw authorities after the Supreme Court decision of 1954 that outlawed racially segregated schools. These schools remained shut for several years, during which the children of nonpoor white families went to new private schools, and the children of the poor whites and of most blacks were deprived of all schooling. As Mayeske noted, "the children who had never attended school could not even hold a pencil, let alone follow detailed instructions or take a test."

More than that, the educational retardation was of long duration, as shown by comparing the test performances of children whose schooling had been interrupted with that of Negro children in a neighboring county, who were of similar socioeconomic backgrounds but whose schooling had not been denied to them by stubbornly racist county officials. "It was found," Mayeske reported, "that the children whose schooling had been interrupted exhibited severe educational retardation, particularly on tests more closely related to school curricula such as spelling and arithmetic. On an intelligence test the scores of these [educationally deprived] children were 15 to 30 points lower than those of the children in the adjacent county who had continued in school." This also, as it happened, demonstrated anew the enormous cultural (environmental) influence on IQ test scores.

"Clearly," Mayeske commented, "the schools *do* have an important function. Just as clearly, one of the goals in improving the schools must be to increase the influence they have on their students that is *independent* of the students' social background—their educational influence, in short."

Mayeske and his collaborators used the Coleman data to help illuminate some of the areas in which beneficial changes could be made with a view to improving the over-all achievement records of all students. But the Coleman data also underscored the sheer complexity of the American educational problems, and the futility of even dreaming of an all-purpose cure for what ails our national classroom achievement levels. Better teacher training, more parental involvement, more individualized instruction, and dozens of other proposed approaches to improving the school experience are not, and in themselves never can be, enough. To Mayeske, "there does not seem to be any *single* variable . . . by which we can transform the achievement levels of lower-achieving schools so that they can 'catch up' with the higher-achieving schools in a few years. Rather, we are embarking upon a longer voyage into an only partially explored ocean. There is some consensus among the crew as to where we would like to go, but no one is quite sure of the best way to get there."

Certain facts were, however, made even more definitive by the massive Coleman data. One of these immutable realities was that no school exists in an environmental or a historical vacuum. "For both students and teachers," Mayeske observed, "the American [educational] system reflects the structure of American society. It therefore tends to perpetuate and even further increase the differential learning experience that students bring to the educational setting by virtue of their birth."

The original Coleman data helped document the extremely complex but measurable links between affluence and high marks; poverty and low marks; family aspirations and expectations and student achievement. At the APA meetings in 1971, Dr. Mayeske opened his presentation of the Coleman data by showing a table summarizing the relationships between race-ethnic groups and classroom achievement scores that had led people such as Ardrey and Jensen and Shockley to cite them as proof positive of the mental inferiority of Negroes and other nonwhite minority people. These data showed that in classroom achievement test scores "whites attain the highest score, with Orientals following them by about 4 points," and that "approximately 5 to 7 points below them lie the Indians, Mexicans [Chicanos] and Negroes with the Puerto Ricans following these groups by another 4 points." These data have, since 1966, been flaunted by eugenicists and other amateurs in genetics as "scientific evidence" of white intellectual superiority.

However, when subjected to the cold tests of regression or statistical analysis, it turned out that "for all practical purposes, all of the differences among students in their academic achievement that are associated with their racial-ethnic group membership can be explained," Mayeske reported, "by factors that are *primarily social in nature and origin*" (italics added).

After careful examination and testing of the data, the Mayeske group had to conclude that "we cannot make any generalizations about the 'independent effect' of membership in a particular racial-ethnic group on achievement, for this membership is almost completely confounded with a number of social conditions." This concept was very much in harmony with the findings of various studies by pediatricians and psychologists the world over, which have shown—over the past fifty years and more—that poor whites, as well as poor blacks, Orientals, and Spanish-speaking groups subjected to the chronic hunger, diseases, and housing of poverty, suffer the same sociobiological impairments to mental development that they mutually suffer to their physical health.

Of the five most important sets of environmental variables affecting learning, the Mayeske study revealed that Racial-Ethnic group membership (RETH) was "the fourth *lowest* in explanatory power." The other sets of variables, each with greater effects on learning, were Home Background (HB), which includes Socioeconomic Status (SES) and Family Structures (FSS); Family Process (PRCS), which includes family attitudes, aspirations, values, and expectations of education; Area of Residence (AREA), which weights regional as well as urban, suburban, and rural factors; and Aggregate School Outcomes (SO), the five student-body variables of achievement and the motivational levels of the students one goes to school with.

All of these variables are to one degree or another interrelated. However, of the total variation in achievement associated with these sets of variables, about 27 percent is totally unrelated to a student's racial-ethnic group. And behind the cold numbers fed into the HEW computers—the classroom test scores and the family income data, the digital symbols of family aspirations and teacher expectations—there were also the historical realities from which all of these data derived.

Therefore, the final version of the Mayeske report on racial-economic variances in school achievement, in its discussion of family background, observes that "because of social practices in the United States that resemble a caste structure based on skin color, the racial-ethnic group to which a student's family belongs *will* affect their socioeconomic status, which *will* in turn affect family structure. Family *structure* may in turn affect socioeconomic status (e.g., there may be an extended family with several wage earners, father absence with or without child support payments, etc.)." (Italics added.)

Thus, in examining achievement scores, Mayeske noted that the procedures by which different socioeconomic and racial-ethnic groups arrive at such widely varying test scores "reflect inequities imposed by society at large, at least as of 1965 [when the Coleman data were collected], by virtue of one's social class and racial-ethnic group membership."

Family background, Mayeske wrote, plays an important role in the development of achievement, "not only through the social and economic well-being of the family but through the values its members hold with regard to education, and the activities that parents and parental surrogates engage in with their children to make these values operational. For whites we saw that such values and activities outweighed social and economic well-being as a factor in achievement. Analyses for other ethnic groups, however, showed that their depressed social and economic well-being, relative to that of whites, had detrimental effects on achievement, and that these effects were more difficult to overcome through what we called 'educationally-related child-rearing practices.'" Such practices range from reading to preschool children to simply "talking with them frequently about their school work and about things that happened in school." (Personal communication, G. W. Mayeske.)

Even in broken homes, the Mayeske group found that "it was not so much the mere presence or absence of key family members that affected a child's achievement level as it was the expectations and aspirations that parents or parental surrogates had for the child and the supporting [educational] activities in which they engaged." Translated into the realities of daily life—where blacks, Chicanos, and other minority groups have been bludgeoned by society not to court disaster by entertaining *uppity* dreams of rising above their born lowly stations—this means that minority group poor families have less hope of upward mobility via education for their children than even equally poor white families. If there is very little room at the top

Racial-Ethnic Group Achievement Means Adjusted for Social Background Conditions

"... the differences among the various racial-ethnic groups in their achievement levels approach zero as more and more considerations related to differences in their respective social conditions are taken into account." From Mayeske, 1971.

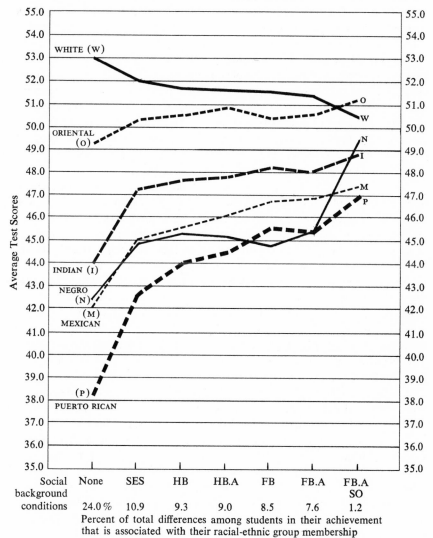

Social background conditions	None	SES	HB	HB.A	FB	FB.A	FB.A SO
	24.0%	10.9	9.3	9.0	8.5	7.6	1.2

Percent of total differences among students in their achievement that is associated with their racial-ethnic group membership

SES = Socioeconomic status

HB = Home background; this set of variables includes both socioeconomic status and family structure

HB.A = Family attitudes about their ability to influence the course of their lives and the extent to which education benefits them

FB = Family background; educational, social, etc.

FB.A = Family background and area (of residence: regional; urban; rural, etc.)

FB.A/SO = Quality and achievement-motivational mix of school attended for the children of the poor, most of this spare patch is reserved for whites and WASPs.

The conclusions of the massive computer-assisted analysis of the Coleman data that Mayeske presented before the APA meetings showed that while, in the sixth-grade students singled out for purposes of demonstration, the maximum weight that could be accounted for by their membership in any of six racial-ethnic groups and by their classroom achievement scores was 24 percent, this percentage dropped to only 1.2 percent "after a variety of social condition variables have been accounted for." These included "the social and economic well-being of the family; the presence or absence of key family members; the student's and parents' aspirations for his schooling; their beliefs about how he might benefit from an education; the activities they engaged in to support these aspirations; one's [geographical] region of residence; and the achievement and motivational levels of the students one goes to school with."

Far from supporting in any way the popular racist and pseudogenetic interpretations of the Coleman data, Dr. Mayeske's study showed, as he told the APA members, that in terms of school achievement scores, "the differences among the various racial-ethnic groups . . . approach zero as more and more considerations related to differences in their respective social conditions are taken into account." In fact, when Dr. Mayeske's team adjusted the records of racial-ethnic group achievement for social background conditions, the six-point spread between the classroom test scores of the whites and those of the next-highest group, the Orientals, was reduced to a one-point differential—*in favor of the nonwhites.* The whites dropped from test averages of 53 to 51, while the Orientals climbed from 49 to just over 51. The Negroes, incidentally, rose from 42 to just under 50, closing the gap between blacks and whites to a statistically insignificant figure.

When Professor Mercer and Dr. Mayeske completed the presentation of their unrelated papers, many of the psychologists attending the meetings felt that they had just witnessed the dawn of a new era. The chairman of the panel, Dr. Edward J. Cassavantes, a psychologist on the staff of the U.S. Commission on Civil Rights, observed that "these papers refute Shockley and Jensen. This new evidence from two separate and independent studies is the strongest ever presented documenting that environmental and social factors affect test scores. Many social scientists have always felt there were no basic intelligence differences between racial and ethnic groups, but until now there has been an absence of scientific data."

Dr. Mayeske observed, after the formal presentation of his paper, that at the start he and his collaborators "intended to study the effect of race on test scores—and ended by studying the effects of racism on test scores." What also impressed many of the scientists present at this panel was the fact that Mercer and Mayeske confined themselves to only those measurable social and behavioral factors that came within their respective fields of professional competence. They did not mention the genetic and environmental biological factors known or presumed to be involved in the inheritance and development of human intellectual capacities. They had been quite content to leave such matters as the biology and chemistry of brain growth and development of mental capacity to the biologists, and the child psychiatrists and experimental psychologists with special training and ongoing scientific experience in those precise areas—just as they left the genetics of human behavioral traits, such as "intelligence," to the geneticists.

For all practical purposes, such as the formulation of government policies, Mercer and Mayeske had clearly reduced the question of heredity *versus* environment in the achievement of high IQ and academic test scores to—at the very least—a quite meaningless and time-wasting speculation. After all, if IQ and classroom test scores could indeed, as the California sociologist and the Washington psychologist demonstrated, be improved by the simple procedures of improving the socioeconomic and cultural environments of the tested children—so that in each racial-ethnic group the low scores climbed to *predictable* levels matching those of the higher-scoring groups as socioeconomic status and related sociocultural levels rose—then the entire question of the very existence or nonexistence of the single genes, or polygenic combination of genes, for high or low IQ test scores had been reduced, in effect, to the status of a non-question.

The presentation of both of these important papers on human mentality was, or should have been, a glowing moment in the history of American science, an episode to be recalled by all who were present as the day that the child-destroying pseudo-data of the new scientific racism were demolished in broad view of America's gathered behavioral scientists by the sharp cutting edge of the replicable data organized and presented by scientists of not one but two disciplines—sociology and psychology.

There was only one slight flaw in this pretty picture: the people who should have been listening were not listening. Far from having become *Kulturhelden* for their parallel feats of slaying the dragons of ignorance, eugenics, and the other forms of the new scientific racism, Mercer and Mayeske were to be ignored and relegated to the ranks of non-people by the same electronic and printed media that had made folk heroes of Jensen, Shockley, and Herrnstein. In our culture, scientific heroes are not made by their deeds but by what the greater society chooses to make of their work.

WHAT IS INTELLIGENCE—AND WHY TEST IT?

While the Jensens and the Herrnsteins were grabbing the headlines with their revival of the eugenics and IQ testing concepts of the World War I era, many serious life and behaviorial scientists were wrestling with the basic problems that had concerned such pioneers of mental testing as the great Alfred Binet. As we saw earlier, Binet went to his grave in 1911 certain only that he could *not* define intelligence itself. His early American imitators, Goddard and Terman, were certain they knew exactly what intelligence was, as well as that it was biologically a unitary trait, controlled by the single gene for this "unit character."

Like Van Evrie in 1868, Goddard and Terman were also certain that the ability to handle what they and Van Evrie termed "abstract" thought was the inherited unitary trait that distinguished the born smart from the born stupid. In his classic of 1916, *The Measurement of Intelligence,* Terman wrote: "An individual is intelligent in proportion as he is able to carry on abstract thinking." Terman's associate in the official evaluation of the Army intelligence tests of World War I, Dr. Edwin G. Boring—later professor of psychology at Harvard—had an even simpler definition of intelligence: "Intelligence is what the tests test."[2]

By 1929, their colleague and collaborator in the testing of nearly two million World War I recruits, Professor Carl C. Brigham, not only had lost his faith in the racial interpretations of these intelligence tests but, as he made plain in his famous retraction of 1930, no longer felt that intelligence was even a unitary trait.

Brigham's successor as a popularizer of the racial interpretations of IQ test scores, Arthur R. Jensen, declared on page 19 of his *HER* eugenics tract that "if we must define it [intelligence] in so many words, it is probably best thought of as a capacity for abstract reasoning and problem solving."

Professor David Wechsler, the author of the WISC IQ test that produced such divergent results in different languages in California, is less taken with the abstraction called "abstract thought." His definition of intelligence reads: "Intelligence is the aggregate or global capacity of the individual to act purposefully, to think rationally and to deal effectively with the environment."

Two highly important modern authors, whose thoughts about the nature of intelligence and the reasons for IQ testing are probably far more representative of serious scientific thinking than are the writings and statements of Jensen and his disciples, have been—like Mercer and Mayeske—ignored by the same communications media that made Jensen, Shockley, and Herrnstein famous. They are the psychiatrist Leon Eisenberg, chief of psychiatry at Massachusetts General Hospital and professor of psychiatry at the Harvard University Medical School, and the research psychologist Michael Lewis, director of the Infant Laboratory at the Center for Psychological Studies, Educational Testing Service, in Princeton, N.J. Were their writings as publicized as those of the Jensens of our culture, it is quite probable that the vast and measurable harm now being inflicted upon hundreds of thousands of children by the golem-like administration and interpretation of IQ tests so prevalent in our nation could have been prevented, and the battering and misbranding of poor children—as exposed in California by both Jane Mercer and the California State Department of Education—would not continue to inflict human and social losses all over the nation.[3]

Professor Eisenberg touched on both the abuses of IQ testing and the nature of intelligence itself in a perceptive chapter, "Clinical Considerations in the Psychiatric Evaluation of Intelligence," written for *Psychopathology of Mental Development.* Dr. Eisenberg warned that "the abuse of intelligence testing stems from the assumption that the clinician can employ the IQ as a useful and practical measure without troubling himself about the problems of definition and methodology. . . . Indeed, clinicians behave as if there were an innate quality termed 'intelligence,' measurable by IQ tests, and as if 'intelligence,' and hence the IQ, were constant."

Dr. Eisenberg observed that the tests "to measure 'intelligence' have narrowed their focus on what might be termed 'abstract' as opposed to 'practical' intelligence, to the derogation of the latter." That is, he explained, the IQ tests measure one's ability to solve a mathematical equation or write a literary review but not the ability to repair an engine or manage a household. And, observing that our available IQ tests ignored "extra-academic indices of life success in choosing the criteria against which test validity is to be determined," Dr. Eisenberg asked:

> By what eternal scale of values can it be said that a competent physicist is better able to "act purposefully, to think rationally and to deal effectively with his environment" than a competent cabinet-maker, politician, farmer, housewife, or, for that matter, con artist? . . . Suffice it to say, we have developed measures which assess but one aspect of intelligent behavior—*one, to be sure, that does matter,* but only one: and yet, we speak grandly of intelligence which, from an evolutionary viewpoint, must have to do with much broader aspects of adaptive capacity.

Dr. Eisenberg called attention to the extraordinary differences in intelligence "found within members of a single social class and of a particular family within that class." This statement, of course, should come as no surprise to any parent who has raised more than one child, and thereby discovered what all good behavioral scientists have long known and measured—that the differences between two human beings born of the same parents and reared in an identical environment are often more striking than the similarities.[4] In the light of such differences, Dr. Eisenberg wrote that it therefore seemed "inescapable that there must be biological differences that contribute to this as they do to other phenotypic variations."

But since Leon Eisenberg is a life and behavioral scientist and not an evangelist of eugenics or other racist religions, he can and indeed does add:

> It is a far cry, however, from this statement to the contention that there are *innate* differences in *intelligence.* The behaviors of the neonate and the infant are insufficiently differentiated to permit any measurement of such differences, as is evident from the low correlation between infant test scales [usually Development Quotient, or DQ] and later standard [IQ] measures, except for damaged infants. A precise specification of the anatomic, physiologic, or biochemical basis of higher nervous activity (and methods to assess it in the intact organism) will be required before we can determine the anlage [the primordium, or earliest discernible embryonic form of an organ or other body component] of later behavioral differences. We are still far from this long sought-after theory. In its absence, it is a scholastic exercise, *certainly in no sense a scientific one,* to speculate about hypothetical, unmeasurable, innate differences in an intelligence *that does not yet exist.* [Italics added.]

As these sentences from the pen of a modern life and behavioral scientist demonstrate, more than concepts have changed since 1916. More than acquiring only new ideas, we are now inundated by literally millions of discrete bits of hard fact and functional data on the biology, the chemistry, the physics, and the sociology of human growth and development. Hard and important facts that could never have even been dreamed of before veritable tidal waves of scientific knowledge (some of it the product of new scientific technologies) in biochemistry, virology, and cytology, as well as neurology and psychology, helped us understand *how much we have yet to learn.* For one of the oldest, if not *the* oldest rule of science is that every new discovery of a hitherto unknown aspect of nature creates in its wake questions we could never before have formulated about entities and phenomena we could never before have found or recognized. In the absence of such banks of new knowledge about the nature of human mentality, it was understandable—but *not* forgivable—that nonbiologists, nonchemists, and nongeneticists such as Thorndike, Terman, and McDougall confused their eugenic faith with scientific fact.

Today, because of the vast bodies of knowledge that are accessible to modern life and behavioral scientists, the kind of pro- or anti-hereditarian didacticism that underscored earlier pronouncements by life and social scientists on the meaning and measurement of intelligence, let alone its relative or positive indices of heritability, is no longer understandable—or forgivable. We know too much about the biochemistry of genetics and mentality to fall into the traps of sweeping simplisms, such as those of Shockley and Jensen and Herrnstein. Therefore, the following passage from Dr. Eisenberg's essay speaks as much for the restraint and humility (born of awareness of what we have yet to learn) as for the thinking of modern life and behavioral scientists on the subjects Jensen tackled in his eugenics tract in the *Harvard Educational Review:*

> Whatever the biochemical forerunners which, under favorable conditions, lead to the physiologic basis of intelligent behavior, these metabolic processes are not "intelligence" nor are they *destined* to determine the acquisition of "intelligence" in the absence of intercellular and extraorganismic [that is, environmental] input—no more than the height of the man is to be found anywhere in the zygote. Development is an epigenetic process, moving from stage A to stage B as a consequence of the interaction of the elements in set A with the elements in its environmental surround; these together give rise to the new conditions that specify B. The task is to characterize the essential attributes of A and B, the boundary conditions between them, and the energetics of the transformation.
>
> So much, then, for the research for a non-existent intelligence in zygote, fetus, and neonate. What can be said about the search for an intrinsic intelligence in the older child, an intelligence to be inferred from IQ test measures?
>
> What we measure is intelligent behavior.

This, then, represents the thinking—*and the knowledge*—of modern psychiatry.

Dr. Michael Lewis, in 1973, in an article on the use and misuse of infant intelligence tests, attacked the hoary eugenic dogma of the late Sir Cyril Burt, who in 1934 had written that intelligence "is inherited or at least innate, not due to teaching or training. It is intellectual, not emotional or moral, and remains uninfluenced by industry or zeal."

Lewis observed that, in 1973, Burt's 1934 "view cannot be supported by the data."[5] In the interim, of course, scores of major studies, such as those of Klineberg, Skeels, Skodak, Pasamanick, Drillien, Butler, Harris, Heber, and many other scientists, had shown that there is far more to a child's IQ test scores than the genes that he inherits. "Why then," Lewis asked in 1973, "should this view of intelligence hold such a dominant position in the thinking of contemporary scientists and public alike?" It is a question whose answer demands wide public attention:

> The answer to such a question may be found by considering the function or use of the IQ score in a technological society. The function of the IQ score is and has always been to help stratify society into a hierarchy. The purpose of this hierarchy is to create a division of labor within the culture. That is, to determine who will go to school in the first place, who will get into academic programs that lead to college, etc. These divisions in turn determine the nature of labor the child will perform as an adult. This division of labor, a necessity in a complex society, is then justified by scores on a test designed to produce just such a division. If we cannot make the claim that IQ differences at least in infancy are genetically determined, then we must base them on differences in cultural learning. But these differences, for the sake of the division of labor, are exactly what the IQ tests are intended to produce. The hierarchy of labor is maintained by the genetic myth. The hierarchy produces the test differences and the test differences are used to maintain the hierarchy. Thus, IQ scores have come to replace the caste system or feudal systems which previously had the function of stratifying society. Wherein these latter systems were supported by evoking the Almighty, the present system evokes Mother Nature.[6]

Given a half century of eugenic brainwashing in the American colleges, graduate schools, and teachers' colleges, it is small wonder that the voices of such modern scientists as Lewis, Eisenberg, Mercer, and Mayeske were to be ignored by the media and the government as thoroughly as Joseph Goldberger and his 1914–15 life-enhancing discoveries about the preventable cause of pellagra were ignored by Harvard's president-emeritus Charles W. Eliot in his December 27, 1915, American Association for the Advancement of Science presidential address, "The Fruits, Prospects and Lessons of Recent Biological Science." Nor was it too surprising that the senior advisers to the President of the United States, in the Age of Jensenism, included one of America's foremost apostles of the concepts of Spencer's Social Darwinism, and one of the nation's most passionate supporters of Shockley's strident demands of the National Academy of Sciences for a eugenic inquiry into the nature of "racial" intelligence.

22 The Jukes and the Kallikaks Ride Again:
The New Scientific Racism Speaks in the White House

A minimum living wage is desirable for every man, but the idea of giving every man a wage sufficient to support a family can not be considered eugenic. . . . It is an attempt to make it possible for every man, no matter what his economic or social value, to support a family. Therefore, in so far as it would encourage men of inferior quality to have or increase families, it is unquestionably dysgenic.
—PAUL POPENOE and ROSWELL HILL JOHNSON, in *Applied Eugenics* (rev. ed., 1933)

In the spring of 1971, I was one of the many Americans on the mailing list of William Shockley who received an unsolicited reprint of page 875 of the March 5, 1971, issue of *Science,* the weekly magazine of the American Association for the Advancement of Science. Shockley could hardly be blamed for sending these reprints out to everyone on his mailing list, for it demonstrated that regardless of what the life, behavioral, and social scientists of the National Academy of Sciences had said and done about his claims that Negroes were genetically the mental inferiors of whites, no less a scientific spokesman than Dr. Edward E. David, Jr., sided completely with Shockley in his controversy with his fellow members of the Academy.

It was, of course, Dr. David's privilege—as an individual citizen—to hold that the refusal of the National Academy to spend millions of dollars on the investigation of "Negro intelligence" demanded by Shockley was a "disaster [that] divorces our decision-makers from reality." As an individual, he was even within his rights when he disagreed with Professor Edsall and other NAS members' declaration of 1970 that "Dr. Shockley's proposals are based upon such simplistic notions of race, intelligence, and 'human quality' as to be unworthy of serious consideration by a body of scientists."

"WE MUST NOT PLACE LIMITATIONS ON BIOLOGICAL EXPERIMENTS . . ."

Under the First Amendment of our Constitution, every man not only has the right to say what he believes but also to believe what he says. There is nothing in our Constitution that says that every man must read all he has to read to avoid being ignorant of biology, its history, and its dangers when abused—let alone that he must read the history of experiments by qualified medical and biological scientists between 1933 and 1945 in that eugenic paradise called Nazi Germany. The right to be ignorant is as sacred as the right to be informed.

However, Dr. Edward David was not *any* man. When he made the statement cited by Shockley in 1971, David was the top science adviser to the President of the United States. And when he, as the holder of this

Losing Our Nerve to Experiment?

Edward E. David, Jr., the President's science adviser, believes this country is losing its technological nerve.

David told a science writers seminar last week that the American public is becoming increasingly alienated from rational ways of thought. "There are many evidences that society does not believe that technology can be controlled in a rational way," he said. "Because of that, society is losing its courage to experiment. This trend leads to disaster for it divorces our decision-makers from reality."

David said that "we must not place limitations on biological experiments" despite warnings from such eminent scientists as James D. Watson, Harvard Nobelist, that genetic engineering may lead to test-tube babies and a host of ethical and social problems. David also reiterated his opinion that we should build two prototype supersonic transports (SST's) to determine whether the technical and environmental problems can be overcome so that it becomes feasible to build a fleet of SST's. Finally, he cited the negative reaction given by the National Academy of Sciences to suggestions by Nobelist William Shockley that research should be performed in an effort to identify characteristics peculiar to different races.

"Make no mistake," he said, "a limitation on experimentation in whatever cause is the beginning of a wider suppression. When we fail to experiment, we fail. In failing, we bring the best part of American society as we know it today to a halt.

"Already we see timidity in new undertakings," David continued. "We require overanalysis before we are willing to find out what are the real possibilities. If these trends progress, our society will become dull, stodgy, and altogether stagnant."—P.M.B.

The National Academy of Sciences' rejection of Shockley's demand for a costly study of "racial" differences in intelligence and other behavioral characteristics enraged the senior scientific adviser to President Nixon, Edward E. David, Jr. "We must not place limitations on biological experiment," he declared. To do so, Dr. David warned, will "bring the best part of American society as we know it today to a halt."

immensely important office, was quoted as declaring that "we must not place limitations on biological experiments," he was not merely endorsing Shockley's claims. Although it was quite probably not his intent, with this mindless new cliché Dr. David was speaking directly counter to all of the ethical standards of the life and biomedical sciences—whose investigators and clinicians have always made the distinction between scientific experiment and cold-blooded murder.

No legitimate scientists have ever claimed that biological experiments involving laboratory animals or human beings or whole social classes and races of human beings justify anything and everything such experiments do to the bodies and the dignity of animals or human beings. During the years of the Third and Eugenic Reich, the world's legitimate biologists and scientists denounced as anti-scientific and homicidal the biological experiments that were performed by Nazi doctors, physiologists, and other qualified scientists upon Jews, Nordics, Slavs, Spaniards, and other human beings unfortunate enough to have been captured by the Nordic warriors of Hitler, Goering, and other race purifiers. These "scientific experiments" included the beheading of healthy human beings to make life and death cephalic-index measurements; killing concentration camp TP's (test persons) in simulated oxygen-free atmospheres of 29,400 feet altitude to study the effects of high altitudes on the respiration of Nazi bomber pilots; placing TP's (human) naked on stretchers on the grounds of Dachau and Buchenwald in midwinter, then pouring ice water over them to study what caused death by freezing, and to *experiment* with the physiological mechanisms involved when the unconscious prisoners allowed to live through half of these "experiments" were revived by being placed, naked, between two naked female prisoners of their racial superiors; and, at Ravensbrück and other Nazi institutions of racial purification, sterilizing women with drugs designed to sterilize all dysgenic racial types at low cost to the Vaterland, while other women prisoners were deliberately given gas gangrene wounds to make possible *experiments* in the treatment of such wounds in Nordic fighting men.[1]

Of course Dr. David had *no* intention of retroactively approving the right of Nazi doctors and physiologists to conduct such biological experiments. Of course he probably agrees with the verdicts of the Nuremberg judges who sentenced some of these biological "experimenters" to death.

But therein lies the rub. In March 1971, when he endorsed Shockley's demands of the NAS and attacked the Academy's rejection of these demands as putting a dangerous "limitation on experimentation" on American science, he was speaking as the chief scientific adviser to the President of the world's most powerful nation—and it was his scientific and *moral* responsibility to know enough of the history of *unlimited* experimentation upon human beings to give him some socially more mature insights into the moral implications of the ongoing controversy between Shockley and his scientific peers in the National Academy of Sciences.

Even more unpardonable, however, was the fact that the science adviser to the President of the United States was apparently unaware of the many ongoing and extremely valuable federal agency and federally supported non-governmental longitudinal studies of the physiological and mental growth and development of white, black, rich, poor, north, south, and every other kind of American. Supported largely by federal funds, many fine scientists were already investigating, in serious, scientifically and morally responsible ways, precisely the interacting roles of the genotypical and environmental factors involved in the physiological and intellectual development of white, nonwhite, urban and rural, English-speaking and Hispanic Americans that Shockley said were being ignored by America's scientists.

It happened to have been Dr. David's professional duty to know about these ongoing studies of human development, paid for by federal funds, and conducted in some instances by U.S. Public Health Service life, behavioral, and social scientists who—in the tradition of Stiles and Goldberger—were continuing to demonstrate that, in terms of value per dollar received, the U.S. Public Health Service is the greatest bargain the U.S. taxpayers have received since the purchase of Alaska. So that, as the science adviser to the Chief of State, Dr. David—instead of excoriating the semi-official National Academy of Sciences for not giving in to Shockley's demand for a large-scale study of Negro-white differences in intelligence—should have let the world know what Shockley, *and the Academy itself,* should have known. To wit: not only that such questions were being studied by many serious federally funded scientific studies, but also that *all of these interdisciplinary investigations were producing hard evidence that IQ test score and academic achievement test scores were not determined by race but by the divergent social biologies of poverty and affluence.*[2]

Small wonder that, in a nation where the chief science adviser to the President was a person with the obvious cultural limitations of Dr. David, one of the same President's senior advisers on urban affairs should prove to be Edward Banfield, America's foremost exponent of Herbert Spencer's Social Darwinist brand of scientific racism.

ON THE MENACE OF MINIMUM-WAGE LAWS

The September 1970 issue of *The Atlantic* led off with a five-page article called: "A Theory of the Lower Class—Edward Banfield: The Maverick of Urbanology." Since the article was written by Richard Todd, an associate editor of the venerable Brahmin broadside, it did not need the kind of editorial foreword all the editors joined in giving Herrnstein's IQ essay the following September. The Todd article was preceded by this epigraph:

> The lower-class individual lives in the slum and sees little or no reason to complain. He does not care how dirty or dilapidated his housing is either inside or out, nor does he mind the inadequacy of such public facilities as schools, parks, and libraries: indeed, where such things exist he destroys them by acts of vandalism if he can. Features that make the slum repellent to others actually please him. He finds it satisfying in several ways.
>
> —*The Unheavenly City,* Edward C. Banfield

It was quite fitting that this epigraph, which most effectively summarizes the entire message of the Banfield book, should have been used to open the article. For it was, in reality, an essay review of the book, which, as Todd pointed out in his opening paragraph, "contradicts every received idea about urban problems, including the widespread assumption that they have reached a new point of crisis." Life, Banfield wrote, was improving for all classes of

Americans, but our heightened expectations for the general welfare had out-stripped this rate of progress, and, Todd quoted Banfield as writing, "although things have been getting better absolutely, they have been getting worse *relative to what we think they should be.*"

The admiring Todd then commented: "There is an obvious question: Are ever higher standards not a good thing? Banfield replies quite earnestly that, on balance, no, they probably are not. A 'crisis mentality' breeds actions that have effects contrary to what they intend, and it stirs hopes that cannot be fulfilled. All of this assumes, of course, that there are rather narrow limits to what money, goodwill, and even intelligence can accomplish, and this is precisely Banfield's opinion."

Three other major opinions of Banfield's are reiterated throughout the book. He feels that minimum wages and legislation to enforce them are a prime cause of social and economic troubles in our society; that compulsory high school should be abandoned, and children allowed to leave school at fourteen, since many lower-class children are simply ineducable by nature, and keeping them in school is a waste of the taxes paid by their betters; and, finally, that the quality of a human being is measured by the degree to which he is "present-oriented" or "future-oriented." He also believes, with Jensen, that the blacks are economically and biologically far better off than the standard data, such as the annual *Statistical Abstract of the United States,* would indicate. Finally, there is so little white racism in this country that for Negro leaders to explain to their followers that racial prejudice no longer denied black people any of the social and economic opportunities available to whites would destroy their very reason for being Negro leaders.

Like William Vogt, that earlier scourge of the dark-skinned un-trammeled copulators who made up the poor of India, Banfield fears the effects of modern medicine on the survival of the sex-mad hereditary hewers of wood and drawers of water. In the good old days of fifty and more years ago, "the Malthusian checks of poverty and vice may have been operating so strongly on the lower class of the largest cities as to cause it nearly to die out every generation or two" (p. 213). But that was before our society decided to go in for what the population extremist Paul Ehrlich called "short-sighted programs on death control" which started to interfere with such blessings.

By page 213, Banfield declares that thanks to such public health and clinical advances as better environmental sanitation and miracle drugs, as well as improved hospital care for the indigent, "poverty is no longer a major cause of death in the cities." Therefore, even with its much higher death rates, the lower class is now reproducing itself only too efficiently— and Banfield warns us that until better contraceptives are invented, and superior ways developed to motivate lower class types to stop having lower class children, the lower class will continue to proliferate. Then, if only we could discover a "way to motivate lower-class people to begin birth control before unwanted children have arrived and to use it regularly," then, ah then: "If the lower-class birthrate falls far enough, the decline in the death-rate may be offset, so that the lower class will again fail to replenish itself."

This biological elimination of the poor is the only solution to America's nonexistent urban problems, since, according to Banfield, nothing meaningful could be done about ameliorating a city's most pressing problems as long as it harbored a large lower class. Chronic unemployment would persist, he writes, even if good jobs were made available to all. "Slums may be demolished, but if the housing that replaces them is occupied by the lower class it will shortly be turned into new slums." Family income programs, such as are common in most European democracies; higher welfare payments; better schools; smaller classes—all, he warns, would be of no avail. The lower class progeny who drop out of or even graduate from the adequate schools would still, Banfield predicts, prove to be functional illiterates.

The chief problem, Banfield tells us on page 211, is that among the lowly all human problems reduce themselves to the one fatal flaw of their class: their extreme present orientation that "attaches no value to work, sacrifice, self-improvement, or service to family, friends, or community."

The higher the class, the more future-oriented are its members. This is all spelled out on pages 47 to 54, and it explains just why even people who are of low social status, poorly educated, and bone poor may really be upper-class in nature. The reason, writes Banfield, is very simple: people who are psychologically capable of providing for their future are, by virtue of this very trait, actually innately upper-class. By the same token, the professor adds, rich people of high social position who lack the capacity to plan ahead or keep their impulses under control are, in fact, really lower-class inside. As the Victorian sage Alfred Lord Tennyson, no mean scholar of real class himself, wrote in his merciless exposé of the ostensibly upper-class Lady Clara Vere de Vere: "Kind hearts are more than coronets,/and simple faith than Norman blood."

Where all of this very precise information on the temporal orientations of the various social classes was derived from, and how Banfield came into possession of these data, is never quite explained in the book. Unlike the Lapouges, who measured cephalic indexes, he used no calipers. Unlike the Jensens, he used no IQ tests. It is said, in Cambridge, that the time-orientation measurements were made by a time machine invented by the Wizard of Oz, and deposited under Banfield's pillow by the Tooth Fairy of Louisburg Square, but this of course is an orthodox egalitarian canard.

What is clear is that not since the Norse *Rassenhygienist* Jon Alfred Mjöen, in his 1921 Congress of Eugenics paper "Harmonious and Disharmonious Racecrossings," showed how crossing genetically superior (or upper-class) dolichocephalic Norwegian Nordics with genetically inferior (or lower-class) Lapp brachycephalics resulted in unbalanced offspring with an oversupply of M.B.-type genes for "stealing, lying, and drinking," had so revolutionary—*or so preposterous*—a discovery about the nature and origins of human traits been announced by any other scientist in this century. (Banfield, at that time Harvard professor of urban affairs, is a sociologist.)

Unlike Goddard—the discoverer of the Kallikaks in 1912—Banfield did not suggest that an individual's time orientation was a matter of his biological inheritance. He felt that "the time horizon is a cultural (or subcultural) trait passed on to the individual in early childhood from his group."

But otherwise, Banfield's lower-class sex fiends and slum makers, with their excess proportions of mental disease, were identical in every way to the earlier hereditary hewers of wood and drawers of water who were doomed, by their bad genes, to be feebleminded sex maniacs, burglars, drunkards, and parents of illegitimate little Jukes and Kallikaks.

SOCIAL DARWINISM IN OUR TIME

If many of the things that Banfield has to say in this 1970 book sound familiar, it is probably because they are. Some of Banfield's concepts are quite eclectic indeed, going back to Sumner and Gobineau, Shaler and Spencer, Gilbert and Malthus in the eighteenth and nineteenth centuries. Many of Banfield's concepts were exceedingly common at the turn of this century. For example, in the *American Journal of Sociology* for May 1901, Dr. Charles A. Ellwood, professor of sociology at the University of Missouri, wrote that "it is plainly discernible in the pathological phenomena of the social life [that] the 'instinctive criminal' and the 'hereditary pauper' are such, not because of the contagion of vice, crime, and shiftlessness which certain models in society may furnish, but because inborn tendencies lead them to seek such models for imitation rather than others; because they naturally gravitate to a life of crime or pauperism."

To this paragraph, the author, who was shortly to serve as president of the American Sociological Association, felt impelled to add this footnote: "It is unnecessary [*sic*] to point out that this is practically the unanimous conclusion of all experts engaged in the study of these classes."

Professor Banfield, in short, is a modern practitioner in a long line of Spencerian sociologists who—like their peers among the behavioral, life, and physical scientists—labored long and with notable successes in the grisly effort to put the blame for poverty and all of its individual and social pathologies on the shoulders of the poor alone. For this reason, it might be somewhat helpful to examine the roots of this sociological brotherhood.

It was the Founding Father of scientific racism, Thomas Malthus, who fired the opening salvos against minimum wages as early as 1798, in the very first edition of his *Essay on the Principle of Population,* declaring, as we saw above: "The receipt of five shillings a day, instead of eighteen pence, would make every man fancy himself comparatively rich and able to indulge himself in many hours or days of leisure. This would give a strong and immediate check to productive industry, and in a short time, not only the nation would be poorer, but the lower classes themselves would be much more distressed than when they received only eighteen pence a day."

All of which also showed that Malthus, who published before sociology existed, was some 172 years ahead of Banfield in writing that doubling or tripling the income of the poor would not relieve their squalor and misery.

Two modern schools of sociology stem essentially from the writings and actions of two Englishmen. The first was Jeremy Bentham (1748–1832), born eight years before Malthus; the other was Herbert Spencer (1820–1903). From Bentham, a man of enormous wealth and formidable erudition, graduated as a lawyer but active as a social reformer and philosopher, has come the school of sociology which bases itself on the equalitarian philosophy of the greatest good to the greatest number. From Spencer, born to a family of modest means, largely self-educated, a railroad engineer who ended up as an author and essayist, has come the school of sociology which bases itself on the elitist viewpoint that the world consists of the deserving rich and the undeserving poor.

From Bentham and his disciples were to come social innovations ranging from free education, health and hygienic laws and services, to the Chartist and trade union movements, prison reform, birth control, and industrial safety regulations. From Spencer and his disciples were to come an equal number of social and intellectual innovations, starting with a costly and still flourishing perversion of Darwinian evolution, the increased attention paid to the sciences in public education, and a whole new set of rationales in the name of science for social concepts and customs that were old long before Spencer was born.

Spencer's creation of the philosophical system called Social Darwinism —which was (a) antisocial and (b) anti-Darwinian—introduced to the world the phrase "survival of the fittest." This, despite popular belief, was neither created nor *believed* by Darwin himself.

To Spencer, the Malthusian par excellence, the sufferings of the poor were nature's mechanism for assuring the survival of the fittest—just as the wealth of the rich was nature's means for assuring the propagation of superior types. Thus, whereas Bentham and his followers called for living wages and free education and factory safety, and public sewage, clean-water, and other environmental hygienic improvements for the growing populations of urban poor, to Spencer "the whole effort of nature is to get rid of such, to clear the world of them, and make room for better." It was not overcrowded slum living that produced tuberculosis and other infectious diseases; it was the innate lack of the human will to survive. Therefore Spencer could write, of the victims of slum living and mine and mill accidents: "If they are sufficiently complete to live, they *do* live, and it is well they should live. If they are not sufficiently complete to live, they die, and it is best they should die."

So that when Banfield, in 1970, mourned the possibility that modern health services to the poor might have enabled them to survive "the Malthusian checks of poverty and vice" which formerly caused the urban lower class "nearly to die out every generation or two," all he actually added to Spencer's century-old concepts were the words "lower class" to replace the word "unfit" to describe the victims of the socially and medically avoidable deaths of poverty. Spencer was also a century ahead of Banfield when he wrote that government interference with the processes of natural selection [i.e., health, minimum-wage, education, and welfare legislation] violated the "general truths" of the laws of biology by causing "the artificial preservation of those least able to take care of themselves."

Spencer's great nineteenth-century disciple in America, Yale Professor William Graham Sumner (1840–1910), was, like Malthus, a clergyman who ended up teaching political economy. Sumner's magnum opus, *Folkways,* was an enormous scientistic account of the "natural laws" under which the mores and the values of a people were so ingrained in their biological systems that any attempt to interfere with natural law by social or governmental programs was doomed to failure because it violated the laws of nature itself.

To insure the non-survival of the unfittest, therefore, Sumner violently opposed any laws and actions designed to protect the poor from slum living conditions, from dangerous working conditions, and from the perils of not having enough money to pay for medical and hospital care when ill or injured. To insure the survival of the rich, whom Sumner designated as the fittest by the sole virtue of their wealth, Sumner was equally opposed to the Interstate Commerce Act and all other legislation that proposed to regulate the railroads the nineteenth-century Robber Barons built with federal money and operated under the Vanderbilt slogan: "The public be damned."

To Sumner, "poverty belongs to the struggle for existence, and we are all born into that struggle." He was therefore concerned that things be called by their right names, in the interests of scientific clarity. To him, the economists who "seem to be terrified that distress and misery still remain on earth" and who feel that this continuum of social injustices "bears harshly on the weak" were all guilty of very fuzzy and unscientific thinking. These misguided fools, Sumner wrote in 1879, ". . . do not perceive, furthermore, that if we do not like the survival of the fittest, we have only one possible alternative, and that is the survival of the unfittest. The former is the law of civilization; the latter is the law of anti-civilization. We have our choice beween the two, or we can go on, as in the past, vacillating between the two, but a third plan—the socialist desideratum—a plan for nourishing the unfittest and yet advancing in civilization, no man will ever find."

This selfsame concern for the semantics of social science is quite evident, ninety years later, in Banfield's *The Unheavenly City,* in which he warns (p. 236) that one of the dangers inherent in efforts to keep the lower class in its proper place is that these measures might inadvertently also be directed against the rights and freedoms of future-oriented Americans who are not really members of the real lower class at all. "This danger exists in part because euphemisms—e.g., 'the poor'—have collapsed the necessary distinctions between *the competent and the semi-competent"* (italics added).

Is there also a template, in the history of Spencerian sociology, for the Banfield concept of the time-orientation basis of human classification? A partial answer to this question can be found in an essay by Professor Edward A. Ross, at that time professor of sociology at the University of Nebraska, in the July 1901 *Annals of the American Academy of Political and Social Science.* The entire issue had been devoted to the papers presented at the annual meeting of the Academy, which that year was devoted to a look at America's Race Problems. Professor Ross, who was to emerge in history as one of the most effective spokesmen for the pre-World War I waves of American anti-Semitism and scientific racism, presented the paper "The Causes of Race˙ Superiority." Here, on pages 75 and 76, Ross, in listing and describing the various scientifically measured and confirmed attributes of racial superiority, offered the following:

> For economic greatness perhaps no quality is more important than *foresight* [Ross's italics]. To live from hand to mouth taking no thought of the morrow, is the trait of primitive man generally, and especially of the races in the tropical lands where nature is bounteous, and the strenuous races have not yet made their competition felt. From the Rio Grande to the Rio de la Plata, the laboring masses, largely of Indian breed, are without a compelling vision of the future. The Mexicans, our consuls write us, are "occupied in obtaining food and amusement for the passing hour without either hope or desire for a better future."

These low types, "taking no thought of the morrow" and concerned only about "obtaining food and amusement for the passing hour without either hope or desire for a better future" are amazingly like Banfield's lower-class schnooks whose time-orientation facilities have so atrophied that they attach "no value to work, sacrifice, self-improvement, or service to family, friends, or community," and who finds slum housing "satisfying in several ways."

By now it should be clear what the function of the Spencerian or Social Darwinist tradition of sociology has always been and still remains. It exists to provide new rationalizations, in new clichés and new scientific jargon, for the natural tendency—in the absence of societal intervention in the form of tax-supported social legislation providing for the health, education, and welfare of the greatest number of people in any population—of the rich to get richer and the poor to get poorer. All that really changes, from Malthus and Spencer to Sumner and Ross, Charles B. Davenport and Edward Banfield, is the jargon. The message remains what it was in 1798: them as has, simply hate to pay any taxes, and they will rally to the social and economic support of them as preach sermons or write scientific tracts proving that there is no basis in Divine or Natural Law for one farthing in tax funds to be spent on the alleviation of human misery, or the prevention of the socially and medically preventable diseases and premature deaths of the poor.

HOW LOWER-CLASS TYPES GET TO BE PRESIDENT

Like Spencer and Galton, Banfield believes that the born losers lack the ability to rise above their naturally destined stations in life. On page 212, for example, Banfield declares that, with the exceptions of a few gifted *and black* men such as Frederick Douglass, Malcolm X, and Claude Brown, "there is no direct evidence of there ever having been any upward mobility from the lower class" in America. On pages 58–59, however, Banfield seemed to be somewhat undecided about the upward mobility of the Irish lower classes.

In a segment on the chain of events that caused the grandson of Patrick Kennedy, an Irish laborer who arrived in the "future-oriented atmosphere" of Boston in 1848, to become President of the United States in 1960, Banfield confessed to an inability to explain this phenomenon. Had the heady environment of mid-nineteenth-century Boston, he asks, endowed Patrick Kennedy with a more future-oriented *Weltanschauung?* Or had it, Banfield wonders, enabled the already innate future orientation in Patrick Kennedy's blood to develop an intensity hitherto denied to this fine upper-class trait in famine-stricken Ireland? Or had the future-oriented environment of Boston, Banfield asks, possibly produced both effects? He can only formulate without answering these seminal questions.

Charles Benedict Davenport did far less straddling and dissembling to make the rise of a hillbilly from the Kentucky backwoods to the presidency of the United States conform to the eugenic dogma that the white Anglo-Saxon Appalachian hill country Americans were poor and backward because they were born of genetically lower-class Jukes-Kallikaks type of blood, and that nothing but hereditary paupers and criminals could come out of such feebleminded stock. Thus, in Davenport's 1911 book, *Heredity in Relation to Eugenics* (like Banfield's *Unheavenly City,* a widely used college textbook in its era), on page 250 he asks the question: ". . . do traits never arise *de novo?* If you deny it, how do you account for the presence of great men from obscure origin? For example, Mohammed, Napoleon and Lincoln?"

It was, as Johnny Boyle said to Joxer Daly, a dahrlin' question, and Dr. Davenport proceeded to answer it forthwith:

> . . . in seeking for an explanation of the origin of such [biological] "sports" of which history is full, we must inquire if the *putative* paternity is the real one. Not infrequently a weak woman has had illegitimate children by the wayward scion of a great family. The oft repeated story that Abraham Lincoln was descended on his mother's side from Chief Justice John Marshall of Virginia, whether it has any basis or not, illustrates the possibility of the origin of great traits through two obscure parents.[3]

Banfield's preface to *The Unheavenly City* noted that he drew on history, economics, sociology, political science, and other fields because his book was not a work of social science *per se* but, rather, the attempt of a social scientist to analyze the problems of urban America as they were seen from the standpoints of scholarly research in history, sociology, economics, and other sciences. For a typical example of his use of history, let us examine his statements on pages 56 and 57 concerning the history of immigration and education in this country.

According to Banfield, "after 1840 immigration increased rapidly, the immigrants coming mainly from peasant cultures—first Irish and then, after 1885, southern Italy and eastern European—that were more present-oriented than those of New England, Great Britain, and northern Europe." *How* Banfield knew that the Irish fleeing the famines, the Italians fleeing the poverty of Italy, and the Eastern Europeans—mostly Jews present and future fleeing anti-Semitic persecution—were present-oriented is one of the unresolved mysteries of his book. But then he goes on with more "history."

Where literacy was universal among the "native Americans," according to Banfield, it was rare in the peasant immigrants. The non-Nordic immigrants from what Banfield categorizes as present-oriented cultures showed little zeal for free education and upward socioeconomic mobility. "Even to some sympathetic [*sic*] observers it appeared that many of them would as soon live in hovels and shanties as not. Unlike the native American and the more future-oriented immigrants from England and Northern Europe [i.e., the Nordics], the peasant immigrants seldom patronized mechanics free libraries."

To be sure, Banfield concedes, some native American employers were prejudiced against the non-Nordic peasant immigrants, and refused to hire them. But, of course, one of the primary reasons these non-WASP types rarely became skilled workers who saved for a rainy day was because the future-oriented life style of the successful native Americans was not for them. The Irish and other non-Nordic peasant immigrants were so lacking in the inner-directed drives to get ahead for the future that, in 1852, Massachusetts was compelled to pass what Banfield describes as this nation's first compulsory education bill. "Until then it was taken for granted," he writes, that few human beings would fail to take advantage of free education for themselves and their children.

This passage sums up just about everything that is wrong with Banfield's approach to the needs and dignity and *history* of human beings. In the first place, the drive for compulsory education was so much part of the new immigrant tradition in America that as early as 1647 compulsory education was a reality in Massachusetts and, notes Professor Adolphe E. Meyer in his 1965 *An Educational History of the Western World,* "by 1671 all New England, aside from Rhode Island, had adopted some sort of compulsory education." In the second place, one of the major reasons why the world's disinherited and dispossessed started migrating to America was precisely because in the New World—in contrast to the Old World, where spokesmen for the ruling establishments on the order of Thomas Malthus and Davies Giddy Gilbert, president of the Royal Society, opposed all proposals to educate the poorer classes—the poor and the lowly were encouraged to learn.[4]

It was not because Nordic and Anglo-Saxon types were more future-oriented than Irish or Italian or Eastern European types that, by 1850, according to our own census figures only 10 percent of the white population over twenty years old was illiterate. It was because the educational policies in this nation had been established by the descendants of Europeans who came from classes traditionally denied an education in England, France, Germany, and other nations. Negroes were illiterate in 1850, obviously, not because blacks lacked the time orientation the Banfields seem to think is the key to literacy, but, of course, because they were not taught and also because in most southern states it was a serious crime to teach them the alphabet.

The primary reason why very few peasant immigrants patronized the free mechanics libraries was that they had come from countries where government policies were designed to keep them illiterate, such as Ireland, Italy, and Russia. Those who were literate had not yet learned to read English. Once the people who were to build America's railroads and her sewage systems, bridges, steel industry, and various other health- and wealth-producing resources learned to read and write, they started flocking to the mechanics libraries and the night schools in droves.

Banfield's citing of "sympathetic" observers who reported that the immigrants "would as soon live in hovels and shanties as not" is the kind of know-nothingism that the very history of the United States has been refuting since the first imported bonded servant, achieving his freedom, opened his first shop and hired his first free workers. It is merely another manifestation of the ancient view that the poor are poor and live in hovels because God-Nature planned things that way, and because there is nothing a poor man loves better than a hovel, a shanty, starvation, pellagra, tuberculosis, and freezing to death in winter. It was this essential cruelty in Banfield's approach to the poor that caused the reviewer for *Business Week,* certainly no "bleeding heart" publication, to observe: "Essentially, this book's curiously repellent quality derives not from Banfield's judgments, which have mostly shock value, but from his obtuseness about issues that are frankly moral, not scientific."

WHY THE IRISH RIOTED: THE BANFIELD VERSION

This moral obtuseness leaves a self-revealing trail through all of Banfield's chapters, from the chapter called "Rioting Mainly for Fun and Profit" to his list of twelve *feasible* and logical measures to resolve the urban crises. The main theme of the chapter on urban rioting is that neither the elimination of poverty nor the amelioration of such social end products of racial injustice as unemployment and slum housing would in any way reduce the amount of urban rioting in America for decades to come. "Boys and young men of the lower classes will not cease to 'raise hell' once they have adequate job opportunities, housing, schools and so on. Indeed, by the standards of any former time [such as 1848, during the potato famine in Ireland and chattel slavery in the South] they have these things now."[5]

Innocent dupes of naïve liberal propaganda might believe that the urban riots of the 1960's had to do with conditions such as children bitten by rats in slum tenement cribs, and the huge distances (particularly for poor people without cars) between slum children bitten by rats or felled by lead poisoning and the nearest hospital emergency rooms, but such facts have never fooled Banfield. The real purpose of the riots was to provide fun and profit for "looters and youthful rampagers" who were too present-oriented to work for a living.

In the course of making this case, Banfield again dips into the history of the lower-class immigrant groups, this time the great New York Draft Riots of 1863, which in four July days were to kill or injure over a thousand New Yorkers, a disproportionate number of them blacks. To Banfield, these riots were merely excuses for pillage, with perhaps an element of fury in the hearts of the Irish working class as they contemplated a future in which they would be forced to compete with emancipated black slaves for jobs. It was also possible, Banfield writes, that this atypically future-oriented rage was exacerbated by the anger of the lower-class Irish against what Banfield describes as the "alleged injustices"[6] of Mr. Lincoln's draft law.

There was nothing "alleged" about the provisions in the Civil War draft law under which, as all standard history books of Banfield's student years made crystal clear, "a drafted man might avoid service by providing a substitute or by paying $300."[7]

To the poor, and the Irish were the poorest of New York's white poor, three hundred dollars in 1863 was like three million dollars today. It was more than the poor white Irish lower-class yen to loot a few stores and set fire to Negro orphan asylums that touched off the bloody draft riots of 1863.

Banfield's "ideal solutions" for the problems of our society are not very original. Ideally, he writes, the termination of all further immigration of unskilled types from "present-oriented cultures" would keep the future population level of the lower-class down, while the compulsory surgical sterilization of "chronic delinquents and heads of 'problem families' would before long eliminate"[8] them completely.

However, such immigration policies would, Banfield feels, "create ill-will for the United States abroad." And, in a change of posture, Banfield concedes that "the violation of human rights involved in involuntary sterilization would be an intolerably high price to pay for even the many benefits that would follow from the [gonadal] elimination of the lower class."

Banfield's solutions to the urban problems of America are pure Social Darwinism, starting with the repeal of the minimum wage and "occupational licensure [*sic*] laws that enable labor unions to exercise monopolistic powers." These laws include the National Labor Relations Act of 1935, which affirms labor's rights to organize and bargain collectively and to protect unions and their members from unfair labor practices in violation of these rights. In the nineteenth-century tradition of Spencer and Sumner, Banfield advocates terminating compulsory education at age 14, and eliminating what he dismisses as "rhetoric" that gives people expectations of improved living standards and encourages the presumably black person to "think that 'society' (e.g., 'white racism') not he, is responsible for his ills."[9] The professor also takes a firm stand against "rhetoric" that makes social problems seem more soluble to the lower-class clods than upper-class intellectuals know them to be.

The Banfield prescriptions for what ails us also include one or two modalities that might not have previously been used by Louis XIV. One is the prohibition of live television coverage of riots and other happenings that, in Banfield's opinion, are likely to inspire off-screen riots. Another is to work up a crash program of instruction on birth-control techniques that would be geared to the mental and moral semi-competence of the undeserving poor.

However, since Banfield feels that politics will prevent the enactment of his idealistic proposals, he ends his book on an extremely pessimistic note. But a note, says *The Atlantic,* that must be heard, for "if Banfield is right," then the "noblest efforts of the past thirty years"—that is, social security, minimum-wages-and-hours laws, low-cost housing, increased provisions for public health, education, and welfare—"have been wrong" and "what progress has occurred has been accidental, and only a 180-degree shift in sensibility can begin to save us." In short, it might be time to turn the clock back to the era of Herbert Hoover.

A PRIMER FOR PRESIDENTS

Banfield's book and thoughts are significant for two reasons. The first is that, as of mid-1976, the special college edition of *The Unheavenly City* was in its twelfth printing, and the book was being used in college courses all over the country. This filled the gap left for a generation when, after three editions and one major revision between 1918 and 1933, Popenoe and Johnson's *Applied Eugenics* was, after three decades of intensive use, dropped from most college campuses because of the widespread revulsion against the sanguinary realization of the eugenic dream in Nazi Germany. Now, again, there was a widely used college text that described such socioeconomic artifacts as minimum wages, and such sociobiological "naturefacts" as poor people, from the viewpoint of Sir Francis Galton, Herbert Spencer, and Ivan the Terrible.

The second reason Banfield's book is of such importance in our times is that, although *The Atlantic* associate editor who composed the panegyric to Banfield went out of his way to forget the fact, Banfield was at that time a senior adviser on urban affairs to President Nixon. As *The New York Times* reported in a long story beginning on page 1 of its September 11, 1970, edition: "A year ago, the President appointed a new study group, headed by Dr. Banfield, to make a thorough evaluation" of the Model Cities program the Nixon administration had inherited. The *Times* also noted that Professor Banfield was the principal author of the Banfield Report on this federal program designed to aid the poor—i.e., the lower classes—of the inner cities.

One of the penalties of growing older is that we remember too much of the past. The vision of President Nixon in his Oval Office, with Edward David whispering advice into one ear on biological experimentation and Shockley "genetics," and with Edward Banfield whispering the lowdown on the lower classes and minimum wages into the other, triggered images in my mind of earlier days in the same White House, with other Presidents receiving equally slanted inputs on the genetic perils of non-Nordic immigration and the dangers of disharmonic race mixture.

Nor have the voices of advisers who find their wisdom in the prophets of eugenics and Social Darwinism been silent in the same corridors of national power since the changing of the guard. As I close this chapter, a story clipped from *The New York Times* of December 5, 1973, sits at my right, and under the headline ROCKEFELLER GETS A FORD ACCOLADE is a story that begins: "With Vice-President-designate Gerald R. Ford at his side, praising him as Presidential timber, Governor Rockefeller held the first meeting of his new National Commission on Critical Choices for America and denied that it was a 'stepping stone' toward the White House." The *Times* story ends with:

"After cocktails, lunch and the news conference, Dr. Irving Kristol of

New York University discussed man and his environment, Dr. Edward Teller, the physicist, touched on ecology and energy, and John S. Foster, former research director for the Pentagon, talked about national security."

As I read this sentence, I had the distinct feeling that little had changed relative to man and his cultural environment since the days when President Teddy Roosevelt was issuing ukases against the fecundity of unworthy breeding stocks. And when President Coolidge, after Congress passed the Immigration Act of 1924, proclaimed, "America must be kept American," as he signed the bill to make us *Judenrein*. And when President Franklin Delano Roosevelt accepted the judgments of John B. Trevor and Charles B. Davenport and Harry H. Laughlin in the matter of relaxing or not relaxing the barriers the very same 1924 Immigration Act had raised against sanctuary in America for Jews, Italians, Poles, Hungarians, Greeks, Slavs and other holocaust-threatened species.

The power of truly bad ideas survives their originators for lifetimes without end.

PART FOUR

The Legitimate Sciences' Answer to the Old and New Scientific Racism— Once and for All?

23 What the New Scientific Racism Hides about Human Genetics and Development

I: From Macrocosm to Cytoplasm

> The general spread of the light of science has already laid open to every view the palpable truth, that the mass of mankind has not been born with saddles on their backs, nor a favored few booted and spurred, ready to ride them legitimately, by the Grace of God.
>
> —THOMAS JEFFERSON, June 24, 1826, on the eve of the fiftieth anniversary of the Declaration of Independence
>
> Living organisms are fundamentally alike, even if one kind buzzes around a banana and another kind peels it.
>
> —GEORGE and MURIEL BEADLE, in *The Language of Life: An Introduction to the Science of Genetics* (1966)

The classic example of the philosophy of the new scientific racism in action that we examined at the opening of chapter 17—in the instance of President Nixon signing the $382 million population control bill on the same day that he vetoed the $225 million bill to train more family doctors—is but one example of what happens to the national health and well-being when the dogmas of scientific racism prevail over the logic of the legitimate life, behavioral, and social sciences.

Consider the even more unconscionable matter of what has happened to the great advance in the medical care of all American children that was voted into law by the Congress in 1965, to the cheers of parents, pediatricians, and all concerned citizens aware of how great and as yet tragically unmet human needs would be attended to by this legislative act.[1] At that time, as part of Title XIX, the Medicaid part, of the Social Security Act of 1965, the Congress mandated that by no later than 1975 Medicaid would pay the costs of comprehensive preventive and therapeutic medical care for all children whose parents could not afford to pay the normal costs of such standard medical services.

TEN MILLION CHILDREN ROBBED OF MEDICAL CARE MANDATED BY CONGRESS

By June 8, 1975—midway through the year when this congressionally voted goal should have been achieved—*The New York Times* revealed:

> More than seven years ago Congress passed legislation requiring states to set up, under Medicaid, a health screening and treatment program for poor children. *Fewer than 3 million of the 13 million eligible have received this mandated health care,* and the Federal Government has decided to try to prod the states. Based on the screening already done, at least one quarter [2,500,000] of the untreated children require dental

care, more than one million need correction of eye problems and more than 300,000 need care for hearing defects. [Italics added.]

Untreated eye and ear defects are not only preventable learning handicaps to schoolchildren; they are also involved in thousands of preventable automobile, household, and school accidents suffered by children.

The child health care provisions of the Social Security Act of 1965, like the compulsory smallpox vaccinations of the equally maladministered British Vaccination Act of 1853, are enforceable only to the degree that state and national governments truly desire to turn them into living human health actions rather than dead or moribund entries in the legal codes. We are, it is said, a government of laws, not men—but men and women administer these laws. If these fine comprehensive medical care provisions of the Social Security Act of 1965 were in the social and scientific spirit of Lincoln and Jacobi, Pasteur and Goldberger, their maladministration and/or sabotage were in the equally well-defined scientific racism tradition of Malthus, Galton, Spencer, and Davenport.

The century of the twisting, by scientific racism, of the lexicon of science into a license to deny the full measure of the human benefits of the sciences to entire nations, is marked by more than the vetoes of beneficial public health laws and the sabotage of excellent health care laws passed by Congresses and signed by Presidents. There is also the matter of puffball legislation, such as our current new environmental regulations, that turn out to preserve and protect at least one of the major sources of the environmental health hazards they are supposed to reduce or eliminate.

As we have seen, the health and social pathogens heaved into the environment by our over 131 million registered motor vehicles include not only the fossil fuel pollutants they emit but also the two million people they have killed here since 1900, and the five million people they injure annually, and, among many other health-diminishing pollutants, the noises and non-fuel debris they spew into our ears, our air, and our water. The Nixon administration environmental programs have done nothing about reducing the present ratio of one motor vehicle to every 1.6 individual Americans; nothing about mandating and funding the alternate forms of safe, swift, and nonpollutive mass transportation that can save this nation from the present killing density of motor vehicles. Instead, by setting the possibly unattainable goal of "cleaner" emissions from our burgeoning population of cars and trucks, the environmental "reforms" of the People Pollute school have completely obscured the clinical reality that the saturation of our air with pathogenic fossil fuel emissions is only one of many ways in which our present biologically unsupportable density of motor vehicles degrades the entire human environment.

In the case of these present environmental laws, violence was done not only to the environmental movement started by the Sanitary Reformers of England early in the Industrial Revolution, but also to history. In 1975, Theodore H. White, in his book on the fall of Richard Nixon, wrote that, as of 1970, "of the two major achievements of [Nixon's] new administration—

environmental policy and the successful invasion of Cambodia—the former had been ignored, the latter denounced." Of this environmental policy, White wrote a few pages later that it "would shortly make America the world's leader in environmental management."[2]

What is being denied the nation by such federal and state management of the problems of child, family, and environmental health are all the viable reasons of hope for the improvement of the human possibility that have been created for all humankind by the life, behavioral, and social sciences between 1775 and the present moment.

If the vast series of discoveries of the legitimate sciences provide a viable basis for hope for humankind, when they are denied to all of us because of the lasting effects of past and present propaganda of eugenics, Social Darwinism, and other systems of scientific racism, each instance of the perversion or betrayal or withholding of these scientific benefits to the nation emerges as a costly example of how human hope is killed daily by scientific racism.

Out of the false hypotheses of scientific racism have come such human tragedies as (1) the failure of industrial societies to act early enough and scientifically enough against the pathogenic and carcinogenic environmental by-products of the Industrial Revolution; (2) the Great Pellagra Cover-Up; (3) the Cult of the Nordic and its crowning glory, the Nazi holocaust of 1933–45; (4) the current epidemic of the forced sterilization of the poor and the defenseless in the United States; and (5) the continuing and growing lag between the life-extending discoveries of the legitimate sciences and their swiftest possible application to the greatest number of people.

Out of the confirmed theories of the legitimate sciences have come (1) vaccinations to protect us for life against many of the most devastating infectious diseases in human history, from smallpox and poliomyelitis to diphtheria and German measles; (2) antibiotics to cure us of microbial diseases, such as pneumonia and chronic streptococcal nephritis, which at the start of this century were the first and fifth leading causes of death in America; (3) hard biomedical knowledge that turned once invariably fatal organic diseases, such as diabetes and pernicious anemia, into conditions so medically controllable as to allow their victims to lead productive, self-supporting lives; (4) new knowledge of clinical nutrition that enables us to prevent deficiency and infectious diseases by means of adequate diets; (5) promising new knowledge about the major causes of chronic obstructive pulmonary diseases, heart diseases, cancers, and stroke; and (6) new knowledge about simple but practical environmental reforms that can eliminate many of the known causes of such killing diseases from our present environment. Finally, leading to what might well result in the most important advances in the biomedical sciences since the Cell Theory, the Germ Theory, and the Theory of Evolution, our legitimate sciences have begun to gain new life-prolonging knowledge about the basic mechanisms of heredity and the cellular environment in which they function.

Up until now, in this book, we have concentrated our attention on the words and deeds of the scientific racists who have functioned—only too successfully—to rob us, our children, and their children of the full benefits of the knowledge of the legitimate sciences. In the chapters that follow, we will examine a few areas of the hard knowledge of what we are, biologically, and why we are enabled to grow up to be as human or as inhuman, as strong or as weak, as smart or as stupid as we ultimately turn out to be.

Biology, however, does not act in a social, political or historical vacuum. From gene to cell, from gland to brain, from nerve to muscle, the biological mechanisms of our kind can never act independently of human history; rather, they *interact* with the changing daily environment in which we live, and for all the days of our years.

HIGHER WAGES = LOWER DEATH RATES

The basic biological differences between species—between bacteria and people, birds and mice, fish and kangaroos—reduce themselves to variations on Virchow's maxim, *omnis cellula e cellula:* every cell comes from a cell. No living organisms, from the multicellular Nordic supermen to the unicellular bacteria, can escape the genetic outlines programmed into their cells from the moment of their creation.

However, whether the microscopic or mighty biological products of the genetic information contained in their ancestor's cells will be fertile or infertile, of maximum possible height or dwarfed, short-lived or long-lived, or —in those species with brains, nervous systems, and complex sensory organs—developed to their maximum genetic or genotypical potentials or irreversibly damaged between the moments of conception and birth, is never determined exclusively by the information contained in their ancestral genes.

This is particularly true in our species, which is so socially and culturally dependent. Whether or not a newborn child will be born maimed or whole; grow up healthy or unhealthy; skilled in the arts and sciences; a respecter or corrupter of political office; a high scholastic and economic achiever or a total failure in the quest for knowledge and/or money; a solvent grandparent or—as are millions of children born to the poor of this planet every year— doomed to die during the perinatal, infant, and early childhood stages of the phenotypical development of his inherited genetic potentials, is obviously not nearly as much a matter of the developmental potentials in his genes as it is a matter of the social biology of the environments in which the genomes of newborn children express themselves.

"Genomes," "phenotypes," "genetic expression" are all modern words. In our culture, however, we have yet to improve upon the basic description of the processes involved in the ultimate expression (i.e., the phenotype) of the potentials inherited with the genome (i.e., the full complement of genes in the cells of all living things) by the Victorian-era British novelist Benjamin Disraeli. It was as early as 1845, before our kind knew of the existence of the 100,000-plus genes we inherit at birth, that Disraeli came close to saying it all when he wrote of mankind being divided into two nations—the nation of the well-fed, well-clothed, and well-housed rich and the nation of the malnourished and inadequately clothed and housed poor.

The novelist understood the sociobiological implications of this insight so very well that, thirty years later, as Prime Minister of England, Disraeli told the Parliament: "The public health is the foundation on which repose the happiness and the power of a country. The care of the public health is the first duty of a statesman."

For by 1875, the Prime Minister was a thirty-year veteran of the Health of Towns movement—through whose efforts for sanitary reform Great Britain had installed the clean-water systems that were to end epidemic cholera in England by 1866, a full eighteen years before Koch rediscovered the germ of cholera. Mr. Disraeli was intelligent enough to take his clues about human development and potentials from the nation's biologists and medical scientists, who understood the critical role of a healthy environment in the optimum biological development of each of the greater society's newborn children and their elders.

It is a tribute to the intelligence of Disraeli and his advisers that—as did other British governments—they rejected the scientific racism of the Preformationists or pseudobiological determinists such as Galton and Pearson in favor of the more scientific concepts of England's medical, sanitary, and social reformers, from Edwin Chadwick and John Snow to Sir John Simon and Alexander P. Stewart, who, in contrast to the eugenicists, did not believe that the solution to the sociobiological problems of poverty lay in annihilating the poor. For this reason, whatever the searing inequities of British life that remained in force, the sociobiological policies of England during the century that followed the publication of Galton's *Hereditary Genius* in 1869 were, more often than not, to be based on concepts which ran directly counter to that fountainhead of scientific racism.

The results of these societal decisions to make life more bearable, the environment healthier, hope more viable for increasingly greater numbers of all newborn children in England were neatly summed up in the standard textbook *An Introduction to Social Medicine,* by Professors Thomas McKeown and C. R. Lowe, in three simple graphs. The first of these graphs, on page 15, deals with what is still the most sensitive of all measurements of the total quality of any nation's health: infant mortality.

During these 130 years, the major causes of infant mortality were such infectious diseases as measles, cholera, dysentery, diphtheria, pneumonia, scarlet fever, mumps, and influenza. The germ theory of disease was not developed by Pasteur and Koch until the late 1870's. Koch did not discover the tubercle bacillus until 1882—and an effective vaccine against this germ was not to exist until after 1950. The causal organism of scarlet fever, a major cause of deafness as well as death in children, was not isolated until 1883. Antibiotics against this and other microbial pathogens were not discovered until the era of World War II. Yet the sharp decline in infant mortalities in England and Wales started four decades *before* the medical community had antitoxins to cure the major infectious diseases and vaccines to immunize children against them at the very outset of life.

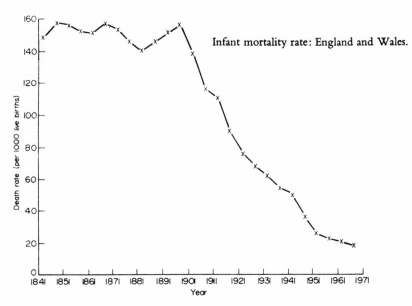

Infant mortality rate: England and Wales.

This decline of infectious disease deaths starting a century *before* doctors had anything other than a good bedside manner with which to fight them was nowhere more dramatic than in the death rates of tuberculosis in England and Wales, as summarized by McKeown and Lowe:

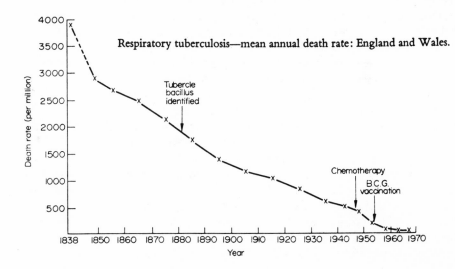

Respiratory tuberculosis—mean annual death rate: England and Wales.

Not, of course, that medical intervention played no role at all in the historic decline in infectious disease mortalities since Galton's prime. Consider the mean annual diphtheria death rates of children under the age of fifteen in England and Wales.

Until the isolation and identification of the *Corynebacterium diphtheriae*,

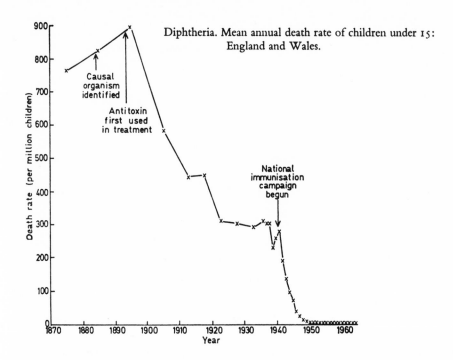

Diphtheria. Mean annual death rate of children under 15: England and Wales.

the germ of diphtheria, in 1884, nothing much could be done about controlling this then major killer of children. Isolation of the causal organism led to the development of the antitoxin that helped cure the disease by neutralizing the toxin, or poison, given off by the diphtheria germs living in the bodies of their human victims. Hygienic measures, from the reduction of overcrowded human dwellings to inside baths and inexpensive laundry facilities, helped reduce the incidence of diphtheria between 1895 and World War II. However, once the vaccine against diphtheria began to be widely used to immunize infants against this killer for life, both the incidence and the mortalities of diphtheria were eliminated from British life.

The mere existence of these clinical data is enough to render scientifically untenable the eugenic dogma that the historically predictable diseases of poverty are caused by the inferior genes of the poor. For this pseudo-scientific nonsense to have any biological validity today, the entire British people would, starting in 1838, have had to undergo not one but literally billions of discrete genetic mutations to alter their so-called hereditary predisposition to tuberculosis and other infectious diseases.

More than that: the downward trend in deaths from infectious diseases in England has also been paralleled in North America and in France, Germany, Holland, Russia, Norway, and other nations on the continental side of the English Channel. Therefore, these "mutations" that "altered" the "hereditary" tendencies of the British poor to die of scores of different bacterial, viral, and other parasitic protozoan and fungal diseases would also have had to occur during these same seven decades on the non-British sides of the waters that separate Nordic from non-Nordic, pure Aryan from diluted Aryan, Protestant from Catholic.

Miraculously sudden and universal genetic mutations were not involved in these declining rates of deaths from infectious diseases. The explanation for the precipitate drops in deaths from these major communicable diseases for reasons that are clearly neither genetic nor medical is neither secret nor mysterious. These huge reductions in both the prevalence of and deaths from the most common microbial, viral, fungal, and other parasitic diseases in Europe and America were due primarily to the over-all improvements in the total biological, socioeconomic, and cultural environment into which all successive generations of the industrialized nations were born since the middle of the last century.

Clean water, inside sinks and baths, sanitary indoor toilets, proper waste-disposal systems, free and adequate universal public health and educational systems, and—probably above all other death-reducing factors—*higher than mere subsistence wages,* were the key elements in reducing the incidence and mortality of such common diseases. Then, as now, the best, the most life-enhancing, and the cheapest way to cope with the infectious and the deficiency diseases of the environment of poverty was to eliminate the filth, the hunger, the slum housing, and the ignorance that all combine to cause such preventable plagues.

This is the answer in macrocosm. Before we depart the subject, however, it might be worth a very brief exploration of the general nature-nurture complexes involved in the biology of the growth and development of a few of our compatriot species on this planet. In fact, that of the simplest of all social orders—the single-cell organisms, the bacteria that cause infectious diseases in man, and the viruses, those non-organismic protein-coated packets of rogue genetic chemicals that plague these bacteria as fatally as they plague other organisms, from potatoes to people.

IS MAN THE GENETIC INFERIOR OF THE PIG?

The single-celled bacteria, all of whose varying species are too small to be seen except under a microscope, are measured in microns. In the metric scale, a micron is $1/1,000$ of a millimeter, or $1/25,000$ of an inch. The most studied of all bacteria, the *Escherichia coli* (*E. coli*), the common bacillus normally found in the human gut, is about 2 microns long and only 1 micron in diameter.

Different species of bacteria have varying shapes and forms. The streptococci (which cause human infections ranging from the familiar "strep throat" to scarlet fever, rheumatic fever, erysipelas, and streptococcal pneumonia) are strung out like chains of beads, 1 micrc￢ wide and from 5 to 10 microns long. The *Chlamydiae,* the weakly infectious bacteria that give people trachoma and psittacosis and the third most prevalent of the venereal diseases, lymphogranuloma venereum, are barely 1 micron long, wide, or round.

Although these bacterial cells are, on the average, about 500 times smaller than the average human cell, the unicellular bacteria manage to share most of the same chemical constituents and metabolic processes common to all compatriot life forms on this planet, from insects and sequoias to paupers and princes. The genes of each bacterial cell, like the genes of every human cell, are composed of the same chemical, deoxyribonucleic acid (DNA). Each bacterial cell, just like each human cell, contains DNA, which in turn causes the cell to produce a second nucleic acid, ribonucleic acid (RNA), which in turn mediates the translation of the genetic information contained in the DNA into the proteins vital to life.

About 70 percent of the total bulk of an *E. coli* cell consists of water, and another 15 percent is made up of the upward of 3,000 different kinds of proteins estimated to exist in these bacteria.[3] About half of the proteins are used to synthesize or assemble the walls, membranes, chromosomes, and other cellular constituents. The rest of the proteins in these and all other living cells act as enzymes, or catalysts that help put into motion one or another of the billions of possible chemical reactions involved in the primary functions of any cell.

The maintenance of cellular life is accomplished by its metabolism, the process of chemically breaking down into simpler compounds the complex and nonliving chemical compounds it takes in from the outside as nutrients.

Most of the compounds vital to the life processes, called the metabolites, are synthesized in the cell by metabolism. A few of the metabolites cannot be synthesized by the cell itself, and must be taken in from the nutrients used by the cells. These external metabolites include certain amino acids and vitamins.

Like the bacteria, we too can break the molecules found in our environment down to some of the metabolites we require—but we are at the same time unable to synthesize certain others. Cows, pigs, and most other animals are born with the biochemical ability to make the enzymes that enable them to break down glucose, a simple sugar derived from dietary starches and sugars, into vitamin C. Human beings of all races, however, are born with a serious genetic flaw: we lack the metabolic ability to, like the pigs, manufacture our own vitamin C.

In order for our species to survive, we have to find in our environment the vitamin C manufactured by other life forms that do not share this genetic defect of our species—and this means all fruits and vegetables. Some plants, such as paprika and cabbages, citrus fruits and tomatoes, are richer in vitamin C than others. They are also more expensive than rice, corn, and other dietary staples of the poor.

Given adequate amounts of vitamin-C-bearing foods, we manage to live healthy lives despite our genetic defect. If we are born into that two thirds of the human race that is poor, and cannot afford to buy fruits, fresh vegetables, and other foods rich in vitamin C, our chances of coping fully with the pathological effects of insufficient vitamin C are not nearly as good as are those of our nonpoor compatriot humans.

Vitamins are, essentially, coenzymes, that is, organic compounds that assist enzymes in effecting the chemical processes of metabolism. Because the same vitamins are used over and over again, in *E. coli* as in Englishmen, only trace amounts of vitamins are required for normal metabolism. The bacterial cells, since they are programmed only by their thousands of genes, never take in more vitamins than they biologically require. Bacteria, however, are not prey to the blandishments of the hucksters and the food faddists, and are thus never subjected to the eye and kidney damage, the bone weakening, the loss of appetite, and other effects of *hyper*vitaminosis A or D. The dangers of taking too many vitamins can be as serious as vitamin deprivation.

Few vitamins are more important to human health and development than vitamin A, which is essential to bone and tooth development, and the formation and maturation of the skin and the digestive, respiratory, and urinary systems. In 1932, Professor George Wald—like Joseph Goldberger the son of immigrant Jewish parents—discovered how essential vitamin A is to human vision. In the poorer nations where most of the world's people live, the World Health Organization (WHO) reported in 1972, "xerophthalmia due to Vitamin A deficiency is the most important single cause of blindness."[4] This is easy to understand, since the common dietary sources of already formed vitamin A are liver, whole milk, butter, cheese, and eggs. We can also, as a species, synthesize our own vitamin A from the provitamin A found in the carotene present in green vegetables and yellow fruits and vegetables, such as carrots.

Supplementary vitamins are often added to bread, cereals, milk, and other foods. A few years ago, in England—when competition for the children's prepared-food markets saw manufacturers adding increasingly spectacular quantities of vitamins A and D as cereal, milk, and canned-food fortifiers—the widespread overconsumption of these coenzymes by otherwise well-nourished children caused virtual epidemics of hypervitaminosis A and D. These epidemics were ended by new government regulations limiting the dosages of these vitamins in common packaged foods.

HOW GENE-ENVIRONMENT NORMS-OF-REACTION CONTROL BIOLOGICAL GROWTH RATES

Since the nutrient requirements of *E. coli* and other bacteria are fairly well known, it is possible to slow down or accelerate their growth or reproduction rate by decreasing or increasing the proportions of such metabolites in their nutrient media. Under perfect nutritional, thermal, and atmospheric conditions, laboratory-grown bacteria can easily double their starting population in only 20 minutes. This means that during the same 24 hours that elapse between the division of most human cells some 72 generations of *E. coli* or other bacteria can be reproduced.

On bare subsistence diets, such as the nutrient formula in which glucose is the sole source of the organic molecules used in *E. coli* metabolism, even at the ideal incubation temperature of 37° C. (98.6° Fahrenheit) it takes 60 minutes to double the original *E. coli* population—or three times as long as it takes properly nourished *E. coli.* On the other hand, *E. coli* will also "regularly grow and divide at temperatures as low as 20° C."[5]

Because they can produce dozens of new generations overnight, *E. coli* and other bacteria are ideal laboratory organisms in which to study hundreds of basic biological processes. Bacteria can be grown in nutrient broths, and they can be raised on flat layers of agar—a gelatine-like material made from seaweeds—to which all known bacterial nutrient formulas can be added. Although the bacteria themselves are too minute to be seen with the naked eye, they become quite visible when billions of them start multiplying in the nutrient-enriched agar in a shallow Petri dish only four inches in diameter as they form "lawns" of various colors, but most often on the dull buff side. Some perfectly harmless bacteria that happen to grow in livelier colors have—because ignorant people did not understand what they beheld—"caused the death of more men than many pathogenic [disease-producing] bacteria."[6]

While the genes of bacterial cells are—*under optimal growth conditions*—capable of directing the doubling time of a given bacterial cell population in 20 minutes, the same cells bearing the same genes would be hard put to double their mass in ten times 20 minutes in less favorable nutritional, climatic, and atmospheric environments. This control of growth rates by the total developmental environment is as true in our multicellular species as it is in the unicellular *E. coli.*

Even though the body of a man can contain up to 5×10^{12} (that is, the sum of five times the number which is otherwise written as the figure 1 followed by 12 zeros) individual cells of thousands of shapes, sizes, and functions, our cells are, like the *E. coli*'s, controlled by the genetic information packed into their DNA. And big as we are, multicellular as we are, intelligent and even moral as we are, our metabolic and growth rates are as inevitably determined by the nutritional, climatic, and atmospheric environments in which we function as biological entities as are those of the bacteria.

The same genes that under one set of environmental growth conditions will cause one child to be short, underweight, and sickly will—in a changed environment—cause his siblings to attain more normal height, weight, and general health. So that while you can never produce bacteria or people without the specific genes for bacteria or people, the quality and the duration of life in the unicellular and multicellular organisms owe as much and often more to the environmental conditions in which these genes express themselves than to the epigenetic and phenotypical information packed in the genes themselves.

THE NORM-OF-REACTION IN BACTERIAL CELL WALLS

The cells of mammals and bacteria are bounded by nonrigid permeable membranes. In multicellular organisms, such as people, the cells touch membrane to membrane, as constituents of various body tissues and structures. In man, all of the tissues and organs are surrounded by a wall of strong but flexible skin. In unicellular bacteria, the permeable cytoplasmic membranes are surrounded by a wall of strong, rigid cellulose-like material.

These rigid bacterial cell walls serve, among other life functions, as external skeletons.

It is possible to cause many species of bacteria to grow without these cell walls. All that has to be done is to add minute amounts of various bacteria-inhibiting agents to the flask or dish in which such bacteria are growing. These agents that induce bacteria to grow without forming most or any of the cell walls they are genetically capable of forming include the enzyme lysozyme (the active agent in the white blood cells that engulf and destroy bacteria in our bodies), and the antibiotic penicillin. In large quantities these agents cause the destruction of the growing bacteria. But in tiny, nonclinical, nonlethal (to bacteria) doses, they inhibit the formation of rigid cell walls.

When the bacteria are inhibited from forming the rigid cell walls or exoskeletons that they are genetically capable of growing in environments free of such pollutants, the bacteria continue to grow in wall-defective forms. In some atypical forms, for example, the outer walls are either too thin or too lacking in normal constituents to be rigid. In other atypical and totally cell-wall-free forms, the live bacteria are not even surrounded by defective elements of rigid cell walls.

As long as the same subclinical amounts of antibiotic, or sublethal amounts of lysozyme, are added to each new batch of nutrient media, the bacteria will continue to grow in this so-called spheroplast form. These cell-wall-defective or cell-wall-free phenotypes of bacteria are generally called L-forms, after the Lister Institute of London, where they were discovered in 1935 by the German refugee biologist Emmy Klieneberger-Nobel. They continue to grow and reproduce in these skeletally defective or atypical phenotypes as long as they are grown in the presence of the environmental pollutants that diminish or repress their genetic ability to grow rigid exoskeletal walls.

However, when the L-forms of bacteria are removed from their polluted growth media, gently washed, and then grown in a nutrient media free of added cell-wall-growth inhibitors, the flexible phenotypes quickly start to regain their genetic ability to grow rigid cell walls again. Inside of a generation portions of the rigid cell-wall structures begin to appear on the daughter cells of these L-forms of bacteria. Within a few bacterial generations—or a day or two by our calendar—the formerly cell-wall-defective or cell-wall-free phenotypes of these bacteria are growing as rigid-walled or "classic form" bacteria again.

The evolutionary question now raised is: Which of the two forms of phenotypes is "normal" for bacteria? The rigid-walled "classic" phenotype? Or the flexible, membrane-walled "atypical" or L-form phenotype?

The answer, obviously, is that neither is the *normal* form. In either phenotype, the bacteria happen to be growing in their genetically determined form best suited for life and reproduction in environments polluted or not polluted by chemicals that inhibit cell-wall formation.

THE NORM-OF-REACTION IN THE HUMAN FEMALE PELVIS

Bacteria are no exception to the rules of genetic expression. As the geneticist Jerry Hirsch observed, "Norm of reaction is a developmental reality of biology in plants, animals, and people."[7] Consider, therefore, what the influence of the varying qualities of different growth environments on the exoskeletons of bacteria have to tell us, by way of analogy, about the pelves of people.

The pelvis is, essentially, a ring of bone. The bony pelvis, in the females of our species, is also the rigid portal through which every newborn child emerges from the womb to the outside world. The same genomes bearing the genes that determine the morphological ranges of intrauterine fetal growth also bear the genes that give every potential mother the phenotypical capacity to develop so that in both form and size her pelvis will permit the normal passage of a normal-sized baby without injuring mother or child. That is, of course, if this mother grows up in an environment conducive to normal human skeletal growth.

As early as 1823 Francis Place opened his contraceptive handbill, "To the Married of Both Sexes in Genteel Life," with the declaration:

"Among the many sufferings of married women, as mothers, there are two cases which command the utmost sympathy and commiseration.

"The first arises from constitutional peculiarities, or weaknesses.

"The second from *mal-conformation of the bones of the Pelvis . . .* [it] is always attended with an immediate risk of life. Pregnancy never terminates without intense suffering, seldom without the death of the child, frequently with the death of both mother and child" (italics added).

Long before the Victorian era, it was well known in Europe and America that girls who had suffered rickets (caused by lack of vitamin D, which is found in sunlight, as well as in milk, eggs, and fish) and other bone deformities of poverty as children, grew up to be women whose pelvic arches were nearly always malformed and much too small to permit easy or even normal births. Victorian-era physicians were also well aware of the role that such cephalo-pelvic-disproportion birth injuries played in causing what was then described as "idiocy" in children.[8]

In 1933, Professor Herbert Thoms, of the Yale University School of Medicine, developed a technique of studying the precise forms and dimensions of the pelves of men and women by X-ray techniques. His early studies of white women of different social classes by means of X-ray pelvimetry revealed that the "high incidence of [normal] pelves among the largest women of both groups suggests the possibility that nutritive and other factors . . . make for the attainment of maximum normal growth" in the female pelvis.[9]

By 1950, through the use of Thoms's techniques, the world's anatomists had established that there were four pelvic types in women: round, long oval, flat, and triangular (scutiform). The first two permitted normal births, the other two types were the flattened-out pelvic shapes that caused brain and

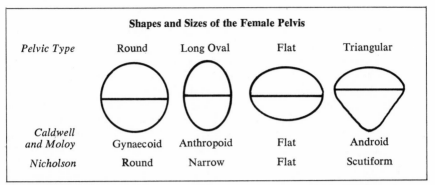

Genetically, all women have the potential for growing round or long oval pelvic brims, each large enough to permit safe passage of infants during the act of birth. However, lifelong malnutrition and other constants of the social biology of poverty waste such healthy genetic potential. These developmental barriers cause the pelvic brims of the world's poorer mothers to be flattened and otherwise malformed, making them far too small to permit their children to be born free of high risks of head, brain, and skeletal injuries during the act of birth.

Redrawn from R. M. Bernard, "The Shape and Size of the Female Pelvis," Transactions, Edinburgh Obstetrical Society Supplement, *Edinburgh Medical Journal,* Vol. LIX, No. 2 (1952), 1–16.

central nervous system injuries to the child and severe to fatal skeletal and tissue injuries to the mother.

At about that time, Professor Robert Bernard, of the University of Aberdeen, in Scotland, undertook a sophisticated study of the size and shape of the female pelvis in a racially but not socially homogeneous group of 200 pregnant women, "100 over 5 ft. 5 ins. and 100 under 5 ft. in height," all of them under thirty years of age and "booked for confinement in the Aberdeen General Hospital."[10]

In Bernard's 200 young Aberdeen mothers, "the incidence of flat pelves among small women . . . approximately one in every three, has been confirmed by further routine X-ray examination of all women under 5 ft. 1 in. in height." So that pelvic malformation was not only an apparent function of total body growth, but also of social class, since in Scotland, as throughout the rest of the world, the lower the social class, the shorter the average height of the people. As Bernard put it:

Since, in fact, a flat pelvis occurs much more frequently among small women, some common factor must be found to account for both pelvic flattening and short stature and this factor is most likely to be environmental.

Even though her parents may also be small, there is no proof that the stature of a small woman is inherited. *Heredity only determines the potentials of an individual, whereas environment determines how far these potentials are achieved* [italics added].

Bernard came to the conclusion that his investigation of this Aberdeen population showed that

> the shape and size of the pelvic brim are closely connected with stature, and suggest that environment, through its influence on stature, is probably the greatest single factor influencing the development and the ultimate shape and size of the pelvis. During childhood and adolescence a good environment encourages optimal growth, and as a result, the pelvic brim reaches its full size and is round or long oval in shape. On the other hand, a poor environment with all it implies—poor diet, poor housing conditions with overcrowding, lack of fresh air and sunshine [source of the vitamin D that helps prevent rickets], and lack of exercise—stunts growth and as a result the pelvic brim is not only smaller than it should be but is flattened, particularly in the posterior segment.

The mental retardation caused by brain injury because of CPD (cephalo-pelvic disproportion) is quite obviously environmental and not hereditary in origin. It also, as the late Dr. Herbert G. Birch noted, mimics true biological inheritance;[11] in the presence of hereditary poverty, CPD is passed down from mothers to children in patterns that can be plotted on a genealogical or pedigree chart. Only after three generations of environmental improvements have passed—that is, when the grandmother (first generation) has had the type of birth-to-first-pregnancy nurture that enables her to give birth to a well-formed girl through a properly proportioned pelvis, and this (second generation) girl is equally well nourished, well housed, and well cared for, so that her pelvis as a pregnant woman presents no brain-injury hazards to her normal children—can the third generation daughters be said to have been born with full equality of opportunity for pelvic and other skeletal development.

Like the oval, round, flat, and triangular forms any human pelvis can take, the rigid, semi-rigid, and wall-free forms in which bacteria can grow and reproduce are phenotypes. That is, the ultimate forms of pelves and bacterial cell walls are the biological expressions of the genes for these morphological traits, as conditioned by the range of reactions most normal in the environment in which they developed.

It is true, of course, that the cause of cell-wall defects in bacteria is not the lack of nutrients but the pollution of their growth environments by antibiotics and other harmful agents. This does not make them unique: the 1972 edition of the *Toxic Substance List,* issued by the federal government's National Institute for Occupational Safety and Health, lists only 13,000 of the *known* chemical substances in our environment that are harmful to man, while also warning that although 13,000 "toxic substances are listed, there are perhaps double that number which have not been included."[12]

They are not all in chemical plants, asbestos mines, and other dangerous occupation sites. More than 60 percent of all presently identified toxic environmental pollutants in America are delivered daily to our lungs, eyes, and other organs by our more than 131 million cars, trucks, and buses. And these do not include the nonchemical, nonphysical pollutants, the noises and the emotional tensions—caused by economic insecurity, racism, war, and other very common causes—that affect the norms-of-reaction of all human beings.

The bacteria are spared the emotional causes of mental and physiological trauma. Small as they are, however, the bacteria are as subject as we are to all forms of virus infections—from the kind that kill at once to the types of virus infection whose major pathological and/or fatal effects are not manifested until years after the virus attacks the cell.

THE PACKETS OF ROGUE GENES CALLED VIRUSES

Where bacteria are measured in microns—each micron is the equivalent of 1/25,000 of an inch—viruses are measured by the angstrom (A.). An angstrom is equal to 1/10,000 of a micron. Unlike the bacteria, which can be seen under any good light microscope, the viruses—which range in size from 200 to 3,000 A.—are too small to be seen under the best of light microscopes. To "see" a virus, it is necessary to use an electron microscope, which has an almost infinitely greater resolving power.

The viruses are not living organisms or even cells. They are particles consisting of a shell (or coat) composed primarily of proteins and, within each shell, a core of nucleic acid. Some viruses have a few enzymes in their cores. Unlike living cells, which all have both DNA and RNA, the viruses have only one kind of nucleic acid; in most viruses this proves to be DNA, but some viruses (including many of the tumor viruses) have only RNA.

The viruses have no metabolism. Whether they are a form of life or one of its building blocks is a question that has no real answers. Two things are biologically certain. The first is that the core of every virus consists of nucleic acid, the stuff of which the genes of life are made. The second is that the viruses replicate by getting their own genes into the cells of any living organism, from bacteria to tulips, gorillas and professors.

Once in the living cells parasitized by viruses, the viral genes take over the metabolism of their host cells. Usually, the infected cells start to assemble and release fully formed virus particles. With certain viruses, however, the viral genes are incorporated into the nucleic acids or chromosomes of the host cells, and for many generations only the virus genes or proviruses are replicated by the infected cells and transmitted to their daughter cells.

Since they are not living cells but "Twilight Zone" particles consisting of inert shells surrounding far from inert nucleic acids, viruses cannot be killed as one kills a living organism. However, their chemical potential can be inactivated or destroyed by various means. These include the antibodies, the interferons and other chemicals of the hosts' immune systems; the white blood cells that engulf and neutralize foreign pathogens; old-fashioned ethyl alcohol, the common laundry bleach sodium hypochlorite (Clorox), and other man-made chemicals; and high heat. We can also be immunized against many disease viruses by effective vaccines.

The viruses that parasitize bacteria were named bacteriophages (literally, eaters of bacteria) by Félix d'Hérelle, their co-discoverer, in 1915. They are commonly called phages, for short, and their manner of replication provided the final experimental proof of the role of DNA in the heredity of all living organisms. Briefly, what happens when a bacteriophage infects a bacterium is that the virus attaches itself to a specific receptor site on the rigid cell wall and transfers its nucleic acid or genetic core chemicals into the cytoplasm—that is, all of the materials except for the nucleus within the bacterial cell. The virus coat, minus its core, now falls away from the cell wall as a hollow "ghost" of the virion, or the complete virus particle.

Inside the cell, the viral genes combine with the cell's genetic chemicals and take over the direction of the metabolism of the bacterial cell—so that now, instead of making new bacterial cells, the bacterium starts to assemble, out of its own nutrients, hundreds of mature bacteriophage particles. Ultimately the infected bacterium lyses or bursts and the newly formed phage virions infect healthy bacteria of the same species in the same environment.

The viruses that attack man and other mammals are very similar to the bacteriophages, consisting as they do of a core of nucleic acid surrounded by a coat consisting largely of proteins. However, mammalian cells do not have rigid cell walls. The receptor sites of the membrane-bounded animal cells are locations to which the viruses cling and invaginate themselves into the cytoplasm as intact, coated virions. Inside the cells, they cause the host cells to produce enzymes that strip the virus coats off the cores bearing the viral genes.

There are three types of animal viruses—the necrotic, which kill the cells; the infectious, which alter them; and the tumor viruses.

Necrotic viruses, such as the polio virus, lyse mammalian cells just as phages destroy bacterial cells. Thus, the polio virus causes paralysis by necrotizing—killing—the spinal cord nerve cells that control the muscles.

The tumor viruses can, in mice, rats, and other experimental animals, cause leukemia (or blood cancers), sarcomas (or solid cancers), and various other kinds of malignant disorders associated with the transformation of healthy cells to cancer cells. While no viruses have yet been proven to cause cancers in people, certain types of viruses that appear in the bodies of patients with cancer bear close resemblance to the viruses cancer researchers now use to induce cancers in animals.

The commonest of the disease-causing viruses are the infectious viruses, which cause a wide spectrum of disorders, ranging from measles and rubella to mumps and smallpox. These viruses do not all kill their host cells, but they do alter them in many ways as the infected cells start synthesizing and shedding new virus particles.

The virus of rubella—German measles—for example, causes the human cells it invades or infects to replicate and shed mature and infective rubella virus particles. But the infected cells go right on reproducing, and when you grow these infected cells in the laboratory they keep on shedding newly formed rubella viruses. When you examine these infected cells under the microscope, they appear to be normal in every way. There is, however, one important difference between healthy human cells grown in laboratory plates or flasks and rubella-shedding cells: the rubella-infected cells take twice as long to replicate as do the healthy cells. This has been described as a "reverse-of-cancer" effect.

When this happens in a developing embryo or fetus, because of a rubella infection in the mother that causes her cells to shed viruses that infect the progeny growing inside her womb, the developing organs and muscular and skeletal systems then predictably grow at only *half* the growth rate for which they are genetically programmed. Rubella-reduced rates of brain, ear, eye, and central nervous system growth—depending on how early in gestation they occur—can cause wide arrays of damage, ranging from irreversible idiocy, total blindness, and immobilizing palsies and scleroses to minimal and more often than not undetected brain, vision, hearing, and nervous system damage in newborn infants. Minimal as these often undetected congenital (that is, associated with birth but not genetic) defects caused by rubella infection *in utero* are, they happen to be enough to keep children from talking, walking, learning, and developing as well as their rubella-free sisters, brothers, and classmates.

The equally ubiquitous cytomegalovirus causes exactly the same kinds of defects and injuries to fetuses and infants as those caused by the rubella virus. In a recent study of a population of mentally retarded English children living at home with their parents, reported in *The Lancet*[13] by four British investigators, it was concluded that "this virus may account for about 10% of all such cases," while the rubella virus and the toxoplasma—a protozoan parasite found in undercooked meats—"were responsible for about 2–3% of all the cases of mental deficiency" in these children.

Some chronic degenerative diseases considered by many doctors to be purely genetic in origin, such as diabetes, have in recent years been shown to be associated with prenatal or early childhood infections by the rubella and mumps viruses. A continuing study of the association of diabetes mellitus with congenital rubella, conducted in 50 young Australian adults with a medical history of congenital rubella infection, showed that "forty-five of the group of 50 have now been tested and reassessed annually, and at the end of 1973, there were 5 with overt diabetes and 13 with latent diabetes, a combined incidence of 40%."

According to Dr. Jill M. Forrest, of the Australian Children's Medical Research Foundation group, "the chronic persistence of the rubella virus probably occurs in the pancreas, as it does in so many other tissues; and that in this site it may damage the islet cell [where the insulin is synthesized in the pancreas] by the inhibition of cell growth . . . or by interference with blood supply or by a direct cytolytic [cell-bursting] action."[14]

A continuing study of the epidemiological association between mumps and diabetes, conducted since 1962 by the Department of Social and Preventive Medicine of the State University at Buffalo, New York, has produced good evidence that "mumps virus may be an etiologic factor in the occurrence of diabetes mellitus among children."[15] As we saw earlier, mumps-virus-induced brain inflammations have been implicated in the causation of mental retardation in children. (See page 27.)

The most common rubella-caused congenital birth defects are, in the order of their prevalence, partial to total deafness, heart damage, and partial to total blindness. Birth defects also common in congenital rubella (and other virus) infections are brain injury with cerebral palsy, mental retardation, and various behavioral disorders. "Rubella infection during pregnancy has been shown to be one cause of autism in children," observed Dr. Louis Z. Cooper, then director of the New York University Medical Center Rubella Project. "So that in these children, at least, the tragedy of this catastrophic mental disease is now shown to occur independently of parental attitudes or other factors."[16] The noted child psychiatrist Dr. Benjamin Pasamanick has found that "not only rubella infections but *all* pre-natal infections that cause organic damage in the developing fetus may also cause autism in children."[17]

The postnatal growth of children with congenitally acquired rubella is slower than that of normal children. Because the infected cells keep shedding new infective virus particles as they replicate, however slowly, most of the children who suffer intrauterine German measles infection acquire more than a single congenital handicap. In rubella infants, deaf *and* blind, spastic *and* deaf, blind *and* mentally retarded combinations of two or more biological defects prove to be the rule rather than the exception. All of which means that, due to a single environmental and patently nongenetic agent called a rubella virus, the offspring of a pair of tall, handsome, intellectually gifted people of any race or social class can be condemned to reach maturity as an undersized, blind, deaf, and mentally retarded human being.

BIOLOGICAL AND SOCIAL VIRUSES

These rubella and other infections that cause physical and mental retardation do not take place in either a historical or a socioeconomic vacuum. Thanks in large measure to research financed by our taxes and carried to fruition by government scientists such as Drs. Harry D. Meyer, Jr., and Paul D. Parkman, rubella infections during pregnancy can now be prevented by new vaccines that provide children with lifelong immunity against the rubella virus long *before* they can become pregnant. Even if exposed to the rubella virus during pregnancy, vaccine-immunized mothers now produce in their immune systems the antibodies needed to destroy any rubella viruses before they reach the womb.

Although the American Academy of Pediatrics recommends that children be immunized against rubella at the age of one year, rubella vaccination is neither mandatory nor universal in this country. According to the Academy, one in three American preschool children is not immunized against preventable infections, and "that means more than five million children one to four years old are unprotected against either polio, measles rubella (German measles), whooping cough, diphtheria or tetanus."

That minority of American families in the socioeconomic classes in which regular preventive and therapeutic pediatric care is afforded to each child from the moment of birth, are the fortunate families in which such immunizations are provided each child at the proper stages of infancy and childhood. In some more enlightened communities, local public health agencies make such vaccines available to the children of the poor. In most states, however, for socioeconomic and societal, rather than genetic or other unavoidable biological reasons, the unborn children of the poor and the undereducated are much more likely to suffer the biological and developmental insult of rubella infection than children conceived at the same moment by affluent parents.

These differential social factors are equally operative in families where children are born with congenital damages caused by rubella and other viral, bacterial, and microbial pathogens. Not every affluent mother is naturally or medically immunized against the rubella virus; prenatal infections are as merciless in the fetuses of the rich as the poor. Yet even here the developmental disadvantages of poverty tip the scales in favor of the nonpoor child afflicted with a congenital birth defect.

This is particularly true of those rubella and other virus infection damages to the eyes, ears, and central nervous system that occur during the last weeks of pregnancy. These late-pregnancy rubella defects are not nearly as apparent as are the gross deformities caused by German measles in the period of gestation. They cause not blindness but defective vision; not deafness but reduced hearing; not cerebral palsy but a barely apparent diminution of motor coordination. Unless such defects are diagnosed and corrected in time, however, they can represent the difference between full and partial comprehension, between passing and failing school marks.[18]

Those barely perceptible damages caused by rubella and other congenital infections to vision and hearing—which trained pediatricians recognize and test for routinely—are in most cases never diagnosed in children without regular pediatric checkups. This means that in affluent families, such defects are detected long before the child enters school, and corrected by glasses, hearing aids, and special training in the seeing and hearing and verbal skills required of a school child. Such fortunate children enter the school system equal in sensory skills to children free of such congenital handicaps.

Poor children, equally handicapped by the same viruses and other congenital infection agents, enter the same schools with their hearing and seeing defects undiagnosed for the want of cradle-to-kindergarten pediatric care. Their learning difficulties, caused by these preventable and/or correctable handicaps, cause them to be categorized as dull and ineducable by their classroom teachers. But worse mislabeling is yet in store for them, for soon they are given their first IQ tests, which "find" that they are lacking in what the IQ testing pundits call the genes for abstract learning—and the problem children are now put squarely on the path to becoming problem adults, welfare clients rather than taxpayers.

As raw statistics, these *in utero* infection-handicapped children—who often wind up warehoused for life in institutions for the mentally retarded and, in at least twenty-one states, subject to compulsory sterilization—will be cited by the simplicists and the statistical wizards of the new scientific racism as proof-positive of the genetic mental inferiority of the poor.

Actually, of course, they are the needless victims of the organ and brain defects inflicted upon their cells before they were even born by the genes of infectious viruses that could have—and should have—been neutralized by the pre-pregnancy vaccinations of their mothers.

LYSOGENIC BACTERIA, CANCER, AND SLOW VIRUS DISEASES

In some species of bacteria, some cells grow in what are called lysogenic phenotypes. These lysogenic bacteria can be invaded by the genes of bacterial viruses. But once the virus genes enter the lysogenic bacterial cells, nothing seems to happen. The viral genomes do not take over the metabolic machinery of the cells, and the cells do not start the self-destructive processes of synthesizing new phage virions. The virus-infected lysogenic cells continue to live and divide as before.

Well, not quite exactly as before. Because inside these lysogenic bacterial cells, the genomes of the viruses are also replicated by the cell. While these immature *proviruses* develop to no further than their pre-virion stage, they do replicate. And the proviruses are handed down to daughter cells and all of their descendants for generations.

It would seem, then, that because these variant strains of bacteria can take in and live at harmony with the genomes of their viruses, they represent a step up the evolutionary ladder. Our species, too, can live in peace and honor with the billions of bacteria in our gut, from the lactobacilli that help us break down milk and milk products to the equally ubiquitous species of enterobacteria whose metabolic by-products include the vitamin B they help us synthesize.

There is, however, more to even so simple a unicellular organism as a bacterium than merely the external form and the growth rate we can observe under the microscope. The chemistry of the cell is too complex a system to experience the addition of a whole new set of foreign genes without suffering any important changes. Consider what happens, for example, to the metabolisms of two fairly familiar species of bacteria, the *Streptococcus pyogenes* and the *Corynebacterium diphtheriae,* whose non-lysogenic or normal strains are completely harmless to man. However, when viral genes invade the lysogenic phenotypes of these species of bacteria, they cause the bacteria to undergo what is known as lysogenic conversion.

Such conversion gives the lysogenic bacteria "the capability of synthesizing one or more chemical constituents which it is unable to make in the absence of the virus." In other words, as the late Professor W. Barry Wood wrote, "the incorporation of the phage DNA into its genetic apparatus so alters the [lysogenic *Corynebacterium diphtheriae*] cell's metabolic machinery as to cause it to produce a new metabolite. In the case of lysogenic diphtheria bacilli which are infected by a specific bacteriophage known as the [beta] phage, the new metabolite formed is the all-important disease-producing toxin [of diphtheria]."[19] In the lysogenic *Streptococcus pyogenes,* the chemical effects of lysogenic conversion include the ability to synthesize the toxin that causes "many major human streptococcal infections, including scarlet fever."[20]

These lysogenic conversions are the least of the changes that occur in lysogenic phenotypes of bacteria. In the laboratory, when colonies of lysogenic bacteria are grown for various studies, it often happens that, without advance warning, the prophages—the seemingly permanently inactive virus genomes handed down for generations of viral somnolence—suddenly start to behave like other virus genomes. The cells now begin to carry out the instructions coded in the newly active virus genes. Inside of a few minutes the cells produce enough mature phage particles to tear themselves apart.

What causes these prophages of viruses that infect lysogenic bacteria to perform like a buried land mine of a forgotten war?

In the laboratory, all one has to do is to put the flasks or the plates in which colonies of lysogenic bacteria are growing under an ultraviolet-ray lamp, or expose it to a few rads of X rays. Or, if such sources of irradiation are not available, to add a little hydrogen peroxide or nitrogen mustard to the culture media. Any of these environmental pollutants will trigger the dormant prophages within the lysogenic cells into furious virus-replicating, host-cell-wrecking action.

As it happens, the radiations and the chemicals that activate the latent genes of the viruses of lysogenic bacteria are all known to be either mutagens or carcinogens or both. Biologists can and do induce various forms of leukemia in mice, rats, guinea pigs, and other animals by subjecting them to excessive irradiation or exposing them to carcinogenic chemicals. They can also induce leukemias by injecting mass doses—quantities far higher than would be normal in nature—of various well-defined C-type virus particles that cause leukemias in animals.

These C-type particles, in their mature forms, can be found in the body tissues of non-leukemic (or pre-leukemic) animals. A good biologist can also determine whether or not an animal is free of the presence of such C-type virus particles *in their mature (virion) form.* However, after leukemias have been induced in such C-type virion-free animals by radiation or chemicals, mature C-type virus particles appear in the tissues of the now leukemic animals.

Such viruses—called, for convenience, radiation viruses or chemical viruses—are very similar to the C-type leukemia viruses. And, like the other C-type viruses found in nature, large enough doses of these radiation and chemical viruses can and do induce the same kinds of leukemia produced by "natural" leukemia viruses in animals. So that we find ourselves with a possible multifactorial model for how cancers are caused in mammals: whether the laboratory animals are made leukemic by radiation or chemicals —two man-made environmental pollutants—or by leukemia viruses, in each instance C-type virus particles appear in the bodies of these animals. And, regardless of how they were caused to grow in their mature forms, these viruses cause animal leukemias indistinguishable from animal leukemias triggered by radiation or chemical carcinogens, such as those found in the exhaust gases of automobiles.

The lysogenic bacteria might also be cytological models for the processes involved in what are now called persistent and slow virus infections. There are, in medicine, various diseases known to be caused by the slow or delayed action of pathogenic viruses. Consider the clinical history of a varicella virus infection, for example.

The varicella or herpes zoster virus is one of the DNA herpes viruses. In childhood, the varicella virus induces a fairly innocuous infection called chicken pox, a self-terminating disease that causes a high fever, an itchy rash, and small raised spots that eventually all disappear. Normally, the immune systems of children who have had chicken pox are programmed by the varicella virus antigens to produce antibodies to the virus for the rest of the child's life, although reinfections with chicken pox a decade or two later are not highly unusual.

Some of the varicella viruses, however, are not neutralized by the antibodies and other anti-viral substances of the children with chicken pox. They seek shelter in such tissues as the nerves leading to the spinal cord, where they remain in a latent or dormant form, much like the inactive viral genes in the cells of lysogenic bacteria. However, many years later, when their hosts are well past middle age, various events—from emotional shocks to the side effects of certain severe diseases—trigger the residual but dormant chicken-pox viruses into their active states. Now, as the no longer dormant varicella viruses start to replicate in nerve and brain cells, they cause the much more painful disease called herpes zoster or shingles. This chicken pox–shingles cycle is a typical example of a persistent or delayed virus effect.

Investigations of slow virus infections by Carleton Gajdusek and his colleagues at the National Institute of Neurological Diseases and Blindness, as well as in other centers of the world, have produced evidence suggesting that many of the chronic degenerative and neurological diseases commonly considered to be genetic or hereditary diseases might well prove to be slow virus diseases.[21]

Retrospective studies in Europe and America, for example, have indicated that Parkinson's disease—shaking palsy—might be a late sequel to the great influenza pandemic of World War I. This could explain why Parkinson's, in our times, seems to be a disease that affects only people old enough to have been child victims of Spanish influenza during the first of our century's wars to end war.

A study conducted at the Birth Defects Institute of the New York State Department of Health revealed that "in examining the epidemic of anencephalus [failure of the brain to grow in the fetus] and spina bifida [a congenital defect of the spinal cord associated with paralysis, brain disorders, and hydrocephalus] which peaked in the years 1929–1932, it is apparent that the parents of these children were alive and affected by the pandemic of influenza in 1918."[22]

Laboratory and clinical studies now suggest that such classic "genetic" diseases as Huntington's chorea and amaurotic familial idiocy (Tay-Sachs disease)—both degenerative diseases that cause insanity and death—and multiple sclerosis, some of the presenile dementias, and amyotrophic lateral sclerosis might all be the final expressions of long dormant viral genes activated by biological changes associated with maturity, trauma, and/or aging. So astute an investigator of the biology of physical and mental development as Dr. Benjamin Pasamanick, Adjunct Professor of Pediatrics at the Albany Medical College of Union University, and until 1976 Associate Commissioner for Research Evaluation in the Division of Mental Retardation and Children's Services, New York State Department of Mental Hygiene, now feels quite strongly that "all chronic degenerative diseases occurring after infancy should now first be suspected of being slow virus conditions."[23]

The work of the world's microbiologists and experimental clinicians on the role and mechanisms of the slow and latent viruses has barely begun. But what we do already know of virus infections in man is enough to enable us to understand how complex, how difficult to survive, are all of the interacting biological, physical, and chemical hazards of our total environment.

Man, however, requires more than merely a biologically safe environment and adequate food, clothing, and health care to enable him to develop into a phenotype as human as his genome enables him to be. These are all important factors in human development, because a sound body is the essential biological prerequisite for a sound mind. But the nonbiological, the purely emotional and cultural requirements for brain growth and mental development must neither be neglected nor atrophied—not in us *thinking* primates, at any rate.

ENRICHMENT, DEPRIVATION, AND GENETIC POTENTIALS

A classic laboratory model for the roles of purely cultural and environmental factors in the mental and the physiological and biochemical development of the brain has been provided in our times by the work of investigators such as Mark R. Rosenzweig, Edward L. Bennett, and Marian Cleeves Diamond at Berkeley, M. W. Fox of Washington University, and other psychologists, neurologists, and biologists around the world. In these studies, it has been shown that rats, mice, and other animals do experience discrete and predictable anatomical and chemical changes in the brain in response to culturally "enriched" and "deprived" environments.

In the Berkeley experiments, rat littermates are separated at birth and raised in varying sets of conditions. In each group, the diets are identical in quality and quantity, and the physical environment—air quality, temperature, noise, and similar factors—is also identical. The basic environmental differences are social. In the three environments used with the Berkeley rats, for example, there is the standard laboratory colony, which consists of three animals kept in a cage built to hold three rats. In the impoverished environment, a single rat is reared alone in a cage of equal quality and comfort. In "the enriched environment 12 rats live together in a large cage furnished with playthings that are changed daily. Food and water are freely available in all three environments."

When the animals raised in environments "intellectually enriched" with various and frequently changed "stimulus objects" (toys) and a full "social life" are sacrificed and autopsied after as little as thirty days, they prove to have developed brains that are larger, heavier, and anatomically and chemically different from the brains of their genetically identical littermates who were raised on identically perfect diets but within socially "deprived environments." The rats raised three to a cage in standard cages, and the rats raised in individual cages in which they were intellectually "impoverished" not only by the absence of toys but even of visual contact with other animals, have—in descending order—smaller and lighter cerebral cortexes of the brain, and significantly smaller quantities of the enzymes crucial to brain activity (acetylcholinesterase, cholinesterase, and hexokinase) than the rats reared in the culturally enriched environments.[24]

In our own species, the recent discovery of a pathological condition called deprivation dwarfism has provided a more clinical model for the effects of nonbiological factors such as happiness and emotional distress on human development. This disorder is found, usually, in children abused, neglected, or abandoned by their parents. In these emotionally abused children, the normal output of growth hormone by their pituitary glands is severely reduced. Children with deprivation dwarfism fail to gain in weight or height even in the presence of adequate food.

In a typical case of deprivation dwarfism reported by Professor Orville Green of Northwestern in 1971, a seven-year-old boy whose mother "had held his hands over a gas flame to punish him for stealing food from the refrigerator" was brought to the hospital and found to have the height of a normal four-year-old. His pituitary gland, which is where the growth hormone is produced, tested out as unresponsive to the normal tests for this gland's functions. In the absence of supplementary growth hormone, "the boy was placed in a foster home, and a remarkable change took place. . . . With no treatment except the loving care given by his foster mother, he grew at an accelerated rate. . . . After two years . . . he had recovered complete pituitary function in a new emotional environment."[25]

The children with deprivation dwarfism are usually (but not always) also sufferers from insufficient food intake, or nutritional deprivation. Pediatric studies around the world, from Malawi to Mississippi, indicate that nutritionally and environmentally induced incidences of hypopituitary dwarfism in poor children vastly outnumber the cases of such underdeveloped phenotypes due primarily to emotional or nonnutritional deprivation.

The pediatrician Myron Winick, now director of the Institute of Human Nutrition at Columbia University, has made a number of studies of the effects of malnutrition on the brains of children and laboratory animals. At a 1972 international conference[26] on malnutrition in fetal and early life and its effects on the nervous system and behavior, Dr. Winick and other leading investigators reviewed the evidence that "cell division normally ceases in the human brain around the end of the first year of life. Proliferation in all three [brain] regions so far studied—cerebrum, cerebellum, and brain stem—stops at the same time. Malnutrition during the first year of life will reduce the rate of cell division and result in fewer cells in the whole brain and all three brain regions."

While there are still questions about the degree of hunger required to produce such reductions in the rate of brain growth, "the fundamental principle that undernutrition during proliferative growth will retard the rate of cell division in the human brain appears well established."[27]

The importance of this early brain growth in humans was also underscored, at the same conference, by Dr. Donald Cheek, professor of pediatrics and head of the Growth Division at Johns Hopkins, and by Professor John Dobbing, of the Department of Child Health, University of Manchester, England. Cheek, citing the studies of Winick and Dobbing, as well as his own, on the percentages of brain-cell DNA and brain weight relative to total body weight, confirmed that "the human has one-third of his eventual number of brain cells at birth," and that "from 15 to 20 weeks postnatally there is a period of maximum cell growth."[28]

Dr. Dobbing suggested that "there is a period of human development extending from the second trimester of gestation well into the second postnatal year, during which the brain appears to have a once-only opportunity to grow properly. *It is at this time especially important that children should grow at a proper rate and under the proper environmental condition.* Among these, *nutrition is central to proper growth, a restriction of which may have lasting behavioral consequences.*" (Italics added.)[29]

Winick described earlier studies in which it had been found that the quantities of a cellular enzyme, DNA polymerase—which is involved in the synthesis of new molecules of DNA—are increased in the cell's cytoplasm preceding the increase in the synthesis of new DNA molecules.

"More recently it has been shown that activity of this enzyme in the liver of growing rats is in part under the control of pituitary growth hormone. Hypophysectomy [surgical removal of the pituitary gland] reduces enzyme activity and growth hormone replacement elevates the [DNA polymerase] activity before any increase in synthesis can be shown."[30]

Earlier Winick had shown that in the brain "the activity of DNA polymerase parallels the rate of cell division during normal growth."[31]

Studies at other centers had suggested to Winick that neonatal malnutrition will reduce the activity of this DNA polymerase on the liver. "Thus, the evidence at this stage indicates that one way by which early malnutrition may curtail the rate of DNA synthesis and perhaps indirectly regulate the distribution of available nucleotides [the molecular building blocks of DNA and RNA] is by reducing the activity of DNA polymerase [an enzyme essential in the cellular chemical processes in which nucleotides are assembled into nucleic acids]."

All of this meant to Winick that in man "malnutrition during proliferative growth will curtail net protein, RNA, and DNA synthesis and result in an organ with a reduced number of cells."[32]

These biochemical sequelae of malnutrition affect all organs of the body, as well as the brain, which is why more mental retardation will always be found in babies of abnormally low birth weights than in children with normal weights at birth. The human brain, because its period of proliferative growth is over at the age of 150 weeks, is the organ most irreversibly harmed by

The Human Brain Attains Most of Its Adult Growth by Age 6

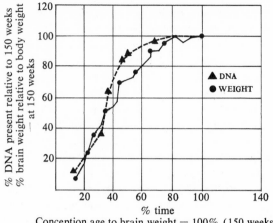

From Pan
American
Health
Organization,
WHO, 1972,
p. 11.

% time
Conception age to brain weight = 100% (150 weeks)

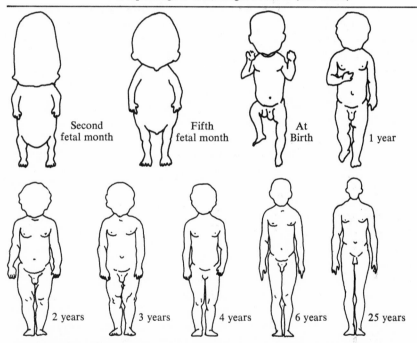

Second
fetal month

Fifth
fetal month

At
Birth

1 year

2 years   3 years   4 years   6 years   25 years

Redrawn from Joseph Handler, *Le Livre de la Santé*, vol. 12, *La Croissance* (Monte Carlo: Les Éditions Rencontre, 1968), pp. 32–33.

As revealed by post-mortem biochemical and morphological studies , the human brain has, at birth, one-third of its adult number of cells. "There is a period of human development," notes the British brain expert John Dobbing, "extending from the second trimester of gestation well into the second post-natal year, during which the brain appears to have a once-only opportunity to grow properly." For the first six years of life, during which the brain achieves 90% of its adult growth, the human head and brain grow much faster than the rest of the body. Professor Dobbing warns that during the pre-natal and post-natal periods of maximum brain development, "nutrition is central to proper [brain] growth."

maternal and child malnutrition. In laboratory studies with animals, Winick observed, "combining maternal [dietary] restriction with postnatal restriction, we can produce even more marked effects on cellular growth of the developing brain. Although in either prenatal or postnatal malnutrition there is an approximately 15% reduction in brain cell number, the combination of both produces a 60% reduction."[33]

Studies of the brains of children who had died of various types of malnutrition suggested to Winick that similar degrees of brain damage are caused in low birth weight babies. Malnutrition alone, however, is not the only cause of mental retardation in the children of the world's poor people. This point was stressed by Cheek and his co-authors, who noted that investigators of psychological and behavioral changes due to undernutrition were "becoming more convinced that environmental deprivation (stimulation) is the key factor. Undoubtedly, altered reaction to stimuli, emotional instability, and withdrawal are characteristic of nutritionally deprived subjects. If lack of stimulation is added to protein deprivation, then mental retardation may be expected."

Lack of stimulation can, of course, as the many case reports on deprivation dwarfism indicate, derive from nonnutritional factors. But "in countries where protein-calories restriction is remarkable, the mothers have a low I.Q.—and the malnourished infant fails to stimulate the mother and vice versa. Hence, a *double insult* to the infant exists. The child grows up in an atmosphere of apathy and rejection, and if recurrent illness does not cause death, the child will show eventual mental impairment." (Italics added.)[34]

In various studies cited by Cheek, it has been shown that "food intake stimulates insulin secretion." Cheek suggested that insulin is as important to human brain growth as the thyroid hormone released by the thyroid gland in response to the stimulation of thyrotropin (a hormone produced by the anterior pituitary gland) is to the chemistry of the body.

THE CYTOPLASM: WHERE THE GENES FUNCTION

What such lines of biological and pediatric research dramatize is that the modern studies of the effects of malnutrition and emotional deprivation on human growth and development have progressed far beyond the observation and measurement of such physiological and emotional insults merely in terms of height, weight, and mental test scores. *Modern* scientific studies of the mental and physical effects of adverse environments on human development now deal essentially with the mechanisms of these retardations at the systemic, organic, cellular, and subcellular or molecular levels.

For example, the hormones secreted by various endocrine organs of the body and delivered to the cells when and as needed are now known to be chemical activators and repressors or deactivators of the genes themselves: so that the diseases and symptoms caused by not enough growth hormones, such as hypopituitarism, or by too many growth hormones, such as hyperpituitarism, are more than disorders that cause dwarfism or giantism. They are also disorders of the cytoplasm, those nonnuclear elements within the cell that constitute the immediate biological environment of every gene.

The cytoplasm consists of everything inside of the living cell except for the cell nucleus. Inside the nucleus are the chromosomes bearing the deoxyribonucleic acid (DNA) of which our genes are made. DNA is found only in the cell nucleus.

The genetic information that controls every act of growth and development, from the synthesis of thousands of highly specific types of proteins to the rigidly fixed calendars of proliferative and normal growth, is all contained in the DNA or the chromosomal nucleic acid. Even the three known types of RNA, the so-called cytoplasmic nucleic acids, are made in the cell nucleus on the templates of the DNA molecules themselves. These molecules of ribosomal, messenger, and transfer RNA (abbreviated as rRNA, mRNA, and tRNA) are the biochemical agents through which the genetic blueprints incorporated into the DNA are translated into new proteins, new enzymes, new metabolites, and ultimately new daughter cells.

The developmental history of every human being is the history of the achievement or non-achievement of those optimal levels of nutritional, chemical, immunological, emotional, and other environmental factors that in the final analysis determine the quality and the quantities of the cytoplasmic RNA through which the genes in their nuclear DNA express themselves.

However, the cytoplasmic nucleic acid, RNA, is far from being all there is to the cytoplasm, or all that makes the hour-to-hour condition of the cytoplasm the permanent controller of both the qualitative and quantitative effects of the multifaceted functions of the genes in the chromosomal DNA.

Within the viscous cellular mass that surrounds and interacts with the chromosomal DNA are, as the diagram (opposite) illustrates, various organelles (a cellular organelle is analogous to organs such as the heart, the kidneys, the pituitary gland, etc., in the body), such as the ribosomes around which most of the RNA molecules in the cytoplasm cluster, but whose genetic and metabolic functions are still not fully known. All of the organelles of each cell are involved in the manufacture of the materials of life itself, the ultimate function of the endless chemical processes described by the collective title of metabolism.

The raw materials which the metabolic processes convert into the cellular proteins and daughter cells are, of course, derived from the food we ingest. But equally essential to the cellular biochemical actions are the various chemicals and trace elements found at different times in the cells. These range from the literally thousands of different kinds of enzymes synthesized in the cytoplasm to the hormones—produced by different glands and organs throughout the body—that pass through the permeable outer membranes of the cells. The hormones not only play essential roles in the repression and derepression (or activation) of the genes; they also help control the synthesis and the performances of the enzymes.

The hormones are, of course, controlled in many ways by emotional and nutritional pluses and minuses. Nutritional and emotional stresses, let alone the side effects of chronic infectious and neurological diseases and dis-

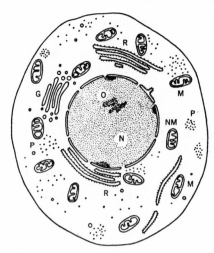

THE WORKING ENVIRONMENT OF OUR GENES

The cytoplasm consists of everything in a living cell outside the nucleus (N). The chromosomes containing the DNA bearing the genes of heredity are inside the nucleus. The nucleic acid RNA, synthesized within the nucleus on the DNA templates, delivers the genes' hereditary information to the cytoplasm. All of the organelles found in the viscous fluid of the cytoplasm are involved in the endless chemical processes termed the metabolism. These organelles—analogous to the heart, lungs, and other organs of the entire body—include the mitochondria (M), the Golgi apparatus (G), and the ribosomes (P) found within and outside the endoplasmic reticulum (R). From E. H. Mercer, *Cells: Their Structure and Function* (N.Y.: Anchor Books, 1967), p. 26.

The more than 100,000 genes we inherit at birth perform their tissue-building and other somatic functions only by interacting with the cytoplasm in the 2000 billion cells we have at birth and their infinitely greater number of descendant cells in mature bodies.

Our genes are not cosmic blueprints for predetermined heights, weights, lifespans, IQ test scores, and bank balances. What we inherit in our genes are, rather, vast ranges of reactions normal to the specific qualities of the environment—the cytoplasm—in which they occur.

The make-up and the volume (i.e., the qualities and the quantities) of the biological end products of the gene-cytoplasm interactions are determined largely by biochemical and quantitative factors in the cytoplasm at the time of these genetic interactions. These factors range from the dimensions of the sheer mass of the cytoplasms in the fetal cells to the specific amounts—at any given time of life—of the gene-activating and gene-repressing hormones and other chemicals produced elsewhere in the body and delivered to the cell cytoplasms via the body's circulating fluids.

While a human fetus will, if it survives the hazards of pregnancy and birth, always develop into a human infant, and not a calf or a starling, the rate and final limits of fetal growth and development are always determined primarily by many interdependent and interacting factors of pregnancy. The nutritional status of the pregnant mother, for example, will control the actual size of the cytoplasmic mass itself. The absence or presence of maternal infections has a critical influence on the nature of the cytoplasm. Regardless of the *potentials* for normal fetal growth and development present in the genes at the moment of conception, such infectious agents as the rubella virus during the nine months of gestation will prevent the fetus from growing normal-sized and -formed brains, or eyes and ears that can see and hear, or central nervous systems that function normally.

After birth, the biopsychological effects of wars, religious and racial segregation, and poverty can result in the human glandular production of abnormally low or dangerously high amounts of hormones and other gene-controlling body chemicals capable of seriously disrupting gene-cytoplasm interactions.

Many of the biological, chemical, and societal factors now known to impair the mass and qualities of the cytoplasms where the genes function—from malnutrition and preventable infections to wars and racism—are obviously not genetic but societal in origin. They are each, therefore, capable of being eliminated by responsible societal intervention. Such all-important factors of the bodily and external (or ambient) environments function, for good or ill, to exactly the same degree in the cell cytoplasms of people of all races, all social classes, and all political systems.

orders, are known to affect the levels of hormone production in the body.[35] So that the qualities of the constant interactions between the cellular enzymes and the hormones are at all times affected by the nature of the total environment—as, indeed, are the quantity and the quality of the biologically essential macromolecules (such as hormones and antibodies) that the cells manufacture for export to other cells, organs, and skeletal systems in the same body.

Just as trace amounts of vitamins the body cannot manufacture for itself are vital to cellular metabolism, so too are trace minerals and metals, such as cobalt, selenium, lithium, magnesium, iron, manganese, copper, vanadium, molybdenum, and zinc. We have barely begun to learn the infinitely complex ways in which such trace elements are involved in the genetic and metabolic cell functions, but we do know that at least one percent of many enzymes consists of one or more of these trace elements, and we know something of some of their functions.

Cobalt, for example, is now known to be essential to the xanthine oxidase enzymes that break down the amino purines and cause them to be excreted from the body as uric acid. The iron-like metal also happens to be an essential component of cyanocobalamin—vitamin B_{12}—essential in the manufacture of red blood cells in the bone marrow. In the absence of the cobalt-containing vitamin B_{12}, the body develops pernicious anemia, a genetic condition that was always fatal until its nature was understood—and its treatment, originally with frequent doses of liver extract rich in the "antipernicious anemia factor" later identified as vitamin B_{12}, and now with monthly injections of the vitamin, permitted its victims to lead perfectly normal lives.[36]

The contributions of the cytoplasm as a whole to the biological interpretations of the master plans of the genes are controlled by organs—and by socioeconomic, chemical, biological, and cultural events—beyond the cell membranes within which the chromosomal and the cytoplasmic nucleic acids carry out the processes of life and its continuation. Consider, again, only the pituitary gland, which Sir Walter Langdon Brown so aptly termed "the conductor of the whole endocrine orchestra."[37]

The Oxford cell biologists John B. Gurdon and H. R. Woodland have reported that "the rupture of the germinal vesicle [oocyte nucleus] leading to the capacity of egg cytoplasm to induce DNA synthesis, is an event that can be brought about *in vitro* and *in vivo* only by pituitary hormone."[38]

Without pituitary hormone, then, there can be no production of DNA itself after the egg has been fertilized. With insufficient amounts of pituitary hormone, the cells cannot synthesize enough DNA to produce phenotypes of normal size. So that, in human beings, a malfunctioning pituitary gland diminishes the functions of every gene in every single one of the body's trillions of cells.

More than hormones and enzymes are involved in the endless interactions of cellular DNA and RNA in the cytoplasm. There are, as previously indicated, the healthy and unhealthy cytoplasmic levels of minerals, vitamins, and other metabolites derived from the diet of the multicellular organism called a human body. There are also the relative presence or absence in or near the cells of viruses, bacterial toxins, cellular parasites, and other pathogenic agents—as well as antibodies, interferons, complement, leukocytes, lysozyme, and other cell-produced immune or host defense agents. There is also the presence or the absence of the viable but inactive genes of the slow viruses, those non-chromosomal episomes of viral DNA.

In a brilliant presentation before the Twelfth International Congress of Genetics in 1969, Gurdon described in great detail the nature of the interactions that take place between the nucleus and the cytoplasm of the cell. He divided these interactions into two categories: (1) the effects of the cell nucleus on the activity of the cytoplasm and (2) the effects of the cell cytoplasm on the activity of the nucleus.[39]

He showed, among other examples of cytoplasmic effects on the gene, how the change induced in the cytoplasm by the "physical or chemical suppression of RNA synthesis by X-irradiation or by Actinomycin-D [an antibiotic] permits normal cleavage [the cell divisions that occur in the zygote immediately after its fertilization] but very quickly inhibits gastrulation, the developmental stage when functional differences between cells first become obvious." Thus, in the absence of the sufficient amounts of cytoplasmic RNA to mediate its genetic information, the cell's nuclear DNA is unable to express itself in a normal manner, and as a result it either fails to divide or it divides and its daughter cells are created as defective cells.

In view of the findings on the hormonal-cytoplasmic effects of nutritional and emotional deprivations alone, Gurdon's own interpretation of the much wider biological and social implications of the elucidation by modern geneticists and cell biologists of the countless effects of the cells' cytoplasm on their genes constitutes a resounding affirmation of the role of *informed* free will in the shaping of human destinies:

"A good understanding of the way cytoplasmic components can control gene activity is of great importance, not only because much of the embryonic development and cell differentiation depends upon processes of this kind, but also because *few other conditions are known in which the activity of the genes can be altered in a predictable way*" (italics added).

If I had to select any single sentence that best speaks for the real scope of modern developmental biology, as well as for a proper understanding of the basic role that modern genetics is now equipped to perform in the elimination of completely preventable physical and mental dwarfism from the entire human race, it would be this concluding sentence from Gurdon's essay, "The Cytoplasmic Control of Gene Activity."[40]

We cannot, because of the great dangers of causing deformity and death, interfere directly with the genes of any human being. Concurrently, we no longer have the moral right, because of what we already know of the physical and mental consequences of failure to intervene, to refrain from doing whatever we possibly can do to improve the nurturant qualities of the cytoplasm in which the genes of every human cell function.

This grave human responsibility—and opportunity—that we all share is underscored by all serious modern studies of the biology of fetal and postnatal growth and development the world over. The American pathologists Richard L. Naeye and William Blanc, for example, in a recent study, "Effects of Maternal Malnutrition on the Human Fetus,"[41] not only found that "the larger brain size in newborn of mothers who were well nourished raises the possibility that fetal brain growth may reach its full genetic potential only under such circumstances of full nutrition"; they also made some highly significant findings in the cells of the liver and the adrenal glands. In each of these glands and organs, they found that the cytoplasmic mass in the cells of children of malnourished mothers was far smaller than in the cells of children of well-nourished mothers.

Not only the cytoplasmic mass but also the number of cells per gland and organ were also considerably lower in the children of undernourished mothers than in the liver and adrenals of the babies of well-nourished mothers. It was only in the diameters of the liver and the adrenal cell nuclei—the sites of the genetic DNA—that the size differences between the children of the well fed and the children of the malnourished were not as striking.

What this and all other modern studies of human development at the intracellular interfaces where the genes interact with the cytoplasm also tell us, of course, is that the venerable words "environmentalist" and "hereditarian" no longer represent scientific values but philosophic or sentimental postures. As Professor Leon Eisenberg of Harvard declared recently, "I trust that we have passed beyond the hoary counterposing of genetic and experiential factors as though they were diametric opposites."[42]

Today we know that regardless of the diameter of the cell nuclei containing the chromosomes, when malnutrition and other environmental factors severely reduce the dimensions of the cytoplasmic mass in the cells of a developing fetus or a newborn child, these *preventable* biological insults also diminish at least as drastically (and most probably even more traumatically) the biological opportunities of the chromosomal DNA—the genes—to express themselves to anywhere near their innate potentials.

It is, patently, high time that we ceased labeling ideas and/or people with such now scientifically meaningless abstractions as "environmentalist" or "hereditarian" and that, instead, we start paying more attention to the pediatricians, the child psychiatrists, the pathologists, and the other serious modern scientists working in human development—who know that there is infinitely more to a human being than his pedigree chart and IQ test score.

24 What the New Scientific Racism Hides about Human Genetics and Development

II: From Chromosomes to Congress

> "Two nations: between whom there is no intercourse and no sympathy; who are as ignorant of each other's habits, thoughts, and feelings, as if they were dwellers in different zones, or inhabitants of different planets; who are formed by a different breeding, are fed by a different food, are ordered by different manners, and are not governed by the same laws."
> "You speak of . . ." said Egremont, hesitatingly.
> "THE RICH AND THE POOR."
> —BENJAMIN DISRAELI, in *Sybil* (1845)

Whatever might have been known about genetics in Galton's time, today we are beginning to understand its mechanics at the cellular and subcellular levels, where the genes of heredity function. Even in terms of our presently limited early knowledge, however, we already know enough about the interactions of gene and human destiny at the primary interface of the cytoplasm to realize that not every genetic burden of our species is necessarily inescapable or beyond therapeutic neutralization or management.

In our own century, some of the brightest pages in the history of modern biomedical sciences have been those bearing the accounts of how totally genetic diseases, such as pernicious anemia; and sex-linked genetic diseases, such as hemophilia; and diseases classified as partially genetic, such as diabetes, have been rendered nonfatal or neutralized. By monthly prophylactic injections of vitamin B_{12} in pernicious anemia. By treatments with adequate amounts of the anti-hemophilic clotting factor that is now produced easily and inexpensively in any blood bank. By injections of animal or fish insulin to supplement the inadequate synthesis of this hormone in the pancreas of a diabetic.

Therefore, the fact that a defect of human physiology or mentality in *individuals* of any and all human races might prove to be wholly or partially genetic in origin is no excuse for branding such people as genetically doomed *and as such* unworthy of any health care, education, or employment. Or forbid them to marry and bear children. Nor does the fact that a disease is *labeled* as genetic, as pellagra was by Davenport and the true believers in eugenics in 1917—or even as diabetes, Huntington's chorea, and other chronic degenerative diseases are classified by most competent and responsible medical scientists today—mean that it might not in due time prove to be a disease of malnutrition, or emotional deprivation, or a slow-acting virus disease, or a systemic reaction to one or more of the thousands of toxic agents new (in some instances very new, since an estimated 500 new chemical pollutants are added to the American environment each year) to the human environment. Such as, for example, the known cases of mental disorders that have been traced to lead poisoning in children. Or the schizo-phrenia-mimicking complex of behavioral aberrations of bromism, the terri-fying mental disorder caused by swallowing too many bromides for head-aches, upset stomachs, and hangovers.

Bromism is a frequent cause of hospitalization with a diagnosis of schizophrenia. It is a misdiagnosis. When correctly diagnosed, bromism can be cured in less than twenty-four hours by drinking liberal doses of sodium chloride—common table salt—in plain water. When diagnosed as schizo-phrenia, bromism is eliminated within a few weeks of hospital life (with or without expensive psychiatric treatment) by the normalization of the body chemical balance on a diet that, of course, is free of Bromo-Seltzers and other common over-the-counter bromides. However, many a case of "schizo-phrenia" that goes into the medical records as "cured" by psychiatric treat-ment "recurs" very quickly and quite often once the patient is released and again starts ingesting bromides.

The medical literature is filled with reports of other conditions mis-diagnosed as "schizophrenia." In 1973, for example, Dr. Ronald R. Fieve, chief of psychiatric research at the New York State Psychiatric Institute, reported at the annual meeting of the American Medical Association that a recent study had demonstrated that over 30 percent of the patients labeled as "schizophrenic" in New York State mental hospitals were found to be, in reality, suffering from depression or manic-depressive psychoses. Thanks to their original misdiagnoses, Dr. Fieve reported, "these patients have under-gone countless sessions of psychotherapy, as well as treatment with electro-shock and multiple tranquilizers and anti-depressants."[1]

If psychiatrists and other behavioral scientists are uncertain of what the diagnostic criteria for schizophrenia as a mental disease entity really are to begin with, many are growing less certain than they were before that schizo-phrenia is a wholly genetically predetermined disease. In recent years, a growing scientific literature on identical twins *discordant* for schizophrenia has suggested that the presumed genetic factors involved in causing this mental disease might not be all that it takes to cause a supposedly genetically susceptible human being to develop schizophrenia.

In 1966, for example, Dr. William Pollin, chief of the Section of Twin and Sibling Studies at the National Institute of Mental Health, and three of his NIH colleagues, published a study of eleven such pairs of *identical* twins (that is, twins developed from a single egg) in which only one twin of each pair developed schizophrenia.[2] What was so revealing here was that in these identical twins discordant for schizophrenia, they found the "re-lationship between lower birth weight and schizophrenia is statistically significant."

To be sure, schizophrenia—like low birth weight, pellagra, tuberculosis, and other preventable diseases of poverty—does, *statistically,* often seem to be a familial or genetic disorder. Why, then, in those pairs of identical twins discordant for schizophrenia, was it the twin with the lower birth weight who generally came down with the mental disorder?

Even if the biochemical changes said to be caused by genetic flaws associated with schizophrenia are eventually pinpointed, "the cause of mental disorder will not be found exclusively in derangements in body chemistry," writes the Harvard psychiatrist Leon Eisenberg. "Patients are people; the mind affects the brain and the body just as brain and body affect the mind. The stresses of life can unhinge the mind by causing emotional changes and the consequent release of hormones."[3]

The growth hormone, synthesized by the pituitary gland, and produced in pathologically low quantities by the emotional and nutritional stresses of deprivation dwarfism, is only one of many hormones that—when produced in quantities too low or too high for human health—cause physical and behavioral disorders. When a thyroid gland produces too little thyroid hormone, for example, a person becomes lethargic and his mind dulled; hyperthyroidism, caused by the glandular overproduction of the same hormone, makes people tense, irritable, and suffer from acute insomnia.

To Professor Eisenberg, "the strong evidence for the role of genetic factors warrants optimism rather than the pessimism evoked in most people by the idea that a disease is hereditary. For if we can tell what persons are at risk from a given disease, we are better able to find ways of intervening in order to prevent it," as, indeed, we today employ medical and dietary techniques of preventing the onset or the physical and/or mental symptoms of such demonstrably genetic diseases as phenylketonuria, galactosemia, and pernicious anemia.

"In the case of schizophrenia, we know only that children born of a schizophrenic parent run 10 times the usual risk of becoming ill; *yet only 1 in 10 is affected.* If we can pick out that one child before he shows symptoms, we can find some way to reduce the likelihood of his getting sick." (Italics added.)[4]

As classroom and IQ test scores have been telling us for years, people who live in urban or rural slums will almost invariably get lower achievement and IQ test scores than the children of more affluent families. Statistically it is easy to use these test score data to "prove" that the poor are *genetically* less intelligent than the nonpoor, and are, in "fact," ineducable. Terman and McDougall did just that during and after World War I. The new scientific racists peddle the same simplisms in our times.

THE INTERACTING ROLES OF THE TOTAL ENVIRONMENT IN CHILD DEVELOPMENT

Biological and behavioral scientists of a more modern scientific orientation, however, have for some decades been studying the evidence that there is more to the mental capacity of a child than his father's socioeconomic status, his parents' skin color, or his racial heritage. Few serious scientists involved in the studies of human development ignore the highly multifactorial roles of the total environment in which a child grows and develops before, during, and after the years in which he is subjected to classroom and IQ tests.

Pediatrician Roy E. Brown and psychologist Florence Halpern are two scientists involved in such holistic studies of human development, in which children are studied as whole human beings in various environments, and not as ciphers with high or middling or low IQ test scores. In 1968, working out of the Tufts University medical center established in one of the poorest rural counties in the poorest state of the Union, Mississippi, Drs. Brown and Halpern conducted studies of physical and mental development in a cohort of 402 poor black children aged 1 week to 36 months. As they noted in a preliminary report on this joint pediatric and psychological investigation:

"*Both* adequate nutrition and proper environmental stimulation are necessary for the optimal development of the growing human organism. Since many persons living in poverty often are deprived of these essential factors, the impact of such deprivations on children's *physical and intellectual growth* seems to us an important area of study" (italics added).[5]

The population they chose to study was, in many ways, an ideal one for such an inquiry into human development:

The black population of Mississippi provides excellent subjects for such evaluation research. Genetically these people are homogeneous. They are involved mainly in agricultural pursuits and a large percentage fall beneath the poverty line as defined by the federal government. Living in homes without heat, running water or sanitary facilities, with as many as eight or ten people occupying two or three rooms, the local people exist in large measure on such staple foods as bread, potatoes, and beans. Conspicuous by their absence are the cultural enrichments and opportunities that are taken for granted in middle class homes. *In particular, where children are concerned, there is generally a dearth of books, toys, pictures and those other playthings which are generally deemed important stimuli for development.*

The psychological portion of the Tufts University Northern Mississippi field study was made with the batteries of developmental tests created by Drs. Arnold Gesell and Catherine S. Amatruda at the Yale University Clinic of Child Development.[6] These psychological tests, which start at the age of one week and continue to the age of five years, are designed to measure and evaluate four basic areas of human growth and development: (1) motor development, (2) adaptive behavior, (3) language, and (4) personal and social behavior. For most of the first year, the Gesell-Amatruda Developmental Quotient (DQ) tests are administered by nonverbal techniques. To this extent—and it is a limited extent—these early infancy and childhood DQ test scores are possibly more revealing of genetic mental capacities than are the language-based IQ tests.

What Brown and Halpern found in the DQ test scores of these poor children in northern Mississippi was summarized succinctly in one paragraph:

These infants aged one week to three months of age showed an average Developmental Quotient (DQ) 17 points *higher* than that of the white population on whom the tests were standardized. Their average DQ was 117. Between 3 months and 6 months the average fell to 101, and remained there until about 15 months of age. A steady decline then began. By 16 months the average DQ was 96 and three months later it was only 90. It then sank slowly to reach 86 at 36 months. Thus, by the time he is 3 years old the Mississippi child's DQ averages are as much as 14 points *below* the norm, in contrast to the 17 point acceleration during the first few months of life. [Italics added.]

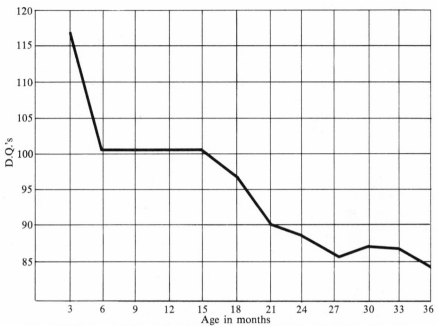

Developmental quotients at three-month intervals of 344 rural Mississippi black children during the period between birth and the end of the third year of life.

From Roy E. Brown and Florence Halpern, "The Variable Pattern of Mental Development of Rural Black Children," *Clinical Pediatrics* (July 1971), 406.

It was in the tests of verbal ability that the children showed the greatest developmental lag. "By the time he is three years old the child is lagging in all areas; the verbal lag is 10 months or more, whereas that in other areas is only 6 months."

Is the verbal area of human development more genetic than the motor, adaptive, and personal and social behavior areas? Or is this lag in verbal development due, as Drs. Brown and Halpern noted in 1968, to the fact that "the parents of these black Mississippi infants and children are themselves rather limited verbally, in the sense that they have rather meager vocabularies and do not easily communicate what they are thinking and feeling."

Behavioral and verbal tests, however, are not the only developmental measures of a growing child. Consider, as Drs. Brown and Halpern did, the clinical measurement (called a hematocrit) of the proportion of red blood cells in the whole blood. This is a somewhat more precise yardstick than an IQ or a DQ test, since it measures totally objective biological cells rather than subjective descriptions and evaluations of behavioral performances. The hematocrit, for example, can reveal by such objective and verifiable measurements whether or not a child is anemic—that is, whether or not the child is suffering from a "reduction below normal in the number of erythrocytes [red blood cells] per cubic millimeter."

What the hematocrits run in the Brown-Halpern study showed was that "more than one-third of the surveyed children can be considered to demonstrate anemia—compared with 7% of a randomly selected population of children aged one to six who represented a geographic and socio-economic cross-section of the United States. However, recent studies of underprivileged children of all races indicate that between 10 and 25 percent are anemic."[7] All of the other measurements of biological development made in the study of this population of poor black children showed parallel adverse developmental and health effects of poverty. The malnutrition that causes anemia breeds scores of other biological and mental disorders.[8]

Nevertheless, for all of the purely biological insults that reflected themselves, ultimately, in the mass and composition of the cytoplasm of each cell in their malnourished bodies, the emotional stresses of being black children in rural Mississippi in 1968 were at least as damaging, and certainly as apparent in their patterns of human development. In *The Tyranny of Hunger,* the book Dr. Brown wrote on his field studies of the social biology and the social psychology of poverty, a long passage in the closing pages tells us more about the reasons for the verbal and intellectual characteristics observed in poor black and nonblack people than do all the IQ test scores in the history of mental testing.

> Again, children with mothers who are harassed by many obligations and who spend 12 or more hours working in the fields or in someone else's kitchen, are not very likely to enjoy the maternal play and verbal exchange that children in more fortunate circumstances receive. In fact, because their mothers are working, these children often know and relate to many "mothers," including older siblings, grandmothers, neighbors and so on. This complicates the developmental progress of the child in verbal as well as in other areas since there is no consistent speech pattern to be followed under such circumstances, and no consistency in the rewards meted out for successful performance. There is only a shifting and confusing world.
>
> The general child-rearing practices typical for this entire population add to the difficulties the child experiences in acquiring speech (as well as other forms of learning). Wanted and loved, regardless of the poverty that exists in the home, the child enjoys much body contact during the early months of life, much playing and fondling by the various members of the household.
>
> However, as soon as the child begins to move about and explore his environment, as soon as he makes any attempt to assert himself, he is vigorously and unequivocally cut down. He is likely to be told he is "bad," "nothing," "worthless," and assured that he will never amount to anything. This attitude on the part of those caring for the child is actually an expression of their love for him. As they see it, if he is to survive in the hostile white world of the Delta he must be discouraged from any kind of independent, self-realizing activities and

from anything that might bring him to the negative attention of the white man. The result of this training is excessive inhibition and repression for the majority of these children. They learn very early to adopt the overly compliant, conforming manner that adults assume whenever they are in contact with outsiders.

In fact, they soon learn to live two lives, one acceptable to "Mr. Charlie," and another simple but freer one to follow when they are with their own people. In order to maintain this dichotomy, much that they think and feel must be kept to themselves. For some children the process becomes automatic, but for some it requires constant alertness and absorbs much energy. In either case, the inhibitions that are placed on the child and that he gradually places on himself do not leave him open to new ideas, but seriously hamper the whole learning process even at the age of two or three. The universal impact of this training soon overshadows whatever individual differences might have existed among these children.[9]

Most of what we read above could, as easily, have been written about the white Anglo-Saxon children of poor British miners and mill workers in the first decades of the Industrial Revolution. Or, for that matter, of the children of the equally white Anglo-Saxon Appalachian and urban poor whites in our own country today. The culture of poverty is color-blind.

HOW TO DOOM UNBORN CHILDREN TO PHYSICAL AND MENTAL RETARDATION

There are more ways to demolish the potentials of a human being than by merely denying him adequate nutrition and adequate social, cultural, educational, and employment opportunities after he is born.

A classic way of really handicapping the child prenatally is to arrange for the mother of the child to be, herself, so malnourished from the moment of her own birth, and so deprived of food and medical care during her pregnancy, that the child is certain to be born weighing less than 2,500 grams (5½ pounds). Children born weighing less than 2,500 grams are, by clinical definition, abnormally low birth weight infants. All premature infants are low birth weight infants, but most low birth weight or immature babies are full-term babies.

The entire spectrum of the medical and mental sequelae of being born weighing less than is considered normal for the average healthy baby is, probably, the central core of modern pediatric physiological, psychiatric, and developmental research. As the late professor of both pediatrics and experimental psychology Herbert G. Birch observed, "serious and detailed consideration of the consequences of low birth weight for later behavioral consequences can properly be said to have been begun by Pasamanick, Knobloch and their colleagues shortly after World War II."[10]

The decades of pioneering studies of the mental and physical consequences of abnormally low birth weight and other biological traumas of poverty by the psychiatrist Benjamin Pasamanick and his wife, the pediatrician Hilda Knobloch, started in New Haven during the closing years of World War II. At that time they were both young research assistants of Arnold Gesell and Catherine S. Amatruda at the Yale University Clinic of Child Development.

Their first major published studies derived from their attempts to find some scientific explanations for the apparently racial differences in IQ test scores of white and black children. They found what the earlier researches of Ballantyne and Boas, Klineberg and Skeels, had led them to suspect: to. wit, that the IQ test score differences in individual white and black children had literally nothing whatsoever to do with race, but were determined entirely by sociobiological variables, the most persistent of which seemed always to be related to the individual child's weight at birth and the postnatal growth curves.

Subsequently, as they pursued their individual careers in various New York hospitals, at the University of Michigan, Johns Hopkins, the University of Illinois, Ohio State University, Chicago Medical School, Columbia University, Mount Sinai School of Medicine, Albany Medical College, and in a number of major state mental health administrative posts around the nation, the Pasamanicks continued studying prenatal, intrauterine, and postnatal aspects of child growth and development in white and nonwhite populations of various social classes. By 1955 they had developed their now famous concept of "the continuum of reproductive casualty," a term describing the permanent sequelae of harmful events during pregnancy and childbirth that damage the fetus or the newborn infant, primarily in the central nervous system. This was stated in four propositions:

> 1. Since prematurity and complications of pregnancy are associated with fetal and neonatal death, usually on the basis of injury to the brain, *there must remain a fraction injured who do not die.* 2. Depending upon the degree and location of the damage, the survivors may develop a series of disorders. These extend from cerebral palsy, epilepsy, and mental deficiency through *all types of behavioral and learning disabilities which are a result of lesser degrees of damage sufficient to disorganize behavioral development* and all thresholds to stress. 3. Further, these abnormalities of pregnancy are associated with certain life experiences, *usually socioeconomically determined,* and consequently 4. they themselves and their resulting neuropsychiatric disorders *are found in greater aggregation in the lower (socioeconomic) strata of our society.* [Italics added.][11]

In all of this planet's human populations, regardless of race or color, it is precisely these brain- and central-nervous-system-injured infants who—as the Pasamanicks have documented in massive scientific detail over three decades—grow up to constitute the majority of the constitutionally handicapped of all societies. They are the children slow to sit, slow to crawl, slow to walk, slow to talk, slow to learn: the monopolists of low IQ and classroom test scores.

Although they have managed to survive, physically, the hazards of *in utero* infection and malnutrition and the traumatic complications of the act of birth, they also continue to pay the terrible costs of having been denied anything remotely resembling equality of opportunity to develop the genetic potentials present in the zygotes, the cells from which each of them develops. Through no fault of their own—*or in their genes*—they have been deprived from the moment of their conception of the infinitely greater freedom from the preventable reproductive casualties enjoyed by the nonpoor children of the same race, the same color, the same culture.

It is impossible in this book to do full justice to the mountains of hard, verifiable, and *ultimately hopeful* evidence that the Pasamanicks have, since 1946, published on the actual physiological and intellectual potentials of all of the children in our species. And on the humanly and socially alterable factors that perform so many clearly defined functions in the physical and mental development of every newborn child. Fortunately, some key segments of their work have been summed up in tables prepared for many of their scientific studies and, as the four tables below demonstrate, these hard data speak for themselves, as objectively as a hematocrit.

The tables I have selected from the body of their work define the dimensions of the brain- and body-affecting barriers to optimum phenotypical expression of the inborn potentials of the human genotype. These hazards to normal development fall into three basic categories.

The first, *abnormalities of complicated pregnancy,* includes everything from prematurity and toxemias and skeletal defects to placental disorders, rubella, nonspecific and chronic infections, and prolonged labor.

The second, *nutrition,* deals with all aspects of adequate and inadequate nutrition in the pregnant mother, the developing fetus, and the developing child.

The third, *sociobiological environment,* in which nutrition is of course an integral element, includes the biology, the economics, and the intellectual resources of the immediate and total environment in which the child grows and develops.

All of the variables in these three categories of human developmental experience, from the infections and toxemias of pregnancy to the corrosive effects of slum or rural-hovel existence, to the paucity or total absence of books and other intellectual stimuli at home, "almost certainly operate multi-factorially, each tending to aggravate the effects of the others. They operate by diverse physiologic paths and require preventive efforts at various levels."[12]

More than this, nearly all of them are subject to intervention and control: the toxemias, the fatal or deforming infections, the skeletal deformities, the individual and societal consequences of biological and intellectual malnutrition proven to prevent millions of potentially self-supporting taxpayers from being anything but marginal earners and welfare clients—*if* they survive the premature births and the perilous infancies of poverty—can quite feasibly be eliminated by societal intervention from the American family experience.

Consider, for example, the matter of premature births, those harbingers of possibly a majority of reproductive casualties. In the cross-sectional study the Pasamanicks made with two Johns Hopkins colleagues of 4,700 premature infants born in Baltimore, Maryland, in 1952, it was shown that black mothers had more and smaller premature babies than did white mothers. In Baltimore, in 1952, the lower the weight of the prematures, the larger the proportion of black mothers. As summed up in two of their tables:

Weight Distribution, by Race, of Premature and Full-Term Control Infants

Weight Group (in grams)	White	Nonwhite	Total
	%	%	%
1,000 or less	0.9	0.5	0.7
1,001–1,500	3.2	6.5	5.0
1,501–2,000	6.7	9.9	8.5
2,001–2,500	37.9	34.8	36.2
Subtotal for premature infants	48.7	51.7	50.4
2,501 or more (control)	51.3	48.3	49.6
Total	100.0	100.0	100.0

Total Deviations from Normal of Premature and Full-Term Control Infants

Weight Group (in grams)	No. of cases	Abnormality, %
1,500 or less	57	50.9
1,501–2,500	44.3	24.5
Subtotal for premature infants, adjusted for weight distribution	—	25.7
2,501 or more (controls)	49.2	12.8

Both tables from Hilda Knobloch et al., "Neuropsychiatric Sequelae of Prematurity," *Journal of the American Medical Association,* CLXI, No. 7 (June 16, 1956), 581–85.

In a sample consisting of 500 single-born premature infants and 492 infants born at full term, the two groups of newborn human beings were matched as to race, socioeconomic status, and other significant factors. (See tables above.) When these children were compared at age 40 weeks, it developed:

"The incidence of [developmental] abnormalities was found to increase as the birth weight group of the infant decreased, so that 50.9% of the infants with a birth weight of less than 1,501 grams [3.3 pounds] had defects ranging from minor neurological damage to severe intellectual deficiency. Some of these also had major visual handicaps."

While the children came from all races and socioeconomic classes, when examined it was shown that, regardless of color or social status, 25.7 percent of the premature infants showed some developmental abnormalities as compared with only 12.8 percent of the full-term babies.

"Analysis of the findings shows that *there is no significant difference between whites and nonwhites in the incidence of neurological and intellectual deficit* when adjustment is made for differences in weight distribution between the races. *The incidence of abnormality increases as the birth weight group of the infant decreases.*" (Italics added.)

However, when the population of premature babies born in Baltimore was tabulated by age and socioeconomic class, it could be seen that among white as well as black mothers, the lower the socioeconomic class (and the fewer the health, nutritional, and cultural resources of their families) and the higher a mother's age, the greater did their chances of having a premature baby become. (See the next table.) Far from being in any way a racial trait, "prematurity is inversely related to economic status."[13] The means of prevention of premature births, therefore, are primarily economic, and not medical.

Prematurity Ratios among First-Born Babies by Socioeconomic Fifths and Age of Mother, Baltimore, Md., 1950–1951

Socioeconomic Fifth	Age of Mother (years)			
	All Ages	*Under 25*	*25–29*	*30 and Over*
Total—White	7.1	7.1	6.2	9.4
Highest	5.6	5.2	4.8	8.5
4	6.9	6.9	5.9	8.9
3	7.0	6.4	7.3	9.7
2	8.2	8.5	7.0	8.8
Lowest	8.4	8.1	7.8	12.5
Nonwhite	14.6	14.1	14.1	21.8

From Rider et al., "Associations Between Premature Birth and Socioeconomic Status," *American Journal of Public Health*, XLV (1955), 1022–28.

In terms of the complications of pregnancy—most of them preventable by proper nutrition, sanitation, and medical care—it turned out that white mothers in the lowest fifth of the socioeconomic scale had nearly three times as many complications of pregnancy as did white mothers at the most affluent socioeconomic levels. But the nonwhite mothers, whose poverty was the most severe of all the groups observed, had more than three times as many complications of pregnancy as did the poorest whites, and ten times as many as did the most affluent of the white mothers. This finding suggested to the Pasamanicks and their collaborator, Dr. Abraham Lilienfeld, the theoretical possibility that "prematurity rates increase exponentially below certain socioeconomic thresholds."[14]

The Pasamanicks also observed the life-enhancing effects of nurturant and supportive intervention at various interfaces of human development and the total growth environment. A study of the world's scientific literature documented their certainty that, for example, prematurity could be controlled by medical and nonmedical societal intervention. While it was true, they noted, that in India the prematurity rates rose from 9 percent in the highest or most affluent socioeconomic class to 31 percent in that nation's poorest classes, it was equally evident that "in some of the Scandinavian countries [where the biology of poverty has been pretty well eliminated from the greater society] the incidence of prematurity has been reduced to 3%, and in one clinic in Copenhagen, with additional attention being paid to prenatal care, it has even been reduced to well below 2%."

They reported on how, in many nations, intervention in the form of nutritional protein and vitamin supplements had reduced the rates of toxemias of pregnancy and other complications in various populations that

Complications of Pregnancy among Nonwhite and White Upper and Lower Economic Fifth Mothers of Children Living Through the Neonatal Period and Not Known to Have a Neuropsychiatric Disorder

	White Upper Economic Fifth		White Lower Economic Fifth		Nonwhite	
Total number of cases in each category upon whom hospital records are available	159		185		459	
	No.	*%*	*No.*	*%*	*No.*	*%*
Toxemias of pregnancy	8	5.0	17	9.2	84	18.3
Bleeding during pregnancy	0	0.0	2	1.1	16	3.5
Nonpuerperal complications*	0	0.0	7	3.8	127	27.7
Miscellaneous puerperal complications†	0	0.0	1	0.5	5	1.1
Total complications of pregnancy	8	5.0	27	14.6	232	50.6

* Nonpuerperal complications include diseases associated with, but not related to, pregnancy.
† Miscellaneous puerperal complications include pyelitis, other genitourinary diseases, and hydramnios.

From Pasamanick et al., "Socioeconomic Status and Some Precursors of Neuropsychiatric Disorder," *American Journal of Orthopsychiatry,* XXVI (1956), 594–601.

usually suffer staggering rates of prematurity and low full-term birth weights:

"With full supplementation of protein and vitamins, the prematurity rate was 3%. If proteins alone were given, it was 4.3%, if vitamins alone, 5.6%, and in the absence of any dietary supplement, 6.4%. If the mother was underweight at the start of pregnancy and had a less-than-average weight gain, or a weight loss, the incidence of prematurity was 23.8%. *This was reduced to less than 2%* by protein and vitamin supplement in mothers with good pregravid [pre-pregnancy] nutrition" (italics added).[15]

As biological and behavioral scientists, the Pasamanicks could, therefore, "view determinants of behavior as consisting of interactions among the biological and sociocultural factors of human existence. The biological determinants serve primarily to establish the psychological limits and the floor of [genetic] potential in the organism." Like Jacobi, Binet, Ballantyne, Boas, Klineberg, Skeels, and other honored predecessors in the study of the biological and intellectual development of human beings, the Pasamanicks could also conclude:

"The sociocultural factors, on the other hand, are like the soil in which the plant is nurtured. By enrichment or impoverishment human behavioral potential can be made to blossom or wither, to achieve its [genetic] limits or to fall far short of them."[16]

With Gurdon, at Oxford, and Alfred Mirsky, at Rockefeller University, the Pasamanicks opted for giving the genes of all newborn children the qualities of cellular, systemic, and social environments most conducive to having the most desirable human phenotypes develop from these genes. At mid-passage, summarizing their lifelong studies to 1966, they wrote:

The findings point to the overwhelming importance of the factors of prenatal and maternal health, preschool stimulation and later educational effort which are the major foci in the anti-poverty programs for children today. These programs should be geared up to the elimination and modification of such results of poverty and deprivation as malnutrition, infection and other forms of stress, prenatally in the mother and postnatally in the child. In addition, it seems apparent to us that psychosocial deprivation, faulty stimulation and inadequate education in childhood require fully as much attention, if not more, in preventive programs.[17]

Nor were they talking of untried medical and social measures, of visionary but unpractical and costly experiments: "It would seem apparent from examination of the data . . . that a good deal of reproductive casualty is *immediately preventible* and that much of the remainder could in time be prevented if we begin to plan and institute our preventive measures as soon as possible."[18]

VISION, IQ TEST SCORES, AND MADNESS

It is no longer news, to the world's serious life and behavioral scientists, that whatever the potential differences packed into the genes inherited by all newborn children, it remains the general nature of the total growth environment—from the cytoplasm of the cell, which is the sole functional environment of all genes, to the greater society—that determines the general ranges of reaction of human genes to human history. The sheer numbers and complexities of the factors involved in the individual's gene-organism interactions make the correlational analyses of Galton and his modern disciples offensive to serious life and behavioral scientists. The *only* scientific difference between the false correlations Goldberger denounced when the Davenports used them to "prove" that the hunger disease pellagra was the product of inferior genes, and the equally simplistic correlations the Jensens and their ilk derive from the IQ test scores and the per capita annual incomes of the poor (and disproportionately black) people, is that, in 1916, the false correlations dealt with a physiological disorder.[19]

The data recently issued by the World Health Organization on the prevalence of blindness constitute a devastating commentary on the continued use of false correlations. WHO estimates that total blindness afflicts between 10 and 15 million people on this planet. (The health authorities of India, which alone now has an estimated 10 million blind people, put the world's blind at 20 million.) The survey of WHO found: "Blindness rates are in general around 200 per 100,000 population in America and Europe. In less developed [that is, poor] countries, especially those in Africa and Asia, the rates are considerably higher and reach values about 1,000 per 100,000."[20]

If the genes involved in the development and maintenance of vision were all there were to the causes of blindness, this would mean that the largely nonwhite peoples of Asia and Africa are at least five times as prone by *heredity* to becoming totally blind as are the largely white peoples of America and Europe. But, says this same WHO study, "half to two-thirds of cases of blindness could have been prevented if they had been detected and treated in time."

These preventable types of blindness are, reported the WHO medical scientists, those total and partial losses in vision caused by bacteria, viruses, and other parasites that, in affluent nations, are all subject to control by many societal actions that keep them from infecting people's eyes. These include clean water supplies, proper sewage and garbage systems, and other standard elements of environmental sanitation; proper nutrition that keeps the antibody, complement, interferon, leukocyte, lysozyme, and other elements of our inherited host defense or immune systems in proper working order; early and adequate medical care, including the proper uses of antibiotics and other drugs that kill the agents of eye infections; immunizing vaccinations and other prophylactic services of public health agencies. Next to glaucoma, cataracts, and accidents, "when the cause [of blindness] is known, trachoma ranks highest" in the world.

The agent of trachoma is a *Chlamydia,* a "weakly infectious"[21] bacterium that, in societies where even the poor have access daily to running water and soap, is rarely able to infect a human eye. WHO estimates "that there are from 400 to 500 *million* cases of trachoma in the world, and of these 125 million are in India alone. In most of the developing countries, trachoma and associated infections are still the main causes of ocular pathology, which, when complicated by severe sequelae such as entropion [inversion of the eyelid] and trichiasis [ingrowing eyelashes], may lead to partial or total loss of vision."

In America, the last time I checked on the reported cases of trachoma, in 1970, they added up to exactly 455 cases annually.

Next to preventable infections, such as trachoma, measles, syphilis, gonorrhea, and fetal rubella, there are the common forms of blindness caused by malnutrition, such as those due to vitamin-A deficiency. Where there is inadequate protein-calorie intake, particularly in growing children, there is widespread nutritional anemia. Where there is widespread nutritional anemia, there are huge numbers of eyes diminished or wholly blinded by deficiencies in vitamin A and other nutrients directly or indirectly essential to normal vision.

Thus, on the same day in 1973 that Dr. Dharam Batta Vaidya, the health minister of Uttar Pradash, India's poorest state, told a conference of eye surgeons that in his state "blindness caused by infectious diseases is on the decline, but that blindness caused by malnutrition is on the increase, particularly among children and expectant mothers," the Indian Council of Medical Research reported that "one child in two in India's population suffers from nutritional anemia." The survey also showed that "50 million children one to six years old are affected by protein-calorie malnutrition."[22]

Blindness—the total loss of functional vision—is only one aspect of the continuum of visual casualties our species endures. Between blindness and the lesser vision losses caused by environmental factors are such visual disorders as myopia (nearsightedness), astigmatism, and strabismus (squint). The Pasamanicks have found strabismus, due largely to muscle paralysis or imbalance, to be a common symptom of the syndrome of minimal brain damage in infants. They also found two significant facts about strabismus in our urban societies: it is a precursor to, or found to be associated with, mental diseases, and it is clearly derived from the pathological factors of poverty.

All of these visual disorders, from mild myopia and slight astigmatism to serious losses of visual acuity, have fairly predictable affects on such cultural activities as reading and learning, and on such economic activities as making a living. What is true of the proportions of blindness and vision deficiencies and disorders caused by preventable disorders is, in somewhat equal measure, applicable to the continuum of partial to total hearing losses caused by preventable environmental factors, from prenatal infections and malnutrition to the noises produced by our present American levels of motor-vehicle traffic.

The question of whether or not the poor and largely nonwhite poor people of Asia, Africa, Latin America, and Mississippi, who today suffer the world's highest proportions of blindness and lesser visual deficiencies, would —in the absence of malnutrition and pathogens like the microbial agent of trachoma—also prove to be *genetically* more prone to blindness than are the nonpoor and white people of the industrial nations (or for that matter, the affluent people of their own nations) cannot be answered under present circumstances. However, if the poor who suffer most of the blindness and eye disorders on our planet were given, *for at least three generations,* opportunities equal to our own to be free of the known environmental causes of blindness, then this might be considered a legitimate scientific question.

In an America where the nineteenth-century scourge of trachoma among poor Nordic whites has been spontaneously reduced to a twentieth-century statistical asterisk of less than 500 cases per year, I can also be pardoned for suggesting that equality of opportunity to develop their presently underdeveloped genetic capacities for good eyes (and ears) would also render irrelevant such presently pseudobiological questions about the poverty-ridden nonwhite majority of our species.

What is measured by IQ and classroom achievement tests is—in both instances—the end product of information received by eyes and ears and stored and processed by brains equally dependent upon nutrition, sanitation, relative freedom from stresses, and other nongenetic factors involved in their growth and development. The present gaps between the IQ and classroom test scores of the world's poor children and those of the children of the planet's *minority* of nonpoor people (of all races), therefore, are as much the product of the nutritional, infectious, and other biological aspects of the total growth and development environment as they could ever possibly be of the as yet ephemeral gene or genes for "intelligence" and "educability."

Even if such totally hypothetical genes should, to the utter amazement of the world's geneticists, prove to be realities, it would still remain the nature of the total society, the total growth environment, that in the end controls the general range of the physical, mental, and moral development of the average human being. For, starting in the cytoplasm of the cell that surrounds each gene, the nature of the total environment in which any gene expresses itself biologically determines in major measure the range of reactions within which the genetic potentials of the gene are ultimately translated into such biological organs as brains, central nervous systems, and endocrine glands, let alone into whole human beings.

A TALE OF ONE CITY

Many of the gene-enriching elements of the total environment are purely social in nature. These include adequate nutrition, sanitation, housing, education, and freedom from racism, wars, fear of wars, oppression, noise, insecurity, and other stresses of organized societies, and they are all products of the intelligent use of a nation's natural and intellectual resources. As we have just seen, most of the diseases that cause blindness, for example, are prevented by the sheer equality of opportunity to provide running water, soap, and adequate nutrition to one's children. Other diseases, once endemic in this country and still widespread in poorer nations, have been virtually eliminated by the universal utilization of specific immunizations. Some once common killing diseases, on the other hand, have been eliminated by the combination of social and medical resources.

Diphtheria, to cite a famous example, was—like scarlet fever and measles—also reduced in prevalence by hygienic and nutritional advances in the industrial nations prior to the availability of the anti-diphtheria germ vaccine developed on the eve of World War I. Nevertheless, as McKeown and Lowe note, in addition to these purely environmental anti-diphtheria measures, "it seems probable that immunization has had more effect on the control of this disease than of any other, with the exception of smallpox." McKeown and Lowe comment that recent evidence for England and Wales "suggest that the risk of an attack of the disease is about six times greater, and the risk of a fatal attack ten times greater, in those not immunized than in the immunized."[23]

Thus, a society which chooses *not* to eliminate all diseases preventable by available immunizations also elects to continue the risks of life-wrecking contagions at their pre-vaccine levels. This was brought home with some force, recently, in what may be called "A Tale of One City."

The one city is Texarkana, a municipality of 50,000 people bisected by the Texas-Arkansas state line. Between June 1970 and January 1971, the city suffered an epidemic of 633 cases of measles. The children in the Arkansas segment of the city, where more than 95 percent of the children aged one to nine years had previously been vaccinated against measles by their county health department, suffered only 27 of these cases, giving them an attack rate of 4.3 measles cases per 1,000 children. The children on the Texas side of the city, where there had been no matching community vaccination effort, and where only 57 percent of the children had received natural plus vaccine-induced immunity, suffered 539 of the cases in the one-to-nine age group, for an attack rate of 105.9 measles cases per 1,000.

Measles is no minor disease. Its effect upon vision, for example, is particularly acute. According to a standard medical textbook, keratitis, the inflammation of the cornea of the eye, "and corneal ulcerations are often found" in measles. These lesions usually "heal without scarring, but severe bacterial keratitis and blindness sometimes occur."

Measles encephalitis, another serious viral complication of this virus disease, is "encountered in about 1 case in every 1,000." The mortality rate of measles encephalitis—that is, inflammation of the brain—"ranges from 10 to 30 percent in different epidemics; approximately 40 percent of survivors show permanent sequelae of mental retardation, personality changes, and behavioral disorders. In addition, electroencephalographic abnormalities have been noted in 50 percent of children with measles who have no *clinical* signs of encephalitis." This suggests that the subclinical complications of measles encephalitis might be widespread in many "mild" cases of measles.[24]

The protection the immunized children of the Arkansas portion of Texarkana received against measles will, undoubtedly, be reflected in their IQ and classroom test scores when compared to those of the Texarkana children who suffered medically *preventable* attacks of measles.

Texarkana is, of course, only one city of 50,000 people. But, in his report on the tiny city's measles epidemic, Dr. Philip J. Landrigan, of the U.S. Public Health Center for Disease Control in Atlanta, concluded by noting:

> The difficulties in vaccine delivery encountered in Texarkana resemble those seen in many areas of the United States, *particularly among the low income groups.* The 1970 U.S. Immunization Survey indicates that only 40.7% of children aged 1 to 13 years in urban poverty areas have received measles vaccine, while more than 54% of the children in the "non-poverty" metropolitan areas have been vaccinated. The experience in Arkansas indicates, however, that *a community immunization campaign can deliver sufficient vaccine to all segments of a population to prevent an epidemic.* [Italics added.][25]

The clinical results of the measles epidemic in Texarkana leave us with questions whose answers help define the real biological, as against the pseudobiological, problems of our society. For, given the development of vaccines to immunize all infants against measles, what was it, really, that left 6,350 children between one and nine on the Arkansas side of the city immunized against measles, and their ethnically homogeneous 4,835 peers from the Texas side unimmunized?

The vaccine? Yes, but only in part. In these more protected children, immunity against the measles virus was actually the product of the superior human value system of the Arkansas county, just as susceptibility to the measles virus was a product of the inferior value system of the Texas county.

A TALE OF TWO NATIONS

Texarkana proved to be, in microcosm, an excellent model of the two nations in which it sits. These are two nations with a common geography, and a common name—the United States of America.

Disraeli called these nations the nations of the poor and the rich. But Arkansas is poorer than Texas. Perhaps we should, therefore, think of these nations as (1) the nation of the unconcerned local governments and (2) the nation of the concerned local governments. On the other hand, both nations —the medically deprived and the medically protected—are parts of the same federal government. And in America it is the national government whose values are the model for all of its state, county, and municipal components.

By 1971, Dr. Dorothy Horstmann, in an editorial, "Lagging Immunity of Our Children," in the *New England Journal of Medicine,* observed that the most recent of the national immunization surveys made by the U.S. Public Health Service "reveals a steady fall in the percentage of children immunized against poliomyelitis and measles, particularly among the urban poor. The explanation for these unfortunate trends undoubtedly involves many factors, including the expiration in 1968 of the federal Vaccination Assistance Act, which had provided funds to state and local health departments for the purchase of vaccines."

Two years later, in the continuing absence of a national Vaccination Assistance Act, the federal government, according to the Center for Disease Control (CDC), was growing concerned about the "alarming trend of *declining* immunization levels among preschool children against polio, measles, rubella, pertussis (whooping cough), and diphtheria." According to the same CDC statement, "national surveys indicate that approximately 5 million of the nearly 14 million 1 to 4 year old children are unprotected against either polio, measles, rubella, pertussis, or diphtheria. As an example, immunization levels for polio have dropped to a low of 63 percent in 1972 from a high of 88 percent in 1964 [the year before the major escalation of American intervention in Indochina began]."

Without a meaningful federal vaccination program to provide the vaccines needed to protect these five million children, the federal Department of Health, Education, and Welfare, in cooperation with various professional and voluntary health societies, launched a series of annual public relations campaigns in 1973 and 1974. These publicity campaigns, under the slogan "Immunization Action Month," consisted of "calling upon the people of the United States to observe the month with appropriate action." The goal of the crusade, said the Department, was "to motivate parents to check the immunization status of their children through *family physicians* and public health clinics" (italics added).

Motivation is, of course, the cliché that makes Madison Avenue tick. It is not, however, a viable public health alternative to a federal Vaccination Assistance Act that guarantees the provision of both the vaccines and the medical and paramedical people to administer them to the already vaccination-motivated poor states, poor counties, and poor cities where the five million preschool children of perfectly motivated poor parents—who have never been able to afford family physicians—remain unimmunized against preventable killing and maiming diseases.

By the end of 1975, as government mimeographs continued to churn out mountains of new press releases aimed at motivating all parents—the poor, the medically indigent lower middle classes, the rich—to have their children vaccinated during the first years of life, *The New York Times* reported that "Dr. John J. Witte, director of the immunization division for the [U.S. government's] Center for Disease Control in Atlanta said that the proposed Federal budget for the fiscal year 1976 for immunization assistance to states *was only half what was spent last year*"[26] (italics added).

Can we really believe that, in the absence of an epidemiologically adequate national vaccination program, the unprotected children of the poor have as equal an opportunity to develop their physical and intellectual potentials as do the adequately immunized children of the nonpoor? Can we continue to permit a Congress ignorant of the elementary rudiments of social medicine to do nothing about giving every newborn child protection more realistic than a public relations campaign against the medically preventable fatal and deforming infectious diseases of our environment?

These are far from academic questions. Behind them sit growing mountains of biological, chemical, and physical evidence proclaiming that today—as never before in the history of our species—we possess the knowledge, the techniques, and the economic resources that can truly liberate millions of our children from the preventable diseases that keep them from developing to anything near the full potentials present in their genes at birth.

Since the start of the Industrial Revolution, the dawn of our modern era, many once inevitable barriers to proper physical and mental development have been either eliminated or made completely preventable by our sciences, our economic advances, our changing human value systems. These barriers to optimum development of our genetic potentials include:

1. All of the infectious diseases, such as diphtheria, measles, polio, whooping cough, tetanus, and rubella, for which vaccines already exist.

2. All of the environmentally preventable cases of tuberculosis, rheumatic heart disease, hookworm, malaria, trachoma, influenza, pneumonia, infant diarrhea, pellagra, malnutrition, vitamin and mineral deficiencies—and the continuum of physical and mental disorders directly attributable to the by now insane overuse and misuse of automotive transportation and of fossil fuels in industry and housing, and the careless use of the more than 25,000 known toxic chemicals, drugs, and other substances produced by our industries and now proven to cause human diseases ranging from respiratory and endocrine disorders to heart diseases and cancers.

3. All of the preventable brain-damaging and body-crippling complications of fetal life and pregnancy—the "continuum of reproductive casualty," from low birth weight to cephalo-pelvic disproportion.

4. All of the cultural deficiencies—each a societal artifact and not a genetic effect—that result in mental retardation and intellectual underdevelopment, and contribute to the low IQ and classroom test scores of poor children. These societal deficiencies include: the near or total absence of free and effective pre-kindergarten nursery schools for all children, and daycare centers for the children of working mothers; the absence of books, toys, and other aids to self-development in the homes of the poor; and the tragic prevalence of underbudgeted, understaffed, and educationally backward schools, run by undereducated supervisors and teachers, in precisely those communities where better schools *can* make a difference.

These barriers to healthy human development are our real problems—and our historic opportunities—in the husbandry of our actual and priceless genetic resources. In most instances, these are problems that can be solved by knowledge and techniques already well within our nation's scientific, social, and economic resources.

As with the now measurable successes or failures to provide mandatory and free immunizations against the preventable infectious diseases of childhood, these human developmental problems—from infectious and other poverty diseases to cultural anemia—are not genetic but societal in origin. They are therefore also societal, and not genetic, in solution.

To hold otherwise is not only to exacerbate their already great and costly effects and complications. It is also to render them ultimately insoluble.

This is why the current revival of the eugenics cult, with its moth-eaten IQ Big Lies, and its resurrection of the "meritocracy" myths of Francis Galton and Edward M. East, and the more exotic forms of scientific racism based on the allegedly different "time orientations" of the rich and the poor, are the problems of the greater society as a whole—and not merely of the poor blacks, the poor Amerindians, the poor Chicanos, or even the white Anglo-Saxon poor, *who happen to outnumber the nonwhite, non-English-speaking poor by ratios of three to one.*

From lowered industrial productivity to soaring welfare costs—let alone the personal and tax-dollar costs of the crimes of desperation and violence that have always, historically, been committed by poor people deprived of human hope—the ultimate economic and somatic penalties of unrelieved poverty are always paid by the affluent and the nonpoor in proportions greater than any society can long endure.

25 The Viable Scientific Bases for Human Hope: The Hope Scientific Racism Threatens to Kill

> It seems obvious that under our present-day conditions there are still countless infants born with biological constitutions and potentialities for development well within the normal range who will become mentally retarded and noncontributing members of society *unless appropriate intervention occurs.* It is suggested by the findings of this study and others published in the last 20 years that sufficient knowledge is available to design programs of *intervention* to counteract the devastating effects of poverty, sociocultural deprivation, and maternal deprivation.
> [Italics added.]
>
> —HAROLD M. SKEELS, in *Adult Status of Children with Contrasting Early Life Experiences* (1966)

Just as scientific racism is distinctly a product of the Industrial Revolution, two of the most useful constructs of the modern life and behavioral sciences—longitudinal studies and intervention programs—are uniquely the products of twentieth century advances in the complex of legitimate sciences devoted to the study and the betterment of humankind.

Longitudinal studies are investigations in depth, generally in large segments of a population, and chart the interactions of these people and whole complexes of biological, social, physical, and cultural variables. Usually, they cover long periods of time, ranging from months to many years.

An outstanding example of such modern longitudinal investigations is the British National Child Development Study which has been following the physical, mental, cultural, and social development of nearly 16,000 children born in the same week of March 1958 in England, Scotland, and Wales.[1]

Because of the great costs of conducting such studies and storing their findings, the vast majority of the significant longitudinal studies of human development in our times have had to be mounted and supported by national governments.

Intervention experiments and programs, such as the one designed by Professor Rick Heber and his colleagues in the current Milwaukee Project, are also, because of their costs and nature, originated or financially supported by government agencies. The *concept* of societal intervention in human health is not new: the construction of sanitary water and sewage systems, and free mass vaccinations against smallpox, diphtheria, polio, and other medically preventable infectious diseases, were classic instances of planned benign social intervention in public health. The concept of free and compulsory universal education was a classic instance of social intervention in mass mental development.

The philosophy of our modern intervention programs was well stated by Harold M. Skeels in his final report on his intervention in the lives of white Nordic orphans in Iowa:

"It is suggested by the findings of this study and others published in the last 20 years that sufficient knowledge is available to design programs of intervention to counteract the devastating effects of poverty, sociocultural deprivation, and maternal deprivation."[2]

A measure of how profoundly both longitudinal and intervention studies symbolize the current directions of all of our public health sciences became clear to me in 1972, when I attended the hundredth anniversary meeting of the American Public Health Association (APHA) meetings in Atlantic City. By my own tally, fully 130 of the 530 papers presented at this centenary meeting were longitudinal studies of various aspects of public health.

They included long-range studies of the levels of lead in the blood of urban, semi-urban, and inner-city children; the effects of preventable diseases of children on IQ test scores; water-quality management; the early detection of cancers; drug abuse; the long-term effects of vaccination programs; epidemiological determinants of respiratory diseases in children; black-lung disease in coal miners; venereal diseases in Vietnam War veterans; the epidemiologies of cancer, stroke, and low birth weight; social class and race differences in hypertension incidences; the threshold of fetal safety in X rays of pregnant women; the health benefits and complications of abortions; the epidemiology and prevention of Down's syndrome (mongolism); cystic fibrosis; prenatal care and its absence; maternal education as it relates to child health; poverty and chronic diseases in nine urban areas; the accuracy of clinical laboratory testing; and various other serious and, for the most part, soluble health problems.

Since it costs far more to prevent and/or treat than to measure and quantify the health hazards described in longitudinal studies, there were only —by my own far from scientific count—about a dozen progress reports on what could be considered modern intervention projects in human development. Four of these intervention programs dealt with the training of parents to help in the cognitive development of their socially deprived and/or handicapped children. Two dealt with the counseling of families with sickle-cell anemia problems. One dealt with medical and nonmedical interventions in school drug-abuse tragedies. One dealt with an imaginative experiment aimed at eliminating various health hazards from slum tenements. Another dealt with the effects of sending nutritionists to the homes of poor people to teach them how to make the best nutritional use of their meager food dollars.

What struck me quite forcefully at this meeting, as I went through the program and heard some of the presentations, was the fact that the hundredth anniversary of the APHA coincided nearly exactly with the hundredth anniversary of the publication, in 1869, of Galton's *Hereditary Genius* and the birth of the eugenics movement. For a century, the APHA and its European precursors, such as the Health of Towns movement, had been dedicated to studying the causes of, and liberating our entire species from, the handicaps and early deaths of the preventable infectious, contagious, respiratory, deficiency, accidental, and mental diseases and disorders of industrialized nations. During the same century, the eugenics movement and all other denominations of scientific racism had been (and remain) equally dedicated to the proposition that such afflictions as poverty, pellagra, tuberculosis, illiteracy, cholera, high infant-mortality rates, and their invariably concomitant low IQ and achievement test scores were (and continue to be) the end results of hereditary (genetic) defects—and not of the preventable sequelae of the sociobiological or environmental hazards of poverty.

The enormous factual and moral gaps between scientific racism and scientific method are demonstrated in the structure and findings of three of the most important longitudinal studies of our times, and two of our most significant intervention projects. The longitudinal studies include the above-mentioned National Children's Bureau study of 16,000 British children; the lesser known and ongoing National Health Examination Survey initiated by the U.S. Public Health Service under the authority of the National Health Survey Act of 1956; and the ongoing Collaborative Perinatal Study on cerebral palsy, mental retardation, and other neurological and sensory disorders of infancy and childhood, conducted since March 1966 by the Perinatal Research Branch of the National Institute of Neurological Diseases and Stroke (NINDS; formerly National Institute of Neurological Diseases and Blindness, NINDB) in collaboration with twelve university medical centers and institutions.

The National Health Survey obtained "data through direct examinations, tests, and measurements of samples of the U.S. population," of all ages and social classes in every geographical region.

TOWARD THE PREVENTION OF CONGENITAL BIRTH DEFECTS

The perinatal study of the NINDS has, since 1961, followed 55,908 pregnant mothers and their children with the stated purpose of increasing "the likelihood of the birth of healthy babies free from disease and impairment, and capable of optimal physical and intellectual development. The achievement of this goal depends upon the enlightened and widespread application of measures to prevent perinatal mortality and the *continuum of reproductive wastage* which includes mental retardation, congenital malformation, cerebral palsy, and handicapping neurosensory defect" (italics added).[3]

As can be seen from the above, the NINDS perinatal study is the direct scientific and intellectual offspring of the decades of painstaking studies of what they termed the "continuum of reproductive casualty" by Drs. Benjamin Pasamanick and Hilda Knobloch and their various associates since 1945.[4] In fact, one of their long-time collaborators, Dr. Abraham M. Lilienfeld, chairman of the Department of Epidemiology at Johns Hopkins, is one of the senior advisers of the study and a member of its Perinatal Research Committee.

In *The Women and Their Pregnancies,* the 1972 report of the NINDS study, the authors declared: "The heavy losses in productivity and the extent of human suffering associated with the high rates of perinatal mortality are not the only cause for concern. Of far more importance to the individual, the family, and the community is the 'continuum of fetal insult' manifested by congenital malformation, cerebral palsy, mental retardation, deafness, blindness, and other neurosensory defects. . . . Estimates are that approximately *20 million individuals in the United States have handicaps or defects which fall within this general category"* (italics added).[5]

One quarter of these handicapped Americans suffered from the mental retardation and cerebral damages "which may represent in part" the continuum of preventable reproductive casualty most prevalent among the poorest social classes. To these 5,000,000 mentally retarded and 550,000 cerebral palsy victims, the NINDS study added 1,500,000 people with epilepsy, 345,000 blind, and 760,000 deaf.

Since all of these British and American longitudinal studies of human development were scientific inquiries and not eugenic rituals, they looked to more than IQ and classroom achievement scores, clinical diagnoses, and race as the *sole* units of human measurement. In each of these longitudinal studies, the socioeconomic and other environmental factors associated with these indices were studied with equal objectivity. None of these major studies set out to prove or disprove any "hereditarian" or "environmental" or even "orthodox egalitarian" dogmas. They were and remain dedicated to only one basic purpose, and that is the objective scientific quest for some viable data about why we are as smart or stupid, as healthy or as frail, as long-lived or as prematurely dead as we are.

Consider, for example, the NINDS perinatal study of the continuum of reproductive casualty known to play a major role in the etiology of such

A Comparison of Family Incomes in the Study Population with State Populations* by Race and Institution

Institution[6]	Percent under $2,000				Percent $8,000 or Over			
	White		*Negro*		*White*		*Negro*	
	Study	State	Study	State	Study	State	Study	State
BO	4.8	4.0	3.5	19.9	10.5	33.7	18.2	12.5
BU	1.5	3.9	4.0	22.1	48.0	30.5	40.0	9.7
CH	—	5.5	35.6	29.6	—	28.8	0.1	4.6
CO	2.2	4.4	5.2	15.9	5.3	35.2	8.1	11.6
JH	10.1	3.8	13.4	18.7	5.0	31.9	6.0	10.4
VA	15.4	3.3	32.8	25.1	2.7	33.2	0.9	6.1
MN	9.4	3.1	8.3	16.5	6.8	34.7	16.7	14.9
NY	6.6	4.4	13.6	15.9	1.2	35.2	1.1	11.6
OR	25.8	(5.1)†	31.7	(5.1)†	2.2	(30.8)†	0.7	(30.8)†
PA	16.8	3.6	20.7	18.6	0.8	33.6	1.0	10.4
PR	15.4	(5.9)†	24.5	(5.9)†	3.6	(21.7)†	1.8	(21.7)†
TN	42.8	5.0	30.4	33.0	4.8	26.5	0.6	3.0

* Source: United States census of 1960.
† Percents in parentheses are for white and Negro combined.

From *The Women and Their Pregnancies*, 1972.

As these family income data show, the family incomes of the white women in the study were lower than the average white family incomes in their respective states. The Negro mothers in the study, on the other hand, had higher family incomes than the average black families in their states. Even in this study group, however, the family incomes of the white mothers were higher than those of the black women. The average white mothers in the study population had completed more years of school than had the average black mothers. However, the black mothers in the study population were far better educated than the average black women in their states.

neurological entities as the 5,000,000 cases of mental retardation and the 1,500,000 cases of epilepsy in this nation today. This NINDS study deals not only with the known *medical* problems of pregnancy and their pathological complications, but also with the environmental and social conditions of our society—from the number of cigarettes smoked by pregnant women during their nine months of pregnancy to the race, family income, and education of the 46 percent of the mothers in the study who are white, the 46.2 percent who are black, the 6.3 percent who are Puerto Rican, and the 1 percent who are Oriental. Among these mothers, as among the general population of the nation, the black mothers in every one of the collaborating hospitals from Boston and New York to Baltimore and Memphis had far lower family incomes than the white women in the study.

The outcomes of each pregnancy were predicted by these socioeconomic indexes with terrifying accuracy:

Birthweight and Perinatal Mortality Associated with Conditions and Complications of Pregnancy

Outcomes	Outcomes by Race					
	White			*Negro*		
	All Cases	Number	Rate	All Cases	Number	Rate
Births	19,048			20,167		
Perinatal deaths		668	35.07		845	41.90
Stillbirths		415	21.79		457	22.66
Fresh stillbirths		200	10.50		246	12.20
Livebirths	18,633			19,710		
Neonatal deaths		253	13.58		388	19.69
Livebirths with known birthweight	18,481			19,504		
Birthweight under 2,501 gm.		**1,319**	**71.37**		**2,617**	**134.18**
Mean birthweight		3,272			3,039	
One year exams	14,662			17,123		
Neurologically abnormal at 1 year		253	17.26		274	16.00

What these clinical indices show is that, in the outcomes of pregnancies as in the outcomes of IQ and academic achievement tests, the entire spectrum of conditions of family and community life constitutes major influences in the physical and mental health of human beings. Consider, for example, the postnatal implications of the raw numbers in the table showing that the percentage of birth weights under the danger threshold of 2,501 grams is nearly twice as high among black mothers as among whites. This scientific information not only helps explain *and predict* the higher perinatal and postnatal death rates of black infants in our culture; it also helps predict that among the survivors of these low birth weight groups the rates of nonlethal but handicapping brain, nervous system, and other organic defects associated with the complications of pregnancy and birth must inevitably be proportionately much higher among black infants than among white infants. And these congenital barriers to learning will all be products of clearly nongenetic and purely socioeconomic or environmental causes.

FAMILY INCOME, IQ TEST SCORES, AND RACE

The National Health Survey (NHS), the massive longitudinal study of the total health of the American people, quite properly considers the mental aspects of human health to be as important as the physiological. Since 1960, therefore, under the direction of Dr. Dale B. Harris, co-author of the revised Goodenough-Harris Drawing Test, the NHS has been measuring what the study designates as the "intellectual maturity of children" with this standard mental test.[7] Two other standardized psychological tests have also been used in this segment of the huge health audit, and their results combined with those of the Goodenough-Harris tests.

As Dr. Harris observed in the second and largest report of this portion of the National Health Survey,[8] an assessment of school achievement (classroom test scores) "was also included because of its relation to both the intellectual status and the social and emotional adjustment of the child."

The psychological measurements of the health of the nation's children were, therefore, not only *not* based on any *single* type of IQ test, but even the matter of intelligence itself was projected in a modern manner that was, scientifically, more than a half century ahead of the archaic notions being broadcast by that noisy minority of psychologists to which individuals such as Jensen and Herrnstein belong. As Dr. Harris and Jean Roberts, of the Division of Health Examination Statistics, wrote on page 19 of their 1972 progress report:

> . . . psychological theory with respect to the nature of intelligence has been inadequately developed. So-called intelligence tests have been developed empirically, based largely on measurements of available populations under various conditions and not on theory derived from experimentally generated data. Thus theoretical discussions are based largely on inference from observations made under uncontrolled conditions. It is entirely possible that intelligence is so complex a phenomenon and a product of so many variables that a completely definitive answer, suitable to a positivistic science, will continue to be elusive.[9]

Not only, in short, are we still in the dark about the very nature of intelligence per se. There are, as yet, no tests that can measure this *still undefined* phenomenon or complex of behavioral traits as positively, as definitively as a thermometer measures body or room temperatures, or as a speedometer measures the speed of a moving motor vehicle. Moreover, they reported:

> It is quite well recognized by psychologists that "intelligence" tests do *not* measure a pure, innate [genetic] potential. While some psychologists insist that biologically derived components undoubtedly inhere in abilities, as in many human characteristics, they understand that we measure only manifested, *developed* abilities or capacities. *Therefore, [intelligence] test performance inevitably reflects differential educational experiences and different familiarity with "test taking."* [Italics added.][10]

The tables in which the NHS data were summarized reflect the insights of *modern* psychology about the relationships of socioeconomic as well as race and sex variables to test results. As we can see in their table, the Goodenough-Harris standard test scores of white children are as reflective of family incomes and social status as are those of black children:

Average standard scores for white and Negro children on the Goodenough-Harris Drawing Test, by annual family income, age, and sex, with standard errors for total averages: United States, 1963–65

Age and sex	Annual family income					
	Less than $3,000	$3,000–$4,999	$5,000–$6,999	$7,000–$9,999	$10,000–$14,999	$15,000 or more
	White standard score					
Both sexes 6–11 years	94.4	98.8	100.8	102.2	103.8	102.4
Boys						
6–11 years	95.4	99.5	101.7	102.1	103.5	103.2
6 years	95.1	98.5	103.5	100.9	102.8	101.8
7 years	97.1	100.9	102.0	99.9	106.1	103.1
8 years	96.5	100.6	98.0	102.9	103.6	101.4
9 years	93.6	100.7	102.6	102.4	102.4	103.7
10 years	93.8	99.5	102.9	102.8	102.4	104.5
11 years	96.4	97.3	101.4	103.4	103.7	105.3
Girls						
6–11 years	93.5	98.1	99.7	102.2	104.1	101.3
6 years	92.4	98.4	99.5	103.5	103.9	97.5
7 years	92.1	96.8	99.1	102.1	104.7	106.1
8 years	91.3	99.8	100.3	101.0	104.0	98.0
9 years	94.6	99.4	99.2	101.2	105.1	103.0
10 years	94.7	98.4	101.1	101.7	102.9	101.5
11 years	95.2	95.9	99.3	104.0	104.0	103.3
	Standard error					
Both sexes 6–11 years	1.15	0.81	0.49	0.53	0.64	1.38
Boys 6–11 years	1.24	1.10	0.60	0.77	1.10	1.51
Girls 6–11 years	1.27	0.76	0.56	0.53	0.95	1.73
	Negro standard score					
Both sexes 6–11 years	90.4	96.2	99.4	101.4	98.7	—
Boys						
6–11 years	90.6	96.3	99.3	101.2	95.0	—
6 years	86.9	96.1	100.5	100.8	96.6	—
7 years	89.1	90.2	102.5	110.0	—	—
8 years	93.7	92.5	96.0	100.1	78.0	—
9 years	87.4	100.9	95.6	98.5	—	—
10 years	93.2	99.6	97.2	88.8	101.3	—
11 years	93.0	98.2	104.5	106.3	—	—
Girls						
6–11 years	90.2	96.0	99.5	101.6	103.5	—
6 years	89.4	99.6	106.5	101.3	—	—
7 years	90.9	95.3	103.9	99.9	105.0	—
8 years	91.7	96.9	97.1	110.4	100.4	—
9 years	90.7	97.4	100.8	98.7	—	—
10 years	89.4	93.8	92.9	100.0	—	—
11 years	89.9	90.8	99.2	103.0	—	—
	Standard error					
Both sexes 6–11 years	1.57	0.90	1.97	1.65	8.45	—
Boys 6–11 years	2.69	1.50	2.15	3.09	11.85	—
Girls 6–11 years	1.44	0.96	2.47	3.02	32.98	—

From Harris and Roberts, 1972.

Therefore, in presenting their study, *Intellectual Maturity of Children,* the NHS authors added the subtitle *Demographic and Socioeconomic Factors.* And they concluded, in words that precisely paraphrase the conclusions of Mayeske and Coleman:

> On this test [Goodenough-Harris], while white children performed better than Negro children, the racial differential is lower than on the other two standardized psychological instruments used in the [NHS] survey—the Wechsler Intelligence Scale for Children and the Wide Range Achievement Test—*and is reduced to a negligible amount when the effects of differences in parents' education and family income are controlled* [italics added].[11]

In short, the findings of each of the major American studies of human biological and mental development supported the basic conclusion arrived at as far back as 1892 by Franz Boas. To wit, that when the developmental effects of parents' income, education, and social class are measured and weighted along with the medical and mental test results, the differences in the intellectual and physiological development of children are reduced to nearly or exactly zero.

COMPENSATORY INTERVENTION: THE WEAPON OF CONCERNED SCIENTISTS

This basic truth was dramatized in both of the major American intervention experiments I have selected as equally significant examples of modern (as contrasted to Galtonian) uses of the data and the methods of science in the cause of human betterment.

The first of these intervention experiments, the Infant Education Research Project, was started in 1965, by Dr. Earl S. Schaefer, then at the National Institute of Mental Health, and Msgr. Paul H. Furfey, research associate in the Bureau of Social Research, Catholic University of America, in Washington.

In this project, two groups of Negro boys from poverty-level homes in Washington, D.C., in which the mother's formal education was less than twelve years and/or she had a work history of holding only menial or semiskilled jobs, were followed for a number of years, starting at the time they were 15 months old. In the intervention or experimental group, for the next 21 months the children were taught in their own homes for one hour a day, five days a week, by specially trained college-graduate tutors. Tutoring consisted of playing games and reading or telling stories, activities carefully designed to develop verbal and cognitive skills.

These tutoring sessions made lavish use of drawing materials, books, toys, and educational games in the home—as well as trips to firehouses, zoos, and other points of fun and interest. The tutors also took the experimental group of poor black children to the study's center for birthday parties, as well as for periodic intelligence tests of various categories.

The children in the control group received no tutoring, but did take the same intelligence tests. After 21 months, the results showed how useful this early intervention had been, as summarized in the tables below:

Mean IQ Scores at Intervals during Infant Education Project		
Age, in Months	Experimental	Control
14	105	108
21	97	90
27	101	90
36	106	89

Mean Scores on Various Tests at 36 Months, Infant Education Project		
Test	Experimental (n, 28)	Control (n, 30)
Peabody	87.11	76.23
Johns Hopkins	11.61	6.60
Aaronson-Schaefer	13.43	12.40

When the boys were three years old, the tutoring of the experimental group was ended. From that time on, both groups developed without any intervention other than periodic intelligence testing. Within one year, the average IQ test scores of the previously tutored children dropped by over 6 points, from a mean of 106 at the age of 36 months to 99.6 at the age of 48 months. Over the following three years, these were the results:

Mean IQ Scores at Intervals Following End of Tutoring		
Age, in months	Experimental (Tutored between ages 14 to 36 months)	Control (Untutored between ages 14 to 36 months)
48	99.6	90.3
60	97.7	92.3
72	99.8	96.5

To Dr. Schaefer and his NIMH associate, Dr. May Aaronson, this post-intervention drop in the IQ test score of the experimental group of under-privileged children did not mean, as it had in related Head Start compensatory education settings to Jensen, that "compensatory education has been tried and it apparently has failed." Investigators such as Schaefer and his collaborators were far too experienced to reach any such simplistic and hasty conclusions. They were too well aware of the well-known fact that the earlier findings of research studies conducted by many competent American, British, Swiss, and other psychologists had shown that:

> . . . low-socioeconomic-status infants do not have low mental test scores prior to 15 months of age. At 21 months the control [i.e., the untutored] group had a mean IQ of 90, which remained near that level through four years of age. This finding, that without intervention a rela-

tively low but stable level of mental-test scores was established by 21 months of age, suggests that *intellectual stimulation should begin prior to that time.* The tutored [i.e., experimental] group dropped slightly below norms by 21 months—mean IQ = 97—*but with continued tutoring climbed to a mean Stanford-Binet IQ of 106 by 36 months of age,* a conclusion that was supported by the tutors' reports that some of the infants showed signs of early deprivation at the time tutoring began. [Italics added.]

The fact that the IQ test scores of the previously tutored children dropped after the educational intervention in their development ended, did not mean, to Schaefer and his associates, that the Negro children were genetically less intelligent than white children. It suggested, rather, that their tutoring intervention had started too late and had been terminated much too soon. Therefore, the findings from this experiment, as well as from earlier studies by Clarke (1960), Klineberg (1935), and Lee (1957),

> . . . suggest a need to shift our emphasis from the need for early education to the need for *early and continued education in the family as well as in the school.* . . . Education is a process that begins at birth and continues until death. . . . *The family is the major educational institution in our society.* Statistical data indicate that schools do not change the rate of intellectual functioning that is developed by the family, despite the fact that *disadvantaged children have higher intellectual potential than is being developed by the current environments. Support for the child's education in the family should supplement the current emphasis on the school.* [Italics added.]

Far from calling (as did Terman in 1916, Cannon in 1922, and Jensen in 1969) for the establishment of schools designed to train black and other poor children with the predictable low IQ test scores of poverty to become humble hewers of wood and drawers of water, Schaefer and Aaronson concluded that "the short-term and long-term data of the project suggest the need for a *comprehensive system of education that goes beyond the current emphasis upon academic education* to include pre-academic, para-academic, and post-academic education in the family and community" (italics added).

HOW TO RAISE THE IQ TEST SCORES OF DEPRIVED CHILDREN

In 1961, when Dr. Rick Heber, one of the nation's ranking experts on mental retardation, launched what is today known as the Milwaukee Project, he was not thinking in terms of organizing a "program of intervention to counteract the devastating effects of poverty, sociocultural deprivation, and maternal deprivation." Heber, a professor of education and child psychology at the University of Wisconsin, saw the project as "a longitudinal study to determine whether 'cultural-familial' or 'socioeconomic' mental retardation can be prevented through a program of family intervention beginning in early infancy."[12]

The "cultural-familial" mentally retarded are those 80 percent of

America's six million retarded people in whom "there is no identifiable pathology of the central nervous system" and/or other apparent physical defects caused by prenatal or early infancy diseases and disorders. In practice, these physically healthy people are individuals whose IQ test scores average 75 and below. While, Heber observed, "the basic cause of this type of retardation is unknown, factors of inheritance have been implicated."

Before World War II, Heber noted, the quasi-genetic concepts of eugenics "were used to support the view that 'cultural-familial' mental retardation was a direct function of hereditary determination." However, by 1960, "the complex interactions involved in the expression of genetically-related behavioral characteristics [were] becoming more obvious." In this scientifically more sophisticated atmosphere, Heber obtained a federal grant to help set up and run the study formally entitled "Rehabilitation of Families at Risk for Mental Retardation."

The 1960 census tract data led Heber and his associates to a Milwaukee area in the lowest category of both income and education and "in the highest category in terms of population density per living unit, percent housing rated as dilapidated, and unemployment." Most of the residents of this neighborhood were blacks, the majority of them migrants from even more acute southern rural and urban poverty. Out of this population, the Milwaukee Project investigators induced the parents of forty as yet unborn children to volunteer their cooperation. The average IQ test score of the future mothers of these forty families was under 75 in the Wechsler Adult Intelligence Scale (WAIS).

Before their children were born, the mothers were divided by random selection into two groups—the experimental and the control. The experimental mothers and their infants were given courses of rehabilitation and developmental intervention. The lives of the mothers and children of the control group were not altered in any way, except that they were observed and tested at various times.

With the cooperation of two large Milwaukee nursing homes, the mothers in the experimental group were given job training, and academic education in the reading and arithmetic skills required for well-paying jobs as nursing assistants, dietary aides, hospital housekeepers, and laundry workers. During their pre-vocational training, the experimental mothers were also given classes in child care, home economics, interpersonal relations, and community-oriented social studies. "As the mothers become successfully employed, the maternal program shifted to increased emphasis on training in general care of family and home, budgeting, nutrition and food preparation, family hygiene and mother's role in child growth and development."

After the forty children of both groups were born, the twenty children of the control group led the same lives led by all other children in their slum neighborhood. The direct intervention program for the twenty children of the experimental group, however, started when they were only a few weeks old. At that time, the paraprofessional workers who were to serve as their personal "teachers" for the next six years began to visit them at home, where they helped bathe and feed them, and gained their confidence.

Starting at the age of three months, the twenty individual "teachers" called for the infants every morning and took them to the Infant Education Center set up in a nearby school facility. From then on, for the next six years, the infant education program was conducted from nine thirty to four every day, five days of each week, twelve months of each year.

At the center, the children not only received at least one nourishing meal and a snack every day, but also had naps in comfortable cribs and youth beds, and access to literally hundreds of a wide variety of educational toys, books, drawing and writing materials, musical instruments, sports equipment, and other developmental tools. The project's infant educational program was focused "heavily on language and cognitive skills and on maintaining a positive learning environment for the children."

A typical daily schedule for these preschool slum children, conducted in various special rooms devoted to learning and play, is described in the table, on page 34 of the project's 1972 report, that appears opposite.

All but one of the "teachers" at the center were paraprofessionals. They were chosen, according to Heber, because they were "language facile, affectionate people who had had some experience with infants or young children. The majority of the 'teachers' resided in the same general neighborhood as the children, thus sharing a similar cultural milieu. The 'teachers' ' educational experiences ranged from 10th grade to a master's degree."

To the children, the "teachers" performed the role that is provided by educated nonworking mothers in affluent families—that of a trusted, affectionate adult who is the source of games, stories, toys, books, and essentially educational growth experiences. By education and experience, these "teachers" were better equipped than the biological mothers of the children to provide the cultural and developmental factors essential to achieving high IQ test scores.

Intervention also included, as it had in the Skeels population of white Nordic Iowa orphans before World War II, professional supervision of their physical development. Medical examinations of the Milwaukee children were conducted by the staffs of the Children's Hospital and the Marquette University Dental School. As the project's progress report notes:

"The nutrition and medical services provided the Experimental children were incomparably different than those for Controls. Experimental children were given lunch at the Infant Education Center and were referred for medical and dental care when problems were identified by staff at the Infant Education Center. In addition, the Experimental mothers were instructed in nutrition and general health care. As a consequence, any differences in the developmental status of the two groups could conceivably be a function of this factor rather than the modification of the experiential environment."

Periodically, over the next six years, the children of the experimental and the control groups were given identical IQ and other psychological tests. The results, summed up in the graph below speak for themselves:

In the experimental group, the children of poor slum mothers with IQ test scores of 75 or below had a mean IQ test score of 126. Some of them achieved IQ test scores of 140, and all of them scored above 100.

The children in the control group racked up scores ranging from 65

A Day of Benign Intervention in the Lives of Deprived Children

Time	Language	Reading	Math/Problem solving	Free flowing	Choice of special activity
9:30–10:00	◀	●	●	○	
10:00–10:30	◀	●	●	○	
10:30–10:45				○	
10:45–11:15	◀	●	●	○	
11:15–11:45	◀	●	●	○	11:30–12:00 "Sesame Street"
11:45–12:00	LUNCH			○	
12:00–12:30	LUNCH				
12:30–1:45	NAP				
1:45–2:00	SNACK				
2:00–2:30	●	◀	◀	○	Music
2:30–3:00	●	◀	◀	○	Music Story reading
3:00–3:30	●	◀	◀	○	Story reading
3:30–4:00				○	Gymnasium/Outside

The symbols indicate that at a particular time children are:

● = in a small group learning area

○ = in a free flowing room

◀ = this subject area teacher is either helping children individually in his/her area or is available in the free flowing room to extend activities in his/her subject

For five days a week during their first six years of life, starting when they were around three months old, the twenty children of the experimental group spent their daylight hours in the learning center of the Milwaukee Project. Here these children ate, slept, played, and enjoyed the happy preschool learning experiences common in the homes of educated and nurturant American middle-class and upper-class parents. The daily activity schedule shown here summarizes the highlights of a typical day in the lives of these very poor inner-city Milwaukee black children.

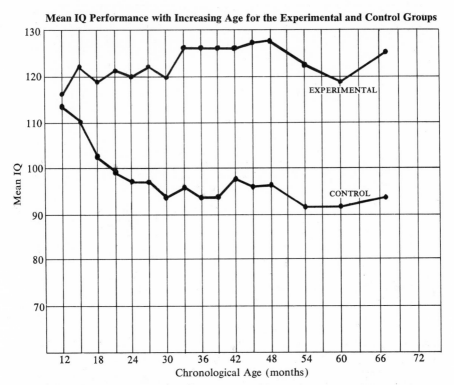

The 22-point difference between the mean IQ test score of 126 in the children in the experimental group, and the mean IQ test score of 94 achieved by the equally matched children of the control group, underscores the value of early and sustained intervention by a concerned community into the lives of deprived children.

From Heber et al., *Rehabilitation of Families at Risk for Mental Retardation: December 1972 Progress Report*, p. 49a.

to 105, with a mean of 94. Even in this group, greater familiarity with IQ testing probably resulted in scores higher than those of their less test-experienced mothers.

The Milwaukee Project is probably one of the most important studies in social biology and social psychology to be conducted since Goldberger, in 1914, showed pellagra to be a nonhereditary vitamin deficiency disease of poverty, and Skeels and his Iowa collaborators, a generation later, showed that merely taking orphaned children from a sterile custodial institution and placing them with affectionate and nurturant families could raise their IQ test scores over those of their parents by similar proportions.

However, where Skeels had worked with orphaned or unwanted and abandoned children, the Heber team worked with *wanted* children loved by their mothers. Far from taking the children from their own families, Heber left them in the same environment in which their mothers had achieved IQ test scores at or below the mental retardation levels.

What the Milwaukee Project did was to intervene in this environment of the poverty that breeds low hematocrits and low IQ test scores and to provide some—but far from all—of the major requirements for the optimal expression of the genetic endowment of any newborn child on this planet. The factors added to the inner-city slum environment of the twenty randomly selected experimental children included:

(1) At least one (and often two) nourishing meals a day. (2) Preventive and therapeutic medical and dental care from the first weeks of life to age six. (3) One-to-one relationships for five days of every week of the year with intelligent, affectionate, moderately to superbly educated adults, carried on in (4) a vermin-free, well-lighted, well-heated and well-cooled, attractive play-learning center, rich in all of the animate and inanimate play and learning resources normally available only in the homes and neighborhoods of affluent, educated, concerned, and nurturant families.

Therefore, more than the IQ-testing potentials in the genes of the twenty inner-city experimental children and their twenty peer controls was tested in the Milwaukee Project. What was also being tested was the value system of the greater society as it affects the growth and development needs of millions of children daily.

What Skeels had inadvertently stumbled upon while carrying out his official duties as a state of Iowa psychologist involved in the welfare of the state's Depression-era orphan wards, Heber now tested and proved in an elegantly designed control study.

Whereas Skeels had shown that the IQ test scores of children of parents with low IQ test scores could be raised to levels averaging 28 points above those of their mothers by merely placing them for adoption in affectionate families, Heber demonstrated that the IQ test scores of the children of poor inner-city slum parents, with IQ test scores at the mental retardation level, could average 36 points above those of their mothers by merely adding a few unsophisticated improvements to the inner-city environment in which they had been born and still lived.

Skeels showed that the ability to achieve higher IQ test scores than one's mother was most probably not a genetically transmissible trait.

Heber, however, showed that to even suggest, after 1972, that the array of cultural and behavioral traits measured by IQ testing were largely or wholly determined by genetic transmission of the yet-to-be-found and very hypothetical genes for these multiple traits was, scientifically, no longer permissible.

16,000 BRITISH CHILDREN FROM BIRTH TO SEVEN

What Heber and his co-workers found in forty very poor urban black children was more than amply validated by the findings made during the first seven years of the lives of nearly 16,000 children who were born to English, Scottish, and Welsh families during one week in March 1958—and who have been observed since then by the pediatricians, psychologists, educators, and other participants in the continuing National Child Development Study. This

British longitudinal study, like the Coleman inquiry, examined most of the standard social conditions in the family of each child. However, the British scientists went an important step further: on the sound principle that even before a child becomes a social being he or she is, first and foremost, a biological entity, they also gathered and analyzed some extremely vital medical and dental data. Brain damage at birth, as the English orthopedic surgeon Little demonstrated in 1853, can be as devastating to a child's mental development (and IQ test scores) as is the total lack of intellectual and cultural stimulation during the five formative preschool years.

The principal authors of the latest volume of the findings of this British study,[13] the educational psychologist Ronald Davie, the pediatrician Neville Butler, and the statistician Harvey Goldstein—calling upon the words of Professor R. M. Titmuss—observed:

"We delude ourselves if we think we can equalize the social distribution of life chances by expanding educational opportunities while millions of children live in slums without baths, decent lavatories, leisure facilities, room to explore and space to dream."[14]

The nearly 16,000 children of the longitudinal study cohort were born long after Great Britain's socially advanced family assistance (cash subsidies for each child) programs guaranteed them against the ravages of prenatal and postnatal malnutrition that is the lot of millions of their less fortunate white and nonwhite contemporaries in those few industrial countries— including the United States—which do not have such universal family assistance programs. But man does not live by bread alone. His children still need "room to explore and space to dream," let alone access to the three "basic" amenities referred to in the graphs below—hot water, bathroom, and indoor lavatory. And, as the British investigators learned and documented, such amenities—or the lack of them—are indeed reflected in the quality of their academic achievements.

There is little in these data that would surprise any *modern* life or behavioral scientist. As the British authors noted, "it is not difficult to think of plausible reasons why there should be a relationship between overcrowding (at home) and poor performance at school." They found, in fact, that "the effect of overcrowding is equivalent to two or three months retardation in reading age. . . . The effect of absence—or shared use [with other families] —of all basic amenities is equivalent to about 9 months retardation in reading age."

What was surprising, in an England where local well-baby clinics are operated free of charge by the government to examine, advise, and immunize children and older people against preventable bacterial and virus diseases, was the discovery that the lower the social class, the greater were the proportions of children who were "deprived of the protection that is theirs by right."

Even more surprising, in a Britain where dental care is also free to every individual under the National Health Service, was the fact that only "12 percent of children had healthy dentition and 10 percent had poor dental health." In this specific public health index, while the rich had relatively

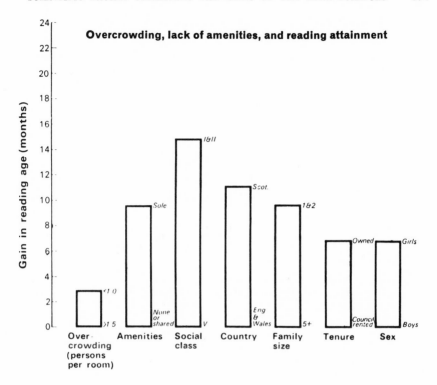

Overcrowding, lack of amenities, and reading attainment

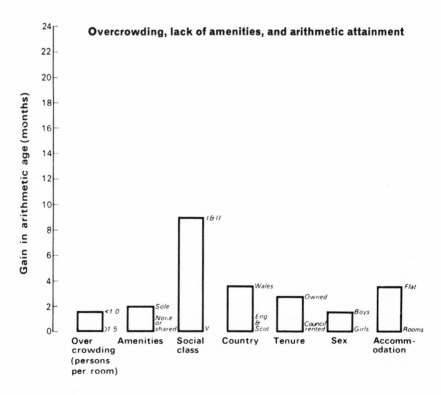

Overcrowding, lack of amenities, and arithmetic attainment

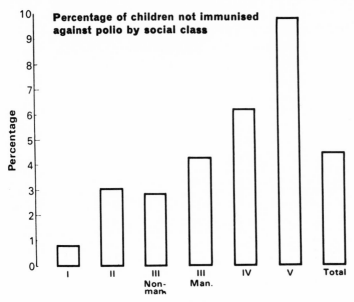

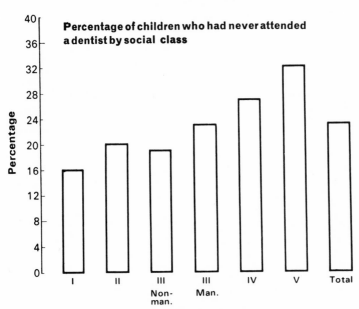

KEY

I	Higher professional
II	Other professional and technical
III (non-manual)	Other non-manual occupations
III (manual)	Skilled manual
IV	Semi-skilled manual
V	Unskilled manual

healthier teeth than the poor, the divergences were not as great as they were in levels of immunizations.

Since they are scientists and not eugenicists, the authors of the British study did not conclude, from these health data, that the ethnic English, Scottish, and Welsh parents were genetically incapable of taking their children to the dentist. But they did have a few nonracist hypotheses to explain this nearly universal failure of the parents of the United Kingdom to give their children adequate dental care—and they did make a most interesting and practical suggestion about what could be done to encourage the parents of the realm to utilize free dental facilities, as we shall see presently.

Consider, first, two paragraphs from their concluding chapter. Notice, in the first paragraph, how their conclusions directly parallel those of their American colleagues Schaefer, Furfey, and Aaronson as well as those of Coleman, Mayeske, and their U.S. Office of Education colleagues:

> One of the most striking features which emerges from the results we have presented is the very marked differences between children from different [social] circumstances, which are already apparent by the age of seven. For example, the estimated gap in terms of reading performance of the most and the least [socially] advantaged children . . . was over four years. Furthermore, *the most potent factors were seen to be located in the home environment.* The most obvious implications to be drawn from this are, first, that *equality of educational opportunity cannot be achieved solely by improving our educational institutions.* Secondly, *"for the culturally under-privileged child, enriching or compensatory education needs to be provided during the pre-school years and continued during the school years"* [here they were quoting from *11,000 Seven Year Olds,* 1969, the study's first report]. Wiseman (1970) too stresses that *"environmental deprivation bears most heavily on the earliest years of childhood; before ever the child reaches the infant school he is already handicapped in the vital area of language development.* If nothing else is done to remedy this, he arrives at school unable to take advantage of the opportunities offered,˙the deficit becomes cumulative, and he is permanently lodged in the ranks of the slow learners." [Italics added.][15]

Instead of excoriating the genes or the ethics or even the time orientation of the lowest-social-class parents "who have most need of statutory [health] services but tend to use these services least," so that "at seven their children are poorly adjusted at school, with bad teeth and showing signs of delayed development in . . . bladder control, speech and physical coordination," the authors wrote:

> There is no evidence to suggest that parents in these social groups have any less concern for their children's welfare. The answer must surely be that either the statutory services are not in the general area seen by these parents as being relevant to their children's welfare; or else there are barriers, physical and psychological, to their attending. Certainly, the mothers most often go out to work; and they have more children in the family to care for. The [health] services may not always be well sited for them.[16]

A TALE OF TWO HUMAN VALUE SYSTEMS

This report, *From Birth to Seven,* was delivered to the British government, and released to the general public, in June 1972. On December 9 of the same year, Anthony Lewis reported from London in *The New York Times* that, in Parliament:

"Edward Heath's Conservative Government proposed, to much applause, a program to provide public [free] nursery schools by 1981 for all the 3- and 4-year olds whose families want nursery school education for them."

Their government, in short, had read their report, and sensibly accepted the advice of the concerned scientists responsible for this longitudinal (and still continuing) study of child development.

There is no doubt in the minds of any of these pediatricians and child psychologists that the new free preschool nursery centers in the United Kingdom will, if need be, also include "toddler clinics" for the free health care and guidance of all the children and their parents. Or that the public nursery schools will be merely the first step in the United Kingdom on the road to the kind of "comprehensive system of education that goes beyond the current emphasis upon academic education" that Schaefer and Aaronson feel is so urgently needed in the United States.

The reaction of our own government to the recommendations of our pediatricians and child psychiatrists and psychologists in our times has been somewhat different from the response of government in England, Denmark, France, and other industrial nations.

Our government research data has, however, always been as lucid as those of England's. A half century ago, for example, the *Journal of the American Medical Association (JAMA)* reported that the U.S. Department of Labor had just completed a study in Baltimore that showed the relationships between the industrial employment of mothers and infant mortality. Working mothers had higher rates of stillbirths, and "the infant mortality rate during the first month of life was 77.3 for each thousand among babies of mothers employed, or nearly twice the rate, 39.9, among the babies of mothers not employed."[17]

Nor was this the start of the interest of our federal health, welfare, and labor agencies in the sociobiology of infant mortality, and the medical and *social* ways of reducing it. *JAMA*'s short notice ended with these still highly pertinent sentences:

> Previous reports of the [U.S.] Children's Bureau of infant mortality have shown a definite connection between income and infant mortality. *As the income increases, the infant death rate decreases.* This Baltimore study, based on a larger group than the previous studies, permits a closer analysis of the single factor of employment of mothers. The importance of this factor may be realized from the fact that *even within the same income groups the mortality rate is higher for babies whose mothers are employed outside the home.* [Italics added.]

For most of the years of this century, public health workers in various national administrations have managed to introduce legislative proposals to extend the vast benefits of the Industrial Revolution and the medical advances of two centuries to working mothers, poor mothers, and their children. Many—but, alas, far from all—of these medically logical proposals have been enacted into law and made the bases of life-enhancing social programs. If our infant-mortality rates have been reduced from 27.7 per 1,000 live births in 1920 to 16.5 in 1974, these efforts to extend such advances as immunizations to many (alas, not all) children in general, and to outlaw child labor (at least in mines and factories, if not in the cotton fields), must be credited with a major role in achieving such improvements in the total quality of American life; a healthy child has made a head start toward becoming a healthy mother.

One of the most far-reaching of these continuing attempts by public health workers and concerned legislators was the bill passed by the U.S. Congress in 1971 "to provide for every child a full and fair opportunity to reach his full [genetic] potential" by, among other measures, the immediate establishment of a national system of comprehensive child development and child care centers for the children of working mothers. As various public health, labor, religious, and civic groups hailed the passage of this bill, it was sent to the White House for the President's signature.

The President's message to the Congress concerning this bill agreed that "we cannot and will not ignore the challenge to do more for America's children in their all-important early years." Nevertheless, observing that "neither the immediate need nor the desirability of a national child development program of this character has been demonstrated," President Nixon vetoed the entire bill. Not only was the program wasteful of federal funds in proposing "the expenditure of $2 billion in a program whose effectiveness has yet to be demonstrated," but it also, according to the President, duplicated his own evanescent *proposals* for day-care centers. He reminded the Congress that one of the objectives of his own welfare program was "to bring the family together," and that "this child development program appears to move in precisely the opposite direction."

Exactly how clean, safe, and intellectually stimulating child day-care centers, run by trained professionals in child health and education, would shatter American family unity was not spelled out in this veto message. But the President did add that "a good public policy requires that we enhance rather than diminish both parental authority and parental involvement with children—particularly in those decisive early years when social attitudes and a conscience are formed and religious and moral principles are first inculcated."

All of which meant, in real life, that the vast majority of working mothers who were, prior to the passage of the bill, without access to inner-city or rural-slum-region day-care centers staffed by professionally trained women, would be forced to continue to leave their children with aged relatives, young neighbors' children, and other available (and not always willing and able) substitutes for "parental involvement with children." It also meant that, as this veto message was followed by major cuts in the levels of previous federal support for local child day-care centers, many educated mothers stopped working and went on public welfare in order to assure their children of safe and proper care during their "decisive early years."

Two years later, in June 1973, the same President proposed cuts of $1.2 billion in various existing public health programs that had already been weakened and restricted by the decades of war in Indochina. These cuts would eliminate many medical residencies and most postgraduate clinical training programs, cause the elimination of outpatient clinics in hospitals in many large American cities, and phase out or curtail much needed maternal and child health programs in New York, New Haven, Newark, Hartford, Buffalo, and many, many other cities between Maine and California. As Dr. Franklin M. Foote, the Commissioner of Health for Connecticut, told *The New York Times,* the new federal cutbacks in aid to existing local health programs were "going to hurt the things we're doing in the field of prevention and community prevention programs, and two programs in our bigger cities aimed at low-income and disadvantaged groups are going to be curtailed."

The cuts, he said, would hurt children particularly. And, he noted: "If it happens to be your child who gets a case of measles because of lack of immunization, and he becomes mentally retarded, you would consider it a serious thing."[18]

Historically, such vetoes of new programs, and cutbacks in our ongoing governmental programs to improve the basic qualities of human life and to widen the opportunities to develop our human genetic endowments, always add up to a continuation of the misguided compulsions to preserve the social biology of what the father of scientific racism, Thomas Malthus, had called the "necessary stimulus to industry"—poverty.

If this vested and often unthinking interest in the preservation of poverty is the iceberg that blocks and freezes hopes for human betterment, we also have to realize that the current controversies over schools, busing, housing, and compensatory or early education and its "failure" to effect permanent increases in IQ test scores in two easy lessons, are merely the tip of this ancient iceberg.

In terms of actual human developmental needs, better schools—*important and needed as they are*—are not nearly as necessary to our society as are comprehensive social programs for meeting the biological, cultural, and emotional imperatives of human development from long before the actual moment of conception to and beyond the first day of school. Few periods of human life are more critical than the years between birth and kindergarten, for those are the years when our children and grandchildren are most amenable to the development of cognitive facilities—and least resistant to the biological and mental damages of hostile environments that make the low IQ and achievement test scores of the children of the poor and other socially deprived families of all races biologically, psychologically, and culturally inevitable.

It is precisely because the doctrines of scientific racism can and do have some of their major adverse effects on the developing bodies and brains of millions of infants and young children in America that it has become a clear and pressing danger to the entire national welfare. The basic mechanism by which scientific racism acts to inhibit this society from making full and proper use of the life- and money-saving findings and methods of the world's real sciences is the false labeling of the children of 40 million people—three quarters of them as white as Arthur Jensen and Francis Galton—as genetically inferior. As innately ineducable. As condemned, by their inferior or dysgenic biological endowments, to be more prone than are their "genetically superior" betters to the various infectious, endocrine, neurological, nutritional, cardiopulmonary, and neoplastic diseases of our species. As carriers of such inferior genes for intelligence, morality, and health that, if allowed to breed unchecked, would threaten the heirs of us superior types with "genetic enslavement."

Therefore, by classic Galtonian or correlational logic, to spend another dime of taxpayers' dollars on day-care centers for the children of working mothers; nursery schools; better grade and high schools; better and wider medical and health care; better job training; better standards of wages, hours, and work safety; better water, sewage, and other sanitation needs; and widely increased support of basic and applied research in the biological, medical, behavioral, and environmental sciences—is simply to waste the sacred tax dollars of the prudent superior people on fatuous and futile efforts to appease the knee-jerk liberals and the orthodox egalitarians. For eugenics claims to prove by correlations and equations that inferior subhumans are doomed, *by heredity,* to be beyond human redemption or rehabilitation.

This is the real thrust of the IQ test score and meritocracy and time-orientation cant of the new scientific racism.

Unfortunately, because the tenets of the new scientific racism are accepted as genuine science by far too many of the leaders in law, government, and industry who make our laws, manage our mass media, and teach the teachers of our children and grandchildren, the findings of our life and behavioral scientists—from Jacobi and Boas to Skeels and Klineberg, Pasamanick and Knobloch, Naeye and Harris, Eisenberg and Heber—are now as wasted by the greater society as, a half century and more ago, were the findings of the microbiologists and the parasitologists and the epidemiologists that such common scourges of mankind as tuberculosis, "white southern shiftlessness," and pellagra were not the hereditary fate of people of inferior breeding stock but *preventable* diseases of the filth, overcrowding, shoelessness, and undernutrition of poverty. Paupers are made—not born.

The controversies aroused by the proponents and opponents of the new scientific racism are not academic arguments between two legitimate schools of scientific thought. They are, rather, battles in a tragic war between scientific racism and scientific method, in which the stakes are nothing less than the health and welfare of well over 212 million living Americans—and of their children and grandchildren yet to be born.

26 What the Geneticists Have Long Proposed for Humankind: The Antidote for the Poisons of Scientific Racism

> After all, children are the future makers and owners of the world; and their personal, intellectual and moral condition is going to determine whether the globe we live on is going to be Cossack or republican.
>
> —ABRAHAM JACOBI (1830–1919), first professor of pediatrics in the United States[1]

> Once in a lecture, a busybody in the audience asked me, "Professor, has science discovered any way to improve human intelligence?" "Yes, of course," I said. "Feed kids better!"
>
> —GEORGE WALD, Nobel laureate in physiology and medicine, 1972[2]

All of the current outpouring of solemn warnings—that unless we do something drastic about checking the birth rates of the people with low IQ test scores we will be plunging this nation into something that used to be called "racial degeneration" and is now called "genetic enslavement"—has come primarily from educational psychologists, pigeon psychologists, and electrical engineers. The one thing, aside from their blind faith in the validity of IQ test scores as measurements of innate human worth and potentials, that the spokesmen of the new scientific racism have in common is that none of them are geneticists.

Genetics is, by definition, that part of *biology* dealing with heredity and variation. The biological scientists who work in the discipline of genetics are, before anything else, biologists. None of the foremost spokesmen of the new scientific racism are biologists either.

As it happens, the biologists who, by education, training, and experience, are qualified as professional geneticists have for over thirty-five years been trying to tell us what we, as a nation and as a world, have to do about deriving the optimum benefits packed into the genotypes of every newborn child. That is, how to make certain that whatever genetic potentials for physical, mental, and moral worth are present in the zygote—the newly fertilized egg—of every newly conceived child are developed to their fullest phenotypical possibilities. The world's leading geneticists even put their proposals for the maximum development of all human genetic possibilities into a very plain-spoken manifesto as far back as August 1939 at the Seventh International Genetics Congress in Edinburgh. Unfortunately, World War II broke out less than a week after the Congress ended, and the very practical advice contained in the geneticists' manifesto was quickly forgotten.

Since the proposals of the geneticists are even more practical now than they were on the eve of World War II, it is time that they be revived and put to use in the service of as many living and yet to be born human beings as possible. First, however, before examining the text of the geneticists' blueprint for hope for the entire human condition, it is necessary to pay our final respects to the revival of the IQ test score mystique that, along with the population bomb hysteria, helped inaugurate the new scientific racism.

THE IQ IS DEAD! LONG LIVE THE EEG?

As we have seen earlier, without the pre-World War I civilian IQ testing of Jews, Italians, and other non-Nordic European immigrants by IQ test pioneer Herbert H. Goddard, and the mass testing of Jewish, Italian, Polish, Hungarian, and other immigrant military recruits with the Army versions of these so-called intelligence tests written and administered by Goddard, Yerkes, Terman, and other eugenicists, the old American scientific racism could never have won the greatest of its national legislative triumphs. That victory, of course, was the successful crusade for the Immigration Act of 1924, in which the very low intelligence test scores of the Jewish, Italian, Polish, Hungarian, and other non-English-speaking immigrants were used to devastating effect by the official Expert Eugenics Agent of the House Committee on Immigration and Naturalization, Harry H. Laughlin, as the ultimate "scientific" grounds for barring the further admission to this country of Jews, Italians, and "other genetically inferior" types.

Anti-Semitism, of course, existed long before Goddard introduced IQ testing to America in 1908; the Immigration Restriction League had been formed by Hall and Ward and other scientific racists as early as the spring of 1894. Eugenicists such as Professor Samuel J. Holmes, who sounded the nineteenth-century Immigration Restriction League warning in his college textbook *Studies in Evolution and Eugenics* (1923) that instead of the good Nordics "who made up the bulk of our immigration before 1880, we have been receiving hordes of Poles, Southern Italians, Greeks, Russians, *especially Russian Jews,* Hungarians, Slovaks, and other Southern Europeans —stocks less closely related to us by blood than the Northern Europeans and less readily imbued with the spirit of our institutions," did not really need the results of the Binet-like Army intelligence tests to make him clamor for closing the gates to non-Nordic immigration. In making the point on page 211 that "the greatest permanent danger . . . lies in the likelihood of receiving stocks of inferior intelligence" and that "some of our racial immigration is of low racial value," Holmes did add a footnote asserting: "This suspicion has been *strengthened* by the results of the mental tests applied to the recruits for the United States Army in the late war." (Italics added.)

The IQ tests, in short, had nothing to do with causing Holmes's anti-Semitism; they merely reinforced it. Anti-Semitism is still very much with us, but in 1975 the IQ test scores of Jewish schoolchildren and their parents could no longer be cited as scientific proof that the Jews are by heredity too stupid to be entitled to the privilege of American citizenship. Times have changed since, during the years 1912 to 1917, the IQ tests administered by

Goddard and his field workers to Jewish immigrants at Ellis Island not only "proved" the bulk of them to be hereditary mental retardates with minds of the "moron grade" but also showed that intellectually "these people cannot deal with abstractions." They had also changed since 1923, when Princeton psychology professor Carl C. Brigham, in his *A Study of American Intelligence,* an analysis of the Goddard-Terman-Yerkes-written World War I Army intelligence tests, solemnly reported that the Army Alpha and Beta test scores of World War I "disprove the popular belief that the Jew is highly intelligent."

By 1973, the Jensen disciple Dwight J. Ingle was, on page 58 of his eugenics tract, *Who Should Have Children?,* citing the statistic that "Jewish children in the United States have an average IQ about eight points above the national average" as "scientific" proof of the racial inheritance of high intelligence.

In that same year, two events were duly recorded in our newspapers: (1) that great pressures were being put on the government of the Soviet Union to permit Russian Jews to emigrate to Israel (see Taylor, 1976); (2) that the birth rate in the United States fell to the replacement level of 2.0 children per family—well below the 2.1 replacement level at which we reached zero population growth. Coincidence or no, it was at that time that Carl Pope, the Zero Population Growth, Inc., lobbyist in Washington, announced a shift in targets. The gonads of the hitherto untrammeled copulators having been trammeled to levels that would have delighted even William Vogt, the ZPG now turned a full historical circle to aim their guns at the original targets of the pioneer of the twentieth-century population explosion hysteria, E. M. East. In terms that would have delighted—but not surprised—East and Davenport, Pope told a writer for *The New York Times Magazine* (September 16, 1973) that the final solution for our population problem, now that the native American fertility rate had reached ZPG, was *Immigration Restriction.* As Pope put it: "Immigration is a sentimental symbol whose day is long past. We could take in 100,000 immigrants [instead of our present quota of 400,000] and still serve that symbol."

What gave this opening of the second front of the ZPG crusade an added *déjà vu* flavor was, of course, the fact that the Nordics-only national origins quotas written into our U.S. Immigration Act by the old scientific racists in 1924 had, by 1965, been finally dropped—and the barriers lowered against Jews, Italians, and other non-Aryans, including Asians and Africans.

A year later, when, with the backing of two of the world's superpowers, and more than a few of its major oil companies, huge majorities of member states voted again and again in the United Nations for policies and measures designed to crush the Jewish state in Israel, and thereby deny that present sanctuary to Russian Jews who manage to get out of the socialist motherland, it was becoming clear that a post-Auschwitz holocaust of the near future might create new humane pressures for permitting the immigration of its targeted victims. More and more spokesmen for the new scientific racism began to talk of the need for new immigration restrictions—not, of course, because of the low IQ test scores of the Jews, but because of the "population explosion" and the "environmental crisis."

Thus, on Sunday, October 27, 1974, a full-page ad, headed "Declaration on Population and Food," placed by the Environmental Fund in *The New York Times* and signed by such veterans of the population control crusade as Garrett Hardin and William C. Paddock, not only called for Paddock's triage ploy abroad and Draconian population controls at home, but also declared: "Obviously, immigration should also be brought into balance with emigration immediately. This can easily be done with legislation."

So that, as the sheer force of scientific logic dooms the IQ test score myths which helped the eugenics movement win the 1924 Immigration Exclusion Act, it would be a mistake to believe that the new scientific racism will lower itself into the grave already awaiting IQ testing. The scientific racism that managed to integrate Galton's eugenics, and Spencer's Social Darwinism, and Davenport's pseudobiology into the value systems of the statesmen who framed and administered all of our health, education, and other social policies for most of this century, established its legislative beachheads *without* the aid of its later IQ testing weapons. The same scientific racism also managed to survive such traumas as the loss of its genocidal cephalic-index weapon at the hands of Boas' 18,000 anthropometric measurements before World War I. Malthus and his successors have been nothing if not adaptable to new opportunities, and new techniques, of selling the same old package.

A case in point is Hans J. Eysenck, England's foremost disciple of Jensen, Shuey, and their school. Professor Eysenck's book *The IQ Argument* (1971) not only defends the Jensen-Shockley-Herrnstein case for the 80-percent-plus genetic origin of all IQ test scores. He goes a vital step further: on page 53 he gives us a preview of the new brain wave tests that will beyond all doubt be marshaled to fill the breach when the IQ tests are abandoned as obsolete and socially harmful.

In his acknowledgments, Eysenck said that both of these very objective brain wave readings had originally been published in an article by John Ertl in *Nature* in 1970. This was, indeed, true; the article by John Ertl, "Brain Response Correlates of Psychometric Intelligence," had been published in *Nature*. However, far from validating the accuracy of IQ test scores of individuals, let alone the existence of racial differences in intelligence, all of Ertl's work with his system of measuring the neural efficiency of the human brain by means of measuring certain evoked potentials by the standard electroencephalograph (EEG) had served to completely refute both concepts. According to John Ertl, his brain wave measurements have, to date, proven :

[1] My brain wave readings disagreed with about 40% of the lower IQ test score readings—and with less than 2% of the higher IQ test score results.

[2] No single parameter of the electrical activity of the brain that is measured by the EEG has been shown to be different in different

Will Electronic Brain Wave Tests Succeed IQ Brainwashing?

7a "Brain waves" resulting from a sudden stimulus (light). The waves are the "evoked potentials" recorded on an electroencephalograph, and the score used is the wave length of the first four waves to occur (E^1 to E^4). Note that the wave lengths are shortest for the brightest, longest for the dullest subject. In other words, transmission of information is quick for bright, slow for dull subjects.

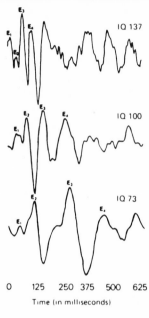

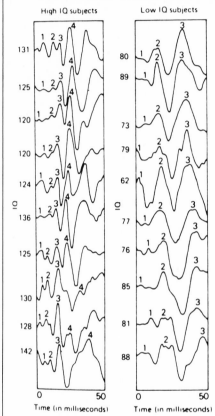

7b Evoked potentials from ten bright and ten dull subjects. Note the quick waves in the former, and the slow waves in the latter.

races. On the other hand, several brain wave studies made with the EEG have shown a very high degree of correlation between the brain wave activity of identical twins. In fact, if you plotted the brain wave activity of identical twins, you would come up with the same correlation as Shockley has shown for identical twins and IQ test scores. So from this one can conclude that, first, EEG parameters are sensitive enough to detect similarities between twins, and, second, you can therefore assume that electroencephalographs are also sensitive enough to detect qualitative differences between racial groups. There have been several serious studies by competent investigators of just this area of human brain development, and thus far no one has succeeded in finding any brain wave differences between races.[3]

The EEG brain wave readings do, however, show that children who have been malnourished from the moment of conception, and who have as infants and children suffered from vitamin deficiencies, hormone imbalances, and other developmental pathologies of poverty, do quite measurably prove to have slower evoked responses—that is, lower neural efficiency scores— than do the well-fed, well-housed, well-doctored children of affluent parents. For this reason, a switch-over from mass IQ testing to mass brain wave screening in schoolchildren would serve the same social and political functions for the scientific racists as IQ testing does now—and at about 2 percent of the present cost of IQ testing. Like the IQ tests, the neural efficiency brain wave tests would—in the hands of the modern eugenicists—continue to "prove" the biological existence of what E. J. Lidbetter, of London, described at the Second International Congress of Eugenics in 1921 as "a definite race of chronic pauper stocks" with genetically low intelligence quotients.

In short, in the hands of what Walter Lippmann, writing of the original IQ tests in 1922, called "muddle-headed and prejudiced men," the switchover from the mass "intelligence" screening by IQ tests to the infinitely cheaper brain wave screening of evoked potentials is capable of mislabeling as mentally inferior and retarded whole generations of infants yet to be born —just as, since 1911, the older IQ tests have falsely condemned entire generations of perfectly normal children to the human and socioeconomic fates reserved for the mentally retarded.

Since I knew, from my own reading of Ertl's scientific papers, that Eysenck's use of his work did not represent Ertl's thinking, I visited the inventor of the neural efficiency system cited by the British Jensen disciple to tape his views, with his permission, on a little Sony that sat between us on the table of his Ottawa living room.

WHAT BRAIN WAVE TESTS *REALLY* SHOW ABOUT IQ TESTING

Because, for now obvious legal reasons, the IQ tests are—within the next few years—probably going to be abandoned in most American school systems, it is quite likely that some form of brain wave tests, such as those developed by Ertl, then professor of psychology and director of the University of Ottawa Center of Cybernetic Studies, will be employed to replace them in many state and city educational systems. Since, in the hands of scientific racists, the brain wave tests can cause as much human and social harm as the benignly conceived IQ tests they supplant, it might be worth reading what Ertl himself had to say on the subject. The questions that follow are by me; the answers by Ertl, as transcribed from our taped colloquy.

First, what exactly do the brain wave tests measure?

They measure neural efficiency, which I define as the rate of information transfer within parts of the brain. The mechanism for measuring this depends upon the electrical activity of the brain, which is then analyzed by a computer. The system measures gross electrical activity related to something real in the individual cells of the brain.

Is that what some people call intelligence?

The relationship of this concept of neural efficiency to intelligence as defined by psychologists is something else again. First of all, I want to get out of the circular reasoning trap that intelligence is a score on an IQ test. And the way I've attempted to do this is to show that neural efficiency is related to medical or biological factors. What we are measuring is, after all, *biological* electrical activity.

For example, just as a deficiency of thyroid hormone diminishes the mental awareness or acuity of a human being, the level of neural efficiency measured by brain waves is reduced in people with thyroid hormone deficiency. Just as deficiencies of certain vitamins and the presence of alcohol, tranquilizing drugs, and anesthetics in the body systems cause a reduction of mental acuity, they also cause reductions of the neural efficiency measured by EEG readings.

How is the neural efficiency test administered to get the kind of brain wave tracings Eysenck republished in his book?

Standard EEG electrodes are attached to the scalp with a conductive paste. The leads of these electrodes are then passed through a small brain wave amplifier, which in turn feeds its signals into a computer. The person being tested is then exposed to about a hundred flashes of light, in bursts of an average rate of one flash per second. The whole test takes about two minutes.

What happens to the information about the responses these lights evoke?

I'll trace the pathway for you. The light is perceived by the eye, and the information is then transmitted to the reciprocal cortex in the back of the head. The transmission time for this is very short, a few thousandths of a second. Now, having received this input, the brain decides, what am I going to do with it? Is the brain going to order a blink, or a jerk of the hand, or simply ignore the whole signal? Whatever happens, processing takes place, and this processing takes time. The shorter this time, the better the neural efficiency of this brain. It is real—but it is all very simple, and I am sure that it does not explain one tenth of what the complex organ called the brain is all about. But it does explain something.

As Ertl envisioned the system, he saw it as a clinical tool for determining, quickly, such things as whether a brain is organically damaged (by accident, sickness, or heredity), and whether a patient operated upon for removal of a brain tumor is or is not making a full recovery from this surgery. As he tested it on large groups of children and adults, however, it also told him many things about the gross inadequacies of many standard intelligence tests.

To arrive at a patient population whose low IQ test scores truly reflected brains that were mentally less capable of learning than other brains, the IQ tests were of little or no use at all. It was extremely hard to find people who got low IQ test scores on all three of the standard IQ tests Ertl used—"a child who achieved 75 on the WISC might get 108 on the Cattell, and 92 on the Revised Standard Binet, and so forth."

The children who scored low marks on all three IQ tests were, as a rule, suffering from obvious eye, ear, or clinical deficiencies that had predictable adverse effects upon their mental and neural efficiency. In short, they showed "all the physical symptoms of slowness that you imagine are there," Ertl said. So "there were not too many disagreements between the IQ test scores and the neural efficiency test scores in these poor sick children. But in large random samples of children of different social classes and races, and we have examined thousands of children in Canada and the United States, the divergence between the IQ test scores and the neural efficiency scores was as high as 40 percent."

Tests run with his system, by himself and other investigators, have suggested to Dr. Ertl that while, until now, low birth weight was the most accurate predictor of future low IQ and classroom test scores, evoked brain wave potential tests of newborn infants were possibly even more accurate in predicting future learning lags. And, he hopes, such early diagnoses of future learning difficulties could be used to start intervening with nutritional and other necessities to correct the brain growth and development deficiencies pinpointed by these early brain wave tests. What the evoked potential tests measure, in short, is the level of neural efficiency a human brain of given biological health is capable of maintaining.

Ertl is also fearful that, like Binet's well-intentioned IQ tests, his own neural efficiency tests will be used not to help children who can be helped, but to hurt children who do not deserve to be misbranded by the willful abuse of his work. "There is no question," he said, "but that poor nutrition affects the neural efficiency score just as it affects the IQ test score, just as does prenatal and postnatal brain and other organ damage caused by infectious diseases. But mass screening with brain wave techniques will probably not be preceded by mass medical and neurological examinations—any more than mass IQ testing has, up until now, been preceded or accompanied by mass clinical screening for anemia, eye and ear acuities, and neurological damage in the same children."[4]

When we spoke, in the autumn of 1972, Dr. Ertl feared that the growing inflation would widen the gap between the $12.50 it then cost to administer and process an IQ test to the average schoolchild in America and the thirty cents it cost to make a neural efficiency test by his system. He was afraid that for purely economic reasons state and city school systems would abandon IQ testing in favor of the much less costly EEG brain wave tests.

However, costs were not the only factors that might cause the early abandonment of IQ testing in government-operated school systems and other institutions. On Thanksgiving Day, 1974, a federal judge in San Francisco gave thousands of black, Chicano, and other ethnic minority families a second good reason to celebrate this day henceforth. As reported the next day by *The New York Times:*

A Federal judge says I.Q. tests must not be given to black children in California if the tests do not reflect the cultural background of the students.

United States District Judge Robert F. Peckham ruled two years ago that San Francisco schools could not use the tests to determine whether black children should be placed in classes for the educable mentally retarded.

His order yesterday extended this ruling statewide and he said the tests could not be given to black children at all. He ruled in a suit brought by the N.A.A.C.P. Legal Defense and Educational Fund, Inc., and other groups. Three black psychologists held a news conference about the ruling and estimated that 75 per cent of the black children in classes for the retarded did not belong there. Dr. William Pierce said I.Q. tests were "based on white middle-class standards."

It would be nice, as an intellectual, to say that this long overdue action was taken because of the fine scientific research conducted by American behavioral, biological, and medical scientists from Wallin in 1913 and Arlitt in 1921 to Skeels and Klineberg in the 1930's and Pasamanick and Knobloch, Mercer, Mayeske, and Heber after World War II. But this would be only half the story. The other half was implicit in the new demography, born of the New Enclosures; and the new politics, born of the civil rights explosion touched off by *Brown* v. *Board of Education* in 1954. Twenty years later, black men held the office of mayor in Detroit, Los Angeles, and other large cities; thousands of other black men and women held offices ranging from U.S. senator to sheriff.

At the Watergate conspiracy trial of the former Attorney General of the United States in 1974, eight black people sat on the jury of twelve hearing this case. And, two days before the federal judge in San Francisco outlawed the IQ testing of minority group children in California, the first character witness presented by former Attorney General John Mitchell was a black housemaid for the Mitchell family. Like the poll taxes nullified by the Supreme Court, the equally regressive IQ testing agent of socioeconomic polarization was doomed.

The blacks were still a long way from matching, in income and health, the Irish who had also found in political power the secret of escaping from hereditary poverty in America. But they were on their way, and once they made it, the white poor—who even now outnumber the black poor by more than three to one—would again be, as they were in the days of Malthus and Galton, the primary victims of scientific racism in America. For these threatened white Americans, then, the proposals of the world's geneticists at their international congress in Edinburgh become particularly vital.

THE FORGOTTEN GENETICISTS' MANIFESTO OF 1939

An insight into how to cope with the clear and pressing dangers to the welfare of our children and our society implicit in the current revival of scientific racism was provided in a lecture delivered, ironically at the invitation of Charles Benedict Davenport, by the geneticist Hermann J. Muller at the Biological Laboratory of the Brooklyn Institution on July 25, 1921. In this lecture Muller related in some detail the history and nature of the scientific experiments that, between 1910 and 1912, had destroyed the unit character (one gene = one trait) dogmas on which the transmission genetics concepts of eugenics were (*and remain*) based. (See pages 191–92.)

Muller concluded his lecture with a passage that speaks for the real spirit of the best of our scientists, past and present. He had been describing his own efforts (ultimately to succeed and win him his Nobel prize) to induce artificial mutations in genes, and had confidently predicted that these efforts to manipulate the genetic materials of fruit flies by altering the environments in which they grew "will furnish further data on the nature of genes and mutations." And, having concluded his report on various genetic experiments, Muller closed with:

> It will be a far call from this, perhaps, to an analysis of the gene, but who can say what may not be possible in the history of science? There is a long future ahead of genetics, and ahead of the Drosophila [fruit fly] work. And I have seen in the space of only one year a student—Sturtevant—when only 19 years old—first in the world [to] get it into his head that it might be possible to make a real *map* of the inside of a *chromosome,* and then proceed actually to do it. Just one boy, in one year, could do that. What then cannot be done by the friendly cooperation of many people in the course of time? [Muller's italics.]

Nearly twenty years later, in August 1939, Muller was one of the signers of the Geneticists' Manifesto drawn up at the Seventh International Genetics Congress in Edinburgh. From it we can construct a very feasible, very practical, totally nondestructive, and, above all else, very *scientific* program with which to answer the pseudo-genetic but socially dangerous propaganda claims of the new scientific racism by, indeed, making possible "the friendly cooperation of many people" in a massive application of the findings of human genetics and all other sciences dealing with human heredity, gestation, birth, physical and mental development, and human well-being.

Nothing that has been discovered in genetics or biology in general since this historic Geneticists' Manifesto was signed in 1939 has altered any of its scientific premises. On the contrary, all of the great longitudinal and intervention projects since 1939 have served only to bolster its concepts, and to lend added weight and urgency to each of its recommendations.

Presented only a few days before the outbreak of World War II, the Geneticists' Manifesto was quickly forgotten once the leaders of "the Nordic Movement in Germany," hailed in the *Eugenical News*[5] as the noble deliverers of "Nordic race-hygiene," sent their dolichocephalic Wehrmacht and their Nordic Luftwaffe and their Aryan concentration camp sadists into action against the dysgenic populations of Europe and the American armed forces. It is, perhaps, high time that the basic postulates of the Geneticists' Manifesto of August 1939 were put to the use of mankind. Particularly the three key postulates incorporated in the following passages:

. . . the effective genetic improvement of mankind is dependent upon major changes in social conditions, and correlative changes in human attitudes. *In the first place there can be no valid basis for estimating and comparing the intrinsic worth of different individuals without economic and social conditions which provide approximately equal opportunities for all members of society* instead of stratifying them from birth into *classes with widely different privileges.*

The second major hindrance to genetic improvement lies in the economic and political conditions which foster antagonism between different peoples, nations and "races." The removal of race prejudices and of *the unscientific doctrine that good or bad genes are the monopoly of particular peoples or of persons with features of a given kind* will not be possible, however, before the conditions which make for war and exploitation have been eliminated. . . .

Thirdly, it cannot be expected that the raising of children will be influenced actively by considerations of the worth of future generations unless parents in general have a *very considerable economic security and unless they are extended such adequate economic, medical, educational and other aids in the bearing and rearing of each additional child* that the having of more children does not overburden either of them. *As the woman is more especially affected by child bearing and rearing she must be given special protection* to ensure that her reproductive duties do not interfere too greatly with her opportunities to participate in the life and work of the community at large.[6] [Italics added.]

The geneticists, in 1939, were of course writing of the necessity to create the kind of social and economic conditions that would provide "adequate economic, medical, educational and other aids" involved in the bearing and rearing of children, and "special protection" for the women of all of this planet's nations. In due candor, particularly after our costly decades of massive military intervention in Asia, we no longer can afford to think of making all these gene development advantages possible for the people of the entire world. Especially not after we ourselves have, in these past two decades, fallen so far behind other industrial nations in the same areas of human health and development.

A MODEST PROPOSAL

What, however, would be wrong with applying the letter and the spirit of the advice of the world's geneticists in the entire state of Mississippi? This nation has everything to gain, and literally nothing to lose, by in effect turning Mississippi into our national laboratory of human betterment.

There are two priceless treasures no society can long afford to waste. The first is human life. The second is human knowledge. This is particularly true in all situations where human lives are being wasted for failure to apply the available knowledge of how to prevent such deaths. This knowledge, primarily biosocial, represents real hope for mankind. There is no justification for any society to permit its stubborn minority of scientific racists to continue to denigrate and diminish this hope.

The findings and the advice of our competent and serious life and behavioral scientists—the heirs to the great scientific tradition of Jacobi and Boas, Wilder and Klineberg, Myerson and Skeels, Stiles and Goldberger—speak to us today in the new lexicon of longitudinal studies, and planned intervention between every newborn child and "the devastating effects of poverty, sociocultural deprivation, and maternal deprivation."[7] They speak to us not with Pecksniffian rhetoric but with millions of discrete biological, chemical, physical data—scientifically collected, stored, measured, and analyzed. Hard, objective, unbiased facts, such as the treasure house of socioeconomic and sociocultural data gathered by the great Coleman study of equality of educational opportunity in 650,000 children in all fifty states of the Union. And the matching studies of the National Health Survey of all Americans, and the NINDS perinatal study, and the British National Child Development Study.

We also have the hard data gathered in various intentional and unintentional intervention projects over two centuries, from the building of clean-water systems in great cities to the initiation of compulsory vaccination programs in entire nations. Thanks to the longitudinal studies of our times, we *know* what causes vast numbers of our children and mothers to develop the degrees of physical and mental strengths and weaknesses they possess. Thanks to intervention experiments in this nation, we also know considerably more than we ever dreamed of knowing about what happens to children and parents when they are spared the biological, emotional, and cultural traumas of poverty and racism. Our entire national history since 1776 is, after all, one of reaping the unlimited benefits of extending at least some of the blessings of equality of opportunity to the children of servants as well as masters.

What I propose for Mississippi is, therefore, not only good science but also as American as Abraham Lincoln. Essentially, it boils down to a vast longitudinal study of what would follow if—through massive medical, nutritional, educational, and social intervention—we made it possible, *really* possible, for every newborn child, every family in Mississippi (regardless of skin color), to enjoy at least two generations of genuine equality of opportunity to develop their inborn or genetic potentials to their fullest ranges. For equality of opportunity goes far beyond the grudging right, at the age of six, to start attending the superior schools of the white children. And, particularly in Mississippi, equality of opportunity is as scarce in the average white family as among the blacks. In all ethnic groups, throughout the entire world, equality of opportunity includes:

1. Equality of opportunity to be protected, *in utero,* against the brain-damaging, body-deforming, preventable viral, bacterial, and fungal infections, from rubella and cytomegalovirus infections to syphilis and toxoplasmosis.

2. Equality of opportunity to be born to a mother whose own lifelong freedom from hunger has spared her the malnutrition-caused metabolic disorders of the liver, the placenta, and other organs that, over generations of poverty, result in cross-generational *preventable*

sequelae such as epilepsy, mental retardation, cerebral palsy, minimal brain dysfunction, and, frequently, infant deaths.

3. Equality of opportunity to have, from conception, enough to eat so as to meet the body's nutritional requirements for proper brain and bodily growth and development for the first nine months of prenatal life and for at least the first six years of postnatal life—when the brain achieves most of its mature weight and bulk.

4. Equality of opportunity to be immunized, *beginning during the first year of life,* against polio, diphtheria, measles, and other eye-damaging, ear-damaging, brain-damaging, and paralyzing infectious diseases of childhood, according to the medically wise schedule recommended by the American Academy of Pediatrics. This schedule, now enjoyed by only a minority of America's children, calls for immunizations to be completed before the age of six years—the time at which, in those thirty-three states with erratically enforced mandatory immunization laws, most poor children are supposed to receive their first immunizations upon entering public school.

5. Equality of opportunity to enjoy—*as do all of the children in most other industrialized nations*—statutory programs of free cradle-to-grave preventive and therapeutic medical care. This includes regular periodic examinations for the detection, diagnosis, and correction of vision, hearing, and other defects and abnormalities of the skeletal, endocrine, and other body systems; preventive and corrective dental care; and, where needed, psychological guidance and psychiatric care.

6. Equality of opportunity to be educated in clean and safely constructed schools by professionally competent teachers; to go to and from such schools without being forced to run gauntlets of obscenity-screaming, rock-throwing, club-wielding adults; to be trained for any trade, educated for any profession, for which the individual has both the aspirations and the mental capacities required; to compete on equal terms for employment in such trades or professions upon completion of one's training, regardless of the color of his skin or the religion and/or national origins of his great-grandparents.

7. Equality of opportunity to raise one's children in a family environment where toys, drawing materials, books, and other educational materials are always on hand; where the levels of verbal communication stimulate the learning processes; where community and family sanitation, family living space, and family food resources offer each growing child the maximum protection against the chronic, *preventable* respiratory and deficiency diseases that cut down on individual attention spans and classroom attendance records.

8. Equality, finally, of the opportunity to be conceived, to be born, and to grow up in developmental environments free of the *preventable* adverse genetic, somatic, and mental effects of the chemical, physical, noise, and psychic pollutants of industrialized societies, most of whose political leaders have yet to grasp the moral necessity—made

manifest as far back as 1775 by Sir Percivall Pott—for mandating the *pre-engineering* of the biological safety as well as the functional efficiency of all mines and motor vehicles, power plants and pesticides, factories and food additives.

To talk of such genuine equality of opportunity to develop and employ the gifts that are innately one's own does not in any way mean the same thing as saying that every human being is born equal in every single respect. To suggest such "equality" would be to claim that all human beings, given such equality of opportunity to develop their zygotic potentials, would end up *identical* in height, weight, longevity, IQ test score, and bank account. This pseudo-egalitarian concept of equality of opportunity is an absurdity invented for self-serving purposes by the scientific racists, a stuffed horse fabricated only to be flayed on the rialto.

In human, as in other forms of life, *diversity* is the basis of individual reality, of natural selection, of evolution, of survival. It would have been a very short-lived human species, indeed, if our entire species had all been members of one "pure race," genetically identical in thought, myths, physiology, and temperament.

The function of this proposed compensatory intervention project, then, is certainly not to make all Mississippi children alike, be they white or black. Two centuries of endless, grinding, pulverizing poverty have already created enough of *that* kind of phenotypical sameness in that socially deprived state. On the contrary, the genetic function of the massive intervention in Mississippi is precisely to liberate the genes of all of Mississippi's children from the kind of leveling *uniformity* imposed by poverty—a continuum of genetic damage that is bad enough in whites, and even more repressive when, for the nearly 37 percent of the state's black people, it is imposed by the hammer of race prejudice on the anvil of socioeconomic deprivation.

The primary reason for suggesting a statewide intervention program in Mississippi, rather than any other state, is self-evident. For over half a century, Mississippi has been our poorest state. According to *Hunger—1973,* a report issued by the U.S. Senate Select Committee on Nutrition and Human Needs, no other state in the Union is as pathologically poor as Mississippi, where 34.4 percent of the state's people live in poverty.

Mississippi's infant-mortality rates per 1,000 live births—17.0 for white and 35.6 for black babies—are by far the highest in the nation. Connecticut, the state with the highest per capita income in the nation, has infant-mortality rates of 13.3 for whites and 24.9 for Negroes. The death rate of children under five years of age stands at 773 per 100,000 population in Mississippi and 399 in Connecticut. Mississippi's 1974–75 annual expenditure per public elementary and secondary school pupil was the lowest in the nation—$921, 36 percent or $510 below the U.S. average of $1,431. (This average expenditure per pupil ranges from New York's $2,241 and the District of Columbia's $1,957 to the annual expenditures per pupil of $960 and $933 reported by Kentucky and Alabama.)[8]

SAVE OUR CHILDREN—AND WE SAVE OUR WORLD

Without going into the details of any plans, what I propose is simply that we let the nation's pediatricians, obstetricians, family physicians, clinical nutritionists, epidemiologists, and all other health professionals tell us what is needed—in the way of immunizations, nutritional supplements and guidance, civic and family sanitary facilities and services, and other interventions—to reduce and/or eliminate the essentially biological causes of Mississippi's scandalously high infant-mortality and child death rates.

Under the supervision of our health professionals, let us then proceed without further delay to do everything that their scientific training and clinical experience convince them has to be done to prevent the continuum of reproductive casualty and fatal and debilitating infectious and deficiency diseases of infancy and early childhood. Physically and *genetically,* there is no reason why the 2,324,000 white and nonwhite people of a bucolic Mississippi, whose density of population is only 49.1 people per square mile, should not have infant-mortality rates as low as the world's most densely populated industrialized nations, Japan and the Netherlands. Japan, with a population density of 757 people per square mile, has an infant-mortality rate of only 11.3 per 1,000 live births; the Netherlands, with a population density of 855 people per square mile, has an infant-mortality rate of only 11.0.

At the same time as we intervene between the tender bodies of all newborn Mississippi infants and the biological hazards to their normal growth and development, we must also call in our better child psychiatrists and psychologists, educational specialists, cultural anthropologists, and speech therapists to plan and carry out massive programs of in-home and community interventions between developing children and the verbal, cultural, and psychological barriers to the optimum development of the genetic mental potentials of the white and black children of Mississippi. As the authors of *From Birth to Seven* remind us, "before even the [environmentally deprived] child reaches the infant school he is already handicapped in this vital area of language development."[9] In Mississippi, as in England, Scotland, and Wales, the ability of white and black children to learn how to read starting at age six is, after all, considerably enhanced when the lexicon of the school primers happens to be the same as the lexicon of verbal communication in a child's family and preschool cultural growth environment.

Operationally, it is perfectly feasible to apply all of the IQ-test-score-boosting discoveries of Heber's Milwaukee Project and Schaefer's Infant Education Project—let alone the invaluable Coleman study of what goes into high and low achievement test scores—to every newborn child, to every family, in Mississippi. And at a net cost in tax dollars of considerably less than the more than $400 billion we have spent and mortgaged[10] to date on our Indochina wars and the satrap dictators they enriched, since it costs far less to salvage a baby's brain in Mississippi than to bomb a baby's brain in Vietnam. We not only have the money to save Mississippi, but we also are rich in the superbly trained and motivated teachers, psychologists, social workers, and therapists who would be willing to leave more lucrative posts to live and work with both the children and the parents of our poorest state as tutors, guides, and living bridges to the language and culture of the twentieth century.

Like the medical and paramedical workers involved in delivering their types of intervention in the development of all Mississippi children, the educational, behavioral, and social science professionals would also collect and process the hard data on the learning and other behavioral effects of their segments of the intervention in the total quality of life in Mississippi. These behavioral findings can also be applied without undue delay to the same problems of children in California, New York, and other affluent states.

Naturally, as part of the total intervention, decent schools and preschool nursery centers would have to be built throughout the state. These new buildings not only would serve as schools, but would also contain well-baby clinics, immunization and other preventive medical clinics, ambulatory medical care clinics, and adult education centers where professionals would teach parents everything from the care of sick children to how to read and write. The schools could also be used as the base headquarters for circulating libraries of books, toys, phonograph records, reproductions of art works, and other artifacts that would have to be made available, on loan, to every Mississippi family.

Since, as a function of the intervention program, the parents as well as the children would have to be educated and trained for skilled jobs at which they could earn enough money to raise children properly, vast new work projects would have to be set up as part of the intervention. Not all of these work projects would necessarily have to be financed by the program; a good number of industrialists would leap at the chance to capitalize on the availability of skilled and healthy work forces so near the sources of their raw materials.

These work projects could serve many purposes—from generating income for the entire state in the form of profits from the sale of various goods and services, to demonstrating that the poor white people of the South, libeled by the scientific racists as being born *genetically* "the shiftless, ignorant, and worthless class of anti-social whites of the South" (a libel accepted as scientific fact by Oliver Wendell Holmes, Jr., and other educated Americans for over half a century), are as willing to work and as capable of being trained for skilled trades as are any other Americans.

Mississippi would also be an ideal spot in which to conduct large-scale pilot programs in the growing and processing of various types of conventional and novel protein foods.

Two relatively new sources of conventional protein foods—pond-grown fresh-water fish and soybeans—are already important sources of farm income in Mississippi. At least two sources of novel or unconventional forms of food protein, for the enrichment of human or animal foods, could easily be produced in huge and commercially profitable quantities in Mississippi. These are (1) leaf-protein concentrate (LPC) and (2) single-cell protein (SCP). Leaf-protein concentrates are precipitated by hot-water treatment of the juices extracted by crushing the leaves of trees; the waste leaves of beets, sugar cane, cotton, and other farm plants; alfalfa; water hyacinths and other water weeds. Single-cell proteins are produced by culturing microorganisms in petroleum wastes, alcohols, and food-processing wastes in fermentation reactors. In his excellent new book, *The Nutrition Factor* (1973), Alan Berg writes that the production of single-cell proteins "requires so little space that half a square mile could produce 10 percent of the world's protein supply. And SCP raises few waste disposal problems because nearly all its constituents are used."

The alcohols in which SCP can be produced also happen to constitute an inexhaustible font of renewable fuel resources for transportation, industry, and electric power dynamos. As the American Society for Microbiology reminded us in 1975:

"As long as the sun shines, plants will grow and will provide us not only with food and fiber but also with an inexhaustible supply of raw materials for alcohol production. Much plant material ends up as waste in the form of paper and paper products, sawdust, stubble on the farm, garbage and sewage in the city. These waste materials have a high content of carbohydrates, especially cellulose, which are readily degraded and converted to alcohol by microorganisms."[11]

Too many of us tend to forget that Pasteur's great work on the role of yeasts in the fermentation of alcohols was done nearly two decades before he turned his attention to the role of other microorganisms, such as bacteria and what we now know as viruses, in causing animal and human diseases. Let us not forget what the American Society for Microbiology, in the same statement, said about the current potentials of microorganisms in this regard:

"The microbiological and biochemical knowledge and talent needed for the research and development of 'microbe power' to create new fuel sources from renewable and inexhaustible natural resources is available. Greater public support, as well as that of government and industry, is needed to translate this knowledge into production."[12]

Of equal importance, Mississippi could also be used to test new patterns of urban and rural transportation designed to free every family from dependence upon the dominant one person–one car patterns of daily transportation that have made our air unsafe to breathe, our water unsafe to drink, have put our lives at annual risk of 56,000 traffic accident fatalities and over 5 million traffic accident injuries, and now constitute our major cause of accidental crippling. Such biologically more intelligent patterns of daily transportation would also end our mounting "energy crises" and make us immune to the extortions and the crude blackmail ploys of various oil-rich countries and their multinational cartel partners.

The late E. Richard Weinerman, professor of medicine and public health at Yale University, declared, "I think of the automobile (although I love mine for the pleasure it gives me) as Public Health Enemy Number One in this country. My solution is not to find a less dangerous fuel for it to burn, but to find a different system of inner-city transportation."[13] Where electric buses and trolleys, powered by electricity produced in plants where fuel is burned in properly low-pollution power plants and in new generations of more efficient and cleaner turbines operated on "power gas" produced from coal or oil,[14] can help make inner-city transportation safer for human life, the prospects of making life better by the development of superior systems of inter-urban and regional transportation are fine moral reasons for extensive experimentation.

Scientific and engineering developments in such fields as air-cushion vehicles and "vehicles that 'fly' a foot or so above a guideway, lifted and propelled by electromagnetic forces"[15] now present good evidence that while the People Pollute population-bombers were crying for Clean Cars, the age of long-distance wheeled transportation might well have been starting to end. Certainly such experimental systems could well be tested in our Mississippi experiment in gene-environment interactions—not only for what they can teach us about our real genetic potentialities but also for what they might contribute to the advancement of the decentralization of American industry.

Every good public health experiment must also have a good control population. The *Statistical Abstract of the United States* offers an ideal candidate: the adjacent state of Arkansas—the sixth-poorest state in the Union, equally southern, with a population of 2,062,000 people, of whom 352,400 (18.3 percent) are black. During the period of the intervention project (or, at least, for the first ten years) Arkansas is to be left strictly to its own present devices. But every medical examination, every battery of mental tests—including, of course, the Iowa (classroom) Achievement Tests and, if they are still taken seriously, the IQ tests in standard use in forty-eight other states[16] —used in Mississippi on the intervention experimental population would be administered to the children and the parents of Arkansas, the control state, at the same time.

As the ongoing NCDS study in England, Scotland, and Wales has already demonstrated, the educational results of this massive intervention would make themselves apparent inside of only two generations. If the British findings on their children of all social classes have any bearing on the children of Mississippi, the occupation (or social station) of neither the paternal nor the maternal grandparents of any child has any effect upon that child's academic achievement levels. The three R's, apparently, are no more genetic than is the ability to speak any of the Aryan languages.

Naturally, as the results of the intervention program in Mississippi would become known to the nation, their publication might cause Arkansas and other states to demand identical health, education, and job training and opportunity interventions for their populations. The nation that spent and mortgaged at least $400 billion in a futile effort to bomb Vietnam, Cambodia, and Laos "back to the Stone Age," and another $30 billion on landing on the sterile moon, could very well afford to meet these demands for biological and sociocultural interventions here in the living populations of Americans. The money these salvaged Americans returned to the nation in taxes on incomes they could never have had without such interventions in their lives would more than equal the tax-dollar costs of the compensatory interventions.

None of this is to say that the proposed intervention program in Mississippi—and, if demanded by their citizens, in other states—would turn every barefoot child in every urban slum and rural hovel into a Nobel laureate or a tycoon. Skeels did not turn the orphaned and mentally retarded (by IQ test score criteria) Iowa children into professors of electrical engineering and bank presidents when he and his colleagues placed them for adoption in loving, nurturant families. But this simple and human act of intervention in their lives *did* enable these young candidates for nearly certain lifetimes of institutional and ambulatory consumption of welfare funds to mature, instead, as self-respecting, self-supporting, and tax*paying* citizens.

Perhaps, had the Iowa programs of nutritional, medical, nursing, domiciliary, and above all else psychological interventions been started while the children were still in their mothers' wombs, they might conceivably have risen to even higher stations in life. It is quite likely that, in such individuals, the benign intervention had come too late in their development to undo *all* of the damage done to them by maternal malnutrition during pregnancy, or emotional deprivation between the moments of birth and adoption. Let alone from the sea of preventable troubles in which poor children find themselves from birth: early, perinatal malnutrition; early infections by the agents of measles, diphtheria, whooping cough, tetanus, mumps, and rubella; early and chronic cases of the classic diseases and disorders of poverty, the respiratory and deficiency diseases, pneumonia, diarrhea, and the behavioral disorders now shown to be associated with nutritional anemia.[17]

This we will never know. What we do know, however, is that an employed, self-sufficient taxpayer and parent—however modest his or her calling—is still of far greater value to the greater society than a lifetime inmate of an institution for the mentally retarded or a chronic and involuntary public welfare client. It was certainly not their genes that originally retarded the low-IQ-test-scoring orphan and unwanted children whom Skeels and the other Iowa psychologists reclaimed for the ranks of the lifelong taxpayers; there is no evidence at all that it was their genes that kept them from rising even greater in the socioeconomic scale than they did thanks to the healing interventions in their previously shattered lives.

What happened to the twelve children salvaged by benign intervention in Iowa was paralleled, in their own lifetimes, by what happened to an entire generation of poor Americans of all races after World War II. Lest we forget, it was only under a G.I. Bill of Rights that had never existed prior to Pearl Harbor that all American ex-servicemen were guaranteed college or trade school admission, the costs of studying, and living subsidies by our government.

The right to a college or technical education under the G.I. Bill of Rights did not depend upon IQ test scores, or social class, or race or skin color, but on the serviceman's personal history and duration of military service. Education, for these men, became one of their rights of citizenship, and not a privilege bestowed by the accident of birth upon the children of economically solvent families.

The result was that literally thousands of ex-servicemen from poor families—whose IQ test scores invariably reflected their inherited social status and *not* their genetic mental capacities—did go to college and trade schools at the expense of a grateful nation. And they did emerge, most of them, as professionals, skilled craftsmen, businessmen, and managerial and craft employees of far higher social class and income levels than the Galtons and the Davenports, the McDougalls and the Easts and the rest of the scientific racists of their childhood had ever conceded they had the genetic capacities to achieve.

As we saw earlier, it was the cross-generational effects of the educational benefits of the G.I. Bill of Rights of 1945 that were responsible for the "miracle" of Windsor Hills—which witnessed the children of the nearly all-black Windsor Hills Elementary School achieve the highest IQ test scores in the entire Los Angeles, California, school system in 1969.

It should be plain that such massive planned interventions as this proposal for Mississippi advocates—much as they would affect the cellular and broader milieus in which human genes express themselves—are not to be classed as *genetic* interventions. We have neither the moral nor the scientific right to meddle, either by alteration or selection, with the genes of human beings—much as we do have the moral obligation to apply scientific findings on genetic mechanisms to improving the social, biological, cellular, and subcellular *environments in which genes are created and function.*

I make this statement because of a question with which Professors I. I. Gottesman and L. Erlenmeyer-Kimling opened an article in *Social Biology*[18] in 1971:

"Who is minding the quality of the human gene pool? Hardly anybody, it seems, except for a large handful of eugenically-minded scientists, some of whom are organized under the flag of the American Eugenics Society and other national eugenics societies."

In terms of such attentions—the legislative codes of twenty-one of our fifty states still have valid laws calling for the compulsory *eugenic* sterilization of human beings with low IQ test scores and related stigmata of "social inadequacy" passed under the leadership of the same "American Eugenics Society and other national eugenics societies"—the best thing, the *noblest* thing the eugenicists could do for "the human gene pool" is to stop minding it, now and forever.

If they are really concerned about the quality of the human gene pool, they might do something about bringing to an end the current epidemic of literally hundreds of thousands of compulsory eugenic sterilizations yearly of white and nonwhite Americans who committed the "crime" of having been born to poor parents.

The people of this nation—and this world—have in this century paid about as high a price as any species deserves to pay for gene watches mounted by race hygienists past and present. The human gene pool is too valuable a resource to be subjected to further doses of eugenical aggressions. A half century of cruel eugenical interventions into the value systems of the people who make public policies in the fields of health, education, and welfare—and the people who elect them to their posts—have robbed the American people of the advanced social programs and benefits enjoyed by the people of Sweden, the Netherlands, England, France, Denmark, Japan, and all of the other industrial nations with lower infant-mortality and higher longevity rates than we have here.

We have far better things to do with our gene pools than to turn them over to the gelders, the stigmatizers, and the would-be breeders of supermen. We are not slaughterhouse turkeys. We are people.

As members of the human race, it is high time that we started to find out exactly how valuable are the actual—but for the most part never fully realized—genetic possibilities present at the moment of creation in the zygote of every newborn child. It is time to demonstrate to ourselves, and to the world, the vast social and economic benefits that would be inherent in the act of making the entire human race as fully, as truly *human* as we know it to be genetically capable of becoming.

If, as a nation, we do not respond to the imperatives of liberating all human genes from the *preventable* biological, physical, chemical, emotional, and sociocultural hazards that now prevent all but a few of them from ever developing their full genetic range of expression, then the costs to the nation of such capitulation to the false dogmas of scientific racism are quite predictable. The continued neglect of the majority of the productive, intellectual and moral potentials inherent in millions of the newborn children now denied the equality of opportunity for optimum epigenetic or postnatal development must—in the not very distant future—inevitably accelerate our present rates of social, economic, and political decline as measured by the world's social, economic, and health achievements.

Let us, therefore, unleash our pediatricians and their scientific collaborators in Mississippi. They might just be better able to show us the ways out of our spreading and deepening morass than are the politicians and the generals whose leadership stranded us in it.

NOTES AND REFERENCES

PREFACE

1. Flexner, 1910, p. 20.
2. Patrick Buchanan, " 'The IQ Myth' Second Showing Still Leaves Skeptics," *TV Guide,* August 2, 1975, pp. A3–A4.
3. Hague, 1914, IV, 567.
4. Chase, 1972, pp. 284–87.
5. Flexner, 1910, p. 20.
6. *Ibid.,* p. 19.

CHAPTER 1

1. Quoted from William Shannon, "Mr. Nixon's Revenge," *The New York Times,* October 12, 1971, Op-Ed page.
2. U.N. Demographic Yearbook, 1974. See also the table, "National Wealth and Its Effects," *The New York Times,* September 28, 1975, Sec. 4, p. 3.
3. The original discovery of the cholera bacillus had been made in 1854 by the Florentine physician Filippo Pacini (1812–83). For years, at various scientific congresses, Pacini presented ample proof that he had discovered the causative agent of cholera. His evidence was ignored. Pacini lived long enough to witness the birth of the germ theory of disease as a result of the work of Pasteur and Koch between, roughly, 1875 and 1885. He died one year before Koch's rediscovery of the cholera bacillus. In 1965, the International Commission on Bacteriological Nomenclature recommended that the preferred name of the cholera bacillus should henceforth be *Vibrio cholera* Pacini 1854.
4. See also Duell, 1942, pp. 219–29.
5. William Shockley, "Dysgenics—A Social-Problem Reality Evaded by the Illusion of Infinite Plasticity of Human Intelligence?" Paper read before annual meeting, American Psychological Association, September 7, 1971; see p. 8.
6. "Vaccination in Ireland," unsigned editorial, *British Medical Journal,* February 24, 1877, p. 236.
7. Hofstadter, 1959, pp. 44–47.
8. Both quotes are from Niswander and Gordon, 1972, p. 2.
9. Galton, 1883, pp. 24–25.
10. I. F. Henderson and W. D. Henderson (eds.), *A Dictionary of Scientific Terms.* Seventh Edition by J. H. Kenneth (Princeton: D. Van Nostrand Co., 1960), pp. 74, 148.
11. Pearson, 1914–40, III (1930), 220.
12. Francis Galton, "The Possible Improvement of the Human Breed under the Existing Conditions of Law and Sentiment," the second Huxley Lecture of the Anthropological Institute, 1901. Reprinted in *The Popular Science Monthly,* LX (1901), 218–33, see p. 221.
13. Galton, 1869, p. 327.
14. Pearson, 1914–40, II (1924), 209.

15. Pearson, 1914–40, III (1930), 218–20.

16. Laughlin, 1922, pp. 446–47.

17. Robitscher, 1973: see particularly Tables 1 and 2 in appendices and chapter by Professor George Tarjan (Chapter 2, "Some Thoughts on Eugenic Sterilization"). See also Pickens, 1968: Chapter 6, "Sterilization: The Search for Purity in Mind and Body," for a historical survey linking forced surgical sterilization in the United States to the influence of the eugenics and Social Darwinist movements. See also: George Tarjan, "Sex: A Tri-Polar Conflict in Mental Retardation," paper read before the symposium *Choices of Our Conscience,* sponsored by the Joseph P. Kennedy, Jr., Foundation on Human Rights, Retardation and Research, 1971.

18. United States District Judge Gerhard A. Gesell, Opinion in *Relf* v. *Weinberger et al.:* civil actions Nos. 73–1557, 74–243, U.S. District Court for the District of Columbia, March 15, 1974.

19. *Ibid.*

20. By 1975, the National Center for Health Statistics' preliminary estimates suggested that the annual rate of hysterectomies alone had climbed to 690,000 of which 12,000 would end in death for the women who had them.

21. Joann Rodgers, "Rush to Surgery," *The New York Times Magazine,* September 21, 1975, p. 40.

22. *Ibid.,* p. 34.

23. *Ibid.,* p. 40.

24. Dr. Wood's report, kindness of the Association for Voluntary Sterilization, Inc., New York; McGarrah quotation from "Voluntary Female Sterilization: Abuses, Risks and Guidelines," in *Hastings Center Report,* June 1974, pp. 5–7 (adapted from Wolfe and McGarrah, 1973). In the same article McGarrah presented the following table on increases in the number of sterilization procedures and hysterectomies in the two-year interval between July 1968 and July 1970, at the Women's Hospital of the Los Angeles County Medical Center:

Surgical Procedure	Increase
Elective tubal ligation	470%
Tubal ligation after delivery	151
Elective hysterectomy	742

25. "More in U.S. Rely on Sterilization: About 1 in 6 U.S. Families in Main Child-Bearing Years Choosing Operation," news story signed by Harold M. Schmeck, Jr., *The New York Times,* March 25, 1974.

26. In Hall, 1970, "The Obstetrician's View," pp. 85–95. See p. 93 for "Package Deal." See also Krauss, 1975.

27. See AP unsigned news story, "H.E.W. Bars Full Matching Aid for the Poor in Abortion Cases," *The New York Times,* December 9, 1974. See also Sheila M. Rothman, "Funding Sterilization and Abortion for the Poor," *The New York Times,* February 22, 1975, Op-Ed page.

28. Second International Congress of Eugenics, 1921. Lidbetter: I, 391–97; Laughlin: II, 286–91.

29. Third International Congress of Eugenics, 1934, pp. 364–68.

30. *Ibid.,* pp. 201–09.

31. *Ibid.* Sadler: pp. 193–200; Gosney: pp. 369–71.

32. Morton A. Silver, "Birth Control and the Private Physician," *Family Planning Perspectives,* IV, No. 2 (1972), 42–46.

33. In 1972, Richard Friske, an officer of the German Luftwaffe in World War II and now a conservative Republican member of the Michigan state legislature, made public a less sexist variant of this proposal. Representative Friske, who had also been an organizer for the John Birch Society and the George Wallace

for President crusade, issued a release to the press calling for "curbing the growth of the drone population" to be found "in America's slums." At the same time, in the classic eugenics tradition established by Galton a century earlier, Friske also called for the encouragement of increased birth rates among "the educated, propertied Americans." See "Controls Asked on Births by Poor; Michigan Legislator Urges Larger Affluent Families," *The New York Times,* May 7, 1972.

34. "The Public Speaks," *Philadelphia Inquirer,* September 23, 1971.

CHAPTER 2

1. Director of the University of Wisconsin Mental Retardation Center and research director of the President's Panel on Mental Retardation during the Kennedy administration.
2. Stevens and Heber, 1964, p. 1.
3. Adams, 1971, pp. 8–9.
4. Holmes and Moser, 1972, introduction, p. 1.
5. Crome and Stern, 1972, pp. 112–13.
6. Eichenwald, 1966, p. 25.
7. *Ibid.*
8. Jack Tizard, "Can the Brain Catch Up?" *World Health,* February–March 1974, pp. 10–14.
9. National Center for Health Statistics, 1974, HRA No. 74–1219–1, p. 180.
10. Center for Disease Control, USPHS, Atlanta. U.S. Immunization Survey, 1973.
11. U.S. Senate, Select Committee on Nutrition and Human Needs, *To Save the Children,* January 1974, p. 47.
12. Conley, 1973, pp. 48–49.
13. *Ibid.,* p. 48.
14. *Ibid.,* p. 323.
15. Reed and Reed, 1965, p. 1.
16. Gottesman in Glass, 1968, p. 62; Jensen in Jensen, 1969, pp. 92–93.
17. Reed and Reed, 1965, p. 2.
18. Frankwood E. Williams, biographical essay on Walter Elmore Fernald, *Encyclopædia of the Social Sciences* (1st ed., 1935), pp. 184–85.
19. Quotations taken from pp. 12–13, 41, and 43 of the AMA's *Mental Retardation* handbook.
20. The term "feeblemindedness," which is no longer used clinically, was in the late nineteenth and early twentieth centuries one of the many terms used to describe mental retardation. Other more or less interchangeable terms included "idiocy," "imbecility," "moronity," "borderline mental deficiency," "hysteria," and "insanity." For contemporary accounts of the training and diagnostic competence of the eugenics field workers trained by Davenport, Goddard, and other early American eugenicists, see chapter 6. These field workers were famous, in their day, for being able to make clinical diagnoses of such purportedly hereditary conditions as "pauperism," "feeblemindedness," "criminality," and "shiftlessness" *at a single trained glance.* They were equally expert, according to their mentors, in diagnosing diseases ranging from syphilis to dementia praecox in the ancestors of today's feebleminded members of the degenerate classes, even in ancestors who died before the American Revolution. See also chapter 7 for how Goddard's Kallikak-study field workers needed only "a glance" to establish a deaf boy's low mentality, while the boy's father "showed by his face [*sic*] that he had only a child's mentality."
21. Reed and Reed, pp. 8, 10.
22. Personal communication, Richard E. Wexler.
23. Charles Benedict Davenport, "Euthenics and Eugenics," *The Popular Science Monthly,* January 1911, pp. 16–20.
24. *Ibid.,* p. 17.

25. Stevens and Heber, 1974, p. 2.
26. Robert W. Friedrichs, "The Impact of Social Factors upon Scientific Judgment: The 'Jensen Thesis' as Appraised by Members of the American Psychological Association," paper presented before the American Sociological Association's annual meeting, New Orleans, 1972. Reprinted in the *Journal of Negro Education,* Fall 1973.
27. Max Seham, *Blacks and American Medical Care* (Minneapolis: University of Minnesota Press, 1973), p. 14.
28. *Ibid.,* p. 14.
29. Milton Alter, "Is Multiple Sclerosis an Age-Dependent Response to Measles?" *The Lancet,* February 28, 1976, pp. 456–57. See also unsigned leading article, "Multiple Sclerosis," *The Lancet,* January 17, 1976, pp. 129–31.

CHAPTER 3

1. "Annals of Eugenics Portrait Series, No. 1," *Annals of Eugenics: A Journal for the Scientific Study of Racial Problems,* issued by the Francis Galton Laboratory for National Eugenics, University of London, Vol. I (1925–26) (Cambridge: Cambridge University Press, 1925).
2. Davies, 1955, pp. 38–39.
3. McDougall, 1923, p. 47.
4. "New I.Q. Test Excites Educators," *McCall's,* June 1972, p. 51.
5. *Science,* October 12, 1973, p. 115.
6. Archives of the Eugenics Record Office, American Philosophical Society Library, Philadelphia.
7. Chase, 1972, p. 242.
8. Finer, 1952, p. 233.
9. *Ibid.,* p. 237.
10. Letter to Lord John Russell, 1838. Quoted in Longmate, 1974, p. 62.
11. In some countries, such as England, Sweden, Holland, and Japan, such environmental reforms as the Victorian Sanitary Reformers advocated have been achieved without social revolution. In others, such as Czarist Russia and semi-feudal China, social revolution became the necessary precondition for such legislative programs of preventive medicine as universal and free vaccinations against serious infectious diseases, and cradle-to-grave free medical and nursing care for entire populations. The universal free vaccination laws of modern China, obviously, are no more Marxist or Maoist than the older and identical vaccination laws of modern England are Tory or Liberal or Labour. Neither viruses nor vaccines have any political ideologies or tropisms.
12. Finer, 1952, pp. 238–39.
13. Stewart and Jenkins, 1969, p. 1.
14. *Ibid.,* p. 70.
15. Signet edition (1961), pp. 115–16. Dickens shared the revulsion of most socially concerned English people to the role of Malthus and his teachings in the rationalizations of the needless physical and mental suffering endured by the factory "hands" and miners who actually produced the products on which the new industrial classes of factory owners and mine owners grew wealthy in Manchester and other "Coketowns." He, therefore, in painting the portrait of Thomas Gradgrind, the archetypal Coketowner (and Member of Parliament), added to the Gradgrind family "two younger Gradgrinds" named Adam Smith and Malthus (p. 29). This also happens to be their only appearance in *Hard Times.* For more on Dickens and Manchester, see Marcus, 1974, pp. 30–32, 40–41.
16. Chase, 1972, pp. 182–84.
17. *The New York Times,* June 18, 1975, p. 1.
18. Stewart and Jenkins, 1969, p. 69.
19. See pages 275–81, *Health United States 1975,* the first annual report to Con-

gress by the Health Resource Administration and the National Center for Health Statistics (January 1976; DHEW Publication No. [HRA] 76-1232).

CHAPTER 4

1. Julius Paul, "The Return of Punitive Sterilization Proposals: Current Attacks on Illegitimacy and the AFDC Program," *Law and Society Review,* August 1968, pp. 77–106.
2. "Science and the Race Problem: A Report of the AAAS Committee on Science in the Promotion of Human Welfare," *Science,* November 1, 1963, pp. 558–61. See also statement by Otto Klineberg and seventeen other social scientists, most of them members of the American Psychological Association. "Does Race Really Make a Difference in Intelligence?" *U.S. News and World Report,* October 26, 1956, pp. 74–76.
3. Mead, Dobzhansky, Tobach, and Light, 1968, pp. 122–31.
4. Quoted from Gossett, 1963, p. 109.
5. In Harris E. Starr, *William Graham Sumner* (New York, 1925); quoted from Gossett, 1963, p. 154
6. Quoted from Higham, 1963, p. 42.
7. See, for example, Baltzell, 1964; Gossett, 1963; Higham, 1963; Kamin, 1974; Morse, 1968; Solomon, 1956.
8. Osborne, 1970, p. 32.
9. *Ibid.,* p. 32.
10. Dean and Cole, 1967, p. 68. For effects of continuing Agricultural Revolution in English life, see Blythe, 1969, table entitled "Domesday for 1936 and 1966: The Second Agricultural Revolution," pp. 29–30.
11. Ashton, 1964, p. 5.
12. Osborne, 1970, p. 37.
13. Finer, 1952, p. 215.
14. Chambers and Mingay, 1966, p. 120.
15. Dean and Cole, 1967, p. 62.
16. Longmate, 1974, p. 34.
17. Chambers and Mingay, 1966, p. 119.
18. Wilson, 1954, p. 139. See also p. 235 for account of how, at the start of the Civil War in 1861, the 85-million-bushel grain harvest of 1840 was doubled, and how by 1890 "the American wheat crop was doubled again to become one-sixth of the entire world's production."
19. Williams, 1969, p. 102.
20. Folke Dovring, "Soybeans," *Scientific American,* February 1974, pp. 14–21.
21. *Statistical Abstract of the United States,* 1971, Table 953, p. 586.
22. More than a decade later, in a collage of extracts entitled "Prophets Revisited," the Malthus passage that inspired this doggerel was reprinted in a special box on the last page of *The Other Side,* the newsletter of The Environmental Fund of Washington, D.C., in its issue of May 1975. Hardin is one of the six directors of this group.
23. Longmate, 1974, p. 44.
24. The full text of this handbill can be found in Himes, 1963, pp. 214–17; Pike, 1966, pp. 300–03; Hardin, 1964, pp. 192–93. See also Fryer, 1965, Chapters 5 and 6, the first on Place, the second on Carlile.
25. For some reasons that have had literally nothing to do with the writings and philosophy of Thomas Robert Malthus, post-Malthusian authors have persisted in terming birth control *the Malthusian solution* to what they see as the population problems of the nineteenth and twentieth centuries. A classic instance of this misreading of Malthus was provided by Margaret Sanger, who on page 125 of her *Autobiography* (1938) made the flat statement that "the father of family limitation was Thomas Robert Malthus." Malthus was, of course, the first major enemy of the movement to enable poor people to

limit the size of their families by the use of contraceptive techniques and devices.

CHAPTER 5

1. *Dorland's Illustrated Medical Dictionary* (24th ed.; Philadelphia: W. B. Saunders Company, 1968).
2. Needham, 1956, pp. 183 *et passim.*
3. Gobineau, 1915.
4. The elitist Gobineau's fear of democracy was well founded, since it did threaten all of the royalist concepts of the natural inequality of races. This was underscored in the letter Thomas Jefferson sent on June 24, 1826, to the organizers of the celebration of the fiftieth anniversary of the Declaration of Independence, which was held on the same day, July 4, 1826, as Jefferson's death at the age of eighty-three. Wrote Jefferson: "The general spread of the light of science has already laid open to every view the palpable truth, that the mass of mankind has not been born with saddles on their backs, nor a favored few booted and spurred, ready to ride them legitimately, by the Grace of God." Jefferson apparently understood as early as 1826 the basic differences between the legitimate sciences, with their search for truth, and the pseudosciences, with their continuing quest for scientific-sounding rationales for poverty and inequality.
5. Heiden, 1944, pp. 245–46.
6. Stocking, 1968, pp. 60–62.
7. Baker, 1974, p. 48.
8. Hankins, 1926, p. 128. Hankins, after quoting the statement Lapouge had made "in 1887 and 1889," added: "After the Great [1914–18] War he [Lapouge] pointed to this statement as a prophecy whose fulfilment had exceeded his worst apprehensions!" See also Klineberg, 1935, p. 7, and Benedict, 1943, p. 3.
9. Gossett, 1963, p. 77.
10. *Ibid.*
11. Galton, 1869 (quotations from final revised edition, 1914), pp. 327–28.
12. Francis Galton, "Hereditary Improvement," *Fraser's Magazine,* January 1873, p. 118. It was a cardinal tenet of the eugenics and other scientific racist dogmas that only the blacks and Orientals were lower on the evolutionary scale than the Irish.
13. Pearson, 1914–40, II (1924), 208–09.
14. Galton, 1869, p. 12.
15. "Hereditary Improvement," p. 118.
16. *Ibid.,* pp. 119,123.
17. Galton took a dim view of urban life, since he felt that it tended to render sterile the sexuality of the naturally gifted types and to favor the fecundity of "the less suitable strains" of Englishmen. He failed, completely, to understand that the great decline in the live-birth rates of the affluent classes was caused by the sharp reductions in urban infant, child, and maternal deaths that all resulted from the health effects of the Agricultural and the Industrial revolutions and the great Sanitary Movement. Clean water and hygienic sewage systems; cheap and washable cotton cloth for underwear and bed linens; vaccination against smallpox; and aseptic obstetrical care, let alone adequate protein-calorie intake levels, were all social gains that were first enjoyed by the affluent and upper middle classes. As a result, their falling birth rates were the first to reflect these sociobiological advances.
18. "Hereditary Improvement," p. 129.
19. Richard Herrnstein, "I.Q.," *The Atlantic Monthly,* September 1971, pp. 43–64.
20. Jensen, 1969, p. 95.
21. Blacker, 1952, p. 104.

22. "Hereditary Improvement," p. 117.

23. Stewart and Jenkins, 1969.

24. Reprinted in *The Popular Science Monthly,* LX (1901), 218–33.

25. *Ibid.,* pp. 223, 231, 233.

26. Galton, 1869, p. 346.

27. Higham, 1963; see first four chapters.

28. That this error also carried over into the twentieth century was amusingly demonstrated in 1955 in the blurb on the jacket of Richard Hofstadter's *Social Darwinism in American Thought.* Although Professor Hofstadter pointed out on page 39 that Spencer had "coined the expression 'survival of the fittest,' " the blurb began: "The centenary of the publication of Charles Darwin's *The Origin of Species* is a startling reminder of the tremendous impact its ideas have made upon our *total* culture, and of how widely its principal themes, *survival of the fittest, natural selection, and evolution,* have become assimilated" (blurbist's emphasis).

29. *We know* nothing of the sort. Races per se have no such characteristics. As the anthropologist Marvin Harris of Columbia University has observed, today in the Mongoloid (Oriental) race of man we can find every kind of culture and society in human history—from primitive forest tribes to capitalist industrial nations like Japan with close to 100 percent literacy; to nearly 100 percent illiterate fascist dictatorships; to communist and industrializing nations such as North Korea on the border of capitalist and industrial South Korea; let alone theocratic, military, and political dictatorships; colonies and semi-colonies of European and other non-Mongoloid powers; and the world's largest nation, Communist China. Racially, all of the peoples of these vastly diverse cultures are *identical.* None are in any way *typical* of their race, any more than any of them are *untypical.* (Personal communication, Marvin Harris.)

30. David G. Croly and George Wakeman, *Miscegenation: The Theory of the Blending of the Races, Applied to the American White Man and Negro* (New York, 1864), pp. 29–30; and Joseph LeConte, "The Race Problem in the South," bulletin, Brooklyn Ethical Association, No. 28 (1893), p. 360. Both quoted from John S. Haller, Jr., 1971, p. 162.

31. Solomon, 1956, p. 64.

32. Francis Amasa Walker, *The Atlantic Monthly,* June 1896, pp. 822–29.

33. In 1908, in the fifth volume of his *History of the American People,* Wilson wrote (pp. 212–13) that the census of 1890 revealed that while "immigrants poured in steadily as before," there had also been "an alteration of stock which students of affairs marked with uneasiness. Throughout the century men of sturdy stocks of the north of Europe [i.e., Nordics] had made up the main strain of foreign blood which was every year added to the vital working force of the country, or else men of the Latin-Gallic stocks of France and northern Italy; but now there came multitudes of men of the lowest class from the south of Italy and men of the meaner sort out of Hungary and Poland, men out of the ranks where there was neither skill nor energy nor any initiative of quick intelligence; and they came in numbers which increased from year to year, as if the countries of the south of Europe were disburdening themselves of the more sordid and hapless elements of their population, the men whose standards of life and work were such as American workmen had never dreamed of before." This passage was, of course, a clear paraphrasing of Walker's words.

Four years later, when Wilson became the Democratic candidate for the presidency of the U.S., his Republican opponents reprinted these words in English and in other languages and distributed them to naturalized voters and their American-born children. Wilson not only had to eat these words in public, but he also worked furiously to prove his lifelong admiration for southern Italians, Hungarians and Poles of all sorts, and other non-Nordics. See also Higham, 1963, p. 190.

CHAPTER 6

1. Pearson, 1910–40, III (1930), 613.
2. *Boston Evening Transcript,* quoted from Chaplin, 1959, p. 24.
3. Chaplin, 1959, p. 26. See Ray Allen Billington, "The Burning of the Charlestown Convent," *The New England Quarterly,* March 1937, pp. 4–24.
4. Solomon, 1956, p. 88.
5. *Ibid.,* p. 112.
6. Sidney Cohen, *The Drug Dilemma.*
7. David F. Musto, *The American Disease: Origins of Narcotic Control* (New Haven: Yale University Press, 1973), p. 5.
8. Quoted from *The Educational Test: Its Futility and Harmfulness as a Measure to Regulate Immigration* (New York: National Liberal Immigration League, 1912).
9. Scholars familiar with the history of scientific racism have wondered why Jewish scientists, such as Boas and Myerson, ever had anything to do with the clearly anti-Semitic eugenics movement. Boas, as a matter of fact, had long since denounced the eugenics movement when he inserted a clause into his will in 1924 leaving his vast collection of anthropometric measurements, notes, and other data to the library of the Eugenics Record Office, headed by Davenport. This was because, with its own collection of height, weight, and other body measurements, the Eugenics Record Office was as of 1924 America's largest archive of the raw data of physical anthropology. Four years later, however, when Columbia University, where Boas was professor and chairman of the Department of Anthropology, started its own Department of Physical Anthropology, Boas informed Davenport that he had revoked the 1924 will and had now left his anthropometric data to Columbia University. Not long ago, I asked an aging American writer who is Jewish and, when younger, had known Boas how people like Boas and Myerson could have any dealings with anti-Semites such as Davenport and Osborn. He answered with a question of his own: "If we Jews had to choose to work only with Gentiles certified to be a hundred percent free of anti-Semitism, who could we ever really work with?"
10. In his 1935 book, *Out of the Night,* Muller had proposed preserving the sperm of great men for fertilizing mothers of lesser "genetic qualifications" in future generations and even future centuries. At the 1966 Congress on Human Genetics, held in Chicago six months before Muller died, he renewed this proposal. However, as "Professor Jérôme Lejeune of the Paris Faculty of Medicine pointed out, in the first draft of his paper Dr. Muller had suggested including among potential donors for his sperm bank several French scientists, and in the political field, Stalin and Lenin. In the second version of the speech, after the Khrushchev period, Stalin's name had disappeared." (*UNESCO Features,* No. 662, 1974, p. 27.)
11. Oscar Riddle, National Academy of Sciences, *Biographical Memoirs,* XXV (1949), 75–110. See also biographical sketch by E. C. McDowell, a co-worker of Davenport's, in *Bios,* XVII (1946), 1–50. Riddle's *Memoir* is based primarily on McDowell's *Bios* article.
12. H. H. Goddard, "In Defense of the Kallikak Study," *Science,* XCV, No. 2475 (1942), 574.
13. Myerson, 1925, pp. 77–80.
14. M. Haller, 1963, p. 72.
15. These quotes and data on the early history of state compulsory eugenic sterilization laws are all from Laughlin, 1922.

16. Dight, "an eccentric Minneapolis physician [who] as a result of great frugality for a time lived in a tree," accumulated a small fortune, which he left to the University of Minnesota to establish the Dight Institute for Human Genetics. (M. Haller, 1963.)
17. Garrett Hardin, "Parenthood: Right or Privilege?" editorial in *Science,* CLXIX (1970), 427.
18. The British geneticist J. B. S. Haldane, in his 1938 book *Heredity and Politics,* observed (p. 17) that under the terms of this law, the poet John Milton would have been subject to forced sterilization under clause 7 (Blind); the composer Ludwig von Beethoven would have been subject to compulsory sterilization under clause 8 (Deaf); and that Jesus Christ would have been subjected to eugenic sterilization under clause 10 (Dependent, including orphans, ne'er-do-wells, the homeless, tramps, and paupers)—since, *by Laughlin's definitions of social inadequacy,* Jesus was homeless, a tramp, and a pauper.
19. Landman, 1932, p. 259.
20. Adolf Hitler, *Mein Kampf,* 1925. Translation quoted from Blacker, 1952, p. 44, and English-language edition of *Mein Kampf* (New York: Reynal & Hitchcock, 1941), p. 608.
21. Hans Harmsen, "The German Sterilization Act of 1933: Gesetz zur Verhütung erbkranken Nachwuches," *Eugenics Review,* January 1955.

CHAPTER 7
1. Mrs. P. F. Hall, 1922, pp. 1 *et passim.*
2. Robert DeCourcy Ward, "The Restriction of Immigration," *North American Review,* 1904, pp. 226–37.
3. Fairchild considered himself to be a disciple of Hall's as early as his university days prior to World War I. As a young professor and author, Fairchild worked very closely with Hall in advancing the cause of ending non-Nordic immigration. See, for example, Hall's letter to Charles Evans Hughes, July 29, 1916; Hall to William S. Sadler, April 12, 1918; and Fairchild's letter to Hall, dated January 27, 1919, concerning the political infighting involved in his efforts to carry on the IRL crusade in Washington. All in the archives of the Immigration Restriction League, Houghton Library, Harvard University.
4. Goddard, 1912, pp. 116–17.
5. See, for example, Richard Todd, "A Theory of the Lower Class: Edward Banfield, the Maverick of Urbanology," *The Atlantic Monthly,* September 1970, pp. 51–55.
6. McKeown and Lowe, 1974, p. 86. A case in point is the modern turkey bred solely to provide maximum amounts of meat for the $500-million-dollar-a-year turkey industry. Whereas the turkeys found in their original wild state by Columbus et al. weighed from twelve to fifteen pounds, by 1972, according to Nadine Brozan ("Pilgrims Wouldn't Recognize Today's Thanksgiving Turkey," *The New York Times,* November 23, 1972, p. 48), "technology and genetics have altered almost everything about a turkey from the color of its feathers" to the twenty-five pounds of meat on the skeleton of the average commercially bred turkey.

    Genetic engineering has given the modern raised-for-slaughter commercial turkey characteristics extremely rare in natural strains of turkeys. Those characteristics include "white feathers instead of bronze to eliminate a residue of pigment in the meat, broader breasts, shorter legs. Of course, progress has its price—*while his wild ancestor could fly at about 50 miles per hour, the domestic turkey is lucky if he can get his top-heavy body off the ground*" (italics added).

    The modern turkey, triumph of genetic engineering that he is, has lost more than the ability to fly from danger. Thanks to the genetic "engineers," the commercial turkey's enormous weight and grotesque proportions have deprived him of the only pleasure left to the poor—sex. The modern turkey "intended for the marketplace is born out of artificial insemination. Because of biological progress [*sic!*], natural mating has been virtually eliminated. The breeder tom (some turkeys are grown exclusively for breeding purposes) has grown so large, weighing about 32 pounds, that he would harm the hen, 20 to 24 pounds, if permitted to follow his natural instincts."

However profitable to the slaughterhouses these manipulations of the genetic potentials of wild turkeys have made their tame ancestors, they have also turned these once swift-flying, self-reliant, and self-replicating turkeys into evolutionary cripples completely at the mercy of an alien species—man—for food, shelter, and even reproduction.

7. Unit character "genetics" and its wholesale or cheaper-by-the-dozen variant—"Polygenic Inheritance"—have, of course, remained the core dogma of eugenics and its contemporary true believers to this very day. See, for example, Reed and Reed, 1965, p. 75, and Jensen, 1969, pp. 32–33, 70.

8. Charles B. Davenport, *The Feebly Inhibited: Nomadism, or the Wandering Impulse, with Special Reference to Heredity: Inheritance of Temperament,* Publication No. 236, and *Naval Officers: Their Heredity and Development,* Publication No. 259 (Washington, D.C.: The Carnegie Institution of Washington, 1915, 1919).

9. Like other original thinkers, Davenport was at times unduly pessimistic. During the last dozen years of his corporeal life on earth, Davenport knew that the Nordic eugenics activist Adolf Hitler employed on a mass scale the "infinitely superior" weapon of capital punishment against Nordic and non-Nordic race polluters whose inferior genes threatened the purity of the Aryan racial gene pools of the Third Reich. He even defended the Nazi pledge to exterminate the Jews; see *Eugenical News,* September–October 1932, p. 117, editorial note to the article "The Nordic Movement in Germany," by the Nazi propagandist Karl Holler.

10. Most of the letters to and from Madison Grant are from one and often more than one of three primary sources: the archives of the Immigration Restriction League at the Houghton Library, Harvard University; the archives of the Eugenics Record Office at the library of the American Philosophical Society; and the papers of Henry Fairfield Osborn in the archives of the library of the American Museum of Natural History. In some instances, the original and copies of a single letter are to be found in all three archives.

11. Hellman (1969) quotes the ornithologist Robert Cushman Murphy as revealing, in private conversation, that: "Osborn followed some of Madison Grant's prejudices, but he was very evasive about them. He wouldn't see the press when they sent Jewish reporters to question him on some of his published statements. He delegated this to Dr. [William K.] Gregory." Dr. Gregory's services to Osborn continued after the death of his chief in 1935. By 1937, when Gregory wrote the Osborn essay for the *National Academy of Sciences Biographical Memoirs* (XIX [1938], 53–119), the many anti-Semitic, anti-Catholic, and anti-democratic causes and societies in which Osborn and Grant had shared leadership over three decades had, because of their intellectual and even working links with the German Nazi movement, at last become less than respectable to growing numbers of American scientists. Gregory could hardly omit any mention of Madison Grant in his obituary essay on Osborn, but he could and did confine it to a single mention on page 81: "Ably seconding his friends Madison Grant and John C. Merriam, he [Osborn] was one of the founders of the *Save The Redwoods League* . . ."

12. Franz Boas, "War Anthropology: Peoples at War," reviews of *The Passing of the Great Race,* by Madison Grant, and *Long Heads and Round Heads, or What's the Matter With Germany?,* by W. S. Sadler, *American Journal of Physical Anthropology,* I (1918), 363.

13. Frederick Adams Woods, book review, *Science,* October 25, 1918, pp. 419–20.
14. A footnote on page 598 of the 1941 edition of *Mein Kampf,* by the ten prominent American scholars who annotated the translation, reads: "Possibly a reference is made here to American protagonists of racial science, notably Lothrop Stoddard and Madison Grant. Some of this group had argued that the 'barbarism' of the Germans during the War (1914–18) had been due to the fact that the best elements of the German population had been annihilated during the Thirty Years' War."

    See also "The Nordic Movement in Germany," translated from the German of Karl Holler, *Eugenical News,* September–October 1932, which describes the role of non-Germans—from Gobineau and Chamberlain to Ripley and Lapouge and "the books of the two Americans, Madison Grant and Lothrop Stoddard, [which] were translated [into German] and found a large circle of readers"—in awakening "in Germany the Nordic thought, that is, the movement for the preservation and increase of the Nordic race." The same article went on to say that "it was especially the party of Adolf Hitler, 'The National Socialistic German Worker Party,' which took up the Nordic idea in its program and publicly subscribed to it."
15. Hitler, 1941, pp. 597–600.
16. Probably the most famous passage in the book, this sermonette on the dangers of race mixing was, of course, merely a nonmathematical statement of the old (and by 1916 thoroughly disproven by the world's *professional* geneticists) Galton-Pearson "law of regression to the hereditary mean."

CHAPTER 8

1. Franz Boas, "Rudolf Virchow's Anthropological Work," *Science,* September 19, 1902, p. 443.
2. Like Van Evrie, Jensen (1969) looked to the anatomy of the human brain to explain what he described as the Level II, or abstract thinking, ability of white children, writing that *"certain neural structures* must also be available for Level II abilities to develop, and these are conceived of as being different from *the neural structure* [*sic*] *underlying Level I"* (italics added).
3. In the year the American Civil War ended (1865), Huxley wrote, in his essay "Emancipation—Black and White," that "no rational man, cognizant of the *facts,* believes the average negro is equal, still less the superior, of the average white man . . . it is simply incredible that, when all his disabilities [i.e., slavery] are removed, and our *prognathous* relative has a fair field and no favor, as well as no oppressor, he will be able to compete successfully with his *bigger-brained and smaller-jawed* rival, in a contest which is carried on by thoughts and not by bites. The highest places in the hierarchy of civilization will assuredly not be within reach of our dusky cousins." (Italics added.)
4. See J. H. Comstock, "Burt Green Wilder," *Science,* May 22, 1925, pp. 531–33; and Edward H. Beardsley, "The American Scientist as Social Activist: Franz Boas, Burt G. Wilder and the Cause of Racial Justice," *Isis,* LXIV (1973), 50–66.
5. J. Haller, 1971, pp. 84–85; and Beardsley, 1973, p. 51.
6. Burt G. Wilder, book review, *Science,* XLII (1915), 768. See Beardsley, 1971, for account of how "ultimately Wilder had to settle for publication of a much shorter review than he wanted. *Science* editor James McKeen Cattell not only found the article long but also thought it improper to 'take up in *Science* the question of social relations, etc. of the Negro.'" Cattell, who had studied under Galton, was a foremost spokesman for Galton's views in America.
7. Wilder, 1909, p. 22.
8. Bean's Article, "Some Racial Peculiarities of the Negro Brain," was published in the *American Journal of Anatomy,* V (1906), 353–432. His popular version

of his "discoveries" ran in the September and October 1906 issues of the middlebrow *Century Magazine,* pp. 778–84, 947–53. Mall's article, "On Several Anatomical Characteristics of the Human Brain Said to Vary According to Age and Sex, with Special Reference to the Frontal Lobes," ran in the *American Journal of Anatomy,* IX (1909), 1–32.

9. It was quite possibly because of the pre-World War I work of Wilder, Mall, and other anatomists that, in 1969, when Arthur R. Jensen helped revive the Van Evrie-Bean-Shufeldt types of brain anatomy myths, he studiously refrained from claiming that such anatomical differences between the brains of white people and black people actually exist. Instead, Jensen suggested that "certain neural structures *must* [italics added] also be available" to make the brains of white people so much more capable of handling analytical and "abstract" mental problems than those of black people. See Jensen, 1969, p. 114.

10. Franz Boas, "Review of William Z. Ripley, *The Races of Europe,*" *Science,* September 1, 1899, pp. 292–96. Reprinted in Boas, 1940, pp. 155–59.

11. Higham, 1963, pp. 154–56.

12. Franz Boas, "The Growth of Children," *Science,* May 6 and 20, 1892, and December 23, 1892, pp. 256–57, 281–82, and 531–52. See also Boas, 1940, pp. 103–30.

13. Stocking, 1968, p. 174.

14. See correspondence between John C. Phillips and his brother William, then Assistant Secretary of State, dated January 15, 1915, re reference of Jenks for job of IRL propaganda agent; also correspondence between Jenks and Hall, Lee, and other IRL leaders on his propaganda results for IRL, 1915. IRL archive, Houghton Library, Harvard. See particularly undated letter, John C. Phillips to Hall, saying: "On consultation with Mr. [Joseph] Lee this A.M. I closed with Jenks for $300.00 a month." *The Immigration Problem,* which Jenks wrote in collaboration with W. Jett Lauck in 1912, was for some years a widely used college textbook and was cited by the Imperial Wizard of the Knights of the Ku Klux Klan, Hiram Evans, as proof of the Klan claim that non-Nordic—particularly Jewish and Italian—immigration was causing racial deterioration in America. See Evans, 1921, p. 8.

15. In the same letter, Patten informed the leaders of the Immigration Restriction League that he and South Carolina Congressmen Hayes and Lever had set in motion "a peach of a plot to completely discredit the cuss O.S.S. [Oscar S. Straus, Secretary of Commerce and Labor] at the White House, and it seems to me it will impress W.H.T. [incoming President William Howard Taft] with the dire need of reorganizing the immigration bureau and disgust him and every good citizen with 'The Jew in the Cabinet.' " Straus, a member of the family that owned the Macy's department store, had been appointed to the Cabinet by President Theodore Roosevelt in 1906 in a successful effort to win Jewish votes for the New York Republican party. Author of a scholarly biography of Roger Williams, twice appointed Ambassador to Turkey by Presidents Cleveland and McKinley, Straus earned the enmity of the Immigration Restriction League on two counts. In addition to being a Jew, as Secretary of Commerce and Labor Straus was also in charge of the U.S. Immigration Service—where his humane administration of the immigration laws rankled the immigrant baiters.

16. Boas, 1940, pp. 61–65.

17. Bernice Kaplan, "Environment and Human Plasticity," *American Anthropologist,* LVI (1954), 780–800, reviews post-Boas literature. See also William W. Greulich, "A Comparison of the Physical Growth and Development of American-born and Native Japanese Children," *American Journal of Physical Anthropology,* XV (1957), 489–515, and Greulich, "Growth of

Children of Same Race under Different Environmental Conditions," *Science,* March 7, 1958, 515–16.

18. Hooton, 1937, pp. 208, 211.
19. Baker, 1974, pp. 201–02. In the same book, Baker writes (p. 59) of the Nordicists that "only one . . . Lapouge, strongly condemns the Jews. Treitschke is moderately anti-Jewish; Chamberlain, Grant, and Stoddard *mildly so;* Gobineau is equivocal. The rest show little or no interest in the Jewish problem. . . . With the exception of Lapouge it is impossible [*sic*] to imagine any of these men participating in or condoning actual cruelty to Jews, though some of them would limit their immigration.' " (Italics added.)
20. The achievement of "more favorable conditions during the period of growth" can, of course, be accomplished by methods other than immigration. In our own times after the flood control, agricultural, and health care programs of the government formed by the Chinese Revolution of 1949 proceeded to end the famines that had periodically plagued China for more than a century, and to make medical and/or trained paramedical care available to all families, some changes seem to have occurred. Professor Jay M. Arena, past president of the American Academy of Pediatrics, who visited China with a delegation of American Medical Association specialists in 1974, noted that "Chinese children are now growing up to be taller and stronger. The seven-year-olds are reported to be 3–5 kgs. [6½ to 11 pounds] heavier and more than 10 cm. [3.94 inches] taller than those in 'pre-liberation' days." ("China's Children," *Nutrition To-Day,* September–October 1975, pp. 20–25.)
21. W. Johannsen, "Heredity in Populations and Pure Lines," in Peters, 1959, pp. 20–26.
22. Cavalli-Sforza and Bodmer, 1971, p. 523. A simple experiment that can be performed in any home or apartment will demonstrate the role of clearly environmental factors in the phenotypical variations of the biological or physiological potentials present in any genome or genotype.

This experiment calls for the removal, and the planting and rearing under varying experimental conditions, of the six peas (seeds) generally contained in a single pea pod. Each of these peas are, of course, phenotypes grown from a single genotype.

Plant each of these peas in an individual pot of identical soil, and number each pot from 1 to 6. Grow the pea in Pot 1 under optimum growth conditions—that is, with a maximum of sunlight, daily watering, and even a little added fertilizer or other soil nutrients. Grow the pea in Pot 2 with the same amount of sunlight, but provide it with no fertilizer or soil nutrients, and water it only once a week. Grow the pea in Pot 3 in a dark closet, in soil nourished by fertilizers or soil nutrients, water it daily, but give it only one hour of sunlight daily. Grow the pea in Pot 4 in the same closet, give it one hour of sunshine on alternate days only, deprive it of all fertilizers and soil nutrients, and water it only once weekly. Grow the pea in Pot 5 in the same dark closet, but never take it out of the closet, and water it only once monthly. Grow the pea in Pot 6 in the same dark closet, but provide it with no sunlight whatsoever, and give it neither fertilizer, nutrients, nor water from the moment it is placed in the dark closet.

Those of the six pea plants that survive the varied environmental insults, such as partial or total lack of light, water, and nourishment—probably the pea plants in Pots 1, 2, and 3—will each be very different phenotypes, in terms of height, color, weight, and vigor, of the same ancestral genotype. Genetically, however, these markedly variant phenotypes will be, of course, identical.
23. See G. H. Hardy, "Mendelian Proportions in a Mixed Population," in Peters, 1959, pp. 60–62.
24. Kenneth M. Ludmerer, "American Geneticists and the Eugenics Movement:

1905–1935," *Journal of the History of Biology*, II, No. 2 (1969), 337–62; see p. 347.

25. Walter S. Sutton, "The Chromosomes in Heredity," in Peters, 1959, pp. 27–41.

26. T. H. Morgan, "Sex Limited Inheritance in Drosophila," in Peters, 1959, pp. 63–66.

27. On July 25, 1921, Muller delivered a lecture at the Biological Laboratory of the Brooklyn Institute, Cold Spring Harbor, New York, then part of the Davenport scientific empire. Davenport himself suggested the title for this lecture: "An Episode in Science." The typed transcript of the entire talk was locked away in the Davenport Cold Spring Harbor files with the hand-written notation on page 1: "Keep until Dr. Muller's authorized release or after 20 years." I found and Xeroxed this transcript in the collection of Davenport's letters and papers now at the library of the American Philosophical Society in Philadelphia, in 1971, and with their gracious permission these excerpts are here reprinted—after a lag of more than fifty years.

28. Morgan to Henry Fairfield Osborn, January 9, 1914, in Osborn papers, library of the American Museum of Natural History.

29. See Davenport-Grant correspondence, March–April 1917, in which Davenport advised Grant to delete references to unit characters from first (1916) edition in his second edition (1918). See also letter from Grant to Davenport, dated November 20, 1917, thanking Davenport for "making some favorable suggestions in regard to my book" and assuring him that "I am also omitting references to 'unit' characters." Archives of the Eugenics Record Office, American Philosophical Society Library.

30. This account of Stiles and his work in hookworm disease is derived from both his own memoir, "Early History, in Part Esoteric, of the Hookworm (Uncinariasis) Campaign in Our Southern United States" in the *Journal of Parasitology*, Vol. XXV, No. 4 (August 1939), and Greer Williams' excellent popular book on the work of the Rockefeller Foundation, *The Plague Killers* (New York: Scribner's, 1969). The latter is one of those fine books that should be but, alas, are not mandatory reading in all general education courses in our high schools and colleges. It is our failure to utilize such inexpensive resources that has helped turn the United States into a nation governed by illiterates in human biology and social medicine.

31. During World War I, according to the National Academy of Sciences, "a comparison was made at Camp Travis between the intelligence scores of 632 white men and 130 negroes who were infected with hookworm and the scores of 5,615 white men and 2,877 negroes who were free from the disease." The noninfected whites who qualified for the Alpha intelligence test, for literates, achieved a mean Alpha intelligence test score of 118.50; the whites with hookworm disease had a mean Alpha score of 94.38. Among the blacks who took the Alpha tests, the noninfected recruits had a mean Alpha score of 40.82, while the blacks with hookworm had a mean score of 34.86. The infected whites had a mean Beta test score of 45.38, as against the noninfected white mean of 53.26. For the same Beta test, the infected blacks had a mean score of 22.4, as against the mean of 26.00 for the noninfected blacks. While conceding that, obviously, white and black recruits suffering from hookworm disease had, as Stiles had predicted, lower intelligence test scores than white and black soldiers free of this environmental infection, the authors of this report also felt that "the results may mean that hookworm disease has a more serious effect on whites than on negroes." (Yerkes, 1921, pp. 809–10.)

32. G. Williams, 1969, pp. 43–44.

33. "Hereditary Improvement," *Fraser's Magazine*, January 1873, p. 118. Here Galton wrote that "in every malarious country, the traveler is pained by the sight of the miserable individuals who inhabit it. These have the pre-eminent gift of being able to survive fever, and therefore by the law of economy of

structure [*sic*], are apt to be deficient in every quality less useful to the exceptional circumstances of their life."

CHAPTER 9

1. Major, 1947, pp. 607–14.
2. *Ibid.*
3. Stimson, 1938, p. 39.
4. *Report of the Pellagra Commission of the State of Illinois, November, 1911* (Springfield, Ill.: Illinois State Journal Co., 1912).
5. Parsons, 1943, p. 222.
6. Joseph Goldberger, "Typhus Fever: A Brief Note on Its Prevention," *Public Health Reports,* XXIX, No. 18 (May 1, 1914).
7. Lash, 1971, pp. 177–78.
8. Joseph Goldberger, "The Etiology of Pellagra. The Significance of Certain Epidemiological Observations with Respect Thereto," *Public Health Reports,* XXIX, No. 26 (1914), 1683–86. See also Joseph Goldberger, "The Cause and Prevention of Pellagra," *Public Health Reports,* XXIX, No. 37 (1914), 2354–57.
9. After Goldberger's death in 1928 the P-P factor was renamed Vitamin G in his honor. It has since proven to be niacin, one of the vitamins in the B complex. Niacin is produced in our bodies from protein-bearing foods, after ingestion, by the conversion of tryptophan, one of the amino acids present in proteins. It is also found already synthesized in green vegetables, peanuts, and whole-grain cereals.
10. Parsons, 1943, pp. 305–06. See also Joseph Goldberger, G. A. Wheeler, and Edgar Sydenstricker, "A Study of the Relation of Family Income and Other Economic Factors to Pellagra Incidence in Seven Cotton-Mill Villages of South Carolina in 1916," *Public Health Reports,* XXXV (November 12, 1920), 2673–714. Also, by the same authors, "A Study of the Relation of Diet to Pellagra Incidence in Seven Textile-Mill Communities of South Carolina," *Public Health Reports,* XXXV (March 18, 1920), 648–713.
11. *Proceedings of the National Association for the Study of Pellagra; Journal of the American Medical Association,* December 4, 1915, p. 2028.
12. *Public Health Reports,* XXXI, No. 46 (November 17, 1916); also published in *Southern Medical Journal,* Vol. X, No. 4 (April 1, 1917).
13. De Kruif, 1928, pp. 359–61.
14. Parsons, 1943, p. 307.
15. Joseph Goldberger and G. A. Wheeler, "Experimental Pellagra in the Human Subject Brought About by a Restricted Diet," *Public Health Reports,* XXX (November 12, 1915), 3336. See also Joseph Goldberger, "Pellagra: Causation and a Method of Prevention. A Summary of Some of the Recent Studies of the U.S. Public Health Service," *Journal of the American Medical Association,* February 12, 1916, pp. 471–76.
16. Parsons, 1943, p. 308.
17. *Ibid.*
18. *Ibid.*
19. *Proceedings of the National Association for the Study of Pellagra, 1915; Journal of the American Medical Association,* December 11, 1915, p. 2115.
20. Nam Hollow was the natural habitat of the Nams, the Jukes-Kallikaks type of family Davenport had described in Eugenics Record Office Memoir No. 2, *The Nam Family: A Study in Cacogenics,* 1912. "Cottages" was the term the Vanderbilts and other American nabobs applied to their multimillion-dollar summer palaces on the cliffs of Newport.
21. Flexner, 1910, p. x.
22. Parsons, 1943, p. 327.

23. Henry Pelouze de Forest, "Peanut Worms and Pellagra," *West Virginia Medical Journal,* May 1933.

24. De Forest's search of the literature for a demonstration of the true cause of pellagra was not to be matched until 1972, when Professor Richard Herrnstein of Harvard searched the literature and found that most of the experts in the field agree that IQ scores are 80 percent genetic. See chapter 20, below.

25. Stimson, 1938, p. 42.

26. In 1943, the War Food Administration—on the advice of military and civilian doctors—ordered the enrichment of white bread flour with an array of specific vitamins and minerals. The mandated vitamins included the B-complex, whose B-2 component contains niacin (nicotinic acid). This nutritional fortification of bread flour and other grains, such as corn meal, has remained mandatory under post-war federal regulations to the present time. Thus, while well over 25 million Americans still live in poverty, the addition of niacin—Goldberger's pellagra-preventing factor—to the staple foods of the diet of the poor has acted to pretty well eliminate pellagra as an endemic disease in the United States.

   Pellagra is, of course, still endemic in those poor nations where the vast majority of the world's population now lives—and food-processing and distribution complexes such as are common in industrialized nations, and average family incomes large enough to purchase fortified bread, flour, and cereals, simply do not exist. A recent World Health Organization study in Africa, for example, reveals that pellagra "has been commonly observed, especially among adults, in some countries of south-eastern Africa. It has been closely studied in Lesotho, where it was found to be a major cause of impaired working efficiency and also of psychiatric disorders. After therapy with nicotinic acid [niacin], however, it was possible to discharge 50% of the subjects in the mental hospital." Quoted from K. V. Bailey, "Malnutrition in the African Region." *WHO Chronicle* 29 (1975), pp. 354–364. See p. 361.

27. John Dobbing, "Lasting Deficits and Distortions of the Adult Brain Following Infantile Undernutrition," Pan American Health Organization, WHO, 1972, p. 22.

CHAPTER 10

1. Stoddard, 1922, p. 69.

2. McDougall, 1921, p. 162. Technically, since the Army mental tests of World War I were graded by letters from A to E, and not by numerical IQ scores—that is, by the numbers arrived at by dividing the so-called mental age (MA) of the testee by his chronological age (CA) and then multiplying this quotient by 100—the Army tests were not IQ tests but intelligence tests. However, since during the past five decades these World War I Army Alpha and Army Beta mental tests have been universally described as IQ tests, and since the types of questions asked on the civilian IQ tests that were also written by Goddard, Terman, Yerkes, and other authors of the Army mental tests were quite identical in form, structure, and proposed function to the Goddard-Binet, the Stanford-Binet, the Point Scale, and other American IQ test questions, I have used the terms "IQ" and "intelligence" and "mental" interchangeably to characterize the Army Alpha and Beta tests of 1917–19.

3. Yerkes, 1921.

4. Klineberg, 1935(a), 1935(b). Also Otto Klineberg, "A Study of Psychological Differences Between 'Racial' and National Groups in Europe," *Archives of Psychology* (Columbia University), 1931.

5. Alfred Binet and Victor Henri, "La Psychologie individuelle," *Année Psychologique,* 1895, 2, 411–65.

6. Hunt, 1964, pp. 210–11.

7. *The Psychological Review,* II (1895), 475–86.
8. McDougall, 1921, pp. 68–71.
9. Read D. Tuddenham, in Postman, 1962, p. 490.
10. H. H. Goddard, "The Binet Tests in Relation to Immigration," *Journal of Psycho-Asthenics,* December 1913, pp. 109–10.
11. Terman, 1916.
12. *Ibid.,* p. 115.
13. *Ibid.,* pp. 91–92. This was published by Terman fifty-three years before Professor Arthur R. Jensen published his *Harvard Educational Review* article, "How Much Can We Boost IQ and Scholastic Achievement?" In that article, Jensen—like Terman neither a biologist nor a geneticist nor an anthropologist —asked, "Is there a danger that current welfare policies, *unaided by eugenic foresight,* could lead to the *genetic enslavement* of our population? The possible consequences of our failure seriously to study these questions [i.e., the fecundity and IQ test scores of black people] may well be viewed by future generations as our society's greatest injustice to Negro Americans" (italics added). (Jensen, 1969, p. 95.)
14. Hunt, 1961, pp. 42–44.
15. Quoted from Hunt, 1961, p. 13.
16. *Ibid.*
17. Terman, 1925–59, II: "The Early Mental Traits of Three Hundred Geniuses," by Catherine M. Cox, 1926. See also Lewis M. Terman, "Psychological Approaches to the Biography of Genius," *Science,* October 4, 1940, p. 293.
18. *Science,* October 4, 1940, p. 294.
19. Terman, 1916, p. 79.
20. Wallin, 1955, pp. 118–19.
21. *Ibid.,* p. 119.
22. Yerkes, Bridges, and Hardwick, 1915.
23. *Proceedings,* 24th Annual Meeting of the American Psychological Association, *The Psychological Bulletin,* February 15, 1916.
24. Lewis M. Terman, "The Use of Intelligence Tests in the Army," *The Psychological Bulletin,* XV, No. 6 (1918), 177–85.
25. Yoakum and Yerkes, 1920, p. 43.
26. Terman, "The Use of Intelligence Tests in the Army," p. 178.
27. Official U.S. Army *Examiner's Guide,* quoted from Yoakum and Yerkes, 1920, p. 55.
28. The Sicilian child who drew in the missing crucifix *common on houses in his native cultures* was, of course, marked wrong, since the answer the authors of the test demanded was a chimney. Similarly, at the Training School in Vineland, New Jersey, and in the immigrants' barracks at Ellis Island, when the people tested by Goddard defined a table as "something to eat on," or a fork "it is to eat with," or a horse "is to ride," and so forth, Goddard found unacceptable such "use" definitions—*which of course showed that the New Jersey native and Ellis Island immigrant testees knew exactly what tables, horses, and forks actually are*—and Dr. Goddard cited such answers by native and immigrant children as scientific proof that "these people cannot deal with abstractions." What they could not deal with, of course, were IQ testers with a eugenic bias. (Goddard, 1912, 1913, 1917.)

CHAPTER 11

1. Popenoe and Johnson, 1918, 1920, 1927, 1933; in 1933 edition, p. 334.
2. Prescott Hall was one of the founders and leaders of the Immigration Restriction League. General Walker's essay "Restriction of Immigration," in the June 1896 issue of *The Atlantic Monthly,* was the core "scientific" basis of the League's propaganda.
3. *Historical Statistics of the United States, Colonial Times to 1957.*

4. McDougall, 1921, p. 195. See also the Terman IQ test scale on page 236 by which thousands of people—many of them of perfectly normal intelligence —were at that time being classified as "indisputably feeble-minded."

5. *Ibid.,* p. 195.

6. *Ibid.,* p. 194.

7. *Ibid.,* p. 161.

8. *Ibid.,* pp. 54–55.

9. As the Republican presidential candidate in 1920, Senator Warren G. Harding told a southern election rally that "whoever will take the time to read and ponder Mr. Lothrop Stoddard's book on 'The Rising Tide of Color' . . . must realize that our race problem here in the United States is only a phase of a race issue the whole world confronts. Surely we will gain nothing by blinking the facts. That is not the American way of approaching such issues." (*The New York Times,* September 15, 1920; quoted from Gosset, 1963.) An equally fervent admirer of Stoddard's plea for white supremacy was the British sexologist and eugenicist Havelock Ellis, whose three-page review of the book ("The World's Racial Problem," *Birth Control Review,* October 1920, pp. 14–16) hailed Stoddard as a thinker whose "conclusions are, after all, fundamentally in harmony with those of sober and judicial observers in Europe."

10. Stoddard, 1922, p. 30.

11. "The Great American Myth," lead editorial, *The Saturday Evening Post,* May 7, 1921, p. 20.

12. Stoddard, 1922, pp. 30–31.

13. Brigham, pp. 190–92.

14. *Ibid.,* p. 192. This, of course, was the eugenic and polygenic myth of "Selective Migration."

15. *Ibid.,* pp. 110–11.

16. McDougall, 1921, p. 135.

17. Brigham, pp. 208–09, 210.

CHAPTER 12

1. In foreword to Brigham, 1923, p. vii.

2. Osborn, 1923.

3. Ross, 1914, p. 155.

4. *Ibid.,* p. 154.

5. *Ibid.,* p. 145.

6. This and all other quotes on pages 242–87 from papers read at the Congress are from the two-volume proceedings of the Second International Congress of Eugenics, published in 1923 (see Bibliography).

7. *Ibid.,* I, 218–25.

8. *Ibid.,* II, 1–6.

9. The aging anti-Semite and cephalic-index enthusiast made at least one convert in America, for in March 1925 Mrs. Margaret Sanger brought him back to address the Sixth International Malthusian and Birth Control Conference in New York. As Sanger described the visit of "Dr. G. O. Lapouge, a French eugenist," the committee sent to meet his ship found him on the pier, where this "inconspicuous, desolate man [sat] on top of his luggage, reading, waiting patiently for someone to come for him—so unimportant-looking that no one would have suspected him of being a renowned scientist [*sic*]. The next morning the Hotel McAlpin, where the convention was to be held, called me to report that Dr. Lapouge had been severely burned, and an interpreter was needed. Dr. Drysdale hurried off to find this poor little man of seventy in excruciating pain . . . without understanding how to regulate the shower he had stood under it and turned on the hot water. The skin fairly peeled

off his chest. Nevertheless, bandaged and oiled, he undauntedly attended all the sessions." (Sanger, 1938.)

10. Second International Congress of Eugenics, 1923, I, 391–97.
11. Lidbetter, 1933, pp. 15–16, 23–24.
12. Second International Congress of Eugenics, 1923, II, 62–63.
13. *Ibid.,* II, 373–76.
14. *Ibid.,* II, 286–91.
15. *Ibid.,* II, 402–06.
16. *Ibid.,* I, 20–28.
17. *Ibid.,* II, 430–31.
18. *Ibid.,* II, 41–61.
19. Ross, 1914, p. 148.
20. *Ibid.,* pp. 164–65.
21. Baltzell, 1964, pp. 210–12, and Gossett, 1963, pp. 372–73.
22. Brigham, 1923, p. 210.
23. Mann, 1969, p. 187.
24. "Biological Aspects of Immigration," statement of Harry H. Laughlin, Hearings, Committee on Immigration and Naturalization, House of Representatives, April 16–17, 1920.
25. "Analysis of America's Modern Melting Pot," statement of Harry H. Laughlin, Hearings, Committee on Immigration and Naturalization, House of Representatives, November 21, 1922.
26. "Restriction of Immigration," Hearings, Committee on Immigration and Naturalization, House of Representatives, on H.R. 5, H.R. 101, and H.R. 561, Serial 1-A, December 1923 and January 1924, pp. 608–20.
27. *Ibid.,* pp. 570–71.
28. Stoddard, 1922, p. 69; Brigham, 1923, pp. 124–25.
29. Evans, 1923, pp. 13–15.
30. Letter from Davenport to Grant, dated April 7, 1925, in Archives of the Eugenics Record Office, American Philosophical Society Library.

CHAPTER 13

1. See, for example, Margaret Wooster Curti, "The Intelligence of Delinquents in the Light of Recent Research," *The Scientific Monthly,* 1926, pp. 132–38.
2. Garrett Hardin, "The Competitive Exclusion Principle," *Science,* April 29, 1960, and Garrett Hardin, "Equality, One World, and Science," *Western Destiny,* XI, No. 4 (1966), 17. Here Hardin, amplifying on his 1960 *Science* essay, declared: *"All men are, by nature, unequal* [Hardin's italics]—this is the uncensored truth of our century. We are as afraid of admitting the consequences of this truth as the Victorians were of the consequences of admitting that men are animals." Since truths must be recognized, Hardin added, "we must acknowledge it and go ahead to explore its implications. To do so is an act of faith in science, faith in the future, faith in the essential goodness of truth."
3. Hellman, 1968, pp. 196–97.
4. Louis Z. Cooper, "Rubella in the U.S.A.: The Problems," *Scandinavian Journal of Infectious Diseases,* Supplementum 6: *Rubella Vaccination, A Symposium Held in Stockholm, Sweden, on October 15, 1971,* 1972, pp. 10–11.
5. Quotes following are from Harry H. Laughlin, "The Legal Status of Eugenical Sterilization," Supplement to the Annual Report of the Municipal Court of Chicago, 1930, pp. 16–19.
6. Felix Frankfurter, *Mr. Justice Holmes and the Supreme Court* (Cambridge: Harvard University Press, 1961).

7. Howe, 1953, II, 937–38.
8. *Ibid.,* pp. 939–41.
9. Rosenberg, 1962, p. 135, n. 4.
10. In 1910, Galton was so moved by an article, "The Scope of Eugenics," by H. J. Laski in the July issue of the *Westminster Review* that he wrote a letter to the author and asked if they could meet. The article had declared, among other things:

"The time is surely coming in our history when society will look upon the production of a weakling as a crime against itself. When we remember that the highest duty is parenthood, it is surely only right to ask that the parents have no serious heritable trait. As Galton has so finely said, we must hold the eugenic ideal of parenthood with the fervour of a new religion."

Galton had never heard of H. J. Laski, and was understandably astounded to discover that the author was, as Galton wrote to his niece Milly, "a school boy at Manchester, aged 17. It is long since I have been so much astonished. The lad has probably a great future before him and he will make his mark if he sticks to Eugenics, which he says has been his passion for two years. I as yet know nothing more about him, but hope to learn."

What Galton soon learned was that Laski was the son of a Manchester textile man named Nathan Laski, who was not only Jewish but a leader of the Manchester Jewish community. A month later, Galton wrote to Milly that "my wonderful boy Jew, Laski by name, came here with his brother to tea." The wonderful boy Jew was so stimulated by this meeting with the anti-Semitic founder of eugenics that, in 1911, he spent the first half of the year studying eugenics under Karl Pearson at University College, London, before entering New College, Oxford. See Pearson, 1914–40, III (1930), 606, 608, 609.

In 1927, the year of the Holmes decision upholding the eugenic and sterilization dogmas of Galton and Pearson, a leading American eugenicist, Edward M. East, in his book *Heredity and Human Affairs,* cited a 1913 memoir, "On the Correlation of Fertility with Social Value," of which Laski, along with Pearson, had been one of the six co-authors, as proof that it was folly to think that the bad conditions of poverty "can be changed entirely by raising wages" (pp. 255–56).

As the letter to Holmes revealed, Laski had not abandoned his passion for Galton's eugenics when he added to them the socialism of Marx, Pearson, and the other Fabians.

11. Howe, 1953, II, 941–42.
12. *Ibid.,* pp. 964–65.
13. Henry H. Goddard, "Feeblemindedness: A Question of Definition," *Journal of Psycho-Asthenics,* XXXIII (1928), 219–27.
14. Carl C. Brigham, "Intelligence Tests of Immigrant Groups," *The Psychological Review,* XXXVII (1930), 158–65.

CHAPTER 14

1. Quoted from Gossett, 1963, p. 372.
2. Rosenberg, 1962, p. 62.
3. Charles Benedict Davenport, "Euthenics and Eugenics," *The Popular Science Monthly,* January 1911, p. 19. The nation's scientific racists never quite abandoned their dogma that tuberculosis, that classic infectious disease of poverty, is a genetic or hereditary disease. As late as the fourth edition of *Applied Eugenics* in 1933, Popenoe and Johnson solemnly wrote: "The death rate from tuberculosis in the United States has been declining for many years. It was particularly high in the first few decades after 1840. The Irish were at that time emigrating in such numbers that within a generation about two-fifths of the entire population of Ireland had moved to the United States. As a group they had and still have a notably high death-rate from tuberculosis,

and their presence here seems to be associated with the great increase in the death-rate in 1840–1850. Since then the rate has been decreasing, as the more susceptible [*sic*] strains died out."

4. Osborn to Davenport, November 26, 1921, Archives of the Eugenics Record Office, American Philosophical Society Library.
5. John Hope Franklin, *From Slavery to Freedom: A History of Negro Americans* (3rd ed.; New York: Alfred A. Knopf, 1966), pp. 474–84; Samuel Eliot Morison, *The Oxford History of the American People* (New York: Oxford University Press, 1965), pp. 884–85.
6. Archives of the Eugenics Record Office, American Philosophical Society Library.
7. Third International Congress of Eugenics, 1934, pp. 193–200.
8. *Ibid.,* pp. 201–09.
9. *Ibid.,* pp. 364–68.
10. *Ibid.,* pp. 369–71.
11. *Ibid.,* pp. 138–44.
12. Muller himself, impatient to share the delights of the new, scientific, and just society, was soon to journey to the Soviet Union. There, in the socialist homeland, he would learn to his horror that Mendel, and Muller's mentor and friend Thomas Hunt Morgan, were bourgeois idealists and betrayers of the working class, and that Mendelian and chromosomal genetics were anti-socialist, let alone anti-materialist. More than that, Stalin's concepts of genetics and heredity were not those of the geneticists but of the Lamarckian Party hack Lysenko—pseudoscientific concepts much closer to those of reactionary amateurs in biology, such as the equally fervent Lamarckians Herbert Spencer and William McDougall, than to the classless and objective concepts based on the experimental laboratory experiments and observations of serious scientific workers, including Mendel, Johannsen, Boas, Boveri, Sutton, Morgan, Sturtevant, Bridges—and Muller himself.

It was a Muller humiliated by the abuse of pygmies, and shattered by disillusion, who finally fled the land of "socialist genetics." At that he was lucky: he still had an American passport, and thus avoided dying of pneumonia in a heatless Soviet jail as did Nikolai I. Vavilov, the great Soviet plant geneticist, who had professed the same anti-Lamarckian "heresies" known elsewhere in the world as modern genetics.

For Muller's attack on Lysenko—written as both a legitimate geneticist and a committed Marxist—see Graham, 1972, pp. 451–69. Muller's article, "Lenin's Doctrines in Relation to Genetics," appears on pages 453–69. For an account by a Soviet biologist of the Lysenko controversy, see Medvedev, 1969. For the biological thoughts of Lysenko himself—concepts that, in modern times, were to help turn the Soviet Union into an importer rather than a major exporter of wheat and other grains—see Lysenko, 1948. To the extent that Lysenko's 1948 book demonstrates his enormous ability to pervert the realities of the life sciences to fit and "prove" his own half-baked pseudoscientific concepts—and to then convince the rulers of a huge 20th century superpower that his notions are the one true science of biology—this means that Lysenko was, beyond all doubt, the Soviet Charles Benedict Davenport.
13. Estimates of Ku Klux Klan membership during the 1920's range from Higham's figure of three million (Higham, 1963, p. 297) to the figure of four to five million stated in the *Columbia Encyclopedia* (3rd ed., 1963) and also by Arnold Foster and Benjamin Epstein, *Report on the Ku Klux Klan* (New York: Anti-Defamation League, 1965).
14. George D. Stoddard, "The IQ: Its Ups and Downs," *Educational Record, Supplement,* January 1939, p. 46.
15. Harold M. Skeels, "Mental Development of Children in Foster Homes," *Journal of Consulting Psychology,* II, No. 2 (1938), 33–41.

16. Stoddard, *op. cit.,* p. 48.
17. *Ibid.*
18. Skeels, 1966. See also Marie Skodak, "The Mental Development of Adopted Children Whose True Mothers Are Feebleminded," *Child Development* IX (1938), 303–08; Marie Skodak, and Harold M. Skeels, "A Final Follow-Up Study of One Hundred Adopted Children," *Journal of Genetic Psychology* LXXV (1949), 85–125; H. S. Skeels, and H. B. Dye, "A Study of the Effects of Differential Stimulation on Mentally Retarded Children," *Proceedings of the American Association on Mental Deficiency,* XLIV (1939), 114–36; Harold M. Skeels, Ruth Updegraff, Beth L. Wellman, and H. M. Williams, "A Study of Environmental Stimulation: An Orphanage Preschool Project," *University of Iowa Studies in Child Welfare,* XV, No. 4 (1938); and Beth L. Wellman, "Our Changing Concepts of Intelligence," *Journal of Consulting Psychology,* II (1938), 97–107.
19. Skeels, 1966, p. 3.
20. *Ibid.* See Table 8, p. 38; Table 9, p. 38; Table 12, p. 44; Table 13, p. 46; and Summary and Implications, pp. 54–57.

   In November 1974, Mr. Robert Ferguson, who was born in Iowa on June 14, 1934, and renounced by his mother on that date, was interviewed by Andrew H. Malcolm for *The New York Times.* Ferguson, a contemporary of the orphaned and abandoned Iowa children Skeels had worked with, was raised in state institutions. First he lived in a state orphanage, cared for by adults whose "faces changed with each shift." Later he was sent to an Iowa state institution for the mentally retarded. As he told Malcolm, there "we sat at a table all day with our arms folded. If you stood up without raising your hand for permission, someone hit you." As he outgrew this and other institutions he was sent to others, equally sterile emotionally and mentally. He was interviewed by the *Times* because, at the age of forty, he had spent most of his adult years in different prisons for a variety of petty crimes, and now —upon his release from the Iowa State Penitentiary after his latest prison term—he confessed that, having spent thirty-nine of his forty years "inside state walls," he feared to live in freedom outside of the walls. See "For One Convict, 'Freedom' Is Another Word for 'Fear,' " *The New York Times,* November 20, 1974.
21. Abraham Jacobi (1830–1919) and his wife, Mary Putnam Jacobi (1842–1906), were two of the most useful physicians of their era. Jacobi's 1873 paper, "Infant Diet," presented before the Public Health Association of New York, was reissued in 1874, "revised, enlarged and adopted to popular use by Mary Putnam Jacobi, M.D.," and again revised and reprinted in 1880. It was, for many years, a basic book for both pediatricians and the families of small children. Mary Jacobi, daughter of the publisher George Palmer Putnam and the first woman to attend the Ecole Médicine in Paris, was a noted neurologist and professor of medicine at the New York Post-Graduate Medical School. In 1911, when Jacobi was eighty-one, he was elected president of the American Medical Association. His presidential address, in 1912, "Infant Mortality," advocated postnatal assistance for overworked mothers, and asked: "Which of the poor is not overworked?" He added that the time had come to think of social programs of maternal and child health care in more civilized terms: "the term *charity* should be supplanted by *responsibility.* It is useless to call [maternal and child health care] socialism or communism." (Jacobi's emphasis.) At a time when President Theodore Roosevelt and the eugenicists were denouncing birth control as "race suicide," Jacobi, in the same address, said that in families too poor to be able to provide the necessities of health to all of their children, "to limit the number of children, even the healthy ones, is perhaps more than merely excusable." (Truax, 1952, pp. 172, 248.)

22. Truax, 1952, p. 140.
23. Skeels, 1966, p. 56.

CHAPTER 15

1. See "Memorandum to Miss Wycoff," October 26, 1933, by R. V. Coleman, Charles Scribner's Sons Archives, Princeton University Library.
2. *Rassenkunde des deutschen Volkes.* The proper translation of *Rassenkunde* is "Raceology"—which is not to be confused with the legitimate science of ethnology, which is by definition "the science which treats of races and peoples, and of their relations to one another, their distinctive physical and other characteristics, etc." (OED). Raceology is not a science.
3. Hellman, 1969, p. 194.
4. See Shirer, 1960, pp. 263–422.
5. Stoddard, 1940, p. 187.
6. Frick, as a young police officer in Munich, became a spy in police headquarters in 1923 for Adolf Hitler. Frick was one of the Big Five of the Nazi Party, along with Hitler, Goering, Goebbels, and Strasser. He was the first of the Big Five to hold major political office, in the province of Thuringia— where, at Hitler's orders, he had a chair in "raceology" created for Hans Günther in 1930. He was appointed Minister of the Interior when Hitler became Germany's Nazi dictator in 1933 and held this post until a few months before the international war crimes tribunal at Nuremberg sentenced him to death. Frick was hanged on the same day as were Hitler's Nordic philosopher, Alfred Rosenberg, and Julius Streicher, the whip-carrying sadist and pornographer and publisher of the smut sheet *Der Stürmer,* whose profits had financed the Nordic Nazi movement in its infancy. Darré, Minister of Agriculture, was an Argentine-born German agronomist who, writes Wilhelm Shirer, was forced to quit his pre-Hitler-era jobs in the Agriculture Ministries of Prussia and the Reich because of his political convictions. He retired to his Rhineland home in 1929, and there wrote a book, *The Peasantry as the Life Source of the Nordic Race.* Rudolf Hess brought Darré to Hitler, who commissioned him to draw up the Nazi agricultural program. Under the Darré-written Hereditary Farm Law of September 29, 1933, "only an Aryan German citizen who could prove the purity of his blood back to 1800" (Shirer, *op. cit.*) could own a farm in the new pure-blooded Nordic Vaterland of Streicher, Rosenberg, and Hitler's Big Five.
7. What Stoddard here describes as "the Jewish problem" was, of course, *the German non-Jewish problem* of how to annihilate the Jews physically. To the German and non-German Jews trapped in Nazi Germany or the territories the *Wehrmacht* seized, "the Jewish problem" was, rather, one of trying to make certain that their physical elimination from the Third Reich was via emigration to non-Nazi countries and not by annihilation at Auschwitz, Buchenwald, Maidenek, and other Nazi centers of applied eugenics.
8. To the members of more than one major American book club, Stoddard's use of Günther's *rassenkundlich* phrase "the ideal type" and his quotation of Günther's own explanation of this pseudogenetic construct might well have sounded a familiar, contemporary note. Perhaps what it brought to mind was a 1966 English translation of *Das Sogenannte Bose: Zur Naturgeschichte die Aggression,* by one of Hans Günther's fellow toilers in Hitler's scientific vineyards, Dr. Konrad Lorenz. In this book (issued in English as *On Aggression*), Lorenz writes, in Chapter 11:

   "The concept 'normal' is one of the most difficult things to define in the whole of biology . . . by normal we understand not the average form taken from all the single cases observed, but rather the *type* [Lorenz's italics] constructed by evolution which for obvious reasons is seldom to be found in a pure form; nevertheless we need this purely ideal conception of a type in order to be able to conceive the deviations from it. The zoology textbook cannot do more than describe a perfectly intact, ideal butterfly as the representative of its species, a butterfly that never exists exactly in this form because, of all the specimens found in collections, every one is in some way malformed or damaged.

"We are equally unable to assess the ideal construction of 'normal' behavior in the Graylag Goose or in any other species, a behavior which would occur only if absolutely no interference had worked on the animal and which exists no more than does the *ideal type* of butterfly [italics added]. People of insight *see* [Lorenz's italics] the ideal type of a structure or behavior, *that is, they are able to separate the essentials of type from the background of little accidental imperfections*" (italics added).

The American geneticist Professor James C. King, of the New York University School of Medicine, cited this passage from *On Aggression* as "a graphic description of the notion of the Platonic type in its most virulent form."

Dr. King went on to declare, on page 7 of his excellent monograph for the lay reader, *The Biology of Race,* that: "This mystical concept of a perfect model, never completely actualized in the crude material world, *is much closer to theology than to science.* It contrasts sharply with the ideal of a modal *phenotype* envisaged by those who think in terms of the biological species." (Italics added.)

9. Quoted from Mosse, 1968.
10. Lorenz, 1966, pp. 194–95.
11. Grunberger, 1971, p. 365.
12. Duell, 1942, p. 221.
13. Duell, 1942, p. 224.
14. Stoddard, 1940, p. 192.
15. Hans Harmsen, "The German Sterilization Act of 1933: Gesetz zur Verhüting erbkranken Nachtwuches," *Eugenics Review,* January 1955, p. 227.
16. Stoddard, 1940, p. 196.
17. In Archives of Eugenics Record Office, American Philosophical Society Library.
18. Kenneth M. Ludmerer, "American Genetics and the Eugenics Movement: 1905–1935," *Journal of the History of Biology* II, No. 2 (1969), 358.
19. M. Haller, 1963, p. 179.
20. Morse, 1968.
21. Evans, 1923, p. 22.
22. Morse, 1968, pp. 172–73. Laughlin's career as an expert in human eugenics came to an end on Pearl Harbor day—December 7, 1941—when America's entrance into World War II rendered superfluous his very active propaganda and congressional lobbying efforts to keep Jewish, Italian, and other non-Aryan refugees from sanctuary in the United States. As C. B. Davenport revealed in the obituary he wrote after Laughlin died in 1943, Laughlin's last years "were devoted to a study of the inheritance of racing capacities in thoroughbreds." Like Laughlin, Davenport remained steadfast in the faith, and concluded his obituary (*Science,* February 26, 1943) with: "Some of Laughlin's conclusions and their applications in legislation [i.e., compulsory sterilization and immigration restriction laws] were opposed by those committed to a different social philosophy, founded on a less thorough analysis of facts. One cannot but feel that a generation or two hence Laughlin's work, in helping bring about restricted immigration and thus the prevention of our country from the clash of opposing ideals and instincts found in the more diverse racial or geographical groups, will be the more widely appreciated as our population tends toward greater homogeneity."
23. The other eugenicists who served as president of the AAAS included David

Starr-Jordan (1909), Charles W. Eliot (1914), J. McKeen Cattell (1924), Henry Fairfield Osborn (1928), and Edward G. Conklin (1936). In 1927, Conklin's name headed the list of signers of a "Memorial on Immigration Quotas," sent to the President and the Congress, urging "the extension of the quota system to all countries of North and South America . . . in which the population is not predominantly of the white race." The co-signers of this white-supremacy memorial included such veteran eugenics die-hards as Robert DeCourcy Ward, Edward M. East, Edward A. Ross, Madison Grant, Henry Fairfield Osborn, Robert M. Yerkes, Henry Pratt Fairchild, C. C. Little, Charles B. Davenport, and Harry H. Laughlin. (*Eugenical News,* 1927, pp. 27–28.)

24. In 1913, in a lecture to his Columbia students, Thorndike declared: "Long before a child begins his schooling, or a man his work at trade or profession, or a woman her management of a home—*long indeed before they are born*— their superiority or inferiority to others of the same environmental advantages is determined by the constitution of the germs and ova from which they spring, and which, at the start of their individual lives, they *are*." (See "Eugenics: With Special Reference to Intellect and Character," by Professor Edward L. Thorndike, a lecture given at Columbia University in March 1913, in *The Popular Science Monthly,* 1913, pp. 125–38.)

25. Galton, 1883, pp. 65–66. Galton, of course, was not the first literate ignoramus to ascribe vicious madness to epilepsy, or to incorrectly label it as a genetic disorder. As Peter Wingate (1972) reminds us: "Hippocrates' treatise *On the Sacred Disease,* probably written about 400 B.C., tried to debunk the superstition indicated by the title and [to] present epilepsy as a disease like any other, a result of natural causes. Yet even now many people are uneasy in the presence of epileptics, and not only in primitive societies."

26. Aaron Capper, "The Fate and Development of the Immature and of the Premature Child. A Clinical Study. Review of the Literature and Study of Cerebral Hemorrhage in the New-Born Infant," *American Journal of Diseases of Children,* 1928, pp. 443–91.

27. *Ibid.,* p. 479.

28. Arthur Jensen, "Race and the Genetics of Intelligence: A Reply to Lewontin," *Bulletin of the Atomic Scientists,* May 1970, p. 17. Lewontin's article "Race and Intelligence," to which Jensen here replied, appeared in the March 1970 *Bulletin of the Atomic Scientists.* It was subsequently reprinted in Senna, 1973, and in Block and Dworkin, 1976. Lewontin, one of the nation's leading population geneticists, had in his article expressed very low opinions of the "genetics" of both Thorndike and Jensen.

29. Grant and Davison, 1930, pp. 1–3.

30. *Ibid.,* pp. 86–98.

31. *Ibid.,* pp. 225–29.

32. *Ibid.,* pp. 230–36.

CHAPTER 16

1. Otto Klineberg, "A Study of Psychological Differences Between 'Racial' and National Groups in Europe," *Archives of Psychology* (Columbia University), 1931; see pp. 27–37. See also Klineberg, 1935(a) and 1935(b).

2. Klineberg, 1935(a) and 1935(b).

3. Chase, 1972, pp. 137–39.

4. Oswald T. Avery, Colin M. Macleod, and Macklyn McCarty, "Studies on the Chemical Nature of the Substance Inducing Transformation of Pneumococcal Types: Induction of Transformation by a Deoxyribonucleic Acid Fraction Isolated from Pneumococcus Type III," *Journal of Experimental Medicine,* LXXIX (1944), 137–58. Reprinted in Peters, 1959, pp. 173–92. See also Alfred E. Mirsky, "Chromosomes and Nuclear Proteins," F. F. Nord and

C. H. Werkman, eds., *Advances in Enzymology and Related Subjects of Biochemistry* (New York: Interscience Publishers, 1943), III, 1–34.

5. Brigham, 1923, p. 210.
6. Kennedy, 1970, p. 119.
7. In the same presidential address, Davenport said: "It will be a long time before we can improve practically on nature's method of race improvement —a high birth rate and a high death rate."
8. Burch and Pendell, 1947, pp. 44 *et passim.*
9. Hardin, 1949; 1951 edition.
10. J. B. S. Haldane, "The Interaction of Nature and Nurture," *Annals of Eugenics,* XIII (1946), 197–205.
11. The University of Illinois geneticist and psychologist Jerry Hirsch, in his now classic "Behavior-Genetic Analysis and Its Biosocial Consequences" (*Seminars in Psychiatry,* Vol. II, No. 1, 1970), estimated that even in "the extreme, but unrealistic, case of complete environmental homogeneity"—i.e., if the entire population were born and raised in an identical environment—"the heritability [of traits] value would approach unity, because only genetic variation would be present. Don't forget that even under the most simplifying assumptions, there are over 70 trillion potential human genotypes—no two of us show the same genotype no matter how many ancestors we happen to have in common."
12. Hardin, 1951, p. 620.
13. See Chase, 1972, pp. 173–74, 238–41.
14. Fairfield Osborn, 1948, pp. 25–26.
15. *Ibid.,* pp. 192–93.
16. *Ibid.,* p. 41.
17. *Ibid.,* p. 42.
18. *Ibid.,* p. 202.
19. Vogt, 1948.
20. *Ibid.,* p. 48. The well-known custom celebrities have of rarely reading books for which they write forewords is quite possibly underscored by these attacks on the medical profession and its social effects. Bernard Baruch, who signed a foreword to Vogt's book, was the son of a South Carolina physician and a lifelong defender and supporter of medical research and advanced medical care. The Baruch foreword to the Vogt book speaks only of the importance of improving the environment in which people live, and makes no mention of Vogt's savage attacks on clinical and public health medicine.
21. *Ibid.,* p. 228.
22. Kingsley Davis, "Population," *Scientific American,* September 1963, p. 62. See also Chase, 1972, appendix, Table 2.
23. Rae Goodell, *The Visible Scientists,* doctoral dissertation, Stanford University, January 1975. (Now in press: Boston: Little, Brown & Co.)
24. Glenn Fowler, "Hugh Moore, Industrialist, Dies; Birth Control Crusader Was 85," *The New York Times,* November 26, 1972.
25. Lader, 1971, p. 6.
26. *Ibid.,* pp. 35–36.
27. *Ibid.,* pp. 38–39.
28. *Ibid.,* pp. 79–81.
29. World Health Organization, "The Effect on Man of Deterioration of the Environment," *WHO Chronicle,* XXVIII (1974), 549–53.
30. Christensen, 1972.
31. Chase, 1972, pp. 178–79.
32. Snell, 1974, p. A-31. Here the Senate report reveals that "contrary to popular belief, the Pacific Electric, not the automobile, was responsible for the [Los Angeles] area's geographical development. First constructed in 1911, it established traditions of suburban living long before the automobile had arrived."

33. New York Environmental Protection Agency, *The Car Is Anti-City*. (New York, 1971).

34. Insurance Information Institute, "Insurance Facts 1973," *Yearbook* (New York: Insurance Information Institute, 1973), p. 50.

35. National Highway Traffic Safety Administration, "A Message to My Patients," folder (Washington, 1975).

36. Commoner, 1971, p. 168.

37. *Statistical Abstract of the United States, 1974.*

38. Prepublication data, U.S. Department of Commerce. See also *Statistical Abstract of the United States, 1975.*

39. Snell, 1974, pp. A-28–A-29, A-2.

40. *Ibid.,* p. 2.

41. *Ibid.,* p. A-32.

42. *Ibid.,* p. A-29.

43. *Statistical Abstract of the United States, 1971.*

44. See John Cairns, "The Cancer Problem," *Scientific American,* November 1975, pp. 64–78. See particularly the table "Cigarette Smoking and Lung Cancer, England and Wales, 1900–1975," on p. 72 of the article.

45. WHO Expert Committee on Smoking and Its Effects on Health, "Smoking and Disease: The Evidence Reviewed," *WHO Chronicle,* XXIX (1975), 402–08.

46. Chase, 1972, pp. 351–52; see also "London Cleans the Thames," *Newsweek,* May 1, 1972, p. 123.

47. *Science,* CLXII (1968), 1243–48.

48. Paddock, 1967, p. 22.

49. *Ibid.,* p. 248.

50. *Ibid.,* p. 222.

51. *Ibid.,* p. 224.

52. S. Wortman, "Agriculture in China," *Scientific American,* June 1975, p. 16.

53. Paddock, 1967, p. 160.

54. *WHO Chronicle,* XXIX (1975), 171–73. See also Victor W. Sidel, "Some Observations on the Health Services in the People's Republic of China," *International Journal of Health Services,* II (1968), pp. 385–95. Victor W. Sidel, "Medical Personnel and Their Training," in *Medicine and Public Health in the People's Republic of China* (Bethesda, Md.: Fogarty International Center, National Institutes of Health, 1972), USDHEW Publication No. NIH 72–67. Victor W. Sidel and Ruth Sidel, "The Delivery of Medical Care in China," *Scientific American,* April 1974, pp. 19–27.

55. There was also an important historical lesson in the post-1949 mainland Chinese experience, and it was that—just as the Agricultural Revolution was started and carried to successful increases in European and English food production some three quarters of a century before the birth of the United States of America—the fate of the poor and the hungry of the world is not quite as subject to the whims and caprices of triage-minded American population extremists as they would like to have us believe. China not only had no help from the United States in its generation of self-directed change from endemic famine to the conquest of hunger (and the diseases and deaths associated with hunger) but even had, to put it in the gentlest terms, the active and deep-rooted enmity of the government and the power establishments of the United States during the years when this historic change took place. The idea of triage, of a Pax Americana in which Washington decides what nations shall live and what nations shall starve to death, might do wonders for the psychological needs of self-designated population experts such as Vogt, the Paddocks, Hardin, and Paul R. Ehrlich—but it can hardly stop China's neighbors from seeking to solve its food and hunger-disease problems as

China did, should any American government decide to apply the genocidal principles of triage to India and other Asian nations.

56. In May 1975, the year of the Paddocks' Doomsday prediction, *Science* brought out a special Food Issue, in which the food-population realities of this planet were analyzed by agricultural, nutritional, chemical, economic, and social scientists working in the fields of food and food-related problems of health and population. Not one of these working scientists agreed with the Paddocks and the other pop Malthusians. Nor, for that matter, did the food-famine-population realities of the planet c. 1975. The gist of what, based on their own fields of knowledge and experience, each of the *Science* Food Issue authors had to report was, perhaps, typified by the conclusions drawn by the Michigan State University professor of horticulture S. H. Wittwer in his article, "Food Production: Technology and the Resource Base," an impassioned "plea for instituting a massive program in agricultural science and technology." Wittwer concluded:

"Despite a growing population and increasing demands of that population for improved diets, it appears that the world is not close to universal famine. There is enough food now to feed the world's hungry. That people are malnourished or starving is *a question of distribution, delivery, and economics, not agricultural limits* [italics added]. The problem is putting the food where the people are and providing an income so that they can buy it." (*Science*, May 9, 1975, p. 584.)

In 1975, as in 1798, when Malthus published his *Essay on the Principle of Population*, intelligent scientists understood very well the differences between the lack of enough money to buy sufficient food and the lack of enough land and other natural resources with which to grow sufficient food.

57. Philip Hauser, "What to Do as Population Explodes, Implodes, and Displodes." *Smithsonian*, December 1970, pp. 20–25.
58. Ehrlich, 1968, p. 16.
59. *Ibid.*, p. 166.
60. *Ibid.*, p. 195.
61. Ibid., p. 140.
62. Franz Boas, "Fallacies of Racial Inferiority," *Current History*, February 1927, p. 679.
63. On the other hand, since birth rates always follow death rates down *or up*, raising the death rates to their 1900 levels just might not bring about Ehrlich's dream of a U.S. population fixed at 150,000,000.

CHAPTER 17

1. Stephen Viederman, "Values, Ethics and Population Education," *Hastings Center Report*, June 1973, pp. 6–8.
2. H. Curtis Wood, Jr., "The Changing Trends in Voluntary Sterilization," *Contemporary Ob/Gyn*, I, No. 4 (1973), p. 39.
3. Viederman, *op. cit.*, p. 7.
4. "President Signs Birth Curb Bill; Programs Are Expanded—Nixon Kills Measure on More Family Doctors," AP dispatch in *The New York Times*, December 27, 1970.
5. For details and effects of the Vietnam war cutbacks between 1965 and 1970, see Chase, 1972, pp. 91, 120, 145, 158–61, 250–58, 293–95, 300–01, 319–20. For continuing details and effects of cutbacks and retrenchments since 1970, consult *The New York Times Index* in categories dealing with U.S. federal budgets, medical education, medical research, health care, preventive medicine, child health, etc.
6. See Lenneberg, 1967.
7. The tragic death, in a New York slum tenement, of one of these women of septicemia caused by a self-induced abortion in the hot summer of 1912,

was to play an important role in the formation of the American birth control movement—by the revolutionary socialist nurse who attended her. The dead woman's name was Sadie Sachs; the nurse's name was Margaret Sanger. For years, until Mrs. Sanger's conversion to eugenics, the cry of the movement was "Remember Sadie Sachs!" See, for example, Douglas, 1970, pp. 32–36, 40, 103, 127, 186. See also Kennedy, 1970, pp. 16–17, 81, 275. Finally, see *Margaret Sanger: An Autobiography* (New York: W. W. Norton, & Co., 1938), p. 92.

8. That this was happening in California and other states as well as New York was reported in "Legal Abortion in the U.S.A.," in *The Lancet* (September 18, 1971), by Malcolm Potts and B. N. Branch, who wrote: "A significant part of the [U.S.] abortions now being performed would previously have taken place illegally."

9. Health Services Administration of the City of New York, *Report on First 18 Months of Legal Abortions,* February 20, 1972.

10. Nicholas M. Nelson, "Sacrifice of the Unfortunate," *New England Journal of Medicine* CCLXXXV, No. 14 (1971), 807–08.

11. Grier, 1971; see, particularly, pp. 11–18.

12. Donald J. Bogue, "Demographic Aspects of Maternity and Infant Care in the Year 2001," paper read at the Plenary Session on "Maternity and Infant Care in the Year 2001" at the annual meeting of the American College of Obstetricians and Gynecologists, San Francisco, May 6, 1971.

13. *Statistical Abstract of the United States, 1971.*

14. Commoner, 1971, p. 158.

15. These figures were based on the raw data of the national census of 1970. Subsequently, spreading unemployment and rising inflation have reduced family incomes and lowered the purchasing powers of the dollar. The latest available Bureau of the Census report, "Money Income and Poverty Status of Families and Persons in the United States, 1974" (Series P-60, No. 99, issued July 1975), showed that, in terms of constant 1974 dollars, "there were about 24.3 million persons below the poverty level in 1974 ($5,038 for a non-farm family of four) constituting 12 percent of the U.S. population. . . . The general downturn in the economy coupled with the substantial inflation which occurred between 1973 and 1974 contributed to a decline in the real income for both white and black families."

   The median 1974 income for all white families was $12,836; for all families of "Negro and other races" it was $8,265. Some 11.1 percent of the nation's 49,454,000 white families had incomes below the official poverty line. Among the families of "Negro and other races," some 29.5 percent of the nation's 6,262,000 black, Hispanic, Amerindian, Oriental, and other non-white families had incomes below the government's official poverty level. This means that, as of the present writing, the 5,489,394 white families with incomes below the poverty level outnumber the 1,815,980 below-poverty-level families of "Negro and other races" by more than three to one.

16. "European Birth Rate Dropping," AP dispatch in the *New York Post,* December 27, 1971, p. 18.

17. Rodney Angove, "France Wants Babies," *N.Y. Post,* October 27, 1971, p. 27.

18. Kasturi Rangan, "Surplus of Grain Reported in India: Breakthrough May Lead to Export of 8 Million Tons," *N.Y. Times,* April 29, 1972, p. 4. See also "India Achieves the Impossible: It Has Food to Spare," *N.Y. Times,* June 21, 1976, p. 4.

19. Jaime Zipper, in "Population, the U.S. Problem, the World Crisis," special supplement, *The New York Times,* April 30, 1972, Sect. 12, p. 21.

20. Tillman Durdin, "China's Changing Society Seems to Cut Birth Rates," *The New York Times,* April 21, 1970.

21. Victor K. McElheny, "Aspen Technology Conference Ends in Chaos," *Science,* September 18, 1970, p. 1187.

22. Borgstrom, 1973, writes: "Few realize that among the 2.5 billion who constitute the Hungry World [most of Asia, Latin America, and Africa] more than half are below 18 years of age. Yet more than half of all infants die before the age of five. The Hungry World currently counts around one billion children, 650 million of whom will never reach adulthood."

23. Dusko Doder, "Population: A Change of Attitude," *New York Post,* August 27, 1974.

24. John D. Rockefeller 3d's action might also have put the quietus on the population extremists and cocktail-party demographers who have until now invariably dismissed the Demographic Transition of our professional demographers as a figment of Marxist propaganda.

CHAPTER 18

1. Terman, 1925–59, IV: *The Gifted Child Grows Up* (1947), pp. 14–15.
2. Karpinos, 1968, p. 18.
3. Terman, *op. cit.,* p. 16.
4. *Ibid.,* p. 16. This, of course, was in 1921, one world war and two presidential wars ago, before the inflation caused by all wars cheapened the American dollar. In 1929, the year of our greatest post-World War I prosperity, the median family income for the entire United States was only $2,335. By 1969 the official poverty level established by the federal government was $4,900 or under for a family of four, but these dollars had far less buying power.
5. See Kotz, 1969, pp. 195–96. See also National Center for Health Statistics, National Health Survey, *Preliminary Findings of the First Health and Nutrition Examination Survey, 1971–1972;* also Series 10, Data from National Health Interview Survey; Series 11, Data from Health Examination Survey; Series 20, Data on Mortality; Series 21, Data on Mortality, Marriage and Divorce; and Series 22, Data from the National Natality and Mortality Surveys.
6. Charles Upton Lowe, "Statement by Charles Upton Lowe, M.D., Chairman, Committee on Nutrition, American Academy of Pediatrics," Committee on Agriculture, House of Representatives, April 24, 1969.
7. *U.S. Senate, To Save the Children,* 1974, p. 47.
8. R. P. Devadas et al., "Protein Intake and Mental Abilities of Selected Pre-School Children," *Indian Journal of Nutrition and Dietetics,* VIII (September 1971), 235.
9. Cecil Mary Drillien, "School Disposal and Performance for Children of Different Birthweight Born 1953–1960," *Archives of Diseases of Children,* XLIV (1969), 562–70.
10. Mark Golden et al., "Social-Class Differentiation in Cognitive Development among Black Preschool Children," *Child Development,* XLII, No. 1 (1971), 37–45. The table that appears on p. 438 was condensed and adapted from the originally published table.
11. *Supplement to Health of the Army, 1969;* Bellas and Vinyard, 1971.
12. Karpinos, 1968, pp. 25–30.
13. National Health Survey, Series 11, No. 13, *Hypertension and Hypertensive Heart Disease in Adults, United States, 1960–1962.* See also table CD.III.23, p. 481, which shows the prevalence of hypertension in persons between 17 and 44 years of age in the United States to be 34.4 per 1,000 persons among whites and 62.3 per 1,000 among "all other" people, *Health United States 1975,* first annual report to Congress, DHEW, HRA No. 76–1232, January 1976.
14. Bellas and Vinyard, 1971.
15. Conley, 1973, pp. 232–33.

CHAPTER 19

1. Mary E. Mebane [Liza], "What a Segregated School Is: The South Has Always Bused Children to Achieve a Racial Pattern," *The New York Times,* March 15, 1972, Op-Ed page.

2. *Why Should I Belong to the Citizens Council?* (undated pamphlet) (Jackson, Miss.: Citizens Council).

3. *What the Citizens Council Is Doing* (undated pamphlet) (Jackson, Miss.: Citizens Council).

4. See pp. 446–47.

5. See, for example, Ada Hart Arlitt, "On the Need for Caution in Establishing Race Norms," *Journal of Applied Psychology,* May 1921, pp. 179–83; Wayne Dennis, "Goodenough Scores, Art Experience, and Modernization," *Journal of Social Psychology,* LXVII (1966), 211–28; Fred Brown, "An Experimental and Critical Study of the Intelligence of Negro and White Kindergarten Children," *Journal of Genetic Psychology,* LXV (1944), 161–75; Klineberg, 1935(a) and 1935(b); George D. Stoddard, 1943, also "The IQ: Its Ups and Downs," *Educational Record Supplement,* XX (1939), 44–57; Lester R. Wheeler, "A Comparative Study of East Tennessee Mountain Children," *Journal of Educational Psychology,* XXXIII (1942), 321–34; the work of Skeels and the Iowa psychologists described in chapter 14, this book, and literally scores of equally well-documented scientific studies well known to professional clinical and research psychologists.

6. Shuey, 1966, p. 317.

7. *Ibid.,* p. 521.

8. George, 1962.

9. *Ibid.,* p. 79.

10. Predictably enough, while the nineteenth-century liberal socialism ascribed to Boas was used by George as another reason to discredit Boas' scientific anthropometry, George completely ignored the fact that Fairchild was a twentieth-century Marxist socialist. Fairchild's 1940 book, *Economics for the Millions,* was described, quite accurately, as "a critique of our capitalist economy from a Marxist point of view" in *The New York Times.* For some years before his death, Fairchild had served as national secretary of the National Council of Soviet-American Friendship. (See obituary, *The New York Times,* October 3, 1956.)

    In Fairchild's book *Race and Nationality,* the source of this attack on the viability of Boas' head-form or cephalic-index data, described by Fairchild on p. 105 as "two careful scholars, G. M. Morant and Otto Samson," was quite revealing. Geoffrey M. Morant, a craniologist and eugenicist who wrote for Pearson's two journals, *Biometrika* and the *Annals of Eugenics,* in the years between World Wars I and II—and was, in fact, the assistant editor of *Biometrika* under Karl Pearson—was co-author with Samson of "An Examination of Investigations by Dr. Maurice Fishberg and Professor Franz Boas Dealing with Measurements of Jews in New York," which appeared in *Biometrika* in June 1936. This article was the sole basis of Fairchild's "refutation" of the Boas cranial measurements. Its chief argument was made on page 30, in its snide comment that before one could accept the validity of Boas' cranial measurements one would need what Boas had not produced: "an indubitable proof that environment was solely responsible for such an observed difference would require *a guarantee that all the mothers and fathers* [sic] of the children *were of pure Jewish descent"* (italics added). This ugly slur on the morality of the immigrant Jewish mothers of the families in the Boas studies had been invented, c. 1910, by scientific and gut racists who had no better answer to Boas' hard anthropometric data. It was recently revived by John R. Baker in a book published in England and New York in 1974 (Baker, 1974, pp. 201–02).

11. In his *The Melting Pot Mistake* (1926) Fairchild—then the secretary-treasurer of the American Eugenics Society, Inc.—cast aspersions on the Jews as being of the inferior "new immigration" that threatened the "racial dilution" of the real American or "old immigration" of Nordic stock (pp. 212 *ff.*), and celebrated the passage of the infamous Immigration Act of 1924 as an act that "may justly be regarded as one of the most influential and far-reaching pieces of legislation ever enacted in human history" because it would reduce to a token minimum "the racial dilution of American stock" (p. 228). In his *Race and Nationality,* published in 1947—only two years after the Nazi holocaust in which six million Jews, many of them infants and minor children, were slaughtered solely because they had been born to Jewish parents—Fairchild devoted a full chapter, "The Jews" (pp. 137–61), to the argument that the Jews were as responsible as their anti-Semitic assassins for the sheer existence of anti-Semitism. In his words (p. 147), ". . . it cannot be ignored that anti-Semitism is a two-sided affair." He explained, at the end of the chapter, that he had avoided the "shameful story" of centuries of anti-Semitic persecutions and pogroms because it was not "the purpose of this volume to voice criticism in any direction."

12. Putnam, 1961.

13. Of the seven times Boas is cited in the index, Putnam refers to Boas being a Jew by the euphemism "minority group" three times. The word "Jew" is never mentioned in connection with Boas or his "minority group" students. In our own times the euphemism "Zionist" is used as Putnam used "minority group" in Moscow, Havana, Uganda, and the speakers' rostrums at the United Nations. Recall also the euphemisms "Alpine Slav," invented by Carl C. Brigham in 1923; and "cosmopolite," created by the propaganda organs of the Soviet Union in the 1950's.

14. Putnam, 1961, p. 67.

15. James S. Coleman et al., *Equality of Educational Opportunity* (Washington, D.C.: U.S. Government Printing Office, 1966), pp. 9, 12.

16. Ardrey, 1970, p. 63.

17. *Ibid.,* p. 63.

18. James S. Coleman, personal communication.

19. Seham, 1973, p. 38.

20. *Pediatrics,* XLIII, No. 1 (1969), 129.

CHAPTER 20

1. Chase, 1972, p. 237.

2. Miller, 1971, p. 86, Table V-26. The author, Herman P. Miller, is chief of the Population Division at the U.S. Bureau of the Census.

3. Terman, 1925–59, IV (1947), 16.

4. Jerry Hirsch, "Behavior-Genetic Analysis and Its Biosocial Consequences," *Seminars in Psychiatry,* II, No. 1 (1970), 97.

5. Original source of this statement: Jensen, 1969, p. 95. See also John Neary, "A Scientist's Variations on a Disturbing Racial Theme: Psychologist Arthur Jensen Believes Blacks Are Born with Lower IQ's Than Whites—And the Furor Goes On," *Life,* June 12, 1970, p. 65. See also "Can Negroes Learn the Way Whites Do? Findings of a Top Authority," *U.S. News and World Report,* March 10, 1969, pp. 48–51.

6. In the same issue of *Environment, Heredity, and Intelligence* (Harvard Educational Review Reprint Series No. 2, 1969) in which Jensen's article appeared, there also appeared articles analyzing it by Hunt, Kagan, Crow, Bereiter, Elkind, Cronbach, and Brazziel. For some of Professor Lederberg's earliest comments, see *Life,* June 12, 1970, p. 64.

7. Jensen was not alone among America's pop experts on the social biology of human mental development in his refusal to accept the statistical data of the U.S. Department of Health, Education, and Welfare's National Nutrition

Survey and other studies of the severity and extent of gross nutritional deprivation in the United States. There was also, for example, the social scientist Professor Daniel P. Moynihan, then on leave from Harvard to serve as an adviser on domestic problems to President Nixon. In 1969, when Dr. Arnold Schaefer, the director of the federal government's National Nutrition Survey, informed the U.S. Senate Select Committee on Nutrition and Human Needs that "our studies to date clearly indicate that there is malnutrition—and in our opinion it occurs in an unexpectedly large proportion of our sample population," Dr. Moynihan was singularly unimpressed. According to *Washington Post* reporter Nick Kotz, when Dr. Arthur Burns, Nixon's economic adviser, expressed serious doubts about the implications of Schaefer's testimony, since it suggested that hunger in America was due to lack of money rather than to the ignorance of the hungry, "Pat Moynihan joined forces with Burns in downgrading the hunger issue in a hurried memo to the President, *expressing doubt as to whether there was in this country any malnutrition serious enough to cause brain damage in infants"* (italics added). Quoted from Kotz, 1969, p. 212.

8. "Mental Tests for 10 Million Americans—What They Show: Some Startling Patterns of the Mental Ability of U.S. Youths—Measured by Race, Region, and Education—Emerge from the Results of a Nationwide Quiz," *U.S. News and World Report,* October 17, 1966, pp. 78–79.

9. Charles E. Greene, "Medical Care for Underprivileged Populations," *New England Journal of Medicine,* CCLXXXII, No. 21 (1970), 1187–93.

10. Professor Moynihan's "benign neglect" memo can hardly be given sole credit for the Nixon administration's open record of deliberately—and with government propaganda agencies trumpeting the facts—exsanguinating and terminating nearly all civil rights and other equality of opportunity programs mandated under the Eisenhower, Johnson, and Kennedy administrations. The Nixon posture on the open neglect of black civil rights and socioeconomic and sociocultural aspirations was based solely on his "Southern strategy" for re-election in 1972. The role that Moynihan's "benign neglect" memorandum played in the shattering of social programs aimed at redressing some of the legitimate black and other minority group grievances was to make many liberals, particularly in academic and editorial circles, as happy with Nixon's treatment of the blacks as were the white Citizens Councils of Mississippi and other southern states. It apparently helped achieve this, as witnessed by the many academic liberals who hastened to sign the full-page advertisements supporting the re-election of Nixon in 1972.

11. Richard L. Naeye et al., "Relation of Poverty and Race to Birth Weight and Organ and Cell Structure in the Newborn," *Pediatric Research,* V (1971), 17–22. See also Birch and Gussow, 1970; Committee on Maternal Nutrition, National Academy of Sciences, 1970; Conley, 1973; Davie, Butler, and Goldstein, 1972; Hardy, 1971; Kotz, 1969; La Rocco, 1968; Lichtenberg and Norton, 1970; Nutrition Foundation, 1969; Pan American Health Organization, 1972; Pasamanick and Knobloch, 1966; Scrimshaw and Gordon, 1968. In medical journal literature, see also Josef Warkany et al., "Intrauterine Growth Retardation," *American Journal of Diseases of Children,* CII (1961), 249–79; Richard L. Naeye, "Malnutrition: Probable Causes of Fetal Retardation," *Archives of Pathology,* LXXIX (1965), 284–91; James D. Harmeling and Marshall B. Jones, "Birth Weights of High School Dropouts," *American Journal of Orthopsychiatry* XXXVIII (1968), 63–66; P. A. Harper and G. Weiner, "Sequelae of Low Birth Weight," *Annual Review of Medicine,* 1965, pp. 405–20; Paul A. Harper, Liselotte K. Fischer, and Rowland V. Rider, "Neurological and Intellectual Status of Prematures at Three to Five Years of Age," *Journal of Pediatrics,* LV, No. 6 (1959) 679–90; and Howard V. Meredith, "Body Weight at Birth of Viable Human

Infants: A Worldwide Comparative Treatise," *Human Biology,* 1970, pp. 217–64.

12. L. Willerman and J. A. Churchill, "Intelligence and Birth Weight in Identical Twins," *Child Development,* XXXVIII (1967), 623–29. Babson, Kangas et al., "Growth and Development of Twins of Dissimilar Size at Birth," *Pediatrics,* March 1964, pp. 327–33. Sandra Scarr, "Effects of Birth Weight on Later Intelligence," *Social Biology,* XVI, No. 4 (1969), 6249–56. See also Richard L. Naeye, Kurt Bernischke, et al., "Intrauterine Growth of Twins as Estimated from Liveborn Birth-Weight Data," *Pediatrics,* March 1968, pp. 409–16.

13. See also William Walter Greulich, "Growth of Children of the Same Race under Different Environmental Conditions," *Science,* March 7, 1958, pp. 515–16; L. J. Filer, "Anemia: The USDA Today—Is It Free of Public Health Nutrition Problems?" *American Journal of Public Health,* February 1969, pp. 327–38; Robert M. Malina, "Growth and Physical Performance of American Negro and White Children: A Comparative Survey of Differences in Body Size, Proportions and Composition, Skeletal Maturation, and Various Motor Performances," *Clinical Pediatrics,* VIII, No. 8 (1969), 476–83; Salvador Armendares et al., "Chromosome Abnormalities in Severe Protein Calorie Malnutrition," *Nature,* CCXXXII (1971), 271–73; M. H. Brenner, "Fetal, Infant and Maternal Mortality during Periods of Economic Instability," *International Journal of Health Services,* III, No. 3 (1973), 145–59; Thomas E. Webb and Frank A. Oski, "The Effect of Iron Deficiency Anemia on Scholastic Achievement, Behavioral Stability and Perceptual Sensitivity of Adolescents," paper presented May 17, 1973, to the annual convention of the American Pediatric Society; and V. Ramalingaswami, "Nutrition, Cell Biology and Human Development," *WHO Chronicle,* XXIX (1975), 306–12.

14. Letter mailed by Shockley to members of the National Academy of Sciences, April 16, 1970, on the letterhead of the Foundation for Research and Education on Eugenics and Dysgenics, with a covering letter by Shockley.

15. On August 5, 1926, in "The Function of Sterilization," an address delivered at Vassar College, Mrs. Sanger, the founder of the American birth control movement, noting that while the anti-Semitic and anti-Italian racial quotas of the U.S. Immigration Act of 1924 had "already taken certain steps to control the quality of our population through drastic immigration laws . . . while we close our gates to the so-called 'undesirables' from other countries, we make no attempt to cut down the rapid multiplication of the unfit and undesirable at home."

Like Harvard's psychology professor McDougall in 1921 and the *Atlantic Monthly* author Cornelia James Cannon in 1922, Mrs. Sanger had been shaken by the "revelation" that, according to the Army intelligence test scores of World War I, the average American adult male had the mentality of a thirteen-year-old child. "When we view the political situation and realize that a moron's vote is as good as an intelligent, educated, and thinking [*sic*] citizen, we may well pause and ask ourselves—'Is America really safe for democracy?' "

To Mrs. Sanger, the remedy was gonadal: "It now remains for the United States government to set a sensible example to the world by offering a bonus or a yearly pension to all obviously unfit parents to allow themselves to be sterilized by a harmless and scientific [*sic*] means. . . . There is only one reply to a request for a higher birth rate among the intelligent, and that is to ask the government to *first* take off the burdens of the insane and feebleminded from your backs. Sterilization for these is the remedy." (Quoted from *Birth Control Review,* October 1926, p. 299.)

16. Van Evrie, 1868, p. 80.

17. March 7, 1966, p. 74.

18. Jameson G. Campaigne, "Personally Speaking: Genetics and Intelligence— $ Facts Needed," *Indianapolis Star,* August 24, 1969, Sect. 2, p. 2.
19. Erik Olaf-Hansen, personal communication.
20. Constance Holden, "R. J. Herrnstein: The Perils of Expounding Meritocracy," *Science,* July 6, 1973, pp. 36–39; see p. 37.
21. Quoted from Frederick Ausabel, Jon Beckwith, and Kaaren Janssen, "The Politics of Genetic Engineering: Who Decides Who's Defective," *Psychology Today,* June 1974, p. 38.
22. Since this judgment could not possibly be made by comparing *The Origin of Species* with Galton's *Hereditary Genius,* or the theory of variation, natural selection, and evolution with Galton's eugenics, it is possibly based on the fact that in Volume II of Terman's *Genetic Studies of Genius* Galton's IQ is given as being at least 200, while Darwin, along with Leonardo da Vinci, Napoleon, and Kant was given an IQ score of only 135 by Terman and his collaborators. See above, pp. 236–37.
23. David Layzer, "Science or Superstition? A Physical Scientist Looks at the IQ Controversy," *Cognition,* I (1972), 226. See also David Layzer, "Is There Any Real Evidence That IQ Test Scores Are Heritable?" *Scientific American,* July 1975, pp. 126–28. See also M. W. Feldman and R. C. Lewontin, "The Heritability Hang-up," *Science,* 190 (1975), 1163–68.

CHAPTER 21

1. John T. Chandler and John Plakos, "Spanish-Speaking Pupils Classified as Educable Mentally Retarded," California State Department of Education, 1970.
2. Edwin G. Boring, "Intelligence as the Tests Test It," *New Republic,* June 6, 1923, p. 35.
3. See Mercer, 1972.
4. Long before Alfred Binet's discovery of the different learning patterns of his own two natural daughters, Madeline and Alice, led to the start of experimental child psychology (see Wolf, 1973; Chapter 3, "Experimental Psychology: Its 'Fatherhood' in France"), a father and a mother in the Garden of Eden had discovered that their natural sons, Cain and Abel, who were of identical genetic endowments, had very different behavioral patterns (see Genesis 3:4).
5. See L. J. Eaves and J. L. Jinks, "Insignificance of Evidence for Differences in Heritability of IQ Between Races and Social Classes," *Nature,* CCXL (1972), 84–88. The authors are geneticists at the University of Birmingham, England. In the same issue, on page 69, see editorial, "How Much of IQ Is Inherited?" See also David Layzer, "Heritability Analyses of IQ Scores: Science or Numerology?" *Science,* March 29, 1974, pp. 1259–66. Layzer concludes: "1. Published analyses of IQ data provide no support whatever for Jensen's thesis that inequalities in cognitive performance are due largely to genetic differences. . . . 2. Under prevailing social conditions, no valid inferences can be drawn from IQ data concerning systematic genetic differences among races or socioeconomic groups. Research along present lines directed toward this end—whatever its ethical status—is scientifically worthless. 3. Since there are no suitable data for estimating the narrow heritability of IQ, it seems pointless to speculate about the prospects for a hereditary meritocracy based on IQ."
6. Michael Lewis, "Infant Intelligence Tests: Their Use and Misuse," *Human Development,* XVI (1973), 108–18; see pp. 113–14. See also Michael Lewis and Harry McGurk, "Evaluation of Infant Intelligence: Infant Intelligence Scores—True or False?" *Science,* CLXXVIII (1972), 1174–77.

CHAPTER 22

1. Shirer, 1960 (Fawcett edition, 1962), pp. 1274–88, segment titled "The Medical Experiments."
2. Much of this research was reviewed in a 1970 monograph of the federal government's National Institute of Mental Health, *Cognitive and Mental Development in the First Five Years of Life: A Review of Recent Research,* by Philip Lichtenberg and Dolores G. Norton.
3. This slur against Lincoln pained at least one of Davenport's closest friends, the paleontologist Henry Fairfield Osborn, Davenport's sponsor for membership in the National Academy of Sciences. On May 21, 1912, in a letter welcoming Davenport to membership in the Academy, Osborn asked: "May I not beg you in another edition of your Eugenics to omit the insinuation regarding the parentage of Lincoln? There may be something in it, but why throw an aspersion for which there is not historic foundation? Similarly I keenly resent Haeckel's insinuation regarding the parentage of Christ."

    Davenport answered, on May 24, 1912, and cited Nicolay's article on Lincoln in the latest *Encyclopædia Britannica,* XVI, 703, in which Lincoln's mother, Nancy Hanks, "is said to have been an illegitimate daughter of one Lucy Hanks." Because he had the *Britannica* as his authority, Davenport added, he "made almost no use of a series of copies of affidavits of fellow townspeople and contemporaries of Nancy Hanks, obtained shortly after Lincoln's death, and deposited at this [Eugenics Record] Office." He concluded that he "felt it a duty, from the standpoint of both biology and eugenics," to let his statement about Lincoln's paternity stand. Copies of correspondence in both the Archives of the Eugenics Record Office, American Philosophical Society, and the Osborn papers, American Museum of Natural History Library.
4. When, on February 19, 1807, Samuel Whitbread, heir to the great brewing fortune and Member of Parliament, brought in a proposed reformed poor-law bill which called, first, for "the establishment of a free educational system," this proposal was "keenly criticised in the press by Malthus, Bone, Bowles and others" (*Oxford Dictionary of National Biography,* VXXI, 26). When the portions of the bill dealing primarily with education were presented as a separate bill, Davies Giddy Gilbert, who was in 1827 to become president of the Royal Society, denounced it in ringing terms: "However specious in theory the project might be, of giving education to the labouring classes of the poor, it would in effect be found to be prejudicial to their morals and happiness; it would teach them to despise their lot in life, instead of making them good servants in agriculture, and other laborious employments to which their ranks in society had destined them; instead of teaching them subordination it would render them factious and refractory, as was evident in the manufacturing counties; it would enable them to read seditious pamphlets, vicious books, and publications against Christianity; it would render them insolent to their superiors; and in a few years the result would be that the legislature would find it necessary to direct the strong arm of power towards them . . ." Quoted from Bronowski, 1969, p. 149. See also Carlo M. Cipolla, *Literacy and Development in the West* (Baltimore: Penguin Books, 1969), pp. 65–70.
5. Banfield, 1970, p. 205
6. *Ibid.,* p. 192.
7. Hockett and Schlesinger, 1940, I, 733–34.
8. *Ibid.,* p. 244.
9. *Ibid.,* pp. 245–46.

CHAPTER 23

1. See Chase, 1972, p. 310.
2. Five years after the start of Nixon's program which according to Mr. White

would "make America the world's leader in environmental management," *The New York Times,* in an editorial, "Polluter No. 1: The Car" (May 18, 1975), observed: "Transportation—particularly the private automobile, taxis and trucks—contributes 97 per cent of all carbon monoxide in the city's air, 51 per cent of hydrocarbons, 28 per cent of nitrogen oxides, 19 per cent of particulates and significant quantities of toxic metals. . . . motor vehicle congestion and pollution have continued to increase . . . Downtown traffic is still rising at a rate of about 2.5 per cent a year. Levels of carbon monoxide creep relentlessly upward, despite emission controls installed in late-model cars." For a somewhat more realistic insight into the functions of Mr. Nixon's ecological crusades, see Schell, 1976, p. 143.

3. Watson, 1970, pp. 100–01.
4. WHO/SEARO Consultants, "The Prevention of Xerophthalmia," *WHO Chronicle,* XXIV, No. 1 (1972), 28. Dr. Halfdan Mahler, Director-General of WHO, writes that "twelve cents will buy enough vitamin A to protect a child from xeropthalmia for a year." *World Health,* February–March 1976, p. 3.
5. Watson, 1970, p. 74.
6. The enteric bacillus *Serratia marcescens,* a bacterium commonly found growing harmlessly in the human gut in its pigmented strains, grows out in very red colonies. While the nonpigmented strains of *Serratia* are serious pathogens which cause septicemia, urinary-tract infections, osteomyelitis, and other killing, crippling, and debilitating diseases, the pigmented *Serratia* is, as a pathogen, quite harmless to man. However, because of its trait of growing on solid nutrient media in fairly vivid shades of red, *Serratia marcescens* is possibly responsible for more deaths in the history of civilization than many other pathogens.

To bacteria, a slice of bread, a Eucharistic wafer, a dish of cold porridge, a slice of potato, a pot of white cottage cheese are all as nutritious as a block of nutrient agar in a laboratory Petri dish. Goughran (1969) writes that as far back as 332 B.C., when these red colonies were discovered flourishing on bread in the armies of Alexander the Great during the siege of Tyre in Lebanon, they were considered to be "miraculous blood." Prior to the Christian Era, the appearances of such "miraculous blood" were considered omens of the gods that predicted victories or defeats for Greek, Roman, Etruscan, and other Mediterrranean Basin armies.

During the Christian millennia it was determined by Church authorities as early as 1169 A.D., in Alsen, Denmark, that the red colonies of *Serratia marcescens* that appeared on the Eucharistic bread in the damp church was, indeed, the change of the Host into flesh and blood. From here to the discovery by Europe's leading scholars that the Jews were responsible for causing the blood to appear by defiling the Eucharistic bread was only a century into the future of the Dark Ages. In 1290 A.D. the appearance of *Serratia* colonies or the Blood of the Lord on the Sacramental wafer in Paris led to a pogrom in which, writes Goughran, "Jews were burned or slaughtered by the thousands." From then until sometime after 1823, when three Italian investigators proved that the holy blood was actually a "microscopic fungus" (the word "bacteria" was yet to be coined), the social history of pigmented *Serratia marcescens* was an unbroken chain of pogroms in Germany, Spain, Austria, and Poland. Scheurlen (1896) is cited by Goughran as having estimated that this harmless gut bacterium "caused the death of more men than many pathogenic bacteria."

R. S. and M. R. Breed (1924) reported that in Naples, Italy, "as recently as 1910 or 1911 when a *'bleeding host'* was found in one of the churches," the Naples newspapers reported excited disturbances that "caused many ignorant people to seek the protection of the priests as did the similar outbreak [of *Serratia* colonies on polenta, or corn-meal mush] in northern Italy."

Quotes from Eugene R. L. Goughran, "From Superstition to Science: The History of a Bacterium," *Transactions of the New York Academy of Science,* Series II, XXXI, No. 1 (January 1969), 3–24, and R. S. Breed and Margaret E. Breed, "The Type Species of the Genus *Serratia,* Commonly Known as *Bacillus Prodigiosus," Journal of Bacteriology,* IX (1924), 545–57.

7. Hirsch, 1970, p. 101.

8. Editorial, "The Obstetric Aspects of Idiocy," *British Medical Journal,* February 24, 1877, pp. 235–36.

9. W. W. Greulich, H. Thoms, and R. C. Twaddle, "A Study of Pelvis Type and Its Relationship to Body Build in White Women," *Journal of the American Medical Association,* CXII (1939), 485–92.

10. Robert M. Bernard, "The Shape and Size of the Female Pelvis," *Edinburgh Medical Journal,* II (1952), Supplement, *Transactions of the Edinburgh Obstetrical Society,* pp. 1–16.

11. Dr. Herbert G. Birch, personal communication. See also: Birch and Gussow, 1970, pp. 115–20.

12. Christensen, 1972, p. xi.

13. H. Stern, S. D. Elek, J. C. Booth, and D. G. Fleck, "Microbial Causes of Mental Retardation: The Role of Prenatal Infections with Cytomegalovirus, Rubella Virus, and Toxoplasma," *The Lancet,* August 30, 1969, p. 7618. See also Thomas H. Weller, "The Cytomegaloviruses: Ubiquitous Agents with Protean Clinical Manifestations," *New England Journal of Medicine,* CCLXXXV, Nos. 4–5 (1971), 203–14, 267–74; and James B. Hanshaw, "Congenital Cytomegalovirus Infection: A Fifteen Year Perspective," *The Journal of Infectious Diseases,* CXXIII, No. 5 (1971), 555–61.

14. Jill M. Forrest, personal communication. See also Jill M. Forrest, Margaret A. Menser, and J. A. Burges, "High Frequency of Diabetes Mellitus in Young Adults with Congenital Rubella," *The Lancet,* August 14, 1971, pp. 332–34; Jill M. Forrest, Margaret A. Menser, and J. D. Harley, "Diabetes Mellitus and Congenital Rubella," *Pediatrics,* XLIV, No. 3 (1969), 445–47.

15. Harry A. Sultz, Benjamin A. Hart, Maria Zielezny, and Edward R. Schlesinger, "Mumps and Diabetes: An Epidemiologic Association," paper read before the 102nd Annual Meeting of the American Public Health Association, New Orleans, 1974.

16. Louis Z. Cooper, personal communication.

17. Benjamin Pasamanick, personal communication.

18. Janet B. Hardy, "Rubella as a Teratogen," *Birth Defects: Original Article Series* V, Vol. VII, No. 1 (February 1971) (The National Foundation); and Janet B. Hardy, George H. McCracken, Mary Ruth Gilkeson, and John I. Sever, "Adverse Fetal Outcome Following Maternal Rubella *After* the First Trimester of Pregnancy," *Journal of the American Medical Association,* CCVII, No. 13 (1969), 2414–20.

19. Wood, 1961, p. 63.

20. Stanier et al., 1970, p. 788.

21. Gajdusek et al., 1965. See also D. Carleton Gajdusek, "Kuru and Creutzfeldt-Jacob Disease: Experimental Models of Neurological-Inflammatory Degenerative Slow Virus Disease of the Central Nervous System," *Annals of Clinical Research,* V (1973), 254–61. See also Jean L. Marx, "Slow Viruses: Role in Persistent Disease," *Science,* June 29, 1973, pp. 1351–54, and July 6, 1973, pp. 44–45. See also Milton Alter, "Is Multiple Sclerosis an Age-Dependent Response to Measles?", *The Lancet,* February 28, 1976, pp. 456–57.

22. Dwight T. Janerich, "Relation Between the Influenza Pandemic and the Epidemic of Neurological Malformations," *The Lancet,* June 5, 1971, p. 1165. See also, by the same author: "Influenza and Neuro-tube Defects," *The Lancet,* September 4, 1971, pp. 551–52.

23. Benjamin Pasamanick, personal communication.

24. Mark R. Rosenzweig, "Effects of Environment on Development of Brain and of Behavior," in Tobach, Aronson, and Shaw, 1971, pp. 303–42. See also Mark R. Rosenzweig, Edward L. Bennett, and Marian Cleeves Diamond, "Brain Changes in Response to Experience," *Scientific American,* February 1972, pp. 22–29. See also M. W. Fox "Neurobehavioral Development and the Genotype-Environment Interaction," *The Quarterly Review of Biology,* XLV, No. 2 (1970), 131–47.

25. Orville C. Green, "Sizing Up the Small Child," *Postgraduate Medicine,* October 1971, pp. 103–09. See also G. F. Powell, J. A. Brasel, and R. M. Blizzard, "Emotional Deprivation and Growth Retardation Simulating Idiopathic Hypopituitarism. Part 1. Clinical Evaluation of the Syndrome; Part 2. Endocrinologic Evaluation of the Syndrome," *New England Journal of Medicine,* CCLXXVI, No. 23 (1967), 1271–82; and Lytt I. Gardner, "Deprivation Dwarfism," *Scientific American,* July 1972, pp. 76–82.

26. Seminar on Malnutrition in Early Life and Subsequent Mental Development, Pan American Health Organization and World Health Organization, Mona, Jamaica, January 10–14, 1972 (*PAHO Scientific Publication* No. 251, 1972).

27. Pan American Health Organization, 1972, p. 25 (Cheek, Dobbing, Winick, Birch, et al.). The rate of growth during the proliferative growth phase is important for many reasons, not the least of which is that the chronology of proliferative growth is quite finite. Thus, in skeletal proliferative growth, which ends shortly after adolescence, dwarfism caused by hypopituitarism can be corrected by injections of supplementary growth hormone *only* during this proliferative growth period. Once the chronological period for skeletal proliferative growth has ended, supplementary growth hormone has no effect at all on human growth. The proliferative growth period in the human brain ends by the time a child is seven years old, at which time the human brain reaches about 90 percent of its adult weight. The weight added to the human brain after that represents few if any new cells. The *chronology* or calendar of proliferative growth is wholly genetically determined, but the *rate* of proliferative growth is considerably determined by the nutritive, emotional, chemical, physical, and various other elements of the total developmental environment during the proliferative growth period.

28. Pan American Health Organization, 1972, p. 7.

29. *Ibid.,* pp. 21–2.

30. *Ibid.,* p. 26.

31. *Ibid.,* p. 26.

32. *Ibid.,* p. 28.

33. *Ibid.,* p. 29.

34. Donald B. Cheek, "What's New and Important in Brain Growth Research?" *Medical Tribune,* November 15, 1972, p. 4.

35. A. E. Axelrod, "Immune Processes in Vitamin Deficiency States," *American Journal of Clinical Nutrition,* XXIV (1971), 265–71.

36. The fact that such doses of vitamin B_{12} help people with pernicious anemia to lead normal, healthy lives does *not* mean that pumping equally large doses of the same vitamin complex into the bodies of healthy people without pernicious anemia will produce anything except streams of urine rich in not needed and therefore nonutilized vitamin B_{12}.

37. Butler, 1959, p. 84. See also Nelson L. Levy and Abner Louis Notkins, "Viral Infections and Diseases of the Endocrine System," *Journal of Infectious Diseases,* CXXIV, No. 1 (1971), 94–103.

38. J. B. Gurdon and H. R. Woodland, "The Cytoplasmic Control of Nuclear Activity in Animal Development," *Biological Reviews,* XLIII (1968), 233–67; see p. 235.

39. J. B. Gurdon, "Nucleo-Cytoplasmic Interactions during Cell Differentiation," *Proceedings XII International Congress of Genetics,* III (1969), 191–203.

40. J. B. Gurdon, "The Cytoplasmic Control of Gene Activity," *Endeavour,* XXV (1966), 95–99; see p. 99. See also J. B. Gurdon, "The Autonomy of Nuclear Activity in Multicellular Organisms," *Symposia of the Society for Experimental Biology,* XXIV (1970), 369–78.
41. Richard L. Naeye, William A. Blanc, and Cheryl Paul, "Effects of Maternal Malnutrition on the Human Fetus," *Pediatrics,* October 1973; and Richard L. Naeye and William A. Blanc, "Relation of Poverty and Race to Antenatal Infection," *New England Journal of Medicine,* CCLXXXIII, No. 11 (1970), 555–59.
42. Leon Eisenberg, "The Future of Psychiatry," *The Lancet,* December 15, 1973, p. 1373.

CHAPTER 24

1. Commenting on southern hospital data on black mental patients, Pasamanick observed that "if the tabulations for manic-depressive psychoses over the years are studied, it is found that in Virginia a proportionate fall has occurred as schizophrenia rose. In 1940 there was 2.5 times as much manic-depressive psychosis as schizophrenia. When the two rates are summed, the total rate is invariant across the years, from 28.5 per 100,000 in 1920 to 27.2 in 1955. *What has apparently occurred is a change in style of diagnosis rather than an increase in schizophrenia*" (italics added). *Journal of the National Medical Association* LVI, No. 1 (January 1964), 6–17.
2. W. Pollin et al., "Life History Differences in Identical Twins Discordant for Schizophrenia," *American Journal of Orthopsychiatry,* XXXVI (1966), 492–509.
3. Leon Eisenberg, "The Pace of Progress," *World Health,* October 1974, pp. 4–7.
4. *Ibid.,* p. 7.
5. Roy E. Brown and Florence Halpern, "The Variable Pattern of Mental Development of Rural Black Children. Results and Interpretation of Results of Studies on Mississippi Children Aged One Week to Three Years by the Gesell Developmental Scales," *Clinical Pediatrics,* X, No. 7 (1971), 404–09.
6. Gesell and Amatruda, 1947.
7. Brown and Halpern, 1971.
8. See also Roy E. Brown, "Decreased Brain Weight in Malnutrition and Its Implications," *The East African Medical Journal,* November 1965, pp. 584–95; and Roy E. Brown, "Organ Weight in Malnutrition with Special Reference to Brain Weight," *Developmental Medicine and Child Neurology,* VIII (1966), 512–22.
9. See also Halpern, 1973.
10. Birch and Gussow, 1971, chapter 3.
11. Pasamanick and Knobloch, 1966, p. 7.
12. *Ibid.,* p. 22.
13. Rowland V. Rider, Matthew Tabach, and Hilda Knobloch, "Associations Between Premature Birth and Socioeconomic Status," *American Journal of Public Health,* XLV (1955), 1022–28.
14. Benjamin Pasamanick, Hilda Knobloch, and Abraham Lilienfeld, "Socioeconomic Status and Some Precursors of Neuropsychiatric Disorder," *American Journal of Orthopsychiatry,* XXVI (1956), 594–601.
15. Knobloch and Pasamanick, 1966, pp. 40–41.
16. *Ibid.,* p. 41.
17. *Ibid.,* p. 42. By 1966, the soaring dollar costs of the war in Indochina acted to limit the federal investment in such anti-poverty programs. After Jensen's 1969 blast at compensatory education in the *Harvard Educational Review,* with its assurances that it had failed because the stupidity of poor children was in their genes and *not* "the result of social, economic, and educational

deprivation and discrimination," the war hawks in the executive and legislative branches of government seized upon the writings of Jensen and followers like Herrnstein as proof that "even the liberal intellectuals" agreed that the sequelae of poverty were due to bad genes and not to bad social conditions—and therefore Head Start and other anti-poverty programs were a sheer waste of tax dollars at a time when billions of dollars were still needed for the bombs and napalm required to win the hearts and minds of the peasants of Vietnam, Cambodia, and Laos. As our military expenditures in Indochina were escalated, most of the compensatory anti-poverty programs were either drastically curtailed or killed entirely.

18. Pasamanick and Knobloch, 1966, p. 23.
19. See Lita Furby, "Implications of Within-Group Heritabilities for Sources of Between-Group Differences: IQ and Racial Differences," *Developmental Psychology,* IX, No. 1 (1973), 28–37. An excellent study of the gene-environment norms of reaction and interaction in the development of behavioral traits, such as the ability to achieve high or low IQ test scores. See, particularly, pp. 35–36.
20. G. B. Bietti et al., "Prevention of Blindness," *WHO Chronicle,* XXIV, No. 1 (1973), 21.
21. Moulder, 1964, p. 13.
22. "Half of Blind Are Indian" and "Nutritional Anemia in India," unsigned news articles, *Medical Tribune,* February 14, 1973.
23. McKeown and Lowe, 1974, pp. 92–93.
24. Robert R. Wagner, in *Harrison's Principles of Internal Medicine* (5th ed., 1966), p. 1727.
25. Philip J. Landrigan, "Epidemic Measles in a Divided City," *Journal of the American Medical Association,* CCXXI (1972), 567–70.
26. "Children Termed in Disease Peril; Experts Say Many in U.S. Lack Immunization," *The New York Times,* October 2, 1975, p. 42.

CHAPTER 25

1. Davie, Butler, and Goldstein, 1972.
2. Skeels, 1966, p. 56.
3. Niswander and Gordon, 1972, p. 2.
4. Pasamanick objects to the term "reproductive wastage" because, as in the case of the term "human refuse" Galton used to describe English emigrants, it suggests that certain classes of people—such as emigrants from England and the handicapped—are worthless to society and to themselves.
5. Serious physicians have for well over a century been aware of the role of complications of pregnancy and birth in such disabilities. It was in 1862 that the English orthopedic surgeon William John Little published his classic study, "The Influence of Abnormal Parturition, Difficult Labors, Premature Births, and Asphyxia Neonatorum on the Mental Health and Physical Condition of the Child. Especially in Relation to Deformities." It was Little who first described Little's disease, cerebral spastic limb paralysis, in his "Treatise on Deformities" in 1853. Little himself was the victim of a congenital birth defect, a club foot.
6. *Key to Institution Abbreviations: BO:* Boston Hospital for Women and Children's Medical Center, Harvard Medical School, Boston; *BU:* Children's Hospital, State University of New York Medical School, Buffalo; *CH:* Charity Hospital, Tulane University, School of Medicine and Medical Center, New Orleans; *CO:* Columbia-Presbyterian Medical Center, Columbia University College of Physicians and Surgeons; *JH:* Johns Hopkins Hospital, Johns Hopkins University School of Medicine, Baltimore; *VA:* Medical College of Virginia, Richmond; *MN:* University of Minnesota Hospitals, Minneapolis; *NY:* Metropolitan Hospital, New York Medical College, New York; *OR:*

University of Oregon, Medical Center, Portland; *PA:* Pennsylvania Hospital and Children's Hospital of Philadelphia, University of Pennsylvania, Philadelphia; *PR:* Child Study Center, Brown University, Providence; *TN:* Gailor Hospital, University of Tennessee College of Medicine, Memphis.

7. The original Goodenough tests were conceived as "culture-free" tests of intelligence by Florence Goodenough in 1926. While the original and the revised (1963) Goodenough-Harris tests are nonverbal and less culture-bound than the Stanford-Binet and Wechsler types of mental tests, they are nevertheless very far from being culture-free.

   Professor Wayne Dennis, in a series of studies of Goodenough IQ test scores achieved by children of various cultures in America, Latin America, Japan, Iran, Lebanon, and Africa, found that—among many other purely environmental and cultural factors—the art experiences of children played *predictable* roles in their IQ test scores. Thus, among Jewish children in Brooklyn, the religious (i.e., cultural) injunction against drawing living people (i.e., graven images) resulted in far lower IQ test scores in children of Orthodox Jewish families, and proportionately higher IQ test scores in Jewish children of less Orthodox or agnostic families. Hopi Indian children— morons according to their Goddard and Terman Binet test scales—"were found to test higher than American whites" on the Goodenough Draw-a-Man IQ tests. See Wayne Dennis, "Goodenough Scores, Art Experience, and Modernization," *Journal of Social Psychology* LXVIII (1966), 211–28. See, particularly, his Table 1, pp. 216–17.

8. Dale B. Harris and Jean Roberts, "Intellectual Maturity of Children: Demo-Series 11, No. 116, 1972. See also Jean Roberts and Arnold Engel, "Family Background, Early Development, and Intelligence of Children 6–11 Years," National Health Survey, Series 11, No. 142, 1974.

9. Harris and Roberts, *op. cit.,* p. 19.

10. *Ibid.,* p. 20.

11. *Ibid.,* p. 1.

12. Heber et al., 1972.

13. Davie, Butler, and Goldstein, 1972.

14. This important point was overlooked by many well-meaning Americans in 1954, when the U.S. Supreme Court unanimously held classroom and all other forms of racial segregation to be illegal.

   In making its ruling, the Supreme Court did not do so in the belief that by ending classroom segregation it would in one swift stroke turn disadvantaged black children into Rhodes scholars. The high court outlawed racial segregation in schools because it was immoral and in violation of nearly every clause in the U.S. Constitution dealing with human rights and human dignity.

   I happen to agree with Father Theodore Hesburgh, president of Notre Dame University, who suggested that the federal government raze to the ground all schools in the dual (and traditionally, in terms of state expenditures per capita, far inferior to the parallel school systems maintained by the same states for white children) school systems maintained for black children. Once this was done, the children who formerly attended these deliberately inferior schools were to be bused to the nearest schools to their homes.

   Busing was far from unknown in most states and cities prior to 1954. It served two functions: (1) throughout the country, as small local schools were consolidated into larger centralized schools, buses transported the children and youths who lived far from the new schools; (2) in the South, busing was also used to help enforce local and state laws against integration, so that children were not only bused to school, but also bused to either white or black schools regardless of how close they lived to schools of the wrong "color."

   When Dr. James S. Coleman, in 1975, suggested that busing was doing more harm than good, and had caused the flight of middle-class whites from the larger *but not the smaller cities* to the suburbs, this led many people to say that the author of the Coleman Report of 1966 had now changed his mind about the need for integrated education. Coleman denied this implication, and I believe he means this sincerely.

I think, however, that Coleman is in serious error in blaming busing alone for the continuing flight of people who could afford to make the move from the larger cities to the less crowded and greener suburbs. The flight from the larger industrialized cities began not in 1954 in America, but in 1854 in Manchester, Birmingham, and other British and European cities, and was then followed by a similar flight from New York, Chicago, Pittsburgh, and other industrializing and industrialized cities starting shortly after the American Civil War. The American hegira from the larger cities reached its frenzied peak shortly after World War II, when the wartime freeze on housing construction caused housing shortages that exacerbated the original cause for this flight from the industrial big cities. That cause, of course, was the dense clouds of smoke from the industrial furnaces that, in time, was to be joined by dozens of other by-products of industrial, chemical, transportation, and electric power furnaces and engines and processes which added to the burdens of chemical, physical, noise, and stress pollution that have, since the dawn of the Industrial Revolution, made living in the larger cities increasingly hazardous to human physiological and mental health.

There is little if any hard evidence that educational integration per se materially advanced the timetables of the white *and the black* families that had long been planning to quit the increasingly polluted big cities. A recent opinion study among metropolitan householders by a M.I.T.-Harvard team headed by David Birch of M.I.T. reported: "Only 23 percent expressed a desire to live in any metropolis; 13 percent wanted to live in a small city, 20 percent in a town or a village, and 29 percent wanted rural isolation." Quoted from Gurney Breckenfeld, "Is the One-Family House Becoming a Fossil? Far from It," *Fortune*, April 1976, p. 88.

15. Davis, Butler, and Goldstein, p. 190.
16. *Ibid.,* p. 192.
17. "Infant Mortality and Employed Mothers," unsigned article in the *Journal of the American Medical Association,* March 31, 1923, p. 948.
18. Quoted from Martin Tolchin, "Health Officials Say Nixon's Cuts Would Hurt Poor: Proposed Reductions in Aid Are Called Threat to Vital Medical Services Here," *The New York Times,* June 24, 1973, pp. 1, 20.

CHAPTER 26

1. Quoted from Truax, 1952.
2. In foreword to Chase, 1972, p. xiv.
3. John Ertl, personal communication.
4. This relationship between poor nutrition and abnormal electroencephalograph (EEG) measurements is well known to all serious neurologists and other medical scientists. In a position paper prepared for the U.S. Senate by a subcommittee on Nutrition, Brain Development and Behavior of the National Academy of Sciences, the subcommittee reported: "In man and lower animals, the normal EEG patterns and their developmental sequence have been shown to become abnormal in chronic and acute states of malnutrition suggesting alterations in normal cellular electrical activity. The abnormal electroencephalographic tracings seen in malnourished subjects generally improve toward normal with adequate nutritional therapy." From Select Committee on Nutrition and Human Needs, U.S. Senate, *National Nutrition Policy: Nutrition, Health And Development* (Washington, D.C.: U.S. Government Printing Office, June 1974), p. 5.

5. On the eve of Hitler's coming to power, Davenport and Laughlin, in the September–October 1932 issue of the *Eugenical News,* published "The Nordic Movement in Germany," translated from the German of K. Holler, which promised that the Nazis would satisfy the world's desire for "Nordic race-hygiene [eugenics]" and also "a political system in the Nordic sense." In an editorial footnote, the editors of the *Eugenical News* wrote: "Without race tolerance, and paradoxical as it may seem, without race pride, practical eugenics cannot be very successful. It is the business of each race and family to take stock of its own qualities, to establish its own ideals, and to breed [*sic*] toward such ideals." By "race tolerance," of course, they meant tolerance for the concentration camps where the Nazis would shortly begin to establish their own Nordic race ideals.

6. "Men and Mice at Edinburgh: Reports from the Genetics Congress. The Geneticist's Manifesto," *Journal of Heredity,* XXX, No. 9 (1939), 371–73.

7. Skeels, 1966, p. 56.

8. These figures are from the *Statistical Abstract of the United States,* 1975; and the National Center for Education Statistics, 1976.

9. Davie, Butler, and Goldstein, 1972, p. 190.

10. Between June 27, 1950, when President Harry Truman sent a thirty-five-man Military Assistance Advisory Group to teach "our" Vietnamese how to handle their American weapons (June 27, 1950, was also the day Truman ordered General MacArthur to dispatch our troops to Korea for the first of our era's presidential wars in Asia), to the time the last of our military forces were withdrawn from Vietnam in 1973, an estimated $150 billion had been spent on the war in Vietnam, Laos and Cambodia. During the Indochina war the government estimated its ultimate costs would come to $352 billion if "that war would end by July 30, 1970." The bulk of these expenditures would be for veterans' benefits and interest on war loans, the last payment coming in the year 2045. However, American military involvement in Indochina did not end until more than a year after the Paris Agreements of January 1973. After the killing ended, the government changed its estimates of the ultimate costs from $352 billion to "Unknown." The economist Robert Lekachman put the costs to the year 2000 as at least $400 billion. In 1973, another economist, Tom Riddel, estimated that the ultimate dollar costs would total about $676 billion (i.e., 676,000 million). *Statistical Abstract of the United States:* 1971, p. 243; 1975, p. 317. *Congressional Record:* March 21, 1975, p. S4642; May 14, 1975, p. S8152.

11. Berg, 1973, p. 133. According to a news article in *The New York Times,* November 10, 1973, single-cell protein plants were already in operation in Scotland and southern France, where they produced, respectively, 4,000 tons and 20,000 tons of SCP a year, and a 100,000-ton production plant was under construction in Sardinia. All of these were being joined by a network of SCP plants in the Soviet Union, with capacities estimated at between 50,000 and 240,000 tons of single-cell proteins each.

12. Carl Lamanna and Guenther Stotzky, "Alcohol Fuel: The Inexhaustible Font," letter to *The New York Times* on behalf of the Committee on Environmental Microbiology of the American Society for Microbiology, *The New York Times,* May 7, 1975. The sunshine required to grow the plants and plant products needed for alcohol production is the same renewable source of energy responsible scientists such as Farrington Daniels and Barry Commoner describe as the safe, inexpensive, non-polluting and eminently practical alternative to the costly and environmentally degrading fossil and nuclear fuels on which we are now forced to rely for light, heat, transportation, and industrial power. See chapter 6, "The Sun," in Commoner, 1976.

13. *Yale Reports,* No. 492, December 1, 1968.

A half-dozen years after Dr. Weinerman indicted the automobile as the nation's Public Health Enemy Number One, the fuel crisis of 1974 led to sharp decreases in gasoline sales and automobile daily transportation. To conserve gasoline, speed limits in all states were reduced to 55 miles per hour. Various studies by federal, state, and research center agencies all showed that this decline in daily automobile use and the lowering of speed limits led to improvements in American health. In September 1975, a team of University of California epidemiologists published the results of their study of first-quarter mortality rates in 1970–1973 and 1974 in urban San Francisco County and less-urbanized and adjacent Alameda County. They showed that, in 1974, there was a 13.4 percent decrease in deaths from all causes in San Francisco County and a 7.7 percent drop in Alameda County. Deaths from cardiovascular diseases fell by 16.7 percent in San Francisco County and 11.2 percent in Alameda County. Chronic lung diseases including chronic bronchitis, asthma, and emphysema—which are equally related to vehicular exhaust fumes—fell by 32.9 percent in San Francisco County and by 38.0 percent in Alameda County during the fuel crisis of 1974.

The authors of the study wrote that "analysis of weather and air stability data, relevant pollutant levels and the pattern of influenza and pneumonia deaths (sometimes thought to influence cardiorespiratory deaths) support the hypothesis that a decrease in vehicular exhaust fumes would have a beneficial effect on health." From Stephen M. Brown et al., "Effect on Mortality of the 1974 Fuel Crisis," *Nature,* September 25, 1975, pp. 306–07.

14. Arthur M. Squires, "Clean Power from Dirty Fuels," *Scientific American,* October 1972, pp. 26–35.

15. Henry H. Kolm and Richard D. Thornton, "Electromagnetic Flight: The future of high-speed ground transportation may well lie not with wheeled trains but with vehicles that 'fly' a foot or so above a guideway, lifted and propelled by electromagnetic forces," *Scientific American,* October 1973, pp. 17 *et passim.*

16. As of this writing, IQ testing has been eliminated by the school systems of California, the District of Columbia, and New York City, and has been reduced drastically in St. Paul, Minnesota, and Philadelphia. With the current opposition to further IQ testing in Massachusetts being led by Massachusetts State Education Secretary Paul Parks, it is quite probable that the days of IQ testing in that state's school system are now numbered. (See *Newsweek,* March 22, 1976, p. 49.) Since IQ tests do not measure such genetic factors as are involved in test performances but do measure the experiential effects of differential health, cultural, and social conditions, these standard IQ tests, particularly as they are still used in most of the nation, can help evaluate the mental sequelae of the massive environmental intervention and its absence in two fairly matched and contiguous southern states, the poorest and the sixth-poorest states in the land.

17. Thomas E. Webb and Frank A. Oski, "The Effect of Iron Deficiency Anemia on Scholastic Achievement, Behavioral Stability and Perceptual Sensitivity of Adolescents," paper presented to the annual convention of the American Pediatric Society, May 17, 1973.

18. I. S. Gottesman and L. Erlenmeyer-Kimling, foreword to "Supplement," *Social Biology,* XVIII (1971), S-1. *Social Biology,* formerly the *Eugenics Quarterly,* and before that *Eugenical News & Current Record of Human Genetics and Race Hygiene,* the quarterly publication of the American Eugenics Society, is the direct and lineal descendant of the *Eugenical News,* founded in 1916 by Charles Benedict Davenport and Harry Hamilton Laughlin and edited by them until 1939. When used as a synonym for "eugenics," the term "social biology" ranks second only to Spencer's nineteenth-century construct, "Social Darwinism," as a howling misnomer, since eugenics is, historically, both antisocial and antibiological.

GLOSSARY OF SCIENTIFIC TERMS

ANTHROPOMETRY: The branch of anthropology and biology dealing with the proportional measurements of the human body, its parts and powers.

ANTIBODY: A protein produced in the human body in response to the presence of "foreign" or "non-self" antigens. Antigens are proteins present in body tissues, as well as in the walls of viruses, bacteria, protozoa, fungi, and other microbial pathogens. All antibodies are specific only to different individual antigens. Chemically, antibodies neutralize antigens by combining with and inactivating them. (See VACCINES.) Our inborn host defense or immune systems against such causes of infectious diseases also include the leukocytes, lysozyme, interferon, and complement.

ASTHENIA: Weakness; lack or loss of strength and energy; debility.

BACTERIUM (pl. BACTERIA): The bacteria are small, single-cell microorganisms related to the plant kingdom but lacking in chlorophyl. They contain both types of nucleic acids, DNA and RNA, but lack the rod-shaped chromosomes present in the cells of higher forms of life. Some bacteria are pathogens (i.e., causes of disease) in man and other mammals.

BRACHYCEPHALIC: Having a relatively broad or round skull, with a cephalic index of over 80.

CACOGENIC: See DYSGENIC.

CEPHALIC INDEX: One hundred times the ratio of the maximum breadth of the skull to the maximum length (front to back) of the skull.

CHROMOSOMES: Rod-shaped bodies, made of the nuclear protein chromatin, found in the nucleus of all but the most primitive cells (such as the bacteria). The genes of heredity are carried, in linear order, within the chromosomes.

COMPLEMENT: A group of substances found in the blood that speed the destruction of certain bacteria and other cells in the presence of specific complement-fixing antibodies.

CONGENITAL: Physiological and mental conditions existing at birth, of hereditary or nongenetic causation. For example: skin and eye colors are hereditary; congenital blindness and mental retardation are usually caused by such nonhereditary factors as viral and other maternal infections during pregnancy, maternal malnutrition, and brain damage suffered during the act of birth.

CRANIOMETRY: The measurement of skulls and facial bones.

CYANOCOBALOMIN: Vitamin B_{12}; the antianemia factor found in liver.

CYTOMEGALOVIRUS (CMV): An ubiquitous virus that causes mild infections in adults but is a major cause of congenital birth defects in brain and body when a CMV infection in the pregnant mother allows the virus to infect the developing fetus in her womb.

CYTOPLASM: Everything inside of the living cell except the nucleus.

DEMOGRAPHY: The science dealing with the study of human populations, their growth and death rates, distribution, and other changes.

DEOXYRIBONUCLEIC ACID (DNA): The nucleic acid that comprises the genetic material, found in the nucleus of the cell in all forms of life.

DOLICHOCEPHALIC: Long headed, with a cephalic index of under 75.

DYSCRASIA: An abnormal or pathological condition.

DYSGENIC (or CACOGENIC): Word coined by Francis Galton to categorize races or strains tending toward "racial degeneration." The opposite of eugenic (q.v.).

ECOSYSTEM: The total environment in which a community of organisms live and interact with one another.

ENCEPHALOPATHY: Any degenerative disease of the brain.

ENDOCRINE GLANDS: Ductless glands that synthesize and release hormones, usually into the bloodstream, to act on other parts of the body.

ENZYME: A protein that accelerates or helps produce chemical changes, such as those involved in metabolism.

EPIDEMIOLOGY: The science dealing with the various single or interacting factors that cause infections and other physiological and mental diseases in human communities.

EPIGENESIS: The gene-programmed development of the body and its various component structures during the embryonic stages of growth.

EPISOME: One of two groups of nonchromosomal genetic elements. One class of episomes is independent of the genome and is replicated in cell cytoplasms; the other type of episomes replicate by integrating themselves into the chromosomes of living cells.

ERYTHROCYTE: A red blood cell.

ETIOLOGY: The sum of knowledge regarding the cause or causes of a disease.

EUGENICS: a scientistic (pseudoscientific) cult started in nineteenth-century England by Francis Galton "to give the more suitable races or strains of blood a better chance of prevailing speedily over the less suitable." Galton saw all human physical and mental traits as due to either superior (eugenic) or inferior (dysgenic) blood, and preached that the "race" was in peril because most people were, *by biological inheritance,* "ineffectives" and "mediocrities."

GAMETE (GERM CELL): A mature reproductive cell. The nuclei of male and female gametes fuse (constituting fertilization) to form a zygote and thereby initiate the creation and development of a new individual organism.

GENE: The basic unit of genetic information, or heredity. Genes are transmitted to the progeny via the chromosomes carried in the parental gametes.

GENETICS: The science dealing with the biological, chemical, and physical realities of heredity and variation in living organisms. Genetics is the opposite of eugenics.

GENOME: The complete set of hereditary factors contained in the genetic (inherited) constitution of an organism.

GENOTYPE: The genetic constitution of an individual, exclusive of all nonhereditary factors involved in his growth and development.

GERM PLASM: The hereditary factors in the cell; an obsolete term superseded by the lexicon of modern genetics.

HEMATOCRIT: A centrifuge that separates the cells and other elements of the blood; used in estimating the volume of red blood cells in the diagnosis of anemia.

HEREDITARY: Characteristics biologically transmissible from parents to offspring.

HETEROZYGOUS: Possessing different alleles (one of two or more genetic characters) in regard to a given hereditary trait.

HOMOZYGOUS: Possessing an identical pair of alleles in regard to a specific or to all genetically transmissible characters.

HORMONES: Chemicals secreted by the endocrine glands, usually into the bloodstream, that act as regulators of various body functions, such as growth, metabolism, and reproduction.

IMMUNOLOGY: The science dealing with the mechanisms of natural and induced immunity to infectious diseases; with induced sensitivity to innate and foreign substances; and with allergy.

INTERFERON: Small proteins, produced by cells in response to virus infection, that act to inhibit or interfere with the parasitic actions of viruses in cells.

LEUKOCYTE: White blood cell.

LYSOZYME: An antibacterial enzyme present in white blood cells, tears, saliva, egg white, and various body fluids.

METABOLISM: The sum of all the physical and chemical bodily processes by which living cells and organisms are formed, maintained, and supplied with energy.

METABOLITE: A substance that takes part in a metabolic process. Most metabolites are produced by the body; others, essential to metabolism, must be taken in from the environment. The latter group of essential metabolites includes certain vitamins and amino acids present in various foods.

NIACIN (NICOTINIC ACID): One of the vitamins in the vitamin B-2 complex; the PP (pellagra-preventive) factor.

NUCLEIC ACID: A family of long chain molecules found in the chromosomes and cells of all living things, and also in viruses. See also DEOXYRIBONUCLEIC ACID and RIBONUCLEIC ACID.

OOCYTE: An immature or prefertilized ovum.

OVUM: A female germ cell; a mature egg cell. The female gamete or sex cell.

PERTUSSIS: Whooping cough.

PHENOTYPE: The end product, at any given stage of fetal or postnatal life, of the perpetual interactions between the array of potentialities present in an individual's genome or genetic inheritance and the continuum of nutritional, physiological, chemical, physical, emotional, cultural, and other environmental factors involved in the growth and development of that individual.

PHENYLKETONURIA: A congenital defective metabolism of phenylanaline, the amino acid essential for growth in infants. It retards mental development in infants.

PHLOGISTON: An imaginary element which, in the eighteenth century, was widely believed to be the essential principle of fire, and to cause combustion and be given off by anything burning.

PHRENOLOGY: Early nineteenth-century scientistic system of "analyzing" the character and intelligence of a person by feeling and studying the shape and protuberances of the skull.

PHYSIOGNOMY: An early precursor of phrenology, in which the evaluation of human traits was made by studying the contours of an individual's face.

POLYGENIC: In *genetics,* pertaining to or influenced by several different genes.

POLYGENIC INHERITANCE: In *eugenics,* the concept that all traits and characteristics, from intelligence and health to the capacity to amass fortunes or become paupers, are controlled—*independently of environmental factors*—by combinations of specific genes. See also UNIT CHARACTER INHERITANCE.

PREFORMATION: The concept, widely accepted prior to the development of modern cell theory and embryology, that the ovum or spermatozoon of each animal (including the human species) contained a miniature adult—which, during and after gestation, grew to full-size adult form. The opposite of epigenesis (*q.v.*).

PROTOZOA: Microscopic single-cell organisms, less primitive than the bacteria. Some protozoa, such as the malaria parasite, cause human diseases.

RIBONUCLEIC ACID (RNA): The nucleic acid, synthesized from DNA in the cell nucleus, that carries the genetic information from the nucleus to the cytoplasm of the cell.

RIBOSOMES: Minute granules of protein and RNA found in the cytoplasm of cells. They are believed to be the main sites of protein synthesis—that is, the carrying-out of the genetic directions transferred from the DNA in the nucleus by the RNA molecules.

SCARLATINA: Scarlet fever.

SCIENCE: Any branch of the systematized knowledge of nature and the physical world in which information and concepts are derived from observation, analysis, and experimentation in order to determine the nature or principle of what is being studied. Scientific disciplines examined in this book include anthropology, biology, demography, embryology, epidemiology, genetics, immunology, pathology, pediatric medicine, psychiatry, and psychology.

SCIENTISTIC: Using the language, symbols, findings, and other attributes of the legitimate sciences to advance unproven preconceptions and dogmas; pseudoscientific. Scientistic cults examined in this book include Malthusianism, eugenics, phrenology, craniology, Teutonism (Nordicism), and Spencer's "Social Darwinism."

SOCIAL BIOLOGY: The continuum of interacting social, biological, historical, agricultural, economic, political, behavioral, biochemical, and biophysical factors involved in the qualities and the duration of human life.

SOCIOBIOLOGICAL: Pertaining to one or more of the interacting complexes involved in social biology.

SPERMATOZOON (*pl.* SPERMATOZOA): The male gamete or sex cell. Sperm; semen.

TROPISM: The tendency of an organism to react in a certain way to a specific stimulus, such as the tendency of plants to grow toward light.

TRYPTOPHAN: An amino acid present in proteins from which it is set free by digestive processes in the body. It is essential for infant growth and nitrogen balance in adults. It is also a metabolic precursor of niacin, which explains why meats, fish, eggs, dairy products, and other protein foods can both prevent and cure pellagra.

TUBAL LIGATION: The surgical cutting and tying-off of the Fallopian tubes of the uterus to cause sexual sterilization in women.

UNIT CHARACTER INHERITANCE: The eugenic concept that every single physical, behavioral, economic, and moral trait in human beings is the product of a specific gene for each specific trait—

and that these traits are genetically transmitted and expressed independently of all environmental factors, such as nutrition, infectious diseases, affluence, and poverty. Unit character inheritance is the intellectual precursor of polygenic inheritance (*q.v.*).

VACCINE: Any microbial preparation made of live, attenuated, or inactivated microbial pathogens or their toxoids that confers limited or lasting immunity against specific infectious diseases upon inoculation or ingestion. Vaccines that immunize infants and adults against diseases include those for smallpox, diphtheria, pertussis, tetanus, poliomyelitis, measles, rubella (German measles), influenza, mumps, and tuberculosis.

VASECTOMY: The surgical cutting or blocking of each vas deferens (the two tubes that carry spermatozoa from the testes to the penis), performed to cause sexual sterilization in men.

VIRUS: A submicroscopic particle, which is invisible under the light microscope but can be observed by electron microscopy, consisting of a core of either DNA or RNA within a coat of protein or of protein and lipid. Although consisting almost wholly of genetic chemicals, viruses have no metabolism of their own, and can replicate only by parasitizing bacterial, plant, and animal cells. Diseases caused by viruses range from potato, cereal, and other food-plant blights to smallpox, poliomyelitis, measles, rubella (German measles), yellow fever, influenza, mumps, and the common cold.

ZYGOTE: The fertilized ovum.

Derived from: I. F. and W. D. Henderson and J. H. Kenneth, *A Dictionary of Scientific Terms* (7th ed., 1960); *Webster's New World Dictionary of the English Language* (College Edition, 1960); *The Encyclopedia of the Biological Sciences* (Peter Gray ed., 1961); E. C. Graham, *The Basic Dictionary of Science* (1965); M. Abercrombie et al., *A Dictionary of Biology* (1966); *Dorland's Illustrated Medical Dictionary* (24th ed., 1968); J. D. Watson, *Molecular Biology of the Gene* (2nd ed., 1970); Robert W. and Isabel Leader, *Dictionary of Comparative Pathology and Experimental Biology* (1971); *The Oxford English Dictionary* (Compact Edition, 1971); *Stedman's Medical Dictionary* (22nd ed., 1972); P. Wingate, *The Penguin Medical Encyclopedia* (1972); and J. D. Ebert et al., *Biology* (1973).

SELECTED BIBLIOGRAPHY

ADAMS, MARGARET. *Mental Retardation and Its Social Dimensions.* New York: Columbia University Press, 1971.

AINSWORTH, MARY D. *See* Bowlby, John.

ALLAND, ALEXANDER, JR. *Evolution and Human Behavior.* Garden City, N.Y.: The Natural History, 1967.

————. *Human Diversity.* New York: Columbia University Press, 1971.

————. *The Human Imperative.* New York: Columbia University Press, 1972.

AMERICAN ACADEMY OF PEDIATRICS. *Report of the Committee on Infectious Diseases.* 17th ed.; Evanston, Ill.: American Academy of Pediatrics, 1974.

AMERICAN MEDICAL ASSOCIATION. *Mental Retardation: A Handbook for the Primary Physician.* Chicago: American Medical Association, 1965.

ANASTASI, ANNE. *Differential Psychology.* 3rd ed.; New York: The Macmillan Company, 1958.

ANFINSEN, CHRISTIAN B. *The Molecular Basis of Evolution.* New York: John Wiley & Sons, 1959.

ANGLE, CAROL A., and EDGAR A. BERING, JR. *Physical Trauma as an Etiological Agent in Mental Retardation.* Proceedings of a Conference on the Etiology of Mental Retardation, 1968. Bethesda, Md.: National Institute of Neurological Diseases and Stroke, 1970.

ARDREY, ROBERT. *The Social Contract: A Personal Inquiry into the Evolutionary Sources of Order and Disorder.* New York: Atheneum Publishers, 1970. (Paperback: Dell Publishing Co., 1971.)

ARENA, JAY M., ed. *Davison's Compleat Pediatrician.* 9th ed.; Philadelphia: Lea & Febiger, 1969.

ASHTON, T. S. *The Industrial Revolution, 1760–1830.* Rev. ed.; New York: Galaxy Books, Oxford University Press, 1964.

AVERY, GILLIAN. *Victorian People in Life and Literature.* New York: Holt, Rinehart and Winston, 1970.

BAKER, JOHN R. *Race.* New York: Oxford University Press, 1974.

BALLANTYNE, J. W. *Manual of Antenatal Pathology and Hygiene.* Edinburgh: William Green & Son, 1904.

BALTZELL, E. DIGBY. *The Protestant Establishment: Aristocracy and Caste in America.* New York: Vintage Books, 1964.

BANFIELD, EDWARD C. *The Unheavenly City: The Nature and Future of Our Urban Crisis.* Boston: Little, Brown and Company, 1970.

BARNETT, HENRY L., and ARNOLD EINHORN, eds. *Pediatrics.* 15th ed.; New York: Appleton-Century-Crofts, 1972.

BARRY, J. M. *Molecular Biology: Genes and the Chemical Control of Living Cells.* Englewood Cliffs, N.J.: Prentice-Hall, 1964.

BATCHELDOR, ALAN B. *The Economics of Poverty.* New York: John Wiley & Sons, 1966.

BATESON, WILLIAM. *Mendel's Principles of Heredity.* Cambridge, Eng.: Cambridge University Press, 1909.

BAXTER, PAUL, and BASIL SANSOM, eds. *Race and Social Difference: Selected Readings.* Baltimore: Penguin Books, 1972.

BEADLE, GEORGE, and MURIEL BEADLE. *The Language of Life: An Introduction to the Science of Genetics.* Garden City, N.Y.: Doubleday & Company, 1966.

BEESON, PAUL B., and WALSH McDERMOTT, eds. *Cecil-Loeb Textbook of Medicine.* 13th ed.; Philadelphia: W. B. Saunders Company, 1971.

BELLAS, JOSEPH J., and JOHN H. VINYARD, JR. *Selective Service Registrants Disqualified for Medical Reasons When Examined for Induction at Armed Forces Examining and Induction Stations: 1958–1970.* Washington, D.C.: Office of the Surgeon General, Department of the Army, 1971.

BENEDICT, RUTH. *Race: Science and Politics.* Rev. ed.; New York: The Viking Press, 1943.

BERG, ALAN, portions with ROBERT J. MUSCAT. *The Nutrition Factor: Its Role in National Development.* Washington, D.C.: The Brookings Institution, 1973.

BERGMAN, PETER M., and staff under MORT N. BERGMAN, eds. *The Chronological History of the Negro in America.* New York: Mentor Books, New American Library, 1969.

BERGSMA, DANIEL, and SAUL KRUGMAN, eds. *Intrauterine Infections.* Papers presented at the Symposium on Intrauterine Infections, New York, 1968. New York: The National Foundation, 1968.

BEST, CHARLES, and NORMAN BURKE TAYLOR, eds. *The Physiological Basis of Medical Practice.* 7th ed.; Baltimore: The Williams & Wilkins Company, 1961.

BINET, ALFRED, and THÉODORE SIMON. *The Development of Intelligence in Children.* Trans. from the French by Elizabeth S. Kite. Baltimore: The Williams & Wilkins Company, 1916.

BIRCH, HERBERT G., and JOAN DYE GUSSOW. *Disadvantaged Children: Health, Nutrition and School Failure.* New York: Harcourt, Brace and World, 1970.

BLACKER, C. P. *Eugenics, Galton and After.* London: Gerald Duckworth & Co., 1952.

BLOCK, N. J., and GERALD DWORKIN, eds. *The IQ Controversy.* New York: Pantheon, 1976.

BLYTHE, RONALD. *Akenfield: Portrait of an English Village.* New York: Pantheon Books, 1969.

BOAS, FRANZ. *The Mind of Primitive Man.* New York: The Macmillan Company, 1911; rev. ed., 1938.

———. *Race, Language, and Culture* (collected scientific papers, 1887–1939). New York: The Macmillan Company, 1940. (Paperback: The Free Press, 1966.)

———. *Race and Democratic Society.* New York: J. J. Augustin Publisher, 1945.

BONNER, JOHN TYLER. *The Ideas of Biology.* New York: Harper & Brothers, 1962.

BORGSTROM, GEORG. *The Food and People Dilemma.* North Scituate, Mass.: Duxbury Press, 1973.

BOWLBY, JOHN. *Maternal Care and Mental Health.* In same volume: Mary D. Ainsworth et al., *Deprivation of Maternal Care.* World Health Organization studies. New York: Schocken Books, 1953.

BRACHET, JEAN, and ALFRED E. MIRSKY, eds. *The Cell: Biochemistry, Physiology, Morphology.* 3 vols.; New York: Academic Press, 1959.

BRENNER, JOSEPH, ROBERT COLES, ALAN MERMAN, et al. *Hungry Children: A Report.* Atlanta: Southern Regional Council, 1967.

BRIGHAM, CARL C. *A Study of American Intelligence.* Foreword by Robert M. Yerkes. Princeton: Princeton University Press, 1923.

BRONOWSKI, J. *William Blake and the Age of Revolution.* New York: Harper Colophon Books, 1969.

BROWN, ROY E. *Fertility, Malnutrition and Mortality.* Master's thesis, University of North Carolina, 1967.

——— *The Tyranny of Hunger,* Foreword by Florence Halpern. New York: Springer, 1976. Read in manuscript, 1972.

BUCHSBAUM, RALPH. *Animals Without Backbones: An Introduction to the Invertebrates.* Rev. ed.; Chicago: University of Chicago Press, 1948.

BUEL, RONALD A. *Dead End: The Automobile in Mass Transportation.* Englewood Cliffs, N.J.: Prentice-Hall, 1972. (Paperback: Penguin Books, 1973.)

BULLOCH, WILLIAM. *The History of Bacteriology.* London: Oxford University Press, 1938.

BURCH, GUY IRVING, and ELMER PENDELL. *Human Breeding and Survival: Population Roads to Peace or War.* Foreword by Walter B. Pitkin. New York: Penguin Books, 1947.

BURDETTE, WALTER J., ed. *Viruses Inducing Cancer: Implications for Therapy.* Salt Lake City: University of Utah Press, 1966.

BURNET, F. M., and W. M. STANLEY, eds. *The Viruses: Biochemical, Biological and Biophysical Properties.* 3 vols.; New York: Academic Press, 1959.

BURNET, MACFARLANE, and DAVID O. WHITE. *Natural History of Infectious Disease.* 4th ed.; Cambridge, Eng.: Cambridge University Press, 1972.

BUTLER, J. A. V. *Inside the Living Cell.* New York: Basic Books, 1959.

CANCRO, ROBERT, ed. *Intelligence: Genetic and Environmental Influences.* New York: Grune & Stratton, 1971.

CARTWRIGHT, FREDERICK F. *Disease and History.* New York: Thomas Y. Crowell Company, 1972.

CAVALLI-SFORZA, L. L., and W. F. BODMER. *The Genetics of Human Populations.* San Francisco: W. H. Freeman and Company, 1971.

CHAMBERS, J. D., and G. E. MINGAY. *The Agricultural Revolution, 1750–1880.* London: Batsford, 1966.

CHAPLIN, J. P. *Rumor, Fear and the Madness of Crowds.* New York: Ballantine Books, 1959.

CHARGAFF, ERWIN. *Essays on Nucleic Acids.* New York: Elsevier Publishing Company, 1963.

CHASE, ALLAN. *The Biological Imperatives: Health, Politics, and Human Survival.* New York: Holt, Rinehart and Winston, 1972. (Paperback: Penguin Books, 1973.)

CHASTEEN, EDGAR R. *The Case for Compulsory Birth Control.* Englewood Cliffs, N.J.: Prentice-Hall, 1971.

CHRISTENSEN, HERBERT E., ed. *The Toxic Substances List, 1972 Edition.* U.S. Public Health Service Publication No. HSM 72-10265. Rockville, Md.: National Institute for Occupational Safety and Health, 1972.

COHEN, SIDNEY. *The Drug Dilemma.* New York: McGraw-Hill, 1968.

COLES, ROBERT. *The Desegregation of Southern Schools. A Psychiatric Study.* New York: Anti-Defamation League of B'nai B'rith, 1963.

————. *Children of Crisis: A Study of Courage and Fear* (Vol. I of a three-volume study). *Migrants, Sharecroppers, Mountaineers* (Vol. II). *The South Goes North* (Vol. III). Boston: Little, Brown and Company, 1967, 1969, 1971.

COMMITTEE ON MATERNAL NUTRITION, NATIONAL RESEARCH COUNCIL–NATIONAL ACADEMY OF SCIENCES. *Maternal Nutrition and the Course of Pregnancy.* Washington, D.C.: National Academy of Sciences, 1970.

COMMONER, BARRY. *The Closing Circle: Nature, Man, and Technology.* New York: Alfred A. Knopf, 1971.

————. *The Poverty of Power: Energy and the Economic Crisis.* New York: Alfred A. Knopf, 1976.

COMMONS, JOHN R. *Race and Immigration in America.* New York: The Macmillan Company, 1908.

CONLEY, RONALD W. *The Economics of Mental Retardation.* Baltimore: Johns Hopkins University Press, 1973.

CROME, LEON, and JAN STERN. *Pathology of Mental Retardation.* 2nd ed.; Baltimore: The Williams & Wilkins Company, 1972.

DAHLBERG, GUNNAR. *Race, Reason, and Rubbish.* Trans. from the Swedish by Lancelot Hogben. New York: Columbia University Press, 1942.

DAVENPORT, CHARLES BENEDICT. *Heredity in Relation to Eugenics.* New York: Henry Holt and Company, 1911.

DAVENPORT, F. GARVIN, JR. *The Myth of Southern History: Historical Consciousness in Twentieth-Century Southern Literature.* Nashville: Vanderbilt University Press, 1970.

DAVIDOWICZ, LUCY S. *The War Against the Jews, 1933–1945.* New York: Holt, Rinehart and Winston, 1975.

DAVIE, RONALD, NEVILLE BUTLER, and HARVEY GOLDSTEIN. *From Birth to Seven: A Report of the National Child Development Study* (1958 Cohort). London: Longmans, in association with the National Children's Bureau, 1972.

DAVIES, JACK. *Human Developmental Anatomy.* New York: The Ronald Press, 1963.

DAVIES, JOHN D. *Phrenology, Fad and Science: A 19th-Century American Crusade.* New Haven: Yale University Press, 1955.

DAVIES, STANLEY POWELL, with KATHERINE G. ECOB. *The Mentally Retarded in Society.* New York: Columbia University Press, 1959.

DAVIS, JOHN A., and JOHN DOBBING, eds. *Scientific Foundations of Paediatrics.* Philadelphia: W. B. Saunders Company, 1972.

DEANE, PHYLLIS. *The First Industrial Revolution.* Cambridge, Eng.: Cambridge University Press, 1965.

———— and W. A. COLE. *British Economic Growth 1688–1959.* 2nd ed.; Cambridge, Eng.: Cambridge University Press, 1967.

DE KRUIF, PAUL. *Hunger Fighters.* New York: Harcourt, Brace and Company, 1928.

DENNIS, WAYNE, ed. *Readings in the History of Psychology.* New York: Appleton-Century-Crofts, 1948.

DEUTSCH, MARTIN. *Happenings on the Way Back to the Forum: Social Science, IQ, and Race Differences Revisited.* Reprinted from *Harvard Educational Review,* XXXIX, No. 3 (Summer 1969). New York: Anti-Defamation League of B'nai B'rith, 1969.

————, IRWIN KATZ, and ARTHUR R. JENSEN, eds. *Social Class, Race, and Psychological Development.* Preface by Ernest R. Hilgard. New York: Holt, Rinehart and Winston, 1968.

DICKENS, CHARLES. *Hard Times.* London: 1854; New York: Signet, New American Library, 1961.

DOUGLAS, EMILY TAFT. *Margaret Sanger: Pioneer of the Future.* New York: Holt, Rinehart and Winston, 1970.

DUBOS, RENÉ, and JEAN DUBOS. *The White Plague: Tuberculosis, Man and*

Society. Boston: Little, Brown and Company, 1952.

DUELL, WALLACE. *People under Hitler.* New York: Harcourt, Brace and Company, 1942.

DUNBAR, ANTHONY. *The Will to Survive.* Foreword by Senator Charles H. Percy. Atlanta: Southern Regional Council, 1969.

EAST, EDWARD M. *Heredity and Human Affairs.* New York: Charles Scribner's Sons, 1927.

EBERT, JAMES D., ARIEL G. LOEWY, RICHARD S. MILLER, and HOWARD A. SCHNEIDERMAN. *Biology.* New York: Holt, Rinehart and Winston, 1973.

EHRLICH, PAUL R. *The Population Bomb.* Foreword by David Brower. New York: A Sierra Club–Ballantine Book, 1968.

EICHENWALD, HEINZ F., ed. *The Prevention of Mental Retardation Through Control of Infectious Diseases.* U.S. Public Health Service Document No. 1692. Bethesda, Md.: National Institute of Child Health and Human Development, 1966.

EVANS, H. W. *The Menace of Modern Immigration. An Address Delivered on the Occasion of Klan Day at the State Fair of Texas at Dallas, October 24, 1923.* Atlanta: Knights of the Ku Klux Klan, 1923.

EYSENCK, H. J. *The IQ Argument: Race, Intelligence and Education.* New York: The Library Press, 1971.

FAIRCHILD, HENRY PRATT. *The Melting Pot Mistake.* Boston: Little, Brown and Company, 1926.

———. *Race and Nationality.* New York: The Ronald Press, 1947.

FINER, S. E. *The Life and Times of Sir Edwin Chadwick.* London: Methuen & Co., 1952.

FIRST INTERNATIONAL EUGENICS CONGRESS (London, 1912). Scientific papers and appendices. London: The Eugenics Education Society, 1912–13.

FLEXNER, ABRAHAM. *Medical Education in the United States and Canada: A Report to the Carnegie Foundation for the Advancement of Teaching.* Introduction by Henry S. Pritchett. New York: The Carnegie Foundation, 1910.

FLEXNER, SIMON, and JAMES THOMAS FLEXNER. *William Henry Welch and the Heroic Age of American Medicine.* New York: The Viking Press, 1941.

FRAENKEL-CONRAT, HEINZ. *Design and Function at the Threshold of Life: The Viruses.* New York: Academic Press, 1962.

FRANKLIN, JOHN HOPE. *From Slavery to Freedom: A History of Negro Americans.* 3rd ed.; New York: Vintage Books, 1969.

FRIED, MORTON H., ed. *Readings in Anthropology.* 2 vols., 2nd ed.; New York: Thomas Y. Crowell Company, 1968.

FRYER, PETER. *The Birth Controllers.* London: Martin Secker & Warburg, 1965.

GAJDUSEK, D. CARLETON, CLARENCE J. GIBBS, and MICHAEL ALPERS. *Slow, Latent and Temperate Virus Infections.* U.S. Public Health Service Publication No. 1378. Bethesda, Md.: National Institute of Neurological Diseases and Blindness, 1965.

GALTON, FRANCIS. *Hereditary Genius.* London: Macmillan, 1869.

———. *Inquiries into Human Faculty.* London: Macmillan, 1883.

———. *Memories of My Life.* London: Methuen & Co., 1908.

GARRETT, HENRY E. *General Psychology.* New York: American Book Company, 1955; 2nd rev. ed., with Hubert Bonner, 1961.

———. Four monographs: (1) *How Classroom Desegregation Will Work,* (2) *Heredity: The Cause of Racial Differences in Intelligence,* (3) *The Relative Intelligence of Whites and Negroes: The Armed Forces Tests,* (4) *Breeding Down.* Richmond, Va.: The Patrick Henry Press, undated but c. 1954–67.

GEORGE, WESLEY CRITZ. *The Biology of the Race Problem: A Study Prepared by Commission of the Governor of Alabama.* Birmingham, Ala.: State of Alabama, 1962; repr. by The Patrick Henry Press, Richmond, Va., undated.

GERARD, R. W. *Unresting Cells.* New York: Harper & Brothers, 1940. (Paperback: Harper Torchbooks, 1961.)

GESELL, ARNOLD, and CATHERINE S. AMATRUDA. *Developmental Diagnosis, Normal and Abnormal Child Development: Clinical Methods and Pediatric Applications.* New York: Paul B. Hoeber, 1941; rev. ed., 1947.

GLASS, DAVID C. ed. *Biology and Behavior Genetics.* Proceedings of a conference under the auspices of the Russell Sage Foundation, the Social Science Research Council, and the Rockefeller University. New York: The Rockefeller University Press and the Russell Sage Foundation, 1968.

GOBINEAU, JOSEPH ARTHUR DE. *The Inequality of Human Races.* Trans. from the French editions of 1853–55 by A. Collins. London: Heinemann, 1915.

GODDARD, HENRY HERBERT. *The Kallikak Family: A Study in the Heredity of Feeble-Mindedness.* New York: The Macmillan Company, 1912.

———. *Feeble-Mindedness: Its Causes and Consequences.* New York: The Macmillan Company, 1914.

GOLDSBY, RICHARD A. *Race and Races.* New York: The Macmillan Company, 1971.

GOSSETT, THOMAS W. *Race: The History of an Idea in America.* Dallas: Southern Methodist University Press, 1963.

GOULD, CHARLES WINTHROP. *America, A Family Affair.* New York: Charles Scribner's Sons, 1920; rev. ed., 1922.

GRAHAM, LOREN R. *Science and Philosophy in the Soviet Union.* New York: Alfred A. Knopf, 1972.

GRANT, MADISON. *The Passing of the Great Race.* Preface by Henry Fairfield Osborn. New York: Charles Scribner's Sons, 1916; new edition, revised and amplified, with a new preface by Osborn, 1918.

——— and CHARLES STEWART DAVISON, comps. and eds. *The Alien in Our Midst or "Selling Our Birthright for a Mess of Pottage": The Written Views of a Number of Americans (Present and Former) on Immigration and Its Results.* New York: The Galton Publishing Company, 1930.

GRIER, GEORGE. *The Baby Bust.* Washington, D.C.: The Washington Center for Metropolitan Studies, 1971.

GRUNBERGER, RICHARD. *The 12-Year-Reich: A Social History of Nazi Germany, 1933–1945.* New York: Holt, Rinehart and Winston, 1971. (Paperback: Ballantine Books, 1972.)

GUZE, LUCIEN B., ed. *Microbial Protoplasts, Spheroplasts and L-Forms.* Baltimore: The Williams & Wilkins Company, 1968.

HAGUE, W. GRANT, M.D. *The Eugenic Marriage: A Personal Guide to the New Science of Better Living and Better Babies.* 4 vols.; New York: The Review of Reviews Company, 1914.

HALDANE, J. B. S. *Heredity and Politics.* New York: W. W. Norton & Company, 1938.

HALL, P. F. *Immigration and Its Effects on the United States.* New York: Henry Holt and Company, 1906.

———. *Eugenics, Ethics and Immigration.* IRL Publication No. 51. Boston: Immigration Restriction League, 1909.

HALL, MRS. PRESCOTT F. *Immigration and Other Interests of Prescott Farnsworth Hall.* Foreword by Madison Grant. New York: The Knickerbocker Press, 1922.

HALL, ROBERT, ed. *Abortion in a Changing World.* Proceedings of an International Conference, 1968. 2 vols.; New York: Columbia University Press, 1970.

HALLER, JOHN S., JR. *Outcasts from Evolution: Scientific Attitudes of Racial Inferiority, 1859–1900.* Urbana: University of Illinois Press, 1971.

HALLER, MARK M. *Eugenics: Hereditarian Attitudes in American Thought.* New Brunswick, N.J.: Rutgers University Press, 1963.

HALPERN, FLORENCE. *Survival: Black/White.* New York: Pergamon Press, 1973.

HANDLER, PHILIP, ed. *Biology and the Future of Man.* New York: Oxford University Press, 1970.

HANKINS, FRANK HAMILTON. *The Racial Basis of Civilization: A Critique of the Nordic Doctrine.* New York and London: Alfred A. Knopf, 1926.

HARDIN, GARRETT. *Biology: Its Human Implications.* San Francisco: W. H. Freeman and Company, 1949.

———, ed. *Population, Evolution, and Birth Control: A Collage of Controversial Ideas.* San Francisco: W. H. Freeman and Company, 1964, 1969.

HARDY, JANET B., ed. *Factors Affecting the Growth and Development of Children.* Proceedings of a symposium convened by the Johns Hopkins Collaborative Perinatal Project. Baltimore: Johns Hopkins University Press, 1971.

HARRIS, DALE B., and JEAN ROBERTS. "Intellectual Maturity of Children: Demographic and Socioeconomic Factors, United States." National Health Survey, Series 11, No. 116, 1972.

HARRIS, MARVIN. *The Rise of Anthropological Theory: A History of Theories of Culture.* New York: Thomas Y. Crowell Company, 1968.

———. *Culture, Man and Nature: An Introduction to General Anthropology.* New York: Thomas Y. Crowell Company, 1971.

Harvard Educational Review. Science, Heritability, and IQ. Reprint Series No. 4. Articles by Richard J. Light and Paul V. Smith; Arthur L. Srinchcombe; F. S. Fehr; Thomas J. Cottle; Martin

Deutsch. Cambridge: *Harvard Educational Review*, 1969.

———. *Challenging the Myths: The Schools, the Blacks, and the Poor.* Reprint Series No. 5. Articles by Charles A. Valentine; Susan S. Stodolsky and Gerald Lesser; Ray C. Rist; Stephen S. Baratz and Joan C. Baratz; Annie Stein. Cambridge: *Harvard Educational Review*, 1971.

HAWRYLEWICZ, E. J., and BACON F. CHOW, eds. *The Relationship of Perinatal Malnutrition to Brain Development.* Special issue, *Nutrition Reports International*, Vol. IV, No. 5 (November 1971). Los Altos, Calif.: Geron-X, Inc., 1971.

HEBER, RICK, et al. *Rehabilitation of Families at Risk for Mental Retardation: December 1972 Progress Report.* Madison, Wisc.: Rehabilitation Research and Training Center in Mental Retardation, University of Wisconsin, 1972.

HEIDEN, KONRAD. *Der Fuehrer: Hitler's Rise to Power.* Trans. from the German by Ralph Manheim. Boston: Houghton Mifflin Company, 1944.

HELLMAN, GEOFFREY. *Bankers, Bones and Beetles: The First Century of the American Museum of Natural History.* Garden City, N.Y.: The Natural History Press, 1969.

HERRNSTEIN, RICHARD. *I.Q. in the Meritocracy.* Boston: Little, Brown and Company, 1973.

——— and EDWIN G. BORING, eds. *A Sourcebook in the History of Psychology.* Cambridge: Harvard University Press, 1965.

HERSKOVITS, MELVILLE J. *Franz Boas: The Science of Man in the Making.* New York: Charles Scribner's Sons, 1953.

HIGHAM, JOHN. *Strangers in the Land: Patterns of American Nativism 1860–1924.* New Brunswick, N.J.: Rutgers University Press, 1955. (Paperback: rev. college ed.; New York: Atheneum Publishers, 1963.)

HIMES, NORMAN E. *Medical History of Contraception.* New York: Gamut Press, 1963.

HIRSCH, JERRY. "Behavior-Genetic Analysis and Its Biosocial Consequences." *Seminars in Psychiatry*, II, No. 2 (1970), 89–105.

HITLER, ADOLF. *Mein Kampf.* Complete and unabridged trans., fully annotated by John Chamberlain et al. New York: Reynal & Hitchcock, 1941.

HOCKETT, HOMER CAREY, and ARTHUR MEIER SCHLESINGER. *Political and Social Growth of the American People.* Vol. I, 3rd ed.; New York: The Macmillan Company, 1940.

HOFSTADTER, RICHARD. *Social Darwinism in American Thought.* Rev. ed.; New York: George Braziller, 1959.

HOLMES, LEWIS B., HUGO W. MOSER, et al. *Mental Retardation: An Atlas of Associated Physical Abnormalities.* New York: The Macmillan Company, 1972.

HOLMES, SAMUEL J. *The Trend of the Race: A Study of Present Tendencies in the Biological Development of Civilized Mankind.* New York: Harcourt, Brace and Company, 1921.

———. *Studies in Evolution and Eugenics.* New York: Harcourt, Brace and Company, 1923.

———. *An Introduction to General Biology: Life and Evolution.* New York: Harcourt, Brace and Company, 1926.

HOOTON, EARNEST ALBERT. *Apes, Men, and Morons.* New York: G. P. Putnam's Sons, 1937.

HOWE, MARK DE WOLFE, ed. *The Holmes-Laski Letters: The Correspondence of Mr. Justice Holmes and Harold J. Laski.* Vol. II. Cambridge: Harvard University Press, 1953.

HUNT, J. MCVICKER. *Intelligence and Experience.* New York: The Ronald Press, 1961.

———. *The Psychological Basis for Using Pre-School Enrichment as an Antidote for Cultural Deprivation.* Urbana, Ill.: Reprinted from *The Merrill-Palmer Quarterly of Behavior and Development*, Vol. X, No. 3 (July 1964).

———. *The Challenge of Incompetence and Poverty: Papers on the Role of Early Education.* Urbana: University of Illinois Press, 1969.

HURLEY, RODGER. *Poverty and Mental Retardation: A Causal Relationship.* Foreword by Senator Edward M. Kennedy. New York: Vintage Books, 1969.

INGLE, DWIGHT J. *Who Should Have Children? An Environmental and Genetic Approach.* Indianapolis: The Bobbs-Merrill Company, 1973.

INSURANCE INFORMATION INSTITUTE. *Insurance Facts, 1973.* New York, 1973.

INTERNATIONAL ASSOCIATION FOR THE ADVANCEMENT OF ETHNOLOGY AND EUGENICS, NEW YORK. Reprint series: (1) *The Inheritance of Mental Ability*, by Cyril Burt. Reprint No. 11. Reprinted from *American Psychologist*, Vol. XII, No. 1 (January 1958). (2) *A Review: Klineberg's Chapter on Race and Psy-*

chology, by Henry E. Garrett. Reprint No. 1. Reprinted from *The Mankind Quarterly,* Vol. I, No. 1 (July 1960). (3) *The Emergence of Racial Genetics,* by Ruggles R. Gates. Reprint No. 6. Reprinted from *The Mankind Quarterly,* Vol. I, No. 1 (July 1960). (4) *The S.P.S.S.I. and Racial Differences,* by Henry E. Garrett. Reprint No. 9. Reprinted from *American Psychologist,* Vol. XVII, No. 5 (May 1962).

JELLIFFE, DERRICK B. *Child Nutrition in Developing Countries.* Public Health Service Publication No. 1822. Washington, D.C.: Agency for International Development, 1968.

JENKS, JEREMIAH, and W. JETT LAUCK. *The Immigration Problem.* New York: Funk and Wagnall's, 1912.

JENNINGS, H. S. *The Biological Basis of Human Nature.* New York: W. W. Norton & Company, 1930.

JENSEN, ARTHUR R. *Environment, Heredity and Intelligence,* in Reprint Series No. 2, *Harvard Educational Review.* Containing Jensen's "How Much Can We Boost IQ and Scholastic Achievement?" and discussion by Jerome S. Kagan, J. McVicker Hunt, James F. Crow, Carl Bereiter, David Elkind, Lee J. Cronbach, William F. Brazziel, and Arthur R. Jensen. Cambridge: Harvard Educational Review, 1969.

———. *Genetics and Education.* New York: Harper & Row, 1972.

———. *Educability and Group Differences.* New York: Harper & Row, 1973.

JEVONS, F. R. *The Biochemical Approach to Life.* Foreword by Frederick Sanger. London: George Allen & Unwin, 1964.

KAMIN, LEON. *The Science and Politics of I.Q.* Potomac, Md.: L. Erlbaum Associates, 1974.

KARIER, CLARENCE, ed. *Shaping the American Educational State.* New York: The Free Press, 1975.

KARPINOS, BERNARD D. (1) *Results of the Examination of Youths for Military Service, 1967: Supplement to Health of the Army.* (2) *Results of the Examination of Youths for Military Service, 1968: Supplement to Health of the Army.* (3) (unsigned) *Results of the Examination of Youths for Military Service, 1969 and 1970: Supplement to Health of the Army.* Washington, D.C.: Medical Statistics Agency, Office of the Surgeon General, Department of the Army, December 1968, June 1969, October 1971.

KENNEDY, DAVID M. *Birth Control in America: The Career of Margaret Sanger.* New Haven: Yale University Press, 1970.

KESSNER, DAVID M., et al. *Contrasts in Health Status.* Vol. I: *Infant Death: An Analysis by Maternal Risk and Health Care.* Vol. II: *A Strategy for Evaluating Health Service.* Vol. III: *Assessment of Medical Care for Children.* Washington, D.C.: Institute of Medicine, National Academy of Sciences, 1973–74.

KING, JAMES C. *The Biology of Race.* New York: Harcourt Brace Jovanovich, 1971.

KLEMENT, ALFRED W., JR., et al. *Estimates of Ionizing Radiation Doses in the United States 1960–2000.* Rockville, Md.: U.S. Environmental Protection Agency, 1972.

KLINEBERG, OTTO. *Race Differences.* New York: Harper & Brothers, 1935 (a).

———. *Negro Intelligence and Selective Migration.* New York: Columbia University Press, 1935 (b).

KOTZ, NICK. *Let Them Eat Promises: The Politics of Hunger in America.* Introduction by Senator George S. McGovern. Englewood Cliffs, N.J.: Prentice-Hall, 1969.

KRAUSS, ELISSA. *Hospital Survey on Sterilization Policies. Reproductive Freedom Project.* New York: American Civil Liberties Union, 1975.

KUYPER, CH. M. A. *The Organization of Cellular Activity.* Amsterdam and New York: Elsevier Publishing Company, 1962.

LADER, LAWRENCE. *The Margaret Sanger Story, and the Fight for Birth Control.* Garden City, N.Y.: Doubleday & Company, 1955.

———. *Breeding Ourselves to Death.* Foreword by Paul R. Ehrlich. New York: Ballantine Books, 1971.

LANDMAN, J. H. *Human Sterilization: The History of the Sexual Sterilization Movement.* New York: The Macmillian Company, 1932.

LA ROCCO, AUGUST, et al. *Poverty and Health in the United States: A Bibliography with Abstracts. With Two Supplements.* New York: Medical and Health Research Association, 1968.

LASH, JOSEPH P. *Eleanor and Franklin: The Story of Their Relationship, Based on Eleanor Roosevelt's Private Papers.* New York: W. W. Norton & Company, 1971.

LAUGHLIN, HARRY HAMILTON. *Eugenical*

Sterilization in the United States. Chicago: Psychopathic Laboratory of the Municipal Court of Chicago, 1922. Rev. ed.; New Haven: American Eugenics Society, 1926.

LAUGHLIN, W. S., and R. H. OSBORNE, eds. *Human Variation and Origins: An Introduction to Human Biology and Evolution.* Readings from *Scientific American.* San Francisco: W. H. Freeman and Company, 1967.

LEAVITT, HELEN. *Superhighway-Superhoax.* Garden City, N.Y.: Doubleday & Company, 1970. (Paperback: Ballantine Books, 1971.)

LEE, JOSEPH. *Constructive and Preventive Philanthropy.* Introduction by Jacob A. Riis. New York: The Macmillan Company, 1910.

LENNEBERG, ERIC H. *Biological Foundations of Language.* Appendices by Noam Chomsky and Otto Marx. New York: John Wiley & Sons, 1967.

LEWIS, ANTHONY, et al. *The Second American Revolution: A First-Hand Account of the Struggle for Civil Rights.* London: Faber and Faber, 1966. (Originally published by Random House as *Portrait of a Decade.*)

LEVY, HILTON B., ed. *The Biochemistry of Viruses.* New York: Marcel Dekker, 1969.

LICHTENBERG, PHILIP, and DOLORES G. NORTON. *Cognitive and Mental Development in the First Five Years of Life: A Review of Recent Research.* U.S. Public Health Service Publication No. 2057. Rockville, Md.: National Institute of Mental Health, 1970.

LIDBETTER, E. J. *Heredity and the Social Problem Group.* London: Edward Arnold & Co., 1933.

LOGAN, RAYFORD W. *The Betrayal of the Negro: From Rutherford B. Hayes to Woodrow Wilson.* Rev. ed.; New York: Collier Books, 1965.

LONG, ESMOND R. *A History of Pathology.* New York: Dover Publications, 1965. (1st ed.; Baltimore: The Williams & Wilkins Company, 1928.)

LONGMATE, NORMAN. *The Workhouse.* London: Temple Smith, 1974.

LORENZ, KONRAD. *On Aggression.* Trans. from the German by Marjorie Kerr Wilson. New York: Harcourt, Brace and World, 1966. (Paperback: Bantam Books, 1967.)

LOVETT, WILLIAM, and JOHN COLLINS. *Chartism: A New Organization of the People.* Facsimile of original edition of 1840. Introduction by Asa Briggs. New York: Humanities Press, 1969.

LUNG PROGRAM, NATIONAL HEART AND LUNG INSTITUTE. *Respiratory Diseases: Task Force Report on Problems, Research Approaches, Needs.* DHEW Publication No. (NIH) 73-432. Washington, D.C.: National Heart and Lung Institute, 1972.

LYSENKO, TROFIM. *The Science of Biology Today.* New York: International Publishers, 1948.

MAJOR, RALPH H. *Classic Descriptions of Disease, with Biographical Sketches of the Authors.* 3rd. ed.; Springfield, Ill.: Charles C Thomas, 1947.

MALTHUS, THOMAS. *An Essay on the Principle of Population, As It Affects the Future Improvement of Society.* London, 1798. Rev. eds.: 1803, 1806, 1807, 1817, 1826.

MANN, ARTHUR. *La Guardia: A Fighter Against His Times, 1882–1933.* Philadelphia: J. B. Lippincott Company, 1959. (Paperback: University of Chicago Press, 1969.)

MANTOUX, PAUL. *The Industrial Revolution in the Eighteenth Century: An Outline of the Beginnings of the Modern Factory System in England.* Preface by T. S. Ashton. Rev. ed.; New York: Harper & Row, 1961.

MARCUS, STEVEN. *Engels, Manchester and the Working Class.* New York: Random House, 1974.

MATTMAN, LIDA. *Cell Wall Deficient Forms.* Cleveland: CRC Press, 1974.

MAYESKE, GEORGE W. *On the Explanation of Racial-Ethnic Group Differences in Achievement Test Scores.* Washington, D.C.: Office of Education, U.S. Department of Health, Education, and Welfare, 1971.

——— et al. *A Study of Our Nation's Schools: A Working Paper.* Washington, D.C.: Office of Education, U.S. Department of Health, Education, and Welfare, 1970–71.

MAYHEW, HENRY. *Mayhew's London: Being Selections from 'London Labour and the London Poor' by Henry Mayhew.* (First published in 1851.) Ed. by Peter Quennell. London: Spring Books, 1969.

McDOUGALL, WILLIAM. *Is America Safe for Democracy?* New York: Charles Scribner's Sons, 1921.

McELROY, WILLIAM D. *Cellular Physiology and Biochemistry.* Englewood Cliffs, N.J.: Prentice-Hall, 1961.

McKEOWN, THOMAS, and C. R. LOWE. *An Introduction to Social Medicine.* 2nd ed.; Oxford, Eng.: Blackwell Scientific Publications, 1974.

McPhee, John. *The Pine Barrens*. New York: Farrar, Straus & Giroux, 1968.

Mead, Margaret, Theodosius Dobzhansky, Ethel Tobach, and Robert E. Light, eds. *Science and the Concept of Race*. Scientific papers of a symposium at the December 1966 meetings of the American Association for the Advancement of Science. New York: Columbia University Press, 1968 (paperback ed., 1969).

Medical Clinics of North America. (1) Jules Harris, guest ed. *Symposium on Clinical Immunology*, 56/2. (2) Raymond V. Randall, guest ed. *Symposium on Endocrine Disorders*, 56/4. (3) Julian Katz and Donald Kaye, guest eds. *Symposium on Changing Concepts of Disease*, 57/4. (4) Philip L. Lerner, Martin C. McHenry, and Emanuel Wolinsky, guest eds. *Symposium on Infectious Diseases*, 58/3. Philadelphia: W. B. Saunders Company, March 1972, July 1972, July 1973, May 1974.

Medvedev, Zhores A. *The Rise and Fall of. T. D. Lysenko*. Trans. from the Russian by I. Michael Lerner. New York: Columbia University Press, 1969.

Mercer, E. H. *Cells: Their Structure and Function*. Rev. ed.; Garden City, N.Y.: Anchor Books, 1967.

Mercer, Jane. *Labeling the Mentally Retarded*. Berkeley: University of California Press, 1972.

Miller, Herman P. *Rich Man, Poor Man*. New York: Thomas Y. Crowell Company, 1971.

Moore, John A. *Heredity and Development*. New York: Oxford University Press, 1963.

Morison, Samuel Eliot. *The Oxford History of the American People*. New York: Oxford Univesrity Press, 1965.

Morse, Arthur. *While Six Million Died: A Chronicle of American Apathy*. New York: Random House, 1968.

Mosse, George L., ed. *Nazi Culture: Intellectual, Cultural, and Social Life in the Third Reich*. Trans. from the German by Salvator Attanasio and others. New York: Grosset & Dunlap, 1968.

Moulder, James W. *The Psittacosis Group as Bacteria*. New York: John Wiley & Sons, 1964.

Muller, Hermann. *Out of the Night*. New York: Vanguard Press, 1935.

Myerson, Abraham. *The Inheritance of Mental Disease*. Baltimore: The Williams & Wilkins Company, 1925.

———, et al. *Eugenical Sterilization: A Reorientation of the Problems*. By the Committee of the American Neurological Association for the Investigation of Eugenical Sterilization. New York: The Macmillan Company, 1936.

National Center for Health Statistics. *First Health and Nutrition Examination Survey, United States, 1971–72: Preliminary Findings on the Dietary Intake and Biochemical Levels of Various Nutrients in a Probability Sample of U.S. Population 1–74 Years of Age, by Age, Sex, Race, and Income Level 1971–72*. DHEW Publication No. (HRA) 74-1219-1. Rockville, Md.: National Center for Health Statistics, 1974.

———. *Data from the Health Interview Survey: Series 10, Nos. 1 through 78;* and *Data from the Health Examination Survey, Series 11, Nos. 1 through 122;* and *Data on Mortality, Series 20, 1 through 13;* and *Data from the National Natality and Mortality Surveys, Series 22, Nos. 1 through 14*. For individual titles, see *Current Listing and Topical Index to the Vital Health and Statistics Series, 1962–1974*. DHEW Publication No. (HRA) 76–1301 (rev.). Rockville, Md.: National Center for Health Statistics, October 1975.

Needham, Joseph. *A History of Embryology*. New York: Abelard-Schuman, 1959.

Nelson, Waldo E., ed. *Textbook of Pediatrics*. 8th ed.; Philadelphia: W. B. Saunders Company, 1964.

Niswander, Kenneth R., and Myron Gordon, eds. *The Women and Their Pregnancies: The Collaborative Perinatal Study of the National Institute of Neurological Diseases and Stroke*. Philadelphia: W. B. Saunders Company, 1972.

Nutrition Foundation, Inc. *Food, Science, and Society*. Papers from a 1968 symposium of the same name. New York: The Nutrition Foundation, Inc., 1969.

Osborn, Fairfield. *Our Plundered Planet*. Boston: Little, Brown and Company, 1948.

Osborn, Frederick. *The Future of Human Heredity: An Introduction to Eugenics in Modern Society*. Foreword by Theodosius Dobzhansky. New York: Weybright and Talley, 1968.

Osborn, Henry Fairfield. *The Approach to the Immigration Problem Through Science*. A paper presented before the National Immigration Conference. New York: National Industrial Conference

Board, 1923. (Reprinted as a pamphlet by the author.)

OSBORNE, JOHN W. *The Silent Revolution: The Industrial Revolution in England as a Source of Cultural Change.* New York: Charles Scribner's Sons, 1970.

OSER, JACOB. *Must Men Starve? The Malthusian Controversy.* New York: Abelard-Schuman, 1957.

OSLER, SONIA F., and ROBERT E. COOKE, eds. *The Biosocial Basis of Mental Retardation.* Baltimore: Johns Hopkins University Press, 1965.

OWEN, ROBERT. *A New View of Society* (first published in 1813/14) and *Report to the County of Lanark* (first published in 1821). Ed. with an introduction by V. A. C. Gatrell. Baltimore: Penguin Books, 1969.

PADDOCK, WILLIAM, and PAUL PADDOCK. *Famine—1975! America's Decision: Who Will Survive?* Boston: Little, Brown and Company, 1967.

PAN AMERICAN HEALTH ORGANIZATION, WHO. *Nutrition, the Nervous System, and Behavior.* Proceedings of the Seminar on Malnutrition, in Early Life and Subsequent Mental Development, 1972. Scientific Publication No. 251. Washington, D.C.: Pan American Health Organization, 1972.

PANOS, CHARLES, ed. *A Microbial Enigma: Mycoplasma and Bacterial L-Forms.* Cleveland: The World Publishing Company, 1967.

PARSONS, ROBERT PERCIVAL. *Trail to Light: A Biography of Joseph Goldberger.* Indianapolis: The Bobbs-Merrill Company, 1943.

PASAMANICK, BENJAMIN, and HILDA KNOBLOCH. (1) *Retrospective Studies on the Epidemiology of Reproductive Casualty: Old and New.* (2) *Prospective Studies on the Epidemiology of Reproductive Casualty: Methods, Findings, and Some Implications.* Chicago: Reprinted from *The Merrill-Palmer Quarterly of Behavior and Development,* Vol. XII, No. 1 (1966).

PAWLEY, WALTER H. *Possibilities of Increasing World Food Production.* Rome: Food and Agriculture Organization of the United Nations, 1963.

PEARSON, KARL. *Life, Letters and Labours of Francis Galton.* 4 vols.; Cambridge, Eng.: Cambridge University Press, 1914–40.

PETERS, JAMES A., ed. *Classic Papers in Genetics.* Englewood Cliffs, N.J.: Prentice-Hall, 1959.

PETERSEN, WILLIAM. *Population.* 2nd ed.; New York: The Macmillan Company, 1969.

PETERSON, J. *Early Conceptions and Tests of Intelligence.* New York: Harcourt, Brace and Company, 1925.

PETTIGREW, THOMAS F. *Negro American Intelligence.* New York: Anti-Defamation League of B'nai B'rith, 1969.

PIAGET, JEAN. *The Origins of Intelligence in Children.* Trans. from the French by Marjorie Worden. New York: International Universities Press, 1952.

———. *Structuralism.* Trans. from the French and ed. by Chaninah Maschler. New York: Basic Books, 1970.

PICKENS, D. P. *Eugenics and the Progressives.* Nashville: Vanderbilt University Press, 1968.

PIKE E. ROYSTON, ed. *Hard Times: Human Documents of the Industrial Revolution.* New York: Frederick A. Praeger, 1966.

PIRIE, N. W. *Food Resources, Conventional and Novel.* Baltimore: Penguin Books, 1969.

PLATT, B. S., and R. J. C. STEWART. "Reversible and Irreversible Effects of Protein-Calorie Deficiency on the Central Nervous System of Animals and Man." Reprinted from *World Review of Nutrition and Dietetics,* Vol. XIII. Basel, Switz.: S. Karger, 1971.

POPENOE, PAUL, and ROSWELL HILL JOHNSON. *Applied Eugenics.* New York: The Macmillan Company, 1918; rev. ed., 1933.

POSTMAN, LEO, ed. *Psychology in the Making: Histories of Selected Research Problems.* New York: Alfred A. Knopf, 1962.

PRESTON, SAMUEL H. *Older Male Mortality and Cigarette Smoking: A Demographic Analysis.* Population Monograph Series, No. 7. Berkeley: Institute of International Studies, University of California, 1970.

PUTNAM, CARLETON. *Race and Reason: A Yankee View.* Washington, D.C.: Public Affairs Press, 1961.

RACE BETTERMENT FOUNDATION. Official proceedings: Vol. I, *Proceedings of the First National Conference on Race Betterment,* January 8–12, 1914. Vol. II, *Second National Conference on Race Betterment, 1915: Held in San Francisco, California, in connection with the Panama-Pacific International Exposition.* Vol. III, *Proceedings of the Third National Conference on Race Betterment.* January 2–6, 1928, Battle Creek, Mich.: The Race Betterment Foundation, 1914, 1915, 1928.

REED, ELIZABETH W., and SHELDON C. REED. *Mental Retardation: A Family Study.* Philadelphia: W. B. Saunders Company, 1965.

RICHARDSON, KEN, DAVID SPEARS, and MARTIN RICHARDS, eds. *Race and Intelligence: The Fallacies Behind the Race-IQ Controversy.* Baltimore: Penguin Books, 1972.

ROBERTS, KENNETH L. *Why Europe Leaves Home.* Indianapolis: The Bobbs-Merrill Company, 1922.

ROBITSCHER, JONAS, comp. and ed. *Eugenic Sterilization.* Springfield, Ill.: Charles C Thomas, 1973.

ROSE, STEVEN. *The Conscious Brain.* New York: Alfred A. Knopf, 1973.

ROSENBERG, CHARLES E. *The Cholera Years: The United States in 1832, 1849, and 1866.* Chicago: University of Chicago Press, 1962.

ROSS, EDWARD A. *The Old World in the New.* New York: The Century Company, 1914.

ROSTAND, JEAN. *Error and Deception in Science: Essays on Biological Aspects of Life.* Trans. from the French by A. J. Pomerans. New York: Basic Books, 1960.

RUCH, THEODORE C., and JOHN F. FULTON, eds. *Medical Physiology and Biophysics.* 18th ed.; Philadelphia: W. B. Saunders Company, 1960.

RUTSTEIN, DAVID D. *The Coming Revolution in Medicine.* Cambridge: MIT Press, 1967.

SADLER, W. S. *Long Heads and Round Heads or What's the Matter with Germany.* Chicago: A. C. McClurg & Co., 1918.

SANDAY, PEGGY R. *A Diffusion Model for the Study of the Cultural Determinants of Differential Intelligence, 1973: Final Report Submitted to National Office of Education, US DHEW.* Unpublished. Copies available from Educational Resources Information Center, Bethesda, Md.

————. *On the Causes of IQ Differences Between Groups and Implications for Social Policy.* Philadelphia: Reprinted from *Human Organization,* journal of the Society for Applied Anthropology, Vol. XXXI, No. 4 (1972).

SANGER, MARGARET. *My Fight for Birth Control.* New York: Farrar and Rinehart, 1931.

————. *Margaret Sanger, An Autobiography.* New York: W. W. Norton & Company, 1938.

SARVIS, BETTY, and HYMAN RODMAN. *The Abortion Controversy.* New York: Columbia University Press, 1973.

Scientific American, readings from. *Altered States of Awareness.* Introductions by Timothy J. Teyler. San Francisco: W. H. Freeman and Company, 1972.

————. *The Nature and Nurture of Behavior: Developmental Psychobiology.* Introductions by Ward C. Greenough. San Francisco: W. H. Freeman and Company, 1973.

SCHELL, JONATHAN. *The Time of Illusion.* New York: Alfred A. Knopf, 1976.

SCRIMSHAW, NEVIN S., and JOHN E. GORDON, eds. *Malnutrition, Learning and Behavior.* Cambridge: MIT Press, 1968.

SECOND INTERNATIONAL CONGRESS OF EUGENICS (New York, 1921). Scientific papers. Vol. I: *Eugenics, Genetics and the Family.* Vol. II: *Eugenics in Race and State.* Baltimore: The Williams & Wilkins Company, 1923.

SEHAM, MAX. *Blacks and American Medical Care.* Minneapolis: University of Minnesota Press, 1973.

SELIGMAN, BEN B., ed. *Aspects of Poverty.* New York: Thomas Y. Crowell Company, 1968.

SENNA, CARL, ed. *The Fallacy of I.Q.* New York: The Third Press, 1973.

SEXTON, PATRICIA CAYO. *Education and Income: Inequalities in Our Public Schools.* Foreword by Kenneth B. Clark. New York: The Viking Press, 1961 (paperback ed., 1964).

SHIRER, WILLIAM L. *The Rise and Fall of the Third Reich: A History of Nazi Germany.* New York: Simon and Schuster, 1960. (Paperback: Fawcett Crest Books, 1962.)

SHUEY, AUDREY M. *The Testing of Negro Intelligence.* 2nd ed.; New York: Social Science Press, 1966.

SHUFELDT, R. W. *America's Greatest Problem: The Negro.* Philadelphia: F. A. Davis Company, 1915.

SKEELS, HAROLD M. *Adult Status of Children with Contrasting Early Life Experiences.* Monograph of the Society for Research in Child Development, Serial No. 105. Chicago: University of Chicago Press, 1966.

SNELL, BRADFORD C. *American Ground Transport: A Proposal for Restructuring the Automobile, Truck, Bus and Rail Industries,* in Part 4A—Appendix to Part 4, Hearings before the Subcommittee on Antitrust and Monopoly of the Committee on the Judiciary, U.S. Senate, 93rd Congress, 2nd Session, on S.1167. 1974.

SOLOMON, BARBARA MILLER. *Ancestors and Immigrants: A Changing New England Tradition.* Cambridge: Harvard University Press, 1956.

SORENSON, JAMES R. *Social and Psychological Aspects of Applied Human Genetics: A Bibliography.* DHEW Publication No. (NIH) 73-412. Bethesda, Md.: Fogarty International Center, 1973.

SPENCER, HERBERT. *The Man versus the State. With Four Essays on Politics and Society.* (First published 1884, 1892.) Ed. with an introduction by Donald Macrae. Baltimore: Penguin Books, 1969.

STAKMAN, E. C., RICHARD BRADFIELD, and PAUL C. MANGELSDORF. *Campaigns Against Hunger.* Cambridge: The Belknap Press of Harvard University Press, 1967.

STANIER, ROGER Y., MICHAEL DOUDOROFF, and EDWARD A. ADELBERG. *The Microbial World.* 3rd ed., Englewood Cliffs, N.J.: Prentice-Hall, 1970.

STANLEY, WENDELL M., and EVANS G. VALENS. *Viruses and the Nature of Life.* New York: E. P. Dutton & Co., 1965.

STENT, GUNTHER S. *Molecular Biology of the Bacterial Viruses.* San Francisco: W. H. Freeman and Company, 1963.

STEVENS, HARVEY A., and RICK HEBER, eds. *Mental Retardation: A Review of Research.* Chicago: University of Chicago Press, 1964.

STEWART, ALEXANDER P., and EDWARD JENKINS. *The Medical and Legal Aspects of Sanitary Reform.* Facsimile of original edition published in London by Robert Hardwicke in 1867. Introduction by M. W. Flinn. New York: Humanities Press, 1969.

STIMSON, A. M. *A Brief History of Bacteriological Investigations of the United States Public Health Service: Supplement No. 141 to the Public Health Reports.* Washington, D.C.: U.S. Government Printing Office, 1938.

STOCKING, GEORGE W., JR. *Race, Culture and Evolution: Essays in the History of Anthropology.* New York: The Free Press, 1968.

STODDARD, GEORGE D. *The Meaning of Intelligence.* New York: The Macmillan Company, 1943.

STODDARD, LOTHROP. *The Rising Tide of Color Against White World-Supremacy.* New York: Charles Scribner's Sons, 1920.

―――. *The Revolt Against Civilization: The Menace of the Under-Man.* New York: Charles Scribner's Sons, 1922.

―――. *Into the Darkness: Nazi Germany Today.* New York: Duell, Sloan & Pearce, 1940.

SWANSON, CARL P. *The Cell.* Englewood Cliffs, N.J.: Prentice-Hall, 1960.

TANNAHILL, REAY. *Food in History.* New York: Stein & Day, 1973.

TAYLOR, TELFORD. *Courts of Terror: Soviet Criminal Justice and Jewish Emigration.* New York: Alfred A. Knopf, 1976.

TERMAN, LEWIS M. *The Measurement of Intelligence: An Explanation of and a Complete Guide for the Use of the Stanford Revision and Extension of the Binet-Simon Intelligence Scale.* Boston: Houghton Mifflin Company, 1916.

―――. *The Intelligence of School Children: How Children Differ in Ability, the Uses of Mental Tests in School Grading, and the Proper Education of Exceptional Children.* Boston: Houghton Mifflin Company, 1919.

――― et al. *Genetic Studies of Genius.* 5 vols.; Stanford, Calif.: Stanford University Press, 1925–59.

TERRIS, MILTON, ed. *Goldberger on Pellagra.* Baton Rouge: Louisiana State University Press, 1964.

THIRD INTERNATIONAL CONGRESS OF EUGENICS (New York, 1932). *A Decade of Progress in Eugenics: Scientific Papers of the Congress.* Baltimore: The Williams & Wilkins Company, 1934.

THOMAS, LEWIS. *The Lives of a Cell: Notes of a Biology Watcher.* New York: The Viking Press, 1974.

THOMPSON, JAMES S., and MARGARET W. THOMPSON. *Genetics in Medicine.* Philadelphia: W. B. Saunders Company, 1966.

THORNDIKE, EDWARD L. *Human Nature and the Social Order.* New York: The Macmillan Company, 1940.

TOBACH, ETHEL, LESTER R. ARONSON, and EVELYN SHAW. *The Biopsychology of Development.* New York: Academic Press, 1971.

TRUAX, RHODA. *The Doctors Jacobi.* Boston: Little, Brown and Company, 1952.

TUMIN, MELVIN M., ed. *Race and Intelligence: A Scientific Evaluation.* New York: Anti-Defamation League of B'nai B'rith, 1963.

U.S. DEPARTMENT OF AGRICULTURE. *Household Food Consumption Survey, 1965–66.* Report No. 6: *Dietary Levels of Households in the United States;* Report No. 11: *Food and Nutrient Intake of Individuals in the United States.*

Washington, D.C.: USDA/Agricultural Research Service, 1969, 1972.

———. *Food: Consumption, Prices, Expenditures.* Agricultural Economic Report No. 138. Washington, D.C.: USDA/Economic Research Service, 1968. Also Supplement for 1970, issued 1971.

U.S. DEPARTMENT OF COMMERCE. *Statistical Abstract of the United States.* Issued annually. Also *Historical Statistics of the United States, Colonial Times to 1957,* and Supplement to same, *Continuation to 1962 and Revisions.*

U.S. PUBLIC HEALTH SERVICE. *Report of the Committee on Environmental Health Problems.* Public Health Service Publication No. 908. Washington, D.C.: U.S. Department of Health, Education, and Welfare, 1962.

U.S. SENATE, STAFF OF THE SELECT COMMITTEE ON NUTRITION AND HUMAN NEEDS. *Hunger—1973* (May 1973) and *To Save the Children: Nutritional Intervention Through Supplemental Feeding* (January 1974). Washington, D.C.: U.S. Government Printing Office, 1973, 1974.

VAN EVRIE, J. H. *White Supremacy and Negro Subordination, or Negroes a Subordinate Race, and (So-Called) Slavery Its Normal Condition.* 2nd ed.; New York: Van Evrie, Horton & Co., 1868.

VOGT, WILLIAM. *Road to Survival.* Introduction by Bernard M. Baruch. New York: William Sloane Associates, 1948.

———. *People! Challenge to Survival.* New York: William Sloane Associates, 1960.

WADDINGTON, C. H. *New Patterns in Genetics and Development.* New York: Columbia University Press, 1962.

WALLIN, J. E. WALLACE. *The Mental Health of the School Child: The Psycho-Educational Clinic in Relation to Child Welfare.* New Haven: Yale University Press, 1914.

———. *Problems of Subnormality.* Yonkers: World Book Company, 1917.

———. *The Odyssey of a Psychologist.* Wilmington, Del.: privately published, 1955.

WATSON, ERNEST H., and GEORGE H. LOWREY. *Growth and Development of Children.* 3rd ed.; Chicago: The Year Book Publishers, 1958.

WATSON, JAMES D. *Molecular Biology of the Gene.* 2nd ed.; Menlo Park, Calif.: W. A. Benjamin, 1970.

WEIDEL, WOLFHARD. *Virus.* Ann Arbor:

University of Michigan Press, 1959.

WHITNEY, LEON F. *The Case for Sterilization.* New York: Frederick A. Stokes Co., 1934.

WILDER, BURT G. *The Brain of the American Negro.* Reprinted from the Proceedings of the First National Negro Congress, New York, 1909. New York: National Negro Committee, 1909.

WILLIAMS, GREER. *Virus Hunters.* New York: Alfred A. Knopf, 1959.

———. *The Plague Killers: Untold Stories of Three Great Campaigns Against Disease* (yellow fever, malaria, and hookworm). New York: Charles Scribner's Sons, 1969.

WILLIAMS, WILLIAM APPLEMAN. *The Roots of the Modern American Empire: A Study of the Growth and Shaping of Social Consciousness in a Marketplace Society.* New York: Random House, 1969.

WILSON, EDMUND B. *The Cell in Development and Heredity.* 3rd ed.; New York: The Macmillan Company, 1928.

WILSON, MITCHELL. *American Science and Invention: A Pictorial History.* New York: Simon and Schuster, 1954.

WINGATE, PETER. *The Penguin Medical Encyclopedia.* Harmondsworth, Eng.: Penguin Books, 1972.

WINTROBE, MAXWELL M., ed. *Harrison's Principles of Internal Medicine.* 6th ed.; New York: McGraw-Hill Book Company, 1970.

WOLF, THETA H. *Alfred Binet.* Chicago: University of Chicago Press, 1973.

WOLFE, SIDNEY, and ROBERT E. McGARRAH, eds. *Health Research Group Study on Surgical Sterilization.* Washington, D.C.: Health Research Group (funded by Public Citizen, Inc.), 1973.

WOOD, W. BARRY, JR. *From Miasmas to Molecules.* New York: Columbia University Press, 1961.

WORLD HEALTH ORGANIZATION. *Health Hazards of the Human Environment.* Prepared by 100 specialists in 15 countries. Geneva: WHO, 1972.

YERKES, ROBERT M., ed. *Psychological Examining in the U.S. Army.* Memoir XV, National Academy of Sciences. Washington, D.C.: National Academy of Sciences, 1921.

———, JAMES W. BRIDGES, and ROSE S. HARDWICK. *A Point Scale for Measuring Mental Ability.* Baltimore: Warwick & York, 1915.

YOAKUM, C. S., and R. M. YERKES. *Army Mental Tests.* New York: Henry Holt and Company, 1920.

INDEX

Note: Page numbers in *italics* refer to illustrative material; t. to tabular matter; n. to notes. The following abbreviations have been used throughout the index in the interests of conserving space: AFQT (Armed Forces Qualification Test); CPD (cephalo-pelvic disproportion); IRL (Immigration Restriction League); NINDS (National Institution of Neurological Diseases and Stroke); PP (Pellagra Preventive) factor.

Aaronson, May, 589–90, 599, 600
abortion: and drop in maternal mortality after legislation, 411; forced sterilization as condition of, 18–19, 42; N.Y.C. statistics, 411; repeal of laws *vs.*, 411
AFQT (Armed Forces Qualification Test) scores, 432–3; Bellas-Vinyard study of, 443–45, 446; Chase's study of, 439–42; Conley on inverse relationship in, 446; Garrett's use of, 446–7; inverse relationship between disqualification rates for mental and medical reasons, 439–47; Jensen's use of, 433–4, 472; and new scientific racism, 433, 466–7, 472; Shockley's use of, 483, 487
Agnew, Spiro: Shockley on, 482
Agricultural Revolution, 72–3, 75–6; and decline in British death rate, 73; and improved food-producing capacities, 72, 365, 377
agriculture: Malthus and myth about food production and population growth, 75–7, 82, 83, 85, 365, 369
agriculture, U.S.: decrease in farm population, 417; costly machinery makes small farms obsolete, 417; production, 75–6, 415, 417; soybean production, 75–6
Aldrich, Thomas B., 111; and the Boston Irish (1892), 111
Alien in Our Midst, The (ed. by Grant and Davison), 357–60
Allied Patriotic Societies of New York: statement before Committee on Immigration, 295–7, 323
American Academy of Pediatrics, 31, 63, 546
American Association for the Advancement of Science (AAAS), 648–9 n.; response to new scientific racists, 461; warning *vs.* revival of scientific racism, 68
American Association on Mental Deficiency: on mental retardation, 24–5, 34
American Breeder's Association, 15, 114; *see also* Eugenics Section of ABA
American Coalition of Patriotic Societies, 357; campaign to keep Congress from relaxing Immigration Act barriers, 353; indicted for sedition (1942), 367
American Eugenics Society, 118, 164, 287, 366
American Federation of Labor, 271, 358
American Medicine: accepts Bean's claims about racial brain differences, 179–80
American Psychological Association, 493; attacks new scientific racism, 461; censures use of IQ tests, 242
Ammon, Otto: and cephalic index, 94–5
anemia, 566; pernicious, 364, 558; and vitamin B_{12}, 558, 663 n.
Annals of Eugenics (ed. by Pearson), 9, 82–3, 308

anti-Catholicism: burning of Boston convent, 111; Grant and, 164; and immigration quotas, 11; and IRL, 113; Osborn and, 164, 634 n.; *see also* Irish
anti-poverty programs, 664–5 n.; Pasamanicks on, 572–3
anti-Semitism, 9, 11, 605; Baker on Nordicists', 637 n.; Chamberlain and, 91; Fairchild and, 656 n.; Gobineau and, 90; Grant and, 164; and Immigration Act of 1924, 9, 11, 44, 50, 52; and IRL, 113, 182, 636 n.; Osborn and, 164, 634 n.; reinforced with IQ tests, 605–6; Ross and, 275–6, 288; *see also* Jews
Applied Eugenics (Popenoe and Johnson), 252–3, 257, 308, 510, 524
Ardrey, Robert: claims by, about Coleman Report, and Coleman's reply, 463; his *Social Contract,* 394
Aristogenics, 119, 132
Arlitt, Ada H., 302, 452; "On the Need for Caution in Establishing Race Norms," 302
Army Alpha test, 243, 244–6; sample test, 245; testing procedures, 243, 244
Army Beta test, 243, 246–8; testing procedures, 243, 246–7
Army intelligence tests, xx, 9, 11, 50, 52, 226–8, 243–51; Cannon article based on, 261–4, 489; Brigham's analysis, 264–73, 296, 298; findings of, 226, 227, 249–51, 252; Holmes's analysis, 261; Klineberg's analysis, 227–8 and t.; McDougall's analysis, 257; "prove" three of four Americans to be uneducable, 226, 249, 252; Stoddard's analysis, 226, 259–61, 296; Yoakum and Yerkes on, 249; *see also* Army Alpha test; Army Beta test
Army Mental Tests (Yoakum and Yerkes), 249
Association for Voluntary Sterilization, Inc., 17, 18, 383, 406, 626 n.
autism, in children, 546

Bache, R. Meade: mental tests, 229–30
bacteria, 535–6, 537–9; lysogenic, 548–9; nature of, 535–6, 537–8; *Serriatia marcescens* as cause of pogroms, 661 n.
Baker, John R.: on Nordicists' anti-Semitism, 637 n.; slurs about parentage of children in Boas' studies, 187
Banfield, Edward C., 8–9; for abandonment of compulsory high school, 514, 523; Nixon adviser on urban affairs, 524; *vs.* aid to the poor, 515, 516, 517; *vs.* minimum wages and legislation, 514, 523; on birth control, 523; on born losers, 519–20; on slum population, 2, 150, 513, 516, 517; on solutions to urban problems, 523; and time-orientation basis of human classification, 149, 514, 515–

A NOTE ABOUT THE AUTHOR

ALLAN CHASE, writer and independent scholar, was born in New York in 1913. *The Legacy of Malthus* is the second of a series of books on social biology. The first volume, *The Biological Imperatives,* was published in 1972. He is now at work on the third volume.

A NOTE ON THE TYPE

The text of this book was set on the Linotype in a face called Times Roman, designed by Stanley Morison for *The Times* (London) and first introduced by that newspaper in 1932.

Among typographers and designers of the twentieth century, Stanley Morison has been a strong forming influence, as a typographical adviser to the English Monotype Corporation, as a director of two distinguished English publishing houses and as a writer of sensibility, erudition, and keen practical sense.

This book was composed by American Book–Stratford Press, Brattleboro, Vermont. Printed and bound by R. R. Donnelley & Sons, Chicago, Illinois. Typography and binding design by Leon Bolognese